Encyclopedia of Environmental Control Technology

VOLUME 8

Work Area Hazards

ENCYCLOPEDIA OF
ENVIRONMENTAL CONTROL TECHNOLOGY

Volume 1: Thermal Treatment of Hazardous Wastes
Volume 2: Air Pollution Control
Volume 3: Wastewater Treatment Technology
Volume 4: Hazardous Waste Containment and Treatment
Volume 5: Waste Minimization and Recycling
Volume 6: Pollution Reduction and Contaminant Control
Volume 7: High-Hazard Pollutants
Volume 8: Work Area Hazards

Gulf Publishing Company
Houston, London, Paris, Zurich, Tokyo

Encyclopedia of Environmental Control Technology

VOLUME 8

Work Area Hazards

Paul N. Cheremisinoff, Editor

in collaboration with—

J. Abraham
V. E. Archer
F. C. Bagnati
A. Baj
P. G. Bardin
C. Bolognesi
R. Boyko
C. Chaikittiporn
N. P. Cheremisinoff
M. Dahiqvist
V. L. David
N. Gavrushenko
T. G. Grant
M. A. Hamilton
S. Ibrahim
T. Kawakami
L. Kolachalam
F. Merlo
G. N. Mhatre
P. Mizerek
G. Nano
J. P. Neglia
V. Nivsargikar
P. Osiri
K. Patel
R. A. Serio
G. Sesana
F. Smith
R. B. Trattner
Y. I. Tsen
U. Ulfvarson
S. V. Van Eeden
C. M. Williams

Encyclopedia of Environmental Control Technology

VOLUME 8

Work Area Hazards

Gulf Publishing Company
Book Division
P.O. Box 2608, Houston, Texas 77252-2608

10 9 8 7 6 5 4 3 2 1

Library of Congress Cataloging-in-Publication Data
Work area hazards / Paul N. Cheremisinoff, editor ; in collaboration with J. Abraham . . . [et al.].
p. cm. — (Encyclopedia of environmental control technology ; v. 8)
Includes bibliographical references and index.
ISBN 0-87201-304-9 (acid-free paper)
1. Hazardous substances—Management. I. Cheremisinoff, Paul N. II. Abraham, J. III. Series.
TD191.5.E5 1989 vol. 8
[T55.3.H3]
628.5 s—dc20
[604.7]
94-27853
CIP

ISBN 0-87201-238-7 (Series)

Printed on Acid-Free Paper (∞)

CONTENTS

CONTRIBUTORS TO THIS VOLUME

J. Abraham, South East Water Pollution Control Plant, Philadelphia Water Department, Philadelphia, PA 19116, USA

V. E. Archer, Rocky Mountain Center for Occupational and Environmental Health, University of Utah Medical Center, Salt Lake City, UT 84112, USA

F. C. Bagnati, Strulowitz & Spear, Randolph, NJ 07869, USA

A. Baj, Health and Safety Institute (UOOML), USSL 63, Piazza Benefattori, Desio (Mi) Italy

P. G. Bardein, Department of Internal Medicine, University of Stellenbosch, Tygerberg Hospital, Cape Town, South Africa

C. Bolognesi, Centro Nazionale per lo Studio dei Tumori di Origine Ambientale, Unit of Toxicologic Evaluation, Istituto Nazionale per la Ricerca sul Cancro, Viale Benedetto XV, 10, 16132, Genova, Italy

R. Boyko, Cycle Chem Corp., Linden, NJ 07036, USA

C. Chaikittiporn, Department of Occupational Health, Faculty of Public Health, Mahidol University, Thailand

N. P. Cheremisinoff, SciTech Technical Services, Inc., Morganville, NJ 07751, USA

P. N. Cheremisinoff, Department of Civil and Environmental Engineering, New Jersey Institute of Technology, Newark, NJ 07102, USA

M. Dahlqvist, Department of Environmental Technology and Work Science, Royal Institute of Technology, S-10044 Stockholm, Sweden

V. L. David, J. M. Huber Corporation, Printing Ink Division, Edison, NJ 08818-4032, USA

N. Gavrushenko, N. J. Transit Headquarters, Newark, NJ 07105-2246, USA

M. A. Hamilton, US Public Health Service, Indian Health Service, Pinetop, AZ 85935, USA

S. Ibrahim, SciTech Technical Services, Inc., Morganville, NJ 07751

T. Kawakami, Division of International Cooperation, The Institute for Science of Labour, Japan

L. Kolachalam, SciTech Technical Services, Inc., Morganville, NJ 07751

F. Merlo, Department of Environmental Epidemiology and Biostatistics, Istituto Nazionale per la Ricerca sul Cancro, Viale Benedetto XV, 10, 16132, Genova, Italy

G. N. Mhatre, H-3, Vishwakutir Shankar Ghanekar Marg Dadar, Bombay-400 028, India

P. Mizerek, Engineering & Environmental Services, Inc., Parsippany, NJ 07054, USA

G. Nano, Istituto di Chimica Fisica Applicata, Politecninco di Milano, via Manicinelli 7, Milano, Italy

J. P. Neglia, Becton-Dickinson and Company, Franklin Lakes, NJ 07417, USA

V. Nivargikar, Mega Management & Testing Services, Inc., Holmdel, NJ 07733, USA

P. Osiri, Department of Occupational Health, Faculty of Public Health, Mahidol University, Thailand

K. Patel, Department of Civil and Environmental Engineering, New Jersey Institute of Technology, Newark, NJ 07102, USA

R. A. Serio, Hunterdon Medical Center, Flemington, NJ 08822-4604, USA

G. Sesana, Health and Safety Institute (PMP), USSL 69, Via Spagliardi 19, Parabiago (Mi) Italy

K. D. Smith, Naval Air Warfare Center, Trenton, NJ 08628, USA

R. B. Trattner, Department of Chemistry, Chemical Engineering and Environmental Science, New Jersey Institute of Technology, Newark, NJ 07102, USA

Y. I. Tsen, Department of Civil & Environmental Engineering, New Jersey Institute of Technology, Newark, NJ 07102, USA

U. Ulfvarson, Department of Environmental Technology and Work Science, Royal Institute of Technology, S-10044 Stockholm, Sweden

S. V. Van Eeden, Department of Internal Medicine, University of Stellenbosch, Tygerberg Hospital, Cape Town, South Africa

C. M. Williams, Clayton Environmental Services, Inc., Edison, NJ 08817, USA

ABOUT THE EDITOR

Paul N. Cheremisinoff, P. E., is Professor of Civil and Environmental Engineering at the New Jersey Institute of Technology. Professor Cheremisinoff has more than 40 years' experience in research, design, and consulting for a wide range of government and industrial organizations. He is author and co-author of numerous papers and books on the environment, energy, and resources, and is a licensed professional engineer, member of Sigma Xi and Tau Beta Pi, and a Diplomate of the American Academy of Environmental Engineers.

PREFACE

The *Encyclopedia of Environmental Control Technology* is a series of volumes focusing on in-depth coverage of selected topics in environmental and industrial pollution control areas. These volumes are intended to provide up-to-date information on technology, research, and future trends in this vast interdisciplinary field.

Volume 8 deals with a broad range of health problems and focuses on hazards in the work area, their risks, and elimination. A broad spectrum of methods exists to deal with such situations and their effects on the workplace environment. Techniques vary widely, and there is no one single control technology to deal with problems resulting from former accepted work area practices.

Many concerns and practices are in response to increasingly stringent regulations. While reduction and elimination at the source have become the desired policy and the goal of government and industry, the correction of existing and past practices is a necessary reality and continues to dominate the safety control field. In practice, activities continue to focus on engineering controls, administrative practices, and personal protective equipment. Many occupationally related hazards address planning and prevention.

Acknowledgement goes to the contributors of this volume, each of whom has given willingly of valuable time and knowledge to ensure the accuracy and timely subject matter in the materials covered as well as those who have previously made this series of volumes possible.

Paul N. Cheremisinoff

CHAPTER 1

WORKPLACE HAZARDS

F. C. BAGNATI

Strulowitz & Spear
375 Route 10
Randolph, NJ 07869, USA

CONTENTS

GENERAL

A hazard is a dangerous condition that can interrupt or interfere with the expected progress of an activity. A chemical presents a health hazard if there is evidence that it may affect the health of exposed personnel. Examples of health hazards are carcinogens, irritants, sensitizers, and corrosives. Physical hazards include flammables, noise, pinch points, and sharp objects.

Toxic compounds are defined as materials that, upon contact with the human body, have the capacity to cause personal injury or illness. Many extremely toxic compounds are used safely in industry without significant risk to health because suitable precautions are taken to limit contact to amounts that have been established as safe.

Distinction between hazard potential and toxicity is important. The toxicity of a substance is its inherent capacity to produce injury, whereas the hazard potential (risk) is the likelihood of injury resulting from use of the substance in the quantity and manner proposed. To evaluate the hazard potential of a particular substance, it is necessary to know not only its toxicity but also many of its physical and chemical properties, as well as the manner and quantity in which the substance is to be used and stored.

Such factors determine how much of a substance enters the body, by what route, how frequently, and for how long—the combination of which can be described simply as the exposure. Thus, depending on the exposure, it is possible to have a high degree of risk associated with a low order of toxicity, a low degree of risk associated with a high order of toxicity, or an intermediate combination.

Hazards can be categorized in many ways, such as:

- How the hazardous material or energy is contained or supplied (compressed-gas cylinders, lecture bottles, cryogenic fluids house gases, vacuum, machinery, tools).
- Special properties (toxic gases, flammable and combustible liquids, reactive chemicals, electricity).
- Miscellaneous (asbestos, housekeeping).

Material Safety Data Sheets (MSDS) should be reviewed carefully for each material used.

CHEMICAL INVENTORIES

An inventory is a special inspection designed to find unused, out-of-date, or excess materials. It is good laboratory practice to inventory combustibles or flammables monthly and all other chemicals annually. Changes in federal and state labeling regulations may affect inventory practices.

MATERIALS SAFETY DATA SHEETS (MSDS)

Every laboratory work area and pilot plant should have an MSDS for each chemical present. Government regulations require that suppliers or manufacturers of any hazardous chemical prepare and provide MSDS for that chemical. The MSDS contains information on the properties and hazards of the chemical, health information, as well as first-aid measures and contacts in case of emergency.

The MSDS is to be requested on the requisition for the first-time purchase of any chemical. A copy of the MSDS, when received, should be appropriately filed.

The manufacturer's MSDS can be sued for a commercial product. If the material is a company product or a variant of a company product described by the generic MSDS, the appropriate MSDS can be used. If there is no existing applicable MSDS, one must be prepared.

For materials not included on the Toxic Substances Control Act (TSCA) inventory, customers shall be informed of this fact when they receive the MSDS. A file of MSDS information should be maintained.

GOVERNMENT NOTIFICATION AND PERMITTING

TSCA Premanufacturing Notice (PMN)

The TSCA inventory requires that a Premanufacturing Notice (PMN) be submitted to the Environmental Protection Agency (EPA) 90 days before manufacturing a new chemical substance. However, the PMN may not be required for the production of chemicals used for scientific experimentation or analysis and product development. Following is a summary of the rules that may apply:

- New materials produced solely for scientific experimentation or analysis on site do not require a PMN. However, personnel working with the material must be advised of pertinent safety and health risks, and appropriate protective measures must be taken.
- New materials produced solely for scientific experimentation or analysis and shipped outside do not require a PMN but must be accompanied by an MSDS. This MSDS shall contain all of the known safety and healthy risks. The receiving organization also shall be advised that the substance is not on the TSCA inventory.
- Sample size is not a factor in determining whether a PMN is required. The technical information and analyses obtained from the tests with the material must be needed for the decision on commercialization. Sale to a commercial user solely for evaluation meets these requirements as long as an MSDS, as described previously, accompanies the sample. However, production of a material for commercial purposes is not normally permissible before a PMN has cleared EPA review.

Permitting for Air Emissions

A permit must be obtained for emissions of vapor or gas to the atmosphere if more than about 2.3 kg/hr (5 lb/hr) of volatile organics or if more than 23 kg/hr (50 lb/hr) of material is processed and air pollution controls are provided. In general, laboratory operations are below these thresholds; pilot plant operations can exceed such guidelines.

Labeling of Chemicals

The objective of chemical labeling systems is straightforward. Hazard information is provided to employees who will or potentially will be exposed to the substance. Commercial labeling of chemicals provides much of the information required for their safe use and disposal. However, the OSHA Hazards Communications Standards (and probably other governmental rules) impose additional requirements on tagging of chemicals.

Materials should be identified with the user's name and hazards in accordance with current labeling practice, especially when transferred to a new container. As a minimum, the labeling shall show the generic identify as on the MSDS, commonly used trade name, chemical group (e.g., corrosive), statement of hazard, date container opened or filled, discard date if required, and the owner's name.

Safety cans and other portable containers of flammable liquids with flash points at or below 26.7°C (80°F) should be painted red and clearly identified with a yellow band or the name of the contents stenciled in yellow. Safety cans used for toxic or carcinogenic materials shall be labeled appropriately. If used for waste collection, safety cans shall have a dated waste disposal tag attached when first put into use. Any unlabeled container shall be handled as a hazardous material unless it can be positively identified safely. Therefore, it is important to keep labels on containers. Labeling of vessels for short-term experiments is not necessary if there is a sign displayed that the vessel is in use. However, such a vessel should be labeled if it will be used for storage following the experiment.

Samples of hazardous materials to be shipped shall be identified as to the nature of the hazard and handling precautions. If an analyst who receives a material is not familiar with it, the sender should be consulted. Filled pressure containers are to be tagged identifying the contents after outage and leak-tightness are confirmed. The person taking the sample shall sing the analysis request tag, chemical name, and physical state of contents, and also complete the hazard handling label.

Flammable/Combustible Liquids Storage

Flammable and combustible liquids are hazardous and must be stored with proper safeguards. The ability to ignite, vaporize, burn, or explode depends on the properties of the liquid.

Because of combustion and fire potential, there are restrictions on the maximum amount of flammable (Class I) and combustible (Class II or Class IIIA) liquids permitted in a single, typical laboratory module (typically 20 ft × 30 ft). Furthermore, only the actual amount needed to work effectively should be in a lab.

Basically, all wet chemistry labs are intended to be Class A, or normal, labs. If a lab cannot meet the Class A requirements, it may be possible to obtain authorization to have one flammable and combustible liquid storage cabinet. If this additional storage is insufficient, the lab director shall declare the laboratory to be a Class B, or special, lab and make sure that the requirements are met.

Flammable or combustible liquids stored in glass containers must be protected by a metal container, even if kept in a cabinet with a vent or a protected by a sprinkler. A paint can or grease pail can serve this purpose. A plastic bucket is not to be used as a container for glass bottles or equipment holding flammable or combustible liquids unless it is inside a metal container. Plastic buckets shall not be used for storing low-molecular-weight organic materials (e.g., solvents).

Class IIIA or non-flammable chemicals may be stored in the cabinets below laboratory fume hoods if the overall lab quantity restrictions are met and if in suitable containers or catch pans. Odorous chemicals may be stored in cabinets if specially packed. An inventory, taken at least monthly and preferable weekly, of the flammable and combustible liquids in each lab, must be posted near the entrance.

Bulk storage of 35-gal and 55-gal drums shall be designated barrel-storage areas meeting OSHA guidelines. A maximum of two 55-gal drums of flammable and combustible liquids is permitted in a pilot plant, excluding drums in use containing working inventory.

OSHA guidelines limit the size and construction materials of containers for flammable and combustible liquids. Class I liquids (and not liquefied flammable gases) kept in a lab must normally be in safety cans. The cans are painted red; those containing liquids with flash points below 26.7°C (80°F) are to be identified with the name of the contents stenciled with a yellow or with a yellow band. The name also should be on the handle where it is less likely to fade from contact with chemicals.

Class I liquids are not to be placed in plastic containers except for half-liter (1-pint) vented squeeze bottled used for rinsing lab equipment. Dispensing and receiving vessels shall be bonded and grounded electrically to prevent ignition from static sparks. Combustible solids shall be stored apart from flammable liquids. Bottle carriers are to be used for unprotected glass containers of flammable and combustible liquids taken outside of laboratories.

Combustible rubber or plastic material shall be separated from other combustibles and stored in an area protected by automatic sprinklers. Storage of materials requiring special handling (e.g., in hot or cold box, dry box, drums, etc.) shall be reviewed.

TRANSFER AND CONTAINMENT OF HYDROCARBONS

Open transfer is any transfer in which vapor released from the liquid being transferred mixes with the atmosphere, possible forming a flammable mixture. Open transfer of limited amounts of hydrocarbons is permitted (e.g., where the waste is being poured into a safety can equipped with an approved flame arrestor). The alternative is closed transfer in which there is no contact of the chemical with the atmosphere.

COMBUSTIBLE SOLIDS

Combustible solids include rubber, plastics, wood, paper, and rags. These materials can contribute to fire hazards if housekeeping is poor or storage is sloppy. A fire involving a combustible solid is a Class A fire. A fire involving both a combustible solid (Class A) and a flammable liquid (Class B) is difficult to extinguish. The extinguishing agents (carbon dioxide, dry chemical) most readily available near a laboratory may put out a Class A fire but leave smoldering cinders that can reignite. However, use of water alone as an extinguisher will not help against a Class B fire and frequently can cause it to spread due to splashing and floating effects. Consequently, it is important to decide quickly which type of fire predominates, then to use the appropriate agent for the predominant type of fire. If a Class A agent (e.g., dry powder) is used first, there should be prompt and careful follow-up with water to quench embers, and then monitoring for possible reignition.

Combustible solids should be stored properly. Non-combustible materials (e.g., aluminum foil) are to be used to cover shelves, bench tops, lab equipment, etc. The storage of significant amounts of polymers or paper, including cardboard boxes, in a work area is strictly forbidden due to the added fire hazard.

COMPRESSED GASES AND CRYOGENS

Compressed gases, stored in pressurized containers, present a major hazard potential. Many are flammable or toxic. Inert gases released from cylinders can dilute oxygen in confined spaces dangerously, as discussed elsewhere. There are strict rules for using compressed gases, the cylinders that contain them, and the piping and fittings for handling them.

It is essential to become knowledgeable about the flammability, toxicity, chemical reactivity, and corrosivity of compressed gases to be used. The conditions of use in the laboratory or pilot plant also are important.

Cryogens (materials that boil at below –100°C [–148°F]) present problems due both to their low storage and flash temperatures, and to the high pressures that can be generated when they vaporize. In addition, escape of excessive liquid nitrogen or helium can cause asphyxiation.

Compressed gases are stored in cylinders either in gaseous form under pressure or as a liquid if the actual vapor pressure is low enough. Cylinders supplying hazardous compressed gases are placed either outside the building on special pads or in a fume hood. Cylinders are stored upright and are to be fastened securely. Piping from cylinders into laboratories or pilot plants is designed and installed according to strict standards.

The size and number of gas cylinders in laboratories should be limited as specified or as listed below, whichever is more restrictive:

- Propane one, size 2L
- Butadiene one, size 4
- Oxygen one, size A
- Nitrogen one, size 1A

Get a lab manager's approval before relocating a cylinder and for proper placement of a new cylinder. All cylinders not in use shall be stored in special outside areas. All standard 1A style gas cylinders are restricted to a maximum allowable working pressure (MAWP) of 1/300 kPa (2,500 psig). However, some vendors may supply high-purity gas blends in high-pressure cylinders at 18,000–19,000 kPa (2,600–2,730 psig). These cylinders are similar to but 10 cm taller than the standard, 142 cm tall 1A gas cylinders. Anyone planning to use or who receives such cylinders shall check the MAWP of all components (e.g., flexible metal hose, tubing) and review the system with the laboratory supervisor.

Employees must become familiar with the features and operating characteristics of cylinder valves before making any connections. Cylinders should be delivered with the valve-protection caps attached **hand tight** only. Any cylin ders received for which the cap or valves cannot be opened by hand effort only are to be returned for replacement. Caps shall be kept on cylinders whenever handled or moved.

Cylinders, except for those handling toxic or radioactive gases, have safety devices to avoid overpressure and possible rupture. Once open, these safety devices cannot be closed. To prevent accidental release of cylinder contents due to excessive temperature and pressure, gas cylinders generally shall not be exposed to temperatures exceeding 55°C (130°F). However, for a new chemical or supplier, the temperature limit of the fusible link should be determined from the supplier's literature. Adjustments to cylinder safety devices shall not be attempted.

It is important to obtain and review the physical and chemical properties of mixed compressed gases before use. Such mixtures may require use of special regulators. Mixed gases must be obtained from a vendor, preparation of mixed gases on site is prohibited, except for recharging self-contained breathing apparatus by authorized personnel.

Users of liquid nitrogen cylinders shall follow proper procedures, as provided by the supplier, which consider the high pressure and low temperature of the liquid nitrogen. Special care must be taken when using cylinders containing corrosive gases, acetylene, hydrogen sulfide, fluorine, and certain doping gases.

The depletion of the contents of cylinders containing gases under pressure (not liquefied gases) is proportional to the reduction of the absolute pressure in the cylinder. The contents of cylinders containing liquefied gases can be determined only by weighing and subtracting the tare weight stamped on the cylinder.

Purchasing

Cylinders are rented from the suppliers. Excess cylinders that are empty or no longer useful can be hazardous in the work area. Consequently, all compressed-gas cylinders no longer in use should be removed promptly from the area. All cylinders are to be marked legibly with either the chemical or trade name of the contents. Color coding is not acceptable.

Storage and Handling

Lecture Bottles

Lecture bottles are a safe means of transporting and using small quantities of gas. Chlorine and hydrogen chloride (HCl) are the most common gases available in lecture bottles. However, the valves and connections are different from those on compressed-gas cylinders.

Specific precautions for the use of lecture bottles follow:

- Only use the gas if the label and tag confirm its identity.
- Label the equipment with the name of the gas being used. This is to avoid cross-contamination. Connections on lecture bottles are the same for all non-corrosive gases.
- Use appropriate tubing for connections. Check for corrosion resistance and pressure rating. Monel tubing is acceptable for chlorine and HCl.
- Pressure-test all lines.
- Secure the bottle in a stand with the valve at the top.
- Install a control valve or regulator before opening the valve on the bottle.
- Close the valve on the bottle when not in use.
- Remove the regulator or control valve before shipping a lecture bottle.
- Determine compatibility before changing gases in a system.

Cryogenic Fluids

The following precautions shall be taken when using or planning to use a cryogenic fluid (cryogen):

- Become familiar with all of the pertinent properties of the material and with accidents that might occur with the particular fluid.
- Provide for sufficient insulation using non-combustible materials.
- Use the proper protective equipment, including insulated gloves, face shields, and body covering.
- Provide for adequate ventilation.
- Use equipment specifically designed for cryogenic service.
- Avoid blocked-in liquid-filled piping or vessels that might rupture at the pressures generated by liquid expansion due to warming.
- Use a closed, pressurized system at above atmospheric pressure to avoid contamination by air.
- Develop precise operating procedures from which deviations are permitted only after approval by a knowledgeable authority.

Cryogens are used and transported in well-insulated containers, commonly the double-walled, evacuated Dewar flask. Small Dewars usually are made of glass; large ones, of metal. Glass Dewars should be wrapped with tape to prevent injury in event of an implosion.

Liquid nitrogen is usually stored in large metal Dewar flasks, although some smaller glass Dewars are used. More volatile materials (e.g., helium) require more insulation to prevent evaporation. Containers shall be kept vertical and protected from mechanical or thermal shocks. They shall be moved using a dolly, not tipped or rolled. OSHA 1910.104 applies to liquid oxygen.

If splashed with a cryogenic fluid, immediately flood the skin or clothing with water from a safety shower or drench hose to prevent freezing the skin.

HOUSE GASES AND VACUUM

House gases available through site utilities piping may include compressed air, nitrogen, argon, helium, low-pressure steam, and hydrogen. To prevent cross-contamination, all connections for house gases must be kept separate. Connections to the low-pressure steam utility shall be insulated to prevent burns.

Use pressure for house nitrogen or air should be no more than 35 kPa (5 psig) unless firmly connected to the lab outlet by screwed or quick-disconnect fittings. Air used for cleaning shall be at no more than 207 kPa (30 psig). House vacuum systems should not be used to collect spills or dispose of gases or vapors. Consequently, a trap should be installed between the vacuum system and any process user. The refrigerant for cold traps shall be chosen to avoid condensation of oxygen. Reduce pressure slowly.

Direct connections to laboratory or pilot units are prohibited. Such connections could be causes of contamination of the utility or overstressing of components operating at lower pressures. Connection with safety valves to block, bleed, and prevent backflow, must be used for utility tie-ins. Air or nitrogen hoses should not be used for cleaning skin unless specific approval is obtained by a laboratory supervisor, who shall identify specific OSHA requirements to be followed.

TOXIC GASES

To avoid health hazards, the following information should be known about each chemical handled, including possible reaction products:

- **Toxicity**—Is the expected level above the health standard, expressed as maximum allowable concentration (Threshold Limit Value [TLV], Permissible Exposure Limit [PEL], or Occupational Exposure Limit [OEL])? The TLV, PEL, or OEL is the maximum concentration to which an individual may be exposed continuously in the working atmosphere for an eight-hour period, repeatedly, without adverse effects.
- **Nature of Effects**—Is there a warning before the danger point is approached? Does intermittent exposure have a cumulative effect? Examples: Ammonia is immediately irritating at low concentrations and, therefore, provides a warning of its presence. Benzene has a rather pleasant odor and gives no indication of its harmful effects.
- **Physical Properties**—Can the expected concentration in the workplace be predicted?
- **Route of Entry**—Is inhalation the only route of concern? Should skin contact and the possibility of ingestion be considered?
- **Precautions**—Should special safety equipment or ventilation be provided?
- **First Aid**—What treatment can be provided in event of exposure?
- **Decontamination**—What should be done in event of a spill? Unknown materials should be treated as hazardous.

The following added precautions should be taken when using or planning to use a toxic gas:

- Use the proper protective equipment, including gloves, face shields, and body covering.
- Provide for adequate ventilation.
- Develop precise operating procedures from which deviations are permitted only after approval by a knowledgeable authority.

Toxic gases (Class 1) are to be used in a hood. They shall not be piped outside the hood or inside the laboratory.

Ventilation shall be monitored by an alarm system that signals any concentration above the TLV, PEL, or EOL of the toxic gas. A trap, check valve, or knockout vessel is to be used if a corrosive or reactive gas is used in a process unit or under the surface of a liquid. Lines handling toxic and flammable gases shall be protected by flow-limiting devices. The device should limit the flow to 1.5 times the normal flow rate.

REACTIVE CHEMICALS

Peroxides are hazardous due to their reactivity. However, the reactivity varies widely with the chemical nature of the compound. These products become less reactive on aging, and some can form unstable degradation products. Know the specific type of peroxide to be used and be thoroughly familiar with the practices for handling as recommended by the manufacturer and other sources.

Organic peroxides act as catalysts. Organic peroxides may be shipped as solids, liquids, or pastes. The most reactive materials are shipped under refrigeration. The pastes are mixtures of peroxides and inert diluents prepared to improve safety characteristics and facilitate dispersion into polymer systems. Liquid peroxides are either pure liquids diluted with a plasticizer or a solvent, or solids dissolved in a plasticizer.

Peroxides may be kept in explosion-proof refrigerators, in ventilated sprinkler-equipped metal cabinets, or on open shelves arranged to be readily wetted by discharge from the room sprinklers. However, some peroxides should not be stored below certain minimum temperatures to prevent crystallization of more hazardous solids.

Some chemicals form peroxides by reaction with oxygen during storage at normal conditions. The hazard is generally less for heavier-molecular-weight compounds.

Solvents or monomers susceptible to autoxidation should be stored in air-free, air-tight amber containers, preferably in the dark. If compatible with the planned use, an inhibitor (e.g., a substituted phenol or aromatic amine) can effectively stop peroxide formation in storage. Distillation and evaporation of peroxide-forming compounds (e.g., unsaturated hydrocarbons, isopropyl ether, easily polymerizable materials) can concentrate the peroxide in the residue. Consequently, these materials shall be tested for peroxides before distillation, such as by using the iodide or ferrous thiocyanate test or by physical means. If peroxide content is found, the solution shall not be distilled unless effectively treated with a reducing agent and/or strong caustic.

Addition of a stable, high-boiling hydrocarbon such as mineral oil can prevent distillation to dryness and limit peroxide concentration in the residue. Air is to be excluded from operations using peroxide-forming chemicals by purging with an inert gas. Inert flushing shall be maintained during cooldown of residual material until ambient temperature is reached. The residue then shall be tested to determine the peroxide content.

Following are requirements for inventorying peroxides and peroxide-forming chemicals:

- All peroxides and peroxide-forming chemicals should be inventoried semi-annually.
- Containers shall be dated when first opened. The discard date and owner's name should be listed.

- A new peroxide container found to have a broken seal or other indication of contact with air should be discarded immediately, following proper procedures, because the extent of exposure to air is unknown.
- These materials should be retained for a maximum period of **six months,** depending on the hazard potential.

Following are some general guidelines for the destruction of peroxides:

- In general, contaminated or overage organic peroxides should be disposed of as hazardous waste.
- Some solid peroxides, chiefly diacyl peroxides, can be destroyed safely by treatment with about ten times their weight of cold 10% caustic solution. Add the peroxide gradually in small portions with sufficient agitation to prevent settling or lump formation. The reaction is mildly exothermic. Dilute the mixture with water as required if it thickens to the point that stirring is difficult. Decomposition should be complete in 24 hours, when no further bubble evolution is observed, or when an acidified sample does not liberate iodide from an acetone solution of sodium iodide.
- An organic peroxide should never be flushed down a laboratory drain.

Pyrophoric Materials

Pyrhophoric materials spontaneously ignite when exposed to air. If diluted, they still are considered pyrophoric if they ignite in air at a concentration of more than 20%. Special training should be provided to anyone who uses pyrophoric materials.

Metal Alkyls

Hazards

The most important hazard of alkyls is that they are generally pyrophoric. Therefore, diluted alkyls should be purchased and used unless no alternative exists to the neat alkyl. Small quantities or solutions of less than about 25% concentration will not burn but will generate smoke and considerable heat.

Organometallic compounds such as lead beryllium and boron alkyls are toxic. Avoid both vapor inhalation and skin contact.

Aluminum alkyls with chain lengths greater than C4 will not burn unless ignited. However, smoking tendency decreases with chain length until at the C_{12} to C_{14} level, where no smoking occurs although heat generation still is evident.

Nickel is a catalyst for alkyl side reactions and decomposition; cobalt and platinum also, to a lesser extent. Therefore, nickel and its alloys should be used only when such side reactions are desirable. Equipment previously used for oxo reactions should be avoided because of possible residual cobalt.

Only qualified personnel shall remove alkyl from its container. Procedures for boron alkyls are different from those for aluminum alkyls. All spills shall be wiped up with absorbent paper before leaving the dry box. The paper should then be placed in a beaker, and heptane or Varsol added before removing the beaker from the dry box for weathering.

First Aid

First aid for contact with alkyls and their solutions is the same as for other corrosive chemicals. Alkyls react with oxygen both in air and body moisture. Immediate flushing with a large volume of water will neutralize residual alkyl, carry away heat, and cool burned tissue. Promptly remove contaminated clothing, including shoes, then shower for five minutes or longer. Eyes, if affected, shall be flushed gently with copious quantities of water for 15 minutes with eyelids held open. After showering/flushing, the exposed person shall go or be taken for medical attention immediately.

Fires with Alkyls

An aluminized protective suit (with hood and gloves) and SCBA should be available in a nearby emergency closet if needed for safe and effective fire fighting.

Fire control with alkyls has two aspects:

- Control and extinguishment of the burning material.
- Control of flammable vapors from the dilute, non-pyrophoric solutions.

Effective control involves use of materials that both blanket and adsorb the liquid. An adequate quantity of an appropriate adsorbent shall be kept readily available—enough to re-extinguish the fire at least two times.

Catalysts, Initiators, Promoters, and Accelerators

The degree of hazard of these materials varies widely, depending on their properties. For this reason, it is essential to know the properties of the material to be used and to follow the recommendations of the manufacturer with regard to storage and handling. Some general precautions for such reactive chemicals are as follows:

- Keep away from all sources of heat or ignition such as open flames, electrical devices, sunlight, naked lights, sparks, and heating equipment. Enforce "no smoking" regulations strictly.
- Keep containers loosely covered at all times to prevent contamination with dust, metals, or other chemicals.
- Store in the same container in which the material is shipped. Do not repackage in odd quantities or different containers.

- Do not mix initiators, catalysts, accelerators, and/or promoters. Violent decomposition or explosion may result.
- Avoid contact of the eyes or skin with these materials.
- Limit the quantities of any catalyst, accelerator, or promoter in a work area to that required for immediate needs when supplied in the original container.
- Use a suitable barrier cream or put on appropriate gloves (e.g., vinyl-impregnated or rubber) before starting work.

Free-radical sources cause mild irritation of the eyes and may be slight skin irritants. Avoid contact with eyes, skin, and clothing. Many have slight to low oral toxicities, but exceptions exist. Therefore, users shall become familiar with manufacturer's health hazard information and follow the precautions given.

Free-radical sources are regulated as hazardous materials. Avoid dusty conditions that can lead to dust explosions because of ignition by static electricity. Use grounded, metal equipment. To avoid buildup of static charges when adding these materials to a flammable solvent, they should first be transferred to an electrically grounded, open metal container or hopper.

Although some of these materials only produce nuisance dusts, some dusts present significant health hazards. Where nuisance dust exposure exceeds 10 mg/m^3 total dust or 5 mg/m^3 respirable dust, wear a dust mask with suitable filter. Consult the American Conference of Governmental Industrial Hygienists (ACGIH) for specific threshold limit values (TLVs).

Do not store these materials in a tightly closed container such as a reactor, a sealed package other than the original vented shipping container, or a sealed storage area. Do not use screw tops. Thermal decomposition of certain materials (e.g., azo compounds) releases gas or vapor that can reach a hazardous pressure if confined. Some free-radical sources have a self-accelerating decomposition temperature, above which thermal decomposition is very rapid.

Do not store in glass. Keep these materials dry. Do not exceed the manufacturer's recommended storage temperature. Lower storage temperatures reduce the tendency of these materials to cake on prolonged storage. Free-radical sources are incompatible with strong oxidizers such as nitric acid.

Do not vacuum spilled material. Avoid friction and heat. If there is a floor drain, flush a contaminated floor with water to the sewer. If free-radical sources are involved in a fire, toxic fumes may be present. Take precautions to avoid exposure to smoke and fumes; self-contained breathing apparatus (SCBA) should be worn when fighting a free-radical fire in any confined area where the fumes cannot be avoided. Do not approach a large fire because there is a risk of rapid decomposition and rupture of a container.

Radiation-Producing Materials or Equipment

Ionizing radiation produced by radioactive materials, X-ray machines, halogen analyzers, light microscopes, and other devices can damage body tissues and organs. Special equipment and procedures are required for the handling and disposal of radioactive materials.

Persons working with or near a source of ionizing radiation must wear a dosimeter badge, and for some activities, a dosimeter ring. Conduct readings on such monitors at least once a month.

Both direct and reflected infrared (IR) and ultraviolet (UV) radiation can cause eye or skin damage. IR light sources include welding, heat lamps, and glassblowing operations. Some UV light sources are welding, the sun, mercury vapor lamps, and ultraviolet lamps. Protective measures include shielding the eyes from the source and reflections. Opaque shields are best for protection against IR and UV light; welder's goggles and shields also are effective. Common lab shielding and safety spectacles provide some UV protection.

Microwaves are electromagnetic radiation ranging from the shortwave radio region to extremely high frequencies. They are harmful because they can cause localized heating of tissues, and a person may not be aware of the exposure. In addition, operation of certain cardiac pacemakers can be inhibited. Use of this type of equipment also can involve other hazards—high-voltage transformers, sealing of submersible UV cells for chemical reactions, etc. The most well-known source of microwaves is the common microwave oven. In addition, nuclear magnetic resonance (NMR) analyzers are a source of magnetically induced radiation. The warning sign on the entrance door should be observed.

Water- and Acid-Sensitive Chemicals

Some materials react with water or water vapor to form flammable or explosive products. Metallic potassium, sodium, calcium, lithium, rubidium, cesium, and their alloys react with water to produce hydrogen. Nitrides, sulfides, selenides, arsenides, phosphides, and metal alkyls react with moisture to produce volatile flammables, spontaneous combustibles, or explosive hydrides. Acid anhydrides, concentrated acids, and alkalis evolve heat when in contact with water.

Contact of water- and acid-sensitive materials with water shall be avoided except under controlled conditions. Such materials cannot be stored in an area protected with an automatic sprinkler system. Water should not be used to fight fires where such materials are stored, unless the hazards are understood and the fire is beyond control by other means. Signs to warn firefighters shall be posted on storage facilities for such materials.

Many materials that are sensitive to water also react with acids or acid mists to generate heat, hydrogen, or other flammable or explosive gases. Contact of acids with concentrated alkalis, arsenic, selenium, tellurium cyanides, and structural alloys also is hazardous. Concentrated acids or alkalis must be stored in separate polyethylene, polypropylene, or stainless-steel trays. Storage is to be below head level to minimize the risk of eye and face contact. Organic materials shall not be stored in the same tray as strong oxidizers or any highly corrosive material.

LASERS

The greatest hazard from lasers is their ability to cause permanent damage to the retina of the eye. Preventive programs and measures are the only effective means to avoid eye exposure to direct or reflected laser beams. entrance to laser labs should be controlled. Persons in the lab shall wear appropriate special eyewear.

STORAGE OF LIQUIDS

Drums are to be stored in compliance with Occupational Safety and Health Administration (OSHA) and National Fire Protection Association (NFPA) requirements. Aisles are to be kept clear of flammables and combustibles; the minimum permissible aisle width is 5 ft. No drums are to be stored in laboratories.

Employees who routinely handle drums shall receive training. Some important precautions are:

- Avoid placing fingers where they may be pinched. Keep fingers, legs, and feet clear. Wear gloves and safety shoes.
- Do not upend heavy barrels and drums without special training.
- Be sure all drums are labeled with the name of the contents and a description of the hazard (e.g., flammable).
- Inspections shall be made daily for leaking drums. The contents of damaged drums should be transferred to new drums, using a funnel and wearing protective clothing. If judged necessary, the leaker shall be placed in an oversize recovery drum.

Drums equipped with dispensing valves and containing flammable liquids require locking pins installed in the valves. When transported by an industrial truck, the drum position is to be attached to prevent accidental valve opening. Drum racks with flammable liquids shall be grounded. The drum and receiving container are to be electrically bonded to the rack.

Safety cans or other portable containers can be used for flammable liquids with flash points below 26.7°C (80°F) received in stock drums. The flash point is the lowest temperature at which a flammable liquid gives off enough vapors to form a flammable mixture with air near the surface of the liquid or within its container and will ignite under standard test conditions.

Examples of low flash compounds typically used are acetone, methyl ethyl ketone, heptane, hexane, and sopropyl alcohol. Although chemicals such as ethylene and propylene have extremely low flash points, safety cans are unsuitable for them because of their high vapor pressures.

Solvent storage cabinets for chemicals are designed and constructed to protect the contents from external fires. Each cabinet is designed to maintain the internal temperature below 3°C (325°F) in a 10-minute fire test. Cabinets with vent openings cannot meet the temperature limit in the fire test. NFPA-30 (1987) states that storage cabinets are not to be vented for fire-protection purposes. However, venting of cabinets is sometimes required for health and safety reasons so the concentration of toxic or noxious vapor does not build up. NFPA recognizes such circumstances, and permits the use of special cabinets designed both for fire resistance and to limit internal vapor concentration.

Flammable chemicals that are neither odorous nor toxic should be stored in non-vented cabinets. Also, shipped materials sealed such that vapors could not escape if the material spilled inside the container can be stored in unvented facilities. Most solids also can be stored without venting. This can be checked using vapor pressure data, if available. It is not good practice to store chemicals on the top of a solvent storage cabinet.

Tires and other materials may be stored if first properly tagged and in a suitable container, if applicable. The tag shall include the name of the material, the owner, the discard date, and any special handling information. **No combustibles or flammables** shall be stored with such materials. Conduct an inventory at least annually. When inventorying material stored in a refrigerator, clean the refrigerator. Remove any material that is no longer of use.

CHEMICAL SAMPLING

Following are some major principles of safe sampling. Specific sampling problems or concerns should be discussed with a supervisor before taking action.

- There should be adequate ventilation where sampling is done.
- Shielding and personal protective equipment should be used where such protection is judged warranted. A face shield is preferred to splash goggles when sampling a system under pressure.
- Withdrawal into a closed system is preferred. Guard against nearby ignition sources, including hot surfaces.
- The equipment should be electrically bonded and grounded to eliminate static accumulation and discharge.
- Samples containing more than 25% gases (ethane or lighter) should not be collected in containers intended for liquids.
- If appreciable cooling or chilling can occur during sampling, use a stainless steel container.
- Whenever possible, hot materials should be cooled before sampling. Warm or corrosive samples should not be placed in non-tempered glassware.
- When very hot material must be discharged through a valve, open and close the valve rapidly so only a small amount flows each time. If the valve is left open, the temperature increase may cause expansion and sticking.

All material submitted for shipment must be accompanied by an MSDS or other indication of hazardous properties for each material to be shipped.

SAFE WASTE DISPOSAL

The individual in charge of a laboratory, pilot plant, or other work area is responsible for ensuring that wastes generated in that area are disposed of in a safe, environmentally acceptable, and legally responsible manner. All supervisors, technicians, researchers, and engineers should become familiar with and follow proper waste-handling and disposal practices.

IDENTIFICATION OF WASTE MATERIALS

No material should be discarded unless its nature is known and it is properly identified if classified as hazardous waste. Individuals responsible for a waste product must identify and determine the physical and chemical characteristics of the material being discarded. If this information is not available as a result of normal work activities, the material must be analyzed by a laboratory.

To provide guidance to the analytical laboratory personnel, every effect should be made to learn something about the unknown material. Possible leads include previous workers in the area, protections in the area, lab notebook number references, etc. Prior to taking a sample, check the condition of the container. If there is no indication that the material has corroded the container or decomposed such as to be hazardous, the container should be placed in a hood, carefully opened, and a sample taken. If the condition of the container is suspicious, consult the lab supervisor.

Together with all the information that can be obtained about the material, the sample should be delivered for analysis. Based on the analytical results, the container should be tagged properly and sent for disposal.

Identification of unknown chemicals can be costly and time consuming. A waste material is identified and characterized by its chemical name, not its commercial name. Where necessary for a mixture of several materials, the major component and all harmful components must be identified. Some sources of information about chemicals are:

- The manufacturer or distributor.
- Materials Safety Data Sheets (MSDS).
- Chemical handbooks.
- Chemical supply company catalogs.
- Industry association manuals.

REGULATED/UNREGULATED WASTE

Regulated waste is material listed on EPA, RCRA, or state hazardous materials inventory or has characteristics of ignitability, corrosivity, reactivity, or extraction procedure toxicity (EP-toxicity). Unregulated waste is material not listed on EPA, RCRA, or state hazardous materials inventory or does not have characteristics of ignitability, corrosivity, reactivity, EP-toxicity.

Unlisted but regulated waste is material not listed on EPA, RCRA, or state hazardous materials inventory which, however, has characteristics of ignitability, corrosivity, reactivity, or EP-toxicity. This includes material of an experimental nature which, in the judgment of the employee responsible, should be sent through regulated disposal channels. Unknown material, as well as uncategorized material, is to be analyzed to determine its status.

For each identified waste, the individual responsible must classify the material as hazardous waste or a non-hazardous waste. In general, this is decided based on the properties of the material.

If a material is known, its relevant properties should be listed. If a material is unknown, it shall be identified before requesting disposal.

Certain chemicals shall be given special handling because of their specific properties and characteristics. Disposal of highly reactive or explosive materials presents environmental problems and is, consequently, costly. Therefore, all laboratory and pilot-plant personnel shall minimize or, if possible, eliminate the purchase, use, and generation of such materials. A material legally must be considered as explosive or highly reactive if it has one or more of the following characteristics:

- It is normally unstable and readily undergoes violent change without without detonating (e.g., pyrophoric or peroxide-producing materials).
- It reacts violently with water.
- It forms potentially explosive mixtures with water.
- When mixed with water, it generates toxic gases, vapors, or fumes in a quantity sufficient to present a danger to human health or the environment.
- It is a cyanide- or sulfide-bearing material that, when exposed to pH conditions between 2 and 12.5, can generate toxic gases, vapors, or fumes in a quantity sufficient to present a danger to human health or the environment.
- It is capable of detonation or explosive reaction if subjected to a strong initiating source or if heated under confinement.
- It is readily capable of detonation, explosive decomposition, or reaction at standard temperature and pressure.
- It is a forbidden explosive (see 40 CFR 261).

ELECTRICITY

Electrical shock has specific effects on humans that are complicated and difficult to predict. The strength of the electrical current is the determining factor. Low voltages can be more dangerous than high voltages, but both can kill.

Electrically bonded equipment, piping, containers, etc. have the same electrical charge. When also grounded, the bonded equipment has no charge. Polarized grounded receptacles are used to assure that a continuous ground is available and that the "hot" lead is directed to the correct part of the electrical device. Therefore, pigtail adapters are not permitted.

An explosion-proof enclosure can withstand an explosion of a gas or vapor that may occur within it and can prevent ignition of flammable gas or vapor surrounding the enclosure by sources of ignition inside.

Electrically hazardous areas are categorized as Class I, Class II, or Class III, depending on the physical properties of the combustible substances that might be present. Class I locations are hazardous because flammable gases or vapors are or may be present. Class II locations are hazardous because combustible dust is or may be present. Class III locations are hazardous because combustible metal dusts are or may be present. NEC Articles 500-2 through 500-5 contain details of the pertinent requirements.

Within each location class, two divisions are recognized, depending on the likelihood that an ignitable atmosphere might be present. These divisions are as follows:

- Division I locations are likely to have the flammables or combustibles present under normal conditions. Such locations can be defined as those where the probability of flammable or combustible atmospheres occurring exceeds 1 in 10,000 hours.
- Division 2 locations are likely to have the flammables or combustibles present only under abnormal conditions such as failure of equipment. Such locations can be defined as those where the probability of flammable or combustible atmospheres occurring is estimated to be between 1 in 10,000 hours and 1 in 1,000,000 hours.

Well-ventilated areas that do not meet the above definitions are unclassified or non-hazardous where general-purpose equipment may be used.

Flammable and combustible substances are placed in one of seven groups (A through G) depending on their behavior in contact with a source of ignition. Unless otherwise specifically stated, chemical fume hoods shall be Class I Div. 2, with consideration that open transfers may take place. Equipment (e.g., vacuum pumps) on or within 2 ft of the floor shall be Class I Div. 2 because of the risk of the presence of heavy flammables. Dry boxes are nitrogen purged so, in principle, general-purpose equipment could be used in them. However, use of equipment classified as Div. 1 or 2 is preferred. General-purpose equipment can be used on benches as long as no flammable chemicals are handled within 3 ft.

Safe procedures for operation, cleaning, etc. should be followed rigorously. Minimize the amount of flammable solvents open to the atmosphere, use negative-pressure vent systems, and avoid unnecessary sources of ignition when flammables are present. Chemical pilot plant areas are normally Class I Div. 2. However, where flammable hydrocarbon emissions may occur, Div. 1

equipment shall be used. In the latter situations, the equipment is situated in a hood or equivalent.

Rotating or reciprocating machinery can be hazardous in several ways. Safeguards are provided to prevent injury from exposure to contact at or with:

- Point of operation (e.g., grinding heel, drill bit).
- Exposed moving parts.
- Power transmission apparatus.
- Mechanical and electrical failures.
- Contact with metal chips, hot metal surfaces, fumes, and splashing of chemicals or lubricants.
- Human failure.

Guards and shields are not to be removed without explicit permission.

When used for flammable materials, ovens and dryers should be equipped with electrical elating elements, controls, motors, and other accessory equipment that will not provide a source of ignition. Mercury glass thermometers should not be used for temperature indication, unless the thermometer is protected from breakage due to mechanical shock. Be sure mercury glass thermometers have an adequate range.

Asbestos is a health hazard; be aware of and report flaking. Asbestos was originally used as insulation in many buildings. Some asbestos may remain in hoods (transite is 35% asbestos) but is safe as long as it remains encapsulated. Drilling into hoods shall not be done because it could release asbestos dust. Special precautions and permits are required for removal of insulation, materials of construction, lab hoods, etc. that may contain asbestos.

CHAPTER 2

PERCEPTION OF RISK: WHY THE DISPARITY?

MARK A. HAMILTON

U.S. Public Health Service
Indian Health Service
Pinetop, AZ 85935, USA

CONTENTS

INTRODUCTION

In a democracy, we have the ability to be involved in the process of making decisions that affect us as a society [1, 2]. These decisions are based in part on how we perceive things to be in the past, present, and future. Our perceptions of risk, then, are important to decisions regarding policy, regulations, recycling stations, waste repositories, and other environmental management issues.

Disclaimer: The views represented in this chapter are those of the author and do not necessarily represent those of the U.S. Public Health Service.

For discussion purposes, three interactive entities can be identified that are often in conflict regarding environmental issues: the lay public, government, and industry. At first glance, the divisions among these three entities appear trichotomous. However, in reality it is very fractionalized. The lay public includes very broad groups, including individual citizenry, community groups (e.g., sporting, charitable, social); formal special-interest organizations (e.g., Sierra Club, National Rifle Association); professional organizations (e.g., American Medical Association, National Environmental Health Association); and educational institutions, to name only a few [3]. Government and industry are equally as fractionalized, represented by other groups such as lawyers, environmental managers, lobbyists, scientists, senators, and congressmen. However, for convenience of discussion, a trichotomous distinction will be made.

These groups and their membership have different interests and values that shape the way issues or risks are perceived. "Values" refers to what is valued by individuals, social groups, government, or industry. Values may have a monetary value, such as marketable goods, or a non-economic value, such as a scenic view [1, 2]. "Interests" may relate to concepts such as control, power, economics, or pleasure. For example, the issues that surround the contamination of a groundwater aquifer by chemicals from a chemical manufacturing plant may go beyond health effects from drinking water. Other issues, concerns, or risks may include property devaluation, the effects to home gardens, health of the household pet, moral obligations for contamination cleanup, safe drinking water as a personal right, or feelings of fear, lack of control, or anger. Or, concerns that the manufacturing plant may be closed for violations, even though it is a major employer in the community. Which becomes the bigger issue, health or loss of employment?

This chapter will discuss how risks are perceived by the lay public, government, and industry by discussing concepts of acceptable risk, scientific risk assessment, and the methods used by the lay public to assess risks. The implications these issues have for environmental management, and strategies that can be used in planning for positive interaction and informed decision making, also will be presented.

The focus of this discussion is on how the lay public views risks. This information is discussed in a generic sense, often referring to "technologies." Bear in mind that technology can refer to a chemical plant, overland transport of hazardous waste, landfills, nuclear energy facilities, coal-fired power plants, or other technological processes that effect health or the environment. Further, this also applies to the specific agents that pose a risk, such as radioactive particles, dioxin, tetrachloroethylene, sulfur dioxide, and a host of others.

ACCEPTABLE RISK: THERE IS LITTLE AGREEMENT

The term "acceptable risk" has been used to characterize those risks society deems acceptable [4]. It has been noted, however, that people do not accept risks, they tolerate risks [5]. Furthermore, people do not choose between risks, they choose between options, each with some level of risk. The selection of a particular option is based on the associated consequences, benefits, facts, and values [6]. In environmental management, the risks and benefits of technology are constantly weighed.

However, the lay public, government, and industry do not agree on the concept of acceptable risk—what degree of a given risk is acceptable, or which risks are more important than others. The three segments of society assess the benefits and risks of technology differently based on their interests and values. Although we are healthier than ever before based on a longer life expectancy, the lay public displays an ever-increasing concern regarding the risks of technology [7]. The lay public argues for a "zero-risk society." Government and industry fear that the concept of zero risk threatens the political and economic foundation of our country [7]. The government may determine that some risks are *de minimis,* or small enough to be ignored [8]. In occupational settings, a risk of death, injury, or disease of 1 in 10,000 may be acceptable given the benefits. On the other hand, a risk of 1 in 1,000,000 may be acceptable in a community setting. Furthermore, risks acceptable today may be unacceptable tomorrow, as evidenced by increasingly stringent environmental regulations.

For industry, the choice between risks and benefits is fundamentally an economic one. The goals for industry are profit, market shares, sales, and influence [9]. Industry will accept some degree of risk if it is profitable. In terms of environmental practices, industry remains relatively secretive. Petulla refers to industry as a mysterious "black box" based on a once-popular toy—a black box that when turned on, extended an arm from its center and turned itself off [10]. With a desire to elicit information from the industrial sector regarding environmental management practices, Petulla asked 60 environmental managers, "Is there some structural obstacle in industry that prevents access to any kind of information that might be beneficial to society?" The explanations offered by industry for maintaining a mysterious posture, in order of importance, were:

1. Risk of bad press and the desire to protect corporate image,
2. The view that anyone not affiliated with the corporation is an adversary,
3. Lack of concern for the environment, and
4. Power.

Risk perception is central to the environmental management issues that divide these three groups. There is a seemingly endless list of examples that illustrate disagreements about technological risk. For example, the experts and lay public disagree about the risks and benefits of nuclear energy and the hazards associated with nuclear waste disposal in Yucca Mountain, Nevada. The Department of Energy sees a nuclear waste repository located in Yucca Mountain as presenting little risk, while the lay public associated it with ideas of "danger," "death," and "pollution" [11, 12]. Disagreements also arise regarding the health risks associated with hazardous waste and radon gas [13, 14, 15].

One of the frequently cited examples of the disagreements between the governmental experts and the lay public is the EPA report "Unfinished Business: A Comparative Assessment of Environmental Problems" [13, 16, 17]. The report lists the environmental concerns most pressing according to scientists and managers at EPA. In 1989, the EPA administrator asked the Science Advisory Board (SAB) to review and evaluate the "Unfinished Business" report. The SAB formed a Relative Risk Reduction Strategy Committee that identified the major environmental risks listed in Table 1. Generally, the SAB report agreed with "Unfinished Business," but added habitat destruction and species extinction. Two of the guiding principles in the SAB's conclusions were that quality of life, human health and the economy are adversely affected by the destruction of natural

Table 1
Risks Identified by the Science Advisory Board

Risks to the Natural Ecology and to Human Welfare
Relatively High Risks
• Habitat Alteration and Destruction
• Species Extinction and Loss of Biological Diversity
• Stratospheric Ozone Depletion
• Global Climate Change
Relatively Medium Risks
• Herbicides/Pesticides
• Toxics, Nutrients, BOD, Turbidity in Surface Water
• Acid Deposition
• Airborne Toxics
Relatively Low Risks
• Oils Spills
• Ground Water Pollution
• Radionuclides
• Acid Runoff to Surface Waters
• Thermal Pollution
Risks to Human Health
Relatively High Risks
• Ambient Air Pollutants
• Worker Exposure to Chemicals in Industry and Agriculture
• Indoor Air Pollution
• Pollutants in Drinking Water

Source: R. Loehr, EPA Journal 17(2)*: 10–11 (1991)*

ecosystems. Secondly, the temporal and spatial dimensions of environmental problems must be considered. Some environmental problems may take a long time to form and rectify, and/or may affect an entire population (e.g., ozone depletion) [16]. When the expert rankings were compared with a 1990 Roper poll (see Table 2) it became clear that the experts and the lay public were not in agreement in terms of which risks were most important [17]. The experts viewed habitat alteration and destruction, species extinction, ozone depletion, and global climate change as high risks to ecology and welfare. Ambient air pollution, worker exposure to chemicals, indoor air pollution, and pollution in drinking water were viewed as high risks to human health. The lay public, while recognizing these as risks, viewed hazardous waste, industrial waste, toxic chemicals,

Table 2
Risks Identified by the Public 1990 Roper Poll

Active Hazardous Waste Sites	67%
Abandoned Hazardous Waste Sites	65%
Water Pollution from Industrial Wastes	63%
Occupational Exposure to Toxic Chemicals	63%
Oil Spills	60%
Destruction to the Ozone Layer	60%
Nuclear Power Plant Accidents	60%
Industrial Accidents Releasing Pollutants	58%
Radiation from Radioactive Wastes	58%
Air Pollution from Factories	56%
Leaking Underground Storage Tanks	55%
Coastal Water Contamination	54%
Solid Waste and Litter	53%
Pesticide Risks to Farm Workers	52%
Water Pollution from Agricultural Runoff	51%
Water Pollution from Sewage Plants	50%
Air Pollution from Sewage Plants	50%
Pesticide Residues in Foods	59%
Greenhouse Effects	48%
Drinking Water Contamination	46%
Destruction of Wetlands	42%
Acid Rain	40%
Water Pollution from City Runoff	35%
Non-Hazardous Waste Sites	31%
Biotechnology	30%
Indoor Air Pollution	22%
Radiation from X-rays	21%
Radon in Homes	17%
Radiation from Microwave Ovens	13%

Source: L. Roberts, Science 249*: 616 (1990).*

oil spills, ozone depletion, and nuclear power as most risky. Pollution in drinking water and indoor air were ranked low on the public's list of concerns. Additionally, it was observed that EPA programs concentrated on the concerns of the public, rather than risks of most concern to the experts [16].

The major reason for the disparity in risk perceptions is that each group uses different methodologies for evaluating risks, and decisions about risk are based on different interests and values. Scientific risk assessment used by government and industry, is often in conflict with public risk assessment used by the lay public [1, 2]. Traditionally, scientific risk assessment has been called "objective" or noted as identifying the "real" risk, whereas public risk assessment has been labeled "subjective" or "irrational" [18]. The more recent arguments, however, note that scientific risk assessment is very subjective and that public risk assessment is anything but irrational [1, 2].

SCIENTIFIC RISK ASSESSMENT

Risk assessment is the process used to evaluate the nature and magnitude of risks to human health or the environment, through toxicity and exposure assessments [8, 19]. Risk analysis and risk management take risk assessment further by adding elements of risk communication and risk reduction [8]. The end result of risk assessment is the quantification of risk in numerical and/or economic terms. For example, the chance of cancer form exposure to chemical A may be 1 in 1,000,000. The experts who practice scientific risk assessment think of risk in these terms [20].

The term "regulatory science" has been used by Rushefsky to characterize risk assessment [21]. It is similar to "normal science" in that it has many characteristics and purposes. It seeks to understand and explain through laws and principles that can be tested experimentally. Regulatory science, however, has unique characteristics. It has direct public policy implications, a public agenda, operates under time pressures, and may have to provide only sufficient information as opposed to absolutes. Therefore, gaps of knowledge filled with assumptions are created.

The Science of Uncertainty

Although one of the objectives of scientific risk assessment is objectivity, it is also very subjective. The need for specific pieces of information and the desire to answer questions, gives it a subjective purpose [1, 6, 22]. The process involves many assumptions and interpretations, and has many uncertainties. There are numerous decision points addressed that are subjectively based on values [8]. For example, the choice of which dose-response model to use is based on a judgment of the degree of desired conservatism, with linear models providing the most conservative estimates [19, 21]. The decision of whether or not to use the maximum tolerated dose (MTD) to elicit a response in the interest of time and money, even though it may overwhelm an animal's defense mechanism and ability to fight disease, is based on a judgment. Or, the decision to weight some data more heavily than others when there is conflicting data is also a judgment based on values.

This is illustrated in a study conducted by Kraus et al. where members of the Society of Toxicology and the lay public were surveyed with regard to beliefs about chemical risks [22]. Among their many findings were that the public and experts see risk assessment differently, and there is disagreement within these groups. For example, both the experts and the public were split on the issue of whether or not animal exposure data is predictive of human health. The fact that the experts were split on this issue illustrates expert disagreement. Additionally, industrial toxicologists viewed chemical risks as generally less risky than their counterparts in academia and government [22]. This shows how interests and values may be commingled with science. Studies have shown that in the face of uncertainty and the formulation of assumptions, expert judgment can give way to the same tools employed by the lay public to conduct public risk assessments [20].

PUBLIC RISK ASSESSMENT

Public risk assessment is different from scientific risk assessment in that it goes beyond statistical probabilities. The lay public uses many different strategies to assess risks. Because different groups have different interests and values, risks are perceived differently. This can result in conflict, where trust and credibility become important aspects of risk perception.

Studies concerning risk perception fall into two basic approaches: revealed preferences and expressed preferences. Simplistically, the revealed-preferences approach makes assumptions about risk perceptions based on what people do. The expressed-preferences approach makes assumptions about risk perceptions based on what people say. The former is based on observation, while the latter is based on direct questioning [23, 24].

Dimensions of Risk

Searching for an approach to measure risks relative to benefits, Starr utilized a revealed-preference approach by comparing fatalities with social benefits, in dollars, for voluntary and involuntary activities. Starr argued that all human activity was either voluntary or involuntary, and that over time an "essentially optimum trade-off of values has been achieved" by society [25]. In other words, society, over time, has determined what is socially acceptable risk and what is not. Voluntary activities were, for a particular individual, selected based upon one's own value system. For example, one chooses to engage in sports that have some degree of risk from injury. In contrast, he also argued that involuntary risks were imposed and those making decisions regarding risk were different from those affected by the risk, noting that war was the most extreme example. Starr's work lead him to conclude that the public is more willing to accept voluntary risks than involuntary risks, the acceptability of risk is proportional to real or imagined benefits, and social acceptance is influenced by the awareness of the benefits [25]. Starr's work provided a foundation for risk perception research and provided a yardstick for making comparisons.

Fischhoff et al. extended Starr's original work, using an expressed-preference approach, by asking people directly what they prefer in terms of risk and benefits. In this study, participants were asked to judge risks and benefits for 30 activities and technologies. They concluded that participants did not view technological risk as being optimally managed by society. Rather, they viewed high risks in the presence of low benefits, and levels of safety for technological activities too risky [26]. Generally, the public views the level of technological safety imposed and practiced by the managers in government and industry as too risky. This was contrary to the assumptions made by Starr [25].

More recently, inspired by the work of Fischhoff et al., Slovic et al. have sought to develop a taxonomy of risk to understand risk perceptions, to predict perceptions and to facilitate informed policy development [26]. Using an expressed-preference approach, participants in three studies made judgments by rating risks based on risk characteristics, acceptability, and risk of dying using a seven-point scale. For example, a risk judged as most voluntary was ascribed a 1 rating, while a risk judged as most involuntary a 7 rating with varying degrees in between. The participants involved four groups: League of Women Voters and their spouses, college students, a community service organization, and a nationwide sample of risk assessment experts. One key finding of the study was that expert perception of risk correlated with annual fatality measurements [26]. In other words, activities with statistically high annual fatalities were perceived as most risky, and activities with low annual fatalities were perceived as least risky. However, the non-experts', or lay public's, judgments of risk did not correlate highly with annual fatality estimates. Their judgments were influenced by a variety of other factors, with two basic dimensions. Factor 1, referred to as dread risk, contained characteristics such as uncontrollability, dread, catastrophic consequences, inequitability, and involuntariness. Factor 2, referred to as unknown risk, contained characteristics that were not observable, unknown to those exposed, had delayed effects, and had risks unknown to science. Of the two factors, dread risk correlated most highly with risk perceptions by the lay public. The higher the score for the factor, the greater its perceived risk and desire for reduction and regulation. The perception of risk is represented by their plotting of hazard scores on a factor-space diagram (see Figure 1). In the upper-right quadrant are the most dreaded and unknown risks, such as nuclear reactor accidents, nuclear fallout, and radioactive waste. Those hazards in the

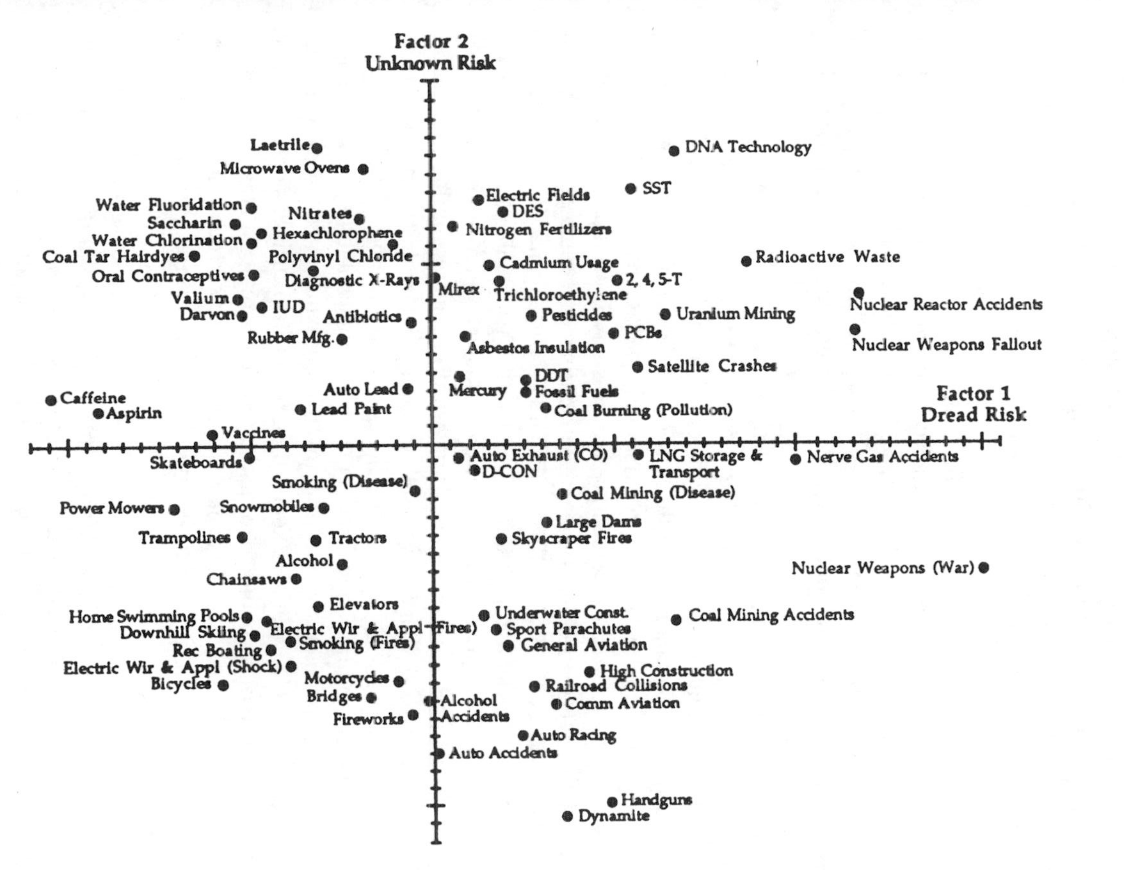

Figure 1. Factor-space diagram. Risks in the upper right quadrant have greater potential to spark higher-order effects. Used with permission of author. Used with permission of author.

most upper-right portion of the quadrant are most dreaded with the greatest perceived risk and desire to be reduced and regulated. Opposite, in the lower-left quadrant, are risks perceived as least risky. These include bicycles, home appliances, and recreational boating. Figure 2 illustrates the different dimensions of risk applied to each technology [20, 26, 27, 28].

The work of Slovic et al. shows that experts in risk assessment base their risk perceptions, and therefore their decisions, on statistical estimates. However, the lay public perceives risk in a broader sense, using a complex system of characteristics to judge hazards and form perceptions. The lay public uses dimensions such as controllability, familiarity, and voluntariness in forming their perceptions. Table 3 summarizes how these dimensions may affect the perceived riskiness of technologies. These dimensions are judged based on individual values and interests and are context specific. Therefore, the table provides only a general guide.

Slovic points out that the factor-space diagram has its limitations and is based on some assumptions, the most important of which is that risks are subjective. Additionally, it is context-specific in terms of the hazard and the questions asked [27, 28]. Others have remarked that the studies were carried out on non-randomly selected subjects and under laboratory conditions [18]. Gardner and Gould claim to have conducted the first expressed-preference study in the country using large samples of the general public to reveal perceptions of technological risks and benefits [18]. The study involved more than 1,000 participants from Connecticut and Arizona. Each was asked 300 questions during personal interviews, mostly regarding six technologies: automobile travel, nuclear weapons, nuclear power, handguns, and industrial chemicals. Item wordings were similar to those of Slovic et al., and participants also used a seven-point scale to rate each technology for regulatory strictness, risks, benefits, and fatalities. The study confirmed that the lay public judges risk, benefits, and acceptability in a complex, multidimensional manner. In addition to characteristics such as catastrophic potential, Gardner and Gould also added dimensions of pleasure, safety, security, and the importance of technology for satisfying human needs. Again, this illustrates the complexity of lay perceptions. Additionally, they found the public sees safety regulations as insufficiently stringent. This is contrary to Starr's "optimum balance" thinking [18].

In an experiment conducted by Fisher et al., participants were asked to list risks of personal concern in their own words [29]. This was different from the studies previously mentioned in that participants were not required to rate a specific set of risks or answer specific questions. The results revealed that accidents (auto, plane, etc.) were at the top of the list, followed by disease and crimes as risks of most concern. Environmental hazards and technological hazards were at the bottom of the list. Other studies have evaluated risk perceptions for a single hazard, with studies of nuclear risks and hazardous materials dominating the literature. Generally, these studies support the notion that the public views risk

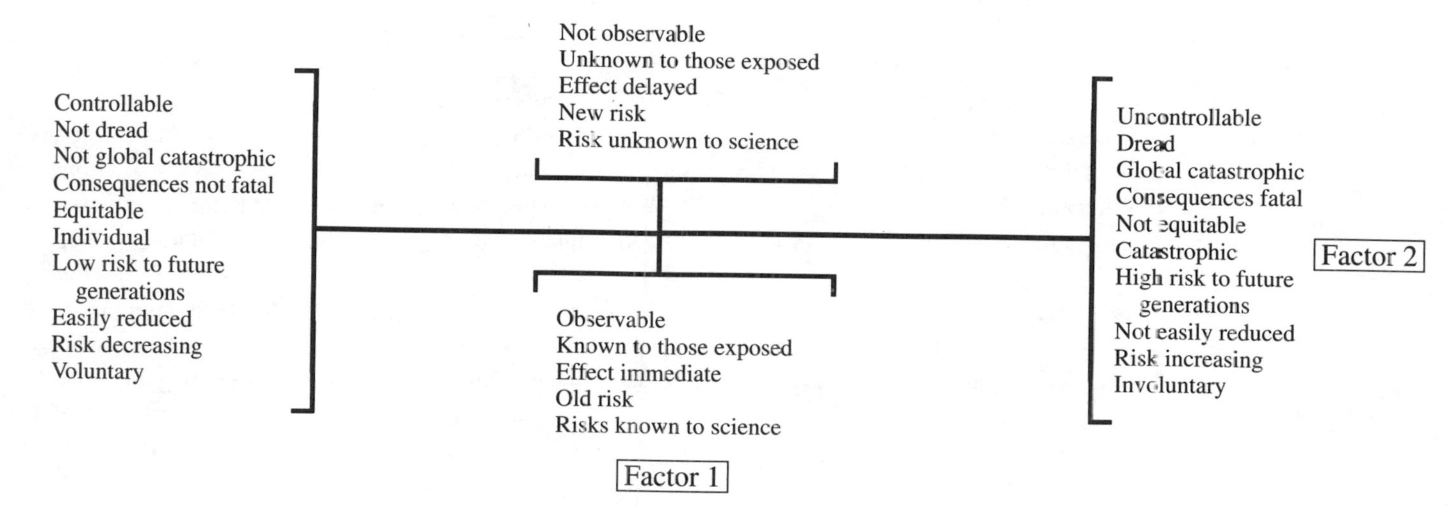

Figure 2. Dimensions of risk applied to technologies. Used with permission of author.

Table 3
Dimensions of Risk and Risk Perception

Dimension	Higher Perceived Risk	Lower Perceived Risk
Severity of Consequences	Large numbers of fatalities or injuries per event	Small number of fatalities or injuries per event
Probability of Occurrence	High probability of occurrence	Low probability of occurrence
Catastrophic Potential	Fatalities or injuries grouped in time or space	Fatalities or injuries distributed randomly in time and space
Reversibility	Irreversible effects	Reversible effects
Delayed Effects	Effects far removed in time	Effects immediate
Impact on Future Generations	Risk to future generations equal or greater	Risks borne by present generation
Impact on Children	Children especially at risk	Risk to adults rather than children
Victim Identity	Victims identifiable	Victims are statistical
Familiarity	Risks unfamiliar	Risks familiar
Understanding	Lack of understanding of risks or causes	Understanding of risks and causes
Scientific Uncertainty	Risks uncertain or unknown to science	Risks well known to science
Dread	Risks associated with fear, terror or anxiety	Risks not dreaded
Voluntariness	Involuntary risk or event	Participate in risk or event voluntarily
Controllability	No personal control over risk or event	Some personal control over risk or event
Clarity of Benefits	Benefits indirect or unclear	Clear benefits
Equity	No direct benefits to those at risk	Those who are at risk are also those who benefit
Trust	Lack of trust in institutions or individuals	Trust in institutions or individuals
Source	Risks from manmade sources	Risks form natural sources
Media Attention	Risks with imagery and repeated frequently	Risks with no imagery or receive little attention

Source: J. J. Cohrssen, V. T. Covello, Risk Analysis: A Guide to Principles and Methods for Analyzing Health and Environmental Risks, *Council on Environmental Quality, 1989*

in a complex fashion based on value judgments. Nuclear energy and waste, and hazardous materials and waste are viewed by the lay public with dread characteristics [11-14, 22, 30-33].

Other dimensions of risk include outrage and apathy. Outrage may be displayed by the lay public when risk assessments are discussed as irrelevant or irrational [2, 15]. Public risk assessments are based on perceptions of voluntariness, controllability, equity, etc. When these factors are ignored by the experts, the public becomes outraged. The relationship of outrage to risk has been characterized by Sandman et al. as.

Risk = Hazard + Outrage [15]

English has taken this further and suggested that outrage occurs "where there is high perceived moral right on the part of the victim, high perceived moral responsibility on the part of the agent, and low perceived fulfillment in terms of risk management" [34]. In other words, when there is a moral obligation by some entity to mitigate risks and it fails to do so, the public becomes outraged. In contrast, when people do not feel threatened, they are apathetic about risk [15].

Judgmental Rules

Psychological and decision research has shown that people use judgmental rules to solve complex problems. These rules are used to make mental tasks simpler. These rules also apply to perceptions about risk and can result in inappropriate perceptions, or bias perceptions. Studies also suggest that experts may rely on these same strategies to solve decision problems [20, 27, 35].

Availability—People tend to judge risks based on their recollection of past events. Risks that occur frequently and have imagery are easier to recall, and tend to be viewed as risky. The media, which will be discussed later, plays an important role in this respect [20, 35].

Reconciling Divergent Opinions About Risk—People tend to base judgments on established beliefs. These beliefs are resistant to change and influence the way contrary evidence is perceived. Information consistent with one's own beliefs is considered reliable, while contrary evidence is dismissed as unreliable. When people do not have strong opinions, information has a major influence on risk perceptions [20, 34, 35].

Desire for Certainty—Uncertainty is denied to reduce anxiety related to a perception. People view the risk as so small and undeserving of attention, or overestimate the risk and make it so large that it must be avoided [20, 35].

Overconfidence—People are overconfident about the decisions they make. This may make for biased reviewing of information and a faulty basis for action [20].

Optimistic Bias—People believe, "It won't happen to me." People may engage in risky behavior, despite being equally at risk [15, 20].

Social Amplification of Risk

Kasperson et al. have introduced the concept of social amplification to describe the societal attenuation of risks [36]. Kasperson et al. use communication theory to explain the social amplification process. An information source sends messages to a transmitter or receiver where they are decoded, and the transmitter decodes and amplifies or attenuates the signal. The transmitter then sends a message to the receiver or next transmitter, and the process repeats. Kasperson et al. go on to explain that risk-event signals are altered by individuals, social groups, institutions, news media, public agencies, and others. As they are processed, they are amplified or attenuated. The social amplification process not only contributes to the disparate perceptions of the lay public and the experts, but also has secondary or tertiary effects. Although higher-order effects usually are neglected by traditional risk analysis, they have far-reaching implications [36]. Slovic describes these effects as "ripples in a pond", where the risk event is at the center with higher-order effects extending to the outer rings. "Signal value" or "signal potential" refers to the potential effect and likelihood of higher-order effects. Slovic comments that signal potential is related to the factor-space diagram (see Figure 1) with risks in the upper-right quadrant having greater potential to spark higher-order impacts [20, 27]. Three Mile Island, Chernobyl, Love Canal, and Times Beach are examples of amplified risks that resulted in higher-order effects, such as property devaluation, stricter regulatory enforcement to related industries, and design or process changes that increased costs to industry and consumers. These effects occupy the outer rings, while the risk event is at the center [35, 36].

As noted in the social amplification process, media can play an important role in the amplification or attenuation of risk. As noted earlier in the judgmental rules discussion, people remember what they see, with imagery and frequency being important to memory recall. The lay public does not have access to technical information, and their experience with technological hazards is primarily through the media [27]. The media have been criticized for biased reporting, exaggerating the news, and misinforming the public. It also has been noted that journalists are not experts but they deal with complex information, operate under deadlines, and compete for stories [35, 37]. Based on a review of evening news abstracts for the three major networks during a 26-month period from 1984 to 1986, Greenberg et al. concluded that risks determined to be most important to scientists received relatively little coverage on the evening news. Instead, stories with prominence, consequence, human interest, and visual impact dominated the news. They found that stories of acute events—such as Bhopal, India (61 stories), and airplane accidents (482 stories)—overshadowed less dramatic stories—such as smoking and tobacco (57 stories)—that clearly have greater health effects on society [37].

Cultural Bias and Political Ideology

Cultural theorists argue that cultural biases influence risk perception. People select and avoid risks based on cultural values and beliefs, and to justify social relations. Cultural theorists recognize three social patterns: hierarchical, egalitarian, and individualist. Hierarchicalists look toward centralized authority, experts, and complex technological solutions to solve environmental problems. They see the benefits of technology to outweigh the risks [38, 39]. They are risk tolerant. In contrast, egalitarians view risks associated with technology to be most serious, especially those associated with the destruction of nature. They view technology as posing many risks with few benefits [38, 39]. They are risk averse. Individualists see risks that affect their individual economic gain and economic freedom to be of greatest concern. Like hierarchical patterns, they also see benefits as outweighing risk. Cultural theorists have found that egalitarians are most likely to perceive the greatest risk from nuclear power [38, 39].

To proponents of political theory, judgments of risk perceptions are based on political ideology rather than facts. Miller notes that there are two opposing forces (although this is oversimplified): technocratic and humanistic. Technocrats subscribe to ideas of centralized government, reliance on experts, mastery of the environment, and the tolerance of risk in the face of rewards. In contrast, humanists look toward a decentralized, participatory decision-making process; distrust of science; harmony with nature, and incorporation of values into the process. These two groups are at odds and in conflict concerning environmental issues [40]. Others also have argued that people with more liberal ideologies will perceive greater risks, while those with conservative ideologies will perceive less risk associated with technology [39].

Conflict, Trust, and Credibility

It is clear that scientific risk assessment and public risk assessment are two different processes that arrive at different conclusions or perceptions about risk. The experts base risk perceptions on statistical information, while the public has a value-based definition. If the disagreement was statistically based, conflict would disappear in light of statistical information. However, this is not the case. Statistical information does not change lay perceptions [27]. Conflicts about risks are inherent because risks are based on values and interests that are different for individuals and segments of society. Therefore, technical expertise or technical information alone will not resolve conflicts based on values and judgments [41]. When there is scientific uncertainty, the values and judgments used at the various decision points may be real issues in conflict, while the science is debated. When there is uncertainty—and there is always some degree of uncertainty—trust in the source of information may become an issue or source of conflict [40, 41]. Four factors can be identified as contributing to conflict:

1. **Differential knowledge**—Expert knowledge of hazards and benefits versus lay public's knowledge of local conditions or practices,
2. **Vested interests**—Those who are at risk from a technology are usually different from those who benefit,
3. **Value differences**—Some may ascribe economic value, while others may see intrinsic value, and
4. **Distrust**—Of institutions and their risk messages [1].

Kasper noted more than ten years ago, "The era of blind trust for the technical experts has ended; the public now often views experts as insensitive and, sometimes, dangerous" [42]. It appears as though nothing has changed during the past decade. Distrust has become institutionalized. The lay public simply does not trust government or industry regarding environmental management issues. There are numerous examples where distrust of the agency, the experts, and the information is an all-important issue with respect to risk perceptions of nuclear technologies and hazardous wastes [11, 13, 22, 28, 33, 38, 41, 43].

Trust and credibility are related issues. Trust is an expectation of truth, reliability, and accuracy. Trust relies on competence, objectivity, fairness, consistency, and faith [44]. Credibility is an attribute based on past performance, trustworthiness, and expertise [1, 44]. To try to decipher the conditions that have contributed to the erosion of trust would be an awesome task. However, a few conditions that have contributed to this situation are worth mentioning. First, expert disagreement has contributed to distrust of government and industry. Generally, you can find opposing expert opinions on any subject. Who is the lay public to believe? Experts become looked upon as self-serving. The lay public does not have the knowledge or resources to ascertain the correct information [1, 43]. Second, scientific uncertainty and conflicting data also contribute to the erosion of trust. In the face of uncertainty, people will trust those sources of information most closely aligned with their philosophy of the environment or technology [45]. Study A says chemical Z causes cancer, while study B says it does not. The public has become confused and distrustful of scientific information. Third, if the lay public, or more specifically those at risk from a given technology, feel they are being ignored and their concerns dismissed, they will become distrustful [2, 43]. Because people base their perceptions on values, distrust will prevail when their values are dismissed as trivial. Fourth, trust in the source of information influences how the information is interpreted. If the EPA is not trusted, its message about chemical risks will not be trusted [1]. Fifth, when people do not trust technical sources of information, they turn to word of mouth, friends, and neighbors for information [1, 3]. Lastly, and perhaps most importantly, once trust is lost it is difficult to regain [1, 43]. Trust is earned, and it is not earned easily or quickly.

Two Examples: Radon Gas and Water Fluoridation

To illustrate many of the points presented thus far, the issues of radon gas and water fluoridation will be briefly discussed. Radon gas has been identified as a major public health concern for the country by the experts [16]. Although risk estimates are based on epidemiological data, there is some degree of uncertainty because these data are based on mining environments. The NIOSH review of epidemiological data found that many of the studies had serious limitations. Further, NIOSH cautioned against applying the mining data to the indoor environment and typical indoor exposures. Although radon presents some degree of risk in indoor environments, indoor home environments are somewhat different from mining environments [46]. However, the EPA has used these data in their risk assessments of indoor radon [47]. In terms of occupational exposure, NIOSH determined that a risk of somewhere between seven and 17 lung cancer deaths per 1,000 miners to be acceptable. This was based on many factors, including the risk associated with typical background exposures and technical feasibility of achieving the standard [46].

The public, on the other hand, does not see radon as a major health concern, and it appears near the bottom of their list of concerns (see Table 2). Studies conducted by Sandman et al. in New Jersey of two groups—one living near the Reading Prong, the other at some distance—found that people were apathetic about radon [15]. Even those in the Reading Prong region, where awareness should be high, were apathetic. Generally, citizens showed "optimistic bias" and underestimated the risk. Reasons given by respondents for not monitoring radon (only 7% had monitored) included not believing they were at risk, they were unsure about the best monitoring methods, and they did not have enough time. Sandman et al. note that there is a social risk to be the first person on your block to test for radon if your neighbors do not view it as a risk [15]. However, the EPA has recommended that people test their homes [48]. The risks of radon also have been disputed, with studies showing no increased cancer risk [49, 50]. In terms of testing devices, studies have shown that some may be unreliable [51, 52]. It is interesting to note that in New Jersey, Maine and Washington, DC, the risk messages used by state and federal agencies, including the EPA's "A Citizens Guide to Radon," failed to motivate citizens to test or mitigate radon [15, 53, 54]. The first EPA guide was criticized for using statistical estimates and risk comparisons that were difficult for the lay public to understand [1]. Could the lay public be distrustful of risk information sources, experts, and the science of radon gas? Because of the uncertainty, are they denying the risk as well as showing optimistic bias?

Water fluoridation is a somewhat different issue. Sixty-one percent of the U.S. population served by community water systems consumes fluoridated water to prevent dental caries [55]. Sodium fluoride is probably the most tested chemical compound of its time. Since 1970, more than 3,700 studies on fluoride and fluoridation have been conducted [56]. There have been numerous studies involving epidemiological data, short-term mutagenicity tests and animal bio-assays. At least 50 epidemiological studies have been conducted, and all failed to show that water fluoridation causes cancer. Short-term mutagenicity tests and animal bioassays have provided conflicting results. Short-term mutagenicity tests used concentrations 2,500 to 70,000 times that found in human blood [57]. Two animal studies were recently conducted, with the study by the National Toxicology Program concluding there was equivocal evidence of carcinogenic activity, a category for uncertain findings with a marginal association [58]. These studies utilized exposures that were 31 to 595 times that of typical human exposures [57]. Because of the conflicting results, the Ad Hoc Subcommittee on Fluoride of the Department of Health and Human Services reviewed the available data and judged that fluoride was safe, effective, and beneficial to society [59]. The greatest risk from water fluoridation is mild dental fluorosis in 10% of the population. However, experts do not consider this to be a public health problem [60]. However, water fluoridation is still a point of conflict for some communities, despite its endorsement by numerous health organizations. Communities still hotly debate the issue and place it on voting referenda [61]. Where response to radon is apathetic, water fluoridation represents the outrage that may be associated with a risk. Opponents argue that water fluoridation is mass medication and violates freedom of choice or is an unnecessary exposure to a chemical. Clearly, the issues are value laden. The issue is not so much one of health as it is one of personal rights.

IMPLICATIONS FOR ENVIRONMENTAL MANAGEMENT

Unfortunately, there is not a cookbook approach for dealing with risk perceptions. There are, however, steps that can be taken to integrate risk perceptions into the planning process to promote positive interaction and informed decision-making.

Attitudinal Surveys

Surveys can be used to elicit a wide variety of information regarding the lay public, or even expert perceptions about technology. The point of the survey is to find out what people think, know, and what are they really concerned about. Examples of questions that can be answered include [6, 23]:

- What are the hidden agendas?
- What issues are at the center of debate?
- What dimensions of risk are involved (equity, controllability, voluntariness, etc.)?
- What steps can be taken to make the project acceptable?
- Are there particular groups opposed to or in favor of the technology?
- Who does the lay public trust and who do they not trust?

Survey design can vary. Surveys can be based on revealed preferences (what people do) or expressed preferences (what people say). Both are based on assumptions. Using both methods provides the most insight because what people do may be different from what people say [23]. Surveys can have open-ended questions that allow for respondents to answer freely with statements of position, or questions can provide alternatives to force decisions. A mixture of these may be possible. Allowing people to answer freely has some attractive qualities, because a respondent can say what is really on his mind. Surveys may be conducted through the mail, personal interview, or at public meetings [6, 23]. Many of the studies discussed earlier ask respondents to rate their views. For example, a question might be framed:

Do you anticipate your property values to be

1	2	3	4	5	6	7
Not Affected At All						Affected A Great Deal

However, surveys are a highly technical process and should be conducted with the assistance of specialists. Questions must be prepared with care and not be too technical or difficult [62]. The way a question is written or worded can have a significant effect on the response. Words, such as "toxic waste" or "carcinogenicity," can evoke imagery, heighten anxiety, or influence people to respond based on other expectations not associated with the issue at hand. Experts also have been found to be influenced by similar wordings [16, 22, 23, 63]. Lastly, survey results should undergo statistical analysis—regression analysis is common—to determine correlations among variables.

Forecasting Risk Perceptions

Four approaches may be used to forecast risk perceptions. First, predictions may be made based on how people may be expected to perceive a given technology. As illustrated by the factor-space diagram (see Figure 1), risks that have dread characteristics, such as nuclear power and DNA technology, are perceived as most risky and requiring stringent technology. One might expect then that other technologies with dread characteristics also may be perceived as risky [27, 28].

Another approach would be to predict perceptions based on cultural or political ideologies. As noted earlier, hierarchical social patterns, technocrats, and conservative ideologies view technology as less risky than egalitarians, humanists, or liberals [38, 39, 40]. The obvious problem here is determining the hierarchical groups from the egalitarians, and so on. Attitudinal surveys would be helpful here.

A third approach would be to forecast risk perceptions based on moral responsibility. When moral rights to health and environment are infringed upon by unfulfillment of moral responsibilities to protect health and environment, perceptions of risk can be expected to be exhibited through outrage or anger [34].

Lastly, signal potential can be used to predict risk perceptions by incorporating a technique used by planners called environmental scanning. Planners have learned that external events have greater effect on an organization than internal events. The process involves screening a variety of sources of information for different criteria. In this case, the criteria would involve cues related to risk perception or signal potential [64, 65, 66]. The purpose of this exercise is to look for information that may shed light on risk perceptions, what people are concerned about, who has the most influence, who is trusted, which groups are supportive, and which groups are not. Additionally, items with signal potential should be sought. The failure of a technology in another state may affect similar technologies across the country. Further, managers should be prepared and plan to address concerns generated by the ripple effect. A committee approach should be used to regularly review the information collected and to evaluate its importance. Generally, reviews of local and national newspapers, demographic data, community profiles, trade magazines and journals, and state and federal documents can provide current information and signal potentials relative to technologies. This should be taken further to include town meetings and meetings of local interest groups [65]. Again, the lay public consists of many publics, all with different interests and values.

Risk Communication

Risk perception has strong implications for risk communication. As shown, experts view risks in terms of statistical probabilities, while the lay public uses a much broader definition. Therefore, risk messages that only communicate statistical probabilities will be ineffective, as illustrated by radon gas risk messages in two states and the District of Columbia [53, 54]. Messages must therefore address the interests and values of their recipients [1]. Attitudinal surveys then also play an important part in forming risk messages. Because the public consists of a wide array of publics with different interests and values, risk messages should be designed to address the different groups. Therefore, one risk message for a large group may be ineffective, whereas several different risk messages designed for different interests and values may have a greater effect [1]. Furthermore, messages that use risk comparison techniques or risk ladders, as the first EPA radon gas pamphlet, also may be ineffective [48]. Such comparisons may compare risks with different benefits, or trivialize the risk by comparing it with familiar, everyday risks perceived differently. For example, compare the risk of exposure to some level of radon gas with the risk of medical X-rays: Clearly, medical X-rays also have a level of benefit not addressed in a risk ladder [1, 2]. Also, because people remember what they see, risk messages should incorporate imagery and frequency which makes things easier to recall. Risk messages also should be augmented with expert advice from trusted sources and supplemented with personal communications [67]. Lastly, risk communication should not be one way. It should be a two-way, interactive process, with an exchange of information taking place [1, 2]. The lay public needs to be involved in the process, and its concerns need to be addressed by the experts. As an interactive process, experts can learn from the public, and vice versa.

Public Involvement

Democracy is supposed to be an interactive process. It has been said that the public has the right to participate in defining the problem, to exchange knowledge, and to determine the future [68]. But traditionally, public involvement has been limited to the public hearing process, where the decisions already have been made, and it is really only an opportunity for the public to air its complaints [2]. Despite our democratic process, there really is not a good forum for public involvement. Often, the courts become the method of public involvement [69]. Or, involvement requires attending a public advisory meeting at some distant place. The public does not have the resources for this type of involvement.

Public involvement needs to do what the term implies: involve the public in the decision-making process. It should be convenient, affordable, and accomplished as early as possible. Because the public has knowledge to share, decisions that are more palatable to all involved will be made [68]. The public can be categorized as the "passive public" (unaware of the issues), the "attentive" public (aware of issues), and the "active" public (trying to affect change) [1]. The latter two likely will be the most vocal. Table 4 lists ways to involve the public in the decision-making process. Group processes should be evenly balanced with competing interests [41]. One approach to this is random selection. Most importantly, public involvement must go beyond the technical information to include the public's deeper concerns, interests, and values [70]. Lastly, if the public is a part of the process and contributes to the final decision, there will be greater acceptability and trust placed in the end result [2, 68, 71, 72].

Inspiring Trust

As discussed earlier, trust is an important aspect to risk perception, risk communication, and environmental management. Trust is required for effective interaction. It is also a social expectation [43]. Table 5 lists factors, attributes, and conditions that inspire trust. If trust has eroded and distrust exists, steps should be taken to rebuild trust. For example, third-party review of programs, plans or projects may help alleviate distrust [32]. Care, of course, must be used to make sure the third party is credible and trustworthy. Seeking assistance and support

Table 4
Forums for Public Involvement

Citizen Advisory Committees (71)	Ombudsman (72)
Citizen Panels (41)	Hot Lines (2)
Citizen Boards (72)	Open Door Policy (71)
Neighborhood Governments (72)	Drop in Hours (2)
Participation on Technical Boards Volunteer Groups (72)	
Goals Programs (72)	Video Presentations (71)
Grievance Committees (72)	Information Booths (71)
Small Group Sessions (41)	Presentations/lectures (41)
One on One Meetings (2)	Newsletters (2)
Attitudinal Surveys (72)	Public Hearings (72)
Citizen Evaluation (71) (72)	Public Advisory Meetings (71)
Citizen Empowerment	Litigation (69)
Facility Tours (71)	
Information and Referral Services (72)	

Sources: See References

Table 5
Factors That Inspire Trust

Be Ethical	Be Open (1)
Maintain Good Reputation (2)	Be Objective (44)
Act Competently (1) (44)	Be Fair (44)
Be Caring (1)	Admit Uncertainty (1)
Be Honest (1)	Admit Errors (1)
Give Consistent Messages (1) (2) (44)	
Show Consistent Behavior (1) (2) (44)	
Show Consistency Between Messages and Behavior (1) (2) (44)	
Listen and Acknowledge Fears, Concerns, Worries (2)	
Practice Unbiased Review of Contradictory Information (2)	
Make Trust a Priority (2)	

Sources: See References

from trustworthy organizations, groups, agencies, or individuals also may be useful; sort of trust by association [2]. Further, acknowledge the lack of trust, try to determine the reason for distrust, and ask what can be done to regain trust [1]. Lastly, be ethical. It is the responsibility of all professionals to act in an ethical manner. Maintaining a high standard of ethics will promote trust.

CONCLUSION

The lay public, government, and industry have different perceptions of risk. Scientists and experts view risk assessment in terms of statistical probabilities, but they also interject values into the process. Public risk assessment involves a complex process, encompassing a much broader definition of risk based on values. Because interests and values are different for everyone, disparity exists in terms of which risks are important or how much of a given risk is acceptable. Generally, the public views technology as imposing a much greater risk to health and safety than do the managers of government and industry. This creates an obvious dilemma for decision- and policy-makers. These different views of risk create conflict, where trust and credibility become important issues. The challenge to the environmental manager is to understand the disparity in risk perceptions and how risks are perceived by different groups. Measures should be taken to gain an understanding of the real issues, using such tools as attitudinal surveys, and involve all those concerned in the process using techniques for public involvement. Lastly, risk communication messages must be more than statements of statistical probabilities. They should incorporate the interests and values of the recipients, and should be supplemented with direct involvement by environmental management professionals.

Research Needs

Much of the information regarding risk perception concerns controversial technologies, such as nuclear energy and waste, and hazardous materials and waste sites. The study of risk perception is still in its infancy, and this is reflected in the literature. These studies are certainly important for establishing a foundation for risk-perception research. However, environmental managers would benefit from a broader information base. Research should be expanded to include, for example, perceptions associated with groundwater contamination incidents, sanitary landfills (because these are much more common then nuclear repositories), occupational safety risk perceptions, and injuries. These studies should focus on real-life situations. A broader information base would promote the understanding of risk perceptions concerning other technologies.

Acknowledgments

The writer would like to gratefully acknowledge Malinda Hamilton for research assistance and administrative support, and Paul Slovic for granting permission to use the factor-space diagram (see Figure 1).

REFERENCES

1. *Improving Risk Communication,* National Research Council, National Academy Press, Washington, D.C., 1989, pp. 1–142.
2. *Improving Dialogue with Communities: A Risk Communication Manual for Government,* Division of Science and Research, New Jersey Department of Environmental Protection and Energy, 1991, pp. 1–83.
3. Canter, L.W., *Environmental Impact Assessment,* McGraw-Hill, New York, 1977, p. 223.
4. Mitchell, J., "The Evolution of Acceptable Risk," *The Environmental Professional, 12*: 114–121 (1990).
5. Kasperson, R.E., Kates, R.W., Hohenemser, C., Hazard Management, *Perilous Progress, Managing the Hazards of Technology,* (R.W. Kates, J.X. Kasperson, C. Hohenemser, eds.), Westview Press, Boulder, 1985, p. 51.
6. Fischhoff, B., Lichtenstein, S., Slovic, P., Derby, S.L., Keeney, R.L., *Acceptable Risk,* Cambridge University Press, Cambridge, 1981, pp. 1–45, 162–171.
7. Slovic, P., "The Legitimacy of Public Perceptions of Risk," *Journal of Pesticide Reform, 10(1)*: 13–15 (1990).
8. Cohrssen, J.J., Covello, V.T., *Risk Analysis: A Guide to Principles and Methods for Analyzing Health and Environmental Risks,* Council of Environmental Quality, 1989, p. 1–54.
9. Bjordal, E.N., "Risk from a Safety Executive Viewpoint," *Risk and Decisions,* (W.T. Singleton, J. Hovden, eds.), John Wiley and Sons, New York, 1987, pp. 41–45.
10. Petulla, J.M., "The Black Box Syndrome," *Environment 26 (1)*: 5, 40 (1984).
11. Slovic, P., Layman, M., Flynn, J. H., "Lessons from Yucca Mountain," *Environment 33(3)*: 7–11, 28–30 (1991).
12. Slovic, P., Flynn, J.H., Layman, M., "Perceived Risk, Trust, and the Politics of Nuclear Waste," *Science 254*: 1603–1607.
13. Kunreuther, H., Patrick, R., "Managing the Risks of Hazardous Waste," *Environment 33(3)*: 13–15 (1991).
14. Hadden, S.G., "Public Perception of Hazardous Waste," *Risk Analysis 11(1)*: 47–57 1991.
15. Sandman, P.M., Weinstein, N.D., Klotz, M.L., "Public Response to the Risk from Geological Radon," *Journal of Communication 37 (3)*: 93–108 1987.
16. U.S. Environmental Protection Agency, "Setting Environmental Priorities: The Debate about Risk," *EPA Journal 17 (2)*: 2–12, 40–43 (1991).

17. Roberts, L., "Counting on Science at EPA," *Science 249*: 616–618 (1990).
18. Gardner, G.T., Gould, L.C., "Public Perceptions of the Risks and Benefits of Technology," *Risk Analysis 9(2)*: 225–242 (1989).
19. *EPA Toxicology Handbook,* Government Institutes, Inc., Rockville, M.D., 1986, p. 2-2, 2-3, 6-16, 6-12.
20. Slovic, P., Fischhoff, B., Lichtenstein, S., "Facts and Fears: Understanding Perceived Risk," *Societal Risk Assessment, How Safe is Safe Enough,* (R.C. Schwing, W.A. Albers, eds.), Plenum Press, New York, 1980, pp. 181–214.
21. Rushefsky, M.E., *Making Cancer Policy,* State University of New York Press, Albany, 1986, pp. 22–27.
22. Kraus, N., Malmfors, T., Slovic, P., "Intuitive Toxicology: Expert and Lay Judgments of Chemical Risks," *Risk Analysis 12(2)*: 215–231 (1992).
23. Mendelsohn, R., Gregory, R., "Managing Environmental Accidents," *Valuing Wildlife Resources in Alaska,* (G. Peterson, C. Swanson, D. McCollum, M. Thomas, eds.), Westview Press, Boulder, in press.
24. Fischhoff, B., Slovic, P., Lichtenstein, S., "Weighing the Risks," *Perilous Progress, Managing the Hazards* (R.W. Kates, C. Hohenemser, J.X. Kasperson, eds.), Westview Press, Boulder, 1985, pp. 265–273.
25. Starr, C., "Social Benefits versus Technological Risk," *Science 165 (3,899)*: 1232–1238 (1969).
26. Fischhoff, B., Slovic, P., Lichtenstein, S., "Characterizing Perceived Risk," *Perilous Progress, Managing the Hazards,* R.W. Kates, C. Hohenemser, J.X. Kasperson, eds.) Westview Press, Boulder, 1985, pp. 91–121.
27. Slovic, P., "Perceptions of Risk: Reflections on the Psychometric Paradigm," *Theories of Risk,* (D. Golding, S. Krimsky, eds.), Praeger, New York, 1992, pp. 118–152.
28. Slovic, P., "Perception of Risk," *Science 236*: 280–285 (1987).
29. Fischer, G.W., Morgan, M.G., Fischhoff, B., Nair, I., Lave, L.B., "What Risks are People Concerned About," *Risk Analysis 11 (2)*: 303–314 (1991).
30. Bord, R.J., O'Connor, R.E., "Determinants of Risk Perceptions of a Hazardous Waste Site," *Risk Analysis 12 (3)*: 411–416 (1992).
31. Morell, J.A., "Community and Individual Reaction to Environmental Hazards: Developing a Measurement Technology," *Environmental Management 11(1)*: 69–76 (1987).
32. Mitchell, J.V., "Perception of Risk and Credibility at Toxic Sites," *Risk Analysis 12(1)*: 19–26 (1992).
33. McClelland, G.H., Schulze, W.D., Hurd, B., "The Effect of Risk Beliefs on Property Values: A Case Study of a Hazardous Waste Site," *Risk Analysis 10(4)*: 485–497 (1990).
34. English, M.R., "Victims, Agents, and Outrage," *The Analysis, Communication, and Perception of Risk,* (B.J. Garrick, W.C. Gekler eds.), Plenum Press, New York, 1991, pp. 199-204.

35. Slovic, P., "Informing and Educating the Public About Risk," *Risk Analysis 6(4)*: 403–415 (1986).
36. Kasperson, R.E., Renn, O., Slovic, P., Brown, H.S., Emel, J., Goble, R., Kasperson, J.X., Ratick, S., "The Social Amplification of Risk: A Conceptual Framework," *Risk Analysis 8(2)*: 177–187 (1988).
37. Greenberg, M.R., Sachsman, D.B., Sandman, P.M., Salomone, K.L., "Network Evening News Coverage of Environmental Risk," *Risk Analysis 9(1)*: 119–126 (1989).
38. Wildavsky, A., Dake, K., "Theories of Risk Perception: Who Fears What and Why?" *Daedalus 119 (4)*: 41–60, (1990).
39. Jenkins-Smith, H.C., Smith, W.K., "Ideology, Culture and Risk Perception," unpublished paper prepared for Annual Meeting of AAAS, (1992).
40. Miller, A., "Ideology and Environmental Risk Management," *The Environmentalist 5(1)*: 21–30 (1985).
41. Renn, O., Webler, T., Johnson, B.B., "Public Participation in Hazard Management: The Use of Citizen Panels in the U.S.," *Risk-Issues in Health and Safety*: 197–225, Summer (1991).
42. Kasper, R.G., "Perceptions of Risk and Their Effects on Decision Making," *Societal Risk Assessment, How Safe is Safe Enough?* (R.C. Schwing, W.A. Albers, eds.), Plenum Press, New York, 1980, p. 78.
43. English, M.R., *Siting Low Level Radioactive Waste Disposal Facilities: The Public Policy Dilema,* Quorum, New York, 1992, pp. 53–73.
44. Renn, O., Levine, D., "Credibility and Trust in Risk Communication," *Communicating Risks to the Public,* (R.E. Kasperson, P.J.M. Stallen, eds.), Kluwer Academic Publishers, Netherlands, 1991, pp. 175–184.
45. Soden, D.L., Conary, J.S., "Trust in Sources of Technical Information Among Environmental Professionals," *The Environmental Professional 13*: 363–369 (1991).
46. *A Recommended Standard for Occupational Exposure to Radon Progeny in Underground Mines,* National Institute for Occupational Safety and Health, DHHS (NIOSH) Publication No. 88-101, 1987, pp. 32–48.
47. *Reducing Radon in Structures,* U.S. Environmental Protection Agency, Washington, D.C., Undated Draft, pp. 1–9.
48. *A Citizen's Guide to Radon, What it is and What to do About it?*, U.S. Environmental Protection Agency, Washington, D.C., 1986.
49. Joyce, C., "Radon and Lung Cancer: The Link is Missing," *New Scientist*: 29 September (1988).
50. Vonstille, W.T. , Sacarello, H.L.A., "Radon and Cancer: Florida Study Finds No Evidence of Increased Risk," *Journal of Environmental Health 53(3)*: 25–28 (1990).
51. Hamilton, M.A., "Radon Reduction: Research Project Targets Homes with Crawl Spaces," *Journal of Environmental Health 53 (3)*: 31 (1990).
52. Field, W., Kross, B.C., "Field Comparison of Several Commercially Available Radon Detectors," *American Journal of Public Health 80(8)*: 926–929 (1990).

53. Johnson, F.R., Luken, R.A., "Radon Risk Information and Voluntary Protection: Evidence from a Natural Experiment," *Risk Analysis 7(1)*: 105 (1987).
54. Doyle, J.K., McClelland, G.H., Schulze, W.D., Elliott, S.R., Russell, G.W, "Protective Responses to Household Risk: A Case Study of Radon Mitigation," *Risk Analysis 11 (1)*; 127–133 (1991).
55. Easley, M.W., "The Status of Community Water Fluoridation in the U.S.," *Public Health Reports 105 (4)*: 348–353 (1990).
56. *Water Fluoridation: A Training Course Manual for Engineers and Technicians,* U.S. Department of Health and Human Services, 1986, pp. 1–12.
57. Hamilton, M.A., "Water Fluoridation: A Risk Assessment Perspective," *Journal of Environmental Health 54 (6)*: 27–31 (1992).
58. *NTP Technical Report on the Toxicology and Carcinogenesis Studies of Sodium Fluoride,* U.S. Department of Health and Human Services, U.S. Public Health Service, National Institutes of Health, Research Triangle Park, N.C., 1990.
59. *Review of Fluoride Benefits and Risks,* Report of the Ad Hoc Subcommittee on Fluoride of the Committee to Coordinate Environmental Health and Related Programs, U.S. Department of Health and Human Services, 1991.
60. Whitford, G.M., "The Physiological and Toxicological Characteristics of Fluoride," *Journal of Dental Research 69*: 539–549, (1990).
61. Hilleman, B., "Fluoridation of Water, Questions About Health Risks and Benefits Remain After More Than Forty Years," *Chemical and Engineering News*: 26–27, (1988).
62. Fischhoff, B., Furby, L., Gregory, R., "Evaluating Voluntary Risks of Injury," *Accident Analysis and Prevention 19 (1)*: 51–62 (1987).
63. Carlo, G., Lee, N.L., Sung, K.G., Pettygrove, S.D., "The Interplay of Science, Values, and Experiences Among Scientists Asked to Evaluate the Hazards of Dioxin, Radon, and Environmental Tobacco Smoke," *Risk Analysis 12(1)*: 37–43 (1992).
64. Slovic, P., "Ripples in a Pond: Forecasting Industrial Crises," *Industrial Crisis Quarterly 1*: 34–43 (1987).
65. Renfro, W.L., Morrison, J.L., "Detecting Signals of Change: The Environmental Scanning Process," *The Futurist*: 49–53 August (1984).
66. Poole, M.L., "Environmental Scanning Is Vital to Strategic Planning," *Educational Leadership 48 (7)*: 40–41 (1991).
67. Weinstein, N.D., Sandman, P.M., Roberts, N.E., "Determinants of Self-Protective Behavior: Home Radon Testing," *Journal of Applied Social Psychology 20 (10)*: 799 (1990).
68. Lynn, F.M., "Public Participation in Risk Management Decisions: The Right to Define, the Right to Know, and the Right to Act," *Risk-Issues in Health and Safety 1(2)*: 95–101 (1990).
69. Carroll, J.D., "Participatory Technology," *Science 171*:647-653 (1971).

70. Koines, A., Cummins, P., Rowe, R.D., "Getting to Yes: Reaching Political Consensus On Local Environmental Priorities," *The Environmental Professional 12*: 154 (1990).
71. Lynn, F.M., "Citizen Involvement In Hazardous Waste Sites: Two North Carolina Success Stories," *Environmental Impact Assessment Review 7(4)*: 347–360 (1987).
72. Lind, A., "The Future of Citizen Involvement," *Futurist 9(6)*: 316–328 (1975).

CHAPTER 3

INDOOR AIR QUALITY OVERVIEW

Paul N. Cheremisinoff
Yuh-Ing Tsen

Department of Civil & Environmental Engineering
New Jersey Institute of Technology
Newark, NJ 07102, USA

CONTENTS

INTRODUCTION

History

Various health complaints related to indoor air quality have been reported to health offices and building owners. A study by the Environmental Protection Agency (EPA) in 1985 indicates that Americans spend about 90% of their time in various indoor environments. Air contaminants sources are both indoors and outdoors. The major types of indoor pollutants are sulfur dioxide, carbon monoxide, carbon dioxide, ozone, nitrogen oxides, hydrocarbon lead, particulates, and pollens. Indoor pollutants can originate from the emissions of building materials, building equipment, maintenance materials, and building inhabitants [1].

Indoor air quality started to get the public's attention during the 1973 and 1978 oil crises because building owners and management teams were paying higher costs to heat and cool their buildings. The consequence of energy conservation often results in problems of inadequate ventilation. The health hazard increases due to the accumulation of pollutants from the building's occupants. The problem of indoor air quality attracted the public's attention when EPA employees complained of headaches, eye irritation, fatigue, and nausea.

Definition

The EPA estimates about 30% of commercial buildings fall into the "sick building syndrome" classification. Health problems associated with buildings are called sick building syndrome, tight building syndrome or building-related illness. However, technically they are different. Sick building syndromes are the most common types of complaints related to indoor air quality, and are associated with building design and characteristics. Unfortunately, a varied terminology coupled with unclear pathophysiologic explanations suggest a multifactorial etiology. Discomfort rather than an acute illness is typical. In contrast to sick building syndrome, building-related illness has a known etiology and distinguishable set of symptoms. These include hypersensitivity disease, infections, dermatitis, reproductive problems, and other health complaints. The sick building syndrome also requires that a building have at least 20% of its occupants' complaints last for more than two weeks, but are relieved when the occupants leave the building.

Overview of Regulations

There is no federal or state regulation on indoor air quality. The EPA presented a proper federal response to the indoor air pollution problem by submitting "The Report to Congress on Indoor Air Quality." The report, required by Title IV of the Superfund Amendments and Reauthorization Act (SARA) of 1986, includes an executive summary and recommendations and three volumes. Volume I, "Federal Programs Addressing Indoor Air Quality," describes indoor air activities and the research agencies. The pollutants under study are radon, asbestos, environmental tobacco smoke (ETS), formaldehyde, chlorinated solvents, and pesticides. Volume II, "Assessment and Control of Indoor Air," addresses pollutant sources and their effects. Volume III, "Indoor Air Research Needs Statement," is a statement of risk-assessment methods, exposure-assessment and modeling needs, source-specific needs, control techniques, and how to build system research, crosscut research, and transfer technology.

SOURCES OF AIR CONTAMINANTS

Energy-conservation efforts that improve the insulation in buildings reduce ventilation, allowing chemicals to accumulate. Office and home air can be extremely dry and can cause an eczema like condition. Office workers may experience fatigue, respiratory problems, headache, and mucous membrane irritation, such as irritation of the eyes, nose, mouth, and throat. Some health complaints required long-term study to determine exposure pollutants. The evaluation of indoor air quality basically includes physical effects, chemical effects, biological effects, and social effects.

Physical Aspect

Physical factors include temperature, humidity, air movement, dust, odors, light, and noise. These factors affect comfort, but may not adversely affect health. An effective temperature is na index of human body response to the combined effects of temperature, humidity, and air movement. ASHRAE 55-1981, "Standard on Thermal Environmental Conditions for Human Occupancy," outlines a range of acceptable temperatures and relative humidity for summer and winter. The specifications are 73° to 79°F and 40% to 60% relative humidity in summer, and 69° to 76°F and 30% to 50% relative humidity in winter. Many ventilation and heating systems do not have the capacity for pollution removal, so pollutants recirculate and often increase.

Good ventilation is required to provide sufficient amounts of outside air. Air movement can keep out noticeable odors and dilute other accumulated pollutants. ASHRAE 62-1989 indicates that the minimum outside air requirement is

15 CFM/person (see Table 1). The required outside air quantities must be delivered to the occupant's breathing zone. Short circulation should be avoided in the design of an air diffuser and air return unit. The ventilation system design should be evaluated carefully. It is a common mistake to locate a makeup air vent close to a source of dirty air exhaust or a garage.

Some studies indicate that air filters are sources of indoor air pollutants. Fiberglass filters can be shredded to submicron-sized fibers and provide nuisance particulates.

Table 1
Outdoor Air Requirements for Ventilation

Application	Estimated Maximum Occupancy Person/1,000 ft^2	Outdoor Air Requirements	
		CFM/person	L/person
Food and Beverage Service			
Dining room	70	20	10
Cafeteria, fast food	100	20	10
Bars, cocktail lounges	100	30	15
Offices			
Office space	7	20	10
Reception areas	60	15	8
Telecommunication centers and data entry areas	60	20	10
Conference rooms	50	20	10
Public Spaces			
Public restrooms		50	25
Smoking lounges	70	60	30
Theaters			
Ticket booths	60	20	10
Lobbies	150	20	10
Auditorium	150	15	8
Transportation			
Waiting rooms	100	15	8
Platforms	100	15	8
Vehicles	150	15	8

Adopted from ASHRAE 62-1989

Chemical Aspect

Thousands of dangerous chemicals can be found in common household products such as insecticides, air fresheners, and fingernail polish. Dry-cleaned clothing can be a source of tetrachloroethylene, a carcinogen. Tables 2, 3, and 4 show the origins of chemical contaminants from building materials, building equipment, and building inhabitants.

Table 2
Building Material Sources of Contamination

Contaminant	Sources
Formaldehyde	Particle board
	Urea-formaldehyde foam insulation
	Pressed wood
	Plywood resins
	Hardwood panelings
	Carpeting
	Upholstery
Asbestos	Draperies
	Filters
	Stove mats
	Floor tile
	Spackling compound
	Furnaces (older)
	Roofing
	Gaskets
	Insulation
	Acoustical material
Organic vapors	Carpet adhesives
	Wool finishes
Radon decay products	Concrete
	Brick
	Stone
	Soil
Man-made mineral fibers	Fiberglass insulation
	Mineral wool insulation

Adopted from Annual American Conference of Governmental Industrial Hygienists, Vol. 10

Table 3
Building Equipment Sources of Contamination

Contaminant	Sources
Ammonia	Microfilm machines Engineering drawing reproduction machines
Ozone	Electrical equipment Electrostatic air cleaners
Carbon monoxide Carbon dioxide Sulfur dioxide Nitrogen dioxide Hydrogen cyanide Particulates Benzo(a)pyrene	Combustion sources including: gas ranges, dryers, water heaters, kerosene heaters, fireplaces and wood stoves, garage
Amines	Humidification equipment
Carbon Methyl alcohol Trinitrofluorene Trinitrofluorenone	Photocopying machines
Methyl alcohol	Spirit duplication machine
Methacrylates	Signature machine
Dust	Video display terminals

Adopted from Annual Conference of Governmental Industrial Hygienists, Vol. 10.

Biological Aspect

Duct work may contain construction debris and even remnants of workers' lunches, and other debris can accumulate over time. Such debris and moisture may provide areas in the system where microorganisms can deposit and multiply to be later entrained in the ventilation air.

Biological agents consist of bacteria, viruses, and fungi, which usually affect health by causing infections or by acting as allergens. Twenty-nine visitors died after attending a convention in a Philadelphia hotel in 1976. This hotel's ventilation system had nurtured bacteria subsequently named Legionnaire's bacteria. Microorganisms pose health effects by causing dermatitis, respiratory problems, allergic reactions, asthma, and hypersensitivity pneumonitis.

Table 4
Building Inhabitants as Sources of Contaminants

Contaminant	Sources
Formaldehyde	Toothpaste Smoking Grocery bags Waxed paper Facial towels Shampoo Cosmetics Medicines
Acetone Butyric acid Ethyl alcohol Methyl alcohol Ammonia Odors	Bioeffluents
Asbestos	Talcum powder Ironing board covers Hot mittens
Acrolein Nicotine CO + 3000 others	Smoking
Dusts and vapors	Personal care products
Vinyl chloride Dusts Vapors	Aerosol spray, propellants, and solvents Cleaning products Hobbies Mothproofing Fire retardants Insecticides
Radon decay products	Ground water Shower
Dusts Vapors	Fertilizers Adhesives
Polychlorinated biphenyls (PCBs)	Carbonless carbon paper

Adopted from Annual American Conference of Governmental Industrial Hygienists, Vol. 10.

Social Aspect

Health complaints may not always be related to indoor air quality. Occupational stress and job dissatisfaction may play an important role in physical illness. Sometimes if one complaint arises about a problem, the occupants may develop an overall negative attitude about their environment before the source of air contaminants can be determined. Therefore, headache and stuffy nose can often be a manifestation of stress. Psychological effects may lead to false identification of the cause of the problem.

ELEMENTS AND HEALTH EFFECTS OF INDOOR AIR POLLUTANTS

Carbon Monoxide

Carbon monoxide (CO) is a common colorless and odorless pollutant resulting from incomplete combustion. The major source of CO emission in the atmosphere is the gasoline-powered internal combustion engine. The chemical can be a fatal poison. The gas can be traced to many sources, including unvented gas appliances and heaters, malfunctioning heating systems, kerosene heaters, and underground or connected garages. ETS is a major source of CO. The gas keeps hemoglobin from binding oxygen and thus causes asphyxiation. Fatigue, headache and chest pain are the result of repeated exposure to a low concentration of CO. Impaired vision and coordination, dizziness, confusion, and death may develop at a high-concentration exposure.

Carbon Dioxide

Carbon dioxide (CO_2) is a colorless and odorless gas. It is an asphyxiating agent. A concentration of 10% can cause unconsciousness and death from oxygen deficiency. The concentration of CO_2 in the blood affects the rate of breathing. CO_2 has been linked to the greenhouse effect. The gas can be released from automobile exhaust, ETS, and inadequately vented fuel heating systems. The gas is heavy and accumulates at low levels in depressions and along the floor.

Nitrogen Oxides

Nitrogen oxides included nitrous oxide (N_2O), nitric oxide (NO), nitrogen dioxide (NO_2), nitrogen trioxide (N_2O_3), nitrogen tetroxide (N_2O_4), nitrogen pentoxide (N_2O_5), nitric acid (HNO_5), and nitrous acid (HNO_2). Of these, nitrogen dioxide is the most important pollutant. It is regulated in the National Ambient Air Quality Standard. The nature of the combustive process varies the concentration of nitrogen oxides. Tobacco smoke is a source of intense exposure to both nitric oxide and nitrogen dioxide.

Inhaling nitrogen oxide gases may cause irritation of eyes and mucous membranes. Prolonged low-level exposure may stain skin and teeth yellowish to brownish. Chronic exposure may cause respiratory dysfunction. Nitric oxide and nitrogen dioxide combine with hemoglobin to form a variety of nitroxyl-hemoglobin complexes and methemoglobin.

Sulfur Dioxide

Sulfur dioxide (SO_2) is a colorless gas with a strong suffocating odor. The major emission source of the gas is fuel combustion. Excess exposure may occur in various industrial processes including ore smelting, coal and fuel-oil combustion, paper manufacturing, and petroleum refining.

Studies of the toxic effects of sulfur dioxide have been conducted on man and animals. The chemical has not been identified as a carcinogen or cocarcinogen by epidemiologic data. However, EPA studies show that short-term acute exposures to a high concentration of sulfur dioxide suggest adverse effects on pulmonary function in man. Absorption of the gas increases from upper to lower airways [3].

Ozone

Ozone (O_3) is a powerful oxidizing agent. It is found naturally in the atmosphere by the action of electrical storms. The concentration of ozone varies in different environments. It is about 0.05 ppm at sea level [4].

The major indoor source of ozone is from electrical equipment and electrostatic air cleaners. Hansen and Andersen have extensively studied ozone in relation to photocopying machines. The indoor ozone concentration is determined by ventilation. It includes the room volume, the number of air vents in the room, the room temperature, and the materials and the nature of surfaces in the room [5].

Ozone is irritating to the eyes and all mucous membranes. Pulmonary edema may occur after exposure has ceased. Ozone is an unstable chemical and is readily decomposed to oxygen. Activated carbon filters are most commonly used in photocopiers. Ozone filters, which do not remove ozone from the air but decompose it to oxygen, also may be used.

Radon

Radon is a naturally occurring radioactive decay product of uranium. A great deal of attention centers around radon (^{222} Rn), which is the first decay product of ^{228}radium. Radon and radon daughters have been linked to lung cancer. The EPA estimates that radon may be attributed to about 5,000 to 20,000 lung cancer deaths a year in the United States. The length of exposure and the concentration of radon are two major factors in developing lung cancer. Smokers may have a higher risk of developing lung cancer induced by radon. The released energy from radon decay may damage lung tissue and lead to lung cancer.

Radon is present in the air and soil. It can leak into the indoor environment through dirt floors, cracks in walls and floors, drains, joints, and water seeping through walls. Radon can be measured by using charcoal containers, alpha-track detectors, and electronic monitors [6]. Results of the measurement of radon decay products and the concentration of radon gas are reported as "working levels" (WL) and "picocuries per liter" (pCi/l), respectively. The continuous exposure level of 4 pCi/L or 0.02 ml has been used by the EPA and the Center for Disease Control as a guidance level for the purpose of further testing and remedial action.

Once identified, the risk of radon can be minimized by engineering controls and practical living methods. The treatment techniques include sealing cracks and other openings in basement floors and installing subslab ventilation. Crawl spaces also should be well-ventilated. Radon-contaminated well water should be treated by aerating or filtering through granulated-activated charcoal.

Table 5 compares risks from radon with risks from smoking and X-rays.

Asbestos

Asbestos is a naturally occurring group of minerals that fracture into millions of fine fibers when crushed. There are three major asbestos minerals: chyrsotile (white fibers), amosite (brown fibers), and crocidolite (blue fibers).

Asbestos possesses a number of physical characteristics that make it useful as thermal insulation and fire-retardant material. It is electrically non-conductive, durable, chemically resistant, and sound absorbent. The properties of asbestos have been recognized since ancient times: The Egyptians used asbestos cloth to prepare bodies for burial. Asbestos has been extensively used as a thermal insulation and a fire retardant since 1940s.

However, lung cancer and mesothelioma have been found to be associated with environmental asbestos exposure. The EPA first listed asbestos as a hazardous air pollutant in 1971.

Chronic exposure to asbestos, via the respiratory system, can contribute to diseases including asbestosis, lung cancer, and mesothelioma. The latency period for asbestos-related diseases can be 10 to 30 years.

A 1979 study indicated that a cigarette smoker who also workers with asbestos is more than 50 times more likely to contract lung cancer than a non-smoking non-asbestos worker (see Table 6).

Formaldehyde

Formaldehyde (HCHO) is a colorless gas with a pungent, suffocating odor. Formaldehyde, a carcinogen, is used industrially as a fungicide and germicide,

Table 5
Radon Risk Evaluation Chart

pCi/L	WL	Estimated number of lung cancer deaths due to radon exposure (per 1,000)	Comparable exposure levels	Comparable risk
200	1	440–770	1000 times average outdoor	More than 60 times non-smoker risk 4 pack-a-day smoker
100	0.5	270–630	100 times average indoor level	20,000 chest x-rays per year
40	0.2	120–380		2 pack-a-day smoker
20	0.1	60–210	100 times average outdoor level	1 pack-a-day smoker
10	0.05	30–120	10 times average indoor level	5 times non-smoker risk
4	0.02	13–50		
2	0.01	7–30	10 times average outdoor level	200 chest x-rays per year
1	0.005	3–13	Average indoor level	Non-smoker risk of dying from lung cancer
0.2	0.001	1–3	Average outdoor level	20 chest x-rays per year

Source: EPA, "A Citizen's Guide to Radon."

Table 6
Asbestos and Smoking

Group	Asbestos Experience	Smoker	Mortality Ratio
Control	No	No	1.00
Asbestos worker	Yes	No	5.17
Control	No	Yes	10.85
Asbestos worker	Yes	Yes	53.24

in disinfectants, and in embalming fluids. The major sources of indoor airborne formaldehyde are furniture, floor underlayment, and ETS. Urea formaldehyde (UF) is mixed with adhesives to bond veneers, particles, and fibers [7]. It has been identified as a potential hazardous source.

Formaldehyde gas may cause severe irritation to the mucous membranes of the respiratory tract and eyes. Repeated exposure to formaldehyde may cause dermatitis either from irritation or allergy. The gas can be removed from the air by an absorptive filter of potassium-permanganate-impregnated alumina pellets or fumigation with ammonia of the entire home.

Exposure to formaldehyde may be reduced by using exterior-grade pressed-wood products that contain phenol resins. Maintaining moderate temperature and low humidity can reduce emission from formaldehyde-containing materials.

Pesticides

Pesticides are the products used to kill household insects, rats, cockroaches, and other pests. Pesticides can be classified in the following groups based on their chemical nature or use: organophosphates, carbamates, chlorinated hydrocarbons, bipyridyls, coumarins; and indandiones, rodenticides, fungicides, herbicides, fumigants, and miscellaneous insecticides [8]. The common adverse effects are irritation of the skin, eyes, and upper-respiratory tract. Prolonged exposure to some chemicals may cause damage to the central nervous system and kidneys.

Some caution should be taken to reduce exposure if pesticide application is required. Chemicals should be mixed or diluted outdoors and used according to the manufacturer's directions. The environment should be well-ventilated after pesticides are used. Unneeded pesticides and unwanted containers should not be stored in or near the home.

Volatile Organic Compounds (VOCs)

The sources of volatile organic compounds (VOCs) include building materials, maintenance materials, and building inhabitants. Building materials include carpet adhesives and wool fabrics. Maintenance materials include varnishes, paints, polishes, and cleaners.

VOCs may pose problems for mucous surfaces in the nose, eyes, and throat. One chemical that has been recognized as a cancer-causing agent is perchloroethylene (tetrachloroethylene), a chemical used in dry cleaning. The best way to reduce exposure is to use these chemicals outdoors. More air should be brought in during and after application if used indoors. Unused containers should be well sealed or discarded.

Lead

Lead has been widely used in the battery industry, the petroleum industry, the ceramics industry, and metal products industry, as well as in paint pigments and insecticides. The battery and petroleum industries are the major consumers of lead. Lead is used in antiknock gasoline additives; about 98% of the airborne lead that has been identified from combustion of gasoline [9].

Lead is only slightly soluble in water. The particle size of lead varies with its physical and chemical state. Regardless of the nature of the dusts, fumes, mists, or vapors, lead can be inhaled and retained in the airways and lungs.

Automobile exhaust is the major source of indoor lead pollutants. Lead is introduced into buildings from inefficiently filtered outdoor air. Toxicity by inhalation of lead may cause pallor, eye irritation, insomnia, and lassitude [6]. Acute exposure to lead can damage the hematopoietic system, kidneys, and nervous system [5].

Respirable Particles

Respirable particles are merely micrometers in diameter. The sources of respirable particles include kerosene heaters, ETS, paint pigments, insecticide dusts, radon, and asbestos. The particles may irritate the eyes, nose, and throat. They may also contribute to respiratory infections, bronchitis, and lung cancer. Figure 1 shows the relative sizes of typical airborne particulates.

Environmental Tobacco Smoke (ETS)

Environmental tobacco smoke (ETS) is a major indoor pollutant. The National Research Council (NRC) in 1985 indicated that passive smoking significantly increases the risk of lung cancer in adults and respiratory illness in children [2]. ETS is composed of irritating gases and carcinogenic tar particles. Inhaling ETS is called "involuntary smoking," "passive smoking," or "second-hand smoking." There are more than 4,700 chemical compounds in cigarettes, such as carbon monoxide, carcinogenic tars, hydrogen cyanide, formaldehyde, and arsenic, 43 of which are recognized carcinogens. The particle size of ETS ranges from

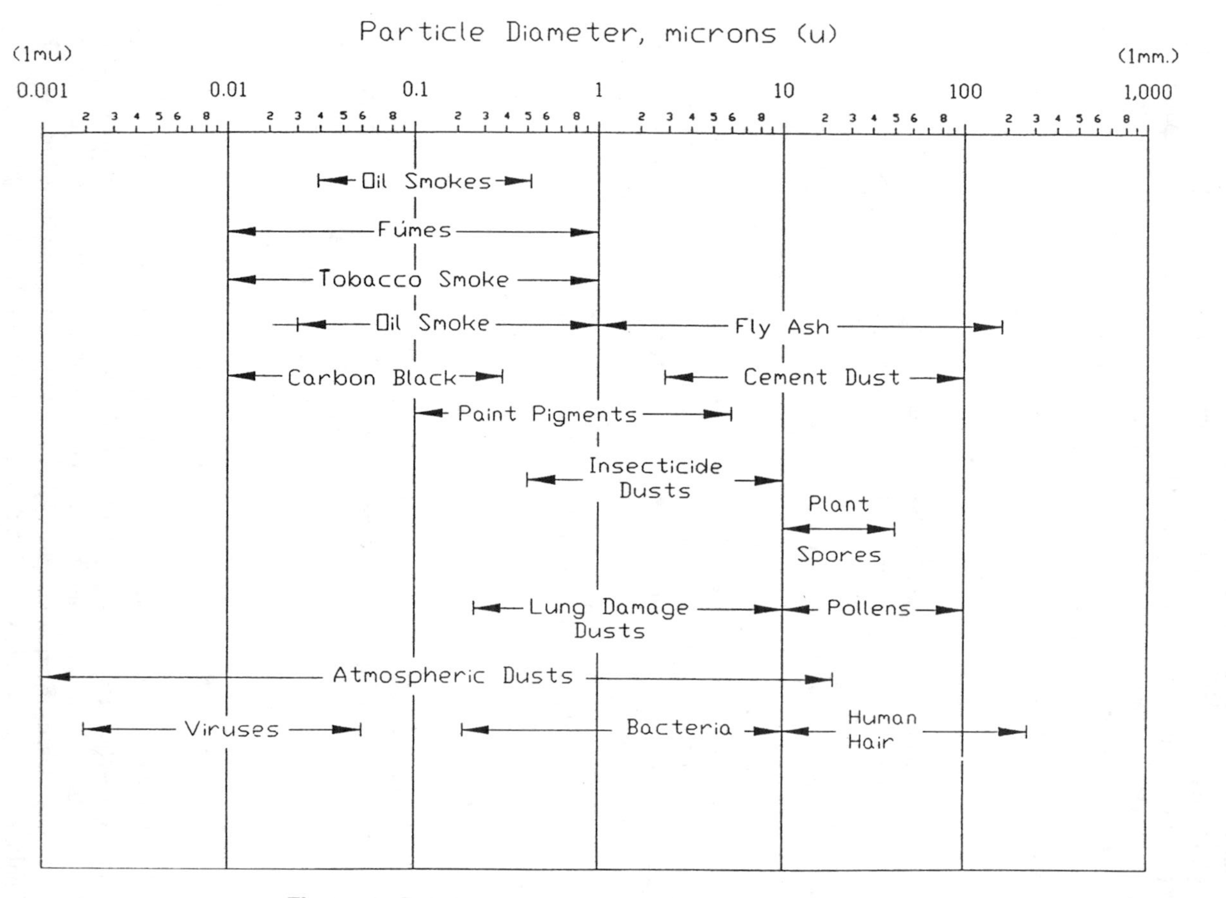

Figure 1. Relative sizes of typical airborne particulates.

approximately 1 to 0.01 microns (see Figure 1). The particles are small enough to be trapped in the upper respiratory tract or on the tracheal or bronchial surfaces, and they can be carried and deposited there directly by inhalation.

ETS is the major source of pollutants that impair health. The most common effect in children from ETS is the development of wheezing, coughing, and sputum. According to 1986 reports by NRC, the risk of lung cancer is about 30% higher for non-smoking spouses of smokers than for non-smoking spouses of nonsmokers. Some studies also showed that ETS has been associated with an increased risk of heart disease.

On April 16, 1988, The Clean Indoor Air Act became law. Smoking is only allowed in designated areas, which should have ventilation systems independent from buildings. The ventilation rate should be increased to dilute the concentration of ETS.

HOW TO CONDUCT AN INDOOR AIR-QUALITY SURVEY

Background Assessment

Background surveys usually can be based on telephone interviews and previous related documents. Telephone interviews can be guided with a questionnaire. Contaminants and their sources are not always obvious; hence, the building's history and a list of symptoms will be useful during investigation. The building information should include details regarding the ventilation system, age of the building, fabric types, the nature and use of the building, and problems with the natural environment. Table 7 may be used as a telephone questionnaire.

Preliminary Assessment

Walk-through evaluations can provide valuable information that may not be obtained during the background assessment. The use of the complained-about area should be defined. NIOSH indicates that inadequate ventilation is to blame in almost 52% of the problems found in indoor air investigations; therefore, the ventilation system is the most important item to investigate. It consists of the location of air intake, the efficiency of makeup air, the distance of air circulation between diffuser and return unit, filters, duct work interiors, and the wet areas of the system.

Personal interviews sometimes are a fast way to pinpoint the characteristic sources of contaminants. Extra caution should be taken because psychological effects may lead to a wrong judgment during the personal interviews.

Environmental monitoring is the most practical method to identify contaminants. Monitorings in each of the complained-about environments should be compared with monitorings of a neutral area. Testing should contain information on temperature and relative humidity.

Table 7
Indoor Air Questionnaire

I. Building Information
1. Building age: ____________________
2. Number of floors: ____________________
   Floor area: ____________ ft^2/floor
3. Any major renovations or operating changes?
   What were they? ____________________
   When did they occur? ____________________
4. Does the building have spray-on insulation? ____________________
5. What type of filtration system is used? ____________________
   How often is it changed? ____________________
6. What type of activities are the building occupants engaged in?
   ____
7. Any new chemical used for building maintenance purpose?
   ____

II. Occupants Information
1. Complaints Yes _____ No _____

_____	too cold	_____	too hot
_____	stale air	_____	noticeable odor
_____	stuffy air	_____	disturbing noise
_____	smoky air	_____	other

2. Health complaints?

_____	eye irritation	_____	headache
_____	fatigue/drowsiness	_____	nausea
_____	skin irritation/itching	_____	sneezing
_____	dizziness	_____	heartburn
_____	chest tightness	_____	back pain

3. When do these problems occur?

_____	morning	_____	daily
_____	afternoon	_____	specific days of the week

4. Symptoms

	symptom clears up 1 hour after leaving work Yes	No	How long
#1 ____________	_____	_____	_____
#2 ____________	_____	_____	_____
#3 ____________	_____	_____	_____
#4 ____________	_____	_____	_____

5. Any health problem or allergies? ____________________
6. Activity
   _____ wear contact lenses
   _____ operate video display terminals at least 10% of the work day
   _____ operate photocopier machine at least 10% of the work day
   _____ use or operate special office machines or equipment
7. Smoke
   _____ yourself _______ others in your immediate work area

8. Age group: ____ 25 or under ____ 25–30
____ 30+ ____ 40+ ____ 50+
9. Job position: ____________________
10. Office number: ____________________
11. Any managerial change? ____ Yes ____ No
12. Any new office equipment? ____ Yes ____ No
What are they? ____________________

Detection tubes are not recommended for use in monitoring because they have approximately a ±25% margin of error. Some portable instruments such as HNU, Photovac, and Foxboro may furnish information on the location of sources of pollutants. Further quantity and quality analysis should follow EPA and NIOSH manuals for chemical identification. Microbiological tests can be guided by U.S. Pharmacopeia XXI.

Follow-up Assessment

An additional site assessment may be required if early surveys are inconclusive. Analytical methods may be conducted on a long-term basis. The chemical and/or microbiological monitoring should be conducted more specifically and extensively. It may not be possible to easily distinguish between building-related symptoms and colds or respiratory infections. More observation may be needed for identification.

EVALUATION

There is no standard of compliance for indoor air quality. The Permissible Exposure Limits (PELs) of the Occupational Safety and Health Administration (OSHA) and Threshold Limit Values (TLVs) of the American Conference of Governmental Industrial Hygienists (ACGIH) are commonly used as chemical exposure references. The physical effects of indoor environments are usually evaluated using the guidelines of the American Society of Heating, Refrigerating and Air Conditioning Engineers, Inc. (ASHRAE).

See Table 8 for evaluation criteria.

POLLUTION CONTROL RECOMMENDATIONS

Removing the source of pollution is the most effective and permanent way to improve indoor air quality, if the pollution source can be identified. For example, a smoking area should have its own ventilation system. When present, water-stained ceiling tile and carpeting should be replaced.

Table 8
Criteria Evaluation

Substance	OSHA-PELs PPM	OSHA-PELs mg/m³	ACGIH-TLVs PPM	ACGIH-TLVs mg/m³
Carbon Monoxide	35	40	50	57
Carbon Dioxide	10,000	18,000	5,000	9,000
Nitrogen Dioxide	—	—	3	5.6
Sulfur Dioxide	2	5	2	5.2
Ozone	0.1	0.2	*C 0.1	*C 0.2
Formaldehyde	1	—	1	1.2
Lead	—	—	—	0.15
Respirable Particles	—	5	—	10
Acetone	750	1,800	750	1,780
Ammonia	—	—	25	17
Beneze	10	—	10	32
Chloroform	2	9.78	10	49
Carbon black	—	3.5	3	3.5
Isobutyl Alcohol	50	150	50	152
Isopropyl Alcohol	400	980	400	983
Tetrachloroethylene	25	170	50	339
Toluene	100	375	100	377

**C: denotes ceiling limit*

Ventilation Adjustment

This could be a cost-effective means of reducing indoor pollutant levels. Additional ventilation should be provided to dilute or exhaust contaminated air. Ventilation time should be prolonged after the occurrence of a pollution episode.

Inadequate ventilation design or installation should be corrected. Operation and maintenance programs have to be conducted properly. Air filters also can be a pollutant source: Particulate filters are not suggested for use in gaseous pollutant removal, but activated charcoal filters are generally effective in removing organic chemicals and particulates.

Public Education

Public education can be the least costly but most effective method of pollution control. It is very important to inform the public, building occupants, management, building maintenance personnel, and others fully regarding the sources and causes of indoor pollutants. Their cooperation can markedly reduce pollutant exposure.

REFERENCES

1. Holt, G.L. *Annual American Conference of Governmental Industrial Hygienists,* Vol. 10, pp. 16-18 (1984).
2. The National Research Council. "Sulfur Dioxide." National Academy of Sciences, Washington, DC, (1978).
3. Windholz, M., et al. *The Merck Index.* Merck & Co., Inc., Rahway, NJ (1983).
4. Hansen, T.B., and Andersen, B. *Ozone and Other Air Pollutants From Photocopying Machines.* American Industrial Hygiene Association (October 1986).
5. Novak, V. "Profits in The Air." *Venture Builder,* p. 74 (December 1988).
6. Smulski, S., Formaldehyde Indoors. *Progressive Builder* (April 1987).
7. "Occupation Diseases—A Guide To Their Recognition," National Institute for Occupational Safety and Health (June 1977).
8. National Academy of Sciences. "Airborne Lead In Perspective." National Research Council, Washington, DC (1972).

CHAPTER 4

VENTILATING INDOOR AIR POLLUTION SOURCES

Valerie Lynn David

J.M. Huber Corporation
Printing Ink Division
333 Thornall Street
Edison, NJ 08818-4032, USA

CONTENTS

INTRODUCTION

Until fairly recently, if someone had been asked an opinion about "air quality," he or she would have spoken about the air outside. Since the Donora episode in Pennsylvania in 1948 and the "Great London Smog" in 1952, outside air quality has become increasingly regulated. The Clean Air Act of 1990 is continuing the trend of regulating air pollution sources and reducing the amounts and types of chemicals allowed into the atmosphere. But it has been only within the past two decades that indoor air quality has become a topic. In the past, it was assumed that air pollution was only outside, and a building offered shelter and protection. As early as 1958, studies indicated that indoor air quality also could pose health problems, but there were few epidemiological studies. In the early 1970's, however, indoor air quality (IAQ) began being researched and investigated in a more direct manner. And by 1974, indoor air quality was a recognized problem, with the EPA reporting, "Indoor generation of air contaminants will assume greater importance as outdoor air quality improves" [1].

Several factors forced IAQ into the limelight. In 1972, the first IAQ conference, titled "Improving Indoor Air Quality," was convened in South Berwick, Maine. The attendees discussed their suspicion that gas stoves generate pollutants indoors, and that IAQ was a more important problem than previously suspected. The energy crisis of 1973 brought about the design of energy-efficient, airtight housing and office buildings. Because of the high costs of heating and cooling, buildings were designed to have significantly less air infiltration, and the American Society of Heating, Refrigeration, and Air-Conditioning Engineers (ASHRAE) recommended a ventilation rate of 5 cfm per person. Infiltration is air leakage through cracks, interstices, ceilings, floors, and walls. The increase in insulation and vapor barrier use led to a decrease in the number of air exchanges per hour and kept moisture within a building. This led to what is now called "tight building syndrome" (TBS). Demand for improved housing with increased living standards, as well as increased labor costs, caused the building industry to switch from traditional materials, i.e., natural woods, to pressed-wood products and fiberboard, which are less expensive. Furniture construction also was changing about the same time, so that plastics, artificial fibers, and composition woods were being used more widely. And household cleaners, insecticides, and personal health-care products, formulated with new synthetic chemicals, were introduced. All of these factors led to the increase in the concentration of contaminants being emitted and staying within a building.

After 1974, people occupying those newly designed energy-efficient office buildings began complaining about "stuffy" and "stale" air. The U.S. National Institute for Occupational Safety and Health (NIOSH) has investigated more than 1,100 IAQ complaints between the late 1970s and 1993. They have determined that 50% of the complaints actually are due to poor ventilation, thermal comfort or humidity; 30% due to some indoor air contaminant such as VOCs, formaldehyde, dusts, and biological agents; 10% due to contaminants from the outside, such as motor vehicle exhaust, pollen, fungi, smoke; and 10% undetermined [2]. If more than 20% of a building's occupants report symptoms and the causes cannot be identified, the phenomenon is referred to as sick building syndrome (SBS). NIOSH also compiled a list of the frequency and types of complaints. They were: 81% eye irritation, 71% dry throat, 67% headache, 53% fatigue, 51% sinus congestion, 38% skin irritation, 33% shortness of breath, 24% cough, 22% dizziness, and 15% nausea [2].

All of these problems can be corrected to some degree by having a well-designed and operating HVAC (heating, ventilating, and air conditioning) system. This chapter describes the strategy for correcting and preventing IAQ problems; defines the nature of the problem; investigates potential causes and sources; compares findings to recognized good practices, standards, and codes; and outlines implementing cost-effective controls. IAQ problems and solutions are addressed for both residential and commercial buildings, with an emphasis on commercial buildings.

DEFINING THE NATURE OF THE PROBLEM

The first step in correcting IAQ problems is to define the nature of the problem. An on-site investigation should include information from the occupants, HVAC and space characterization details, and air testing and monitoring to identify or confirm potential causes and sources. Tables 1 through 8 consist of checklists of information to obtain [2]. Prior to the investigation however, one should be familiar with how HVAC systems operate.

(text continued on page 81)

Table 1
Investigation Checklist

Preliminary steps:

- Meet with someone knowledgeable about the problem, usually a supervisor or the employer.
- Meet with those affected by the problem. Reassure them. Tell them you'll be back to gather more data.
- Conduct a quick walk-through survey to become familiar with the general scope of the building, occupants, and activities.
- Develop a plan for the rest of the investigation.
- Conduct on-site investigations to evaluate suspected problems:
  - Develop and use complaint forms, questionnaires, or interviews to gather more detailed data from those affected.
  - Research the HVAC systems and any other air handlers.
  - Create a diagram of space characterization.
  - Gather information on recent construction, renovations, building materials, and furnishings.
  - Sample and analyze air and water.

If necessary:

- Conduct additional air sampling, medical exams, and other extensive data gathering.
- Summarize your findings and determine all potential causes and solutions.

Conclude the investigation:

- Compare findings and existing conditions to standards, codes, and good practices (e.g., ASHRAE 62).
- Determine and test potential control measures.
- Prepare a report with findings and recommendations.
- Apply controls and conduct followup.

Table 2
Sample Office Questionnaire (#1)

Some people are concerned about the office environment. To investigate these complaints, we would like your cooperation in filling out this questionnaire.

1. Problems. (Please select those applicable to you. Not all complaints have been registered—it is a complete list of all possible complaints.)

☐ Aching joints	☐ Muscle pain	☐ Back pain
☐ Hearing problems	☐ Dizziness	☐ Dry, flaking skin
☐ Discolored skin	☐ Skin itching	☐ Skin irritation
☐ Heartburn	☐ Nausea	☐ Noticeable odors
☐ Sinus trouble	☐ Sneezing	☐ Watering eyes
☐ Congestion	☐ Chest tightness	☐ Eye irritation
☐ Problems wearing contact lenses		☐ Headache
☐ Sore throat	☐ Fatigue	☐ Drowsiness
☐ Sleepiness	☐ Air too dry	☐ Temperature too cold
☐ Too humid	☐ Freq. colds, flu	☐ Temperature too hot
☐ Musty smell	☐ Dirty air	☐ Excessive noise

2. When do these complaints occur?

☐ Morning ☐ Afternoon ☐ All day
☐ Daily ☐ Specific days of week (Specify:)
☐ No noticeable trend or time

3. When do you experience relief of these complaints? ____________________

____

4. Do you have any of the following?

☐ Hay fever ☐ Skin allergies, skin irritation
☐ Other allergies ☐ Colds/flu ☐ Sinus problems

5. Do you smoke tobacco?

☐ No ☐ Yes ☐ Amount ( )

6. Please note your location: ____________________

7. Comments or suggestions: ____________________

8. Name (optional): ____________________

Table 3
Sample Office Questionnaire (#2)

This questionnaire is to determine the cause of complaints expressed by employees. Your cooperation is essential. Your responses will be confidential.

Age: ( ) Sex: ( ) Location ( )

Which of the following symptoms, if any, do/did you have?

Yes	No	Not Sure	
	☐	☐	irritated or watering eyes
	☐	☐	sneezing
	☐	☐	stuffy nose
	☐	☐	post nasal drip
	☐	☐	sore throat
	☐	☐	cough
	☐	☐	difficulty breathing
	☐	☐	headache
	☐	☐	wheezing
	☐	☐	rash
	☐	☐	fever
	☐	☐	vomiting
	☐	☐	diarrhea
	☐	☐	cold symptoms
	☐	☐	sinus trouble

General questions:

☐ Do you know the cause of symptoms?
☐ Do you wear contact lenses?
☐ Do you smoke tobacco?
☐ If yes, how many per day? ( )
☐ Does tobacco smoke bother you at work?
☐ Do you sit near people who smoke?
☐ Do you know the source of the complaints you were experiencing?

Table 4
Quick Walk-Through Survey Worksheet

Name________________________ Date____________________________

Contact______________________ Location_________________________

Phone________________________ ____________________________

Potential Problem	Yes/No	Comments
☐ Employee complaints	______	____________________
☐ Symptoms present	______	____________________
☐ Illness present	______	____________________
☐ Lack of outside air	______	____________________
☐ Inadequate air distribution	______	____________________
☐ Pressure diff between rooms	______	____________________
☐ Air infiltration at perimeters	______	____________________
☐ Detectable odors	______	____________________
☐ Excessive tobacco smoke	______	____________________
☐ Temperature too warm	______	____________________
☐ Temperature too cold	______	____________________
☐ Humidity too high	______	____________________
☐ Humidity too low	______	____________________
☐ Poorly vented heating equipment	______	____________________
☐ Poorly located intakes	______	____________________
☐ Visible mold, slime	______	____________________
☐ Water visible	______	____________________
☐ Water damaged furnishings	______	____________________
☐ Cleaning chemicals	______	____________________
☐ Deteriorated asbestos	______	____________________
☐ Dirty, organic debris	______	____________________
☐ Poor maintenance	______	____________________
☐ Equipment emissions	______	____________________
☐ Improper exhaust ventilation	______	____________________
☐ Poorly located loading docks	______	____________________
☐ Other	______	____________________
☐ Other	______	____________________

Table 5
Building Information Checklist

Building:__________ Contact Person:__________
Address:__________ Telephone:__________
Date: __________ Investigator:__________

Building Description:

▯ Year Built	▯ Owner	▯ Date Occupied
▯ Occupants	▯ No of Floors	▯ Neighborhood type
▯ Construction	▯ Traffic Pattern	▯ Loc. Garages
▯ Emission Sources	▯ Tightness of doors, windows	
▯ Interior Construc.	▯ Interior Layout	▯ Insulation type
▯ HVAC type		

Occupant Space Description

▯ Number of people	▯ Sq.ft/person	▯ Type of activity
▯ Smoking policy	▯ No of smokers	▯ Chemicals present
▯ Cleaning materials	▯ Furnishings	▯ Construction Mat.
▯ Recent Construction	▯ Recent changes	▯ Freestanding fans
▯ Temp., Humidity	▯ Mold/Dirt	▯ Wet surfaces
▯ Adjacent space	▯ Room pressure	▯ Carbon dioxide
▯ Drafts	▯ Stuffiness	▯ Interviews

HVAC Systems

▯ Type of system	▯ Condition	▯ Windows
▯ Type of fuel	▯ Type of Diffuser	▯ Location of Intakes
▯ Location of exhaust	▯ OA provisions	▯ Distribution
▯ Terminal velocities	▯ Noise	▯ Dust/dirt
▯ Economizer cycle	▯ Controls	▯ Zones
▯ Total cfm	▯ Total OA	▯ Heat exchanger
▯ Local Exhaust	▯ Makeup air	▯ Duct type
▯ Water in system	▯ Type humidifier	▯ Restroom exhaust
▯ Air Cleaner type	▯ Air cleaner effic	▯ Person in charge

Table 6
Basic Information Checklist for HVAC Systems

- Plans, drawings, specifications, changes
- The type of system (VAV, constant volume)
- Which rooms have windows that open?
- Location of air intakes
- Location of AHUs
- Percent OA
- How is OA percentage determined?
- How are each of the following controlled (automatic, manual, who, when, how)?
- Fans
  - ☐ OA Damper
  - ☐ RA, SA, fan dampers
  - ☐ Supply terminal dampers
  - ☐ Humidity
  - ☐ Temperature
  - ☐ Air distribution
- Type of filtration
  - ☐ Arrestance/dust spot efficiency
  - ☐ Filter maintenance
- What is the maintenance program for each of the following (frequency, how, who, when)
  - ☐ Fan components
  - ☐ Drive components
  - ☐ Filters
  - ☐ Drain pans, traps, valves, nozzles
  - ☐ Dampers
  - ☐ Controls
- For return air systems (RA)
  - ☐ Location
  - ☐ Plenum or duct
  - ☐ Lining
- For ducts
  - ☐ Type
  - ☐ Insulation (inside, outside, material)
  - ☐ Inspection, cleaning, repair

Table 7
Space Characterization Worksheet

Location ____________________ Date/time ____________________

Name ____________________ Address/phone ____________________

Contact ____________________ Number working in space ____________

Sketch
Show doors, windows, supply and return registers, dimensions; show floor plan or other necessary detail. Use reverse side for more.

L ________ W ________ H ________ Room Volume ____________

T_{DB} ____________ T_{WB} ____________ R.H. ____________

C_{CO2} ____________ C_{other} ____________ C_{other} ____________

T_{MA} ____________ T_{RA} ____________ T_{OA} ____________ %OA ________

C_{SA} ________(CO_2) C_{RA} ________(CO_2) C_{OA} ________(CO_2) %OA ________

Supply volume flow rates (SA)*: Q_1 ________ Q_2 ________ Q_3 ________

Q_4 ________ Q_5 ________ Q_6 ________ $Q_{SAtotal}$ ________

Return volume flow rates (RA)*: Q_1 ________ Q_2 ________ Q_3 ________

Q_4 ________ Q_5 ________ Q_6 ________ $Q_{RAtotal}$ ________

Q_{OA} ____________ $Q_{OA/person}$ ____________ AC/Hr ____________

Terminal (draft) velocities ____________ Location ____________

Pressure relationship with hallways: (+) or (-) Light ________ Noise ________

**Measure supply and return air volumes with doors and windows closed; if doors and windows cannot be shut, measure flow rates through doors and windows.*

Table 8
Testing and Monitoring Equipment

For Measuring	Common Instruments	Typical Suppliers
Air velocity in ducts	Pitot tube and manometer, Velometer	Alnor, TSI, Kurz, Dwyer
Airflow direction	Smoker tube kits	MSA, Draeger, Sensidyne
Airflow at diffusers, return registers	Balometer	Alnor
OA delivery	CO_2 detector tubes or instruments, thermometers	Gastech, Draeger, MSA, Sensidyne
Pressures in duct systems	Pitot tube and manometer	Dwyer
Air distribution	Smoke tubes, soap bubble, neutral-buoyant balloon	MSA, Draeger, Sensidyne
Drafts	Low-velocity velometer	TSI, Kurz
Temperature and humidity	Psychrometer, thermometers	Bachrach, Cole Palmer
Carbon dioxide	Direct reading instruments; indicator tubes	Gastech, Valtronics, B&K; MSA, Draeger, Sensidyne
Carbon monoxide	Direct reading instruments; indicator tubes	Ecolyzer, Interscan, B&K; Draeger, MSA, Sensidyne
Oxides of nitrogen	Direct reading instruments; indicator tubes	Interscan, Ecolyzer; Draeger, MSA, Sensidyne
Ozone	Direct reading instruments; indicator tubes	CSI; Draeber, MSA, Sensidyne
VOCs	Direct reading instruments; indicator tubes	Gilian, MSA, Sipin, B&K; Draeger, MSA, Sensidyne
Bioaerosols	Impactor, culture medium	Anderson
Noise	Sound-level meter	B&K
Radon	Radon monitor	Nuclear associates

(text continued from page 73)

HVAC Operation

Air is composed primarily of nitrogen, oxygen, water vapor, argon, and carbon dioxide. It has a density of 0.075 lbs/ft^3 at standard conditions, specific gravity of 1 for gas or vapors, and a composite molecular weight of approximately 29.0. At standard temperature and pressure (STP), air has a barometric pressure of 29.92 inches Hg, is 68°F, and has 50% relative humidity (RH). The density must be corrected for temperature and pressure, however, when not at standard conditions.

Air flows from high- to low-pressure areas. Atmospheric pressure is recorded as barometric pressure and is measured in inches of mercury. In ventilation systems, there are three types of pressure: static, velocity, and total. Static pressure (SP) is produced by a fan that pushes the air in a ventilation system, is recorded as gauge static pressure, and is measured in inches of water (407 inches water = 29.92 inches Hg). Figure 1 illustrates how the fan has raised the atmospheric pressure, creating 1 in. of positive-gauge static pressure. The U-tube manometer measures this pressure difference [2]. Static pressure is felt

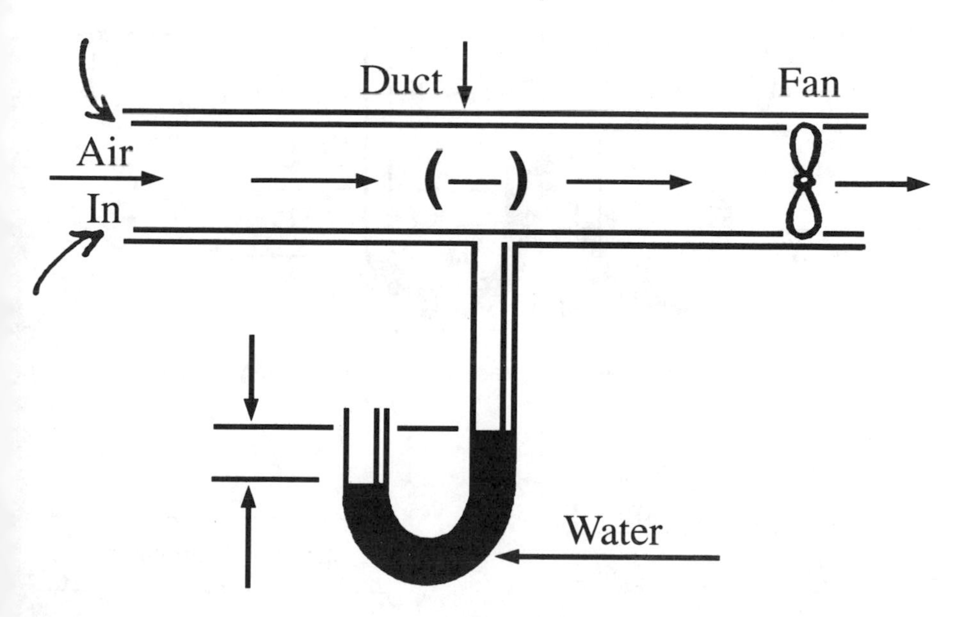

Figure 1. A fan raises atmospheric pressure, creating 1 in. of positive-gauge static pressure.

in all directions within the duct. Negative SP would cause the duct to collapse, while positive SP would cause the duct to expand. Velocity pressure (VP) is the pressure of the air on the manometer probe, and total pressure (TP) is the combination of the static and velocity pressures. The TP, SP and VP values can be positive or negative relative to the fan as illustrated in Figures 2 and 3 [2].

The last two measurements in air ventilation systems are volume flow rate (Q), which equals the velocity (V) × the area (A) in m^3/sec (cfm or scfm for flow at standard conditions, or acfm for flow at actual conditions), and the relationship between velocity (V) and velocity pressure (VP) where $V = 4005\ (VP)^{1/2}$. VP is expressed in inches of water (in. wg) and V is expressed in feet/minute (fpm) [2].

Mechanical air-handling systems (AHSs) distribute air to meet odor, temperature, humidity, and air quality requirements. HVAC engineers talk in terms of zones that define the areas served by the air-handling system. Each zone usually has its own thermostat. An AHS is designed to combine volume flow rate, temperature, humidity, and air quality to produce comfortable working conditions. They require:

- Outside air intakes, plenums, and dusts (OA).
- Filters or scrubbers.

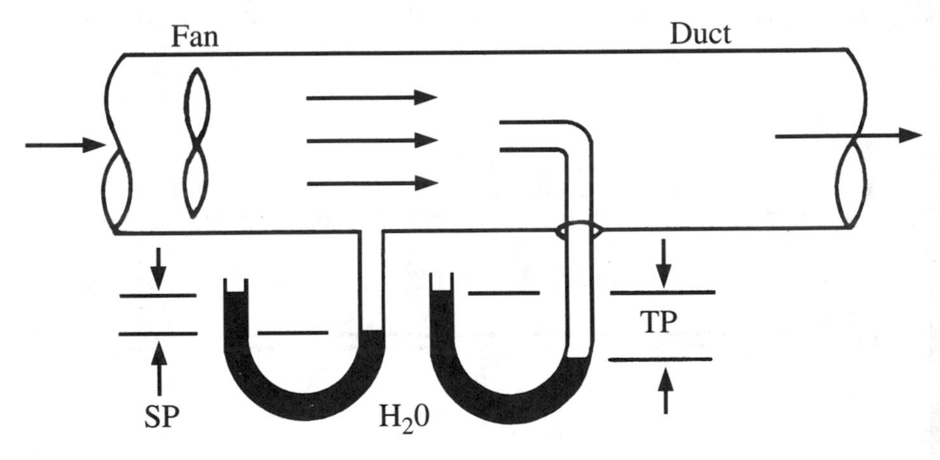

Figure 2. Static pressure (SP), total pressure (TP) and velocity pressure (VP) are relative to the fan.

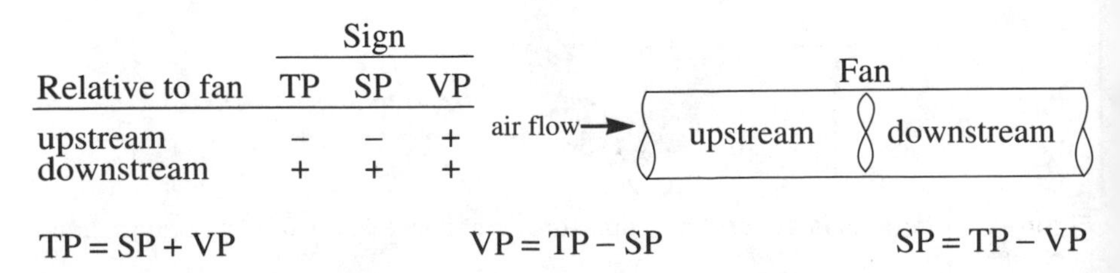

Figure 3. Pressures as they relate to the fan.

- Supply fans (SF).
- Heating and/or cooling coils.
- Humidifying and/or dehumidifying equipment.
- Supply ducts (SA).
- Distribution ducts, boxes, plenums, and registers.
- Dampers.
- Return-air plenums or ducts (RA).
- Return fans (RF).
- Exhaust outlets (EA).
- Controls and instrumentation.

A single-zone constant-volume (SZCV) system incorporating many of these components is illustrated in Figure 4 [2]. The process is as follows:

1. Outside air (OA) comes into the HVAC system through an outdoor air intake.
2. The damper regulates the amount of air admitted.
3. Outside air (OA) mixes with returned air (RA).

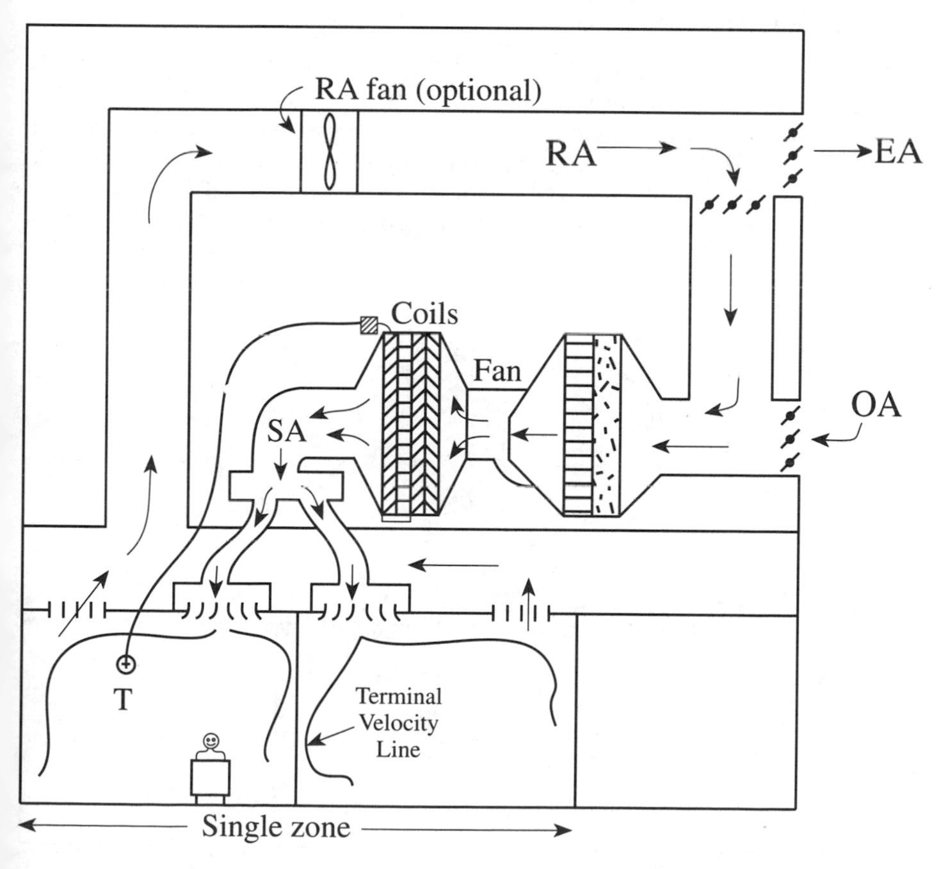

Figure 4. A single-zone constant volume (SZCV) system.

4. Mixed air goes through a prefilter for coarse items and an efficient filter for small particles.
5. Mixed, filtered air goes through a centrifugal fan.
6. Under positive pressure, the air is pushed towards the heating or cooling coils, depending on the temperature of the air and the season.
7. The air passes through a humidifier or dehumidifier.
8. Air is now called supply air (SA) as it travels through the ducts at about 1,000 to 2,000 fpm.
9. Supply air (SA) reaches a distribution box, and then travels through smaller flexible ducts to the supply terminals or diffusers.
10. Supply air (SA) enters the room, and velocity slows to terminal velocity of about 40 to 50 fpm.
11. Supply air (SA) moves towards the return-air register and is now called return air (RA).
12. A return fan (RF) moves the return air (RA) along.
13. Controls determine how much return air (RA) is recirculated and how much is exhausted (EA).

A multiple-zone constant-volume (MZCV) system is illustrated in Figure 5. Multiple thermostats control multiple zones [2].

In addition to constant-volume systems, there are variable air volume systems. A single-zone variable air volume (SZ VAV) system, and a single-zone variable air volume dual-duct (SZ VAV DD) system are illustrated in Figures 6 and 7. The VAV system provides only enough air to maintain the temperature. These

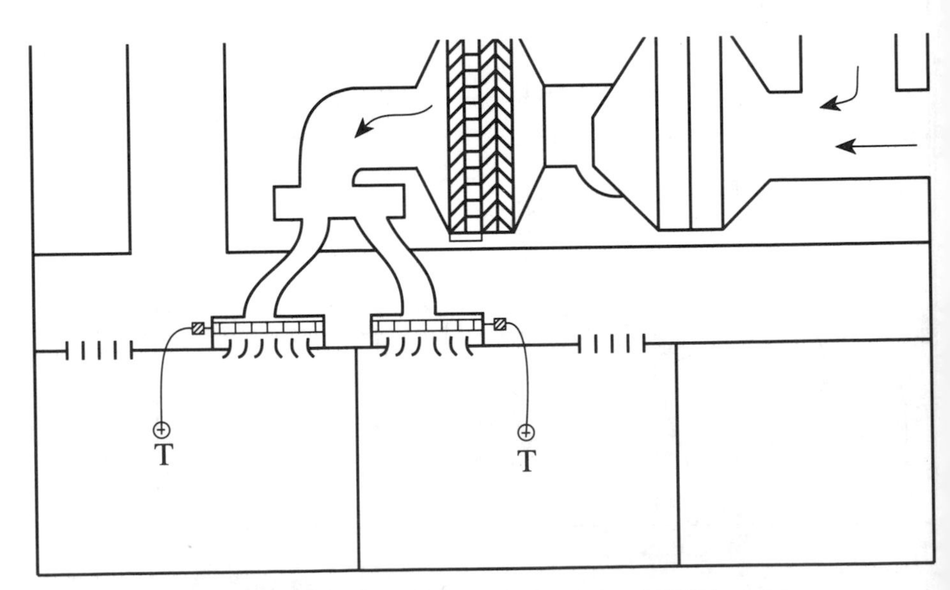

Figure 5. A multiple-zone constant volume (MZCV) system.

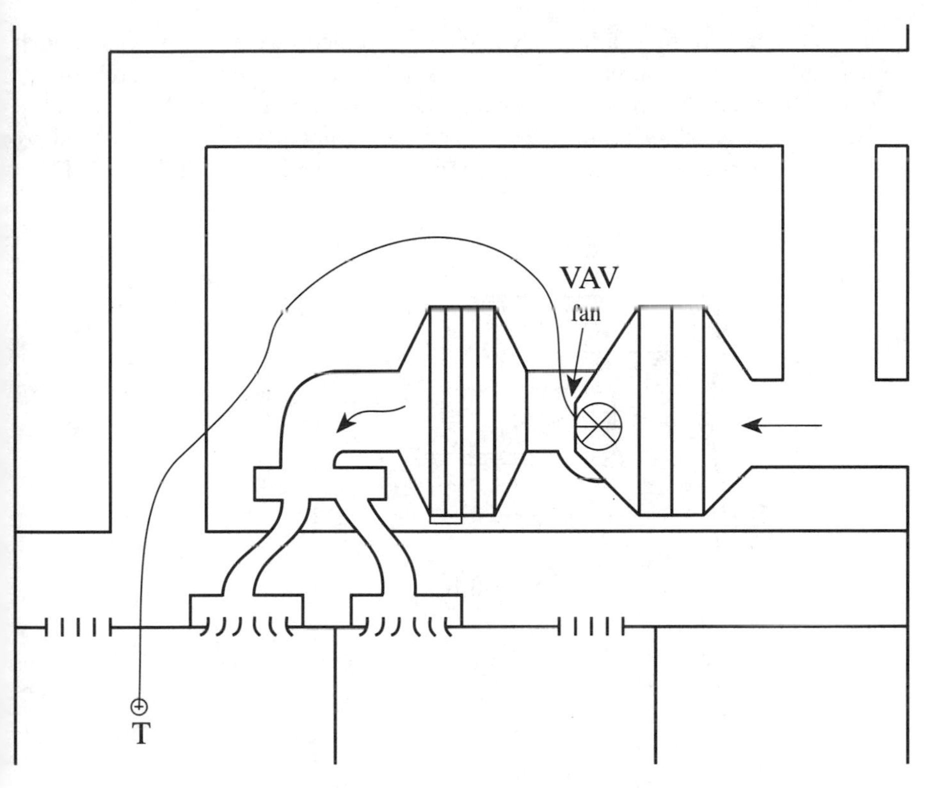

Figure 6. A single-zone variable air volume (SZ VAV) system.

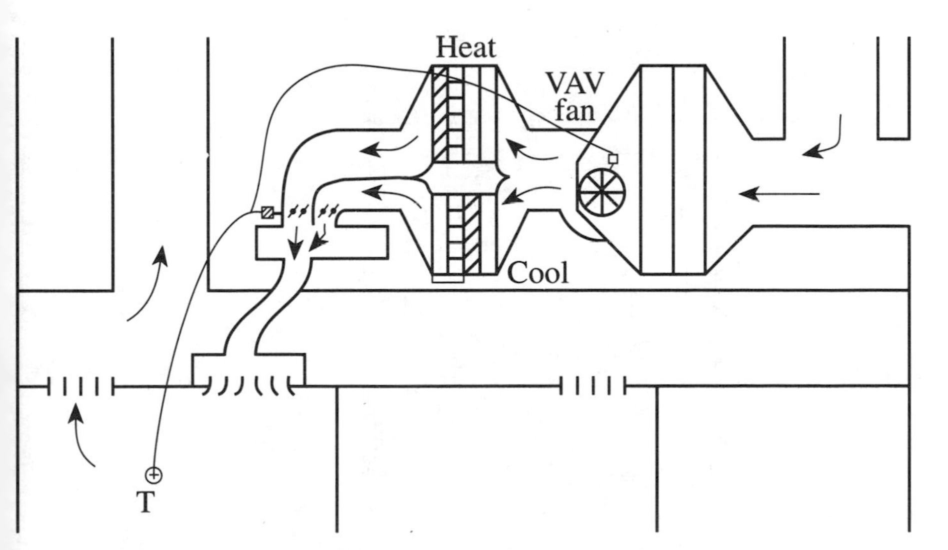

Figure 7. A single-zone variable air volume dual-duct (SZ VAV DD) system.

systems are energy efficient, but they can cause problems by not providing enough outside air. The incorporation of dual ducts solves the minimum air problem by mixing heated and cooled, air and running the fan continuously at the lowest rpm.

Examples of an air-handling system and the symbols and notations used by HVAC engineers are illustrated in Figures 8 and 9 and listed in Table 9 [2].

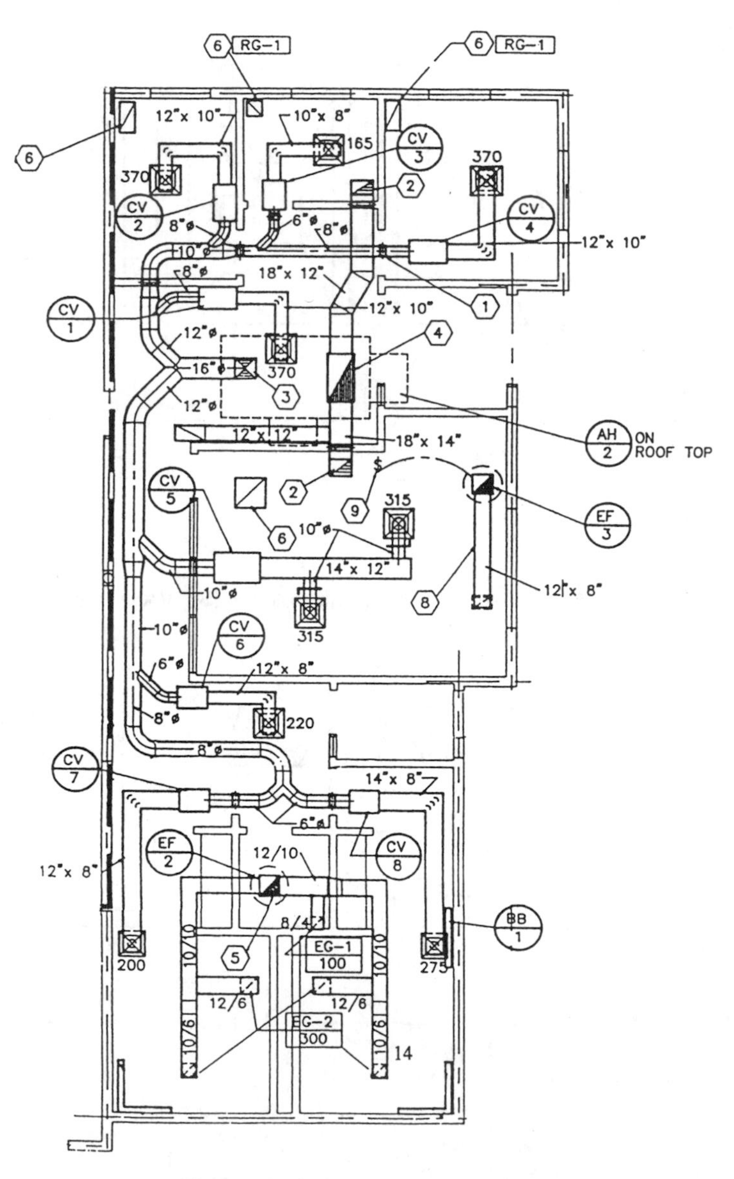

Figure 8. An air-handling system.

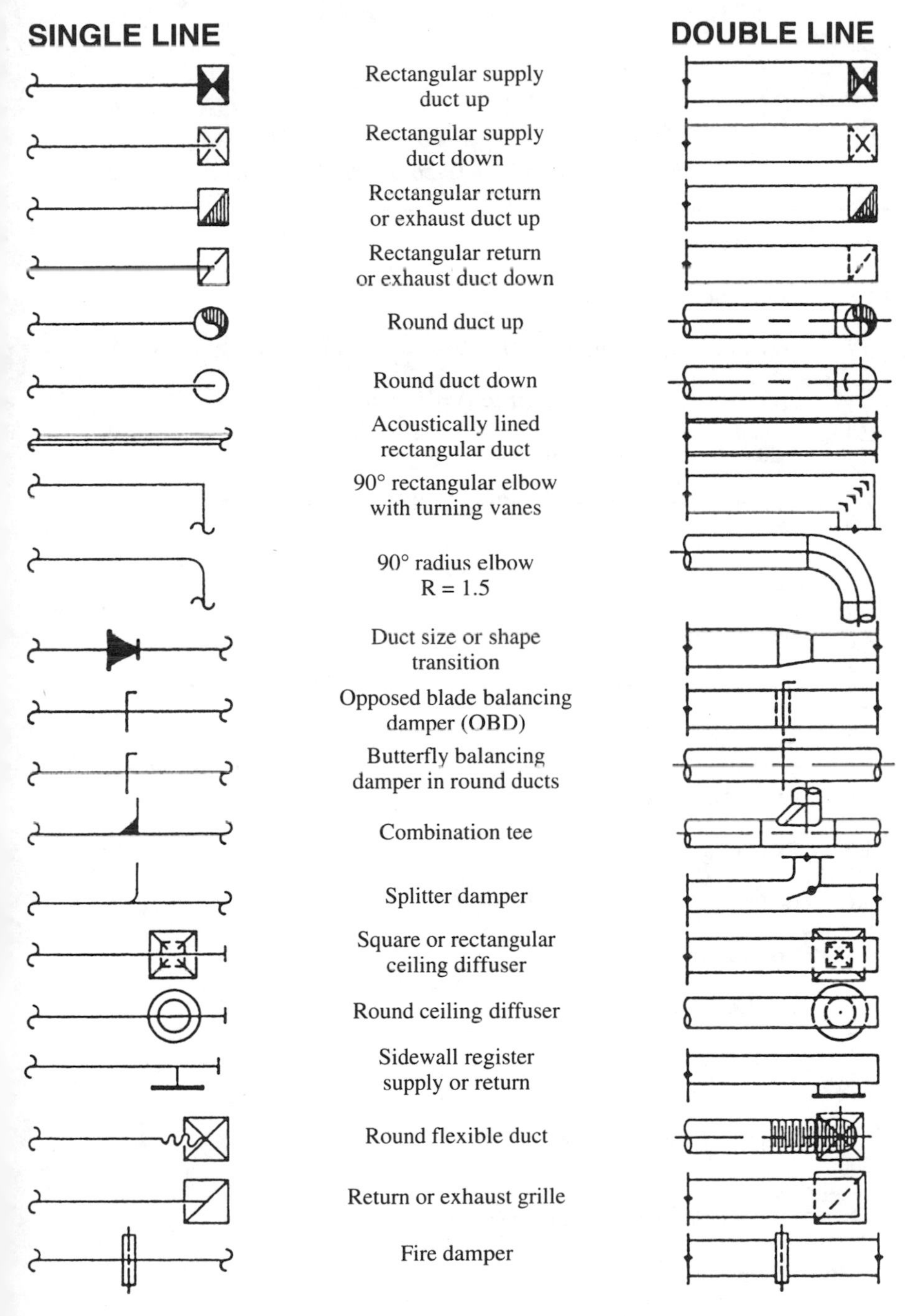

Figure 9. Symbols and notations used by HVAC engineers.

Table 9
Symbols and Notations Used by HVAC Engineers

Symbol	Meaning
⟨1⟩	Fire damper
⟨2⟩	Open-ended return air duct
⟨3⟩	Transition from square AHU outlet to round 16″ supply duct
⟨4⟩	Transition from return duct to AHU inlet
⟨5⟩	Exhaust duct rising to the roof
⟨6⟩	Sound-proofed return air grill to ceiling plenum
⟨7⟩	Unused
⟨8⟩	Duct to exhaust fan on roof
⟨9⟩	Wall switch to control roof-mounted exhaust fan
AH	Air Handler
BB	Base board heater under windows
EF	Exhaust fan
RG	Return grill EG Exhaust grill
CV	Control valve
CD	Ceiling Diffuser
⊠ 200	Ceiling diffuser with cfm noted
◱ 300	Ceiling exhaust grill with cfm noted
12″ x 10″	Duct sizes shown in inches
10″ø	– round duct

POTENTIAL CAUSES AND SOURCES

There are many types of indoor air pollutants. Those that usually affect the IAQ of residential buildings are: asbestos, biologicals, combustion by-products, environmental tobacco smoke, lead, pesticides, radon, and VOCs from both non-formaldehyde and formaldehyde sources. Those that usually effect commercial buildings are: asbestos, biologicals, environmental tobacco smoke, pesticides, and VOCs. The sources and health hazards for each is summarized in Tables 10, 11, and 12. The potential causes and sources contributed by the HVAC system are summarized in Table 13 [2].

(text continued on page 95)

Table 10
Chemical Contaminants and Potential Sources

Formaldehyde	(tobacco smoke, insulation, paneling, furnishings, carpet, stay-pressed cloth, deodorizers, paper products)
Ozone	(photocopiers, electrical equipment, ozone generators, electrostatic air cleaners)
NO_2	(vehicles, gas heating and cooking, industrial processes)
CO_2	(people density, flame operations)
Organic Chemicals	(photocopiers, industrial processes, labs, new and existing, furnishings, building materials, cleaning materials, paint)
Allergens	(pollen, fungi, bacteria, mold, mites, dust, ETS)
Fibers	(insulation, fireproofing, equipment)
SO_2	(industrial processes)
CO_2	(people, open flames, industrial processes, outdoor air, malfunctioning heating equipment, vehicles)
ETS	(tobacco smoke)

Table 11
Emission Factors

Materials	mg/hr-sq meter	VOC	Time of test
Med. density fiberboard	0.7–2.3	HCHO	higher values are for materials
Hardwood plywood paneling	0.06–1.4	"	
Particleboard	0.08–2.0	"	
Urea-formaldehyde foam insulation	0.05–0.8	"	
Softwood plywood	0.01–0.03	"	
Paper products	0.01–0.03	"	
Clothing	0.015–0.02	"	
Plywood	1.0	"	"new"
Silicone caulk	13	TVOC	< 10 hours
	<2	"	10–100 hours
Floor adhesive	220	"	< 10 hours
	< 5	"	10–100 hours
Floor wax	80	"	<10 hours
	<5	"	10–100 hours
Wood stain	10	"	< 10 hours
	< 0.1	"	10–100 hours
Polyurethane wood finish	9	"	< 10 hours
	< 0.1	"	10–100 hours
Floor varnish or lacquer	1	"	< 10 hours
Particle board	0.2	"	2 years old
Chipboard	0.1	"	unknown

Gypsum board	0.03	"	unknown
Wallpaper	0.1	"	unknown
Latex-backed carpet	0.15	4-PC	one week old
	0.08	4-PC	two weeks old
Dry-cleaned clothes	1	PCE	0–1 day
	0.5	PCE	1–2 days

Emissions data for sealants and caulks

Material	Weight loss, fraction of original sample		Hours to dry completely	Primary VOC
	48 hrs	Fully dry		
Butyl rubber	0.05	0.18	430	Aliphatic HC
Acrylic emulsion latex	0.05	0.12	—	TVOC
Silicon	0.02	0.04	480	Xylene
Styrene butadiene rubber	0.14–0.16	0.19–0.35	100–250	Aliph HC, Xylene
Neoprene blend	0.17	0.23	100	MEK, Xylene
Oleoresin	0.01	0.04	2,000	Aliphatic HC
Polysulfide, one-part	0.01	0.06	4,900	Toluene
Acrylic solvent-based	0.03	0.14	1,000	Xylene
Asphaltic, one part	0.01	0.08	4,500	Petrol. HC
Neoprene, one part	0.18	0.33	214	Xylene
Polyurethane, one part	0.01	0.15	8,300	Xylene

Table 12
Properties of Common Solvents, Gases, and Vapors

Solvent/vapor	Molecular Wt MW	Specific Gravity (re: water)	Explosion Limit % LEL	UEL	TLV* (ppm)	IAQ-LV* (ppm)
Acetaldehyde	44.1	0.821	4.0	57	25	0.1 TLV
Acetic Acid	60	1.049	na	na	10	0.1 TLV
Acetone	58.1	0.792	2.7	10.1	750	0.1 TLV
Benzene	78.1	0.879	1.4	7.1	0.1	0.1 TLV
Cresol	108.1	1.05	1.4	na	5	0.1 TLV
Carbon Tet	153.8	1.596	na	na	5	0.1 TLV
Ethyl Acetate	88.1	0.901	2.2	11.4	400	0.1 TLV
Ethyl Alcohol	46.1	0.789	3.3	18.9	1,000	0.1 TLV
Ethylene Glycol	62.1	1.12	3.2	na	50	0.1 TLV
Freon TF solvent	187.4	1.57	na	na	1,000	0.1 TLV
Gasoline	86	0.66	1.1	6.7	300	0.1 TLV
Isopropyl alcohol	60.1	0.785	2.0	11.8	400	0.1 TLV
Kerosene	180	0.80	0.7	5	na	na
Methanol	32	0.792	6.7	36.5	200	0.1 TLV
MEK	72.1	0.806	1.8	11.5	200	0.1 TLV
MIBK	100.1	0.801	na	na	50	0.1 TLV
Mineral spirits	>100	0.80	0.8	5	na	na
Naphtha (solvent)	90	0.7	0.8	5	200	0.1 TLV

Styrene monomer	104.1	0.903	1.3	6.7	50	0.1 TLV
Perchloroethylene	165.8	1.63	na	na	25	0.1 TLV
Toluene	92.1	0.866	1.3	6.7	50	0.1 TLV
1,1,1 TC Ethane	133.4	1.31	7	16	350	0.1 TLV
Trichloroethylene	131.4	1.466	na	na	50	0.1 TLV
Xylene	106.2	0.881	1.0	6.0	100	0.1 TLV

Gas	Molecular Wt MW	Specific Gravity (re: air)	Explosion Limit % LEL	Explosion Limit % UEL	TLV*† (ppm)	IAQ-LV*‡ (ppm)
Air	29	1.0	na	na	na	no odors
Butane	58.1	2.085	1.9	8.4	800	0.1 TLV
Carbon Dioxide	44	1.53	na	na	5,000	1000
Carbon Monoxide	28.1	0.968	12.5	74.2	25	9
Chlorine	70.9	3.21	na	na	0.5	0.1 TLV
Formaldehyde	30	1.08	na	na	0.3	0.1
Hydrogen Chloride	36.5	1.27	na	na	5	0.1 TLV
Hydrogen Sulfide	34.1	1.19	4.3	45.5	10	0.1 TLV
Nitrogen Dioxide	46.0	1.448	na	na	3	0.3
Ozone	48	1.658	na	na	0.1	0.05
Sulfur Dioxide	64.1	2.264	na	na	2	0.14

†"TLV" is 1992-93 ACGIH airborne limits for places of employment.
‡"IAQ-VL" is ASHRAE data or other typical targets for general indoor environments.

Table 13
IAQ Contaminant Sources, by Groups

Major Category	Major Source Type	Contributing Sources (examples)
Outside sources	Contaminated outdoor air	pollen, dust, mold spores industrial pollutants general vehicle exhaust, smoke
	Nearby emission sources	parking garages, lots freeway traffic loading docks, truck parks trash bin odor, debris and garbage building exhaust reentrainment
	Soil gases	radon leakage from HC storage tanks soil contamination (e.g., landfill) pesticides, herbicides
	Standing water	rooftop crawl spaces
Building equipment	HVAC system	leaking refrigerant dirt, dust, debris in ductwork microbiological growth biocides, sealants, water treatment cooling towers
	Other equipment	elevator motors office equipment (e.g., copiers) office cleaning supplies shops, labs, closets, kitchens
Human activities	Personal	smoking perfume, cosmetics body odor cooking
	Housekeeping	cleaning chemicals deodorizers clean activities (e.g., dusty sweeping)
	Maintenance	paint, caulk, adhesive, oil, solvent pesticides cleaning materials
Building furnishings and materials	VOC, ROC emitters	textured surface, carpet, textiles, fleecy particle board, shelving
	Inert aerosol emitters	old asbestos old furnishings
	Microbiological origins	water-damaged furnishings surface condensation standing water (e.g., clogged drains)
Other	Accidents	spills, floods, leaks
	Special use areas	smoking area, exercise room, kitchen

Table 13 (continued)

Major Category	Major Source Type	Contributing Sources (examples)
	Repair and remodeling	new furnishings demolition (dust, bioaerosols) paint, caulk, sealants, adhesives

(text continued from page 89)

By Pollutant

Asbestos

Asbestos is a collective term for a group of mineral silicates used in building materials for insulation and fire retardency. The most commercially important and widely used are chrysotile and amosite. Because the EPA and the CPSC banned several asbestos products in 1973 and manufacturers limited their use, asbestos is most commonly found in older residential and commercial buildings in pipe and furnace insulation materials, fireproofing, thermal and acoustical insulation, friction products, reinforcing materials in cement pipes, shingles, millboard, textured paints and other coating materials, and floor tiles. The presence of friable or damaged asbestos materials does not constitute an imminent hazard to building occupants, however. The degree of risk is determined by fiber concentrations, extent of use, age and condition, location and accessibility, level of occupant activity, potential for disturbance and damage, and potential for high humidity and water damage [3].

Asbestos is hazardous because the fibers are very small, are easily inhaled, and accumulate in the lungs. Asbestos can cause a cancer of the chest and abdominal linings called mesothelioma, and an irreversible lung scarring called asbestosis. Symptoms don't show up until many years after the exposure. Asbestos is a very hazardous material, and any exposure poses some risk. There is no apparent safe level of exposure [3].

Biologicals

Biologicals are composed of microbial agents; allergens; bacteria; mold spores and parts; fungal spores; mildew; viruses; animal dander, hair, excretions, and saliva; dust mites; insects; and pollen. Some biologicals trigger allergic reactions including hypersensitive pneumonitis, allergic rhinitis, and asthma; some transmit infectious diseases such as tuberculosis, influenza, measles and chicken pox; and some release disease-causing toxins [4]. Examples of some illnesses caused by microorganisms are Legionella pneumophila (Legionnaire's disease and Pontiac fever), humidity fever (allergic alveolitis), and aspergillosis [1]. These diseases are directly traced to specific building problems and are called building-related illnesses (BRI). Biologicals grow and proliferate under reduced ventilation and/or increased humidity conditions. In fact, bacteria remain at a constant level and are unaffected by increased ventilation because the more important factors are humidity and "cleanliness" of the air [5].

In residential buildings, reduced ventilation and increased humidity conditions exist mostly in bathrooms, kitchens, basements, and attics, and are spread by passive ventilation. Small room humidifiers also have been identified as sources for microorganisms and antigens. In commercial buildings, the usual breeding media are the cooling tower and evaporative condenser waters; contaminated components of the HVAC system such as ductwork, humidifiers, air washers, and fan coils, or mold- or bacterially infested building materials or furnishings. Contaminated air is actively circulated throughout the building. A common source of contamination is the microbial slime that develops on poorly drained condensation pans in the cooling coil units. The slime dries out, and particles become airborne and enter the supply air ducts. In HVAC systems that employ air washers for cooling, a mixture of outdoor and recirculated air passes through filters and the water spray system. This spray water is collected in a tank, chilled, and recirculated. Microbial slime can grow on the baffle plates and in the sump pump, and can be continuously reintroduced.

Combustion By-Products

Combustion by-products can be produced by unvented kerosene and gas space heaters, wood-burning stoves, fireplaces, gas stoves, vented gas and oil furnaces, and water heaters. The major pollutants released are carbon monoxide, carbon dioxide, nitrogen monoxide, nitrogen dioxide, respirable particulates, aldehydes, and a variety of organic gases. Wood stoves, fireplaces, and unvented kerosene space heaters also emit polycyclic aromatic hydrocarbons (PAHs), and kerosene space heaters can generate sulfur dioxide and acid aerosols. Emissions will depend on the type of heater or burner design, type of fuel, position of the wick, and maintenance. Improperly installed chimneys, flues, and furnace heat exchangers also can introduce combustion products [4]. The effect of air exchange rate (ACH) on emission concentrations of radiant and convective space heaters operated in a 12' × 12' × 8' experimental chamber is illustrated in Figure 10. National Ambient Air Quality Standards (NAAQS), Occupational Safety and Health Association (OSHA), and American Society of Heating, Refrigeration, and Air Conditioning Engineers (ASHRAE) averages are displayed for comparison [3].

Carbon monoxide (CO) is a colorless, odorless gas that interferes with the delivery of oxygen throughout the body. At low concentrations, it can cause fatigue in healthy people and episodes of chest pain in people with chronic heart

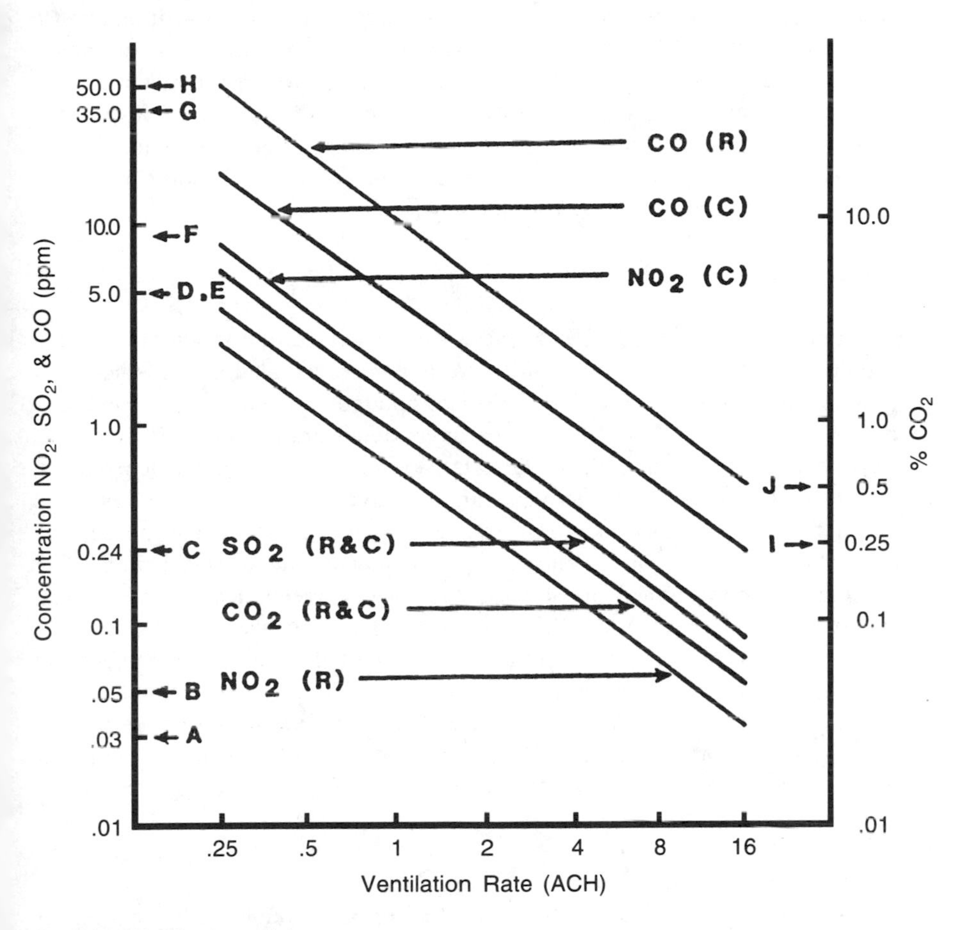

A NAAQS SO_2 Annual Average
B NAAQS NO_2 Annual Average
C NAAQS SO_2 24-hour Average
D OSHA NO_2 Ceiling Value
E OSHA SO_2 8-hour Time-Weighted Average
F NAAQS CO 8-hour Average
G NAAQS CO 1-hour Average
H OSHA CO 8-hour Time-Weighted Average
I ASHRAE CO_2 Guideline
J OSHA CO_2 8-hour Time-Weighted Average

Steady-state concentrations of NO_2, SO_2, CO and CO_2 associated with emissions from radiant (R) and convective (C) space heaters

Figure 10. Steady-state concentrations of NO_2, SO_2, and CO_2 associated with emissions from radiant (R) and convective (C) space heaters.

disease. At higher concentrations, it can cause headaches, dizziness, weakness, nausea, confusion, and disorientation. At very high concentrations, it can cause unconsciousness and death. Nitrogen dioxide (NO2) can irritate mucous membranes and cause shortness of breath after exposure to high concentrations. Continued exposure to low concentrations can increase the risk of respiratory infection. Incomplete combustion also can produce respirable particulates, which can lodge in the lungs, damage lung tissue. Benzopyrene and radon daughters may be attached to respirable particulates [4].

Environmental Tobacco Smoke

Environmental tobacco smoke (ETS) is composed of sidestream smoke (the smoke coming off of the burning end) and the smoke exhaled by the smoker. It is a complex mixture of more than 4,700 compounds, including gases and particles. Some of the more hazardous of the compounds are nicotine, nitrosamines, polycyclic aromatic hydrocarbons (PAHs), carbon monoxide, nitrous oxides, carbon dioxide, formaldehyde, acrolein, and hydrogen cyanide. The carcinogenic potential of tobacco smoke has been associated with the particulate phase, and many of the gas-phase components are potent mucous membrane irritants [3]. According to reports issued in 1986 by the surgeon general and the National Academy of Sciences, ETS is a leading cause of lung cancer in both smokers and non-smokers, and also may contribute to heart disease [4]. A synergistic effect suggested from studies show increased risk of lung cancer in smokers and nonsmokers with indoor radon exposure. Presumably, the radon daughters attached to smoke particulates are selectively deposited at bifurcations of respiratory airways and are highly resistant to dissolution [3].

Lead

Lead can become airborne via lead-based paint being removed by sanding or open flame burning, dust from outside sources being tracked or blown into the building, using lead solder, and previously, from automobile emissions of leaded gasoline. Lead is toxic in both high and low concentrations. It can cause damage to the brain, kidneys, peripheral nervous system, and red blood cells. Children and fetuses are more sensitive to the damaging effects. Effects in children also include delays in physical and mental development, lower IQs, shortened attention spans, and increased behavioral problems [4].

Odor

Organic compounds that are primary contributors to indoor odors are: limonen, α-Pinene, n-Hexanol, 1,3-Xylene, and other terpenes. Odors can act as stressors to initiate complaints of poor IAQ or indicate unwanted organic compounds indoors. No one compound or exposure may cause adverse health effects, but all may be additive to synergistic. The concentrations are usually very low, and the perception will vary with age, sex, susceptibility, and lifestyle. Exposure may increase susceptibility and sensitize some people to other compounds over time [1].

Ozone

Ozone is brought indoors with the makeup air of the HVAC or opening of windows/doors, laser printers, and office and copy machines. Adverse health effects for elevated concentrations of ozone in the outside air include headaches, upper-respiratory problems, and breathing disorders. Safe concentrations have not been determined for indoor air.

Pesticides

According to an EPA survey, 90% of households use pesticides. One study suggested that 80 to 90% of most people's exposure occurred indoors and can be traced to about a dozen sources, such as contaminated soil being blown or tracked indoors, stored containers, and household surfaces collecting and releasing pesticides. Pesticides are used in and around buildings to control insects (insecticides), termites (termiticides), rodents (rodenticides), and fungi (fungicides). Insecticides used indoors include naphthalene, paradichlorobenzene, creosote, pentachlorophenol, sumithrin, methyl demeton, boric acid, diazonin, chlordane, ronnel, chloropyifos, malathion, pyrethrins, and dichlorovos. Antimicrobial or disinfectant pesticides include sodium or calcium hypochlorite, ethanol, pine oil, 2-benzyl-4-chlorophenol, isopropanol, glycolic acid, and 2-phenylphenol [3]. The EPA registers pesticides for use and requires manufacturers to label them. Both the active and inert ingredients in pesticides can cause adverse health effects. Exposure to high levels of cyclodienes (i.e., organic pesticides such as chlordane) produce headaches, dizziness, muscle twitching, weakness, tingling sensations, and nausea. The EPA is concerned that cyclodienes might cause long-term damage to the liver and central nervous system, as well as increased risk of cancer [4].

Radon

Radon (^{222}Rn) is a by-product of the radioactive decay of radium, and has a half-life of 3.8 days. The progeny are electrically charged radioactive daughters. Because they are charged, they can attach themselves to solid and liquid particles that are easily ingested or inhaled. Radon is a naturally occurring noble gas in soil and a number of common rocks and minerals such as granite, schist, and limestone. The usual entry of radon gas into a building is via cracks in the foundation or accumulation in crawl spaces by pressure differential. Once inside the building, the indoor concentration of radon is typically 2 to 10 times that outside, and is approximately inversely proportional to the air exchange rate [1]. Drinking water also can contain radon if the water originated from a contaminated aquifer.

As long as 30 years ago, ^{222}radon and its daughters were recognized as the probable cause of excess lung cancers in uranium miners. Interest then focused on the use of natural building materials with high radium contents and building materials incorporating radium-contaminated by-products such as concrete, sand, gravel, stone, brick, and soil. In 1980, assessments were made of increased lung cancers caused by reduced ventilation, and an excess of 10,000 lung deaths was estimated. More recently, an excess of 1,000 to 20,000 lung cancers per year would likely occur at indoor concentrations of ^{222}radon at 1 pCi/l or 37Bq/m^3 (where 1 Bequerel = 2.702702 E-11 Ci), [1]. Epidemiological studies now suggest a causal role for increased lung cancer risk due to ^{222}radon. Lung cancer rates significantly higher than the national incidence rate are documented for Maine and some counties in Florida [3]. The national average is 1.5 pCi/l, and significant levels in indoor air above the USEPA protection guide of 4 pCi/l have been reported for New Jersey, Pennsylvania, Colorado, New York, and California. A radon risk evaluation chart is shown in Table 14 [2].

VOCs - Nonformaldehyde

After WWII, there was a dramatic proliferation of synthetic organic compounds introduced into buildings, furnishings, and household and personal consumer products. Many of these emit trace to high amounts of VOCs including hydrocarbons, hydrocarbon derivatives, aliphatics, aromatics, alkylbenzenes, ketones, polycyclic aromatics, and chlorinated hydrocarbons. The energy crisis in 1973–74 coupled with breakthroughs in analytical equipment technology allowed the trace amounts of volatile chemicals to be collected, analyzed, and quantified. As early as 1978, total gaseous nonmethane hydrocarbons indoors were found to be higher than outdoors 90% of the time [1]. In most instances, however, the concentrations of organic chemicals found in indoor air are one to two orders of magnitude less than the threshold limit values (TLVs). And because of this, it is difficult to identify any specific VOC as the causal factor of health complaints [3]. In research conducted by the EPA's Total Exposure Assessment Methodology (TEAM), levels of about a dozen common organic pollutants were found to be 2 to 5 times higher inside homes, compared with outside concentrations. The TEAM studies also discovered that the elevated levels persisted in the indoor air long after the activity was completed. Four specific VOCs—benzene, perchloroethylene, paradichlorobenzene, and methylene chloride—were among the most prevalent found by the TEAM studies [4]. In studies of office environments conducted by the Lawrence Berkeley Laboratory, concentrations

Table 14
Radon Risk Evaluation Chart

pCi/l	Estimated Number of Lung Cancer Deaths due to Radon Exposure (out of 1,000)	Comparable Exposure Levels	Comparable Risk
200	440–770	1,000 times average outdoor level	More than 60 times nonsmoker risk
			4 pack/day smoker
100	270–630	100 times average indoor level	2,000 chest X-rays per year
40	120–380		2 pack/day smoker
20	60–210	100 times average outdoor level	1 pack/day smoker
10	30–120	10 times average indoor level	5 times nonsmoker risk
4	13–50		200 chest X-rays per year
2	7–30	10 times average outdoor level	Nonsmoker risk of dying from lung cancer
1	3–13	Average indoor level	20 chest X-rays per year
0.2	1–3	Average outdoor level	

for a variety of aliphatic hydrocarbons, cyclohexane derivatives, alkylated aromatic hydrocarbons, and chlorinated hydrocarbons ranged from 1 to 100 ppb [3]. At present, not much information is known about the detrimental health effects from the levels of VOCs found in homes or office buildings. Some immediate symptoms are mucous membrane irritation, fatigue or lethargy, headaches, dizziness, nausea, visual disorders, and memory impairment. When a significant number of commercial building occupants experience non-specific symptoms that cease after leaving the building, the phenomenon is called sick building syndrome (SBS). Long-term effects are not known for residential or commercial building occupants, even though many of the VOCs are suspected or known to be carcinogenic [4].

VOC - Formaldehyde

Formaldehyde, another post-WWII phenomenon is commonly used to manufacture building materials and household products, and is a by-product of combustion. Sources in the home include ETA; household products; unvented, furl-burning appliances; permanent-press clothing, linens, and fabrics; glues and adhesives; and some paints and coating products. The most significant source in residential and commercial buildings with enough free formaldehyde to significantly contaminate indoor air is usually pressed-wood products containing urea-formaldehyde resin adhesives or urea-formaldehyde foam insulation. Pressed-wood products for interior use include particleboard subflooring, hardwood plywood paneling, medium-density fiberboard, cabinetry, and furniture. Homes with low-level formaldehyde contamination have concentrations of 0.02 to 0.07 ppm, and homes insulated with urea-formaldehyde 1 or more years afterwards have concentrations of 0.03 to 0.13 ppm. Homes with particleboard subflooring 2 to 5 years afterwards have concentrations of 0.06 to 0.30 ppm. Mobile homes have the highest levels, ranging from as high as 1 to 2 ppm for new units, to 0.10 to 0.60 ppm for older models [3]. Formaldehyde emissions generally decrease with time, but when products are new, high indoor temperatures and humidity cause increased emissions as shown in Figure 11 [3].

Urea-formaldehyde resins are chemically unstable. They release free formaldehyde from the volatizable, unreacted formaldehyde trapped in the resin and from hydrolytic decomposition of resin polymer [3]. Formaldehyde is a colorless,

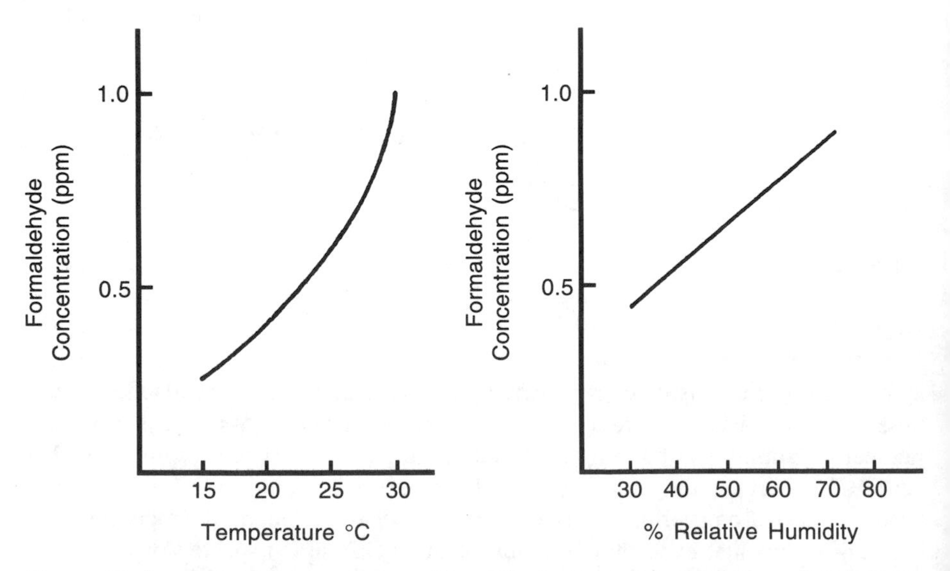

Figure 11. Increases in temperature and humidity increase formaldehyde concentrations.

pungent gas that can cause mucous membrane and skin irritation, nausea, central nervous system depression, and difficulty in breathing in some individuals. There is some evidence that people can develop a sensitivity after exposure. Most formaldehyde complaints share three factors: inadequate ventilation, high population density, and formaldehyde-emitting products that have high emission rates [1]. A causal relationship has not been established between residential formaldehyde contamination and BRI, but one has been established between low-level formaldehyde exposures and a variety of acute irritating symptoms [3]. The first report linking formaldehyde contamination to health problems was published in 1961. During the 1970s and early 1980s, government agencies such as the Consumer Product Safety Commission (CPSC) and the Department of Housing and Urban Development (HUD) began hearing increased complaints. By 1982, the CPSC had received more than 3,000 complaints, and 2,000 of them were related to urea-formaldehyde insulation [1]. Formaldehyde is carcinogenic in animals at high concentrations, but the few epidemiological studies that have been conducted offer conflicting results.

By HVAC System Operation

As mentioned before, air handling systems (AHSs) distribute air to meet odor, temperature, humidity, and air quality requirements. All of these requirements are met when there is adequate ventilation.

Ventilation and Air Movement

Ventilation effectiveness is determined by calculating the mixing factor (K_m), which is the ratio of the amount of air to dilute a contaminant (Q_{actual}), to the ideal amount of air that should reduce it (Q_{ideal}). ASHRAE 62-1989 calculates ventilation effectiveness as the fraction of outdoor air (OA) delivered to the occupied zone. Ventilation effectiveness also requires that the mixed air actually gets to the occupied zone.

Air movement in the interior spaces of a building is governed by several physical elements: the inlet openings, the outlet openings, and the dimensions and divisions of the spaces. The primary principles of air movement are illustrated in Figure 12 [5]. The inlet openings direct the initial pattern of the airflow, the outlet openings regulate the airflow velocity, and the dimensions and divisions modify the pattern and velocity of air movement through the reorganization of pressure differentials and the alteration of airflow inertia [5]. The length, depth, height, and shape of the interior spaces regulate the pattern and velocity of the airflow. This is true for both residential and commercial buildings, but because the windows in residential buildings are operable, it is easier to regulate air movement than in commercial buildings where the air movement is dependent on the HVAC system and placement of partitions. Partitions affect

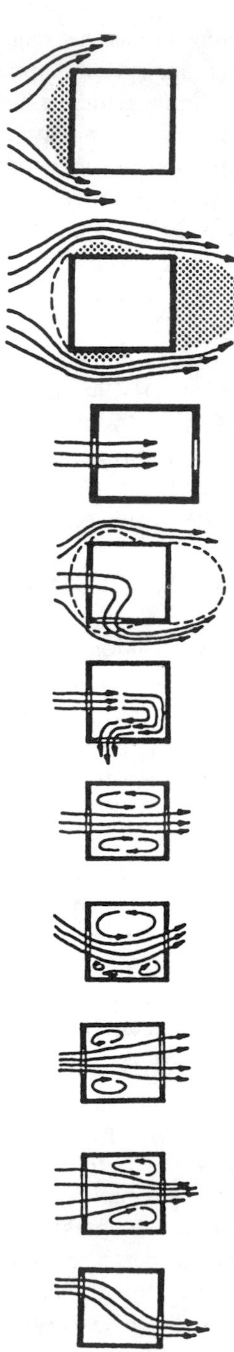

Positive Pressure Areas: An obstacle, such as a house, in the path of air movement will obstruct the airflow and cause it to pile up and slow down until it locates a new path to follow. The affected area is designated as a positive- or high-pressure area. The walls adjacent to the positive-pressure area should contain the inlet opening that allows the airflow to be forced into the building by the pressure exerted on the wall.

Negative Pressure Areas: As the airflow completely surrounds the building, negative-pressure areas, or wind shadows, are created. The size and shape of the area are determined by the configuration and scale of the building. The walls adjacent to the negative pressure areas should contain the outlet openings that permit the airflow to be drawn out of the house by the negative-pressure areas.

Inertia: Air will flow through an opening if another opening is provided and will travel within the building in the same direction as the exterior air movement until the interior airflow encounters an obstruction.

Pressure Differences: The difference between the positive-pressure area and the negative-pressure area helps determine the airflow velocity through the building as well as the exterior airflow velocity, inlet and outlet opening sizes, and directional changes in the airflow.

Directional Changes: A change in the direction of the airflow consumes energy from the airflow and reduces its velocity. The greater the velocity of the airflow when the change of direction occurs, the larger is the loss of energy and velocity.

Optimum Airflow: Inlet openings and outlet openings should be as large as possible to optimize airflow. When the air movement is perpendicular as it encounters the inlet and outlet openings in alignment, the airflow will pass through the building in a narrow stream (*a*). The remainder of the interior space will receive no significant air movement. The result will be maximum airflow in a minimum area. Skewed air movement will result in optimum airflow in a maximum area (*b*).

Maximum Velocity: The maximum velocity of air movement is obtained when the outlet opening is larger than the inlet opening. In that situation, positive pressure is created on the windward face of the building and negative pressure is established inside the building; consequently, increased air movement occurs within the interior spaces (*a*). If the exterior airflow is reversed, positive pressure is created inside the building and negative pressure is established on the leeward face of the building; consequently, increased air movement occurs outside the building (*b*).

Opening Locations: The airflow pattern within a building is determined by the placement of the inlet opening as well as by the initial airflow direction and the location of positive-and negative-pressure areas.

Figure 12. Primary principles of air movement.

air movement patterns and airflow velocity, and should be placed so that airflow is optimized. According to test results, a partition should be located near the outlet opening for optimum air movement through the space, and thorough air movement occurs throughout an entire space when the inlet and outlet openings are not in direct alignment. Figure 13 illustrates these principles [5].

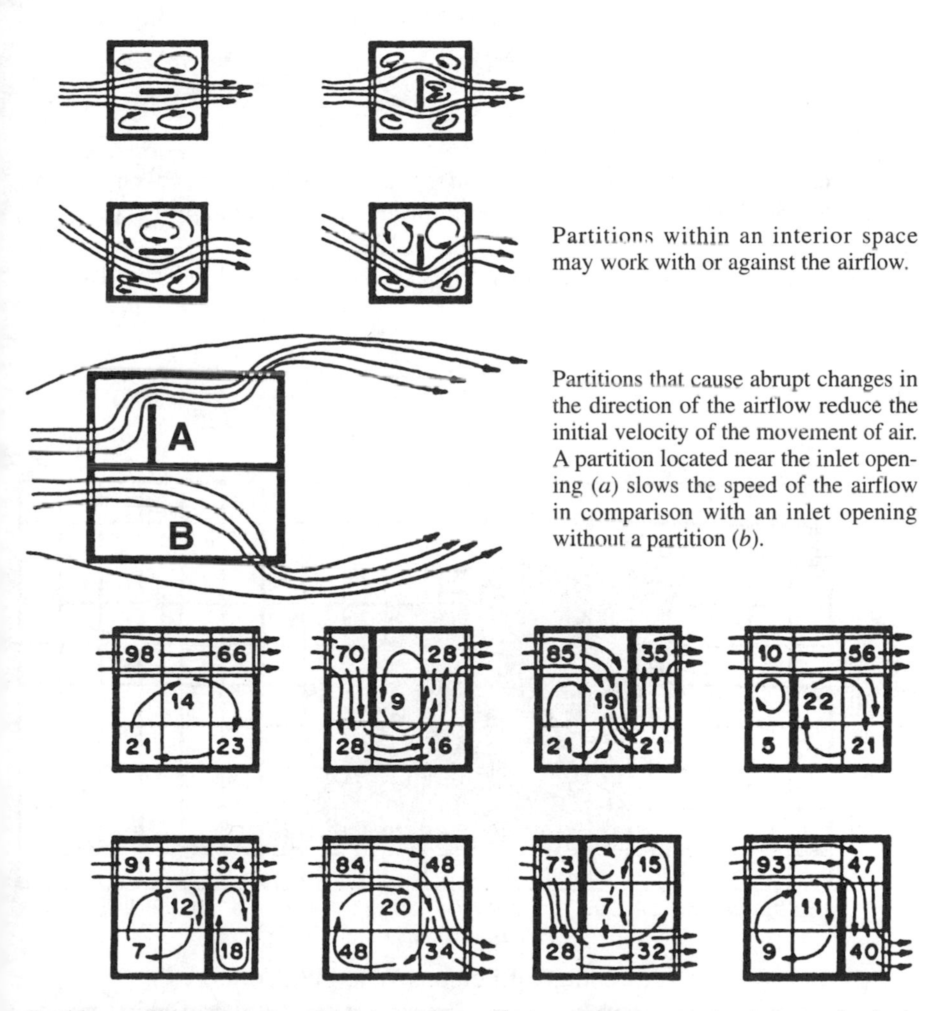

Partitions within an interior space may work with or against the airflow.

Partitions that cause abrupt changes in the direction of the airflow reduce the initial velocity of the movement of air. A partition located near the inlet opening (*a*) slows the speed of the airflow in comparison with an inlet opening without a partition (*b*).

Partitions within interior spaces have various effects on air movement patterns and velocities as the exterior airflow enters the inlet opening perpendicularly. The numbers represent the relative air speeds as percentages of the exterior air velocity.

Figure 13. The effects of partitions on airflow.

Thermal Comfort

Regarding IAQ problems, studies have identified an important relationship between thermal comfort conditions and ventilation. It has been recommended that future studies should be conducted only when the thermal comfort limits were within normal limits (–0.5 < PMV < 0.5), where PMV = personal mean vote set by ASHRAE. The PMV is a seven-point subjective scale ranging from hot to cold. When PMV = 0, 95% of the population will be comfortable. The normal limits of ±5% will therefore satisfy 90%.

Humidity

Another factor that should be determined and adjusted if necessary is the humidity. Based on ASHRAE studies, comfortable relative humidity depends on the dry-bulb temperature as shown in Figure 14 [5].

The most common deficiencies for HVAC systems are listed in Table 15 [20].

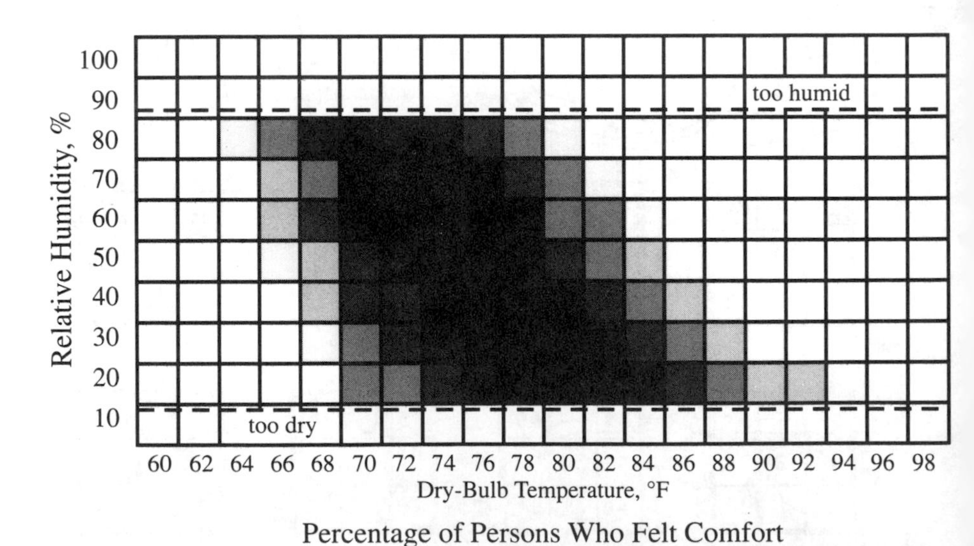

Figure 14. Comfortable relative humidity depends on dry-bulb temperature.

Table 15
Deficiency Checklist for HVAC Systems

Deficiency	Potential causes/Typical problems	Potential corrections
Insufficient total air delivery to occupied space	Inadequate fan capacity	Increase fan speed
	Worn fan blades	Replace/repair wheel
	Faulty fan components	Provide maintenance
	Imbalanced air supply system	Balance air distribution system
	Increased number of occupants	Increase air capacity; redistribute occupants
Insufficient outdoor air (OA) delivered to occupied space	OA dampers set too low	Increase OA; provide fixed minimum OA delivery.
	Imbalanced supply and return systems	Balance systems
	OA damper controls inoperative	Inspect, calibrate, reset controls
	Temperature control capacity insufficient to meet space needs	Increase system capacity
Air distribution within space not adequate; improper; insufficient	Improper supply system balancing	Rebalance
	Poorly operating dampers, boxes	Repair, maintain, inspect boxes and control equipment
	Maladjustment of thermostat/controls	Calibrate thermostats
	Improper location of supply diffuser	Relocate diffusers or occupants
	Diffusers blocked	Remove obstructions
	Office partitions resting on floor	Raise or remove partitions
	Diffusers not attached to supply ducts	Inspect, reattached connecting ductwork
AHU components not operating properly	System does not start before arrival of occupants; shuts down before departure	Reformat controls
	Filters inadequate	Use appropriate filters; install filters in accordance with manufacturers' instructions; change filters on a regular basis.
	Room temperature and humidity controls inoperative	Monitor or calibrate; maintain

STANDARDS AND CODES

As of January 1993, there are no national, mandatory standards for airborne concentrations of IAQ pollutants. The agencies and consensus standards and typical concentrations and concentrations of concern are summarized in Tables 16, 17, and 18 [2].

Other pertinent standards are in Table 19. (See Table 20 for a summary of equations.)

(text continued on page 117)

Table 16
Concentration Guidelines

Contaminant	Concentrations	Time	Source
Asbestos	0.1 fibers/cc	Action level	OSHA
Bacteria	10,000/ml water	—	OSHA
Carbon dioxide	1000 ppm	Continuous	ASHRAE
Carbon monoxide	9 ppm	8 hr	NAAQS
Formaldehyde	0.1 ppm	Continuous	ASHRAE
Fungi	10^6/g dust or 10^3/ml water	—	OSHA
Lead	1.5 μg/m^3	1 year	NAAQS
Nitrogen dioxide	0.05 ppm	1 year	NAAQS
Ozone	0.05 ppm	Continuous	ASHRAE
Particles, <10 m	50 μg/m^3	1 year	NAAQS
	150 μg/m^3	24 hours	NAAQS
Radon	4 pCi/l	1 year	EPA

Table 17
Major Consensus Standards

Topic	Name (often abbreviated)	Organization	Number
General	Ventilation for Acceptable Indoor Air Quality	ASHRAE	ASHRAE 62-1989
	The Ventilation Directory	NCSBCS	Summary of standards
	Various standards for HVAC (to avoid fires)	NFPA	Various
Comfort	Thermal Comfort	ASHRAE	ASHRAE 55-1981
Air filters	Filter testing methods	ASHRAE	ASHRAE 52-1976
	Test Performance of Air Filter Units	UL	ANSI/UL 900-1982
	Air Filter Equipment	ARI	ARI 680-1980

Table 17 (continued)

HVAC/ducts	Rating air flow through outlets	ASHRAE	ASHRAE 70-1972
	Duct standards	SMACNA	SMACNA (various)
	Guideline for the Commissioning of HVAC Systems	ASHRAE	Guideline 1-1989
	Recommended standards for duct cleaning	NADCA	Various standards
	Standards and practice for TAB	NEBB	Various standards
Air mixing	Method of testing ventilation effectiveness (proposed)	ASHRAE	ASHRAE 129P
Fans	Testing fan sound power	ASHRAE	ASHRAE 68-1986
	Standards Handbook	AMCA	AMCA 99-83 + others
Chemical	TLV	ACGIH	TLV Booklet
Bioaerosols	Guidelines for the Assessment of Bioaerosols in the Indoor Environment	ACGIH	ACGIH Publication
Labs	Laboratory Ventilation	ANSI	ANSI Z9.5-1992

Standards Associations

ACGIH	American Conference of Governmental Industrial Hygienists, 6500 Glenway #D-7, Cincinnati, Ohio 45211
ANSI	American National Standards Institute; Contact AIHA (ANSI Z.9 Secretariat) 2700 Prosperity Ave., Suite 250, Farifax, VA 22031
ARI	Air Conditioning Contractors of America, 1228 17th St., NW, Washington, DC 20036
ASHRAE	American Society of Heating, Refrigerating and Air-conditioning Engineers, Inc., 1791 Tullie Circle, NE., Atlanta, GA 30329 (Best source for various other standards related to HVAC system design, operation, and maintenance.)
NADCA	National Air Duct Cleaners Association, 1518 K St NW Suite 503, Washington, DC 20005
NCSBCS	NCSBCS, 505 Huntmar Park Drive, Suite 210, Herndon, VA 22070
NEBB	National Environmental Balancing Bureau, 4201 LaFayette Center Drive, Chantilly, VA 22021
NFPA	National Fire Protection Association, Batterymarch Park, Quincy, MA 02269
SMACNA	1385 Piccard Drive, Rockville, MD 20850
UL	Underwriter's Laboratories, 333 Pfingsten Road, Northbrook, IL 60062

Table 18
Contaminant Concentrations Checklist

Chemical	Typical Concentrations†	Concentration of Concern‡
Carbon dioxide	350–1000 ppm	> 1000 ppm
TVOC	1–2 ppm	Depends on compound
Formaldehyde	0.04–0.1 ppm	> 0.1 ppm
Carbon monoxide	1–5 ppm	> 9 ppm
NO_2	0.03–0.1 ppm	> 0.05 ppm
Ozone	0.01–0.02 ppm	> 0.05 ppm
Particles	< 0.075 mg/m^3 (total) < 0.050 mg/m^3 (PM-10)	> 0.075 mg/m^3 > 0.050 mg/m^3
Bioaerosols	varies with site	varies
Asbestos	< 0.01 fb/cc < 2 μg/m^3	> 0.1 fb/cc
Radon	< 0.5 pCi/lit	> 4 pCi/lit
Odors	None	Any detectable for long periods of time

†Non-industrial environments; concentrations in excess of these values may warrant investigation.
‡ASHRAE 62-1989; WHO Indoor Air Quality Research, *Report 103, 1984; EPA; "Concern" refers to limited health effects and/or comfort.*

Table 19
Selected ASHRAE Standards

ASHRAE 52-1976

Filter rating procedures

Filter Media	Arrestance*	Dust Spot**	DOP***
Fine open foams	70–80	15–30	0
Cellulose mats	80–90	20–35	0
Wool felt	85–90	25–40	5–10
Mats, 5–10 μm, 1/4"	90–95	40–60	15–20
Mats, 3–5 μm, 1/2"	> 95	60–80	30–40

Table 19 (continued)

Mats, 1–4 μm, fibers	> 95	80–90	50–55
Mats, 0.5–2 μm, glass	—	90–98	75–90
Wet-laid glass fibers,			
HEPA	—	—	>95

**measure of a filter's ability to remove coarse dust*
***measure of a filter's ability to remove fine dust,*
****dioctyl phthalate particle measurement of a filter's ability to remove very fine dust*

ASHRAE 55-1981

Relative Humidity = 30–50%

Temperature = winter = 68–74.5°F, summer, = 73–79°F

ASHRAE 62-1989

Ventilation Rate Procedure = minimum OA of 15 cfm/person

= 20 cfm/person in office space where estimated maximum occupancy is 7/1,000 sf

= 60 cfm/person in smoking lounge where estimated maximum occupancy is 70 people/1,000 sf

= 20 cfm/person in conference room where estimated maximum occupancy is 50 people/1,000 sf

Dilution Ventilation Rate

$$Q_{OA} = (q \times K_{eff} \times 10^6)/C_A \text{ in ppm}$$

where Q_{OA} = volume flow rate of dilution air, scfm or m^3/sec

q = volume flow rate of vapor, scfm or m^3/sec

C_A = the acceptable exposure concentration, ppm

K_{eff} = mixing factor to account for incomplete or poor delivery of dilution outside air (OA) to occupants

K_{eff} = 1.0 (wide-open spaces with good supply and return locations, all HVAC equipment working properly)

= 1.1 (conditions not ideal, fans create good mixing)

= 1.2–1.5 (poor placement of supply and return registers. Partitioned offices with generally adequate distribution of registers)

= 1.5–2.0 (crowded office spaces with tight partitions and poor supply of registers)

Table 20
Summary of Dilution Ventilation Equations

Emission rate for evaporated liquid:

$$q = \frac{387 \times \text{lbs evaporated}}{MW \times \text{minutes} \times d}$$

where
q = volume of vapor flow rate in acfm   MW = molecular weight   d = density correction factor

Volume flow rate of dilution air:

$$Qd = \frac{q \times 10^6 \times K_{mixing}}{Ca}$$

where
Qd is volume flow rate of dilution air, acfm   q is volume flow rate of vapor, acfm
Ca is the acceptable exposure concentration, ppm
K is a mixing factor to account for incomplete or poor mixing. Use K = 1.0 to 1.5 for good mixing in occupied space, K = 1.5 to 3 for poor mixing.

Number of air changes per hour in space:

$$N = \frac{Q \times 60}{Vol}$$

where
N = number of air changes per hour   Q = the volume flow rate of air, acfm   VOL = the space volume, cubic feet

Purge and buildup equation:

$$t = -K_m \times \left(\frac{Vol}{Qd}\right) \times Ln \left[\frac{\frac{q \times 10^6}{Qd} - C_t}{\frac{q \times 10^6}{Qd} - C_o}\right]$$

where

q = the generation rate, in acfm
C_t = concentration in air of the emitted substance in ppm, after time t, in minutes
C_o = the initial concentration, in ppm
Q = dilution volume flow rate of air, acfm
VOL = volume of the space, in cu.ft
t = time, in minutes

If the term in brackets is negative, then purge or buildup will not occur completely under the conditions specified.

<u>Where there is no generation (q = 0), the following formula predicts the concentration after a period of dilution.</u>

$$C_t = C_o \times EXP\left[\frac{-Qd}{K_m \times Vol}(\Delta t)\right]$$ EXP is the exponential e

<u>The following formula can be used to estimate the airflow rate through a building from outside wind:</u>

$$Q(cfm) = 88KAV$$

(table continues)

Table 20 (continued)

where
Q = Airflow rate through the building, cfm
A = Upwind or downwind open area, whichever is smaller, sq ft
V = Average wind velocity in miles per hour, mph
K = Angle factor
$K = 0.5$ when the wind is directly perpendicular to the building openings
$K = 0.3$ all other conditions

<u>The following formula can be used to estimate the airflow through a building from temperature differences:</u>

$$Q = 10A\,(H \times \Delta T)^{1/2}$$

where
A = Area of building inlets or outlets, whichever is smaller, in sq ft
H = Height of building between inlet and outlet, ft
ΔT = Difference between average indoor and outdoor temperatures, degrees F

<u>The following formula is used to estimate the amount of outdoor air reaching a space:</u>

$$\%OA = \frac{T_{RA} - T_{MA}}{T_{RA} - T_{OA}} \times 100\text{; also } = \frac{C_{RA} - C_{SA}}{C_{RA} - C_{OA}} \times 100 \qquad \text{also } = \frac{Q_{OA}}{Q_{SA}} \times 100$$

where
T_{RA} = temperature of return air (dry bulb)
T_{MA} = temperature of mixed return and outside air (dry bulb)
T_{OA} = temperature of outdoor air (dry bulb)
C_{RA} = Concentration of carbon dioxide in return air
C_{SA} = Concentration of carbon dioxide in SA,
C_{OA} = Concentration of carbon dioxide in OA
Q_{OA} = volume of outdoor air (cfm)
Q_{SA} = volume of mixed return and outside air, the supply air

Carbon dioxide concentrations can be used to estimate the amount of OA reaching a space:

$$N = \frac{(\ln Ci - \ln Ca)}{\text{hours}} \qquad \text{(works with any tracer gas)}$$

where
N = air exchange per hour, OA
Ci = Concentration of CO_2 at start of test (minus outdoor concentration, about 330 ppm)
Ca = Concentration of CO_2 at end of test (minus outdoor concentration)
hours = time elapsed between start and end of test

(table continues)

Table 20 (continued)

The rate of evaporation (emission rate) from a spilled liquid can be estimated from:

$$q = \frac{0.00813 \times A \times vp \times V^{0.78}}{d \times MW^{0.33}}$$

where
q = rate of release to air, acfm
V = wind speed, ft/min
A = surface area of spilled fluid, sq ft
vp = vapor pressure at ambient temperature, mm Hg
d = density correction factor

The airborne concentration of vapor C at any time t can be estimated from:

$$C = \frac{q \times 10^{6} \times (1 - 10^{Qt/-2.303 \times VOL})}{Q}$$

where
C = airborne concentration, ppm (perfect mixing)
q = emission rate, acfm
Q = airflow rate through room, acfm
VOL = room volume, cubic feet
t = time since initial spill, minutes

(text continued from page 108)

COST-EFFECTIVE CONTROLS

The most cost-effective control is to remove or substitute the source(s) of the IAQ problem. If this cannot be done, the appropriate air pollution control equipment should be chosen, sized, purchased, and installed. The choice will depend on the nature of the contaminant(s), the strength of the emitting sources, whether sources emit continuously or in episodes, the degree of contaminant control required, and cost.

By Pollutant

Asbestos

Because asbestos is a carcinogen, only a trained and certified professional should determine if the residential or commercial building contains asbestos. It is common practice to leave undamaged material alone, periodically inspect it, and prevent damage and deterioration. But if the asbestos fibers are exposed, a trained and certified professional must be consulted to remediate the problem via encapsulation or removal.

Biologicals

The growth and proliferation of biologicals can be reduced by controlling the humidity (30 to 45% for homes); eliminating standing water, water-damaged materials, and water condensation on building materials or wet surfaces; installing exhaust fans to vent kitchens, bathrooms, and clothes driers to the outside; and ventilating attic and crawl spaces. The relative humidity should not exceed 45% because a quantitative relationship has been established between indoor relative humidity and dust mite population [3]. Humidifiers and evaporation trays should be cleaned frequently because they can become breeding grounds for biologicals. Water-damaged carpets and building materials, which can breed mold and bacteria, should be discarded if they cannot be cleaned and dried within 24 hours. Buildings should be kept clean; however, dusting and vacuuming can spread dust mites, pollens, and other allergens. Clean and disinfect basements and operate a dehumidifier, if necessary, to maintain a relative humidity of 30 to 50% [4].

Combustion By-Product

Special precautions should be taken when operating unvented kerosene or gas space heaters. Follow the manufacturer's directions regarding flame adjustments and maintenance. As a precaution, ventilation should be provided by opening a door or window. Exhaust fans should be installed over gas cooking stoves that vent outside. Again, precautions should be taken for proper flame adjustment and maintenance. **Never** use a gas stove to heat your home, and always have the flue open when operating a fireplace. Woodstoves should meet EPA emission standards. Old models should have tight-fitting doors, and EPA and CPSC instruction should be followed when replacing asbestos gaskets. Use only aged or cured wood, and **never** use chemically treated wood. Also, properly maintain chimneys, flues, furnaces, and air-handling systems [4].

Environmental Tobacco Smoke

The most effective method of reducing and eliminating the exposure to ETS is to eliminate smoking. Increased ventilation and segregation of smoking areas do not significantly reduce exposure. Smoking produces large amounts of immediate and residual pollutants that natural or mechanical ventilation systems cannot remove without being cost-prohibitive [4]. If it is not possible to discourage smoking indoors, then smoking should be encouraged outdoors.

Lead

Paint should be tested for lead content prior to any removal attempt. If found, recommendations include covering with wallpaper or another building material, or chemical removal off-site by professionals. If removal is necessary, everyone not involved with the removal should leave the building. Lead dust that is tracked or blown indoors can be reduced with a wet cloth or mop; however, it will remain in carpeting and on furnishings. Increase the ventilation if using lead-based solder or use a no-lead substitute.

Odor

It often is difficult to locate the source of an odor because many of the most offensive odors are at trace concentrations and are insignificant components of materials. It is therefore easier to increase the ventilation than to isolate and substitute the source.

Ozone

Ozone is very difficult to control indoors because it is brought in from the outside and is generated by laser printers, copy machines, and other office equipment. If it is not possible to replace the equipment, ventilation must be increased.

Pesticides

It is illegal to use any pesticide in any manner inconsistent with the directions in its label. Only use pesticides approved for use by the general public, and hire a certified professional to apply the other more hazardous chemicals. Mix, dilute, and apply in well-ventilated areas. Use non-chemical methods of pest control. Keep exposure to moth repellents (paradichlorobezene) to a minimum. This should be used in areas ventilated separately from the building [4].

Radon

Based on the potential dangers, it is recommended that the air be tested, and if ^{222}radon is found to be above the acceptable threshold of 4 pCi/l, the building should be isolated from the soil, cracks and openings sealed in the basement floor, the crawl space ventilated, and a subslab ventilation or a heat recovery ventilator installed. Contaminated drinking water should be treated by aerating or filtering through granulated-activated charcoal [3]. These measures can reduce the radon concentration as much as 90% [5].

VOCs - Nonformaldehyde

There are short-term effects from VOCs, however the long-term effects are not known at this time. Based on the potential dangers, though, prevention and caution should be the policy. Read product labels, and attempt to purchase a product that does not require a warning label. Residential product labels should be read and followed, e.g., "use in well-ventilated areas." If possible, purchase a substitute product in pump rather than aerosol spray form. Properly dispose of partially full or used containers because they can leak. Purchase only the quantities needed. Keep exposure to products containing methylene chloride and benzene to a minimum, and only use in well-ventilated spaces. Common sources of exposure include paint strippers, adhesive removers, aerosol spray paints, pesticide "bombs," ETS, fuels and paint supplies, and automobile emissions. Keep exposure to perchloroethylene from newly dry-cleaned clothing at a minimum by not accepting clothing that has a strong chemical odor. These recommendations mostly refer to household and personal products.

VOC - Formaldehyde

There are short-term effects from formaldehyde; however, the long-term effects are not known at this time. Based on the potential dangers, prevention and caution should be the policy. Whenever possible, read the product labels and attempt to purchase a product that does not contain formaldehyde or urea-formaldehyde resins. To reduce exposure, substitute exterior-grade pressed wood products that contain phenol-formaldehyde resin. Coating the pressed-wood products with polyurethane may reduce emissions, but the coating must cover all surfaces and edges. Ventilation and humidity control will reduce formaldehyde emissions in both residential and commercial buildings [4].

By HVAC System Control

The indoor air pollutants that usually affect commercial buildings are: asbestos, biologicals, environmental tobacco smoke, pesticides, and VOCs. A well-designed, sized and operated HVAC system with good preventive maintenance procedures should not have IAQ problems. If some are discovered, the generally accepted HVAC resolutions are as follows:

Asbestos

Asbestos must be removed or encapsulated by certified professionals. If air samples reveal that asbestos fibers are part of the IAQ problem, it is the responsibility of management to obtain trained professional advice on how to proceed. This IAQ problem cannot be reduced or eliminated with HVAC control.

Biologicals

Biologicals can be reduced or eliminated by controlling moisture, temperature, and filtration.

Environmental Tobacco Smoke

Environmental tobacco smoke can be reduced by dilution ventilation and filtration. Only dedicating a separate ventilation system or banning smoking in the office can eliminate ETS.

Pesticides and VOCs

Pesticides and VOCs can be reduced or eliminated with judicious purchasing, dilution ventilation, filtration, isolation by covering or coating off-gassing materials, emission control, isolation by location, or relocating the air-handling inlet. One potential control is called a "bakeout," where the building is heated to 90°F for one or more days. The increased temperature encourages the volatilization. Formaldehyde does not change appreciably, but other VOCs are reduced 10 to 35% [2].

General Problems and Solutions

Following is a list of common HVAC problems and general solutions [2].

1. Fatigue
   - Check CO_2 levels (exceeding 1,000 ppm suggests too little OA)
2. Temperature too warm or cold
   - Thermostats misadjusted
   - Supply air temperature setting too high or low
   - Too much or too little supply air
   - Supply diffuser blows air directly on occupants
   - Temperature sensor malfunctioning or misplaced
   - HVAC system defective or undersized
   - Building under negative pressure (should be +0.03 to +0.05" w.g.)
3. Air too dry or humid
   - humidity controls nor operating or undersized
4. Air is stuffy
   - Non-delivery or low delivery of air to occupied space
   - Filters overloaded
   - VAV dampers malfunctioning
   - Restrictions in ductwork
   - Ductwork disconnected from supply diffusers
   - Duct leaking
   - Inadequate delivery of OA
5. Air smells
   - Microbiological contamination

Specific HVAC System Components

Dilution Ventilation

Emissions in an office environment can be controlled by judicious purchasing and ventilation (air exchanges and cleanliness of the air). Dilution ventilation is the best choice when [2]:

- Major air contaminants are of relatively low toxicity.
- Contaminant concentrations are not hazardous.
- Smoking is not allowed.
- Emissions sources are difficult or expensive to remove.
- Emission sources are widely dispersed.
- Emissions do not occur close to breathing zones.
- The building is located within a moderate climate.
- OA is of good quality.
- The HVAC system can treat the dilution air.

Dilution ventilation is less effective and more expensive when:

- Air contaminants are highly toxic.
- Contaminant concentrations are hazardous.
- Smoking is allowed.
- The emission sources are easy to remove.
- The emission sources consist of large point sources.
- The emissions source is in immediate vicinity of breathing zones.
- The building is located in a severe climate.
- OA air is more contaminated than inside air.
- The HVAC system is not capable of treating the air.
- There is poor mixing in the occupied zone.
- There will be a need for additional humidification and dehumidification.

If the dilution ventilation choice is applicable, a profile of the contaminant and sources (emission rates), a detailed description of the space (physical parameters of the occupied space), and the acceptable concentration will need to be determined (acceptable exposure based upon some kind of standard such as 10% of TLV) [2]. Dilution is more effective when the dilution air is routed through the occupied zone, the supply air is distributed where it is most effective, returns are located close to the contaminant sources, and auxiliary or freestanding fans are used to enhance mixing [2]. The formula for air changes is:

$$N = (Q \times 60)/Vol$$

where N = number of air changes per hour, ac/hr
Q = the volume flow rate of air, cfm
Vol = the space volume, ft^3

Filtration

ASHRAE rates filters on arrestance—ability to remove coarse dust; dust spot—ability to remove fine dust; and dioctyl phthalate (DOP)—ability to remove very fine dust. Arrestance filters are often dry panels or rolls, dust-spot resistance filters are constructed of envelopes or bags of filter media, and high-efficiency particulate arrestance (HEPA) filters are built into wooden boxes. Most new HVAC systems should be designed to provide 90 to 95% arrestance, and 40 to 60% dust-spot removal. This will reduce the amount of outside dust taken into the system and the recirculation of many microorganisms and larger dust particles. Coarse particulates larger than 2 μm can be removed by sedimentation, electrostatic precipitators, filters, and aqueous sprays. However, these means have a marginal effect on ETS aerosols. To remove ETS aerosols, use a HEPA filter, a charcoal scrubber, or an electrostatic precipitator [2]. If possible, provide a separate ventilation system for a smoking area. If additional or substitute filters are chosen, remember that resistance to flow directly affects fan horsepower, fan class, motor size, electrical usage, and cost. The airflow rates could be reduced as the resistance increases, therefore, the required airflow rates should be designed at the replacement pressure (worst case) of the new filters [2]. Odorous air can be passed through an activated charcoal-type filter for adsorption (air is passed over the non-polar, adsorbing material, odor-causing particles adhere to porous surface area) or air washing (fine mist introduced into the air, odor-causing particles are absorbed by the liquid, collected in a pan, and removed). Activated carbon is most effective for large, less polar organic compounds that boil at normal temperature or higher [6]. Other types of adsorbents used are siliceous (silica gels and molecular sieves that exhibit a greater affinity for polar compounds), metal oxides (activated alumina for polar compounds), and impregnated adsorbents. Air scrubbers and odor masking also can be used. Pollen sizes vary from 10–50 μm [2]. Outside air and RA should be filtered with dust-spot efficiencies of 50% or greater [2].

Choosing the filtration equipment primarily depends on the particle size. Figure 15 is useful when choosing which type to install [7].

Filtering Equipment

The major air-filtering techniques are dry filtering, wet filtering, electrostatic precipitators, air washers, and gas sorbents. Dry-filtering devices come in a wide variety of compositions, efficiencies, and air resistances. Filters separate and collect particulates as the airstream passes through them. Filters are usually accordion pleated, and may be made of cloth or cellulose. Such filters have exceedingly high arrestance, which increases as the interstices clog. They are designed for air velocity of 10 to 50 fpm, depending on the fabric. Airstream moisture has to be controlled because the interstices will clog more rapidly when the particles are damp. Another type of dry filter has fabric layers and collects particles nonuniformly. Viscous-impingement filters are flat, panel filters coated with a viscous substance that acts as an adhesive when the particulates impinge on the fibers. They are characterized by low pressure drop, low cost, and good efficiency on lint but low efficiency on dust. All dry filters have to be replaced periodically as they clog. Some are designed with mechanical scrapers to remove the collected particulates, and others have a continuous belt that scrapes the filter while always exposing a "clean" side to the airstream. These types of filters are often used before high-efficiency filters.

Electrostatic precipitation involves charging a particle and collecting it on an energized plate. The air passes through an ionizer section where the particulates obtain an electrical charge. The air then flows into the collector cells where the charged particulates are drawn to the surface of collector plates by electrical attraction. The particles stick to the adhesive-coated plates. The collector cells are periodically cleaned, and the accumulated dirt is collected. They generally have high efficiencies and low air resistance, but require regular cleaning. There are also continuously sprayed wet electrostic precipitators that utilize electrostatic precipitation and impingement. The airstream, saturated with water, passes

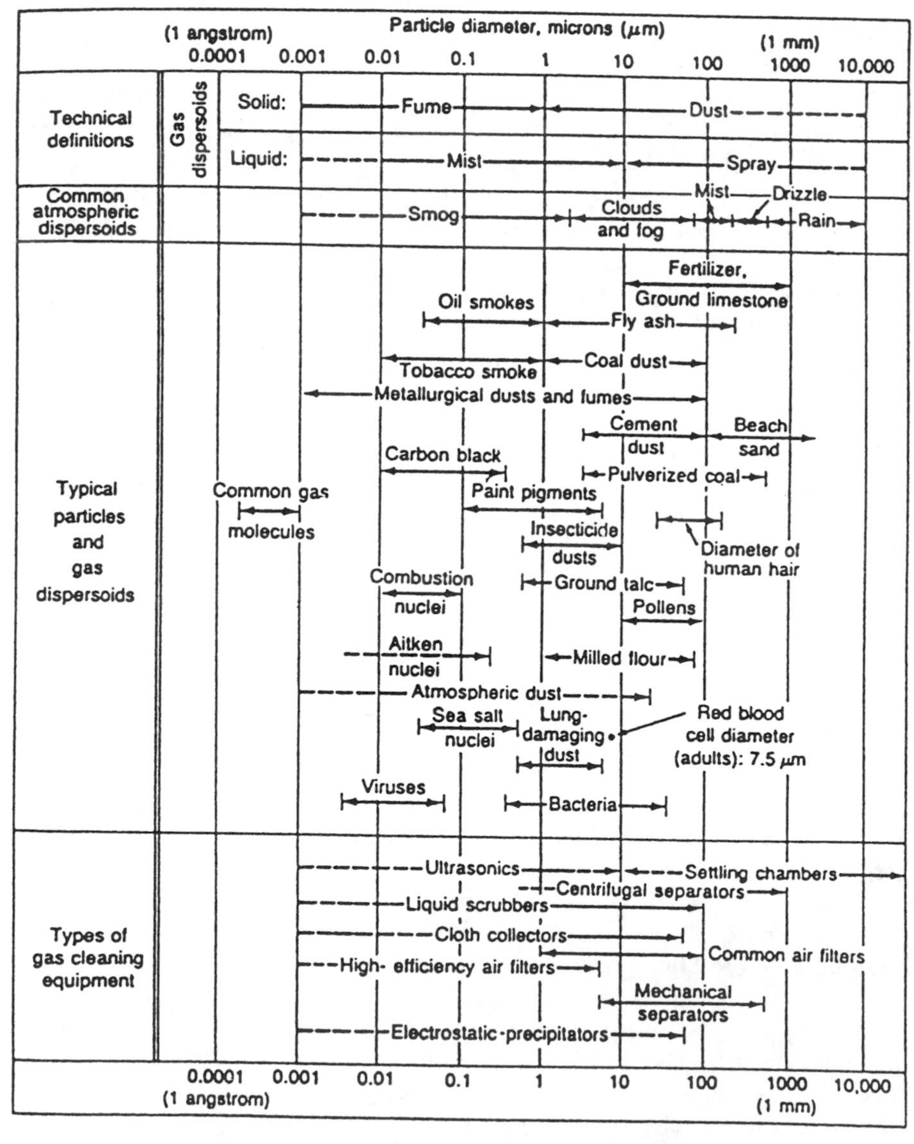

Figure 15. Diameters of typical industrial particles.

through the inlet baffle system, establishing a uniform velocity profile. Scrubbing sprays in the inlet section remove the coarser particulates. The wetted gas then enters the electrostatic field where additional water is sprayed to ensure that sufficient amounts of water are provided for cleaning the internal precipitation surfaces. The water droplets and particulates become charged and migrate towards the collector plates. There they form a continuous water film that goes to a discharge trough. A final set of charged-out baffles removes any remaining droplets and particulates.

Air washers are either the spray type or packed-cell type. The particulate removal efficiency is greater for the packed-cell type and for larger, more wettable particles. The air meets the water spray, and the particles are surrounded by the water. The wetted airstream is then guided by baffles. The droplets impinge on the baffles and fall into the open tank below. The screened water is then recirculated. The most common gas sorbent is activated carbon. It is highly porous, and the gas adheres to the surface area.

Representative diagrams of filtration equipment are shown in Figures 16, 17, 18, 19, and 20.

Moisture and Temperature Control

Humidity and standing water are especially important factors in office buildings where the HVAC system has numerous places where biologicals can flourish

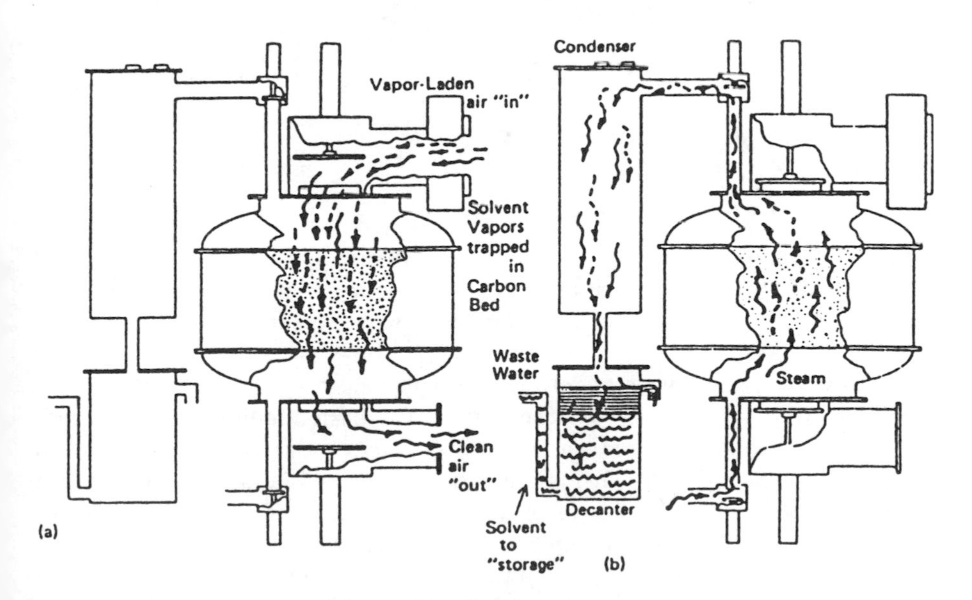

Figure 16. Filtration diagram.

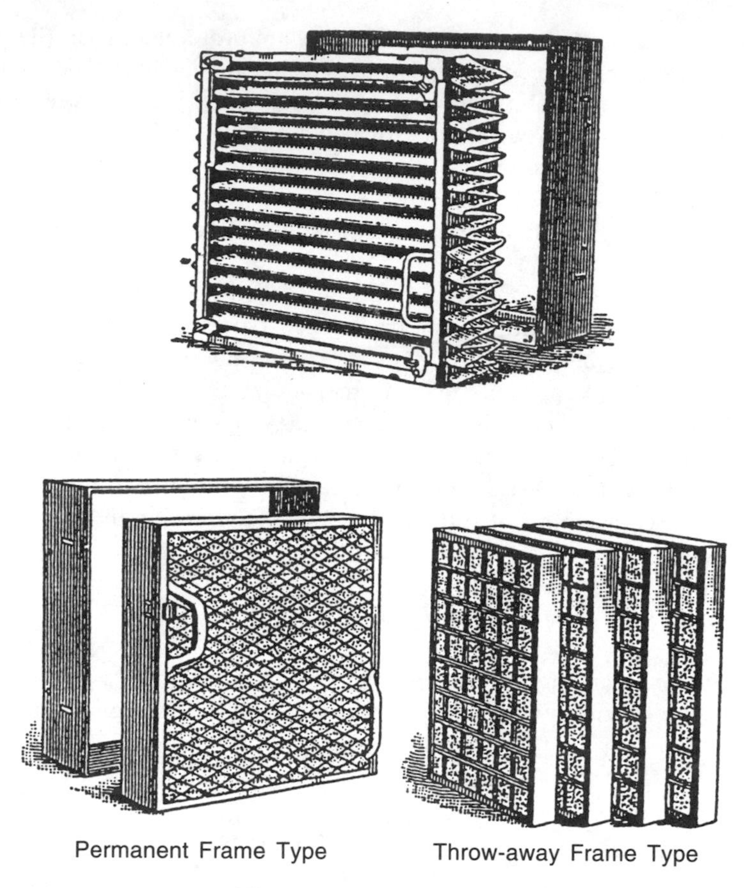

Figure 17. Typical filters.

and subsequently spread throughout the ductwork. Dust mites, pollens, and allergens should be filtered out prior to the air recirculation. These problems can be prevented or at least kept to a minimum with proper maintenance of the HVAC system.

Because microorganisms are always present, look for conditions where they can multiply to unacceptable concentrations. The basic requirements are standing water or water mist/vapor, nutrients for growth, conducive temperatures, and a pathway to reach people. Microbiological agents in air are usually greater than 1 to 2 μm with an upper size of 50 to 100 mm. Filters with a dust-spot efficiency of 50 to 70% should remove many biologicals in the air moving through the filter. A major source of biologicals is stagnant water where bioactivity in HVAC systems can be the source of 40 to 50% of IAQ problems [2].

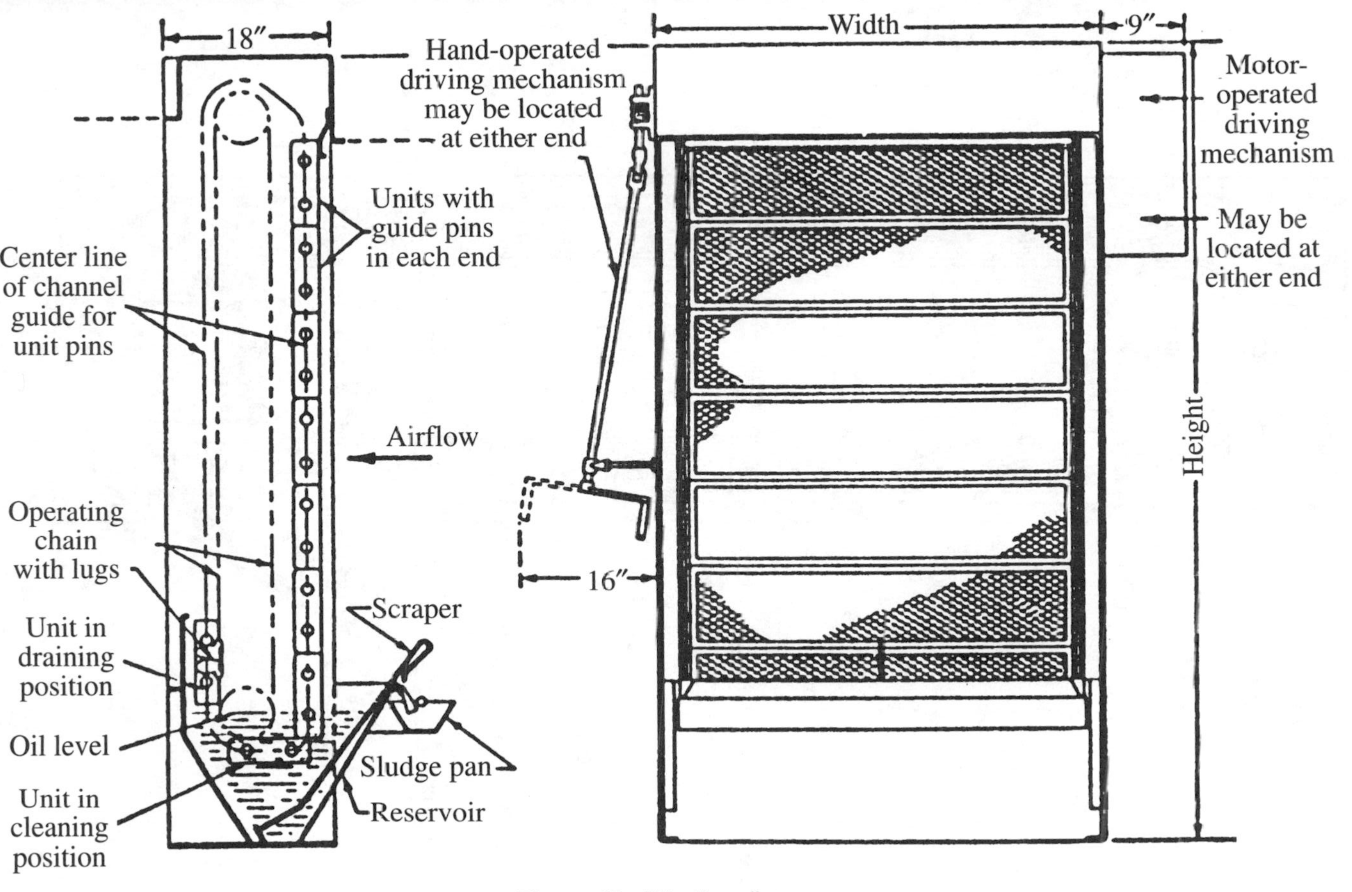

Figure 18. Filtration diagram.

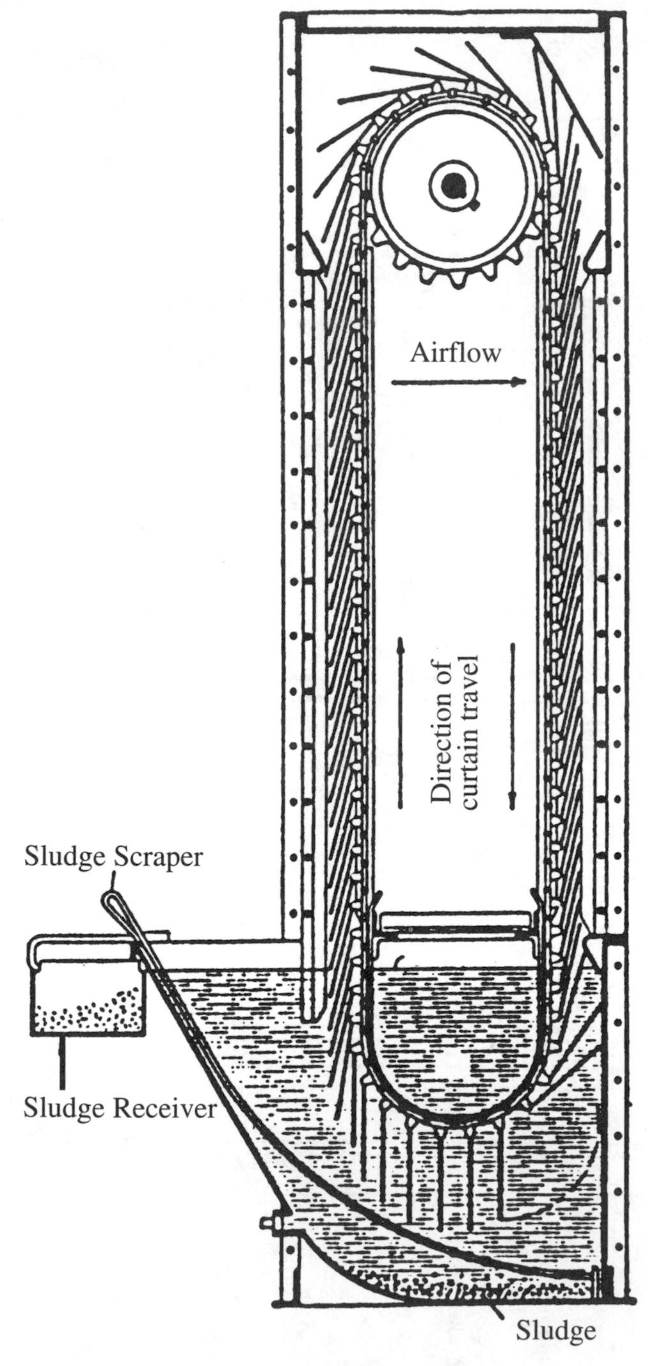

Figure 19. Filtration diagram.

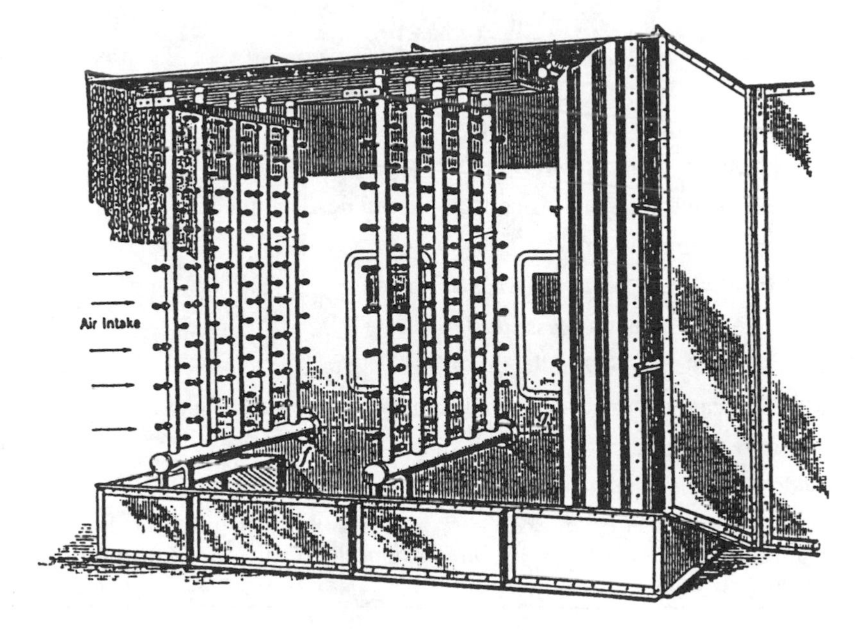

Figure 20. Filtration diagram.

Water temperatures in cooling towers should be reduced to less than 20°C for most of the year by the introduction of a secondary pumped supply. Diligent maintenance and cleaning should be practiced, and biocides should be used. Systems should be designed to avoid sludge pockets and dead legs where bacteria can breed. In hot water systems (HWS), biocides cannot be used, but storage should be at more than 50°C with a nightly lift to 65°C for one hour to kill any bacteria [8]. Use biocides only when the system is off.

Fungi, mold, and mold spores ranges in size from 3 to 200 μm, with most in the 10 to 20 μm range [2]. Control is achieved by removal, isolation, reduction of sources, dilution ventilation, elimination of the reservoir, moisture sources control, elimination of nutrients, and rigorous maintenance [2].

Ducts can become both the source of and pathway for dirt, dust, and biologicals to spread throughout a building. Therefore, ASHRAE strongly suggests keeping dirt, moisture, and high humidity from ductwork [2].

Following is a list of the main components of moisture and temperature. See Table 21 for a checklist for reducing microbial problems [2].

- Eliminate standing water.
- Maintain a RH of 40%-50%.
- Use stream humidifiers.
- Maintain a dew point temperature below 60°F.

Table 21
Checklist for Reducing Microbial Problems

- ☐ Prevent buildup of moisture in occupied spaces.
- ☐ Prevent moisture collection in HVAC components.
- ☐ Remove stagnant water and slime from mechanical equipment.
- ☐ Use steam for humidifying.
- ☐ Avoid use of water sprays in HVAC systems.
- ☐ Maintain relative humidity less than 70%.
- ☐ Use filters with a 50–70% collection-efficiency rating.
- ☐ Find and discard microbial-damaged furnishings and equipment.
- ☐ Remove room humidifiers.
- ☐ Provide preventive maintenance.
- ☐ Provide bird screens on intakes and exhausts (this will prohibit the contamination of the system by bird droppings, feathers, nesting materials, food, and so forth).

- Prevent condensation of water vapor in building spaces.
- Remove water vapor.
- Keep air handling systems water free.
- Clean and disinfect humidifiers and mechanical components.
- Use biocides.
- Do not use cool mist humidifiers.

Relocation of Components

A cooling tower within 25 ft of air intake should be retrofitted with mist eliminators because contaminated water can be reentrained. Contaminants also can be reentrained if the exhaust stack and the intake are not properly positioned. Air intakes should be placed on the lower third of a building, and exhausts on the upper two-thirds or on the roof. Intakes should not be placed near the ground, loading docks, or busy streets. Exhausts should be placed on the predominant downwind side of the building, and uptakes on the upwind side [2]. Good engineering design practices include placing the stack to avoid exhausting into the recirculation cavity (10 ft should be high enough to breach on most roofs), providing sanitary stack heights 10 ft away and 2 ft above the air intake, locating

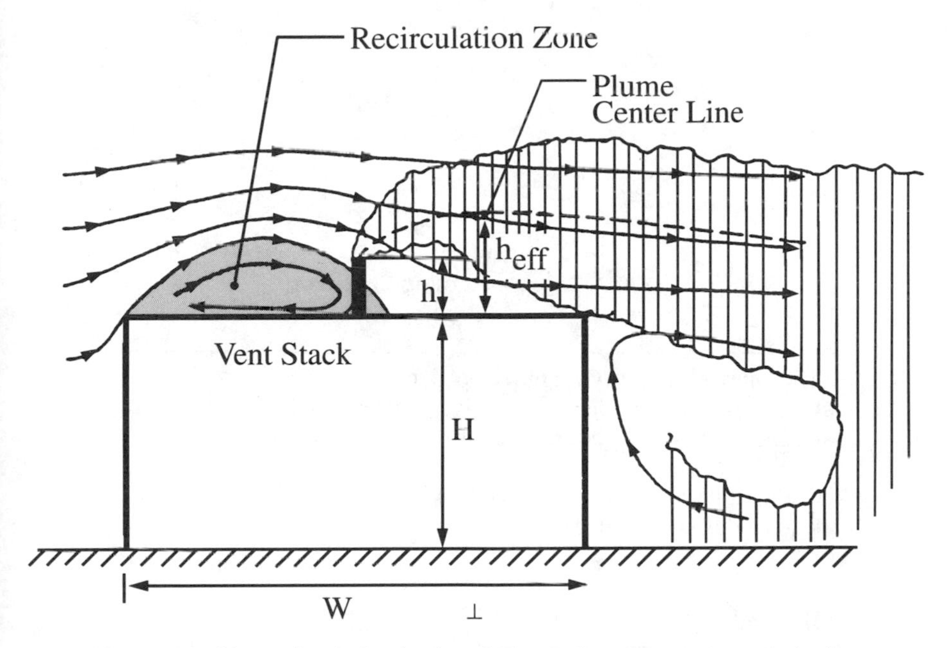

Figure 21. Dispersion behavior in relation to location vents and stacks.

the local exhaust stacks within 50 ft of the roofline. The air intake should be 10 ft tall. Figure 21 is an illustration of dispersion behavior as it applies to location vents and stacks [2].

CONCLUSIONS/RECOMMENDATIONS

It is much simpler to solve IAQ problems in a residential building than in a commercial building for two reasons: The windows are operable, and the occupant usually has much greater control for substituting less- or non-polluting items. In a commercial building, the occupants are dependent on the HVAC system, and they usually have less control of its contents, design, and operation.

The recommendations are to consider alternatives when purchasing items that can potentially contaminate the air. When a substitution is not possible, consider the implications of emissions and the cumulative effects in the building, and try to provide extra ventilation and/or mixing.

Tables 22 through 30 contain several checklists pertinent to commercial buildings.

(text continued on page 142)

Table 22
HVAC Troubleshooting Checklist
Typical Problems

( ) Insufficient outdoor air (OA) introduced to the system
( ) Intake and exhaust dampers inoperative, malfunctioning
( ) Intake and exhaust at improper location

( ) Poor rough filtration—dirt, bugs, pollen in air delivery system
( ) Inadequate dust-spot filtration
( ) Poor system maintenance

( ) Improper balance of distribution system
( ) Distribution dampers at incorrect positions
( ) Building under negative pressure

( ) Terminal diffusers not at correct positions
( ) VAV systems in non-delivery or low-delivery mode
( ) Terminal diffusers not attached to delivery system

( ) Poor distribution or stratification of supply air in occupant space
( ) Draftiness—too much supply air or improper terminal settings
( ) Placement of desks, personnel locations in high-velocity areas

( ) Stuffiness—not enough air delivery or air not delivered properly
( ) Improper pressure differences between rooms—doors hard to open
( ) Temperature extremes—too hot or too cold

( ) Humidity extremes—too dry or too humid
( ) Energy conservation has become No. 1 priority
( ) Settled water in system

( ) Visual evidence of slime or mold

Table 23
Checklist for Preventing and Minimizing IAQ Problems

- ☐ Maintain the HVAC system in top working condition.
- ☐ Provide a written operating and maintenance plan for HVAC systems.
- ☐ Specify building materials with low VOC emissions.

- ☐ Provide appropriate volumes of outside air.
- ☐ Provide good distribution and mixing of supply air.
- ☐ Specify furnishings and materials with low VOC emissions.

- ☐ Restrict smoking to areas with dedicated ventilation systems.
- ☐ Limit open shelving and other uses of pressed wood.
- ☐ Use hard surface flooring and walls where possible.

- ☐ Use lowest temperatures consistent with energy and comfort.
- ☐ Provide relative humidities of 30–50%.
- ☐ Use high-efficiency filters (e.g., ASHRAE dust-spot efficiency 50–70%).

- ☐ Involve and educate occupants.
- ☐ Lower occupant densities.
- ☐ Increase occupant control of environment (e.g., personal fans, more thermostats, involvement in decisions).

- ☐ Minimize exposure for those in stressful jobs.
- ☐ Involve professional assistance in IAQ problems.
- ☐ Monitor systems, air quality.

- ☐ Eliminate standing or stagnant water.
- ☐ Remove contaminated or emitting materials that cannot be controlled.
- ☐ Investigate bakeout for new buildings.

Table 24
Building Owners Documentation and Programs*

- ☐ References and calculations for required supply rate of OA.
- ☐ Methods of measuring/monitoring OA supply.
- ☐ Description of OA control systems.
- ☐ Description of OA control systems for VAV systems.
- ☐ Documentation of temperature/humidity control systems.
- ☐ Filtration descriptions and SOP for use.
- ☐ Filter efficiency documentation.
- ☐ Written preventive maintenance program.
- ☐ Maintenance record keeping.
- ☐ Written Standard Operating Procedures.
- ☐ Plans, drawings, specifications of building (as built/as is).
- ☐ Plans, drawings, specifications of building mechanical systems (as built/as is).
- ☐ Schematic drawings showing locations of all HVAC equipment for non-engineers.
- ☐ Manufacturer's literature for all operating equipment.
- ☐ Testing, balancing, and monitoring records.
- ☐ Building permits, stack permits, other applicable licenses/permits.
- ☐ History of changes to HVAC systems, occupancies.
- ☐ Technical information about control systems.
- ☐ Records of complaints.
- ☐ Records of problems and solutions.
- ☐ Building occupancy records.
- ☐ Other ______________________________
- ☐ Other ______________________________

**Note: To assure IAQ, building operators should maintain these minimum programs, records, and documents.*

Table 25
Occupants Checklist of the Ventilation in a Room

- ☐ Does each room have a source of air?
- ☐ Is air moving through diffusers and return grilles?
- ☐ Are diffusers and grilles open? Blocked?
- ☐ Is the air distributed throughout the space where people are?
- ☐ Do you actually feel air moving?
- ☐ Are there "dead spaces" in the office or room?
- ☐ Do printers, copiers, and other equipment have adequate ventilation?
- ☐ Does the ventilation system operate when people are in the building?
- ☐ Is the air too hot? Too cold? Too humid?
- ☐ Does the air smell bad?
- ☐ Does the air make you uncomfortable? Sick?
- ☐ What do you think contaminates the air?

Table 26
Moisture Control Checklist for Building Occupants and Operators

- ☐ Avoid spilling water at sinks and coffee areas.
- ☐ Clean up spills until dry.
- ☐ Always use exhaust fans in bathrooms, showers, tubs.
- ☐ Avoid the use of room humidifiers.
- ☐ Repair roof leaks.
- ☐ Remove standing water on roofs, in basements, in crawl spaces, and on internal surfaces.
- ☐ Provide adequate drainage of flat roofs.
- ☐ Do not allow water to condense in air handling systems.
- ☐ Do not allow water to condense on perimeter windows and basement walls.

Table 27
Inspection Checklist for HVAC Systems

- Outdoor air intakes
  - ☐ Location
  - ☐ Open
  - ☐ Controllable
  - ☐ Outdoor contaminant sources nearby
  - ☐ Type
  - ☐ Location of exhausts
  - ☐ Predominant wind direction and velocity

- HVAC equipment
  - ☐ Intact
  - ☐ Dry
  - ☐ Clean
  - ☐ Equipment running in accordance with specifications
  - ☐ Filters in place, operating
  - ☐ Slime, mold, dirt, soot removed

- Ductwork
  - ☐ Intact, connected
  - ☐ Dry
  - ☐ Balanced

- Supply air diffusers
  - ☐ Open
  - ☐ Set
  - ☐ Airflow correct
  - ☐ Terminal velocities
  - ☐ Air jet profile
  - ☐ Location
  - ☐ Clean
  - ☐ Quiet

- Return Air Grilles
  - ☐ Location
  - ☐ Air movement
  - ☐ Open
  - ☐ Attached to return system

Table 28
HVAC Equipment Maintenance Checklist

- ■ Cooling towers
  - ☐ Written maintenance and inspection program.
  - ☐ Operated in accordance with manufacturer specifications.
  - ☐ Inspected regularly (monthly, or as required.)
  - ☐ Regular drainage and cleaning (quarterly, or as required).
  - ☐ Treatment of water to control microorganisms, as required.
    - ☐ Follows chemical instructions
    - ☐ Keep records of biocide use—brand, volume, results
  - ☐ Training of employees regarding hazards involved (e.g., *Legionella*)

- ■ Humidifiers
  - ☐ Written maintenance and inspection program.
  - ☐ Inspected weekly during operation.
  - ☐ Cleaned and disinfected, as required.
  - ☐ No visual buildup of mold or slime.
  - ☐ Disinfectants removed before reactivating humidifiers.

- ■ Cooling coils
  - ☐ Written maintenance and inspection program.
  - ☐ Monthly inspection (or, as required) during operation.
  - ☐ Removal of dirt, slime, mold, as required.
  - ☐ Upstream filters operating properly.

- ■ Drain pans, drainage systems
  - ☐ Written maintenance and inspection program.
  - ☐ Monthly inspection (or, as required.)
  - ☐ Drains maintained in free-flowing condition.
  - ☐ No accumulation of stagnant water.
  - ☐ No buildup of slime, mold, or dirt.
  - ☐ Removal of dirt, slime, and mold, as required.
  - ☐ Sample water for microbes, as required.

- ■ Duct and plenum equipment
  - ☐ Written inspection and maintenance program.
  - ☐ Exhaust ducts clear and clean.
  - ☐ Routine inspection of ducts for dirt, debris, microbial growth (e.g., semi-annually).
  - ☐ Provisions for cleanout (e.g., within four feet downstream of duct expansions, supply air openings, or where particulate deposition may occur).
  - ☐ Ductwork attached, and not dented.
  - ☐ Insulation intact, not wet, and no microbial growth.
  - ☐ Ductwork properly balanced.

(table continues)

Table 28 (continued)

- Filtration systems.
  - ☐ Written maintenance, operating, and inspection programs.
  - ☐ Routine inspection.
  - ☐ Provision for measuring pressure drop across filters.
  - ☐ Filter change or cleaning at pre-determined pressure drop.
  - ☐ Filters actually changed at required time.
  - ☐ Filters not wet; no microbial growth.
  - ☐ Proper protective equipment during handling and maintenance.
  - ☐ Change out contaminated filters only during AHU shutdown.

- Dampers
  - ☐ Written inspection and maintenance program.
  - ☐ Periodic inspection.
  - ☐ Control systems operable.
  - ☐ Function, free movement, cleanliness, corrosion.
  - ☐ Periodic testing.
  - ☐ Lubrication, cleaning of linkages.

- Heating coils, heating elements

- When cleaned or washed, all surfaces rinsed with water before use.

- Outdoor air intakes
  - ☐ Written inspection and maintenance program.
  - ☐ Periodic inspection.
  - ☐ Cleaning of debris, dirt, and dead animals.
  - ☐ Located so will not entrain contaminated air.

- Diffusers, terminals, grills, supply and return registers
  - ☐ Written inspection and maintenance program.
  - ☐ Routine inspection (e.g., semi-annually)
  - ☐ Open, unobstructed.
  - ☐ Attached to supply ducts.
  - ☐ Angle of discharge appropriate.
  - ☐ Throw velocities appropriate.
  - ☐ Terminal velocities and locations appropriate.
  - ☐ Provides adequate distribution and mixing.

- VAV system equipment
  - ☐ Written maintenance and inspection program.
  - ☐ Provision for minimum OA air delivery at all times.
  - ☐ Periodically (e.g., semi-annually) calibrated and adjusted.

Table 28 (continued)

- Thermostats
  - ☐ Written inspection and adjustment program
  - ☐ Set for appropriate temperature
  - ☐ Controls inspected and adjusted, as required.
  - ☐ User control defined.

- Outdoor Air supply equipment
  - ☐ Written program.
  - ☐ Meets requirements of ASHRAE 62-1989.
  - ☐ Provision for providing appropriate OA to building and occupied spaces.
  - ☐ Provision for measuring OA delivery to building and spaces.
  - ☐ Periodic (e.g., each season) measurement or calculation of OA.

- Biocide usage
  - ☐ Use only approved materials, chemicals.
  - ☐ Written program.
  - ☐ Appropriate training and education.
  - ☐ AHU's should be shut down during application.
  - ☐ All residual materials should be removed from surfaces before restarting AHU.
  - ☐ Check local codes.
  - ☐ Follow manufacturers recommendations.
  - ☐ Keep records.

- General requirements
  - ☐ All HVAC personnel appropriately trained.
  - ☐ Safety equipment available
  - ☐ Appropriate monitoring equipment available.
  - ☐ Testing and balancing performed by certified personnel.
  - ☐ Standards and codes available for review.
  - ☐ Adherence to standards and codes.
  - ☐ AHUs are provided with access to all components.
  - ☐ IAQ has high priority.
  - ☐ Records are maintained for all maintenance work, unusual conditions, IAQ episodes including names of those affected, equipment cleaning.

- Fans

Table 29
HVAC Fan Maintenance Checklist

- ☐ Fan brand, model, drive, rpm.
- ☐ Motor type; rated hp, sf, rpm, voltage, amperage.
- ☐ Fan curves, manufacturers' literature, installation, and maintenance records.
- ☐ Same fan as on plans, drawings, and specifications.
- ☐ Operating conditions similar to specifications and SOP.
- ☐ Location.
- ☐ Fan assembly intact and clean.*
- ☐ Lubrication of bearings to manufacturers specifications.
- ☐ Minimal vibration; noise acceptable.
- ☐ No pulsations of airflow.
- ☐ Operation point on fan curve.
- ☐ System effect losses.
- ☐ Weekly check of belts and sheaves for wear, tension, alignment.
- ☐ All belts replaced at same time.
- ☐ Pulley sheave size and adjustment proper.
- ☐ All nuts and bolts tight.
- ☐ Motor and fan mounts tight and intact.
- ☐ Vibration isolators, pads, and springs okay.
- ☐ Motor temperature, amperage, and rpm within limits.
- ☐ Electrical connections and insulation okay.
- ☐ Fan direction of rotation okay.
- ☐ Fan wheel mounted correctly.
- ☐ Fan wheel not worn or dirty.
- ☐ Strainers, traps, drains, valves open or clear.
- ☐ Filters and shaft coolers okay.
- ☐ Cutoff at proper location.
- ☐ Duct vibration isolators mounted correctly (outside of duct at fan inlet, inside or outside duct on fan outlet).
- ☐ Inlet and outlet dampers set properly; controls operative.
- ☐ Motor and fan wipe down to remove dirt, oil, and grease.
- ☐ Belt guards appropriate and installed correctly.
- ☐ Stack attached, guyed, and protected from rain.

**Inspection of internal fan components or fan wheel should be performed only when fan is off and locked out.*

Table 30
Providing a Dedicated Ventilation System to a Smoking Area

Step	Action
(1)	Determine the actual boundaries of the proposed smoking area. (This is the total space where air moves, not an area designated on paper.) Estimate the total room volume of the smoking area in cubic feet, V_{space}.
(2)	Estimate the maximum number of persons in the smoking area at any one time, Ps.
(3)	Estimate minimum required fresh outdoor air required ($Q_{OA\text{-}req}$). $Q_{OA\text{-}req} = Ps \times (OA, 60\ cfm/person)$
(4)	Determine the design or actual recirculation percentage at the air handling system, R (e.g., 0.80). Alternatively, determine the design (or actual) OA percentage introduced to the system, N (e.g., 0.20).
(5)	Estimate the total supply air rate in the smoking area, Q_{SA}.
(6)	Estimate the design (or actual) outdoor air supplied to the area: $Q_{OA\text{-}actual} = Q_{SA} \times (100\text{-}R) = Q_{SA} \times N$
(7)	Compare $Q_{OA\text{-}req}$ to $Q_{OA\text{-}actual}$. The latter should equal or exceed the former.
(8)	Estimate room air changes per hour, N: $N = (Q_{SA} \times 60)/V_{space}$ (also, $N = (Q_{OA} \times 60)/V_{space}$ for outside air only)
(9)	Check to see that the return air equals or exceeds supply air to the smoking area.
(10)	Check to see that air from the smoking area is returned directly, not after traveling through other occupied areas.
(11)	Check the mixing potential in the smoking area (walls, fans, etc.)
(12)	Recommend or investigate potential alterations if necessary (e.g., moving partitions, installing fans, moving supply/exhaust registers, closing doors and windows.)
(13)	Check to see if the smoking area is under negative pressure (smoke tube, air flows).
(14)	Establish a comprehensive housekeeping program.

Sample calculation. It is determined that up to five people will occupy a designated smoking lounge, 10' × 10' × 8'. There is a separate, dedicated air supply and exhaust AHU serving the smoking area. $Q_{SA} = 500$ acfm, R = 80% (N = 20%)

(1)	Room volume = 10 × 10 × 8 = 800 cu ft
(2)	Ps = 5 people
(3)	$Q_{OA\text{-}req} = 5 \times 60 = 300$ cfm
(4)	R = 80% (0.80) or N = 20% (0.20)
(5)	$Q_{SA} = 500$ cfm
(6)	$Q_{OA\text{-}actual} = (500)(1.00\text{-}.80) = 100$ cfm (or, 500 × .20 = 100 cfm)
(7)	$Q_{OA\text{-}actual}$ is not sufficient—it should be 300 cfm. (Alterations to the ventilation system will be required to increase the supply for fresh air to the space.)
(8)	$N = (Q_{SA} \times 60)/V_{space} = N = (500 \times 60)/800 = 37.5$ total air changes per hour also, $N = (300 \times 60)/800 = 22.5$ OA ac/hr (Okay)
(9–14)	As necessary. In this case, the recirculation rate R should be reduced to R = 0.40, which will provide 60 cfm per person of outdoor air.

(text continued from page 131)

REFERENCES

1. Gammage, R.B., Kaye, S.V., *Indoor Air and Human Health,* Lewis Publishers, Inc., Chelsea, MI, 1985.
2. Burton, D.J., *IAQ and HVAC Workbook,* IVE, Inc., Bountiful, UT, 1993.
3. Godish, T., *Air Quality,* Lewis Publishers, Inc., Chelsea, MI, 1986.
4. *The Inside Story: A Guide to Indoor Air Quality,* EPA & CPSC, September 1988.
5. Boutet, T.S., *Controlling Air Movement,* McGraw-Hill, New York, NY, 1987.
6. Young, R.A., Ed., Cheremisinoff, P.N., *Air Pollution Control and Design Handbook,* Part 2, Marcel Dekker, Inc., New York, NY, 1977.
7. Licht, W., *Air Pollution Control Engineering,* 2nd ed., Marcel Dekker, Inc., New York, NY, 1988.
8. Sherrott, A.F.C., *Air Conditioning: Impact on the Built Environment,* Nicholas Publishing, New York, NY, 1987.

BIBLIOGRAPHY

Air Pollution Manual–Part II, Control Equipment, American Industrial Hygiene Association, Detroit, MI, 1968.

ASHRAE 1985 Fundamentals Handbook, ASHRAE, Inc., Atlanta, GA, 1985.

ASHRAE Handbook & Product Directory of 1979 Equipment, ASHRAE, New York, NY, 1979.

Burton, D.J., "ASHRAE's Indoor Air Quality Procedure Determines the Outdoor Air Requirement," *Occupational Health and Safety,* Medical Publications, Inc., Waco, TX, February, 1993.

Burton, D.J., "ASHRAE's Indoor Air Quality Procedure Generates the Dilution Ventilation Rate," *Occupational Health and Safety,* Medical Publications, Inc., Waco, TX, April 1993.

Burton, D.J., *Industrial Ventilation Workbook,* 2nd ed., IVE, Inc., Salt Lake City, UT, 1991.

"Energy Conservation in New Building Design," Conservation Paper #43B.

Environmental Control Principles, ASHRAE, Inc., Atlanta, GA, 1985.

Flynn, J.E., Segil, A.W., *Architectural Interior Systems,* Van Nostrand Reinhold Co., New York, NY, 1970.

Gladstone, J., *Air Conditioning,* 2nd ed., Van Nostrand Reinhold Co., New York, NY, 1981.

Jennings, B.H., *Environmental Engineering,* International Textbook Co., Scranton, PA, 1970.

Rafferty, P.J., Ed., *The Industrial Hygienist's Guide to Indoor Air Quality Investigations,* American Industrial Hygiene Association, Fairfax, VA, 1993.

Rowe, W.H., III, *HVAC—Design, Criteria, Options, Selections,* R. S. Means Co., Inc., Kingston, MA, 1988.

Stewart, C.T., Jr., *Air Pollution, Human Health, and Public Policy,* Lexington Books, Lexington, MA, 1979.

Weekes, D.M., Ed., Gammage, R.B., *The Practitioner's Approach to Indoor Air Quality Investigations,* American Industrial Hygiene Association, Akron, OH, 1990.

Young, R.A., Ed., Cheremisinoff, P.N., *Air Pollution Control and Design Handbook,* Part 1, Marcel Dekker, Inc., New York, NY, 1977.

CHAPTER 5

VENTILATING INDOOR AIR POLLUTION AS IT PERTAINS TO HOSPITALS

Ralph A. Serio

Hunterdon Medical Center
2100 Wescott Drive
Flemington, NJ 08822-4604, USA

CONTENTS

INTRODUCTION

As the awareness of and demand for proper indoor air quality increases within a hospital environment, HVAC system designers must develop system solutions that not only respond to energy efficiency but also to high indoor air quality. A healthy indoor environment will provide comfort not only to patients but to support staff and visitors. The final product will be a sense of health and well-being for patients, increased productivity for hospital workers, and comfort for visitors.

This chapter offers a discussion of various areas that affect, enhance, and improve indoor air quality. For existing hospitals, indoor air quality can be improved by developing an indoor air quality profile that addresses the building's original function, whether the building is functioning as designed, what changes in building layout and use have occurred since the original design, what changes may be needed to eliminate or prevent indoor air quality problems, and what new regulatory requirements have been established for various areas of the hospital.

Once an indoor air quality problem surfaces or has been determined, an important function of resolving this indoor air quality problem is investigation. The investigation will include an inventory of outdoor sources, inventory of equipment sources, and a review of building components and furnishings. This investigation also will include the involvement of personnel in the reported areas, housekeeping, shipping and receiving, policy making, facility operation and maintenance, safety officers, and other functions that could affect the indoor air quality in question.

The final part of the paper will deal with the evaluation and inspection of the heat, cooling, ventilation, and filtered outdoor-air system (HVAC). Various parts of system and how they affect and improve the indoor air quality of a building will be discussed.

History

Indoor air quality concerns emerged as an issue in the 1980s primarily as a result of the energy conservation philosophy that evolved from the energy crisis of the 1970s. Hospital renovation was no different in that it, too, adopted the energy conservation philosophy.

To cut down on rising energy costs, most hospitals employed air-handling systems that reduced the amount of outdoor air supply used for heating in the winter and cooling in the summer. The net result is a significant increase in indoor air complaints. Other major factors included the proliferation of poorly designed systems, overcrowding, and alteration of office spaces and changes in building use without appropriate ventilation modifications.

To make matters worse, the 1980s brought new designs of office furniture, carpets computers, hospital equipment, solvents, and cleaning agents that either emitted volatile organic compounds and/or generated high amounts of heat. The final result was that these advancements further taxed the ventilation system of hospitals. Consequently, workers believe hospitals have become a health hazard not only to themselves but also to the patients. They feel vulnerable and susceptible to health-related problems during the time they spend at work. They feel that administration does not care and is not willing to address their concerns about the hospital environment.

Ignoring indoor air quality problems can have expensive and extensive consequences. They not only lead to reduced productivity and absenteeism, which affect revenues, but also to legal expenses from judgments. Moreover, once office workers, patients, and visitors perceive their environment to be hazardous, it can be a monumental task to prove otherwise. Therefore, taking steps to prevent and/or eliminate indoor air problems through a practical operations and maintenance program is a sound investment for hospitals.

INDOOR AIR-QUALITY PROBLEMS

As previously stated, indoor air-quality problems increased dramatically because of energy concerns from the 1970s. The conservation has led to what the industry calls "sick building syndrome." The two most common causes of poor indoor air quality are poor ventilation and poor filtration. Ventilation problems deal with the lack of air distribution and insufficient amounts of outside air. Filtration deals with capturing dust, dirt, and microbes inside the air-handling units and ductwork.

The problem of indoor air quality does not stop here. In fact, there are numerous causes from outside and inside sources (see Table 1). Some indoor sources are high humidity, low humidity, high temperature, low temperature, volatile organic compounds from new furnishings, smoking, solvents, cleaning materials, painting materials, new carpets, light, noise, new curtains, high levels of carbon dioxide, and chemical odors. Some outdoor sources include exhaust from motor vehicles, diesel fumes from loading docks, odors from trash bins, pollen, dust, industrial pollutants, and fungal spores.

Other factors that have increased the number of indoor air quality complaints are the rise in the number of employees working in an office; an increase in the number of repetitive, stressful, and ergonomically demanding jobs; an increase in public sensitivity to health and safety issues; and an increase in the number of interior offices.

All of these problems relate to all industries, but especially to hospitals. It seems that in the past, the vast majority of hospital administrators viewed indoor air quality as a low priority, especially concerning support areas of the hospital. They felt obligated to allot available funds to new medical technology acquisitions. Any remaining funds would be distributed to the support functions. HVAC equipment was put in the support category, and low in priority within that group.

Table 1
Common Causes of Indoor Air Quality Complaints

Ventilation systems	48.3%
Inside contamination (other than smoking)	17.7%
Contamination from outside sources	10.3%
Poor humidity control	4.4%
Contamination from building materials	3.4%
Hypersensitivity (pneumonitis)	3.0%
Cigarette smoking	2.0%
Other or unknown	10.9%

Source: National Institute for Occupational Safety and Health (NIOSH)

Because funds were limited, the maintenance of hospital HVAC systems deteriorated. Plant engineers had to make do with what money they were given. The lack of funds and upgrading for HVAC systems led to increasing complaints from hospital personnel, and eventually patients and visitors.

Because of the increase in indoor complaints, ASHRAE ventilation standards for acceptable indoor air quality were revised in 1989. These standards increased the base minimum ventilation rate from 5 cubic feet of air per minute per person to 15. In developing its new ventilation guidelines, ASHRAE recognized that the previously recommended ventilation rates were insufficient to provide adequate indoor air quality in most modern buildings. See Table 2 for various rates for different industries and hospital functions.

For general office space, ASHRAE recommends a minimum of 20 cfm/person for a maximum of seven people per 1,000 square feet. ASHRAE guidelines also prescribe supply rates for outside air required for acceptable air quality.

The Department of Health and Joint Commission now requires all hospitals to maintain HVAC equipment to provide good indoor air quality. In addition, OSHA requirements also must be met.

Based on ASHRAE standards, there are relatively few sources of indoor air contamination that cannot be least reduced or eliminated through proper design and management of HVAC systems. If the primary factor in the design and operation of a building ventilation system were air quality rather than energy conservation, more emphasis would be given to minimal outside air intake levels, filtration efficiency, and hygiene of air-handling systems.

The first sign of an indoor air quality problem usually occurs when groups of people within a particular area complain of odors. Symptoms commonly

Table 2
Outdoor Air Requirements For Ventilation*

Room Type	Outside Air
Hotel guestroom	30 CFM
Office space	20 CFM
Reception area	15 CFM
Theater	15 CFM
Classroom	15 CFM
Library	15 CFM
Hospital patient room	25 CFM
Operating room	30 CFM

**According to American Society of Heating, Refrigerating and Air Conditioning Engineers (ASHRAE) ventilation standards.*

(CFM = Cubic Feet of Outdoor Air per Minute)

reported are headaches; fatigue; shortness of breath; sinus congestion; cough; sneezing; eye, nose and throat irritation; skin irritation; dizziness; and nausea. Most individuals blame their condition on poor air quality. The actual cause can be widespread, from contaminants in the air to job-related psychosocial problems.

The industry standard to determine if an indoor air quality problem exists in a building is if 20% of the occupants are affected. Therefore, records must be kept to determine if an air-quality study is warranted.

Once complaints are received, it is imperative that serious responses to the indoor air quality complaint be made. Lack of response eventually would lead to mistrust of the staff and patient occupants. Communication is the key: Without open communication, any indoor air-quality problem can become complicated by anxiety, frustration, and distrust, which could delay its resolution. The messages to convey are that management believes it is important to provide a healthy and safe building, that good indoor quality is an essential component of a healthy indoor environment, and that complaints about indoor air quality are taken seriously.

The communication should include the following:

1. What types of complaints management has received.
2. Management's policy on providing a healthy and safe environment, and responding to occupant complaints.
3. What management has done to date.
4. What management plans to do to further investigate and correct the problems.
5. The names and extensions of all hospital personnel involved in the investigation.

The best way to start to evaluate the indoor air quality problem for a hospital or any other industry is to develop an indoor air-quality profile. This profile will enable hospital engineers to identify potential problem areas and to estimate the cost of corrective measures. By implementing this profile, the hospital engineer can predict future indoor air-quality problems that could become critical and cause lack of trust between the staff and administration.

The profile should include the use of the building components and furnishings of the building, mechanical equipment, and occupant population and activities. The next step is to find out if the building is functional as designed. This includes changes in building layout and HVAC systems. The last step is to determine what changes are required to eliminate and/or prevent indoor air-quality problems. Keep in mind any future changes that could cause air-quality problems.

Conduct a walk-through of the affected area. The walk-through will provide an overview of the activities and function of the area, and indicate potential sources of indoor air quality problems.

As part of the walk-through, inspection staff members will be asked about their job responsibilities and whether the indoor air quality problem has surfaced in their area.

The occupant interview should consist of the following:

1. A location in which the person you are interviewing feels comfortable to speak freely.
2. Explanation that the interview is intended to help discover and correct the cause of complaints and that you need their help.
3. Enough time to respond to all questions.
4. Open-ended discussion to allow occupant to expand your hypothesis, or problem.
5. Confidentiality regarding health symptoms.
6. A copy of interview form for each person interviewed.

After the interview, an occupant diary form will be given to each person to record the date and time of symptoms, where and when symptoms appear, and any other information that they feel might be useful. This information then will be compared to hospital operations to see if there is a link to the complaints. See Tables 3, 4, and 5 for occupant interview, diary, and log of activities.

Measurements of the supply, exhaust, temperature, and relative humidity should be taken and compared with the original design. If problems are discovered with any of these items, an investigation of each will be undertaken.

The lack of proper room temperature could be caused by a malfunctioning thermostat, temperature control problems at unit, reheat coil problems, staff use of the thermostat, or change of area layout. Air supply and air exhaust problems

Table 3
Occupant Interview

Building Name: ______________________ File Number: ____________

Address: ____

Occupant Name: ____________________ Work Location: ________________

Completed by: __________________ Title: ______________ Date: __________

Section 4 discusses collecting and interpreting information from occupants.

SYMPTOM PATTERNS

What kind of symptoms or discomfort are you experiencing?

Are you aware of other people with similar symptoms or concerns? Yes ______ No ______

If so, what are their names and locations? ______________________________

____

Table 3 (continued)

Do you have any health conditions that may make you particularly susceptible to environmental problems?

- ❑ contact lenses
- ❑ allergies
- ❑ chronic cardiovascular disease
- ❑ chronic respiratory disease
- ❑ chronic neurological problems
- ❑ undergoing chemotherapy or radiation therapy
- ❑ immune system suppressed by disease or other causes

TIMING PATTERNS

When did your symptoms start?

When are they generally worst?

Do they go away? If so, when?

Have you noticed any other events (such as weather events, temperature or humidity changes, or activities in the building) that tend to occur around the same time as your symptoms?

SPATIAL PATTERNS

Where are you when you experience symptoms or discomfort?

Where do you spend most of your time in the building?

ADDITIONAL INFORMATION

Do you have any observations about building conditions that might need attention or might help explain your symptoms (e.g., temperature, humidity, drafts, stagnant air, odors)?

Have you sought medical attention for your symptoms?

Do you have any other comments?

Table 4
Occupant Diary

Occupant Name: ____________ Title: ____________ Phone: ____________

Location: ____________ File Number : ____________

On the form below, please record each occasion when you experience a symptom of ill-health or discomfort that you think may be linked to an environmental condition in this building.

It is important that you record the time and date and your location within the building as accurately as possible, because that will help to identify conditions (e.g., equipment operation) that may be associated with your problem. Also, please try to describe the severity of your symptoms (e.g., mild, severe) and their duration (the length of time that they persist). Any other observations that you think may help in identifying the cause of the problem should be noted in the "Comments" column. Feel free to attach additional pages or use more than one line for each event if you need more room to record your observations.

Section 6 discusses collecting and interpreting occupant information.

Time/Date	Location	Symptom	Severity/Duration	Comments

could be a sign of clogged reheat coils, clogged filters, dirty ducts, a broken damper, change in use of the area from original design, problems with the air-handling unit, or problems with the exhaust or return unit. Humidity problems could be caused by control problems at the unit. In most older hospitals, no humidity control is provided. When no humidity control is provided, the supplied area will receive 8% to

Table 5
Log of Activities and System Operation

Building Name: ______________ Address: ______________ File Number : ______________

Completed by: ______________ Title: ______________ Phone: ______________

On the form below, please record your observations of the HVAC system operation, maintenance activities, and any other information that you think might be helpful in identifying the cause of IAQ complaints in this building. Please report any other observations (e.g., weather, other associated events) that you think may be important as well.

Feel free to attach additional pages or use more than one line for each event.

Equipment and activities of particular interest:

Air Handler(s): ______________

Exhaust Fan(s): ______________

Other Equipment or Activities: ______________

Date/Time	Day of Week	Equipment Item/Activity	Observations/Comments

15% relative humidity during the heating season. This low relative humidity is below the acceptable range of 30% to 65% recommended by ASHRAE. The end result of low relative humidity is the drying out of sinus and mucous membranes.

When providing humidification improvements to a hospital environment, steam humidifiers should be used. Systems used media other than clean steam must be rigorously maintained to reduce the likelihood of microbiological growth. To insure microbe growth is controlled in a hospital, humidistats should be set at no greater than 45%.

Other types of humidification systems used are:

1. Atomizing.
2. Evaporative - pan.
3. Evaporative - wetted element.

Air supplies are distributed to different parts of the hospital by systems known as air-distribution units. Because the requirements of each area of the hospital are different, there are multiple units. Units that service such areas as the operating room, intensive care unit, and laboratory do not have recirculating units. Their air supply is 100% outside air, and their exhaust goes directly outside. Other areas—such as patient rooms, offices, pharmacies, and general areas—have recirculating capacity. Air recirculation is usually dependent on the outside temperature. In winter and summer, the outside air supply is reduced to a minimum to produce energy savings. To make up for the reduced air supply, a large part of the return air is recirculated. This recirculated air can lead to indoor air quality problems.

As part of measuring the supply and exhaust or return readings in an indoor air quality problem area, the location or lack of supply or exhaust diffuses should be noted. If readings from supply and exhaust are excessive for an area's use, the same complaints and reactions would be received as when not enough air is provided. To measure airflows for supply and exhaust grilles, a machine such as a Balameter must be used.

The last part of the walk-through should identify any chemical pollutants and/or odors. This can be accomplished by making a list of all chemicals, cleaning supplies, and odor-producing activities in the area.

If the indoor air quality problem still cannot be found in the complaint area, a pollutant pathway investigation should be performed. The investigation of the pathway will include doors, operable windows, stairways, elevator shafts, utility chases, ductwork, plenums, and areas serviced by common HVAC controls. The first step in checking pollutant pathways is to check the original design to determine the direction of the air movement intended by the designer. Actual direction of air flow then will be performed by using a micromanometer to measure the magnitude of pressure differences between these areas. Also, the peppermint oil technique can be used to determine air flow. The peppermint oil method is usually done when the areas are unoccupied. A pollutant pathway form, such as Table 6, should be used to record results.

After all indoor sources have been investigated, an inspection should be conducted on outside areas. The investigation should include the area directly outside of the air intakes to the units. The location of any loading docks and stack emissions from a boiler house is of primary importance. Investigate additional

Table 6
Pollution Pathway Form for Investigations

Building Name: ______________________ File Number: ______________

Address: ______________________ Completed by: ______________

This form should be used in combination with a floor plan such as a fire evacuation plan.

Building areas that appear isolated from each other may be connected by airflow passages such as air distribution zones, utility tunnels or chases, party walls, spaces above suspended ceilings (whether or not those spaces are serving as air plenums), elevator shafts, and crawl spaces.

Describe the complaint area in the space below and mark it on your floor plan. Then list rooms or zones connected to the complaint area by airflow pathways. Use the form to record the direction of air flow between the complaint area and the connected rooms/zones, including the date and time. (Airflow patterns generally change over time). Mark the floor plan with arrows or plus (+) and minus (-) signs to map out the airflow patterns you observe, using chemical smoke or a micromanometer. The "Comments" column can be used to note pollutant sources that merit further attention.

Rooms or zones included in the complaint area: ______________________

____

Sections 2, 4 and 6 discuss pollutant pathways and driving forces.

Rooms or Zones Connected to the Complaint Area By Pathways	Use	Pressure Relative to Complaint Area		Comments (e.g., potential pollutant sources)
		+/-	date/time	

sources such as exhaust units, motor vehicle exhaust, construction activity, garbage pickup areas, and other potential sources near the building.

Upon completion of this indoor air quality profile, an evaluation and inspection of the HVAC system that is supplying the indoor air quality problem area should be conducted. The investigation will entail checking all components of the HVAC system or systems that serve the complaint area and surrounding rooms. The goal is to understand the design and operation of the HVAC system well enough to answer the following questions.

1. Are the components that serve the immediate complaint area functioning properly?
2. Are the present uses overloading the HVAC system?
3. Are there ventilation problems?
4. Are airflows too low for units supply area?
5. Are pneumatic controls functioning properly or working according to inappropriate strategies?
6. Are heating and cooling coils operating properly?
7. Are filters installed and properly maintained?
8. Are reheaters installed and properly functioning?
9. Are all preventive maintenance schedules maintained for all unit parts?
10. Are air ducts maintained?
11. Are exhaust or relief fans working properly?
12. Are supply fans working properly?
13. If provided, are humidifiers working properly?
14. Are all thermostats working properly?

As you can see, there are numerous possibilities that could cause indoor air quality problems. In most cases, there are multiple causes that lead to the air quality problems. Following are major areas of concern.

FILTRATION SYSTEMS

One particular area of concern is the filtration system. The air filter should be located after the outdoor-air/return-air mixing chamber and before HVAC equipment. The main purposes of these filters are to reduce air contaminants, and to protect the HVAC equipment from clogging up and restricting the air flow.

Filters came in all types, sizes, and efficiency ratings. They range from low-efficiency filters for lint and dust removal; to medium-efficiency filters for bacteria, pollen, insect, soot, dust and dirt; to high-efficiency filters for exceptionally clean areas. When using a high-efficiency filter, it is recommended to first strain the air through medium-efficiency filters. This procedure will produce the most cost-efficient approach to maximize air quality in HVAC systems.

When upgrading your filtration system from low to medium or high efficiency, the pressure drops must be evaluated so the air volumes to the supply area will not be reduced. Because filter designers have improved on the pressure drops on medium- and high-efficiency filters, older HVAC systems can benefit and provide additional air quality without sacrificing air volumes. Therefore, investigating improving the filtration system in an indoor air-quality problem area is a must. To monitor the pressure drops for filters, a device such as a magnetic hellic gauge should be used. When the pressure drop reaches a certain level established by the filter manufacturer, the clogged filters must be replaced to maintain proper supply air volumes for the unit.

If outdoor odors are problems like stack exhaust or diesel fumes, the indoor environment can be protected by installing a carbon filtration system that greatly minimizes these odors. As previously stated, the pressure drops to air volume supply ratio must be checked when installing in older units.

While filters act to reduce air containments, they also can be contaminant producers. When exposed to cool, damp air, filters can become a source of microbic growth. This is particularly true for high-efficiency filters placed downstream from cooling coils and steam humidifiers. These filters must be monitored stringently and frequently changed.

ECONOMIZER SYSTEMS

Another area of concern is the economizer HVAC system. Although effective in reducing operating costs with respect to cooling and heating building air, economizer designs do not include provisions for meeting the ASHRAE requirements for minimum outdoor air ventilation. Therefore, the controls and systems must be adjusted to comply with and improve on indoor air quality standards wherever possible.

Problems occur when outside temperatures are extremely low or high. The system closes the outside air damper to a minimum and opens the return damper to a maximum. This allows the system to use the warm air in the heating season and the cool air in the cooling season to reduce energy costs. When this happens, the indoor air pollutants such as CO_2, VOCs, and solvents are allowed to recirculate, which eventually increases the level and causes indoor air quality complaints.

HEATING AND COOLING COILS

Heating and cooling coils are placed in the airstream to regulate temperature of the air delivered to the space. Malfunctions of the coil control can result in thermal discomfort. Condensation on underinsulated pipes and leaking in pipe systems often will create moist conditions conducive to growth of molds, fungi, and bacteria. During the cooling season, the cooling coil provides dehumidification as water condenses from the airstream. This condensate collects in the drain pan under the cooling coil, and exits in the deep, sealed trap. If water does not drain properly, bacteria and molds will proliferate unless the pan is cleaned frequently.

During heating season, the lack of high water temperatures in the heating coil can cause the outside air not to be heated sufficiently to maintain thermal comfort. Also, the unit may reduce outside air to meet the conditioning of the outdoor air to meet the required inside temperature. Therefore, heating and cooling coils must be properly maintained, not only for temperature control but for airflow volumes.

DUCTWORK

The feed links that distribute conditioned air throughout a building are called ducts. Ducts are capable of distributing dust and other pollutants, including biological contaminants. Dirt and dust can clog and accumulate on its coils, plenums, and equipment housing, which may lead to contamination or reduction of the air supply.

Therefore, care should be taken on keeping the ductwork as clean as possible. If ductwork requires cleaning, the following procedure will be followed:

1. Any duct cleaning should be scheduled during periods when the building is unoccupied to prevent exposure to chemicals and loosened particles.
2. Negative air pressure that will draw pollutants to a vacuum collection system should be maintained at all times in the duct cleaning area to prevent migration of dust, dirt, and contaminants into occupied areas.
3. The ductwork must be protected.
4. Duct cleaning must include gentle brushing of duct lining to be effective.
5. Only high-efficiency particulate arrestance (HEPA)-filtered vacuum equipment should be used.
6. Only use EPA-registered biocide products.
7. Do not use sealants to cover interior ductwork.
8. Carefully clean coils and drip pans to reduce microbiological pollutants.
9. All water-damaged porous material must be removed from ductwork.
10. Start a preventive maintenance program to keep the system clean.

SUPPLY FANS AND EXHAUST

As part of an HVAC system, fan selection is important. The performance of the fan is expressed in cubic feet per minute (CFM). The size of the fan is determined by the ability to move a given quantity of air at a given resistance or static pressure, and is usually measured in column inches of water.

The final airflow through a system is dependent on the size of the duct, the resistance of the duct, and the velocity of air through the duct. The static pressure in a system is calculated using factors for duct length, speed of air movement, and bends in ductwork system.

Exhaust fans are used in areas to remove odors or contaminants from bathrooms, labs, smoking areas, and areas that do not allow return airflow. For successful confinement and exhaust of these zones, the exhausted areas must be maintained at a lower overall pressure than the surrounding area. Areas where chemical reactions take place will have their own dedicated exhaust systems with enough draw to ensure all harmful gases will be vented. Another important aspect is that the supply to an area always should be slightly greater than the exhaust. This difference will ensure a positive building pressure. The only exception in a hospital is for isolation rooms, which by regulation must be kept under negative pressure so infectious disease will not be spread.

Knowing the fan speed and its relationship to the duct size can help determine if the lack of air supply is caused by renovation, deterioration, or breaks within the duct system. The next step is to check or investigate if the speed of the fan can be increased to create additional volumes of air to help alleviate an indoor air quality problem area. If the fan can be adjusted, make sure ducts can handle increased velocities. If ducts can not handle increased velocity, do not adjust fan speed, because either the ducts will rupture or create a whistling noise, which defeats the purpose of improving indoor air quality.

BOILERS AND COOLING TOWERS

Maintenance of this part of the HVAC should not be overlooked. The boilers are used to supply steam for the heating season and steam for the absorption to produce chilled water for air conditioning. Cooling tower water must be properly treated to control pathogenic bacteria, such as Legionella pneumophila. The chillers must be properly maintained to provide maximum heat transfer to remove heat from the hospital.

MAINTENANCE SCHEDULES

The last part of checking the HVAC system is to review the preventive maintenance records on all equipment. The purpose of this review will provide insight into potential indoor air quality problems caused by lack of maintenance. Examples are cleaning of reheat coils, cleaning of air-handling units, balance of air systems, and overall preventive maintenance of motors and fan systems.

CONCLUSION

The causes of indoor air quality problems are as complex and different as any environmental issue to date. By following the procedures stated in this chapter, hospitals can eliminate or greatly reduce air quality problems in existing hospital buildings. The procedures also can be useful in maintaining new HVAC systems in newly renovated hospitals.

By establishing an indoor air quality profile for your hospital building, your will be able to monitor and plan properly for future renovations to eliminate the possibility of creating indoor air quality problems. In addition by including occupants in affected areas in the investigation process, you will gain trust and acceptance that indoor air quality is important to the administration.

Basic recommendations to reduce indoor air quality problems are:

1. Keep temperature and humidity within desired range.
2. Upgrade filtration system as much as the existing system will allow.
3. Balance supply and exhaust systems, and check as the need arises.
4. Provide the required air volumes of outside air for each use according to ASHRAE standards revised in 1989.
5. Establish and maintain a preventive maintenance program for all parts of the HVAC system.
6. When necessary, provide carbon filtration systems to combat diesel and stack fumes from entering the supply zones of the system.
7. Maintain a clean environment throughout the hospital.
8. Train all HVAC personnel on air quality standards to follow when working on HVAC equipment.
9. When major renovations are performed within a hospital, upgrade the HVAC to meet present standards.

BIBLIOGRAPHY

U.S. EPA, "Building Air Quality: A Guide for Building Owners and Facility Managers," Washington, DC (Sept. 1991).

ASHRAE, Standard 62-1989, "Ventilation for Acceptable Indoor Air Quality," Atlanta, GA (1989).

Burton, Jeff D., *Indoor Air Quality Work Book,* IVE, Inc., Salt Lake City, UT (1991).

National Coalition on Indoor Air Quality, "1992 Indoor Air Quality Conference Work Book," Tampa, FL (1992).

CHAPTER 6

INDOOR AIR QUALITY AUDITS

Colleen M. Williams

Clayton Environmental Services, Inc.
Edison, NJ 08817, USA

Richard B. Trattner

Department of Chemistry, Chemical Engineering and Environmental Science
New Jersey Institute of Technology
Newark, NJ 07102, USA

CONTENTS

INTRODUCTION

The construction and operation of buildings has changed significantly during the past few decades. Some of these changes have resulted in the generation of new types of air contaminants, or indoor air pollutants. These indoor pollutants have been suspected to be related to sick building syndrome and possibly building-related illness. The objectives of this chapter are to present a guide to performing comprehensive qualitative indoor air-quality investigations, descriptions of typical air-quality problems in non-industrial buildings, and mitigation actions that can be used to prevent potential indoor air-quality problems and to improve existing ones.

There is public concern regarding the possible adverse health effects of trace concentrations of air contaminants such as environmental tobacco smoke, radon, volatile organic compounds (e.g., benzene, chlorinated hydrocarbons), and microorganisms in indoor air. This chapter will discuss how to conduct a qualitative walk-through evaluation without performing air or source sampling (quantitative). Several environmental factors, all of which contribute to indoor air quality, will be discussed.

BACKGROUND

Currently, there are no published standards governing indoor air quality in nonindustrial buildings. Facilities such as office buildings, schools, and hospitals likely suffer from poor indoor air quality for the lack of such standards. There are, however, guidelines that can be applied to improve indoor air quality or prevent the accumulation of contaminants that cause poor air quality.

Most people in industrialized countries spend up to 90% of the 24-hour day in the indoor air environment (residence, transport, office). Concern has been raised regarding the potential effects of long-term exposure to indoor air contaminants. The risks associated with breathing indoor air now are thought to exceed the risks associated with pollutants found in the ambient air [1].

Among indoor air pollutants, risk from the exposure to volatile organic compounds (VOCs) may be of similar magnitude as that associated with exposure to chemicals in industrial environments [1]. Microbial aerosols in indoor environments may cause significant health effects, especially among the population predisposed to asthma or allergy. For much of the population, illness caused by respiratory infections is common. One study showed that the incidence of respiratory infection was significantly higher in mechanically ventilated barracks than those barracks naturally ventilated with open windows [2]. Therefore, occupants in mechanically ventilated buildings have a high risk of developing respiratory infections from recirculated microbial agents (viruses) in the indoor air.

Because of public concerns about indoor air quality (IAQ), the American Society of Heating, Refrigeration and Air Conditioning Engineers (ASHRAE) published Standard 62-1973, "Standards for Natural and Mechanical Ventilation." Through years of revisions, the current standard, "Ventilation for Acceptable Indoor Air Quality" (Standard 62-1989), includes recommendations for ventilation requirements for smoking and nonsmoking areas, as well as listing threshold limit values (TLVs) for airborne chemicals such as formaldehyde [3].

One of the most important factors contributing to poor indoor air quality and related concerns is the fuel shortages that caused building owners and engineers to "close up" and design energy-efficient, cost-effective buildings. Buildings are designed with inoperable windows to prevent the loss of valuable conditioned air brought in from outside and passed through the heating, ventilation, and air-conditioning (HVAC) system (Figure 1). Many indoor air quality problems occur because the HVAC system is not bringing in enough outdoor air to dilute contaminants that build up in the occupied space. A qualitative audit of air quality in non-industrial environments can help a building owner, manager, or occupant recognize potential IAQ problems and correct existing IAQ problems.

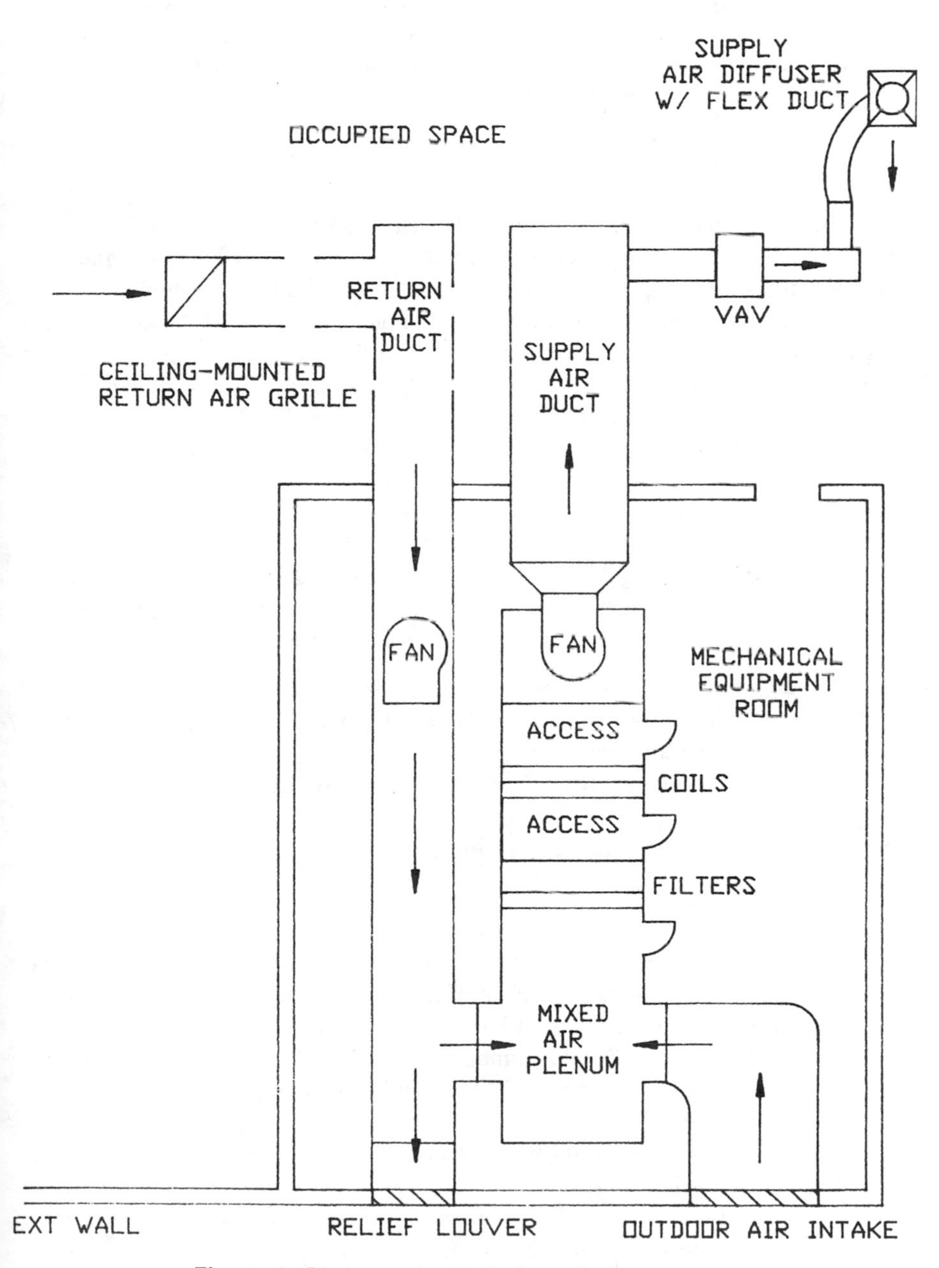

Figure 1. Diagram of a typical ventilation system.

It must be noted, however, that many perceived indoor air quality problems may come from other stressors in the workplace. Ergonomics (chair and desk height, visual display terminals, work schedule), psychological factors (labor management problems, dead-end position with company), and environmental variables (poor lighting, acoustics, vibrational sounds) can elicit complaints from occupants but are not related to indoor air quality.

Sick Building Syndrome

Sick building syndrome (SBS) is defined as an assortment of non-specific, usually sensory complaints including eye, nose, and throat irritation; headache; fatigue; skin irritation; mild neurotoxic symptoms; and odor annoyance [4]. A complaint rate of approximately 20% is needed for the problem to qualify as SBS. Occupants affected by irritation usually report relief when they have breathed outdoor air for a short period of time. Because SBS symptoms are non-specific, they potentially can be associated with one of many contaminant sources in the building, making it difficult to find an exact cause for SBS even in extensively studied buildings.

Building-Related Illness

The term "building-related illness" (BRI) is used to distinguish those instances where occupant health problems resemble a medically identified disease. Examples of BRI and air contaminants involved in disease etiology include:

- Legionnaire's disease—caused by *Legionella* species and serotypes.
- Hypersensitivity pneumonitis, humidifier fever, allergic rhinitis, and asthma—caused microbial allergens.
- Dermatitis—caused by fiberglass fibers and volatile organic compounds (VOCs).
- Cancer—caused by the particulate and gaseous components of environmental tobacco smoke (ETS), some VOCs and radon.

It is possible that the same agents may cause both sick building syndrome and building-related illness. For example, while exposure to VOCs, ETS, and combustion products may cause sick building syndrome, the chronic exposure to these contaminants may cause building-related illness. Recent estimates suggest that approximately 20 to 30% of nonindustrial buildings in the United States are problem buildings, and an estimated 20 to 50 million people are exposed to poor indoor air quality [5].

Three basic approaches should be followed when performing a walk-through evaluation of any facility suspected of being a sick or problem building: Keep in mind the illness and discomfort phenomena experienced by the occupants, investigate HVAC systems in buildings where problems can occur, and consider the types and sources of pollutants that may be present in the building.

INDOOR AIR QUALITY AUDITS

The qualitative IAQ evaluation begins with extensive telephone interviews with the client. Prior to scheduling the walk-through of the building, review telephone conversations and written information provided by the client related to thermal environmental problems, sick building syndrome, building-related illness, ergonomic issues, and potential labor management issues. At this stage of the evaluation, determine if lighting, ergonomic, acoustic, or psychosocial problems may be to blame for occupant complaints. Explore the possibility that labor-management issues are creating employee tension and stress. When health complaints are reviewed, determine if the perceived problem is one of occupant discomfort and annoyance (SBS or thermal discomfort) or building-related illness.

Hypothesize what the likely problems are based on telephone interviews and available documentation. Arrange an initial meeting with the client and be prepared to discuss your hypothesis. Plan for and schedule the walk-through. At the building site, review the nature of the health complaints, visually evaluate environmental factors that may be responsible for complaints, and assess the HVAC system. The objective of the qualitative evaluation is to provide recommendations for remedial actions without requiring expensive contaminant sampling.

Few instruments will be used during the evaluation. Record measurements of relative humidity, temperature, carbon dioxide, and respirable particulates to help verify the hypothesis of the building's problems. It is essential to perform an HVAC evaluation in the analysis of a building. HVAC mechanical equipment components should be visually examined for deficiencies regarding design, operation, and maintenance parameters.

Questionnaire

One of the first steps in initiating an indoor air-quality investigation is to submit an indoor air-quality questionnaire. It is useful in gathering information regarding the types of problems experienced in the indoor environment. It is important that a questionnaire be simple and straightforward to obtain a satisfactory return rate and meaningful results [6].

The consultant issuing the questionnaire should draw a rough floor plan of the complainant area and also should include which individual heating, ventilation, and air-conditioning units serve which zones.

A typical questionnaire should include at least the following:

(1) Employment status:
full time _____
part time _____
temporary _____
other _______________

(2) I have worked in this area since (year) _____, (month) _____, (day) _____
(3) I work in:
a closed office _____
a cubicle _____
an open area shared with others _____
(4) Are any of the following within 30 ft of your work location?
typewriter _____
photocopier _____
video display terminal _____
printer _____
fax machine or blueprint machine _____
other _______________

During the previous week have you experienced the following while working in your area?

(5) Eye symptoms
redness _____
watering _____
burning _____
puffiness _____
dryness _____
irritation _____
blurred vision _____
other _______________
(6) Nasal symptoms
nosebleeds _____
congestion _____
sinus problems _____
sneezing _____
runny nose _____
dry nose _____
other _______________
(7) Throat symptoms
sore throat _____
dry cough _____
other _______________
(8) Are you a contact lens wearer? If so have you experienced problems with the following:
cleaning _____
deposit formation _____
discomfort _____

pain _____
other _______________

(9) Skin problems
dryness _____
flaking _____
rash _____
irritation _____
other _______________

(10) Aches and pains
headaches _____
backache _____
muscle/joint pain _____
other _______________

(11) General complaints
drowsiness _____
dizziness _____
lightheadedness _____
difficulty concentrating _____
other _______________

(12) other symptoms
difficulty breathing _____
digestive disorders _____
reproductive problems _____
other _______________

(13) Were you ever absent because of any health problems that you feel have been aggravated or caused by working in your present location?
If yes, please state health problems: _______________________________
____
____

(14) Have you sought medical attention for the health problems caused by working at your present location? ___________________________

(15) Do you have any known allergies? ___________________________

(16) Are you presently taking prescribed medications for the symptoms you exhibit? ____

(17) Age _____
Sex _____

The following questions help to describe the quality of the office environment. Please note which of these factors disturb you in your work area.

(18) ventilation
temperature _____
humidity _____

air movement _____
other _______________

(19) noise
nearby conversation _____
lighting _____
ventilation _____
office equipment _____
other _______________

(20) lighting
too bright _____
not bright enough _____
glare, flickers _____

(21) If there are odors in your area what do they most resemble:
glue _____
perfume _____
ammonia _____
alcohol _____
vinegar _____
fuel _____
paint _____
musty _____
smoky _____
stale _____
other _______________

Do you have an idea where the smell may originate? _______________
____
____

(22) Are any of the following in your work area?
humidifier _____
air cleaner _____
fan _____
heater _____
desk lamp _____

(23) Do you have any control over your work environment?
ventilation _____
temperature _____
humidity _____
lighting _____

(24) Is there a smoking lounge near your work location? _____
How many feet away? _____

Review the questionnaires for one or several complainant zones to help determine where to begin the indoor air-quality walk-through investigation. There are many factors involved in investigating indoor air-quality matters, and it may be necessary to view each one. The checklists that follow will help uncover deficiencies in building operations and maintenance, and lead to the improvement of the indoor environment under investigation.

Ventilation

Ventilation is the most important factors in indoor air quality. Understanding a building operator's maintenance procedures and HVAC system control strategies is essential to a thorough evaluation. This information is often difficult to obtain because the facilities engineer may be the building owner's representative while the investigator is an agent of the tenant. Therefore, few building operators welcome an audit of their maintenance and operating procedures. This information is essential to acquire a thorough understanding of the HVAC system, and its contribution to poor indoor air quality and must be known. The investigator should not be discouraged in trying to gain this valuable information.

There are several categories under the ventilation umbrellas to address during a walk-through.

Thermal Environmental Parameters

Many occupants have the simple complaint that they are either too hot or too cold at any one time in the office. It is a matter of personal preference. ASHRAE 55-1981 states, "The temperature range which 80% of sedentary office workers will find acceptable is 73 to 78.5°F with 60% relative humidity during summertime (cooling) conditions and 69 to 76° with 30% relative humidity during wintertime (heating) conditions" [7].

During the walk-through, measure the temperature and humidity in several occupied areas to determine whether there is cause for thermal complaints in the building. Use the following checklist to look for other thermal environmental factors.

- Are there portable fans or heaters in the occupied space?
- Is radiant energy (sunlight) entering the building?
- Is there excessive heat gain or loading from electrical equipment?
- Do full-height walls or partitions restrict the movement of air?
- Use smoke tubes to determine if there is stagnation.
- Are thermostats located in the proper place (out of direct sunlight, away from supply airstream)?
- Do air-handling units run continuously in the occupied mode or do they cycle off when the area is thermally satisfied, thus cutting off the flow of supply air?
- Evaluate air motion characteristics such as air velocity in feet per minute coming out of the diffuser. Greater than 30 fpm in winter may mean a draft,

greater than 50 to 150 fpm in summer may mean problems. If air motion is minimal, you have a "stuffy air" situation.

HVAC System Design

Variable Air Volume (VAV) Systems. Most of the buildings in use today utilize what is known as variable air volume (VAV) air-handling units. VAV systems use dampers to constrict airflow to control temperature, as opposed to constant air volume (CAV) systems, which provide constant flow while regulating desired temperatures. Use the following checklist during the walk-through portion of the evaluation. The suggested changes will improve indoor air quality by design.

- Provide damper control to assure a constant amount of outdoor air to be supplied even at the minimum VAV fan setting (approximately 10% outdoor air at all times).
- VAV terminals (thermostats) should have a minimum airflow setting so that supply air is delivered to occupied space even when thermostat is thermally satisfied.
- Do not allow occupants to remove ceiling tiles for long periods of time, as air distributed from supply ducts will short-circuit into ceiling plenum.
- Establish adequate zoning: Do not combine solar exposures on same thermostatic zone. For example, do not combine an eastern exposure with a southern exposure as the solar load may be too high to cool adequately.
- Use slot-type diffusers instead of cone diffusers wherever possible to reduce dumping supply air into occupied space during low-flow conditions.

Return-Air Plenums. Many buildings are designed with false or drop ceilings, which are used to gather return air and deliver it back to the air-handling unit (AHU) to be recirculated, thus saving energy on the reheating or recooling of air.

- Provide adequate openings in full-height walls (perimeter offices) to allow air to return to the main duct, usually in a common interior zone. This can be accomplished by leaving spaces under doors.
- Return-air plenums should be free of asbestos, PVC, gas valves, exposed fiberglass, and construction debris.

Supply air requirements (× cubic feet per minute per square foot) are reduced because approximately 50% of recessed lighting heat load and 70% of roof cooling loads bypass the occupied space.

Heating System. The following are problems that may be encountered upon evaluating the boiler area of the HVAC system.

- Do not use cool-mist humidifiers in the occupied space because harmful microorganisms may amplify in reservoirs and subsequently be delivered to the occupied space on ventilation air.

- Steam used in humidification should not be treated with chemicals such as volatile amines or biocides, which may be potentially harmful if inhaled.
- Combustion air to mechanical rooms must be adequate to supply all necessary equipment with fresh air.
- If humidification is needed, a steam-generating system should be installed in bare (unlined) sheet-metal ducts downstream of air filters.
- If air conditioning systems are shut down at night, contaminants are likely to build up in the occupied space. Also, in temperature locales, an increase in relative humidity will promote the growth of microorganisms (bacteria and fungi).

Makeup-Air and Exhaust-Air Systems

- Exhaust louvers in restrooms, cafeteria, and printing rooms should be separated from outdoor air intake louvers by at least 25 ft to prevent the reentrainment of exhaust into air handling system.
- Environmental tobacco smoke should be discharged directly to the atmosphere, not to ceiling plenums where it will be returned to the air-handling system.
- Photocopier exhaust should be discharged directly to the atmosphere, not to a return-air plenum.
- If chemical or vehicle exhaust is discharged from vertical stacks, the stacks or louvers should have a dispersion velocity, greater than 2,000 fpm.
- If the building contains a parking garage or loading dock, carbon monoxide sensors may be used to activate exhaust systems.
- If normal outside air ventilation is less than building exhaust air, tempered, filtered makeup air should be provided to help keep building pressurization slightly positive relative to the atmosphere. This prevents unfiltered and untempered air from entering the building through doors and open windows.

Ventilation (provision of outdoor air to occupants)

- Provide 20 cubic feet per minute per person (cfm) of outdoor air with proper filtration for office buildings. Refer to ASHRAE 62-1989 for other applications.
- Outdoor air should be distributed evenly to occupants. Do not dump outdoor air in the center of the return-air plenum to be picked up by water-source heat pumps.
- Outdoor air intake should be at least 25 ft from plumbing vents, exhaust fans, boiler flues, relief louvers, cooling towers, and any other potential source of contamination.
- Avoid outdoor air inlet locations where prevailing winds may subject them to cooling tower plume or vehicle exhaust.
- The outdoor air intake should be at least 6 ft from grade level and at least 3 ft above the roof level.

- Outdoor air should be brought in at less than 500 fpm, as anything higher will cause moisture carryover into unit (humidity, rain).
- Outdoor air intakes should be designed with stormproof construction and adequate bird screens.
- Inspect the structural integrity of internal insulation lining supply ductwork. In many instances, duct lining is deteriorated.
- Assure a thorough mix of outdoor air and return air upstream from the cooling coils.
- Provide outdoor air as required to keep building pressure slightly positive relative to the atmosphere.
- Condensate drain lines should be fitted with deep traps to prevent backflow of mechanical equipment room air into the supply air plenum.

HVAC System Maintenance

One of the most important aspects of building ventilation systems is the way they are maintained. Preventive maintenance programs usually are inadequate.

- Use proper filtration based on the presence of certain contaminants.
  A. Use 45%-65% efficiency filters (rated by the ASHRAE atmospheric dust spot test) in all systems to reduce dirt, dust, and spores.
  B. Activated charcoal filters should be installed if contaminants such as volatile organic compounds are present in the occupied space air.
- Filters should be replaced at least six times a year.
- Provide a tight filter for the air handling unit plenum so that all outdoor air is filtered.
- Clean and disinfect cooling coil condensate drain pans once a month during the cooling season. Clean pans with a proprietary biocide when the unit is off and the building is unoccupied. Rinse pans thoroughly with clear water before recommissioning unit.
- Drain pans must be pitched toward the drain outlet, and the outlet connection must be flush with the bottom of the pan.
- Keep floor drain traps wet, especially in air plenums, to prevent the entrainment of sewer gas odors.
- Periodically clean outdoor air intake plenums of dirt, leaves, bird droppings, etc.
- Periodically check outdoor-air and return-air intake dampers to ensure that outdoor air dampers are not completely shut.
- Periodically calibrate thermostats, sensors, and controls.
- Perimeter fan coil units and induction units should be kept free of debris and vacuumed regularly with a high-efficiency, particulate-air-filtered vacuum.
- Rebalance the HVAC system if zones have been altered by renovation.
- The lower-limit volume on variable-air-volume terminals should allow 25% minimum air flow.

- Water-source heat pumps should be run on continuous "fan-on" mode during all hours of occupancy.
- Steam humidifiers should not be treated with biocides or volatile amine additives. Use deionized water in humidifiers.
- Ventilation systems should run continuously during periods of renovation or new occupancy to remove odors from adhesive paints and new furnishings and carpeting.
- Cooling-tower treatment should be provided automatically with electrical feeders. Chemicals should be monitored regularly during the cooling season, including periodic tests for *Legionella* bacteria.

Housekeeping Considerations

Following are several housekeeping factors to consider when performing a walkthrough of any facility.

- Are drain pan outlets properly trapped and pitched and run to waste?
- Are irritants such as paints, adhesives, or volatile chemicals stored in the air plenums of air-handling units?
- What type of lawn chemicals are spread near low outdoor-air intakes?
- Pest control bait should not be placed in perimeter fan coil or induction units enclosures, as particles will become airborne and distributed to the occupied space.
- Do air-handling units provide easily opened access doors large enough for gaining access to clean cooling coils, drain pans, and humidifiers?
- Does the HVAC system provide service clearances for filter maintenance, tube pulls, and water treatment?
- Do not allow rainwater to pool on the roof near air-handling units. Ensure proper drainage through roof drains.

Air Contaminants

It is possible that one or many air contaminants are present in the indoor environment. During a walk-through, it is important to identify likely or potential sources of contamination indoors.

Microbiological Factors

The term "bioaerosol" refers to airborne microbiological particulate matter derived from viruses, bacteria, fungi, protozoa, and their cell-wall components. Although microorganisms are normally present in indoor environments, moisture and substrate will allow microorganisms to proliferate. Certain types of humidifiers, water spray systems, and porous manmade insulation can serve as reservoirs and amplification sites for fungi, bacteria, and protozoa in indoor environments.

Cooling towers and other water-service systems, such as water heaters, can provide water and nutrients for the amplification of harmful microorganisms including *Legionella.* In most cases, the amplification of microbes to atypical concentrations indoors is caused by poor preventive maintenance of these systems [4].

The presence of dirt and moisture invariably translates into fungi and bacterial growth. It is important that leaks or water damage of any kind be immediately removed or dried because the longer a material stays damp, the larger the possibility that microbial growth will occur.

Many people are allergic to molds. When mold is present in the workplace, it is likely that some occupants will develop allergic symptoms including respiratory problems, rash, eye irritation, and cough. Following is a list of possible sources of microbial reservoirs in the workplace.

- Have any floods occurred, or has there been major water incursion from pipe breaks or roof leaks? If so, observe for damp carpeting, wet ceiling tiles, and spray-on fireproofing. If bioaerosols are not readily visible, conduct air sampling to find a source.
- Look for any sources of water or moisture in the air-handling unit, such as biofilm in drain pans, clogged drain pans, moist or damp filters, internal insulation lining plenums, or decaying animals.
- Condensation from cooling coils or cold pipes may introduce moisture onto porous materials (filter media, ceiling tiles), which will subsequently become a microbial amplification site.
- If the air-handling system is equipped with a humidification system, there could be potential problems with the aerolization of microbes from the reservoir water, or the humidity may cause some porous materials to become wet.
- Portable humidifiers in the occupied space can be a source of airborne microorganisms if not properly and meticulously cleaned.
- Excessive humidity in the occupied space (greater than 70%) can promote fungal growth when nutrients, such as dirt and dust, are present.
- Decorative plants, wood bark, wicker, and other organic materials may become amplification sites if they become wet.
- Bioaerosols near the outdoor air intake or an ill-maintained cooling tower may contaminate a building with microorganisms.

Volatile Organic Compounds (VOCs)

Volatile organic compounds are commonly associated with indoor air pollution and SBS. Volatile organic compounds include aromatic hydrocarbons, alkanes, halocarbons, terpenes, alcohols, ketones, and aldehydes. Those VOCs with boiling points above 260°C are known as semivolatile (e.g., pesticides). Compounds with boiling points less than 50–100°C are referred to as very volatile.

These compounds are contained in many substances such as cleaning solvents, paint, adhesives, carpeting, furniture veneers, upholstery, fuel, and other everyday substances including liquid correction fluid, magic markers, and photocopier chemicals. For example, trichloroethane is often found in indoor air because solvents used to clean workstation panels contain this compound. Terpenes are found in air fresheners, cleaners, and polishers in the form of limonene (lemon scent) and pinene (pine scent).

Other major sources of VOCs in buildings include construction materials, maintenance products and combustion materials. Workstation panels, plywood, paints, lacquers, new furnishings, and carpeting emit VOCs in the indoor environment.

Concentrations of these compounds in indoor air can serve as irritants to the mucous membranes and upper-respiratory tract. VOCs are also blamed for fatigue, difficulty concentrating, and unpleasant odors [4].

During the walk-through or pre-walk-through interview, determine which of the following may contribute to occupant indoor air quality complaints.

- Has recent renovation added new sheetrock, paint, carpeting, or furnishings? These can be substantive sources of VOCs in the workplace.
- What kind of cleaning chemicals are used on a regular basis, and are material safety data sheets (MSDSs) available for each one? If so, check the contents of the chemicals to ensure that the least-hazardous products are being used. Overuse of a product can result in high concentrations of VOCs in the workplace.
- Are there special processes that are taking place in the occupied space, such as high-volume printing or photographic developing? If so, are these areas properly exhausted to the outside air, or is VOC-laden air being redistributed to the occupied space with return air?
- How often are pesticides used in the occupied space? Obtain MSDSs on each of the pesticides, and check hazard information.
- If the facility contains laboratory or venting hoods, find out what chemicals are being discharged through hoods. Ensure that outdoor air intakes are a proper distance from the discharge to avoid reentrainment.
- Observe the occupied space for sources and possible sinks for VOCs. Sinks include porous wallcovering and upholstery, fabric-covered workstations, carpeting, draperies, porous ductliners, and air-handling unit filters. Quantitative sampling for VOCs will be necessary if you or the client decide to determine what concentrations are present in the occupied space.
- Look for other outdoor sources of VOCs from loading docks or neighboring facilities.

Combustion Products

Products of combustion—such as carbon monoxide, carbon dioxide, sulfur dioxide, and the nitrogen oxides—may be entrained into outdoor-air intakes of HVAC systems. Exposure to low concentrations of combustion products can cause mucous membrane irritation, headache, dizziness, and lethargy [4]. Clients may want to sample for these compounds to determine concentrations, but identified sources can be removed or reduced without expensive sampling.

- Look for obvious sources of contamination, which include parking garages, parking lots, loading docks, boiler stacks, and furnace flues within 100 feet of outdoor-air intake. Infiltration also may occur through open loading dock doors or windows.
- Negative air pressure in the building or zone near the source of combustion products can cause entrainment.

Environmental Tobacco Smoke (ETS)

Environmental tobacco smoke (ETS) is composed of a complex mixture of chemicals, including gases and respirable particles. ETS contains more than 4,000 chemical compounds, including nicotine, tar, formaldehyde, benzene, styrene, and vinyl chloride. More than 40 carcinogenic compounds are found in ETS [8].

Many facilities have banned smoking indoors. However, such policies have been met with opposition by many in the smoking community. If the facility under investigation allows smoking indoors, it is likely that ETS is causing irritation to some occupants. Common complaints include mucous membrane irritation and headache. Odor annoyance due to ETS is perceived at concentrations much lower than those that cause irritation [9].

- Locate areas where smoking is allowed indoors, and look for a separate exhaust system that discharges directly to the atmosphere.
- In the cafeteria, is there a separate HVAC system for smokers? Such an area should be negatively pressurized relative to other areas to prevent the migration of ETS.
- Keep in mind that many smoking areas, although separated from non-smoking areas, may actually return the smoke-laden air to a common return air plenum to be redistributed to the occupied space.
- Make measurements with an aerosol monitor or respirable particulate meter in smoking zones, non-smoking zones, and in the outdoor air to determine if there is migration of ETS to other areas of the building. Refer to ASHRAE 62-1989 for recommended guidelines [3].
- Good filtration and large amounts of outdoor air to the occupied areas is not the answer to eliminating ETS in the workplace. The best method is to ban smoking indoors.

Particulates Other Than ETS

Other particulates may be generated in or near the occupied space. For example, dust from construction or inadequate housekeeping can be an irritant to the mucous membranes.

- Recent renovation activities such as hanging sheetrock may have generated construction dust. If renovations are taking place in the facility, take care to keep the area negatively pressurized relative to the occupied areas, and avoid tracking construction dust into occupied areas.
- If recent renovations have occurred, it is possible that large amounts of construction dust and debris are located in the occupied space and in the above ceiling return-air plenums. These should be vacuumed immediately with a high-efficiency particulate arrestance (HEPA) filtered vacuum.
- Poor housekeeping may be the cause of dust accumulation in the occupied space. Ask the client about routine vacuuming and dusting schedules.
- Photocopiers and other machines may generate particulate from dry chemicals.
- Look for normal outdoor sources of particulates, including lawn maintenance and construction activity, vehicle traffic, and spores from plants and trees.

CONCLUSION

At the end of the evaluation, be prepared to conduct an exit interview, at which time major observations, conclusions, and preliminary recommendations should be discussed with the building management and employee representatives. Observations of possible contaminant sources made during the qualitative walk-through can help verify preliminary hypotheses of complaint etiology and formulate recommendations for corrective actions.

A report summarizing visual observations and establishing the preliminary hypothesis on complaint etiology should be issued to the client without delay. The most important section of the qualitative evaluation report is the recommendations for improving building performance. Such recommendations usually include a combination of source-control measures and an upgrade of the HVAC system design, operation, and maintenance procedures. If all recommendations are implemented and complaints continue, suggest air sampling for specific contaminants.

REFERENCES

1. Nero, A.V. Jr., *Scientific American,* Vol. 258, p. 42, 1988.
2. Brundage, J.F., et al., *J. Am. Med. Assoc.,* Vol. 259, p. 2108, 1989.
3. American Society of Heating, Refrigeration and Air-Conditioning Engineers (ASHRAE) Standard 62, "Ventilation for Acceptable Indoor Air Quality," Atlanta, 1989.
4. Morey, P.R., and Singh, J., "Indoor Air Quality in Nonindustrial Occupational Environments," *Patty's Industrial Hygiene and Toxicology,* Fourth Edition, Vol. 1, Part A, pp. 531-584.
5. Woods, J.E., *Occupational Medicine: State of the Art Reviews,* Vol. 4, Hanley and Belfus, p. 753-770, Philadelphia, 1989.

6. Guirguis, S., et al., "A Simplified IAQ Questionnaire to Obtain Useful Data for Investigating Sick Building Complaints," *American Industrial Hygiene Journal,* Vol. 52, No. 8, pp. 434-437, August, 1991.
7. American Society of Heating, Refrigeration, and Air-conditioning Engineers, Standard 55, "Thermal Environmental Conditions for Human Occupancy", Atlanta, 1981.
8. United States Environmental Protection Agency, *Indoor Air Facts,* No. 5, ANR-445, 1989.
9. Clausen, G.H., *Proceedings of IAQ 1986,* "Managing Indoor Air Quality for Health and Energy Conservation," ASHRAE, pp. 119-125, Atlanta, 1986.

CHAPTER 7

OSHA AND OSHA REGULATIONS

Nicholas P. Cheremisinoff

SciTech Technical Services, Inc.
Morganville, NJ 07751, USA

Vasantrao Nivargikar

Mega Management & Testing Services, Inc.
Holmdel, NJ 07733, USA

CONTENTS

OVERVIEW

More than 90 million Americans spend their days on the job. Yet, until 1970, no uniform, comprehensive provisions existed for workplace safety and protection against health hazard.

In 1970, Congress considered statistical data that showed:

- Job-related accidents accounted for more than 14,000 worker deaths.
- Nearly 2 1/2 million workers were disabled on the job.
- Ten times as many person-days were lost from job-related disabilities as from strikes.
- Estimated new cases of occupational diseases totaled 300,000.

In terms of lost production and wages, medical expenses, and disabilities compensation, the burden on the nation's commerce was staggering. Human damage cost was beyond calculation. Therefore, a bipartisan Congress passed the Occupational Safety and Health Act (OSHA) of 1970.

Purpose of OSHA

Under the act, OSHA was created within the Department of Labor to:

- Encourage employers and employees to reduce workplace hazards and to implement new or improve existing safety and health programs;
- Provide for research in occupational safety and health to develop innovative ways of dealing with occupational safety and health problems;
- Establish "separate but dependent responsibilities and rights" for employers and employees to achieve better safety and health conditions;
- Maintain a reporting and record-keeping system to monitor job-related injuries and illnesses;
- Establish training programs to increase the number and competence of occupational safety and health personnel;

- Develop mandatory job-safety and health standards and enforce them effectively; and
- Provide for the development, analysis, evaluation, and approval of state occupational safety and health programs.

While OSHA continually reviews and redefines specific standards and practices, its basic purposes remain constant: OSHA strives to implement its Congressional mandate fully and firmly with fairness to all concerned. In all its procedures, from standards development through implementation and enforcement, OSHA guarantees employers and employees the right to be fully informed, to participate actively, and to appeal for actions on all related health and safety items of concern.

What Does the Act Cover?

In general, coverage of the act extends to all employers and their employees in the 50 states, the District of Columbia, Puerto Rico, and all other territories under federal government jurisdiction. Coverage is provided either directly by OSHA or through an OSHA-approved state program.

As defined by the act, an employer is any person engaged in a business affecting commerce and who has employees. The act applies to employers and employees in such varied fields as manufacturing, construction, longshoring, agriculture, law, medicine, charity, disaster relief, organized labor, and private education. Such coverage includes religious groups to the extent that they employ workers for secular purposes.

The following are not covered under the act:

- Self-employed persons;
- Farms at which only immediate members of a family are employed; and
- Workplaces already protected by other federal agencies under other federal statutes.

If another federal agency is authorized to regulate working conditions in a particular industry and it does not do so in specific areas, OSHA standards are applicable to that specific area.

As OSHA develops effective safety and health standards of its own, standards issued under the following laws administered by the Department of Labor are superseded: the Walsh-Healey Act, the Service Contract Act, the Construction Safety Act, the Arts and Humanities Act, and the Longshoremen's and Harbor Workers' Compensation Act.

OSHA STANDARDS

In carrying out its duties, OSHA is responsible for promulgating legally enforceable standards. OSHA standards may require conditions, or the adoption or use of one or more practices, means, methods, or processes reasonably necessary and appropriate to protect workers on the job. It is the employers' responsibility to become familiar with standards applicable to their establishments, and to ensure that employees have and use personal protective gear and equipment when required for safety. Each employee shall comply with all rules and regulations which applicable to his/her own actions and conduct. Where OSHA has not promulgated specific standards, employers are responsible for following the act's general duty clause.

States with OSHA-approved occupational safety and health programs must set standards at least as stringent as the federal standards. Many states adopt standards identical to the federal standards, and make a statement to this effect in their safety and health plans.

Standards Information Source

OSHA standards fall into the following four major categories:

Category 1 — General Industry
Category 2 — Maritime
Category 3 — Construction
Category 4 — Agriculture.

The Federal Register is one of the best sources of information on standards. All OSHA standards are published in this document when adopted, as are all amendments, corrections, insertions, and deletions. The Federal Register is available in many public libraries. Annual subscriptions are available from the Superintendent of Documents, U.S. Government Printing Office, Washington, DC 20402.

Each year, the Office of the Federal Register publishes all current regulations and standards in the Code of Federal Regulations (CFR), available at many libraries and from the Government Printing Office. OSHA's regulations are collected in Title 29 of the CFR, Part 1900-1999.

To assist the public in keeping current with OSHA standards, the OSHA Subscription Service was developed. This service provides all standards, interpretations, regulations, and procedures in easy-to-use loose-leaf form, punched for use in a three-ring binder. All changes and additions are issued for an indefinite period of time. The service is available only from the Superintendent of Documents, and is not available from OSHA or from the Department of Labor. Individual volumes of the OSHA Subscription Service as noted below are available as follows:

Volume I General Industry Standards and Interpretations (includes agriculture)
Volume II Maritime Standards and Interpretations
Volume III Construction Standards and Interpretations
Volume IV Other Regulations and Procedures

Volume V Field Operations Manual
Volume VI Industrial Hygiene Field Operations Manual

Copies of state standards may be obtained from the individual states' appropriate departments.

Standards Development

OSHA can begin standards-setting procedures on its own initiative or in response to petitions from other parties, including the Secretary of Health and Human Services (HHS); the National Institute for Occupational Safety and Health (NIOSH); state and local governments; any nationally recognized standards-producing organization, employer, or labor representatives; or any other interested person.

NIOSH Recommendations

Recommendations for standards may come from NIOSH, established by the act as an agency of the Department of HHS.

NIOSH conducts research on various safety and health problems, provides technical assistance to OSHA and recommends standards for OSHA's adoption. Especially important is NIOSH's investigation of toxic substances and its development of criteria for the use of such substances in the workplace.

While conducting its research, NIOSH may make workplace investigations, gather testimony from employers and employees, and require that employers measure and report employee exposure to potentially hazardous materials. NIOSH also may require employers to provide medical examinations and tests to determine the incidence of occupational illness among employees. When such examinations and tests are required by NIOSH for research purposes, they may be paid for by NIOSH rather than the employer.

Standards Adoption

Once OSHA has developed plans to propose, amend, or delete a standard, it publishes these intentions in the Federal Register as a "Notice of Proposed Rulemaking," or often as an earlier "Advance Notice of Proposed Rulemaking."

An advance notice is used to solicit information that can be used in drafting a proposal. The notice will include the terms of the new rule and provide a specific time (at least 30 days from the date of publication, usually 60 days or more) for the public to respond.

Interested parties who submit written arguments and pertinent evidence may request a public hearing on the proposal, if none has been announced in the notice. When such a hearing is requested, OSHA must schedule one, and must publish, in advance, the time and place for it in the Federal Register.

After the close of the comment period of public hearing, OSHA must publish in the Federal Register the full, final text of any standard amended or adopted and the date it becomes effective, along with an explanation of the standard and the reasons for implementing it. OSHA also may publish a determination that no standard or amendment needs to be issued.

Emergency Temporary Standards

Under certain limited conditions, OSHA is authorized to set emergency temporary standards that take effect immediately and are in effect until superseded by a permanent standards. OSHA must determine that workers are in grave danger from exposure to toxic substances, agents determined to be toxic or physically harmful, or new hazards, and that an emergency standard is needed to protect them. Then, OSHA publishes an emergency temporary standard in the Federal Register, where it also serves as a proposed permanent standard. It is then subject to the usual procedure for adopting a permanent standard except that a final ruling must be made within six months. The validity of an emergency temporary standard may be challenged in an appropriate U.S. Court of Appeals.

Appealing a Standard

No decision on a permanent standard is ever reached without due consideration of the arguments and data received from the public in written submissions and at hearings. However, any person who may be adversely affected by a final or emergency standard may file a petition (within 60 days of the rule's issuance) for judicial review of the standard with the U.S. Court of Appeals for the circuit in which the objector lives or has his or her principal place of business. Filing an appeals petition, however, will not delay the enforcement of a standard, unless the court of appeals specifically orders it.

Variances

Employers may ask OSHA for a variance from a standard or regulation if they cannot fully comply by the effective date, due to shortages of materials, equipment, or professional or technical personnel, or can prove that their facilities or methods of operation provide employee protection "at least as effective as" that required by OSHA.

Employers located in states with local occupational safety and health programs should apply to the state for a variance. However, if an employer operates facilities in states under OSHA jurisdiction, the employer may apply directly to OSHA if state plan allows. OSHA then will work with the states involved to determine if a variance can be granted that will satisfy state and federal OSHA requirements.

Temporary Variance

A temporary variance may be granted to an employer who cannot comply with a standard or regulatory by its effective date, due to unavailability of professional or technical personnel, materials, or equipment, or because the necessary construction or alteration of facilities cannot be completed in time.

The employer must demonstrate to OSHA that all available steps to safeguard employees are being taken in the meantime, and that the employer has mandated a program that complies with the standard or regulation as quickly as possible.

A temporary variance may be granted for the period needed to achieve compliance or for one year, whichever is shorter. It is renewable twice, each time for six months. An application for temporary variance must identify the standard or portion of a standard from which the variance is requested, and the reasons the employer cannot comply with the standard. The employer must document measures to comply with the standard that already have been taken and those that will be taken, including dates.

The employer must certify that workers have been informed of the variance application, that a copy has been given to the employees' authorized representative, and that a summary of the application has been posted wherever notices are normally posted. Employees also must be informed that they have the right to request a hearing on the application.

The temporary variance will not be granted to an employer who simply cannot afford to pay for the necessary alterations, equipment, or personnel.

Permanent Variance

A permanent variance is an alternative to a particular requirement of standard. It may be granted to employers who prove their conditions, practices, means, methods, operations, or processes provide a safe and healthful workplace as effectively as would compliance with the standard. In making a determination, OSHA weighs the employer's evidence and arranges a variance inspection and hearing, where appropriate.

When applying for a permanent variance, the employer must inform employees of the application and of their right to request a hearing. Anytime after six months from the issuance of a permanent variance, the employer or employees may petition OSHA to modify or revoke it. OSHA also may do this of its own accord.

Interim Order

So that an employer may continue to operate under existing conditions until a variance decision, he or she may apply to OSHA for an interim order.

Application for an interim order may bc made at the same time as or after application for variance. Include reasons for the interim order in the application.

If OSHA denies the request, the employer is notified of the reason for denial. If the interim order is granted, the employer and other concerned parties are informed of the order, and the terms of the order are published in the Federal Register. The employer must inform employees of the order by giving a copy to the authorized employees' representative and by posting a copy wherever notices are normally posted.

Experimental Variance

If an employer is participating in an experiment to demonstrate or validate new job safety and health techniques, and the experiment has been approved by either the Secretary of Labor or the Secretary of HHS, a variance may be granted to allow the experiment.

Specific Variances

In addition to temporary, permanent, and experimental variances, the secretary of labor also may find certain variances justified when the national defense is impaired. For further information and assistance in applying for a variance, contact the nearest OSHA office.

Variances are not retroactive. An employer who cited for a standards violation may not seek relief from that citation by applying for a variance. However, an outstanding citation does not prevent an employer from filing a variance application.

Public Petitions

OSHA continually reviews its standards to keep pace with developing and changing industrial technology. Therefore, employers and employees should be aware that, just as they may petition OSHA for the development of standards, they also may petition OSHA for modification or revocation of standards.

RECORD KEEPING AND REPORTING

Before the act became effective, no centralized or systematic method existed for monitoring occupational safety and health problems. Statistics on job injuries and illnesses were collected by some states and by some private organizations; national statistical figures were not based on any reliable projections. The act established a basis for consistent, nationwide procedures including incident reports, a vital requirement for gauging and solving problems.

According to OSHA procedures, those who employ 11 or more people must maintain records of occupational injuries and illnesses as they occur. Employers with 10 or fewer employees are exempt from keeping such records, unless they are selected by the Bureau of Labor Statistics (BLS) to participate in the Annual Survey of Occupational Injuries and Illnesses. The purposes of keeping records are to permit BLS survey material to be compiled, to help define high-hazard industries, and to inform employees of the status of their employer's record.

An occupational injury is any injury such as a cut, fracture, sprain, or amputation that results from a work-related accident or from exposure involving a single incident in the work environment. An occupational illness is any abnormal condition or disorder, other than one resulting from an occupational injury, caused by exposure to environmental factors associated with employment. Included are acute and chronic illnesses or diseases that may be caused by inhalation, absorption, ingestion, or direct contact with toxic substances or harmful agents.

All occupational illnesses must be recorded regardless of severity. All occupational injuries must be recorded if they result in:

- Death (must be recorded regardless of the length of time between the injury and death);
- One or more lost workdays;
- Restriction of work or motion;
- Loss of consciousness;
- Transfer to another job; or
- Medical treatment other than first aid.

If an on-the-job accident occurs that results in the death of an employee or in the hospitalization of five or more employees, all employers, regardless of number of employees, are required to report the accident, in detail, to the nearest OSHA office within 48 hours.

In states with approved plans, employers report such accidents to the state agency responsible for safety and health programs.

Injury and Illness Records

Employers must keep injury and illness records for each establishment. An employer whose employees work in dispersed locations must keep records at each place where employees report for work. In some situations, employees do not report for work at the same place each day. In that case, records must be kept at the place from which they are paid or at the base from which they operate.

Record-keeping Forms

Record-keeping forms are maintained on a calendar-year basis. They are not sent to OSHA or any other agency. They must be maintained for five years at the establishment, and must be available for inspection by representatives of OSHA, HHS, BLS, or the designated state agency. Only two forms, OSHA No. 200 and OSHA No. 101, are needed for record keeping.

1. **OSHA No. 200—Log and Summary of Occupational Injuries and Illnesses:** Each recordable occupational injury and illness must be logged on this form within six working days of the time the employer learns of it. If the log is prepared at a central location by automatic data processing equipment, a copy current within 45 calendar days must be present at all times in the establishment. A substitute for the OSHA No. 200 is acceptable if it is as detailed, easily readable, and understandable as the OSHA No. 200.
2. **OSHA No. 101—Supplementary Record of Occupational Injuries and Illnesses.** The form OSHA NO. 101 contains many more details about each injury or illness. It also must be completed within six working days of the time the employer learns of the work-related injury or illness. A substitute for the OSHA No. 101, such as insurance or workers' compensation forms, may be used if it contains all required information.

Employers selected to participate in the annual statistical survey receive in the mail, soon after the close of the year, form OSHA No. 200S. Each employer selected must complete this report, using form No. 200 as the source of information, and to return it to BLS. Small business employers (employers with 10 or fewer employees) selected for the survey are notified at the beginning of the year and are supplied with a form OSHA No. 200.

Many specific OSHA standards have additional record keeping and reporting requirements.

Posting Requirements

A copy of the totals and information following the fold line of the last page of OSHA No. 200 for the year must be posted at each establishment wherever notices to employees are customarily posted. This copy must be posted no later than February 1 and kept in place until March 1. Even though there were no injuries or illnesses during the year, a zero must be entered on the totals line and the form posted.

Record-keeping Variances

Employers and states wishing to set up a record-keeping system different from the one required by OSHA regulations may apply for a record-keeping variance. Petitions for record-keeping variances must detail and justify the employer's intended procedures and must be submitted to the regional commissioner of BLS for the area in which the workplace is located.

As with applications for variances from standards, an employer filing for a record-keeping variance must give a copy of the application to the employee's authorized representative. The employer also must post a summary of the application wherever notices are normally posted. Employees have 10 working days to submit to BLS their own written data, views, or arguments.

KEEPING EMPLOYEES INFORMED

Employers are responsible for keeping employees informed about OSHA and about the various safety and health matters with which they are involved. Federal OSHA and states with local programs require that each employer post certain materials at a prominent location in the workplace. These include:

- Job Safety and Health Protection workplace posters (OSHA 2203 or state equivalent) informing employees of their rights and responsibilities under the act. Besides displaying the workplace poster, the employer must make copies of the act and copies of relevant OSHA rules and regulations available to employees upon request.
- Summaries of petitions for variances from standards or record-keeping procedures.
- Copies of all OSHA citations for violations of standards. These must remain posted at or near the location of alleged violations for three days or until the violations are abated, whichever is longer.
- Log and Summary of Occupational Injuries and Illnesses (OSHA No. 200). The summary page of the log must be posted no later than February 1 and must remain in place until March 1 of each year.

All employees have the right to examine any records kept by their employers regarding exposure to hazardous materials or the results of medical surveillance.

Occasionally, OSHA standards or NIOSH research activities will require an employer to measure and record employee exposure to potentially harmful substances. Employees have the right to be present, in person or through an authorized representative, during the measuring, as well as to examine records of the results.

Under these substance-specific requirements, each employee or former employee has the right to see his or her examination records, and must be told by the employer if exposure has exceeded the levels set by standards. The employee also must be told what corrective measures are being taken.

In addition to having access to records, employees in manufacturing facilities must be provided information about all of the hazardous chemicals in their work area. Employers are to provide this information by means of labels on containers, material safety data sheets, and training programs.

WORKPLACE INSPECTIONS

To enforce its standards, OSHA is authorized under the act to conduct workplace inspections. Every establishment covered by the act is subject to inspection by OSHA safety and health compliance officers, who are chosen for their knowledge and experience in the occupational safety and health field, and are vigorously trained in OSHA standards and in recognition of safety and health hazards. Similarly, states with their own occupational safety and health programs conduct inspections using qualified safety and health compliance officers.

Under the act, an OSHA compliance officer is authorized to enter the workplace for inspection and investigation. With very few exceptions, inspections are conducted without advance notice. In fact, alerting an employer in advance of an OSHA inspection can bring a fine of up to $1,000 and/or a six-month jail term.

There are, however, special circumstances under which OSHA may give notice to the employer, but even then, such a notice will be less than 24 hours. These special circumstances include:

- Imminent-danger situations that require correction as soon as possible;
- Inspections that must take place after regular business hours or that require special preparation;
- Cases where notice is required to assure that the employer and employee representative or other personnel will be present; and/or
- Situations in which the OSHA area director determines that advance notice would produce a more thorough or effective inspection.

Employers receiving advance notice of an inspection must inform their employees' representative or arrange for OSHA to do so. If an employer refuses to admit an OSHA compliance officer, or if an employer attempts to interfere with the inspection, the act permits appropriate legal action.

Based on a 1978 Supreme Court ruling (Marshall vs. Barlow's, Inc.), OSHA may not conduct warrantless inspections without an employer's consent. It may, however, inspect after acquiring a judicially authorized search warrant based upon administrative probable cause or upon evidence of a violation.

Inspection Priorities

Obviously, not all five million workplaces covered by the act can be inspected immediately. The worst situations need attention first. Therefore, OSHA has established a system of inspection priorities.

Imminent Danger

Imminent-danger situations are given top priority. An imminent danger is any condition where there is reasonable certainty that a danger exists that can be expected to cause death or serious physical harm immediately, or before the danger can be eliminated through normal enforcement procedures.

Serious physical harm is any type of harm that could cause permanent or prolonged damage, or temporary disability as to require in-patient hospital treatment.

OSHA considers that "permanent or prolonged damage" has occurred when, for example, a part of the body is crushed or severed; an arm, legal, or finger is amputated; or sight in one or both eyes is lost. This kind of damage also includes that which renders a part of the body either fundamentally useless or substantially reduced in efficiency on or off the job, for example, bones in a limb shattered so severely that mobility or dexterity will be permanently reduced.

Temporary disability requiring in-patient hospital treatment includes injuries such as simple fractures, concussions, burns, or wounds involving substantial loss of blood and requiring extensive suturing or other healing aids.

Injuries or illnesses that are difficult to observe are classified as serious if they inhibit a person in performing normal functions, cause reduction in physical or mental efficiency, or shorten life.

Health hazards may constitute imminent danger situations when they present a serious and immediate threat to life or health. For a health hazard to be considered an imminent danger, there must be a reasonable expectation that

1. Toxic substances such as dangerous fumes, dusts, or gases are present, and
2. Exposure will cause immediate and irreversible harm to such a degree as to shorten life or cause reduction in physical or mental efficiency, even though the resulting harm is not immediately apparent.

Employees should inform the supervisor or employer immediately if they detect or even suspect an imminent danger situation in the workplace. If the employer takes no action to eliminate the danger, an employee or the authorized employee representative may notify the nearest OSHA office and request an inspection. The request should identify the workplace location, detail the hazard or condition, and include the employee's name, address, and telephone number. Although the employer has the right to see a copy of the complaint if an inspection results, the name of the employee will be withheld if the employee so requests.

The OSHA area director reviews the information and immediately determines whether there is a reasonable basis for the allegation. If it is decided the case has merit, the area director will alert the OSHA regional administrator and the regional solicitor, and assign a compliance officer to conduct an immediate inspection of the workplace.

Upon inspection, if an imminent danger situation is found, the compliance officer will ask the employer to voluntarily abate the hazard and to remove endangered employees from exposure. Should the employer fail to do this, OSHA, through the regional solicitor, may apply to the nearest federal district court for appropriate legal action to correct the situation. Before the OSHA inspector leaves the workplace, he or she will advise all affected employees of the hazard. Should OSHA "arbitrarily or capriciously" decline to bring court action, the affected employees may sue the secretary of labor to compel the secretary to do so. Such action can produce a temporary restraining order—immediate shutdown—of the operation or section of the workplace where the imminent danger exists.

Walking off the job because of potentially unsafe workplace conditions is not ordinarily an employee right. To do so may result in disciplinary action by the employer. However, an employee does have the right to refuse, in good faith, to be exposed to an imminent danger. OSHA rules protect employees from discrimination if:

- Where possible, he or she asked the employer to eliminate the danger, and the employer failed to do so;
- The danger is so imminent that there is not sufficient time to have the danger eliminated through normal enforcement procedures;
- The danger facing the employee is so grave that "a reasonable person" in the same situation would conclude there is a real danger of death or serious physical harm; and
- The employee has no reasonable alternative to refusing to work under these conditions.

Catastrophes and Fatal Accidents

Second priority is given to investigation of fatalities and catastrophes resulting in hospitalization of five or more employees. Such situations must be reported to OSHA by the employer within 48 hours. Investigations will determine if OSHA standards were violated to avoid recurrence of similar accidents.

Employee Complaints

Third priority is given to employee complaints of alleged violation of standards or of unsafe or unhealthful working conditions.

The act gives each employee the right to request an OSHA inspection when the employee feels he or she is in imminent danger from a hazard, or when he or she feels that there is a violation of an OSHA standard that threatens physical well-being.

OSHA will maintain confidentiality if requested, will inform the employee of any action it takes regarding the complaint, and, if requested, will hold an informal review of any decision not to inspect. Just as in situations of imminent danger, the employee's name will be withheld from the employer, if the employee so requests.

Programmed High-Hazard Inspections

Next in priority are programs of inspection aimed at specific high-hazard industries, occupations, or health substances. Industries are selected for inspection on the basis of such factors as the death, injury, and illness incidence rates, and employee exposure to toxic substances. Special emphasis may be regional or national in scope, depending on the distribution of the workplaces involved. Comprehensive safety inspections in manufacturing will be conducted only in those establishments with lost workday injury rates at or above the most recently published BLS national rate for manufacturing. States with their own occupational safety and health programs may use somewhat different systems to identify high-hazard industries for inspection.

Follow-Up Inspections

A follow-up inspection is conducted to determine if the previously cited violations have been corrected. If an employer fails to abate a violation, the compliance officer informs the employer that he/she is subject to "Notification of Failure to Abate" alleged violations, and may propose additional daily penalties while such violation continues.

Inspection Process

Prior to inspection, the compliance officer becomes familiar with as many facts as possible about the workplace, taking into account such things as the history of the establishment, the nature of the business, and the particular standards likely to apply. Preparing for the inspection also involves selecting appropriate equipment for detecting and measuring fumes, gases, toxic substances, noise, etc., among others. Inspector's credentials, opening conference, inspection tour, and closing conference items related to an inspection process are described below.

Inspector's Credentials

An inspection begins when the OSHA compliance officer arrives at the establishment. He or she displays official credentials and asks to meet an appropriate employer representative. Employers should always insist upon seeing the compliance officer's credentials.

An OSHA compliance officer carries U.S. Department of Labor credentials bearing his or her photograph and a serial number that can be verified by phoning the nearest OSHA office. **Anyone who tries to collect a penalty at the time of inspection, or promotes the sale of a product or service at any time, is not an OSHA compliance officer.**

Opening Conference

In the opening conference, the compliance safety and health officer (CSHO) explains why the establishment was selected, and determines if it would be subject to a comprehensive safety inspection based on its workday injury case rate. This rate is the number of lost workday injuries per 100 full-time employees.

The CSHO also will ascertain whether an OSHA-funded consultation program is in progress, or whether the facility is pursuing or has received an inspection exemption.

The compliance officer then explains the purpose of the visit, the scope of the inspection, and the standards that apply. The employer will be given copies of applicable safety and health standards as well as a copy of any employee complaint that may be involved. If the employee has so requested, his or her name will not be revealed.

An authorized employee representative also is given the opportunity to attend the opening conference and to accompany the compliance officer during the inspection. If the employees are represented by a recognized bargaining representative, the union ordinarily will designate the employee representative to accompany the compliance officer. Similarly, if there is a plant safety committee, the employee members of that committee will designate the employee representative (in the absence of a recognized bargaining representative). Where neither employee group exists, the employee representative may be selected by the employees themselves, or the compliance officer will determine if any employee suitably represents the interest of other employees. Under no circumstances may the employer select the employee representative for the inspection.

The act does not require that there be an employee representative for each inspection. However, where there is no authorized employee representative, the compliance officer must consult with a reasonable number of employees concerning safety and health matters in the workplace.

Inspection Tour

After the opening conference, the compliance officer and accompanying representatives proceed through the establishment, inspecting work areas for compliance with OSHA standards.

The route and duration of the inspection are determined by the compliance officer. While talking with employees, the compliance officer makes every effort to minimize any work interruptions. The compliance officer observes conditions, consults with employees, may take photos for record purposes, takes instrument readings, and examines records.

Trade secrets observed by the compliance officer must and will be kept confidential. An inspector who releases confidential information without authorization is subject to a $1,000 fine and/or one year in jail. The employer may require that the employee representative have confidential clearance for any area in question.

Employees are consulted during the inspection tour. The compliance officer may stop and question workers, in private if necessary, about safety and health conditions and practices in their workplaces. Employees are protected under the act from discrimination for exercising their safety and health rights.

Posting and record keeping are checked. The compliance officer will inspect records of deaths, injuries, and illnesses, which the employer is required to keep.

He or she will check to see that a copy of the totals from the last page of OSHA No. 200 has been posted, and that the OSHA workplace poster (OSHA 2203) is prominently displayed. Where records of employee exposure to toxic substances and harmful physical agents have been required, they also are examined.

During the course of the inspection the CSHO will point out to the employer any unsafe or unhealthful working conditions observed. At the same time, the CSHO will discuss possible corrective action, if the employer so desires.

Some apparent violations detected by the compliance officer can be corrected immediately. When they are corrected on the spot, the compliance officer records such corrections to help in judging the employer's good faith in compliance. Even though corrected, however, the apparent violations still may serve as the basis for a citation and/or notice of proposed penalty.

An inspection tour may cover part or all of an establishment, even if the inspection resulted from a specific complaint, fatality, or catastrophe.

Closing Conference

After the inspection tour, a closing conference is held between the compliance officer and the employer or the employer representative. It is a time for free discussion of problems and needs, a time for frank questions and answers.

The compliance officer discusses with the employer all unsafe or unhealthful conditions observed during the inspection, and indicates all apparent violations for which a citation may be issued or recommended. The employer is told of appeal rights. The compliance officer does not indicate any proposed penalties; only the OSHA area director has that authority, which is exercised only after receipt and evaluation of a full report.

During the closing conference, the employer may wish to produce records to show compliance efforts and to provide information that can help OSHA determine how much time may be needed to abate an alleged violation.

When appropriate, more than one closing conference may be held. This is usually necessary when health hazards are being evaluated or when laboratory reports are required. If requested, a closing discussion will be held with the employees or their representative to discuss matters of direct interest to employees. The employees' representative may be present at the closing conference.

The CSHO explains that OSHA area offices are full-service resource centers that provide a number of services such as availability of speakers, handout packages of materials to distribute to interested persons, availability of training and technical materials on safety and health matters, and many other services. The CSHO also explains the requirements of the Hazard Communications Standard, where employers must establish a written, comprehensive hazard communication program that includes provisions for container labeling, material safety data sheets, and an employee training program. It must contain a list of the hazardous chemicals in each work area and the means by which the employer will use to inform employees of the hazards of non-routine tasks.

CITATIONS AND PENALTIES

After the compliance officer reports findings, the area director determines what citations, if any, will be issued, and what penalties, if any, will be proposed.

Citations

Citations inform the employer and employees of the regulations and standards alleged to have been violated and of the proposed length of time set for their abatement. The employer will receive citations and notices of proposed penalties by certified mail. The employer must post a copy of each citation at or near the place a violation occurred for three days or until the violation is abated, whichever is longer.

Penalties

These are the types of violations that may be cited and the proposed penalties:

- **Other Than Serious Violation**—A violation that has a direct relationship to job safety and health but probably would not cause death or serious physical harm. A proposed penalty of up to $1,000 for each violation is discretionary. A penalty for any other than serious violation may be adjusted downward by as much as 80%, depending on the employer's good faith (i.e., demonstrated efforts to comply with the act), history of previous violations, and size of business. When the adjusted penalty amounts to less than $60, no penalty is proposed.
- **Serious Violation**—A violation where there is substantial probability that death or serious physical harm could result, and that the employer knew, or should have known, of the hazard. A mandatory penalty of up to $1,000 for each violation is proposed. A penalty for a serious violation may be adjusted downward, based on the employer's good faith, history of previous violations, the gravity of the alleged violation, and size of business.
- **Willful Violation**—A violation that the employer intentionally and knowingly commits. The employer either knows that what he or she is doing constitutes a violation, or is aware that a hazardous condition existed and made no reasonable effort to eliminate it. Penalties of up to $10,000 may be proposed for each willful violation. A proposed penalty for a willful violation may be adjusted downward, depending on the size of the business and its history of previous violations. Usually, no credit is given for good faith. If an employer is convicted of a willful violation of a standard that has resulted in the death of an employee, the offense is punishable by a court-imposed fine of not more than $10,000, or by imprisonment for up to six months, or both. A second conviction doubles these maximum penalties.

- **Repeated Violation**—A violation of any standard, regulation, rule, or order where, upon reinspection, a substantially similar violation is found. Repeated violations can bring a fine of up to $10,000 for each such violation. To be the basis of a repeat citation, the original citation must be final; a citation under contest may not serve as the basis for a subsequent repeat citation.
- **Failure to Correct Prior Violation**—Failure to correct a prior violation may bring a civil penalty of up to $1,000 for each day the violation continues beyond the prescribed abatement date.

Additional violations for which citations and proposed penalties may be issued are as follows:

- Falsifying records, reports, or applications upon conviction can bring a fine of $10,000 or up to six months in jail or both.
- Violations of posting requirements can bring a civil penalty of up to $1,000.

Assaulting a compliance officer or otherwise resisting, opposing, intimidating, or interfering with a compliance officer in the performance of his or her duties is a criminal offense, subject to a fine of not more than $5,000 and imprisonment for not more than three years.

Citation and penalty procedures may differ somewhat in states with their own occupational safety and health programs.

APPEALS PROCESS

Appeals by Employees

If an inspection was initiated by an employee complaint, the employee or authorized employee representative may request an informal review of any decision not to issue a citation.

Employees may not contest citations, amendments to citations, penalties, or lack of penalties. They may contest the time in the citation for abatement of a hazardous condition. They also may contest an employer's Petition for Modification of Abatement (PMA), which requests an extension of the abatement period. Employees must contest the PMA within 10 working days of its posting or within 10 working days after an authorized employee representative has received a copy.

Within 15 working days of the employer's receipt of the citation, the employee may submit a written objection to OSHA. The OSHA area director forwards the objection to the Occupational Safety and Health Review Commission, which operates independently of OSHA.

Employees may request an informal conference with OSHA to discuss any issues raised by an inspection, citation, notice of proposed penalty, or employer's notice of intention to contest.

Appeals by Employers

When issued a citation or notice of a proposed penalty, an employer may request an informal meeting with OSHA's area director to discuss the case. The area director is authorized to enter into settlement agreements that revise citations and penalties to avoid prolonged legal disputes.

Petition for Modification of Abatement

Upon receiving a citation, the employer must correct the cited hazard by the prescribed date unless he or she contests the citation or abatement date. However, factors beyond the employer's reasonable control may prevent the completion of corrections by that date. In such a situation, the employer who has made a good-faith effort to comply may file for a Petition for Modification of Abatement (PMA) date.

The written petition should specify all steps taken to achieve compliance, the additional time needed to achieve complete compliance, the reasons such additional time is needed, all temporary steps being taken to safeguard employees against the cited hazard during the intervening period, that a copy of the PMA was posted in a conspicuous place at or near each place where a violation occurred, and that the employee representative, if there is one, received a copy of the petition.

Notice of Contest

If the employer decides to contest either the citation, the time set for abatement, or the proposed penalty, he or she has 15 working days from the time the citation and proposed penalty are received in which to notify the OSHA area director in writing. An orally expressed disagreement will not suffice. This written notification is called a "Notice of Contest."

There is no specific format for the Notice of Contest. However, it must clearly identify the employer's basis for filing—the citation, notice of proposed penalty, abatement period, or notification of failure to correct violations.

A copy of the Notice of Contest must be given to the employees' authorized representative. If any affected employees are not represented by a recognized bargaining agent, a copy of the notice must be posted in a prominent location in the workplace, or else served personally upon each unrepresented employee.

Review Procedure

If the written Notice of Contest has been filed within the required 15 working days, the OSHA area director forwards the case to the Occupational Safety and Health Review Commission (OSHRC). The commission is an independent agency not associated with OSHA or the Department of Labor. The commission assigns the case to an administrative law judge.

The judge may investigate and disallow the contest if it is found to be legally invalid, or a hearing may be scheduled for a public place near the employer's workplace. The employer and the employees have the right to participate in the hearing; the OSHRC does not require that they be represented by attorneys.

Once the administrative law judge has ruled, any party to the case may request a further review by OSHRC. Any of the three OSHRC commissioners, at his or her own motion, also may bring a case before the commission for review. Commission rulings may be appealed to the appropriate U.S. Court of Appeals.

Appeals in State Plan

States with their own occupational safety and health programs have systems for review and appeal of citations, penalties, and abatement periods. The procedures are generally similar to federal OSHA's, but cases are heard by a state review board or equivalent authority.

OSHA-APPROVED STATE PROGRAMS

The act encourages states to develop and operate, under OSHA guidance, state job safety and health plans. Once a state plan is approved, OSHA funds up to 50% of the program's operating costs. State plans are required to provide standards and enforcement programs, as well as voluntary compliance activities, at least as effective as the federal program.

To gain OSHA approval as a developmental plan, a state must demonstrate that, within three years, it will provide adequate legislation; standards-setting; enforcement and appeals procedures; public employee protection; a sufficient number of competent enforcement personnel; and training, education, and technical assistance programs. If, at any time during this period or later, it appears that the state is capable of enforcing standards in accordance with the above requirements, OSHA may enter into an "operational status agreement" with the state. OSHA generally limits its enforcement activity to areas not covered by the state in the agreement and ceases all concurrent federal enforcement in the areas covered by the state. Scheduled, accident, and complaint inspections are generally the primary responsibility of the state. OSHA closely monitors state programs.

When all development steps concerning legislation, resources, and other requirements have been completed and approved, OSHA then certifies that a state has the legal, administrative, and enforcement means necessary to operate effectively. This action renders no judgment on how well or poorly a state actually is operating its program, but merely attests to the structural completeness of its program. After this certification, there is a period of at least one year to determine if a state effectively provides safety and health protection. If it is found that the state is operating at an effective level and other requirements including compliance staffing levels are met, final approval of the plan may be granted, and federal authority will cease in those areas over which the state has jurisdiction.

OSHA continues monitoring and evaluating the state program to assure the state maintains its level of effectiveness. If this level should decline, OSHA can begin proceedings to withdraw approval of the program and to reinstituted federal enforcement authority.

Employers and employees should find out if their state operates an OSHA-approved state program and, if so, become familiar with it. State safety and health standards under approval plans must keep pace with federal standards including periodic revisions, and state plans must guarantee employer and employee rights to do so, as does OSHA.

State plans developed for the private sector also must, to the extent permitted by state law, provide coverage for state and local government employees. OSHA rules also permit states to develop plans that cover only state and local government employees; in such cases, private sector employment remains under federal jurisdiction.

Anyone finding inadequacies or other problems in the administration of a state's program may file a complaint about state program administration with the appropriate regional administrator for OSHA. The complainant's name is kept confidential. OSHA investigates all such complaints and, where complaints are found to be valid, requires appropriate corrective action on the part of the state.

EMPLOYER RESPONSIBILITIES AND RIGHTS

Employers have certain responsibilities and rights under the Occupational Safety and Health Act of 1970. Following are checklists of many of these rights and responsibilities. Employer responsibilities and rights in states with their own occupational safety and health programs are generally the same as in states where federal OSHA is in effect.

Responsibilities

As an employer, you must:

- Provide a workplace free of recognized hazards that are causing or are likely to cause death or serious physical harm to employees, and comply with standards, rules, and regulations issued under the act.
- Be familiar with mandatory OSHA standards and make copies available to employees for review upon request.
- Inform all employees about OSHA.
- Examine workplace conditions to make sure they conform to applicable standards.

- Minimize or reduce hazards.
- Make sure employees have and use safe tools and equipment, including appropriate personal protective equipment, and that such equipment is maintained.
- Use color codes, posters, labels, or signs when needed to warn employees of potential hazards.
- Establish or update operating procedures and communicate them so that employees follow safety and health requirements.
- Provide medical examinations when required by OSHA standards.
- Report to the nearest OSHA office within 48 hours any fatal accident or one that results in the hospitalization of five or more employees.
- Keep OSHA-required records of work-related injuries and illnesses, and post a copy of the totals from the last page of OSHA No. 200 during the entire month of February each year. This applies to employers with 11 or more employees.
- At a prominent location within the workplace, post the OSHA poster (OSHA 2203) informing employees of their rights and responsibilities. In states operating OSHA-approved job safety and health programs, the state's equivalent poster and/or OSHA 2203 may be required.
- Provide employees, former employees, and their representatives access to the Log and Summary of Occupational Injuries and Illnesses (OSHA No. 200).
- Cooperate with the OSHA compliance office by furnishing names of authorized employee representatives who may be asked to accompany the compliance officer during an inspection. If there are none, the compliance officer will consult with a reasonable number of employees concerning safety and health in the workplace.
- Not discriminate against employees who properly exercise their rights under the act.
- Post OSHA citations at or near the work site involved. Each citation, or copy thereof, must remain posted until the violation has been abated, or for three working days, whichever is longer.
- Abate cited violations within the prescribed period.

Rights

As an employer, you have the right to:

- Seek advise and off-site consultation as needed by writing, calling, or visiting the nearest OSHA office. OSHA will not inspect merely because an employer requests assistance.
- Be active in your industry association's involvement in job safety and health.
- Request and receive proper identification of the OSHA compliance officer prior to inspection.
- Have an opening and closing conference with the compliance officer.

- File a Notice of Contest with the OSHA area director within 15 working days of receipt of a notice of citation and proposed penalty.
- Apply to OSHA for a temporary variance from a standard if unable to comply because of the unavailability of materials, equipment, or personnel needed to make necessary changes within the required time.
- Apply to OSHA for a permanent variance from a standard if you can furnish proof that your facilities or method of operation provide employee protection at least as effective as that required by the standard.
- Take an active role in developing safety and health standards through participation in OSHA Standards Advisory Committees, through nationally recognized standards-setting organizations, and through evidence and views presented in writing or at hearings.
- Be assured of the confidentiality of any trade secrets observed by an OSHA compliance officer during an inspection.
- Submit a written request to NIOSH for information on whether any substance in your workplace has potentially toxic effects in the concentration being used.

EMPLOYEE RESPONSIBILITIES AND RIGHTS

Although OSHA does not cite employees for violations of their responsibilities, each employee must comply with all occupational safety and health standards, and all rules, regulations, and applicable orders issued under the act. Employee responsibilities and rights in states with local programs are generally the same as for workers in federal OSHA states.

Responsibilities

As an employee, you should:

- Read the OSHA poster at the job site.
- Comply with all applicable OSHA standards.
- Follow all employer safety and health rules and regulations, and wear or use prescribed protective equipment while engaged in work.
- Report hazardous conditions to the supervisor.
- Report any job-related injury or illness to the employer, and seek treatment promptly.
- Cooperate with the OSHA compliance officer conducting an inspection if he or she inquires about safety and health conditions in your workplace.
- Exercise your rights under the act in a responsible manner.

Rights

Employees have a right to seek safety and health on the job without fear of punishment. That right is spelled out in Section 11 (c) of the act.

In accordance with the law, employers are not permitted to punish or discriminate against workers for exercising rights such as:

- Complaining to an employer, union, OSHA, or any other government agency about job safety and health hazards.
- Filing safety or health grievances.
- Participating on a workplace safety and health committee or in union activities concerning job safety and health.
- Participating in OSHA inspections, conferences, hearings, or other OSHA-related activities.

If an employee is exercising these or other OSHA rights, the employer is not allowed to discriminate against that worker in any way, such as through firing, demotion, taking away seniority or other earned benefits, transferring the worker to an undesirable job or shift, or threatening or harassing the worker.

If the employer has knowingly allowed the employee to do something in the past, such as leaving work early, he or she may be violating the law by punishing the worker for doing the same thing following a protest of hazardous conditions. If the employer knows that a number of workers are doing the same thing wrong, he or she cannot legally single out for punishment the worker who has taken part in safety and health activities.

Workers believing they have been punished for exercising safety and health rights must contact the nearest OSHA office within 30 days of the time they learn of the alleged discrimination. A union representative can file the 11(c) complaint for the worker.

The worker does not have to complete any forms. An OSHA staff member will complete the forms, asking what happened and who was involved.

Following a complaint, OSHA conducts an investigation. If an employee has been illegally punished for exercising safety and health rights, OSHA asks the employer to restore that worker's job earning and benefits. If necessary, and if it can prove discrimination, OSHA takes the employer to court. In such cases the worker does not pay any legal fees.

If a state agency has an OSHA-approved state program, employees may file their complaint with either federal OSHA or the state agency under its law.

Additional Rights

As employee, you have the right to:

- Review copies of appropriate OSHA standards, rules, regulations, and requirements that the employer should have available at the workplace.
- Request information from your employer on safety and health hazards in the area, on precautions that may be taken, and on procedures to be followed if an employee is involved in an accident or is exposed to toxic substances.

- Request the OSHA area director to conduct an inspection if you believe hazardous conditions or violations of standards exist in your workplace.
- Have your name withheld from your employer, upon request, to OSHA, if you file a written and signed complaint.
- Be advised of OSHA actions regarding your complaint and have an informal review, if requested, of any decision not to inspect or to issue a citation.
- Have your authorized employee representative accompany the OSHA compliance officer during the inspection tour.
- Respond to questions from the OSHA compliance officer, particularly if there is no authorized employee representative accompanying the compliance officer.
- Observe any monitoring or measuring of hazardous materials and have the right to see these records, as specified under the act.
- Review the Log and Summary of the Occupational Injuries (OSHA No. 200) at a reasonable time and a reasonable manner. An authorized representative also may review the records.
- Request a closing discussion with the compliance officer following an inspection.
- Submit a written request to NIOSH for information on whether any substance in your workplace has potentially toxic effects in the concentration being used, and have your name withheld from your employer if you so request.
- Object to the abatement period set in the citation issued to your employer by writing to the OSHA area director within 15 working days of the issuance of the citation.
- Be notified by your employer if he or she applies for a variance from an OSHA standard, testify at a variance hearing, and appeal the final decision.
- Submit information or comment to OSHA on the issuance, modification, or revocation of OSHA standards, and request a public hearing.

CHAPTER 8

WORKPLACE EMERGENCIES

Nicholas P. Cheremisinoff

SciTech Technical Services, Inc.
Morganville, NJ 07751, USA

Vasantrao Nivargikar

Mega Management & Testing Services, Inc.
Holmdel, NJ 07733, USA

CONTENTS

INTRODUCTION

The importance of an effective workplace safety and health program cannot be overemphasized. There are many benefits from such a program, including increased productivity, improved employee morale, reduced absenteeism and illness, and reduced workers' compensation rates. However, accidents still occur despite efforts to prevent them. Therefore, proper planning for emergencies is necessary to minimize employee injury and property damage.

This chapter details the basic steps needed to handle emergencies in the workplace. These emergencies include accidental releases of toxic gases, chemical spills, fires, explosions, and personal injury. This chapter is intended to assist small businesses without safety and health professionals. It is not intended to serve as an all-inclusive safety program but rather to provide guidelines for planning for emergencies. For companies that already have programs in effect, this chapter can assist in updates and revisions.

PLANNING

The effectiveness of response during emergencies depends on the amount of planning and training. Management must show its support of plant safety programs and the importance of emergency planning. If management is not interested in employee protection and minimizing property loss, little can be done to promote a safe workplace. It is therefore management's responsibility to see that a program is instituted and that it is frequently reviewed and updated. Employee input and support will ensure an effective program. The emergency response plan should be developed locally and should be comprehensive enough to deal with all types of emergencies. When emergency action plans are required by a particular OSHA standard, the plan must be in writing; firms with ten or fewer employees may communicate their plans orally. The plan must include, as a minimum, the following elements:

- Emergency escape procedures and emergency escape route assignments;
- Procedures for employees who remain to perform or shut down critical plant operations before they evacuate;
- Procedures to account for all employees after emergency evacuation has been completed;
- Rescue and medical duties for those employees who are to perform them;
- The preferred means for reporting fires and other emergencies; and
- Names or regular job titles of persons or departments to be contacted for further information or explanation of duties under the plan.

The emergency action plan should address all potential emergencies in the workplace. Therefore, it will be necessary to perform a hazard audit to determine potentially toxic materials and unsafe conditions. For information on chemicals, the manufacturer or supplier can be contacted to obtain Material Safety Data Sheets (MSDSs). These forms discuss the hazards that a chemical may present, list precautions to take when handling, storing or using the substance, and outline emergency and first aid procedures.

The employer should detail the procedures to be taken by those employees who must remain behind to care for essential plant operations until their evacuation becomes absolutely necessary. This may include monitoring power supplies, water supplies, and other essential services that cannot be shut down for every emergency alarm.

Floor plans or workplace maps clearly showing emergency escape routes and safe or refuge areas should be included in the plan. All employees must be told what actions they are to take in emergency situations.

CHAIN OF COMMAND

A chain of command should be established to minimize confusion so employees will have no doubt who has authority for making decisions. Responsible individuals should be selected to coordinate the emergency response team. In larger organizations, there may be a plant coordinator in charge of plantwide operations, public relations, and ensuring that outside aid is called in. Because of the importance of these functions, adequate backup must be arranged so that trained personnel are always available. The duties of the emergency response team coordinator should include:

- Assessing and determining whether the situation requires activating the emergency procedures.
- Directing all efforts in the area, including evacuating personnel and minimizing property loss.
- Ensuring that outside emergency services such as medical aid and local fire departments are called in when necessary.
- Directing the shutdown of plant operations when necessary.

COMMUNICATIONS

During a major emergency involving a fire or explosion, it may be necessary to evacuate offices and manufacturing areas. Normal services—such as electricity, water, and telephones—may be nonexistent. Designate an area to which employees are to report. This area should be equipped for incoming and outgoing calls. The designated area will become an alternate headquarters, and the person in charge should be easily reached there.

Emergency communications equipment (amateur radio systems, public address systems, or portable radio units) should be present for notifying employees of the emergency and for contacting local authorities, such as law enforcement officials, the Red Cross, and the fire department.

A method of communication also is needed to alert employees to the evacuation or to take other action as required in the plan. Alarms should be audible or seen by all people in the plant and should have an auxiliary power supply in the event electricity is effected. The alarm should be distinctive and recognizable as a signal to evacuate the work area or to perform actions designated under the emergency action plan. The employer should explain to each employee the means for reporting emergencies, such as manual pull box alarms, public address systems, or telephones. Emergency phone numbers should be posted on or near telephones, on employees' notice boards, or in other conspicuous locations. The warning plan should be in writing, and management must be sure each employee knows what it means and what action is to be taken.

It may be necessary to notify other key personnel such as the plant manager or physician during off-duty hours. An updated written list should be kept of key personnel listed in order of priority.

Accounting for Personnel

Management will need to know when all personnel have been accounted for. This can be difficult during shift changes or if contractors are on site. A responsible person in the control center should be appointed to account for personnel and to inform police or emergency response team members of those persons believed missing.

EMERGENCY RESPONSE TEAMS

Emergency response teams are the first line of defense in emergencies. Before assigning personnel to these teams, the employer must assure that employees are physically capable of performing the duties that may be assigned to them. Depending on the size of the plant, there may be one or several teams trained in the following areas:

- Use of various types of fire extinguishers,
- First aid, including cardiopulmonary resuscitation (CPR),
- Shutdown procedures,
- Evacuation procedures,
- Chemical spill control procedures,
- Use of self-contained breathing apparatus (SCBA), and
- Search and emergency rescue procedures.

The type and extent of the emergency will depend on the plant operations, and the response will vary according to the type of process, the material handled, the number of employees, and the availability of outside resources. Emergency response teams should be trained in the types of possible emergencies and the emergency actions to be performed. They should be informed about special hazards, such as storage and use of flammable materials, toxic chemicals, radioactive sources, and water-reactive substances to which they may be exposed during fire and other emergencies. It is important to determine when not to intervene. For example, team members must be able to determine if the fire is too large for them to handle or whether search and emergency rescue procedures should be performed. If there is a possibility of members of the emergency response team receiving fatal or incapacitating injuries, they should wait for professional firefighters or other emergency response groups.

Training

Training is important to the effectiveness of an emergency plan. Before implementing an emergency action plan, a sufficient number of persons must be trained to assist in the safe and orderly evacuation of employees. Training for each type of disaster response is necessary so that employees know what actions are required.

In addition to the specialized training for emergency response team members, all employees should be trained in:

- Evacuation plans,
- Alarm systems,
- Reporting procedures,
- Shutdown procedures, and
- Types of potential emergencies.

These training programs should be provided as follows for effective implementation of the emergency plans:

- Initially when the plan is developed,
- For all new employees,
- When new equipment, materials, or processes are introduced,
- When procedures have been updated or revised,
- When exercises show that employee performance must be improved, and
- At least annually.

The emergency control procedures should be written in concise terms and be made available to all personnel. Drills should be held for all personnel, at random intervals at least annually, and an evaluation of performance made immediately by management and employees. When possible, drills should include groups supplying outside services such as fire and police departments. In buildings with several places of employment, the emergency plans should be coordinated with other companies and employees in the building. Finally, the emergency plan should be reviewed periodically and updated to maintain adequate response personnel and program efficiency.

PERSONAL PROTECTION

Effective protection is essential for any person who may be exposed to potentially hazardous substances. In emergency situations, employees may be exposed to a wide variety of hazardous circumstances, including:

- Chemical splashes or contact with toxic materials,
- Falling objects and flying particles,
- Unknown atmospheres that may contain inadequate oxygen to sustain life,
- Toxic gases, vapor, or mists, and
- Fire and electrical hazards.

Safety Equipment

It is extremely important that employees be adequately protected in these situations. Some of the safety equipment that may be used include:

- Safety glasses, goggles, or face shields for eye protection,
- Hard hats and safety shoes for head and foot protection,
- Proper respirators for breathing protection,
- Whole body coverings, gloves, hoods, and boots for body protection from chemicals, and
- Body protection for abnormal environmental conditions such as extreme temperatures.

The equipment selected should be approved jointly by the Mine and Safety and Health Administration (MSHA) and the National Institute for Occupational Safety and Health (NIOSH), or meet the standards set by the American National Standards Institute (ANSI). The choice of proper equipment is not a simple matter. Consult with health and safety professionals before making any purchases. Manufacturers and distributors of health and safety products may be able to answer questions if they have enough information about the potential hazards.

Respiratory Protection

Professional consultation may be needed to determine adequate respiratory protection. Respiratory protection is necessary for toxic atmospheres of dusts, mists, gases, or vapors, and for oxygen-deficient atmospheres.

There are four basic categories of respirators:

1. **Air-purifying devices (filters, gas masks, and chemical cartridges):** These devices remove contaminants from the air but cannot be used in oxygen-deficient atmospheres.
2. **Air-supplied respirators (hose masks, air line respirators):** These devices should not be used in atmospheres that are immediately dangerous to life or health.
3. **Self-contained breathing apparatus:** These devices are required for unknown atmospheres, oxygen-deficient atmospheres or atmospheres dangerous to life or health (positive-pressure type only).

4. **Escape masks:** These devices are required for safety during escape to safe locations.

Before assigning or using respiratory equipment, the following conditions must be met:

- A medical evaluation must be prepared covering safe use and proper care of the equipment, and employees must be trained in these procedures and the use and maintenance of respirators.
- A respirator facepiece fit test must be performed and repeated periodically. Training must provide the employee an opportunity to handle the respirator, have it fitted properly, test its facepiece-to-face seal, wear it in normal air for a familiarity period, and wear it in a test atmosphere.
- A regular maintenance program must be instituted including cleaning, inspection, and testing of all respiratory equipment. Respirators used for emergency response must be inspected after each use and at least monthly to assure that they are in satisfactory working condition. A written record of inspection must be maintained.
- Distribution areas for equipment used in emergencies must be readily accessible to employees.

Self-contained breathing apparatus (SCBA) offers the best protection to employees involved in controlling emergency situations. It should have a minimum service-life rating of 30 minutes. Conditions that require SCBA include:

- Leaking cylinders or containers, smoke from chemical fires, or chemical spills that indicate high potential for exposure to toxic substances.
- Atmospheres with unknown contaminants or unknown contaminant concentrations, confined spaces that may contain toxic substances, or oxygen-deficient atmospheres.

CONFINED SPACE

Emergency situations may involve entering confined spaces to rescue employees overcome by toxic compounds or lack of oxygen. These confined spaces include tanks, vaults, pits, sewers, pipelines, and vessels. Entry into confined spaces can expose the employee to a variety of hazards, including toxic gases, explosive atmospheres, oxygen deficiency, electrical hazards, and hazards created by mixers and impellers that have not been deactivated and locked out. Personnel should never enter a confined space under normal circumstances unless the atmosphere has been tested for adequate oxygen, combustibility, and toxic substances. Conditions in a confined space must be considered immediately dangerous to life and health unless shown otherwise. If a confined space must be entered in an emergency, the following precautions must be followed:

- All lines containing inert, toxic, flammable, or corrosive-materials must be disconnected or valved off before entry.
- All impellers, agitators, or other moving equipment inside the vessel must be locked out.
- Appropriate personal protective equipment must be worn by employees before entering the vessel. Mandatory use of safety belts and harnesses should be stressed.
- Rescue procedures must specifically designed for each entry.

When there is an atmosphere immediately dangerous to life or health, or a situation that could cause injury or illness to an unprotected worker, a trained standby should be present. This person should be assigned a fully charged, positive-pressure, self-contained breathing apparatus with a full facepiece. The standby must maintain unobstructed lifelines and communications with all workers in the confined space, and must be prepared to summon rescue personnel if necessary. The standby should not enter the confined space until adequate assistance is present. While awaiting rescue personnel, the standby may make a rescue attempt using lifelines from outside the confined space.

A more complete description of procedures to follow while working in confined spaces may be found in National Institute for Occupational Safety and Health (NIOSH) Publication Number 80-106, "Criteria for a Recommended Standard . . . Working in Confined Spaces."

MEDICAL ASSISTANCE

In a major emergency, time is a critical factor in minimizing injuries. Most small businesses do not have a formal medical program. However, they are required to have the following medical and first aid services:

- Where the eyes or body of an employee may be exposed to injurious corrosive materials, eyewashes or suitable equipment for quick drenching or flushing must be provided in the work area for immediate emergency use. Employees must be trained to use the equipment.
- The employer must ensure the availability of medical personnel for advice and consultation on employee health matters. This does not mean that health care must be provided, but that medical help will be available if health problems develop in the workplace.

To fulfill the above requirements, the following actions should be considered:

- Survey the medical facilities near the place of business, and make arrangements to handle routine and emergency cases. A written emergency medical procedure should be prepared for handling accidents.
- If the business is located far from medical facilities, at least one and preferably more employees on each shift must be adequately trained to render

first aid. The American Red Cross, some insurance carriers, local safety councils, fire departments, and others may be contacted for this training.
- First aid supplies should be provided for emergency use. This equipment should be ordered through consultation with a physician.
- Sufficient ambulance service should be available to handle any emergency. This requires advance contact with ambulance services to ensure that they become familiar with plant location, access routes, and hospital locations.

SECURITY

During an emergency, it is often necessary to secure the area to prevent unauthorized access and to protect vital records and equipment. An off-limits area must be established by cordoning off the area with ropes and signs. It may be necessary to notify local law enforcement personnel or to employ private security personnel to secure the area and prevent the entry of unauthorized persons.

Certain records also may need to be protected, such as essential accounting files, legal documents, and lists of employees' relatives to be notified in case of emergency. These records may be stored in duplicate outside the plant or in protected secure locations within the plant.

OSHA REQUIREMENTS FOR EMERGENCY RESPONSE

Following is a list of the OSHA requirements pertaining to emergency response. These references refer to appropriate sections of the Occupational Safety and Health Standards (Code of Federal Regulations, Title 29, Part 1910, which are the OSHA General Industry Standards). For additional, applicable, and current information, call the nearest OSHA area office.

Subpart E - Means of egress
- 1910.37 Means of egress
- 1910.38 Employee emergency plans and fire prevention plans
- Appendix to Subpart E - Means of egress

Subpart I - Personal protective equipment
- 1910.132 General requirements - personnel protection
- 1910.133 Eye and face protection
- 1910.134 Respiratory protection
- 1910.135 Occupational head protection
- 1910.136 Occupational foot protection

Subpart K - Medical and first aid
- 1910.151 Medical services and first aid

Subpart L - Fire protection
- 1910.155-156 Fire protection and fire brigades
- 1910.157-163 Fire suppression equipment

1910.164 Fire detection
1910.165 Employee alarm systems
Appendix A–E of Subpart L

Information and Services

Much of the planning and program development for responding to occupational emergencies will require professional assistance. Many public and private agencies provide information and services free or at minimal cost (e.g., federal, state, and local health and labor departments; insurance carriers; nearby universities; etc.). After having exhausted these sources, consider using private specialists.

Workers' compensation insurance carriers probably have safety and health specialists on staff who are familiar with minimum standards and technical information currently available. They may be quite helpful in advising about accident and illness prevention and control.

Trade associations often have technical materials, programs, and industry data available for specific needs.

The Department of Labor, through the Occupational Safety and Health Administration (OSHA), provides information in interpreting the law and on meeting the applicable standards. This information is available free of charge or obligation. The OSHA area office or state plan office nearest to the plant may be contacted for this information.

The Department of Health and Human Services through the National Institute for Occupational Safety and Health (NIOSH) provides printed material relating to employee safety and health in the workplace. Staff from this agency will perform industrial hygiene surveys of plants upon request of employers or employees.

Machine or product manufacturers can be helpful in providing additional information on precautions to take in using their products. Any special problems should be referred to them first.

Professional societies in the safety, industrial hygiene, and medical fields issue publications in the form of journals, pamphlets, and books that may be quite useful (e.g., American Society of Safety Engineers or the Occupational Health Institute). They also can recommend individuals from their societies to serve as consultants.

Local colleges and universities sometimes have industrial hygiene, public health, medical, or other relevant departments with faculty and libraries to assist.

Additional Sources of Information

1. Material safety data sheets (MSDS), guides and manuals from industries. Data sheets on specific substances giving hygienic standards, properties, industrial hygiene practices, specific procedures, and references.
2. AIHA Hygienic Guide Series. American Industrial Hygiene Association, 66 South Miller Road, Akron, OH 44313.

3. ANSI Standards Z37 Series, Acceptable Concentrations of Toxic Dusts and Gases. American National Standards Institute, 1430 Broadway, New York, NY 10018. These guides represent a consensus of interested parties concerning minimum safety requirements for the storage, transportation, and handling of toxic substances. They are intended to aid the manufacturer, the consumer, and the general public.
4. ASTM Standards with Related Material. American Society for Testing Materials, 1916 Race Street, Philadelphia, PA 19103. World's largest source of voluntary consensus standards for materials, products, systems, and services.
5. American National Standards Institute, 1430 Broadway, New York, NY 10018. Coordinates and administers the federated voluntary standardization system in the United States.
6. Factory Insurance Association, 85 Woodland Street, Hartford, CT 06105. Composed of capital stock insurance companies to provide engineering, inspections, and loss adjustment service to industry.
7. Factory Mutual System, 1151 Boston-Providence Turnpike, Norwood, MA 02602. An industrial fire protection, engineering, and inspection bureau established and maintained by mutual fire insurance companies.
8. National Fire Protection Association, 470 Batterymarch Park, Quincy, MA 02269. A clearinghouse for information on fire protection and fire prevention. Nonprofit technical and education organization.
9. Underwriter Laboratories, Inc., 207 East Ohio Street, Chicago, IL 60611. Nonprofit organization whose laboratories publish annual lists of manufacturers whose products proved acceptable under appropriate standards.
10. Medical consultation. Arrange for a local doctor to advise on workplace medical matters. Contact the local Red Cross chapter for assistance in first aid training. If a local chapter cannot be located, write: American National Red Cross, National Headquarters, Safety Programs, 18th and E Streets, N.W., Washington, DC 20006.

STATE OCCUPATIONAL SAFETY AND HEALTH PLANS

The Occupational Safety and Health Act of 1970, under Section 18(b), encourages states to develop and operate job safety and health plans under the approval and monitoring of OSHA. States must set at least as effective standards as OSHA, conduct inspections to enforce those standards (including inspections in response to workplace complaints), cover state and local government employees, and operate occupational safety and health training and education programs. In addition, most states provide on-site consultation to help employers to identify and correct workplace hazards. Such consultation may be provided either under the plan or through a special agreement under section 7(c)(1) of OSHA. Federal OSHA does not conduct enforcement activities in these states, except in very limited circumstances.

SPILL CONTROL

The United States is heavily dependent on chemicals. Everyday, numerous items are used that contain chemicals or that are manufactured from chemicals. Even food is a product of this chemical society. Many of these chemicals present hazards and dangers of which the public is often unaware. An everyday example is "chemicals for lunch":

> Enriched flour, barley malt, ferrous sulfate, niacin, thiamine mononitrate, riboflavin, water, corn syrup, yeast, vegetable oil, salt, wheat, gluten, soy flour, calcium sulfate, calcium stearoyl lactylate, mono- and digliycerides, mono- and dicalcium phosphate, potassium bromate, calcium propionate, beef protein, fiber, hydrogenated soybean oil, blue cheese, white distilled vinegar, sugar, xanthin gum, polysorbate 60, sorbic acid, dried garlic, calcium disodium ethylenediaminetetra-acetic acid, carbonated water, carmel coloring, phosphoric acid, sodium citrate, sodium saccharin, caffeine, sodium benzoate, phenylalanine, citric acid.

This lunch actually was a hamburger, a salad with blue cheese dressing and a diet Pepsi.

Chemicals that seem completely harmless can kill in minutes. The recent disaster in Bhopal, India, associated with Union Carbide is a prime example. The worst aspect concerning exposure and/or production of chemicals is the sudden releases and spills, which are not uncommon.

Spill Categories

Spills can be divided into two distinct categories:

- Oil spills, and
- Hazardous chemical spills.

Extensive research and technology has been developed for responses to spills. Hazardous chemical spills have been receiving emphasis since the passage of the Resource Recovery and Conservation Act of 1976 (RCRA) and the Comprehensive Environmental Response, Compensation and Liability Act (Superfund). Many of the techniques used to treat oil spills may also be used for hazardous chemical spills.

Spill Response

The total quantity of oils and waste oils generated in the United States is estimated to be more than two billion gallons per year. The primary industries producing these oily and metal-containing wastes are petroleum refining, marine transportation, automobile, shipyard, aircraft manufacturing, machine shops, food processing, military installations, textiles, tanning, and fuel handling. Spill response can be divided into four phases: spill prevention, containment, cleanup and recovery, and treatment and disposal.

Spill Prevention

Spill prevention applies to all types of spills, and can be described as the first and simplest approach to spill control. A sound prevention program should include several elements: design, inspection, maintenance, training, and planning. A facility-wide inspection should be instituted to locate possible problem areas. The inspection should look at processes, storage areas/facility drainage, and loading and unloading facilities. Deficiencies should be noted and corrected. Periodic or daily inspections should be conducted to look for possible problems such as leaking or decaying drums, storage tanks, valves, fittings, and pipes. Any problems found should immediately be brought to the attention of supervision and maintenance personnel.

Human error is a major contributing factor to spills and releases. An adequate awareness of spill consequences, preventive measures, and countermeasures will greatly reduce spill occurrences.

Finally, effective planning can reduce the effects of spills by improving response time. Make provisions for spill containment and control equipment, establish chains of command for reporting spills, and train personnel in response techniques.

The federal government requires industries to develop Spill Prevention, Control, and Countermeasures (SPPC) plans and Hazardous Wastes Preparedness, Prevention, and Contingency (PPC) plans. Knowing who to call and what to do can help stop a minor spill from becoming a major catastrophe.

Regardless of how strong a preventive program a facility has, spills and releases still may occur. Oil pollution is a very complex problem that depends mainly on the physical and chemical properties of the oil. Once oil has been spilled, different and sometimes competitive processes will become active. Hydrocarbons tend to evaporate and dissipate in high wind. Other material is subject to oxidation, which is further enhanced by ultraviolet radiation. Strong water movements that include bubbling and, in particular, breaking waves may aerosolize oil from the water surface. Chemical and biological degradation may have similar results, while the formation of water in oil emulsions yields a substance with quite different properties. These changes in oil properties influence the transport and especially the spreading of oil. A combination of factors, such as size of the spill, properties of the oil, and meteorological conditions, determines the fate of oil and other spilled material.

One change in physical properties, emulsification, mainly occurs if the oil layer has sufficient thickness and is promoted by strong agitation of the water surface, such as a heavy rainfall. Emulsions, which may contain up to 80%

water, influence the spread and transport of the spill as well as the removal by natural or artificial means. Tarry lumps eventually form and sink.

Disappearance of visible evidence of an oil spill on the surface of a lake, river, or sea does not coincide with the disappearance of biological damage. Crude and other types of oil products are not readily biodegraded. They settle unaltered to the bottom and are retained in the mud for many months, particularly the more toxic aromatic hydrocarbons that are not detoxified by bacteria. Moreover, oil-laden sediments can move with bottom currents and eventually contaminate unpolluted areas offshore long after the initial accident. Oil concentrates other fat-soluble poisons, such as insecticides and chemical intermediates. When this material bonds with plankton and algae, it is likely to reach edible seafood and subsequently constitute a danger to human health. If films of free oil are present, they can interfere with the natural processes of stream reaeration and photosynthesis, thus procuring serious effects on all living species in the water system.

Containment

Once a spill has been discovered, determine its nature, its size, its direction of travel, and if anyone has been injured. The cause of the spill is irrelevant in the initial response phase; The important step is containment. However, take proper precautions before entering the spill area. Protective clothing can range from gloves to full protective suits, which can either be disposable or reusable. These suits are made of a variety of materials including: Viton, Tyvek, polyvinyl chloride, neoprene, polyurethane, and butyl, nitrile, and chlorinated polyethylene rubber. These materials are all basically resistant to most chemicals and oily wastes. The choice of material is dependent upon the chemical encountered. Once the response personnel are properly protected, they are ready to attack the spill.

The first major spill requirement is to contain it. There are two basic types of products available to contain oil spills: booms and film foaming agents. Booms are designed to contain oil and other chemical spills by providing a physical barrier resembling an underwater curtain, which prevents further migration of the spilled material. Booms are used to concentrate a spill and to push it toward a recovery area.

Booms consist of two components: a draft area taken up by the flotation device and a curtain that extends into the water. The flotation device used in booms can be made of a variety of materials including those that float through the use of air, polyethylene foam, urethane foam, and other similar materials. Sorbent booms contain also absorb the spilled materials. Booms can be disposable or reusable. They usually come in a variety of standard lengths, from 10 ft to 300 ft and can be linked together. Many are not adequately designed or are made of materials that do not float properly. Failure also occur because of incorrect estimates of the pressure against the curtain.

Most booms have subsurface curtains weighted to remain relatively perpendicular to the water's surface. However, problems can exist. For example, the curtain may not remain perpendicular. The curtain must be at a sufficient depth and made of a material not affected by oil. The boom must be adequately reinforced to withstand the forces of wind and water, forces often exceeding 100 lb per linear foot.

The boom also must be balanced to that it will not break apart or tip over under unusual weather conditions. It should have been publicly test-engineered under actual spill conditions. It should be easy to handle, easy to store safely, and deployed quickly and easily with available manpower. Finally, the boom must be the right size for water conditions surrounding the facility.

A second method available of containing floating spills is the use of herders. Herders are film foaming agents that limit a spill's spread by modifying surface tension. Because this technique has not been widely used, its effectiveness is not fully known. Herders are generally applicable to oil spills rather than chemical spills.

Spills of chemicals that sink in water also are problematic. Containment of these types of spills is extremely difficult and expensive but is possible with excavation and dikes. Containment of spills that are soluble or mixable in water can be accomplished with sealed booms that extend to the bottom of the contaminated layer, diversion of the contaminated water to allow for isolation of the mixture, or the use of jelling agents. Booms for these types of spills may contain sorbent material.

Diversion of the contaminated water can be accomplished with earthen dikes and dams, sandbags, and other similar physical devices. For this technique to work, an area must be available to handle the flow. If the flow is small, empty drums or tanks may be adequate. However, for larger flows, temporary ponds must be constructed.

Another technique is the use of jelling agents. These chemicals are designed to immobilize spills. They are generally dry materials. There are several products available, the majority of which contain polyvinyl alcohol and employ polymerization, thickening, or some other chemical technique.

The majority of spills begins on land and works their way into the water if not controlled quickly. Spills of chemicals and oils on land are easier to control provided the materials do not reach the soil but are confined to impervious surfaces such as concrete floors or pavement. There are several devices available to contain small spills from tanks or drums. Magnetic and foam patches can be used to prevent drums and tanks from further leaking. Portable berms also are available to contain leaking drums. These berms consist of rubberized containment ponds with inflatable sides.

Dams can be created around the spill using sorbent pillows, booms, sand, earth, or specially made pillows that solidify when water is added to them. Foamed polyurethane and concrete also can be used as a spill dam. For spills on soil surfaces, trenches can be used to contain the spill, provided the equipment for construction is readily available and the spill is relatively slow moving.

Sorbent compounds can be used to contain as well as clean up spills. Jelling agents can be effective in stopping the spread of spills. Many of these agents act to solidify the spilled materials. Spill-containment kits with jelling agents and sorbent materials recently have been developed for a quick first response.

Cleanup and Recovery

Once a spill has been contained and the source of the spill located and corrected, the major task of cleanup and recovery must begin. Spill cleanup can proceed at the same time as containment, but it's generally the next step in the process. Cleanup can be conducted using a variety of physical, chemical, and biological processes. The actual treatment of spilled materials can be considered part of recovery, or it can be the precursor to disposal.

Physical cleanup techniques include oil skimmers and boats. A wide variety of skimmers and skimming boats are available for oil and other chemicals that float. Skimmers vary in their method of recovery. One device uses centrifugal force to separate oil from water. Other devices use a moving belt or rope designed to pick up the oil. Another method uses wires to collect oil and place it in a reservoir within the device. A fourth method uses the wire-type system, but suctions the oil to a distant reservoir. Suction heads are very efficient and can be used in shallow water along shorelines.

Activated carbon also can be used to treat spills. However, this type of treatment can be expensive. In order to effectively treat a chemical floating on water, it is necessary to transport it to a treatment facility. Booms and surface pumps are normally employed for these tasks, depending upon the size of the spill and the water body. For materials that sink in water, dredging or burial are two alternatives for cleanup. Of the two, dredging is more desirable. Burial of the material on the bottom involves covering the material with an inert material such as clay. The problem with burial is that the spilled material is not removed from the water environment and may resurface at any time. Dredging of contaminated bottom soils removes the contaminant from the water. The soil/chemical mixture then can be dewatered and buried in an approved landfill.

A variety of chemicals and processes are available cleaning up spilled materials. These chemicals may be applied at the spill or at a remote site. Chemical neutralization can be used for acid or caustic spills. Neutralization can be used on a variety of spills in water or in land. Some sorbent pillows used for spill cleanup also contain neutralizing chemicals. A good neutralizer for acid spills is sodium bicarbonate, and for base chemical spills sodium dihydrogen can be effective.

Chemical solidification is still another method of spill cleanup. This type of treatment is particularly effective for hazardous chemicals. Solidification products include cement, silica-based materials, concrete, and various jelling agents. Generally, solidification is conducted after a spill has been cleaned, and the solidified material is placed in a landfill.

For some oil spills, chemical dispersants may be applied. These chemicals reduce the surface tension between the oil and water, and accelerate the natural effects of wind, waves, and currents to disperse the oil. There is a wide variety of dispersants available. However, many of them are potentially toxic, and proper regulatory approval should be obtained prior to use. There are also a wide variety of application methods. Usually, these chemicals are sprayed onto the oil spill. The spray may be applied by hand, airplanes, boats, or helicopters. In addition, there are booms that spray dispersants through discharge nozzles.

Sorbents are available for the cleanup of most chemical spills. Sorbents recover the spilled material either through absorption or adsorption. In adsorption, the sorbent attracts the spilled material and allows it to adhere to the sorbent surface. In absorption, the spilled material penetrates into the pores of the sorbent.

Sorbents can range in composition from inert natural materials such as straw, sand, or peat, to synthetic fibers such as polyethylene or polypropylene. Other sorbents include vermiculite, activated carbon, various silicates, and numerous specialty materials. Sorbents exist for organic wastes, solvents, bases, acids, oil, hydrocarbons, and other types of chemicals. Some sorbents neutralize as well as absorb, and some will absorb vapors as well as liquids. Sorbents may be applied directly on a spill as a dust or may be packaged in booms, pillows, pads, rolls, blankets, etc. There are also sorbents that repel water while collecting the spilled material.

Treatment and Disposal

Once the spill has been cleaned up, it is necessary to properly treat and dispose of the spilled material. Selection of a specific treatment and disposal process depends upon the characteristics of the recovered waste and pollution control regulations. Some materials may be treatable either by using physical, chemical, or biological processes. These processes can be performed on-site or at an approved treatment facility. Mobile units can treat a spill on-site. There are many types of on-site mobile treatment units available that use equipment including:

- Carbon adsorption filters,
- Clarifiers and separators,
- Multimedia filters,
- Polymeric exchange absorbers,
- Equalization/aeration vessels,
- Bioreaction vessels,
- Retention pools, and
- Mixing/transfer units.

CHAPTER 9

HEALTH AND SAFETY AUDITS FOR INDUSTRIAL FACILITIES

Paul N. Cheremisinoff

Department of Civil and Environmental Engineering
New Jersey Institute of Technology
Newark, NJ 07102, USA

Keyur Patel

Department of Civil and Environmental Engineering
New Jersey Institute of Technology
Newark, NJ 07102, USA

CONTENTS

INTRODUCTION

Correcting occupational health hazards is management's responsibility because such hazards can cause compensable illnesses, can impair employees' health to the extent that time is lost from the job, or can negatively affect employees' work capacity. Industrial health and safety issues are regulated by the federal Occupational Health and Safety Administration (OSHA) or state laws. The OSHA of 1970 is a landmark piece of legislation because it made safety and health on the job a matter of law for 4 million American businesses and their 57 million employees.

Employers have certain responsibilities under OSHA. An employer must "furnish to each of his employees employment and a place of employment free from recognized hazards that are causing or likely to cause death or serious physical harm to his employees." OSHA continuously updates old standards and develops new standards. It is very difficult for a manufacturing facility to keep track of and comply with these standards.

Health and safety audits are the best tool to recognize the deficiencies in a company's health and safety programs. Under any regulatory agency, health and safety audits are not mandatory; however, audits are gaining rapid attention because of the company's legal liabilities and heavy penalties by OSHA. Health and safety audit protocol can be divided in three principal phases:

Phase I: Previsit preparations,
Phase II: On-site review, and
Phase III: Health and safety audit evaluation.

Most auditing procedures can be used either by teams contracted to perform health and safety audits at client facilities or by in-plant personnel performing self-audits.

PHASE I: PREVISIT PREPARATION

A. **Size and composition of the audit team**—The size of the team will depend upon the size of the site and the complexity of the potential hazards.
  a. A personal health and safety audit team is composed of:
    1. A team leader with:
      (a) thorough knowledge of the health and safety regulatory requirements.
      (b) good knowledge and understanding of health and safety program management.
      (c) experience in health and safety audits.
    2. A backup person with leadership experience to provide support to the team leader.
    3. A safety specialist with thorough knowledge and understanding of regulatory requirements and the latest safety equipment available in the market.
    4. An industrial hygienist with thorough knowledge and understanding of health hazards in the particular industry and industrial hygiene management programs.
  b. If the site is very large (5,000 or more employees), additional health and safety team members and a well-qualified backup for the team leader are necessary.

B. **Arrangements**—Arrangements for the audit will be coordinated by the team leader or backup person.
  a. **Schedule**—The team leader will contact a company's official to determine a convenient time for the on-site audit when all the operations are active and explain the general activities of the audit.
    1. The duration of the visit will depend upon the size and complexity of the plant.
      (a) The on-site audit should average about two days unless there are more than 1,000 workers or other unusual characteristics exist.
      (b) Chemical plants producing, using, or storing one or more high-risk chemicals will require an additional half day of on-site audit time.

b. **Documentation**—To save time during the review, the team leader should provide the site representative with a suggested list of items to have ready for the team's review before the on-site audit begins.
   1. OSHA log and workers compensation first reports of injury (OSHA 101) for three years
   2. Health and safety training programs
   3. Industrial hygiene sampling reports
   4. Previous health and safety audit reports
   5. Self-inspection reports
   6. Preventive maintenance records
   7. Accident report
   8. Plant safety and health rules
   9. Records of engineering controls
   10. Personnel protective equipment programs
   11. Emergency procedures
   12. Medical records
   13. Safety meeting minutes

C. Advance planning responsibilities for review team members.
   a. **Knowledge of Industry**—The team members should become as familiar as possible with the type of industry and the specific site.
   b. **Detailed review of background information available**—The team members should become as familiar as possible with the provided information well in advance of the on-site audit.
   c. **Information needed**—The team members should review the report format, and make notes concerning the confirmation information and any other data to obtain during the site visit. For chemical plants manufacturing, using, storing, or producing as by-product high-hazard chemicals, obtain documentation and physical evidence that the company has done its own review.
   d. **Checklist**—The team members should carefully review the safety and health program for similar industries and compile a checklist to use during the visit. The training program should be included on that list.
   e. **On-site Questions**—The team leader should carefully review the background information, then compile the list of documents to be reviewed and questions to be asked, provide the list to team members during the strategy meeting, and augment it with questions suggested by the other team members.

D. **Strategy Meeting**—Shortly before the scheduled audit, the team leader will hold a strategy meeting with all the team members to plan the details and to assign specific tasks.
   a. **Timing**—The team leader should carefully plan the time on site.
   b. **Briefing**—The team leader should familiarize team members with on-site procedures and provide a quick review of on-site activities.
   c. **Program Evidence**—The team leader should discuss the various signs of an effective safety and health program.

d. **Assignments**—The team leader should make specific on-site assignments to use each team member's expertise most effectively. The support role of the backup team leader should be clearly spelled out.

PHASE II: ONSITE REVIEW

A. **Initial Interview**—The initial interview with the employer representative and employee representatives, where applicable, will set the stage for the on-site review, letting everyone know what to expect and what assistance will be needed.

a. **Atmosphere**—Team members should establish a cooperative atmosphere.

b. **Discussion Topics**—The team leader should convey the following information.

1. **Schedule**—Briefly cover what activities the audit program will entail and the approximate timing of each stage.
2. **Information**—An employer representative will be needed to answer questions that arise during the audit.
3. **Permission for interviews**—Request permission to conduct private interviews with joint committee members and with randomly selected employees. Interviews of employees are necessary to determine employee involvement in the safety and health program.
4. **Status**—Explain that the team will give the employer representative a brief status report on the progress after the review is completed. The team will discuss its findings with the employer representative to clarify any misunderstanding that may arise.
5. **General information**—Filled out a general information form for the report development and to prepare for the audit. See Table 1 for a suggested form.

(text continued on page 230)

Table 1
General Information Audit

1. What is the name and location of your establishment?

____

____

____

2. Is your establishment independent or part of a larger company, government or private establishment? Please specify and give name of parent company or government agency if applicable.

____

____

____

____

3. Please list the name, address and telephone number of the establishment official who will be the contact person for this survey.

____________________	____________________
name	title
____________________	____________________
	telephone
____________________	
address	

4. How many people are employed at your establishment in each of the following categories:

	Number
Production	______
Maintenance	______
Administration and Office	______
Research and Development	______
Other ____________	______
Total	______

5. What is the nature of your establishment's operation (such as chemical, agricultural, oil refining, etc.)?

____

____

6. What are the major raw materials used in the operation of your establishment?

____

____

7. What are the major types of processes used in the operation of your establishment:

____

____

____

(table continues)

Table 1 (continued)

8. What are the major finished products of your establishment?

____

____

____

9. Indicate to which of the following chemicals there are significant potential exposures on a regular basis in your establishment and the number of employees potentially exposed to each.
(Check each potential exposure)

	Potential Exposure	Number Exposed
a. Halogens	a.______	______
b. Alkaline Materials	b.______	______
c. Toxic Metals	c.______	______
d. Inorganic Compounds of Nitrogen or Carbon	d.______	______
e. Aliphatic Hydrocarbons	e.______	______
f. Alicyclic Hydrocarbons	f.______	______
g. Aromatic Hydrocarbons	g.______	______
h. Aliphatic Halogenated Hydrocarbons	h.______	______
i. Cyclic Halogenated Hydrocarbons	i.______	______
j. Phenols or Phenolic Compounds	j.______	______
k. Alcohols	k.______	______
l. Glycols or Glycol Derivatives	l.______	______
m. Epoxys	m.______	______
n. Ethers	n ______	______
o. Ketones	o ______	______
p. Organic Acids, Anhydrides, Lactones, Acid Halides, Acid Amides, Thioacids	p ______	______
q. Esters	q ______	______
r. Organic Phosphates	r ______	______
s. Aldehydes or Acetates	s ______	______
t. Cyanides or Nitriles	t ______	______
u. Aliphatic or Alicyclic Amines	u ______	______
v. Aliphatic Notro Compounds, Nitrates, Nitrites, Hetercyclic or Miscellaneous Nitrogen Compounds	v ______	______
w. Aromatic Nitro or Amino Compounds	w ______	______
x. Mineral Dusts	x ______	______
y. Organic Dusts	y.______	______
z. Infectious Agents	z.______	______
aa. Resins - Natural or Synthetic	aa ______	______
bb. Other - Indicate Below	bb ______	______

____

____

10. Indicate which of the following physical agents are significant potential hazards on a regular basis in this establishment and the number of employees potentially exposed to each.

		Potential Exposure	Number Exposed
a. Ionizing Radiation	a	________	________
b. Ultraviolet Radiation	b	________	________
c. Laser Sources	c	________	________
d. Abnormal Temperatures	d	________	________
e. Radiant Heat	e.	________	________
f. Microwaves	f.	________	________
g. Vibration	g.	________	________
h. Electrical Hazards	h.	________	________
i. Abnormal Air Pressure	i.	________	________
j. Noise	j.	________	________
k. Other - Indicate below	k.	________	________

____

____

11. How many work shifts does your establishment operate?

a. Day a.________
b. Evening b.________
c. Night c.________
d. Other - Explain d.________

____

____

____

____

12. How many lost-time accidents occurred in your establishment during the last reportable year? 12.________

13. Is your establishment self-insuring or do you belong to a state or federal or private workman's compensation fund?

a. Self-Insuring a.________
b. State Workman's Compensation Fund b.________
c. Federal Workman's Compensation Fund c.________
d. Private Workman's Compensation Fund d.________
e. Other - Explain ________________________ e.________

14. Are your workers unionized?

a. Completely unionized a.________
b. Partially unionized b.________
c. Not unionized c.________

(table continues)

Table 1 (continued)

15. If the answer to question 14 is a or b, please list the name(s) of the union(s).

____

____

____

____

Comments by Respondent

Use this space to clarify any of the questions in this portion of the questionnaire. When doing so, refer to the questions by number.

(text continued from page 226)

B. Document review
  a. Record keeping
    1. **The OSHA 200 Log**—Review the OSHA 200 log for the most recent three-year period to see that the logs have been properly maintained for the entire period. Verify that lost workday case entries are recorded properly by reviewing the company's OSHA 101 forms or their substitute, such as workers compensation or insurance reports of injury. Calculate injury incidence and lost workday rates based on the log.
    2. **Review of health program documentation**—Review documented means of determining potential health hazards, such as a basic industrial hygiene survey or chemical process analysis and records required by relevant health standards. The site health program also must be reviewed to determine that potential and actual exposures are adequately assessed. Check the hazard communication documentation system to ensure that personnel are alert to chemical products at the work site.
    3. **Committee records**—Check minutes and inspection records to verify composition and activities. Committee minutes may provide information concerning the level of employee involvement.
    4. **Complaint records**—Review internal notification files to verify that the system works as described, that cases are well-documented, and that responses seem reasonable and timely.

5. **Chemical process systems documentation**—For chemical plants producing or using high-hazard chemicals, review documents and monitoring systems describing identification of critical failure points, planned redundant protective systems, control system for design or procedure notification, emergency procedures for failure of control systems, procedures for changing back to normal operations after emergencies, and preventive maintenance systems.

b. **Safety and health program**—Document all aspects of the program already in place, such as inspections, accident investigations, emergency procedures, medical records, hazard review, analysis examples, and the personal protective equipment program. For plants producing or using high-hazard chemicals, document project reviews or analyses to identify possible failure points and planned redundant systems.

A suggested documents review checklist is presented in Table 2.

C. Site walk-through

The purpose of the walk-through is to ensure that the safety and health program is operating as described and to ascertain the adequacy of the program

Table 2
Document Review Checklist

	Yes	No	N/A	Comments
A. Injury Records.				
1. Is the log current?	___	___	___	___
2. Are log entries consistent with the 101?	___	___	___	___
3. How does the rate compare with earlier periods and the average for the SIC?	___	___	___	___
4. Are there any trends in the nature of injuries or illnesses which suggest preventive measures are needed?	___	___	___	___
B. Self-Inspections.				
1. Have they been conducted regularly?	___	___	___	___
2. Are records maintained?	___	___	___	___
3. Are hazards identified and abated in a timely manner? How often is the entire site covered?	___	___	___	___
C. Accident/Near-miss Investigations.				
1. Have they been conducted when needed?	___	___	___	___
2. Are the causes identified sufficiently?	___	___	___	___
3. Are appropriate preventive measures taken?	___	___	___	___
D. Handling of Reports of Safety and Health Concerns.				
1. Is a log of reports or some other tracking mechanism maintained?	___	___	___	___

(table continues)

Table 2 (continued)

2. Are reports investigated properly and resolved? ___ ___ ___ ______________
3. Are employees notified of the results of investigations? ___ ___ ___ ______________
4. Are employees satisfied with the outcome of their reports? ___ ___ ___ ______________

E. Employee Training.

1. Are safety/health orientations provided for new employees? ___ ___ ___ ______________
2. Is job hazard prevention training provided on a continuing basis? ___ ___ ___ ______________
3. Is the level of safety and health training adequate to meet the hazards in the workplace? ___ ___ ___ ______________

F. Training For Construction.

1. Are regular "tool box" safety talks held? ___ ___ ___ ______________
2. Is a record maintained of the subject(s) covered and what employees attended? ___ ___ ___ ______________
3. Do subcontractor employees participate in the GC training or hold their own meetings? ___ ___ ___ ______________
4. Are the talks relevant to the phase of construction in progress? ___ ___ ___ ______________
5. Have committee members received adequate training? ___ ___ ___ ______________

G. Hazard Review and Analysis.

1. Are results being used in employee training? ___ ___ ___ ______________
2. Is the company on schedule in conducting any additional reviews planned for? ___ ___ ___ ______________
3. Are the procedures for conducting reviews and analysis satisfactory? ___ ___ ___ ______________

H. Employee Participation.

1. How are employees involved in the safety and health program? ___ ___ ___ ______________
2. Is the participation active and meaningful? ___ ___ ___ ______________

I. Line Accountability.

1. Are supervisors aware of their safety and health responsibilities? ___ ___ ___ ______________
2. Do they implement them appropriately? ___ ___ ___ ______________

J. Safety and Health Program Evaluation.

1. If not already in place, has an evaluation system been developed and commitment made to complete it each year? ___ ___ ___ ______________
2. Is it a system which will enable a full assessment of the strengths and weaknesses of the safety and health program? ___ ___ ___ ______________

K. Health.

1. Is the level of industrial hygiene sampling and/or medical monitoring adequate to meet the potential hazards of the workplace? ___ ___ ___ _______________
2. Are appropriate preventive measures been taken? ___ ___ ___ _______________
3. Is appropriate personal protective equipment available and, where needed, used by employees? ___ ___ ___ _______________

L. Joint Committee Functions (where applicable).

1. Has the committee been meeting regularly? ___ ___ ___ _______________
2. Are minutes maintained and detailed enough to indicate the issues discussed and their resolution? ___ ___ ___ _______________
3. Have a quorum of employee and employer representatives been present? ___ ___ ___ _______________
4. Do committee members participate in inspections? ___ ___ ___ _______________
5. Do committee members participate in or review the findings of accident investigations? ___ ___ ___ _______________
6. Does the committee review complaints and their resolutions? ___ ___ ___ _______________

M. Subcontractor Coverage (where applicable)

1. Are subcontractor foremen/employees aware of the General Contractor/owner participation in Star or Try? ___ ___ ___ _______________
2. Do General Contractor/owner inspections cover hazards created by subcontractor activities? ___ ___ ___ _______________
3. Are these hazards corrected in a timely manner? ___ ___ ___ _______________
4. Are appropriate preventive measures required of the subcontractors by the General Contractor/owner? ___ ___ ___ _______________

---

as implemented. All members of the review team should, if possible, walk through enough of the site to understand the type of work done and to receive a general sense of working conditions. The safety and health specialists must see enough of the site to understand the types of problems that might exist, and to determine that safety and health problems are being addressed systematically by the site safety and health program. Special attention should be given to recurrent problems as identified in previous inspection logs.

a. **Interviews**—During the walk-through, safety and health specialists shall select employees at their workstations and ask them to explain the work procedures and the use and maintenance of any personal protective equipment.

Conduct employee interviews to ascertain the extent of involvement and the awareness of the safety and health program. Suggested questions are listed in Table 3.

b. **Safety specialist review**—The safety specialist shall:

1. Follow the process flow where possible.

(text continued on page 240)

Table 3
Suggested Interview Questions

I. INTERVIEW QUESTIONS FOR SUPERVISORS

Explain your purposes in being at the site.

	Answers
1. How long have you worked here?	____________
2. What other places have you worked?	____________
3. How did the safety and health program(s) compare to this one?	____________
4. When did you become a supervisor?	____________
5. What kinds of hazards are you and/or your employees exposed to?	____________
6. How has management provided protection from those hazards?	____________
7. What do you do when you discover a hazard in your area?	____________
8. What do you do when an employee reports a hazard in your area?	____________
9. Do you provide employee training in safe work procedures? (Is so, please describe.)	____________
10. How often do you use at least the first step of your disciplinary system: What is the most frequent offense?	____________
11. What kind of emergency drills do you run for employees? How often? What is your role in the drill?	____________
12. How are you held accountable for assuring safe and healthful working conditions in your area?	____________
13. (High hazard chemical plants only) Is adequate supervision provided for night and weekend operations?	____________
14. (High hazard chemical plants only) Is maintenance satisfactory, particularly on release prevention equipment?	____________

II. EMPLOYEE INTERVIEW QUESTIONS

These questions are intended for the OSHA reviewer to guide oral employee interviews. State that employee responses shall be kept confidential. Explain your purposes in being at the site and state that responses will not be the sole determinant of the company approval or disapproval.

Answers

A. Background.

1. What is your job here? ____________
2. How long have you worked here? ____________

B. Orientation and Training.

1. Did you receive safety and health training when you began to work here? (Is so, please describe.) ____________
   a. How soon after you began to work did you receive training? ____________
   b. How long did it last? ____________
2. Is you did not get training when you were first hired (or transferred to a new job), have you received any basic safety and health training since that time? (If so, please describe.)
3. Do you receive regular safety and health training? ____________
   a. If so, how often? ____________
   b. How long does it last? ____________
4. Are you aware of company safety rules? ____________
   a. If so, do they seem to cover everything they should? ____________
   b. What happens if an employee disobeys a company safety rule? ____________
5. What is your responsibility in an emergency? When did you last practice it? ____________

C. Hazard Correction.

1. Do you come into contact with any safety hazards? ____________
2. Is so, answer the following questions:
   a. Do the management people responsible for safety understand the hazards associated with your work? ____________
   b. Has management been quick to notice hazards and correct them? ____________
   c. Is you notice a hazard, do you know who in management to contact to get it fixed? ____________
   d. When you do this, do you get quick action? ____________
   e. Do you get appropriate action? ____________
   f. Is management decides no action is needed, is this explained to you? ____________

D. Reports of Safety and Health Problems.

1. Have you ever told your supervisor or other management official about a hazardous condition here? If yes, ask the following questions: (If no, skip to next section.)
   a. What was the condition? ____________
   b. Whom did you notify? ____________
   c. Did you get a response? Is so, was the response satisfactory? ____________
   d. How long did it take to get a response? ____________
   e. Is you did not get a response, did you try again, try someone or somewhere else? (If the latter, describe.) ____________

(table continues)

Table 3 (continued)

2. Have you ever filed a safety and health complaint with OSHA? Is so, how would you compare OSHA's response with company's? ____________

E. Health Program.

1. Do you come into contact with any potentially dangerous chemicals, substances or harmful physical agents such as radiation or noise? Is so, what are they? ____________
a. Do you feel that management has provided enough protection for you? ____________
b. (High hazard chemical plants only) Is maintenance of release prevention equipment satisfactory? ____________
2. Have you ever seen industrial hygiene surveying or monitoring being done in your workplace? ____________
a. Was it just once or are these routine? ____________
b. Is just once, was it in response to a specific problem? Is a specific problem, what was it? ____________
c. Is routine, how often? ____________
3. Has the company had you examined by a physician? Is so, is this done periodically? ____________
a. If routine, how often? ____________
b. If not done periodically, what was the reason for this examination? ____________
c. Did the examination seem thorough? ____________
d. Did the doctor explain what he (she) wash doing and why? ____________
e. Id not, did anyone in management explain? Is so, who? ____________
f. Were the results of the examination explained to you? Id so, who explained them? ____________

F.PPE.

1. Is personal protective equipment (hard hats, goggles, respirators etc.) readily available when needed? ____________
2. Is personal protective equipment is used, is it kept clean and in good repair? ____________
a. Who is responsible for this? ____________
b. What protective equipment have you used? ____________
c. Have you been trained in the use of this equipment? Is so, was the training adequate in your opinion?

G. Safety Committee (where applicable).

1. Are you aware of the committee for safety and health? ____________
2. Is so, please answer the following questions:
a. When did you become aware of the committee? ____________
b. Do you know any of the members? (Is yes, please name the members you know.) ____________
c. Do you know how the employee members were selected? (Is yes, describe.) ____________
d. Have you seen the committee make inspections? Is so, do they appear to be their approach? ____________
e. What other things does the committee do? ____________
f. Would you say the committee is very effective, somewhat effective, or not effective? Why? ____________

H. General.

1. How does this workplace compare to others where you have worked in terms of safety and health? Worse? About the same? Better? Much better? ________
2. (High hazard chemical plants only) Is employee turnover high? ________
   a. If so, why? ________
   b. Also if so, how long goes it take a new employee to learn to work safety alone? ________
3. If your site is approved for this program, OSHA will stop doing routine inspections but will inspect in response to employee complaints, serious accidents or chemical leaks. Under this program, OSHA will come back to evaluate how well things are going as we have done today. How do you feel about that? Do you think it would be OK? ________
4. Is there anything else you think we should know about the safety and health program here? ________

III. INFORMAL EMPLOYEE INTERVIEW TOPICS

<u>Answers</u>

A. Before Approval.

1. Safety and health orientation for new employees. ________
2. On-going safety and health training provided. ________
3. Awareness of the joint committee and its functions (where applicable). ________
4. Safety rules and enforcement. ________
5. Safe work practices. ________
6. Freedom to point out safety or health hazards. ________
7. Awareness of an internal safety and health complaint procedure. ________
8. Responsiveness of management in correcting hazards. ________
9. Emergency procedures. ________
10. Comparison of the safety/health conditions at this this workplace in relation to others. ________

B. After Approval.

1. Questions from list above, as applicable. ________
2. Awareness of VPP program participation -- rights including rights to receive upon request , results of self-inspections or accident investigations. ________
3. Satisfaction with VPP. ________
4. Knowledge of any changes since last OSHA onsite visit. ________

IV. QUESTIONS FOR MAINTENANCE PERSONNEL AT SITES PRODUCING OR USING HIGH HAZARD CHEMICALS

<u>Answers</u>

Preventive Maintenance

1. Is there a scheduled preventive maintenance program? How is it carried out? ________

(table continues)

Table 3 (continued)

2. Does it include:
   a. Critical instrumentation and controls? ____
   b. Pressure relief devices and systems? ____
   c. Metals inspection? ____
   d. Environmental controls, rubbers, filters, etc.? ____
3. Does the design, inspection and maintenance activities include procedures to preclude piping cross-connections between potable water systems and nonpotable systems? ____
4. How are these procedures carried out and how are systems monitored and inspected to find any cross-connections? ____
5. Do maintenance personnel participate in safety committees and other safety functions? ____
6. Is there a priority system for safety/environmental related maintenance items? Is it being followed? ____
7. Does the preventive maintenance program include onsite vehicles, sprinkler systems, detection/alarm equipment, fire protection and emergency equipment? ____
8. Do you have input concerning safety and maintainability for new equipment and machinery purchases? ____
9. Do you have an inventory of spare parts critical to safety and environmental protection? ____

V. EVALUATION INTERVIEW QUESTIONS FOR COMMITTEE MEMBERS

Explain our purpose in being at the site and that responses will not be the sole determinant of company approval or disapproval.

A. General

1. How long have you worked for this company? ____
2. How long have you served on the labor/management committee? ____
3. How are committee members selected? ____
4. How many total committee members are there (for construction or other sites with joint labor-management committees)? ____
   a. Number of management representatives? ____
   b. Number of employee representatives? ____
5. How often does the committee meet? ____
   a. In view of the committee's workload, is this number of meetings: Too many? Just about right? Too few? ____
   b. How are members notified about scheduled meetings? ____
6. How many of the committee members usually attend meetings? All? Most? About half? Less than half? ____
   a. Are members encouraged to attend the meetings? ____
   b. What happens if you miss a meeting? ____
7. Do the safety and health professionals on the committee take the time to explain technical points when they arise? ____

8. What safety and health records does the committee review? __________
9. Does the committee base inspections on this data? __________

B. Inspections.

1. Does the committee do inspections of the workplace? __________
2. Is so, do these inspections cover the whole workplace? Only part? __________
   a. Is only part, how many inspections are needed before the entire workplace has been inspected? __________
   b. Do they only cover each member's own area? If so, how are reports done? __________
3. Do you normally participate in the inspection process? __________
4. How many inspections have you made in the past year? __________
5. Do you consider this an adequate number? __________
6. In terms of keeping the workplace safe, do you consider the inspections very useful? Somewhat useful? Not useful? How would you change or improve them, if you could? __________
7. What role does the committee play in accident investigations? __________
8. Have you seen industrial hygiene inspections at your worksite? __________
   Have you accompanied or participated in any of these inspections? __________
9. Describe the committee's role (if any) in the handling of complaints from workers. __________
10. If the committee overseas the complaint process, do they verify that abatement occurs on valid complaints? __________

C. Training

1. Have you been trained specifically to work on the committee? (If so, describe.) __________
2. Who provided the training? Company? Union? Other? __________
3. Did your training prepare you for committee work? __________
4. Did your training include information on safety hazards? __________
5. Since your initial training have you received supplementary "refresher" training? How often? (Describe briefly.) __________
6. How would you change or improve the training, if you could? __________

D. Communication.

1. Do you think that the committee has had an effect on employee awareness of safety and health problems? (If so, describe.) __________
2. Has the committee made suggestions for safety and/or health improvements? (If yes, give examples.) __________

(table continues)

Table 3 (continued)

3. Was the company responsive to suggestions that the committee offered? (Give examples)	______
4. Is the company does not accept a recommendation, do they explain why? (Give an example.)	______
5. Have there been any disagreements between employees and management about safety and health issues?	______
6. Would you say that the company has been supportive of the time you have spent on committee business?	______
E. Improvements.	
1. Do you think that the committee operations and functions can be improved?	______
2. What else do you think the committee can do to improve safety and health conditions?	______
F. Overall Assessment.	
1. As a whole, how would you characterize the effectiveness of the committee?	______
2. Has the role or amount of activity of committee changed since VPP approval? If so, how?	______
3. Are you pleased with VPP participation? Why or why not?	______

(text continued from page 234)

2. Look for evidence that personal protective and lifesaving equipment, fire safety, signs, signals, barricades, tools, ladders, scaffolding, storage and handling of hazardous materials, machine guarding, power tools, and welding are appropriately managed.

c. Relate the visible problems in the work area to documents concerning work procedures, emergency planning, self-inspection procedures and reports, and complaints or reports of hazards from employees. Make notes concerning program improvements needed to prevent these problems.

d. Make notes about any specific hazards to correct and ensure that a responsible number of management members makes notes as well.

D. **Industrial hygiene review**—The industrial hygienist shall:

a. Follow the process flow where possible.

b. Based on review of monitoring records and material data sheets, recognize potential hazard areas; ventilation, storage, handling and use of toxic materials; emergency equipment, respirator usage and maintenance; and radiation and noise protective measures.

c. For chemical plants producing or using high-hazard chemicals, check process lines to verify documented systems protections and include questions concerning system failure procedures in formal interviews with appropriate operators. Interview maintenance personnel using questions suggested in Table 3.

d. Look for evidence that these hazards are appropriately managed and that no other potential hazards have escaped management's attention.
e. Make notes concerning any health hazard management that needs improvement.
f. Relate these problems to documents concerning safe work practices and training, the respirator program, industrial hygiene sampling and analysis, and hazard communication systems. Make notes concerning program improvements needed to provide management systems to avoid these problems.

PHASE III: SAFETY AND HEALTH AUDIT EVALUATION

At the end of the audit, the team leader will meet with all team members to discuss the apparent serious hazards. Verbally present these hazards to the employer representative and recommend immediate corrective actions. Submit the safety and health audit evaluation report to the employer.

The following report format outlines the subjects to be covered, using questions to elicit written discussions.

A. Introduction
   a. Include the date of evaluation and names of team members.
   b. Indicate the sources of information on which report is based, e.g., review of the on-site documentation, personnel interviewed, the areas toured, and any previous evaluation reports.
   c. Indicate the scope of work for the health and safety audit and who authorized the audit.
   d. Provide a description of the work site and the activities there, including the current number of workers and appropriate standard industrial classification (SIC).

B. Record keeping and injury rates
   a. Do(es) the person(s) responsible for keeping the log vouch for the accuracy of the entries? If not, please explain.
   b. Provide the injury rates for the period since the previous evaluation report, including year-to-date as well as the current coverages of the past three complete calendar years. Compare the rates with the latest Bureau of Labor Statistics industry average.
   c. Explain any foreseen potential problems concerning the rates.
   d. Do monitoring records indicate that required industrial hygiene sampling and record keeping is being done appropriately?
   e. Summarize the quality of injury, industrial hygiene, and medical record-keeping at each site.

C. Safety and health program(s)
   a. The suggested structural requirements

1. Mention any items of the written program that could not be verified. Discuss any missing aspects and reasons for change.
2. Are site inspections covering what they should? Are they as frequent as they should be? Is the documentation of tracking complete? Are written accident investigation reports descriptive? Are preventive actions being taken?
3. Have hazards been adequately tracked and corrected?
4. Is the employee hazard reporting system working efficiently? If not, describe any problems.
5. What is the average number of employee hazard reports per year handled during the past year? During the past three years?
6. Are employees aware of the hazard-reporting system? Are they generally satisfied with the way that it works?
7. Has OSHA conducted any complaint inspections or responded to informal complaints? Did the complainant try the internal system first?
8. Are safety rules, safe work practices, and personal protective equipment requirements adequate? What happens if an employee disobeys one of these? Does the same disciplinary system apply to management?
9. Describe any emergency drills run in the previous year.
10. Only for plants in the chemical industry producing or using high-hazard chemicals, describe the following systems:
    (a) The system that ensures operational processes involving high-risk chemicals are kept within safe bounds during normal operations.
    (b) The system that identifies critical failures points and establishes redundant systems.
    (c) The emergency response system or the emergency close-down/start-up systems.
    (d) Are the system adequate?
11. Is appropriate training begin provided? Do employees understand hazards and how to protect themselves? Do supervisors understand their role in assuring that employees understand and follow protective rules?
12. What evidence have you seen that management has appropriate control over conditions where contract employees are intermingled with regular employees?
13. Has the self-evaluation completed since the previous on-site audit covered all aspects on the safety and health program? Did they include written recommendations? What evidence have you found that the recommendations were responded to? Was the response adequate?

b. General review of safety and health conditions
   1. Does housekeeping appear to be average or better for this type of industry?

2. Are workers using personal protective equipment in areas where it is needed?
3. Based on your tour of the work site, would you characterize the health and safety conditions of this site as above average or below average for this type of industry?
4. Suggest improvements for problem areas.
5. Discuss the effect of any pattern noted in the log review.
6. Provide a summary of health and safety conditions at this site, including employees' perceptions.
7. Assess the adequacy of the safety and health program to provide protection from the potential hazards at this site.

D. Employee awareness and involvement in the safety and health program

a. Atmosphere

1. Was management helpful in providing access for random employee interviews?
2. Did employees appear to be comfortable in talking with you?
3. Were there any factors in the relationship between employees and management that may have influenced their response to you?

b. Awareness

1. Were employees knowledgeable about the health and safety program? Did their overall assessment fit your impressions?
2. Were employees knowledgeable of the provisions, including their rights under the program?
3. Have employees noticed changes in safety and health conditions during the past year? If so, what are they?
4. If a joint committee operates here, were employees knowledgeable about it? Do the employees think that the committee is effective?
5. Describe the ways employees are involved in the safety and health program. How effective is the site employee-involvement program? Outline evidence.

E. Summary

No new information should appear here. Everything stated should be supported by the body of the report.

a. General

Summarize findings in terms of each of the specific areas of the health and safety programs. Summarize the strengths and weaknesses of health and safety programs.

b. Recommendations

Summarize the recommendations to improve the health and safety program, and potential or present health and safety hazards based on the audit's findings.

CHAPTER 10

CHEMICAL HYGIENE AND SAFETY

Paul N. Cheremisinoff

Dept. of Civil and Environmental Engineering
New Jersey Institute of Technology
Newark, NJ 07102, USA

John Neglia

Becton-Dickinson and Company
Franklin Lakes, NJ 07417, USA

CONTENTS

INTRODUCTION

Many companies recognize their responsibility for maintaining a safe and healthy place of business for the protection and well-being of their employees. It is often necessary to develop a chemical hygiene and safety plan in accordance with OSHA's Occupational Exposures to Hazardous Chemicals in Laboratories, 29 Code of Federal Regulations Part 1910.1450. The plan outlines specific workplace practices and procedures to ensure that employees are protected from health hazards associated with chemicals with which they work.

Chemical Hygiene Office - Health/Safety Adviser

- Has overall responsibility for the chemical hygiene plan.
- Interacts with others to implement appropriate chemical hygiene policies and practices.
- Assists in developing effective procedures and work practices.
- Provides input regarding the adequacy of the lab facility relative to its effectiveness for controlling chemical exposures.
- Coordinates the training requirements identified in this plan.
- Assures that annual reviews of the chemical hygiene plan are implemented.

Chemical Hygiene Supervisor - Laboratory Manager

- Is responsible for chemical hygiene in the laboratory.
- Works with laboratory supervisors to develop and implement safe laboratory procedures.
- Monitors procurement, use, and disposal of chemicals used.
- Assures that personnel are effectively trained regarding chemical hazards.
- Ensures that employees know and follow good chemical hygiene practices to minimize exposures to hazardous materials.

- Ensures that exposure controls are effective.
- Ensures that appropriate protective equipment is available and personnel are trained in its use.
- Periodically inspects the facility to assure its adequacy.

Laboratory Employees

- Plan and conduct each operation in accordance with good laboratory practice and establish chemical hygiene procedures.
- Use the controls provided to minimize exposure to chemical hazards.
- Develop and practice good personal chemical hygiene.
- Adhere to all safe practices, general plant safety rules, and laboratory safe practices identified in the chemical hygiene plan.

GENERAL SAFE LABORATORY PRACTICES

To achieve safe conditions for the laboratory and laboratory workers, the following safety procedures must be followed.

Dress Code

Clothing worn by laboratory personnel will protect the body from burns. Shorts, cutoff sleeves, fishnet shirts, and tank tops are not permitted at any time. No employees are permitted to work with any part of their body exposed other than their face and hands, which will be covered with personal protective equipment, when applicable.

Eating

Consumption of food and/or beverages is not permitted in laboratory areas. Glassware or utensils that have been used for laboratory operations never should be used for food or beverages. Laboratory refrigerators and cold rooms are not to be used for food storage.

Emergency Reporting

All emergencies must be reported to an internal contact. All telephones must have stickers bearing the emergency number.

First Aid

In the event of a medical emergency, company safety/health personnel should respond. Basic first aid procedures to be used are as follows until proper medical attention can be obtained.

Eye Contact

Promptly flush eyes at an eyewash station for at least 15 minutes. Be sure to open lids while flushing.

Skin Contact

Promptly flush the affected area with water. Remove all contaminated clothing in a manner that will not cause further contact with the skin. Continue flushing until medical attendants arrive.

Spill Cleanup

Small chemical spills should be cleaned up immediately and disposed of properly. Large spills of hazardous materials should not be handled by laboratory workers. There should be an emergency response team to respond to spills of more than 55 gal.

Hygiene/Housekeeping

Do not use mouth suction to pipette chemicals or to start a siphon; a pipette bulb or an aspirator should be used to provide vacuum.

Wash thoroughly before leaving the laboratory area. Avoid using solvents to wash the skin because they remove the natural protective oils from the skin and can cause irritation and inflammation. In some cases, washing with a solvent might facilitate absorption of toxic chemicals.

The laboratory area should be kept clean and free from obstructions. Cleanup should follow the completion of any operation at the end of the day. Access to exits, emergency equipment and/or controls should never be blocked.

All chemicals and containers must be properly identified and labeled. Unlabeled containers are not permitted in the laboratory. Material safety data sheets (MSDSs) must accompany all laboratory chemicals. Wastes should be deposited in appropriate waste receptacles.

LABORATORY HOODS

Operations involving toxic gases, vapors, aerosols, and dusts must be performed in a hood. Laboratory hoods offer significant employee protection. They prevent toxic or flammable vapors from entering the general laboratory atmosphere, they place a physical barrier between the workers and the chemical environment, and they provide an effective containment device for accidental spills of chemicals.

MATERIAL SAFETY DATA SHEETS (MSDS)

The Occupational Safety and Health Administration (OSHA) requires that all employees maintain a complete and accurate MSDS for each hazardous chemical. Manufacturers/suppliers must supply this information when a material is purchased. These MSDSs must be updated whenever new information becomes available concerning a product's hazards or ways to protect against these hazards.

These MSDSs provide an excellent source of specific information on the chemicals that employees handle. MSDSs should be readily available at workstations. An updated MSDS library should be maintained in the laboratory. The old OSHA-20 MSDSs are quickly being replaced with the new ANSI/CMS standard MSDSs. This sheet has much more information about a chemical and should be the choice whenever possible. It contains 16 sections, as opposed to the OSHA-20 standard 8 sections.

MEDICAL CONSULTATION

A medical department should provide medical attention under various circumstances including:

- New employee physical examinations.
- When an employee develops signs associated with chemical exposure.
- When an occurrence such as a spill, leak, explosion, or other type of emergency occurs in the laboratory resulting in the likelihood of hazardous exposure.
- When exposure monitoring indicates an exposure level triggering medical surveillance, indicated by a routine exposure that exceeds either the action level or the permissible exposure level, in the absence of an action level.
- If respiratory protective equipment will be used.

Consultation with a medical professional should be provided free to the employee, as is any examination or other attention recommended by the consulting physician, at a reasonable time or place. The laboratory standard also requires keeping records of medical consultations and exposure evaluations.

PERSONAL PROTECTIVE EQUIPMENT

Personal protective equipment required to be worn in the laboratory includes:

Gloves

The specific chemical characteristics will determine the glove type to be worn.

Hard Hats

Hard hats are required in designated hard hat areas.

Hearing Protection

Whenever employee noise exposure exceeds the eight-hour time-weighted average of 85 decibels, the employee must wear hearing protection. All high-noise areas contain a posting of hearing protection required to be worn after one minute of equipment operation.

Respirators

When exposures to dust, fumes, mist, radionuclides, gases, and vapors exceed established limits of exposure, respiratory protection is required. If this is necessary, a respiratory protection program must be initiated in accordance with 29 CFR 1910.134.

Safety Glasses

To protect eyes from laboratory hazards, employees must wear safety glasses with side shields. Operations that require improved protection against impact, liquid splash, and other eye hazards will require safety goggles and/or face shields. Contact lenses must not be worn in any work area.

Safety Shoes

Steel-toed safety footwear is mandatory in work areas.

ROUTES OF EXPOSURE

Exposure to hazardous chemicals can be minimized by understanding the common routes of exposure.

Inhalation

Inhalation of toxic vapors, mists, gases, or dusts can produce poisoning by absorption through the mucous membrane of the mouth, throat, and lungs, and can seriously damage these tissues by local action. The degree of injury resulting from inhalation exposure depends upon the toxicity of the material and its solubility in tissue fluids, the depth of respiration, and the amount of blood circulation. To prevent inhalation exposure, adequate ventilation must be supplied.

The American Conference of Governmental Industrial Hygienists (ACGIH) produces annual lists of threshold limit values (TLVs) and short-term exposure limits (STELs) for common chemicals used in laboratories. These values represent conditions in which nearly all workers can be exposed to without adverse health effects.

Ingestion

Many chemicals used in the laboratory are extremely dangerous if they enter the mouth and are swallowed. To prevent entry of toxic chemicals into the mouth, laboratory workers must wash their hands before eating or smoking, immediately after use of any toxic substance, and before leaving the laboratory. Chemicals should not be tasted, nor should pipetting ever be done by mouth. Of course, smoking, eating, and drinking are prohibited in any lab environment and should only be performed in approved areas.

Skin and Eye Absorption

Contact with the skin is a common mode of chemical injury. Chemicals enter the skin through hair follicles, sweat glands, and cuts or abrasions on outer layers of the skin. Skin contact can be prevented by using appropriate protective equipment. Eye contact is a common route of entry when chemicals are splashed. Use proper protective eyewear to avoid eye contact with chemicals.

Injection

Exposure to chemicals through injection is not a common exposure route. However, it is possible through mechanical injury from glass or metal contaminated with chemicals. Safe work practices are the best preventative measure for avoiding exposure through injection. Broken glassware should never be picked up by hand; mechanical devices such as a dust pan and broom or forceps are recommended.

Smoking

Smoking is prohibited in all laboratory areas.

TRAINING

The employer should make provisions for informing and training employees about potential health hazards and measures to protect themselves when working with chemicals in the laboratory environment. Training must comply with OSHA's laboratory Standard (Title 29 CFR 1910.1450), OSHA's Hazard Communication Standard (Title 29 CFR 1910.1200), and state laws.

All new employees should receive instruction, such as audio-visual material and classroom training, within the first month of employment. First-line supervisors should be extensively trained regarding hazards and appropriate protective measures so they will be able to answer employees' questions and provide daily monitoring of safe work practices. As new hazards are introduced, additional training must be provided.

The training program should emphasize these items:

- Summary of the hazard communication standard and the laboratory standard as well as the written training program.
- Chemical and physical properties of hazardous materials (e.g. flash point, reactivity, etc.) and methods that can be used to detect the presence or release of chemicals, including chemicals in unlabeled pipes.
- Physical hazards of chemicals (e.g., potential for fire, explosion, etc.).
- Health hazards, including signs and symptoms of exposure and any medical condition aggravated by exposure to the chemical.
- Procedures to protect against hazards (e.g., personal protective equipment required, proper use and maintenance, work practices, methods to assure proper use and handling of chemical, and procedures for emergency response).
- Work procedures to follow to assure protection when cleaning hazardous chemical spills and leaks.
- Where MSDSs are located, how to read and interpret the information on both labels and MSDSs, and how employees may obtain additional hazard information.
- OSHA's permissible exposure levels (PELs) and American Conference of Governmental Industrial Hygienists' threshold limit values (TLVs).
- OSHA's specific labeling requirements including the name of the chemical substance and health/physical hazards (if any).

WASTE HANDLING AND DISPOSAL

Laboratories are required by federal and state regulations to manage all hazardous waste in a specific manner.

General

All laboratory chemicals must be identified, and all containers with materials designated as hazardous waste should be visibly identified as such. Containers stored outside should be labeled to withstand the elements. Only chemically compatible containers of sufficient strength should be used for waste. The containers should be kept covered at all times and should be arranged for easy access. This will ensure that containers will not be damaged during handling. Care should be taken during all handling to maintain the integrity of the container.

Petroleum Hydrocarbons

Petroleum hydrocarbons may constitute part of the laboratory waste stream. Each type of container used for such substances is disposed of in a different manner.

WORKING WITH SPECIFIC CHEMICAL CLASSES

Corrosives

Acids

General Information. An acid is a substance that yields hydrogen ions when dissolved in water. Acids commonly encountered in laboratory environments include:

- Hydrochloric acid (HCl)
- Nitric acid (HNO_3)
- Hydrofluoric acid (HF)
- Sulfuric acid (H_2SO_4)

Safety Precautions. Use acids under well-ventilated conditions. Prevent all contact of vapor or liquid with skin, eyes, and mucous membranes. Personal protective equipment should include safety glasses, goggles, or a face shield. Contact lenses must NOT be worn. Full-length sleeves and appropriate gloves should be worn. Wear aprons made of acid-resistant material. Use respirators when required. When diluting acids with water, be sure to add the acid to the water, NOT the water to the acid. Be aware of the nearest safety shower and eye wash station before beginning work.

Accident Response. In case of an emergency exposure, dial the local or company emergency telephone number. Have appropriate neutralizers on hand that contain indicators to show when neutralization is complete. Have appropriate containment materials on hand in case of large spills. Notify a supervisor of all exposures or spills.

Transfer and Transport. Use appropriate personal protective equipment as mentioned above, check chemical compatibility of transfer pumps and receiving vessels. When transporting small quantities, use bottle carriers to prevent breakage and to act as a containment vessel should breakage occur.

Storage. Acids should be segregated, preferably in separate cabinets, from active metals, oxidizing acids, organic acids, flammables and combustibles, bases, and chemicals that react with acids to form toxic gases.

Bases/Caustics

General Information. Bases are substances that yield hydroxyl ions when dissolved in water. Bases typically encountered under laboratory conditions include:

- Potassium hydroxide (KOH),
- Sodium hydroxide (NaOH), and
- Ammonia (NH_3).

Safety Precautions. Bases are extremely corrosive and should be handled in a manner similar to acids. Prevent all contact of vapor or liquid with skin, eyes, and mucous membranes. Personal protective equipment should include safety glasses, goggles, or a face shield. Contact lenses must NOT be worn. Full-length sleeves and appropriate gloves should be worn. Wear aprons made of base-resistant material. Wear respirators if necessary. Be aware of the nearest safety shower and eye wash station before beginning work.

Accident Response. In case of an emergency exposure, dial the local or company emergency telephone number. Have appropriate neutralizers on hand that contain indicators to show when neutralization is complete. Have appropriate containment materials on hand in case of large spills. Notify a supervisor of all exposures or spills.

Transfer and Transport. Use appropriate personal protective equipment, and ascertain chemical compatibility of transfer pumps and receiving vessels. When transporting small quantities, use bottle carriers to prevent breakage and to act as a containment vessel should breakage occur.

Storage. Don't store basics or caustics with acids. A reaction would generate large quantities of heat. Because most bases are non-volatile, special storage cabinets are not necessary.

Oxidizers

General Information. These compounds will react violently with flammables and combustibles, and should be separated from these substances as well as from reducing agents. Oxidizers often seen in laboratories include:

- Perchloric acid ($HClO_4$), and
- Chromic acid (CrO_3).

Safety Precautions. Unintentional contact with organic and other oxidizable substances should be avoided. Reaction vessels containing significant quantities of these reagents should be heated using fiberglass mantles or sand baths rather that oil baths. Prevent all contact of vapor or liquid with skin, eyes, and mucous membranes. Personal protective equipment should include safety glasses, goggles, or a face shield. Contact lenses must NOT be worn. Full-length sleeves and appropriate gloves should be worn. Wear aprons of acid-resistant material. Use respirators when required. Be aware of the nearest safety shower and eye wash station before beginning work.

Accident Response. In case of an emergency exposure, dial the local or company emergency telephone number. Have appropriate neutralizers on hand that contain indicators to show when neutralization is complete. Have appropriate containment on hand in case of large spills. Notify a supervisor of all exposures or spills.

Transfer and Transport. Use the appropriate personal protective equipment, and check chemical compatibility of transfer pumps and receiving vessels. When transporting small quantities, use bottle carriers to prevent breakage and act as a containment vessel should breakage occur.

Storage. Oxidizers should not be stored with organics or other oxidizable compounds because they present fire and explosion hazards. They should be stored in glass or other inert containers, preferably unbreakable. Corks and rubber stoppers should NOT be used.

Toxics

Embryotoxins

General Information. Embryotoxins are substances that can act during pregnancy to cause adverse effects on the fetus, including death, malformation, retarded growth, and postnatal functional deficits. These substances also may be called "teratogens." Examples of these substances include organomercurials, lead compounds, and formaldehyde.

Safety Precautions. The period of greatest susceptibility is the first eight to 12 weeks of pregnancy. This includes a period when a woman may not know she is pregnant; therefore, special precautions must be taken at all times. Review proper use of these materials, and review continuing uses annually or when a procedural change is made. Women with childbearing potential should take adequate precautions to guard against spills and splashes. Operations should be carried out using impermeable containers and in adequately ventilated areas. Appropriate personal protective equipment should be worn, especially gloves. Hoods should be inspected for proper functioning before work begins.

Notify the company medical department of pregnancy.

Accident Response. In case of an emergency exposure, dial the local or company emergency telephone number. Notify a supervisor of all exposures or spills.

Transfer and Transport. Use the appropriate precautions whenever transferring or transporting embryotoxins to eliminate exposure.

Storage. Store these properly labeled substances in a well-ventilated area, in unbreakable outer containers.

Allergens

General Information. Allergens are substances that produce skin and lung hypersensitivity. Examples include diazomethane, chromium, nickel, bichromates, formaldehyde, isocyanates, and certain phenols. There is a wide variety of response from one individual to another, so contact should be avoided.

Safety Precautions. Wear suitable gloves and clothing to prevent skin contact with allergens or substances of unknown allergenic potential. Work with adequate ventilation.

Accident Response. In case of an emergency exposure, dial the local or company emergency telephone number. Treat exposure to allergens as you would any chemical. If signs of an allergic reaction appear, get medical attention. Avoid contact with these chemicals when cleaning up spills. Notify a supervisor of any exposures or spill.

Transfer and Transport. Appropriate precautions should be taken to prevent skin contact. Treat with the same respect given to all chemicals.

Storage. No special storage procedures are necessary. Proper labeling should exist, with recognized allergens identified as such.

Chemicals of Chronic or Acute Toxicity

General Information. A chronically toxic substance is one that will cause damage after repeated or long-duration exposure, or will cause damage that becomes apparent a long time following exposure. An acutely toxic substance causes damage as a result of a single or short-duration exposure. Records should be maintained of the amounts of these materials on hand, amounts used, and the names of workers involved.

Safety Precaution. All the general recommendations stated earlier should be followed. In particular, all skin contact should be avoided through the use of gloves and proper apparel. When following procedures that may result in the generation of aerosols or vapors containing the substances, use a properly functioning hood or other suitable containment device. Trap these vapors to prevent their release along with exhaust. Always wash hands and arms immediately after working with these chemicals. Conduct work and mount apparatus above trays made of a chemically resistant material such as polyethylene, or cover work and storage areas with plastic-backed paper. Use and store these substances only in areas of restricted access with special warning signs. Thoroughly decontaminate or incinerate contaminated clothing or shoes. Assure that at least two people are present whenever working with highly toxic compounds or those of unknown toxicity.

Accident Response. In case of an emergency exposure, dial the local or company emergency telephone number. If a major spill occurs outside a hood, leave the area and contact the appropriate company officer. Spills should be cleaned up only by trained individuals wearing appropriate apparel. If a spill involves a toxicologically significant quantity, a full-face supplied-air respirator should be used. Notify a supervisor of any exposures or spills.

Transfer and Transport. Take normal precautions during transfer and transport to avoid exposure and breakage.

Storage. Store breakable containers in chemically resistant trays. Store contaminated waste in labeled impervious containers. For liquid waste, store in glass or plastic bottles half-filled with vermiculite.

Chemicals of High Chronic Toxicity

General Information. These substances include, but are not limited to, certain heavy metals, their derivatives, and potent carcinogens.

Safety Precautions. Use of these chemicals should occur only in a controlled area—a laboratory, or portion of a laboratory or a facility as under an exhaust hood designated for use of highly toxic substances. The controlled area must be conspicuously marked with warning and restricted-access signs. Use of this area is not limited to toxic substances, but all personnel who have access to it must be aware of the substances being used and the necessary precautions.

Protect vacuum pumps against contamination by scrubbers or high-efficiency particulate arrestance (HEPA) filters and vent them into the hood. Decontaminate vacuum pumps or other contaminated equipment, including glassware, in the hood before removing them from the controlled area. Decontaminate the work area before normal work is resumed.

On leaving a controlled area, remove any protective apparel, place it in an appropriate labeled waste container, and thoroughly wash hands, forearms, face, and neck.

Accident Response. In case of an emergency exposure, dial the local or company emergency telephone number. Contingency equipment and materials to minimize exposure of people and property in case of accident should be readily available. Use chemical decontamination whenever possible. If dry powder is used, use wet mops or a vacuum cleaner with a high-efficiency particulate arrestance filter instead of dry sweeping. Ensure that containers of contaminated waste, including washings from contaminated flasks, are transferred from the controlled area in a secondary container under the supervision of authorized personnel.

Transfer and Transport. Conduct all transfers in the controlled area. Make sure all materials are properly labeled before being transported.

Storage. Make certain that all containers have the appropriate identification and warning labels. Store containers of these chemicals only in ventilated limited-access areas in labeled, unbreakable, chemically resistant secondary containers. Keep accurate records of the amounts of these substances stored and used, the dates of use, and the names of all users.

Flammables and Explosives

Flammables

General Information. Flammable substances are those that catch fire readily and burn in air. Examples include acetone, benzene, ethanol, xylene, methanol, and toluene.

Safety Precautions. These substances should only be handled in areas free of ignition sources. They should never be heated by using an open flame. Use a steam, water, oil, or air bath, or a heating mantle. Use adequate ventilation to prevent the formation of flammable mixtures.

Accident Response. In case of an emergency exposure, dial the local or company emergency telephone number. Follow the appropriate first aid procedures. Clean up small spills promptly using the appropriate materials and personal protective equipment. For large spills, evacuate the area and notify the appropriate company officer.

Transfer and Transport. When transferring flammables in metal containers, static-generated sparks should be avoided by bonding and the use of ground straps. Follow all applicable Department of Transportation (DOT) and Environmental Protection Agency (EPA) regulations pertinent to the transport of flammables.

Storage. Store these materials in proper cans as approved by the National Fire Protective Association (NFPA). These substances should be stored in special cabinets designed for this purpose.

Highly Reactive Chemicals and Explosives

General Information. This class of compounds includes peroxide-forming compounds and explosives. These react with oxygen present in the atmosphere to form peroxides. Peroxides are unstable, and there are risks of explosion. The concentration plays an important role, and can change through evaporation and distillation processes. Heat, shock, and friction can create dangerous situations and lead to explosions. Classes that can form peroxides include aldehydes, esthers, most alkenes, and vinyl and vinylidene compounds. Specific chemicals include cyclohexene, cyclooctene, decalin, p-dioxane, diethyl ether, diisopropyl ether, tetrahydrofuran (THF), and tetralin. Aging is a significant factor in the production of peroxides. These compounds often contain additives to prevent the formation of peroxides; however, the addition of additives does not eliminate the hazard, it only delays formation.

Safety Precautions. Explosives should be purchased in small quantities, not stockpiled. Unused peroxides should not be returned to the original container. The sensitivity of most peroxides to shock and heat can be reduced by dilution with inert solvents, such as aliphatic hydrocarbons. Solutions of peroxides in volatile solvents should be handled to prevent evaporation of the solvent, which will increase the peroxide concentration. Metal spatulas should not be used to handle peroxides because metal contamination can lead to explosive decomposition. Ignition sources should not be permitted in the area. Friction, grinding, and other forms of impact should be avoided.

Accident Response. In case of an emergency exposure, dial the local or company emergency telephone number. All spills should be cleaned up immediately. Absorb liquids with vermiculite.

Transfer and Transport. Small quantities of peroxides should be handled to avoid ignition sources, shock, and extreme temperature changes. Follow all applicable DOT and EPA regulations pertinent to transport of these materials.

Storage. Peroxide-formers should be stored in airtight containers in a cool, dry, dark place. Glass containers that have screw tops or glass stoppers should not be used; Polyethylene should be used. To minimize the rate of decomposition, peroxides should be stored at the lowest temperature consistent with their solubility or freezing point, but not lower because they become more sensitive to shock and heat. They should be properly labeled with the receiving date, the opening date, and the date recommended for disposal. These chemicals should not be stored for long periods of time and should be discarded after a given period if not used.

Pyrophorics

General Information & Storage Requirements. These are liquids or solids that spontaneously ignite in air at temperatures less than 130°F. This includes oily rags, dust accumulations, organics mixed with strong oxidizers, and alkali metals such as sodium, potassium, and phosphorus. These substances should be treated with the same respect accorded all chemicals. Their storage requirements, however, require noting. These substances must be stored in inert atmospheres or under kerosene.

Compressed Gases

Compressed gases present the potential for exposure to both chemical and mechanical hazards depending on the particular gas. If the gas in question is flammable, flash points lower than room temperature, compounded by high rates of diffusion, present fire and explosion hazards. There also are reactivity and oxygen displacement considerations. The large amount of potential energy present in the pressure used to compress gas makes for a potential rocket or fragmentation bomb. This creates the need for special handling procedures for compressed gases, the cylinders used to contain them, and the regulators and piping used to control and direct the flow.

General Information

The DOT defines a compressed gas as "any material or mixture having in the container either an absolute pressure greater than 276 kPa (40 lbf/in^2) at 21°C or an absolute pressure greater than 717 kPa (104 lbf/in^2) at 54°C or both, or any liquid flammable material having a Reid vapor pressure greater than 276 kPa (40 lbf/in^2) at 38°C."

Safety Precautions

The contents of any compressed gas should be clearly identified on the cylinder. All gas lines should be clearly labeled to identify the gas being transported. The labels should be color-coded to distinguish hazardous gases. Signs identifying flammable, compressed gases should be clearly posted.

Cylinders should be firmly secured at all times using a clamp and belt or chain. Pressure-release equipment for protecting devices attached to cylinders containing potentially hazardous gases should be vented to a safe place.

Cylinders should be placed in such a way that the cylinder valve is readily accessible at all times. The main cylinder valve should be closed whenever the gas is not in use. This is not only necessary for safety reasons, but also to prevent contamination and corrosion in empty cylinders from the diffusion of air and moisture into the cylinder.

The proper tools should be used on cylinder hardware. Pliers should not be used. Valves should be opened slowly, and it is never necessary to open the main valve all the way. When opening a cylinder containing a toxic gas, stand upwind and to the side. Be aware of the location of fellow workers in case a leak exists.

Do not use the common brass pressure regulators with corrosive gases such as ammonia, boron trifluoride, chlorine, hydrogen chloride, hydrogen sulfide, and sulfur dioxide; special corrosion-resistant regulators should be used. Regulators used with carbon dioxide should have special internal designs and special materials to prevent freeze-up and corrosion problems.

All pressure regulators should be equipped with spring-loaded pressure-relief valves. When used for hazardous gases of any type, these valves should be properly vented. Do not use internal bleed-type regulators.

Sparks and flames should be kept from the area of flammable gas cylinders. All piping, regulators, appliances, and hoses should be kept tightly sealed and in good condition. Equipment used for flammables should not be interchanged with similar equipment used for other gases.

Cylinders should not be emptied less than 172 kPa (25 lbf/in^2) because the residuals may become contaminated if the valve is left open. Empty cylinders should not be refilled; the top valve should be closed, the regulator should be removed, and the valve cap replaced. Mark all empty cylinders "MT" or "EMPTY."

All pressure equipment should be inspected periodically, more often where corrosive or hazardous gases are used.

Storage

Cylinders containing flammables and other hazardous gases should be securely stored in a well-ventilated area. Cylinders of oxygen never should be stored with cylinders containing flammables. Do not store empty and full cylinders in the same area.

Light-Sensitive Chemicals

General Information Storage Requirements

The nature of these compounds is self-explanatory. They should be treated with respect, as all chemicals, and stored in the following manner. These compounds degrade upon exposure to light and should be stored in amber containers.

Water-Reactive Chemicals

General Information/Storage Requirements

These chemicals can lead to the formation of flammable or toxic gases following contact with water. Areas where these compounds are present should be posted in such a way that firefighters are aware of their presence. They should be stored in polyethylene bags, in tightly sealed containers.

ACTIVITIES THAT WARRANT APPROVAL BEFORE IMPLEMENTATION

In the interest of integrating safe laboratory practices, whenever a particular procedure, operation, or activity is to involve the use of carcinogens, reproductive toxins, and/or substances with a high degree of acute toxicity, safety personnel must be informed during the planning stages. It is the responsibility of the project planner to investigate the toxicity of materials of interest and to inform a supervisor in writing of the plans. Prior approval must be given before beginning work with hazardous materials.

Examples of activities and operations requiring prior approval before implementation are given below:

- Any new or radically modified experiment.
- Any switch in chemicals from less to more hazardous.
- Any new activity that involves the use of a chemical falling in the specific chemical classes described earlier.
- Any activity using a toxic substance for an extended period of time. This includes storage as well as handling times. This will limit the need for evaluating operations involving onetime use of a chemical.
- Any activity using a substance that requires special disposal or storage.

DESIGNATED WORK AREAS

Designated work areas should be established for work involving carcinogens, reproductive toxins, and/or substances with a high degree of acute toxicity. Laboratory hoods should be identified throughout the area for such work. Designated hoods may be labeled as follows: THIS HOOD TO BE USED FOR WORK WITH CARCINOGENS, REPRODUCTIVE TOXINS, AND SUBSTANCES WITH A HIGH DEGREE OF ACUTE TOXICITY.

CHAPTER 11

PERSONAL SAMPLING FOR AIR CONTAMINANTS

John P. Neglia

Becton Dickinson and Co.
Franklin Lakes, NJ 07417, USA

CONTENTS

INTRODUCTION

Unnecessary air sampling can tie up laboratory resources and produce delays in reporting results of necessary sampling. Evaluate the potential for employee overexposure through observation and screening samples before any partial or full-shift sampling is conducted. Do not overexpose the employee to gather a sample.

Screening with portable monitors, gravimetric sampling, or detector tubes can be used to evaluate the following:

- Processes, such as electronic soldering.
- Exposures to substances with exceptionally high permissible exposure limits (PELs) in relatively dust-free atmospheres, e.g., ferric oxide and aluminum oxide.
- Intermittent processes with substances without short-term exposure limits (STELs).

- Engineering controls, work practices, or isolation of process.
- The need for protection.

Take a sufficient number of samples to obtain a representative estimate of exposure. Contaminant concentrations vary seasonably, with weather, with production levels, and in a single location or job class.

The number of samples taken depends on the error of measurement and differences in results. If the employer has conducted air sampling and monitoring in the past, review the records.

Bulk samples often are required to assist the laboratory in the proper analysis of field samples. Some contaminants that fall into this category include:

- Silica,
- Portland cement,
- Asbestos,
- Mineral oil and dust,
- Chlorodiphenyl,
- Hydrogenated terphenyls, and
- Chlorinated camphene.

GENERAL SAMPLING PROCEDURES

- Screen the sampling area using detector tubes, if appropriate. Determine the appropriate sampling technique. Prepare and calibrate the equipment and prepare the filter media.
- Select the employee to be sampled and discuss the purpose of the sampling. Inform the employee when and where the equipment will be removed. Stress the importance of not removing or tampering with the sampling equipment. Turn off or remove sampling pumps before an employee leaves a potentially contaminated area (such as when he/she goes to lunch or on a break).
- Instruct the employee to notify the supervisor if the sampler requires temporary removal. Place the sampling equipment on the employee so that it does not interfere with work performance.
- Attach the collection device (filter cassette, charcoal tube, etc.) to the shirt collar or as close as practical to the nose and mouth of the employee. Employee exposure is the exposure that would occur if the employee were not using a respirator.
- The inlet always should be in a downward vertical position to avoid gross contamination. Position the excess tubing so it will not interfere with the employee's work.
- Turn on the pump and record the starting time.
- Observe the pump operation for a short time after starting to make sure it is operating correctly.
- Record the information required.

- Check pump status every two hours. More frequent checks may be necessary with heavy filter loading. Ensure that the sampler is still assembled properly and that the hose has not become pinched or detached from the cassette or the pump. For filters, observe for symmetrical deposition, fingerprints, or large particles, etc. Record the flow rate, if possible.
- Periodically monitor the employee throughout the workday to ensure that sample integrity is maintained and cyclical activities and work practices are identified.
- Take photographs, as appropriate, and detailed notes concerning visible airborne contaminants, work practices, potential interferences, movements, and other conditions to assist in determining appropriate engineering controls.
- Prepare a blank(s) during the sample period for each type of sample collected. For any given analysis, one blank will suffice for up to 20 samples collected. These blanks may include opened but unused charcoal tubes, and so forth.
- Before removing the pump at the end of the sample period, check the flow rate to ensure that the rotameter ball is still at the calibrated mark (if there is a pump rotameter). If the ball is no longer at the mark, record the pump rotameter reading.
- Turn off the pump and record the ending time.
- Remove the collection device from the pump and seal it as soon as possible. The seal should be attached across sample inlet and outlet so that tampering is not possible (see Figures 1 and 2).
- Recalibrate pumps after each day of sampling (before charging).

Figure 1. Improperly sealed cassette allows access to inlet and outlet ports after sample has been taken.

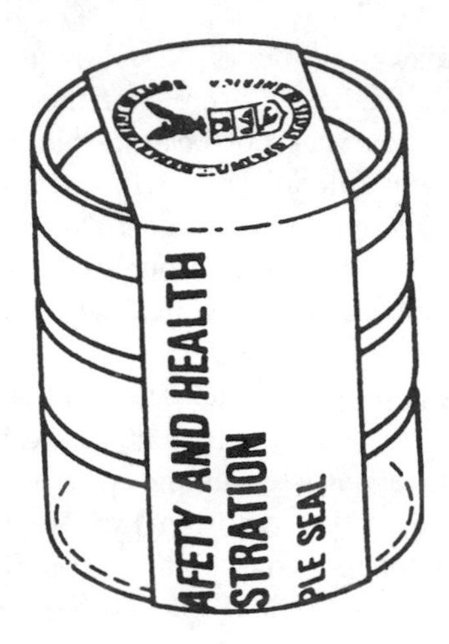

Figure 2. Properly sealed cassette covering inlet and outlet ports provides for sample security.

- For unusual sampling conditions, such as wide temperature and pressure differences from the calibration conditions, seek technical support.

SAMPLING TECHNIQUES

Detector Tubes

- Each pump should be leak-tested before use.
- Calibrate the detector tube pump for proper volume at least quarterly or after 100 tubes.

Total Dust and Metal Fume

- Collect total dust on a pre-weighed, low-ash polyvinyl chloride filter at a flow rate of about 2 liters per minute (lpm), depending on the rate required to prevent overloading.
- Collect metal fumes on a 0.8 micron mixed cellulose ester filter at a flow rate of approximately 1.5 lpm, not to exceed 2.0 lpm. Do **not** collect metal fumes on a low-ash polyvinyl chloride filter.
- Take care to avoid any overloading of the filter, as evidenced by any loose particulate.
- Calibrate personal sampling pumps before and after each day of sampling, using a bubble meter method (electronic or mechanical) or the precision rotameter method (calibrated against a bubble meter).
- Weigh PVC filters before and after taking the sample.

Respirable Dust

- Collect respirable silica dust using a clean cyclone equipped with a weighted low-ash polyvinyl chloride filter at a flow rate of 1.7 lpm.
- Collect silica only as a respirable dust. A bulk sample should be submitted to a laboratory for analysis.
- All filters used shall be pre-weighed and post-weighed.
- Calibration Procedures
  1. Do the calibration at the pressure and temperature where the sampling is to be conducted.
  2. For respirable dust sampling using a cyclone or an open-face filter, set up the calibration apparatus as shown in Figure 3.

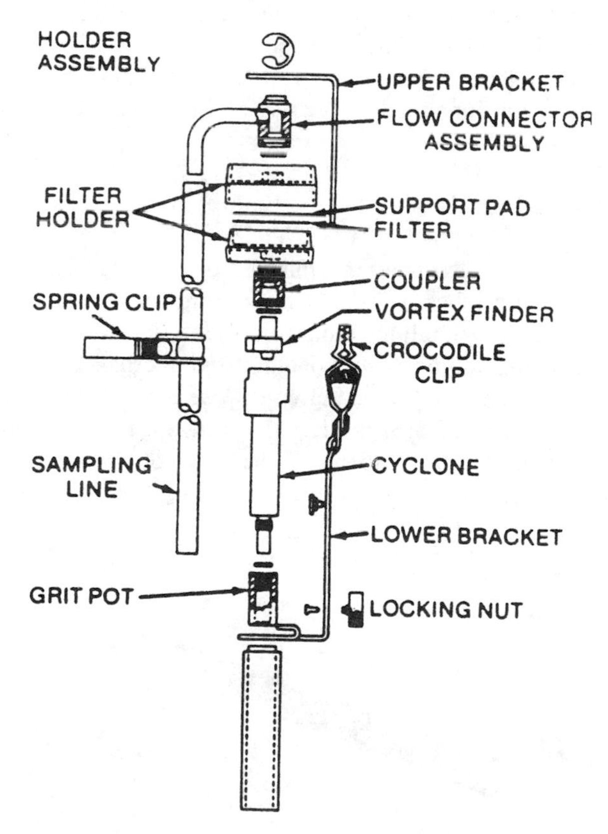

Figure 3. Proper assembly of calibration apparatus for respirable dust sampling using an open-face filter and cyclone.

3. Replace the filter cassette with a 1-liter jar containing the cassette holder assembly and cyclone or the open-face filter cassette. The jar is provided with a special cover.
4. Connect the tubing from the electronic bubble meter to the inlet of the jar.
5. Connect the tubing from the outlet of the cyclone holder assembly or from the filter cassette to the outlet of the jar and then to the sampling pump.
6. Calibrate the pump. The calibration readings must be within 5% of each other.

Cyclone Cleaning

- Clean the cyclone thoroughly after each use to prevent excess wear or damage and to prevent contamination of a sample. Inspect the cyclone after cleaning for signs of wear or damage, such as scoring. Replace the unit if it appears damaged.
- Gently clean the interior, avoid scoring the interior surfaces. Never insert anything into the cyclone during cleaning.
- Leak-test the cyclone at least once a month with regular usage.
- Detailed instructions on cyclone cleaning and leak-testing are available from the manufacturer.

Organic Vapors and Gases

- Organic vapors and gases may be collected on activated charcoal, silica gel, or other adsorption tubes using low-flow pumps.
- Immediately before sampling, break off the ends of the charcoal tube to provide an opening approximately one-half the internal diameter of the tube. Wear eye protection when breaking ends. Use tube holders, if available, to minimize the hazards of broken glass. Do **not** use the charging inlet or the exhaust outlet of the pump to break the ends of the charcoal tubes (see Figure 4).

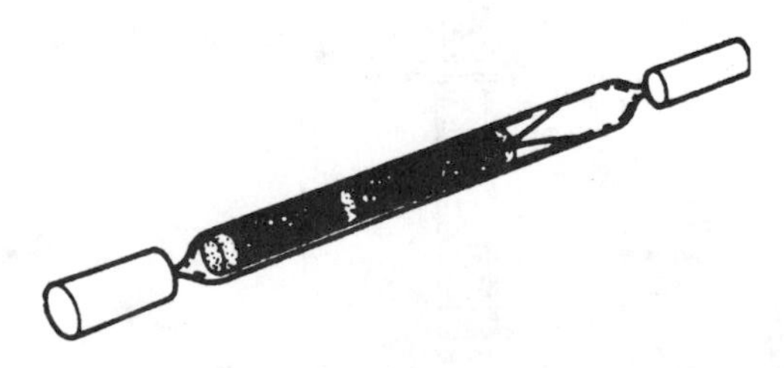

Figure 4. A charcoal of "C" tube with sealed glass ends and NIOSH-approved caps before taking a sample.

- Use the smaller section of the charcoal tube as a backup and position it near the sampling pump. The charcoal tube shall be held or attached in an approximately vertical position with the inlet either up or down during sampling.
- Draw the air to be sampled directly into the inlet of the charcoal tube and not be passed through any hose or tubing before entering the pump tubing.
- Cap the charcoal tube with the supplied plastic caps immediately after sampling and seal as soon as possible (see Figures 5 and 6).
- For other adsorption tubes, follow the same procedures as those for the charcoal tube, with the following exceptions:
  1. Tubes may be furnished with either caps or flame-sealed glass ends. If using the capped version, simply uncap during the sampling period and recap at the end of the sampling period.

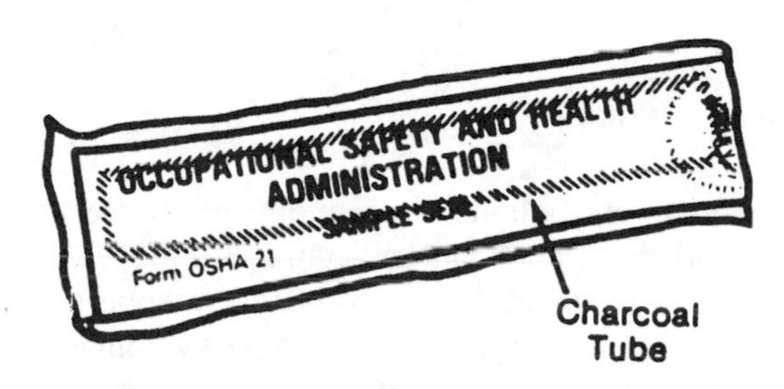

Figure 5. A correctly sealed "C" tube. Sample is completely enclosed in the seal, and no tampering is possible.

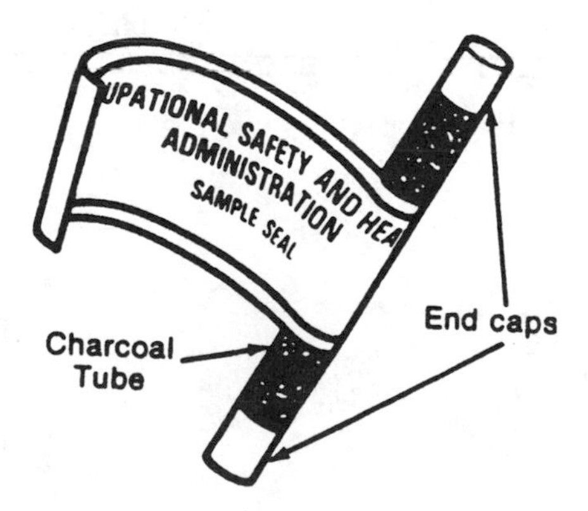

Figure 6. An incorrectly sealed "C" tube. End caps can be removed and sample integrity jeopardized without disturbing seal.

2. The ends of the flame-sealed glass tubes are broken at the beginning of the sampling period and capped at the end of the sampling period. Do not ship with bulk material.

- For organic vapors and gases, low-flow pumps are required.
- Flow rates may have to be lowered or smaller-volume (1/2 the maximum) sorbent tubes used when there is high humidity (above 90%) in the sampling area or relatively high concentrations of other organic vapors.
- Calibration Procedures
  1. Set up the calibration apparatus as shown in Figure 7 replacing the cassette with the solid sorbent tube to be used in the sampling (e.g., charcoal, silica gel, etc.). If a sampling protocol requires the use of two charcoal tubes, then the calibration train must include two charcoal tubes The airflow must be in the direction of the arrow on the tube.
  2. Calibrate the pump.

Midget Impingers/Bubblers

- Take care in preparing bubblers and impingers to see that frits or tips are not damaged and that joints can be securely tightened.
- Rinse the impinger/bubbler, Figure 8, with the appropriate reagent. Then, add the specified amount of this reagent to the impinger flask either in the office or at the sampling location. If flasks containing the reagent are transported, caps must be placed on the impinger stem and side arm. To prevent overflowing, do not add more than 10 ml of liquid to the midget impingers.
- Collect contaminants in an impinger at a maximum flow rate of 1.0 lpm.

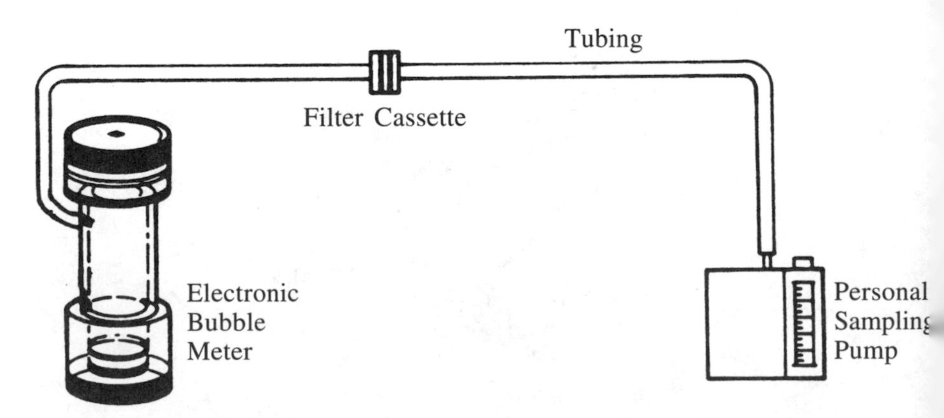

Figure 7. Proper assembly of calibration apparatus using a cassette attached to an electronic bubble meter.

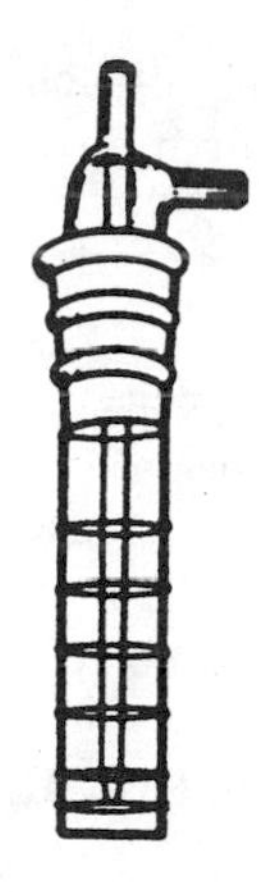

Figure 8. A typical glass bubbler.

- The impinger may either be hand-held by the industrial hygienist or attached to the employee's clothing using an impinger hoister. In either case, it is very important that the impinger does not tilt, causing the reagent to flow down the side arm to the hose and into the pump. (NOTE: Attach a trap using an empty impinger in line to the pump.)
- In some instances, it will be necessary to add additional reagent during the sampling period to prevent the amount of reagent from dropping below one-half of the original amount.
- After sampling, remove the glass stopper and stem from the impinger flask.
- Rinse the absorbing solution adhering to the outside and inside of the stem directly into the impinger flask with a small amount (1 or 2 ml) of the sampling agent. Stopper the flask tightly with the plastic cap provided or pour the contents of the flask into a 20 cc. glass bottle. Rinse the flask with a small amount (1 or 2 ml) of the reagent and pour the rinse solution into the bottle. Tape the cap shut to prevent it from loosening. If electrical tape is used, do not stretch tape because it will contract and loosen the cap.

Calibration Procedure

- Set up the calibration apparatus as shown in Figure 7, replacing the cassette with the impinger/bubbler filled with the amount of liquid reagent specified in the sampling method.
- Connect the tubing from the electronic bubble meter to the inlet of the impinger/bubbler.

- Connect the outlet of the impinger/bubbler to the tubing to the pump.
- Calibrate the pump at a maximum flow rate of 1.0 lpm.

Vapor Badges

- Passive diffusion sorbent badges are useful for screening and monitoring certain chemical exposures, especially vapors and gasses.
- Badges are available to detect mercury, nitrogen oxides, ethylene oxide, formaldehyde, etc.
- Interfering substances should be noted.

SPECIAL SAMPLING PROCEDURES/ASBESTOS

- Collect asbestos on a special 0.8 μm pore size, 25 mm diameter mixed cellulose ester filter, using a backup pad.
- Use fully conductive cassette with conductive extension cowl, Figure 9.
- Sample open face in worker's breathing zone.
- Assure that the bottom joint (between the extension and the conical black piece) of the cassette is sealed tightly with a shrink band of electrical tape. Point the open end of the cassette down to minimize contamination.

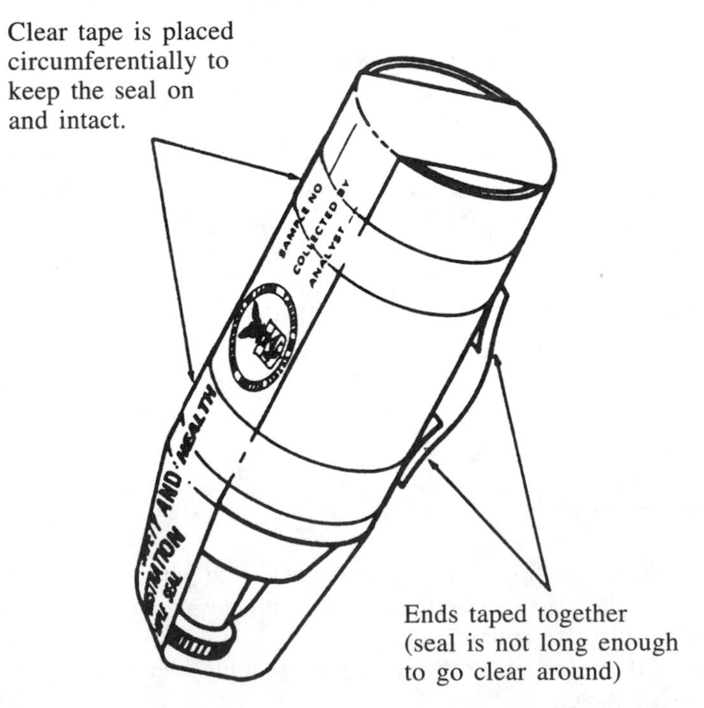

Figure 9. A standard asbestos cassette sealed properly with an OSHA 21 form.

- Use a flow rate in the range of 0.5 to 2.5 lpm. One liter per minute is suggested for general sampling. In an office environment, allow flow rates of up to 2.5 lpm. Calibrate pump before and after sampling.
- Sample for as long a time as possible without overloading (obscuring) the filter.
- Submit at the most 10 blanks, with a minimum in all cases of 2 blanks.
- Where possible, collect a bulk sample of the material suspected to be in the air.
- Mail bulks and air samples separately to avoid cross-contamination. Pack the samples securely to avoid any rattle or shock damage. Do not use expanded polystyrene packing material; Use bubble sheeting as packing. Put identifying paperwork in every package. Do not send samples in plastic bags or in envelopes. Print legibly on all forms.
- Instruct the employee to avoid knocking the cassette and to avoid using a compressed air source that might dislodge the sample.

EQUIPMENT PREPARATION AND CALIBRATION

- Replace alkaline batteries frequently (once a month). Also carry fresh replacement batteries with the equipment.
- Check the rechargeable Ni-Cad batteries **under load** (e.g., turn pump on and check voltage at charging jack) before use.
- Calibrate personal sampling pumps before and after each day of sampling, using either the electronic bubble meter method or the precision rotameter method (calibrated against a bubble meter).

Electronic Flow Calibrators

- These units are high accuracy electronic bubble flowmeters that provide instantaneous airflow readings and a cumulative averaging of multiple samples.
- These calibrators measure the flow rate of gases and present the results as volume per unit of time.
- These calibrators should be used to calibrate all air sampling pumps.
- When a sampling train requires an unusual combination of sampling media (e.g., glass fiber filter proceeding impinger), the same media/devices should be in line during calibration.

Electronic Bubble Meter Method

- Allow the pump to run 5 minutes prior to voltage check and calibration.
- Assemble the polystyrene cassette filter holder, using the appropriate filter for the sampling method. Compress the cassette by using a mechanical press or other means of applying pressure. Use shrink tape around cassette to cover

joints and prevent leakage. If a cassette adaptor is used, care should be taken to ensure that it does not come in contact with the backup pad.

- NOTE: When calibrating with a bubble meter, the use of cassette adaptors can cause moderate to severe pressure drop at high flow rates in the sampling train, which will affect the calibration result. If adaptors are used for sampling, then they should be used when calibrating.
- CAUTION: Nylon adapters can restrict air flow due to plugging over time. Stainless steel adapters are preferred.
- Connect the collection device, tubing, pump and calibration apparatus as shown in Figures 7 or 10.
  1. Inspect all Tygon tubing connections.
  2. Wet the inside of the electronic flow cell with the supplied soap solution by pushing on the button several times.
  3. Turn on the pump and adjust the pump rotameter, if available, to the appropriate flow rate setting.
  4. Press the button on the electronic bubble meter. Visually capture a single bubble and electronically time the bubble. The accompanying printer will automatically record the calibration reading in liters per minute.
  5. Repeat until two reading are within 5%.
  6. While the pump is still running, adjust the pump, if necessary.
  7. Repeat the procedures described above for all pumps to be used for sampling. The same cassette and filter may be used for all calibrations involving the same sampling method.

Precision Rotameter Method

The precision rotameter, Figure 11 is a secondary calibration device. If it is to be used in place of a primary device such as a bubble meter, care must be

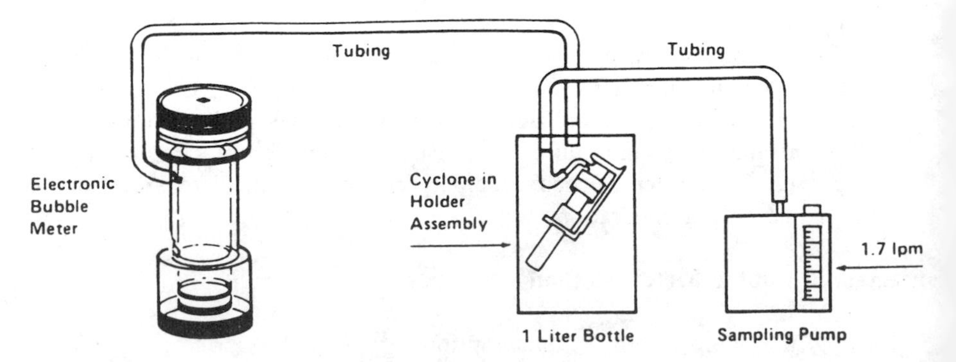

Figure 10. Proper assembly of calibration apparatus using a cyclone attached to an electronic bubble meter.

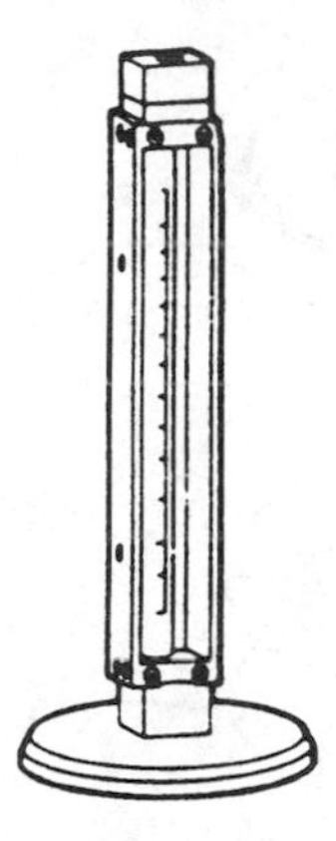

Figure 11. A single-column precision rotameter can be used as a secondary calibration device.

taken to ensure that any introduced error will be minimal and noted. The precision rotameter may be used for calibrating the personal sampling pump in lieu of a bubble meter provided it is:

- Calibrated with an electronic bubble meter or a bubble meter on a regular basis (at least monthly).
- Disassembled, cleaned as necessary, and recalibrated. It should be used with care to avoid dirt and dust contamination which may affect the flow.
- Not used at substantially different temperature and/or pressure from those conditions present when the rotameter was calibrated against the primary source.
- Used such that pressure drop across it is minimal.
- If altitude or temperature at the sampling site are substantially different from the calibration site, it is necessary to calibrate the precision rotameter at the sampling site where the same conditions are present.

FILTER WEIGHING PROCEDURE

The step-by-step procedure for weighing filters depends on the make and model of the balance. Consult the manufacturer's instruction book for directions. In addition, follow these guidelines:

- There shall be **no smoking or eating** in the weighing area. All filters will be handled with tongs or tweezers. **Do not handle the filters with bare hands.**
- Desiccate all filters at least 24 hours before weighing and sampling. Change desiccant before it completely changes color (e.g., before blue desiccant turns all pink). Evacuate desiccator with a sampling or vacuum pump.
- Zero the balance prior to use.

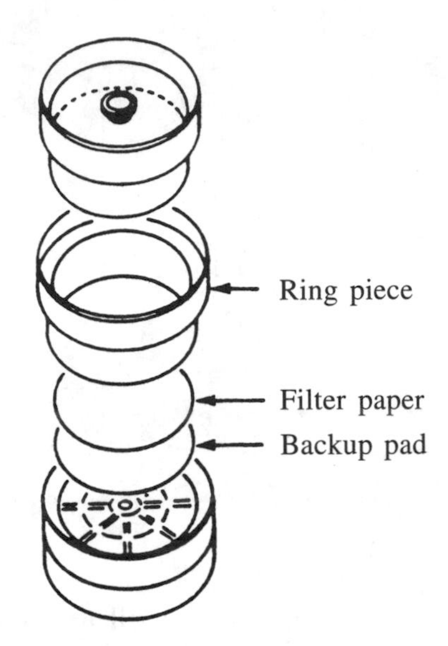

Figure 12. An exploded view of a three-piece cassette showing placement of backup pad.

- Calibrate the balance prior to use and after every 10 samples.
- Immediately prior to placement on the balance, pass all filters over an ionization unit to remove static charges. (Return the unit after 12 months of use to the distributor for disposal.)
- Weigh all filters at least twice.
  - If there is more than 0.005 mg difference in the two weighings, repeat the zero and calibration and reweigh the filter.
  - If there is less than 0.005 mg difference in the two weighings, average the weights for the final weight.
- Record all the appropriate weighing information (in ink) in the weighing log.
- In reassembling the cassette assembly, remember to add the unweighed backup pad (see Figure 12).
- When weighing the filter after sampling, include any loose material from an overloaded filter and cassette.
- NOTE: At all times, take care not to exert downward pressure on the weighing pan(s). Such action may damage the weighing mechanism.

BIBLIOGRAPHY

American Industrial Hygiene Association (AIHA). 1987. *Fundamentals of Analytical Procedures in Industrial Hygiene.* AIHA: Akron, OH.

Hesketh, H.E. 1986. *Fine Particles in Gaseous Media.* Lewis Publishers, Inc.: Chelsea, MI.

Lioy, P.J. 1983. *Air Sampling Instruments for Evaluation of Atmospheric Contaminants.* American Conference of Governmental Industrial Hygienists: Cincinnati.

Lodge, J.P., Jr., Ed. 1988. *Method of Air Sampling and Analysis.* Lewis Publishers, Inc.: Chelsea, MI.

Occupational Safety and Health Administration, U.S. Dept. of Labor. *Chemical Information Manual.* 1987. U.S. Government Printing Office: Washington, D.C.

CHAPTER 12

DETECTOR TUBES AND PUMPS

Randi Boyko

Cycle Chem Corp.
Linden, NJ 07036, USA

CONTENTS

INTRODUCTION

Detector tube pumps are portable equipment that, when used with a variety of commercially available detector tubes, are capable of measuring the concentrations of a wide variety of compounds in industrial atmospheres.

Operation consists of using the pump to draw a known volume of air through a detector tube designed to measure the concentration of the substance of interest. The concentration is determined by a calorimetric change of an indicator in the tube.

Most tubes can be obtained from local suppliers of safety equipment.

Applications/Limitations

Detector tubes and pumps are screening instruments used to measure more than 200 organic and inorganic gases and vapors. Some aerosol levels also can be measured.

Detector tubes of a given brand are to be used only with a pump of the same brand. The tubes are calibrated specifically for the same brand of pump and may give erroneous results if used with a pump of another brand.

A limitation of many detector tubes is the lack of specificity. Many indicators are not highly selective and can cross-react with other compounds. Manufacturer's manuals describe the effects of interfering contaminants.

Another important consideration is sampling time. Detector tubes give only an instantaneous interpretation of environmental hazards. This may be beneficial in potentially dangerous situations or when ceiling exposure determinations are sufficient. When long-term assessment of occupational environments is necessary, short-term detector tube measurements may not reflect time-weighted average levels of the hazardous substances present.

Detector tubes normally have a shelf life at 25°C of one to two years. Refrigeration during storage lengthens the shelf life. Outdated detector tubes (i.e., beyond the printed expiration date) never should be used. See Table 1 for shelf lives of common sampling media.

Performance Data

Specific manufacturer's models of detector tubes are listed in the manufacturer's literature. The specific tubes listed are designed to cover a concentration range that is near the personal exposure limit. Concentration ranges are tube-dependent and can be anywhere from one one-hundredth to several thousand parts per million. The limits of detection depend on the particular detector tube.

Accuracy ranges vary with each detector tube.

Pumps weighing 8 to 11 oz may be handled during operation. Automatic pumps, weighing about 4 lbs, collect a sample using a preset number of pump

Table 1
Shelf-Life of Sampling Media Provided By SLCAL

Sampling Medium	Shelf-Life	Comments
Hydrogen sulfide collecting solutions:		
NaOH	6 months	
$CdSo_4$-STRactan	one year	Solutions are not mixed until just prior to sampling.
Ozone sampling	6 months or whenever solution turns yellow, whichever period is shorter.	
Sodium hydroxide (all normalities)	6 months	
Hydrochloric acid Sulfuric acid Methanol in water	one year	Same for all concentrations of all solutions.
All organic solvents in pure state	4 years	
Bis-chloromethyl ether (BCME) and	2 months	Must be stored in a dark bottle in a refrigerator.
Hydroxylammonium chloride solutions		Should be stored in a refrigerator in a light-protected container.
Hydroxylammonium chloride-Sodium hydroxide mixed solutions (for acetic anhydride, ketene collection)	Stable only 2 hours	Must be prepared fresh just prior to use.
Hydrogen peroxide (0.3N) for sulfur dioxide collection	6 months	Stable if it is protected from light and refrigerated.
Girard Reagent	2 weeks	Store in glassware in the dark.
Folin's Reagent	5 days	Must be stored in a refrigerator.
Passive Monitors		Must be used before the expiration date (if given) printed on the monitor package.
Nitrogen oxides collection tubes		Should be stored in a refrigerator.

strokes. A full stroke for either type of short-term pump has a volume of about 100 cc.

In most cases where only one pump stroke is required, sampling time is about one minute. Determinations requiring more pump strokes take proportionately longer.

Leakage Test

Each day prior to use, perform a pump leakage test by inserting an unopened detector tube into the pump, and attempt to draw in 100 ml of air. After a few minutes, check for pump leakage by examining pump compression for bellows-type pumps or return to resting position for piston-type pumps. Automatic pumps should be tested according to the manufacturer's instructions.

In the event of leakage that cannot be repaired in the field, send the pump for repair.

Record the leakage test on the direct-reading data form.

Calibration Test

Calibrate the detector tube pump for proper volume measurement at least quarterly. Simply connect the pump directly to the bubble meter with a detector tube. Use a detector tube and pump from the same manufacturer.

Wet the inside of the 100 cc bubble meter with soap solution.

For volume calibration, experiment to get the soap bubble even with the 0 ml mark of the burette.

For piston-type pumps, pull the pump handle all the way out (full stroke) and note where the soap bubble stops; for bellows-type pumps, compress the bellows fully; for automatic pumps, program the pump to take a full stroke. For either type pump, the bubble should stop between the 95 cc and 105 cc marks. Allow four minutes for the pump to draw the full amount of air. This time interval varies with the type of detector tube being used in-line with the calibration setup.

Also check the volume for 50 cc (1/2 pump stroke) and 25 cc (1/4 pump stroke), if pertinent. A ±5% error is permissible. If error is greater than ±5% send the pump for repair and recalibration.

Record the calibration information required on the calibration log.

It may be necessary to clean or replace the rubber bung or tube holder if a large number of tubes have been used with the pump.

Additional Information

- **Draeger, Model 31 (bellows)**—When checking the pump for leaks with an unopened tube, the bellows should not be completely expanded for 10 minutes.

- **Draeger, Quantimeter 1000, Model 1 (automatic)**—A battery pack is an integral part of this pump. The pack must be charged prior to initial use. One charge is good for 1,000 pump strokes. During heavy use, it should be recharged daily. If a "U" (undervoltage) message is continuously displayed in the readout window of this pump, the battery pack immediately should be recharged.
- **Matheson-Kitagawa, Model 801 4-400A (piston)**—When checking the pump for leaks with an unopened tube, the pump handle should be pulled back to the 100 ml mark and locked. After two minutes, the handle should be released carefully. It should return to 0 mm or resting position. After taking 100 to 200 samples, the pump should be cleaned and relubricated. This involves removing the piston from the cylinder, removing the inlet and pressure-relief valve from the front end of the pump, cleaning, and relubricating.
- **Mine Safety Appliances, Samplair Pump, Model A, Part No.463998 (piston)**—The pump contains a flow-rate control orifice protected by a plastic filter that periodically needs to be cleaned or replaced. To check the flow rate, the pump is connected to a burette and the piston is withdrawn to the 100 ml position with no tube in the tube holder. After 24 to 26 seconds, 80 ml of air should be admitted to the pump. Every six months the piston should be relubricated with the oil provided.
- **Sensidyne-Gastec, Model 800, Part No.7010657-1 (piston)**—This pump can be checked for leaks as mentioned for the Kitagawa pump; however, the handle should be released after 1 minute. Periodic lubrication of the pump head, the piston gasket, and the piston check valve is needed. Lubrication frequency is determined by use.

Special Considerations

- Detector tubes should be refrigerated when not in use to prolong shelf life.
- Detector tubes should not be used when cold. They should be kept at room temperature or in a shirt pocket for one hour prior to use.
- Lubrication of the piston pump may be required if volume error is greater than 5%.

ELECTRONIC FLOW CALIBRATORS

Description

These units are high accuracy electronic bubble flowmeters that provide instantaneous airflow readings and a cumulative averaging of multiple samples. These calibrators measure the flow rate of gases and report volume per unit of time.

The timer is capable of detecting a soap film at 80 microsecond intervals. This speed allows under steady flow conditions an accuracy of ±0.5% of any display reading. Repeatability is ±0.5% of any display.

The range with different cells is from 1 cc/min to 30 lpm. Battery power will last 8 hours with continuous use. Charge for 16 hours. Electronic flow calibrators can be operated from an A/C charger.

Maintenance of Calibrator

Cleaning before Use

Remove the flow cell and gently flush with tap water. The acrylic flow cell is easily scratched. Wipe with cloth only. Do not allow center tube, where sensors detect soap film, to be scratched or get dirty. **Never** clean with acetone. Use only soap and warm water. When cleaning prior to storage, allow flow cell to air dry. If stubborn residue persists, it is possible to remove the bottom plate. Squirt a few drops of soap into the slot between base and flow cell to ease removal.

Leak-testing

The system should be leak-checked at 6 in. H_2O by connecting a manometer to the outlet boss and evacuate the inlet to 6 in. H_2O. No leakage should be observed.

Verification of Calibration

The calibrator is factory calibrated using a standard from the National Institute of Standards and Technology, formerly called the National Bureau of Standards (NBS). Attempts to verify the calibrator against a glass 1 liter burette should be conducted at 1,000 cc/min for maximum accuracy. The calibrator is linear throughout the entire range.

Shipping/Handling

When transporting, especially by air, it is important that one side of the seal tube connecting the inlet and outlet boss be removed to equalize internal pressure within the calibrator. Do not transport unit with soap solution or storage tubing in place.

Precautions/Warnings

- Avoid using chemical solvents on flow cell, calibrator case, and faceplate. Generally, soap and water will remove any dirt.
- Never pressurize the flow cell at any time with more than 25 in. of water pressure.

- Do not charge batteries for longer than 16 hours.
- Do not leave A/C adapter plugged into calibrator when not in use as this could damage the battery supply.
- Black, close-fitting covers help reduce evaporation of soap in the flow cell when not in use.
- Do not store flow cell for a period of one week or longer with soap. Clean and store dry.
- The calibrator soap is a precisely concentrated and sterilized solution formulated to provide a clean, frictionless soap film bubble over the wide, dynamic range of the calibrator. The sterile nature of the soap is important in the prevention of residue buildup in the flow cell center tube, which could cause inaccurate readings. The use of any other soap is not recommended.

MANUAL BURETTE BUBBLE METER TECHNIQUE

When a sampling train requires an unusual combination of sampling media (e.g., glass fiber filter proceeding impinger), the same media/devices should be in line during calibration. Calibrate personal sampling pumps before and after each day of sampling.

- Allow the pump to run 5 minutes prior to voltage check and calibration.
- Assemble the polystyrene cassette filter holder using the appropriate filter for the sampling method. If a cassette adaptor is used, care should be taken to ensure that it does not come in contact with the backup pad.
- When calibrating with a bubble meter, the use of cassette adaptors can cause moderate to severe pressure drops in the sampling train, which will affect the calibration result. If adaptors are used for sampling, they should be used when calibrating.
- Connect the collection device, tubing, pump and calibration apparatus as shown in Figures 1 and 2.
- Visually inspect all Tygon tubing connections.
- Wet the inside of a 1 liter burette with a soap solution.
- Turn on the pump and adjust the pump rotameter to the appropriate flow rate setting.
- Momentarily submerge the opening of the burette in order to capture a film of soap.
- Draw two or three bubbles up the burette to ensure that the bubbles will compete their run.
- Visually capture a single bubble and time the bubble from 0 to 1000 ml for high-flow pumps or 0 to 100 ml for low-flow pumps.
- The timing accuracy must be within 1 second of the time corresponding to the desired flow rate.
- If the time is not within the range of accuracy, adjust the flow rate and repeat the previous two steps until the correct flow rate is achieved. Perform these steps at least twice, in any event.

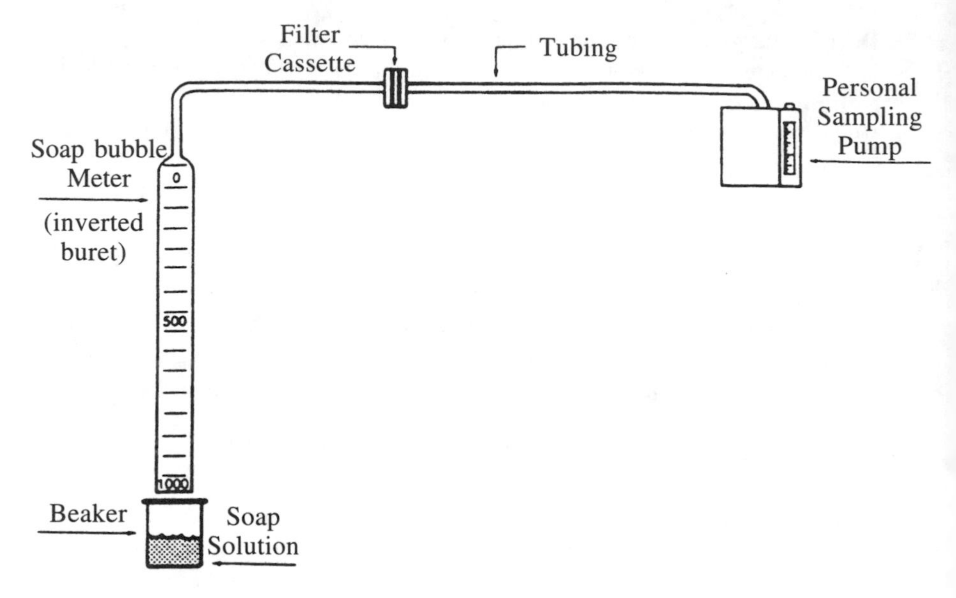

Figure 1. Calibration setup for personal sampling with filter cassette.

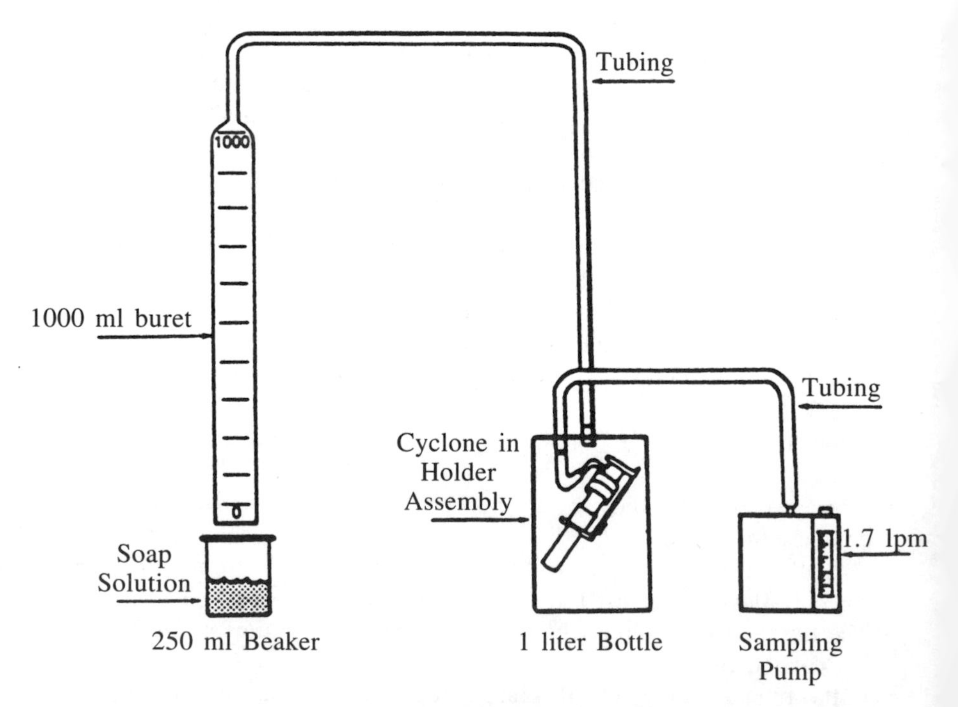

Figure 2. Calibration of cyclone respirable dust sampler using a bubble meter.

- While the pump is still running, mark the pump position of the center of the float in the pump rotameter as a reference.
- Repeat the procedures described above for all pumps to be used for sampling. The same cassette and filter may be used for all calibrations involving the same sampling method.

SAMPLING FOR SPECIAL ANALYSES

Silica Samples Analyzed by X-Ray Diffraction (XRD)

Air Samples

Respirable dust samples are analyzed for quartz and cristobalite by X-ray diffraction (XRD). XRD is the preferred analytical method due to its sensitivity, minimum requirements for sample preparation and ability to identify polymorphs (different crystalline forms) of free silica.

- The analysis of free silica by XRD requires that the particle size distribution of the samples be matched as closely as possible to the standards. This is best accomplished by collecting a respirable sample.
- Respirable dust samples are collected on a tared low ash PVC filter using a 10mm nylon cyclone at a flow rate of 1.7 lpm.
- A sample not collected in this manner is considered a total dust (or nonrespirable sample. Accurate analysis cannot be provided by XRD for such samples.
- If the sample collected is nonrespirable, the laboratory must be advised.
- Quartz and cristobalite are the only polymorphs of free silica presently being analyzed by the laboratory. Tridymite is not currently being analyzed. Samples are analyzed for cristobalite only upon request.
- Quartz (or cristobalite) is identified by its major (primary) X-ray diffraction peak. Because other substances also have peaks at the same position, it is necessary to confirm quartz principally by the presence of secondary and/or tertiary peaks.
- The following major chemicals can interfere with an analysis:
  Aluminum phosphate
  Feldspars (microcline, orthoclase, plagioclase)
  Graphite
  Iron carbide
  Lead sulfate
  Micas (biotite, muscovite)
  Montmorillonite
  Potash
  Sillimanite
  Silver chloride
  Talc

Zircon (zirconium silicate)
Specific additional chemicals should be listed if they are suspected to be present.

- A sample weight and total air volume shall accompany all filter samples. Sample weights of 0.1 to 5.0 mg are acceptable. Sample weights of 0.5 to 3.0 mg are preferred.
- Do not submit a sample(s) unless its weight or the combined weights of all filters representing an individual exposure exceed 0.1 mg.
- If heavy sample loading is noted during the sampling period, it is recommended that the filter cassette be changed to avoid collecting a sample with a weight greater than 5.0 mg.
- If a sample weight exceeds 5.0 mg, another sample of a smaller air volume, whenever possible, should be collected to obtain a sample weight of less than 5.0 mg.
- Laboratory results for air samples are usually reported under one of four categories:
  1. **Percent quartz.** Applicable for a respirable sample in which the amount of quartz in the sample was confirmed.
  2. **Less than or equal to value in units of percent.** Less or equal to values are used when the adjusted eight-hour exposure is found to be less than the PEL, based on the sample's primary diffraction peak. The value reported represents the maximum amount of quartz that may be present. However, the presence of quartz was not confirmed using secondary and/or tertiary peaks in the sample because the sample could not be in violation of the PEL.
  3. **Approximate values in units of percent.** The particle size distribution in a total dust sample is unknown and error in the XRD analysis may be greater than for respirable samples. For total dust samples, an approximate result is given.
  4. **Nondetected.** A sample reported as nondetected indicates that the quantity of quartz present in the sample is not greater than the detection limit of the instrument. The detection limit is usually 10 μg for quartz and 50 μg for cristobalite.
- If less than a full-shift sample was collected, evaluate a nondetected result to determine if adequate sampling was performed.
- If the presence of quartz is suspended, industrial hygienist may want to sample for a longer period of time to increase the sample weights.

Bulk Samples

Bulk samples must be submitted for all silica analyses.

- They have two purposes:

1. For laboratory use only, to confirm the presence of quartz in respirable samples, or to assess the presence of other substances that may interfere in the analyses of respirable samples.
2. To determine the approximate percentage of quartz in the bulk sample.

- A bulk sample submitted "for laboratory use only" must be representative of the airborne free silica content of the work environment sampled; otherwise, it will be of no value.
- The laboratory's order of preference for bulk samples for an evaluation of personal exposure is:
  1. A high volume respirable area sample.
  2. A high volume area sample.
  3. A representative settled dust (rafter) sample.
  4. A bulk sample of the raw material used in the manufacturing process. This is the last choice and least desirable. It should be submitted "for laboratory use only" if there is a possibility of contamination by other materials during the manufacturing process.
- The type of bulk sample submitted to the laboratory should be stated on the OSHA-91 form and cross-referenced to the appropriate air samples.
- A bulk sample analysis for percent quartz will be reported only upon specific request.
- A reported bulk sample analysis for quartz will be semi-quantitative in nature because:

  The XRD analysis procedure requires a thin layer deposition for an accurate analysis.

  The error for bulk samples analyzed by XRD is unknown because the particle size of nonrespirable bulk samples varies from sample to sample.

Samples Analyzed by Inductively Coupled Plasma (ICP)

Metals

Where two or more of the following analytes are requested on the same filter, an ICP analysis may be conducted. However, an industrial hygienist should specify the metals of interest in the event samples cannot be analyzed by the ICP method. A computer printout of the following 13 analytes may be reported:

1. Antimony
2. Beryllium
3. Cadmium
4. Chromium
5. Cobalt
6. Copper
7. Iron

8. Lead
9. Manganese
10. Molybdenum
11. Nickel
12. Vanadium
13. Zinc

Arsenic

Lead, cadmium, copper and iron can be analyzed on the same filter with arsenic, but they are not run without request.

Samples Analyzed by Neutron Activation (NAA) or X-Ray Fluorescence (XRF)

- Filter, wipe, and bulk samples can be qualitatively analyzed by both NAA and XRF. In addition, such samples can be quantitatively analyzed by NAA.
- Requests for NAA or XRF analyses should be preceded by a phone call to determine the extent and value of the analysis.
- Packaging and shipping of such samples should be done in a manner consistent with directions previously given.

SAMPLING FOR SURFACE CONTAMINATION

The terms "wipe sampling," "swipe sampling," and "smear sampling" are all used synonymously to describe the techniques used for assessing surface contamination. However, the term "wipe sampling" will be used in this chapter.

Wipe sampling is most often used to screen for asbestos, lead, other metals, and PCBs. The uses are:

Skin Sampling

- Potential contact with skin irritants may be evaluated by wiping surfaces that may be touched by workers.
- Skin wipes are not recommended for substances that absorb rapidly through the skin. Biological monitoring for these substances or their metabolites, or biological markers, is often the only means of assessing their absorption. Wipe the inside surfaces of protective gear or other surfaces that may contact skin.

Surface Sampling

- Surfaces that may come in contact with food or other materials that are ingested or placed in the mouth (e.g., chewing tobacco, gum, cigarettes) may be wipe sampled (including hands and fingers) to show contamination.

- Contaminated smoking materials may allow the toxic materials, or their combustion products (e.g., lead, mercury), to enter the body via the lungs. Wiping surfaces that smoking materials may touch (e.g., hands and fingers) is useful in evaluating this route of exposure.
- Accumulated toxic materials may become suspended in air, and may contribute to airborne exposures of asbestos, lead, or beryllium. Bulk and wipe samples may aid in determining this possibility.

Personal Protective Equipment Sampling

- Effectiveness of personal protective gear (e.g., gloves, aprons, respirators, etc.) may sometimes be evaluated by wipe sampling the inner surfaces of the protective gear and the protected skin.
- Effectiveness of contamination of surfaces and protective gear may be evaluated by wipe sampling.
- When accompanied by close observation of the operation in question, wipe sampling can help identify sources of contamination and poor work practices.

Evaluation of Sampling Results

- False negative results are possible.
- Use professional judgment on a case-by-case basis when evaluating the significance of positive wipe sampling results.
- Consider the toxicity and the contribution of skin absorption and/or gastrointestinal absorption to the total dose. Other factors to consider when evaluating sample results are the ambient air concentrations and skin irritation.
- Standard reference books list substances that represent a potential for ingestion toxicity, skin absorption, and/or have a hazardous skin effect. This information may be found under the "health" notation. Additional toxicological information concerning chronic skin absorption, dermatitis, etc., should be used to determine if the resulting exposure presents a potential employee hazard (see Bibliography).

GENERAL TECHNIQUE FOR WIPE SAMPLING

Filter Media and Solvents

- Consult a chemical information manual for appropriate filter media and solvents. Dry wipes may be used; solvents are not always necessary but may enhance removal.
- Direct skin wipes should not be taken when high skin absorption of a substance is expected. Under no conditions should any solvent other than distilled water be used on skin, personal protective gear that directly contacts the skin, or surfaces that contact food or tobacco products.

- Generally, there are two types of filters recommended for taking wipe samples:
  Glass fiber filters (GFF) (37 mm) usually are used for materials analyzed by high performance liquid chromatography (HPLC) and often for substances analyzed by gas chromatography (GC).
  Paper filters are generally used for metals, and may be used for anything not analyzed by HPLC. For convenient usage, the Whatman smear tab (or its equivalent) is commonly used.
- Preloading a group of vials with appropriate filters is a convenient method. (The Whatman smear tabs should be inserted with the tab end out.) Always wear clean plastic gloves when handling filters. Gloves should be disposable and should not be powdered.

Procedures

Follow these procedures when wipe samples are taken:

- If multiple samples are to be taken at the work site, prepare a rough sketch of the area(s) or room(s) to be wipe sampled.
- A new set of clean impervious gloves should be used with each individual sample. This avoids contamination of the filter by hands and the subsequent possibility for false positive, and prevents contact with the substance.
- Withdraw the filter from the vial. If a damp wipe sample is desired, moisten the filter with distilled water or other solvent as recommended. **CAUTION:** Skin, personal protective equipment or surfaces that contact food or tobacco products must either be wiped DRY, or wiped with distilled water, never with organic solvents. Skin wipes should not be done for materials with high skin absorption. It is recommended that hands and fingers be the only skin surfaces wiped. Before any skin wipe is taken, explain why you want the sample and ask the employee about possible skin allergies to the chemicals in the sampling filter or media. If the employee refuses, do not force the issue.
- Wipe a section of the surface to be sampled using a template with an opening exactly 100 cm^2.
- For surfaces smaller than 100 cm^2 use a template of the largest size possible. Be sure to document the size of the area wiped. For curved surfaces, the wiped area should be estimated as accurately as possible and then documented.
- Maximum pressure should be applied when wiping.
- To insure that all portions of the partitioned area are wiped, start at the outside edge and progress toward the center making concentric squares of decreasing size.
- If the filter dries out during the wiping procedure, discard the filter, reduce area to be wiped by half, and repeat wiping procedure with a new filter.

- Without allowing the filter to contact any other surface, fold the filter with the exposed side in, then fold it over again. Place the filter in a sample vial, cap the vial, number it, and place a corresponding number at the sample location on the sketch. Include notes with the sketch giving any further description of the sample (e.g., "Fred Employee's respirator, inside"; "Lunch table"; etc.).
- At least one blank filter treated in the same fashion, but without wiping, should be submitted for each sampled area.
- Submit the samples to an analytical laboratory.

SPECIAL TECHNIQUES FOR WIPE SAMPLING

Acids and Bases

When examining surfaces for contamination with strong acids or bases, (e.g., hydrochloric acid and sodium hydroxide), pH paper moistened with water may be used. However, these results should be viewed with caution because of potential interferences.

Direct Reading Instruments

For some types of surface contamination, direct reading instruments may be used, e.g., mercury sniffer for mercury.

Aromatic Amines

Screening may be done to determine the precise areas of carcinogenic aromatic amine contamination. This is an optional procedure.

- Wear clean, disposable impervious gloves. Wipe an area of exactly 100 cm^2 with a sheet of filter paper moistened in the center with five drops of methanol.
- After wiping the sample area, apply three drops of flourescamine to the contaminated area of the filter paper.
- Place a drop of the visualization reagent on an area of the filter paper that has not contacted the surface. This marks a non-sample area or blank on the filter paper adjacent to the test area.
- After a reaction time of six minutes, irradiate the filter paper with ultraviolet light.
- Compare the color development of the contacted area with the non-sample area or blank. A positive reaction will show a discoloration as a yellow color darker than the yellow color of the fluorescamine blank.
- A discoloration indicates surface contamination and possible aromatic amine carcinogen. Repeat a wipe sampling of the contaminated areas using the regular surface contamination procedure.

- The following compounds are some of the suspected carcinogenic agents that can be detected by this screening procedure:

  4,4'-Methylene bis(2-chloroaniline)
  Benzidine
  Napthylamine
  4-Aminobiphenyl

SPECIAL CONSIDERATIONS

- Due to their volatile nature, most organic solvents are not suitable for wipes. If necessary, surface contamination can be judged by other means, (e.g., by use of detector tubes, photoionization analyzers, or other similar instruments).
- Some substances are not stable enough as samples to be wipe sampled reliably.
- Some substances should have solvent added to the vial as soon as the wipe sample is placed in the vial (e.g., Benzidline).
- Do not take surface wipe samples on skin if OSHA or ACGIH shows a "skin" notation, or if the substance is an irritant, causes dermatitis, contact sensitization, or is termed corrosive.

BIBLIOGRAPHY

Adams, R.M. 1983. *Occupational Skin Disease.* New York: Grune and Stratton.

Benezra, C. et al. 1982. *Occupational Contact Dermatitis.* Clinical and Chemical Aspect. Philadelphia: Saunders. 1st ed.

Chaiyuth, C. and L.Levin. A Laboratory Evaluation of Wipe Testing Based on Lead Oxide Surface Contamination. *Am. Ind. Hyg. Assoc. J.* 45:311–317, 1984.

Clayton, G.D. and F.E. Clayton. 1981. *Patty's Industrial Hygiene and Toxicology.* New York: John Wiley & Sons. Vol. II.

Fisher, A.A. 1986. *Contact Dermatitis.* Philadelphia: Lea & Febriger. 3rd ed.

Gellin, G. and H.I. Malbach. 1982. *Occupational and Industrial Dermatology.* Chicago: Year Book Medical Publisher.

Lees, P.S.J. et al. Evidence for Dermal Absorption as the Major Route of Body Entry During Exposure of Transformer Maintenance and Repairman to PCBs. *Am. Ind. Hyg. Assoc. J.* 48:257–264, 1987.

Occupational Safety and Health Administration (OSHA), U.S.Dept. of Labor. 1987. *Chemical Information Manual.* Washington, DC: U.S. Government Printing Office.

CHAPTER 13

PERSONAL PROTECTION EQUIPMENT

Paul N. Cheremisinoff

Department of Civil and Environmental Engineering
New Jersey Institute of Technology
Newark, NJ 07102, USA

CONTENTS

INTRODUCTION

Emergency-response workers face a higher risk of accidents and injuries than the normal, individual employee. Often they must deal with situations that are beyond their control. They may be exposed to a number of substances that are hazardous because of their biological, radiological or chemical characteristics. Thus, personal protection equipment (PPE) will be necessary. Selection of PPE is primarily based on the routes of exposure.

Routes of Exposure

There are four routes by which a substance can enter the body.

Inhalation

For most chemicals in the form of vapors, gases, mists, or particulates, inhalation is the major route of entry. Once inhaled, chemicals are either exhaled or deposited in the respiratory tract. If deposited, damage can occur through direct contact with tissue, or the chemical may diffuse into the blood through the lung-blood interface.

Upon contact with tissue in the upper respiratory tract or lungs, chemicals may cause health effects ranging from simple irritation to severe tissue destruction. Substances absorbed into the blood system are circulated and distributed to organs. Health effects can then occur in the organs sensitive to the toxicant.

Skin or Eye Absorption

Direct contact with the skin (dermal contact) can cause effects that are relatively innocuous, such as redness or mild dermatitis. More severe effects include destruction of skin tissue or other debilitating conditions. Many chemicals also can cross the skin barrier and be absorbed into the blood system.

Once absorbed, they may produce systemic damage to internal organs. The eyes are particularly sensitive to chemicals. Even a short exposure can cause severe effects to the eyes. The substance can be absorbed through the eyes or skin and can be transported to the other parts of the body, causing harmful effects.

Ingestion

Chemicals that inadvertently get into the mouth and are swallowed do not generally harm the gastrointestinal tract unless they are irritating or corrosive.

Chemicals that are insoluble in the fluids of the gastrointestinal tract (stomach, small and large intestines) are generally excreted. Others that are soluble are absorbed through the lining of the gastrointestinal tract. They are then transported by the blood to internal organs where they can cause damage.

Injection

Substances may enter the body if the skin is penetrated or punctured by contaminated objects. Effects can then occur as the substance is circulated in the blood and deposited in the organs.

Therefore, an ensemble of PPE that will provide protection to a level of exposure below established permissible exposure limits should be selected and used.

Respiratory Protection

The respiratory system is able to tolerate exposure to toxic gases, vapors, and particulates, but only to a limited degree. Some chemicals can impair or destroy portions of the respiratory tract, or they may be absorbed directly into the bloodstream from the lungs. Chemicals that enter blood may eventually affect the function of other organs and tissues. The respiratory system can be protected by avoiding or minimizing exposure to harmful substances. Engineering controls such as ventilation help decrease exposure. When these methods are not feasible, respirators provide protection. Certain respirators can filter gases, vapors, and particulates in the ambient atmosphere; other respirators that supply clean air to the wearer also are available. The use of respirators is regulated by the Occupational Safety and Health Administration (OSHA). Regulations stipulate the use of approved respirators, proper selection, and individual fitting of respirators.

RESPIRATORY HAZARDS

The normal atmosphere consists of approximately 75% nitrogen, 21% oxygen, 0.9% inert gases, and 0.04% carbon dioxide. An atmosphere containing toxic contaminants, even at low concentrations, could be a hazard to lungs and body. A concentration large enough to decrease the percentage of oxygen in the air can lead to asphyxiation.

Oxygen Deficiency

The body requires oxygen to live. If the oxygen concentration decreases, the body reacts in various ways. Death occurs rapidly when the concentration decreases to 6%. Physiological effects of oxygen deficiency are not apparent until the concentration decreases to 16%. The various regulations and standards dealing with respirator use recommend that concentrations ranging from 16 to 19.5% be considered indicative of an oxygen deficiency. Such numbers take into account individual physiological responses, errors in measurement, and other safety considerations. In hazardous material response operations, 19.5% oxygen in air is considered the lowest "safe" working concentration.

Aerosols

Aerosol is a term used to describe fine particulates (solid or liquid) suspended in air. Particulates ranging in diameter from 5 to 30 microns are deposited in the nasal and pharyngeal passages. The trachea and smaller conducting tubes collect particulates 1 to 5 microns in diameter. For particulates to diffuse from the bronchioles into alveoli, they must be less than 0.5 microns in diameter. Gravity prevents larger particles from reaching the alveoli. The smallest particulates may never be deposited in the alveoli and may diffuse back into the conducting tubes to be exhaled.

Gaseous Contaminants

Gases and vapors are filtered to some degree on their trip through the respiratory tract. Soluble gases and vapors are absorbed by the conducting tubes en route to the alveoli. Not all will be absorbed, but some will diffuse into the alveoli, where they can be directly absorbed into the bloodstream.

Gaseous contaminants can be chemically classified into:

- **Acidic:** Acids or react with water to form acids.
- **Alkaline:** Bases or react with water to form bases.
- **Organic:** Compounds that contain carbon; may range from methane to chlorinated organic solvents.
- **Organometallic:** Organic compounds containing metals.
- **Inert:** Not chemically reactive.

RESPIRATORY PROTECTION DEVICES

The basic function of a respirator is to reduce the inhalation exposure to airborne contaminants. A respirator provides protection by removing the contaminants from ambient air or by supplying the wearer with an alternate source of clean air. All respiratory apparatus are composed of two main parts: the device that supplies or purifies air, and the face piece that covers the nose and mouth and seals out the contaminants. The first component defines what class of respirator the device is; the second determines the relative measure of protection afforded by the respirator.

Respirator Classifications

Air-Purifying Respirators

Air-purifying respirators (APRs) remove contaminants by passing the breathing air through a purifying element. There are a wide variety of APRs available to protect against specific contaminants, but they all fall into two subclasses: particulate APRs that employ a mechanical filter element, and gas and vapor APRs that utilize chemical sorbents contained in a cartridge or canister.

It is important to realize that there are limitations on the application of APRs. These devices are specific for certain types of contaminants, so the identity of hazardous agents must be known. Because there are maximum concentration limits, the ambient concentration of the contaminant, as well as the maximum use limit (MUL) of the respirator must be known. APRs only clean the air, so the ambient concentration of oxygen must be sufficient (19.5% or greater).

Atmosphere-supplying respirators (ASRs) provide a substitute source of clean breathing air. The respirable air is supplied to the worker from either a stationary source through a long hose or from a portable container. The first type are called supplied-air respirators and the latter are known as self-contained breathing apparatus (SCBA). These devices can be used regardless of the type of airborne contaminant or oxygen concentration. However, the contaminant concentration limits vary for the different types of ASRs, and the wearer must be aware of the limitations of the respirator.

Face Pieces

The protection provided the respirator wearer is a function of how well the face piece (mask) fits. No matter how efficient the purifying element or how clean the supplied air, there is little protection afforded if the respirator mask does not provide a leak-free face piece-to-face seal. Face pieces are available in three basic configurations that relate to their protective capacity.

Quarter Masks

A quarter mask fits over the bridge of the nose along the cheek and across the top of the chin. The head bands that hold the respirator in place are attached at two or four places on the mask (i.e., two- or four-point suspension). Limited protection is expected because the respirator can be easily dislodged, creating a breach in the seal.

Half Masks

A half mask fits over the bridge of the nose, along the cheek, and under the chin. Head bands have a four point suspension. Because they maintain a better seal and are less likely to be dislodged, half masks give greater protection than quarter masks.

Full Face Pieces

A full face piece fits across the forehead down over the temples and cheeks, and under the chin. A full face piece typically has a head harness with a five- or six-point suspension. These masks give the greatest protection because they are held in place more securely and because it is easier to maintain a good seal along the forehead than it is across the top of the nose. An added benefit is the eye protection from the clear lens in the full face piece.

AIR-PURIFYING RESPIRATORS

An air-purifying respirator (APR) is used for respiratory protection when airborne contamination levels exceed acceptable or approved levels. Because the air-purifying respirator removes selected components of the contaminated air rather than supplying clean air, many restrictions exist on their use. The following requirements must be met for use:

- The contaminant must be identified and the air concentration determined. The oxygen content of the atmosphere must be a minimum of 19.5%.
- The maximum use limit (MUL) of the respirator assembly must not be exceeded.
- The maximum use concentration (MUC) of the particular cartridge or canister must not be exceeded.
- The respirator assembly (face piece, cartridges and canister) must be specifically approved for use in that atmosphere.
- The service limit of the cartridges or canisters is not exceeded. The atmosphere does not exceed conditions that make it immediately dangerous to life and health (IDLH).
- The air contaminant possesses adequate warning properties.
- The user has been successfully fit-tested and trained in the use of APRs, and is medically fit for such usage.

Respirator Construction

The complete respirator assembly consists of two major components: the face piece, including all valves, mounting hardware, and lenses; and the air-purifying cartridges or canisters. With some styles of APRs, most notably single-use or disposable units, these two components are combined into a single functioning unit. Most units employ a reusable face piece assembly and disposable cartridges or canisters.

The basic classes of air-purifying respirators follow.

Disposable Dust Respirators

Many disposable cloth and paper respirators are NIOSH approved, but some are not. Those with approvals provide protection against dusts, mists, fumes, and sometimes asbestos. It is very difficult to fit-test and obtain and maintain a good face piece-to-face seal with this type of respirator.

Mouth-bit Respirator

Mouth bit respirators are approved for escape only. The bit is held by a teeth and a clamp is used to close the nostrils. A cartridge-type filter removes the contaminant from the atmosphere. This type of respirator can be used only when the hazard is identified and the respirator is approved for the hazard.

Quarter-mask Respirators

The quarter mask is used with cartridges or cloth filters for toxic and non-toxic dusts with exposure limits (ELs) at or above 0.05 mg/m^3. For dusts with ELs below 0.05 mg/m^3, a more efficient respirator must be used.

Half-mask Respirator

A half-mask respirator fits from under the chin to above the nose. One or two cartridges filter the air and are discarded once the use limits are reached. Where as quarter masks are only approved for dusts, the half masks have cartridges approved for pesticides, organic vapors, dusts, mists, fumes, acid gases, ammonia, and several combinations.

Full-face Respirators

The whole face, including the eyes, is protected by the full-face mask. It gives the ten times protection of the half mask. The full-face mask may be used with twin cartridges, chin-mounted canisters, or chest- or back-mounted canisters. Filters are available for the same materials as for the half mask, plus several more.

Powered Air-purifying Respirators (PAPRs)

PAPRs use a fan or other air movement device to draw or force air through the filtering media (see Figure 1). Thus, they offer more breathing resistance and should maintain a positive pressure in the face piece. They are used with half masks, full-face masks, hoods, and special helmets. It is important that the user keep in mind that the service life of the filter elements may be less than the other APRs because of the higher air flow rate.

Figure 1. Powered air-purifying respirator with high efficiency particulate arrestance (HEPA) filter for asbestos abatement. (Courtesy of International Safety Instruments, Inc., Lawrencevill, GA 30243.)

Face Pieces

The face piece is the means of the sealing the respirator to the wearer. Attached to the face piece is the lens (in the case of the full face piece) and the suspension for holding the mask to the face. An adapter is attached to the cartridge or canister. With the adapter and the mask is an inhalation check valve that prevents the exhaled breath from discharging through the cartridges or canisters. An exhalation valve permits the exhaled breath to be exhausted and prevents air from entering during inhalation. Some respirators provide an air-tight, internal speaking diaphragm. Each respirator manufacturer has a different way of assembling and attaching parts. This prevents hybridizing two different makes into one, which would void its NIOSH approval.

Although many configurations exist, only four types of face piece-element assemblies are recommended as satisfactory by OSHA for use when dealing with hazardous material:

- Half mask with twin cartridges or filters,
- Full-face mask with twin cartridges or filters,
- Full-face mask with chin-mounted canister (gas mask), and
- Full-face mask with harness-mounted canister (gas mask).

In important cases, the best facepiece is to use is the full face piece. It provides eye protection, is easier to fit, and has a protection factor greater than the half mask.

Air-Purifying Elements

Basically, respiratory hazards can be broken down into two classes: particulates and vapors/gases. Particulates are filtered by mechanical means, while vapors and gases are removed by sorbents that react physically or chemically with them. Respirators using a combination of mechanical filters and chemical sorbents effectively will remove both hazards.

Particulate Removing Filters

Particulates can occur as dusts, fumes, or mists. The particle size can range from macroscopic to microscopic, and their toxicological effects can be severe or innocuous.

Mechanical filters are classified according to the protection for which they are approved under scheduled 21C of 30 CFR part 11. Most particulate filters are approved for dusts and/or mists with ELs equal to greater than 0.05mg/m^3. These dusts generally are considered to produce pneumoconiosis and fibrosis. Such filters have an efficiency of 80% to 90% for particles larger than 0.6 microns (10^{-6} m).

Respirators approved for fumes are more efficient, removing 90% to 99% for particle larger than 0.6 microns. This type of respirator is for dusts, mists, and fumes with ELs equal to greater than 0.05 mg/m^3.

Finally, there is a high-efficiency filter, which is 99.97% effective against particles 0.3 microns in diameter and larger. It is approved for dusts, mists, and fumes with ELs less than 0.05 mg/m^3.

Mechanical filters load up with particulates as they are used. As they do, they become more efficient, but also become more difficult to breathe through and should be replaced.

Gas and Vapor-removing Cartridges and Canisters

When selecting a gas or vapor removing element, it must be chosen for protection against a specific type of contaminant. Some of the common types of chemical cartridges and canisters and their OSHA-required colors are listed in 29 CFR 1910.134.

Style and Size

Gas and vapor cartridges or canisters are available in several styles. The physical differences are: size and how they are attached to the face piece. The smallest elements are the cartridges that contain 50 to 200 cc of sorbent and attach directly to face piece, usually in pairs. Chin canisters (gas masks) have a sorbent volume of 250 to 500 cc and attach directly to a full face piece assembly. Larger gas masks or industrial-size canisters contain 1,000–2,000 cc of sorbent, are supported by a harness to the wearer's chest or back, and are connected to the full face piece by a breathing hose.

Service Life

Each sorbent has a finite capacity for removing contaminants. When this limit is reached, the cartridge or canister is said to be saturated. At this point the cleaning element will allow the contaminant to pass through and enter the facepiece. The length of time a cartridge or canister will effectively remove the contaminant is known as the service life of the element. Service life of a cartridge or canister is dependent on several factors: the breathing rate of the wearer, contaminant concentration, and sorption efficiency.

Breathing Rate. If the breathing rate of the user is rapid, the flow rate of the contaminated air drawn through the cartridge is greater than it is at moderate or slow respiration rates. A higher flow rate brings a larger amount of contaminant in contact with the sorbent in a given period of time which, in turn, increases the rate of sorbent saturation and shortens service life.

Contaminant Concentration. The expected service life of an organic vapor cartridge (MUC=1,000 ppm) decreases as ambient contaminant concentration increases. As concentration goes up, the mass flow rate increases, bringing more contaminant in contact with the sorbent in a given period of time. For example, at any constant breathing rate, ten times as much contaminant contacts the element when the concentration is 500 ppm compared to 50 ppm.

Chemical Sorbents. Chemical sorbents vary in their ability to remove the contaminants from air. The efficiency of chemical cartridges to sorb the contaminants from air often is affected by environmental conditions. In particular, air temperatures and humidity may greatly affect sorbent efficiencies. As a rule of thumb, the higher the temperature, the shorter the service life. However, the effects of humidity are much harder to predict. For many contaminants, an increase causes a reduction in sorbent efficiency, but for some water-soluble compounds, efficiency actually increases. Another environmental condition to be considered is the effect of mixtures of contaminants. Little is known of the results of combinations of contaminants on sorption.

Warning Properties. A warning property is a sign that a cartridge or canister in use is beginning to lose its effectiveness. A warning property can be detected as an odor, taste, or irritation. At the first signal, the old cartridge or canister must be replaced with a fresh one. Without an adequate warning property, a respirator's efficiency is dropped without the knowledge of the wearer, ultimately causing a health hazard.

Most substances have warning properties at some concentrations. A warning property detected only at dangerous levels—i.e. greater than its assigned exposure limit—is not considered adequate. An odor, taste, or irritation detected at extremely lower concentrations is also not adequate because the warning begins long before the filter begins lose its effectiveness. In this case, the wearer never will realize when the filter actually becomes ineffective.

The best concentration for a warning property to be first detected is around the established exposure limit. For example, toluene has an odor threshold of 40 ppm and a TLV-TWA of 100 ppm. This is usually considered as an adequate warning property. Conversely, dimethylfarmamide has a TLV-TWA of 10 ppm and an odor threshold of 100 ppm. An odor threshold ten times the TLV is not an adequate property.

If a substance causes rapid olfactory fatigue—that is, the sense of smell is no longer effective, its odor is not an adequate warning property. For example, upon entering an atmosphere containing hydrogen sulfide, the odor is immediately quite noticeable. After a short period of time, however, it is no longer detectable because of olfactory fatigue.

Selection Criteria

The use of an APR is contingent upon a number of criteria. If these criteria are not met, then the then use of an APR is prohibited. Figure 2 illustrates the basic selection criteria in a flow chart.

Oxygen Content

The normal atmosphere contains 21% oxygen. The physiological effect of reduced oxygen is evident at 16%. Based on oxygen content only, the atmosphere must contain a minimum of at least 19.5% oxygen to permit use of an APR. This is a legal requirement. Below 19.5% oxygen, oxygen-supplying respirators must be used.

Identification of Contaminants

It's absolutely imperative that the contaminant(s) be known so that:

- The toxic effects of inhaling the contaminant can be determined.
- Appropriate particulate filters or cartridges/canisters can be chosen.
- It can be determined that the adequate warning properties exist for the contaminant.
- The appropriate face piece be selected. A full-face piece mask is necessary if the contaminant causes eye irritation.

APRs are not to be used to protect against any radioactive gases or vapors and APRs are not be used for any gas or vapor that will react with the sorbent to generate heat.

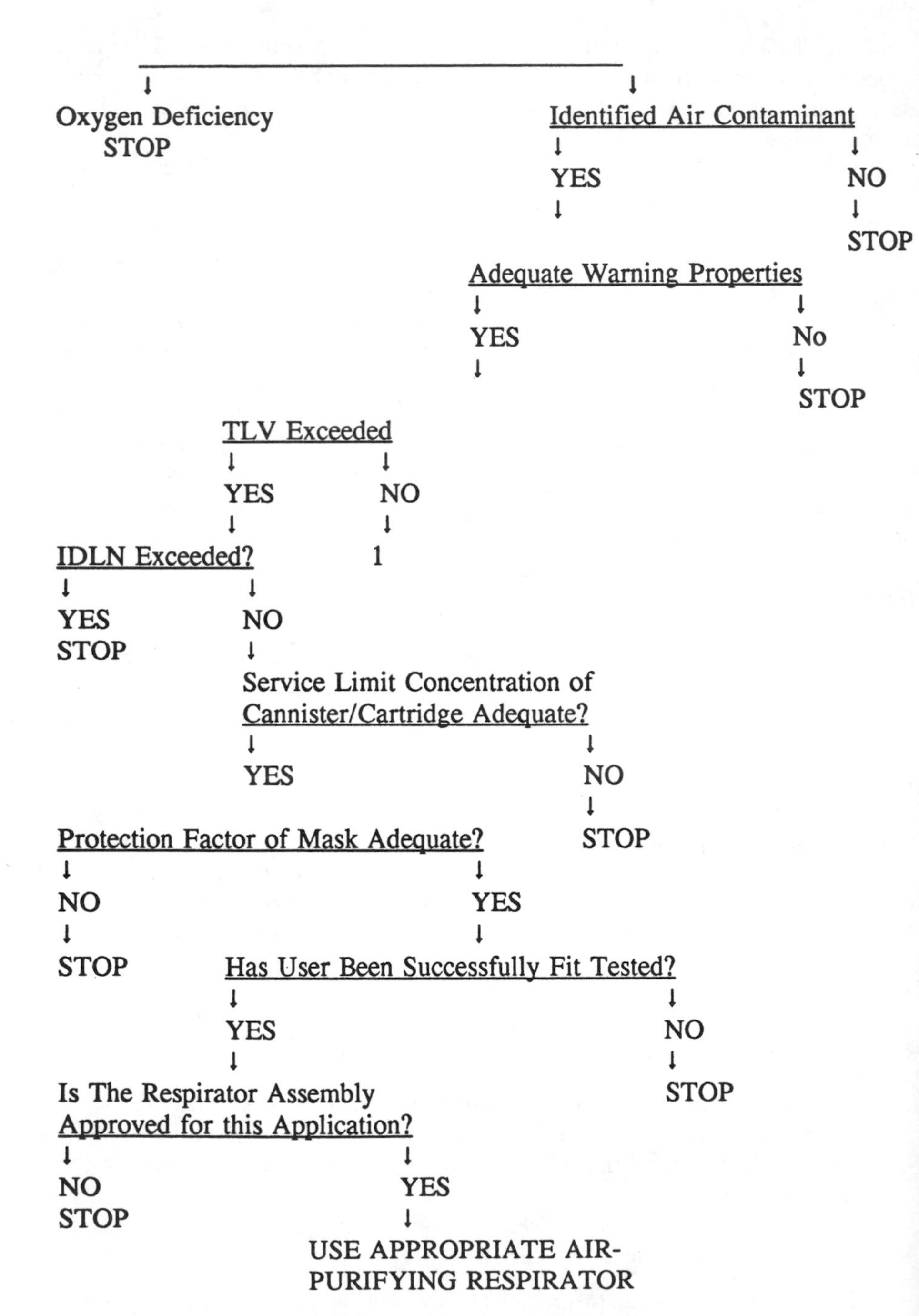

Figure 2. Air-purifying respirator selection criteria flow chart.

Contaminant Concentration Is Known

The maximum permissible air concentration depends on the contaminant and the respirator.

- Concentration must not exceed IDLH;
- Concentration must not exceed Lower Flammable Limit (LFL);
- The MUL of respirator can not be exceeded.
- The maximum use concentration (MUC) of a particular size and cartridge or canister must not be surpassed;
- Expected service life (cartridge/canister efficiency) should be determined, if possible.

Many factors need to be considered such as sorbent efficiency, relative humidity, use temperature, vapor pressure.

Periodic Monitoring of Hazards

Because of the importance of knowing the identity and concentration of contaminants, monitoring of the appropriate work-area equipment must occur at least periodically during the workday. This done to ensure that no significant changes have occurred and the respirators used are adequate for work conditions.

Approved Respirators

The respirator assembly (face piece and air-purifying element) must be approved for protection against the contaminant at the concentration present in the work area. The concentration must not exceed the NIOSH/OSHA designated MUC for that type and size cartridge/canister.

Fit-test and Proper Use

The wearer must pass a qualitative fit-test for the make, model, and size of air-purifying device used. OSHA regulations state: "Every respirator wearer shall receive fitting instructions including demonstration and practice in how the respirator should be worn, how to adjust it, and how to determine it fits properly. Respirators shall not be worn when the conditions prevent a good face seal. Such conditions may be a growth of beard, side burns, a skull cap that projects under the facepiece, or temple pieces on glasses. Also, the absence of one or both dentures can seriously affect the fit of a facepiece. The worker's diligence in observing these factors shall be evaluated by periodic check. To assure proper protection, the face piece fit shall be checked by the wearer each time he puts on the respirator.

Pressure Tests

Place the respirator over the face and draw the straps securely and evenly. The mask should not be so tight as to cause discomfort or a headache.

Negative Pressure Test

This test (and the positive pressure test) should be used only as a very gross determination of fit. The wearer should use this test just before entering the hazard atmosphere. In this test, the user closes of the inlet of the canister, cartridges, or filters by covering with the palm or squeezes the breathing tube so that it does not pass the air; inhales gently so that the face piece collapses slightly; and holds breath for about ten seconds.

If the face piece remains slightly collapsed and no inward leakage is detected, the respirator assembly passes the negative pressure test. If leakage is detected, the straps are readjusted and test procedure is repeated. If leakage is still detected, the respirator cannot be worn.

Although this test is simple, it has severe drawbacks, primarily that the wearer must handle the respirator after it has been positioned on his face. This handling can modify the facepiece seal.

Positive Pressure Test

This test, similar to the negative pressure test, is conducted by closing off the exhalation valve and exhaling gently into the facepiece. The test is considered to be positive if slight positive pressure is built up inside the facepiece without any evidence of outward leakage. For some respirators, this method requires that the wearer remove the exhalation valve cover; this often disturbs the respirator fit even more than does the negative pressure test. Therefore, this test should be used sparingly if it requires removing and replacing a valve cover. The test is easy for respirators whose valve cover has a single small part that can be closed by the palm or a finger.

Care and Cleaning

General Requirements

Any organization using respirators on a routine basis should have a program for their care and cleaning. The purpose of a program is to assure that all respirators are maintained at their original effectiveness. If they are modified in any way, their protection factors may be voided. Usually one person in an organization is trained to inspect, clean, repair, and store respirators.

The program should be based on number of types of respirators, working conditions, and hazards involved. In general a program should include:

- Inspection (including a leak check),
- Cleaning and disinfection,
- Repair, and
- Storage.

Inspection

Inspect respirators after each use. Inspect monthly a respirator that is kept ready for emergency use to assure it will perform satisfactorily.

On air-purifying respirators, thoroughly check all connections for gaskets and "O" rings, and for proper tightness. Check the condition of the face piece and all its parts, connecting air tubes, and headbands. Inspect rubber and elastomer parts for pliability and signs of deterioration.

Maintain a record for each respirator inspection, including date, inspector, and any unusual conditions or findings.

Cleaning and Disinfection

Collect respirators at suitable central location. Train employees required to wear respirators and explain to them that they are responsible for maintaining a clean and sanitized respirator. Clean and disinfect respirators as follows:

- Remove all cartridges, canisters, and filters, plus gaskets or seals not affixed to their seats.
- Remove elastic headbands.
- Remove exhalation cover.
- Remove speaking diaphragm-exhalation valve assembly
- Remove inhalation valve.
- Wash facepiece and breathing tube in cleaner/sanitizer powder mixed with warm water, preferably at 120° to 140° F. Wash components separately from the face mask, as necessary. Remove heavy soil from surfaces with a hand brush.
- Remove all parts from wash waters and rinse in clean, warm water.
- Air-dry parts in a designated clean area.
- Wipe face pieces, valves, and seats with a damp lint-free cloth to remove any remaining moisture.

Repairs

Only a company-trained person with proper tools and replacement parts is permitted to work on regulators. Individuals can make minor repairs following manufacturer's instructions. No one should attempt to replace components or to make adjustments or repairs beyond the manufacturer's recommendations.

Make repairs as follows:

- Disassemble and hand clean the inhalation and exhalation valve assemblies. Exercise care to avoid damaging the rubber diaphragms.
- Replace all faulty or questionable parts or assemblies. Use parts only specifically designed for the particular respirator.
- Reassemble the entire respirator and visually inspect the completed assembly.
- Insert new filters, cartridges, or canisters, as required. Make sure that gaskets or seals are in place and tightly sealed.

Storage

Follow the manufacturer's storage instructions, which are furnished with new respirators or affixed to the lid of the carrying case. In addition, these general instructions may be helpful:

- After respirators have been inspected, cleaned, and repaired, store them to protect against dust, excessive moisture, damaging chemicals, extreme temperatures, and direct sunlight.
- Do not store respirators in clothes lockers, bench drawers, or toolboxes. Place them in wall compartments at workstations or in a work area designated for emergency equipment. Store in original cartons or carrying cases.
- Draw clean respirators from storage for each use. Each unit can be sealed in a plastic bag, placed in a separate box, and tagged for immediate use.

SELF-CONTAINED BREATHING APPARATUS

Respiratory apparatus must frequently be used during response to hazardous materials incidents. If the contaminants are unknown or the requirements for air-purifying respirators cannot be met, then an atmosphere-supplying respirator is required. Several types of atmosphere-supplying respirators are available:

- **Oxygen generating:** One of the oldest respirators is oxygen-generating respirator, which uses a canister of potassium superoxide. This chemical reacts with water to produce oxygen, which replenishes the wearer's exhaled breath. Exhaled CO_2 is removed by a scrubber device containing LiOH. This reoxygenated air is then returned to the wearer. Oxygen-generating respirators have been used for escape purposes in mines. They generally are not used for hazardous materials applications because of the chemical reaction taking place within the respirator.
- **Hose mask:** This type of respirator consists of a face piece attached to a large diameter of hose that supplies clean air from a remote area.

  In units where the wearer breathes the air in, the hose lines can go up to 75 feet. With powered units, the hose length can vary from 50 to 250 ft.

- **Airline Respirator:** The airline respirator is similar to the hose mask, except that breathing-grade air is delivered to the wearer under pressure, either from a compressor of from a bank of compressed air cylinders (see Figure 3). The air may flow continuously, or it may be delivered as the wearer breaths (demands it). The air source must not be depletable, and no more than 300 ft of airline is allowed. An escape device is required for entry into an IDLH atmosphere (see Figure 4).

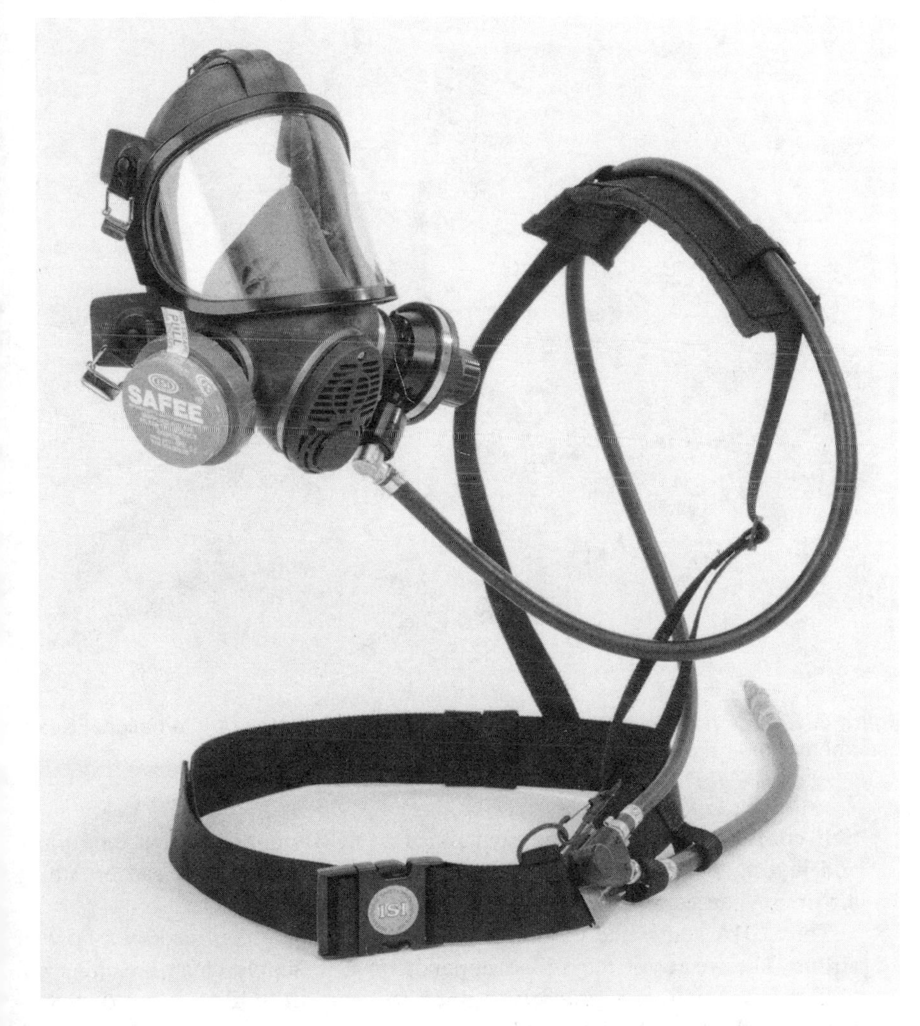

Figure 3. Supplied air line with filter for emergency egress. (Courtesy of International Safety Instruments, Inc., Lawrenceville, GA 30243.)

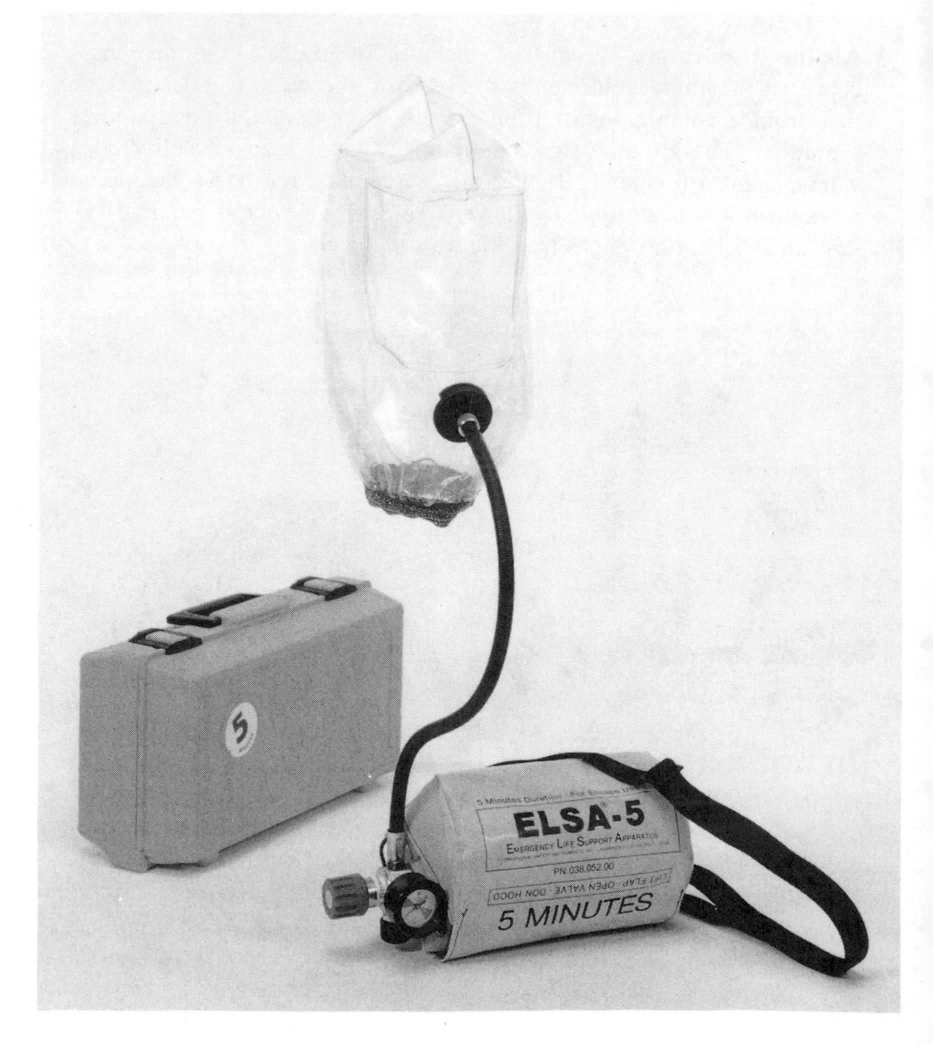

Figure 4. Escape hood with 5 min minute air cylinder. (Courtesy of International Safety Instruments, Inc., Lawrenceville, GA 30243.)

- **Self-contained breathing apparatus:** The SCBA consists of a face piece and regulator mechanism connected to a cylinder of compressed breathing air or oxygen carried by the wearer (see Figure 5).

  The SCBA allows the wearer to work without being confined by a hose or airline. The wearer of the SCBA depends on it to supply clean breathing air.

  If the wearer is not properly trained to wear an SCBA or the device is not properly cared for, then it may fail to provide the protection expected.

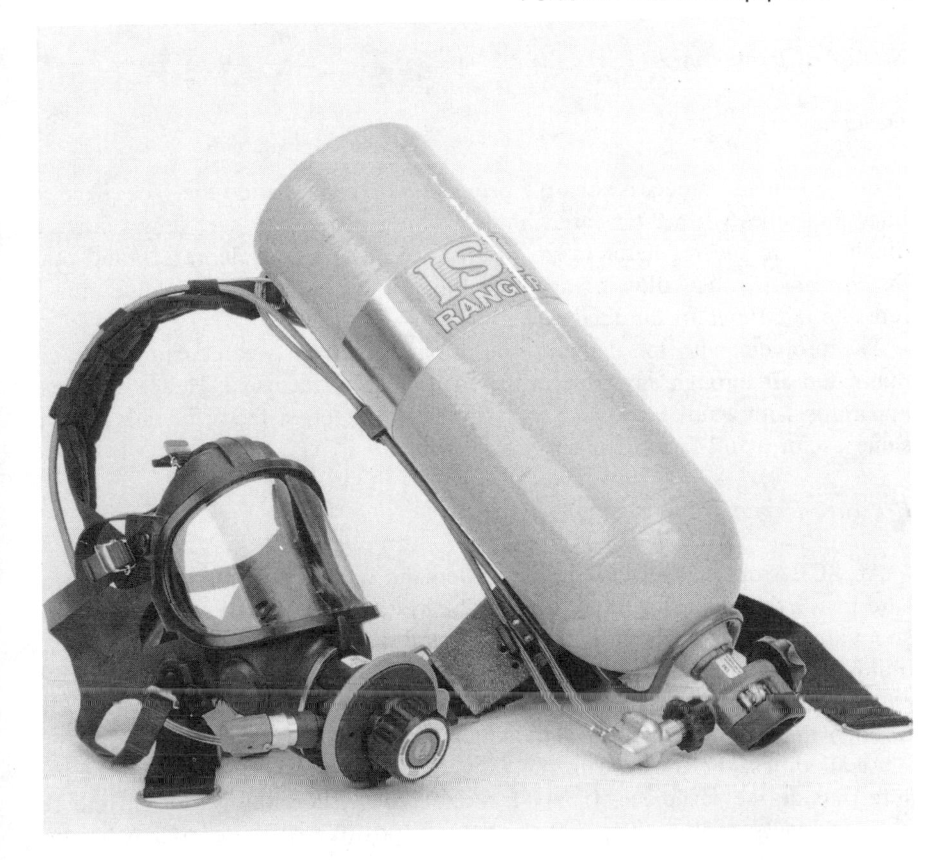

Figure 5. Self-contained breathing apparatus. (Courtesy of International Safety Instruments, Inc., Lawrenceville, GA 30243.)

The user should be completely familiar with the SCBA being worn. Checkout procedures have been developed for inspecting an SCBA prior to use, allowing the user to recognize the potential problems. An individual who checks out the unit is more comfortable and confident wearing it.

There are two types of apparatus: Closed circuit, which use compressed oxygen; and open circuit, which use compressed air. SCBAs operate in one of two modes, demand or pressure demand (positive pressure). The length of time an SCBA operates is based on the air supply. The units available operate from 5 minutes to more than 4 hours.

Pressure demand (positive pressure) is the only approved type of open-circuit SCBA for use in hazardous environments by the USEPA and NFPA.

Modes of Protection

Demand

In the demand mode, a negative pressure is created inside the face piece and breathing tubes when the wearer inhales. This negative pressure draws down the diaphragm in the regulator in an SCBA. The diaphragm depresses and opens the admission valve, allowing air to be inhaled. As long as the negative pressure remains, air flows to the face piece.

The problem with the demand operation is that the wearer can inhale contaminated air through any gaps in the face piece-to-face seal. Hence, a demand apparatus with a full face piece is assigned a protection factor of only 100, the same as for a full-face air-purifying respirator.

Pressure-Demand

An SCBA operating in the pressure-demand mode maintains a positive pressure inside the face piece at all times. The system is designed so that the admission valve remains open until enough pressure is built up to close it. The pressure builds up because the air is prevented from leaving the system until the wearer exhales. Less pressure is required to close the admission valve than is required to open the spring-loaded exhalation valve.

At all times, the pressure in the facepiece is greater than the ambient pressure outside the facepiece. If any leakage occurs, it is outward from the face piece. Because of this, the pressure-demand (positive-pressure) SCBA has been assigned a protection factor minimum of 10,000.

Types of Apparatus

Closed Circuit

The closed circuit SCBA, commonly called the rebreather, was developed especially for oxygen-deficient situations. Because is recycles exhaled breath and carries only a small oxygen supply, the service time can be considerably greater than an open-circuit device, which must carry all of the user's breathing air.

The air for breathing is mixed in a flexible breathing bag. This air is inhaled, deflating the breathing bag. The deflation depresses the admission valve, allowing the oxygen to enter the bag. There it mixes with exhaled breath, from which carbon dioxide has just been removed by passage through a CO_2 scrubber.

Most rebreathers operate in the demand mode. Several rebreathers are designed to provide a positive pressure in the face piece. The approved schedule 13F under 30 CFR part for closed-circuit SCBA makes no provisions for testing demand or pressure-demand rebreathers. The approval schedule was to set up to certify only rebreathers that operate in the demand mode. Thus, rebreathers designed to operate in the positive-pressure mode can be approved strictly as closed-circuit apparatus. Because regulations make no distinction and selection is based on approval criteria, rebreathers designed to maintain a positive pressure can only be considered as a demand-type apparatus. Rebreathers use either compressed or liquid oxygen. To assure the quality of the air to be breathed, the oxygen must be at least medical grade, which meets requirements set by the U.S. Pharmacopeia.

Open Circuit

The open circuit SCBA requires a source of compressed breathing air. The user simply inhales and exhales. The exhaled air is exhausted from the system. Because the air is not recycled, the wearer must carry the full air supply, which limits a unit to the amount of air that the wearer can easily carry. Available SCBA's can last from 5 to 60 minutes. Units with 5 to 15 minute air supplies are only applicable to escape situations.

The air used in open-circuit apparatus must meet or exceed the requirement in the Compressed Gas Association's pamphlet G-7.1, which calls for at least "grade D" air. Grade D air must contain 19.5% to 23.5% oxygen with the balance being predominantly nitrogen. Condensed hydrocarbons are limited to 5 mg/m^3, carbon monoxide to 20 ppm and carbon dioxide to 1,000 ppm. An undesirable odor also is prohibited. Air quality can be checked using an oxygen meter, carbon monoxide meter, and detector tubes.

Components of a Open-Circuit Pressure Demand SCBA

- **Cylinders:** Compressed air is considered a hazardous material. For this reason, any cylinder used with a SCBA must meet the department of transportations (DOT) "general requirements for shipments and packaging" and "shipping container specifications."

  A hydrostatic test must be performed on a cylinder at regular intervals: for steel and aluminum cylinders, every five years, for composite cylinders every three years.

  Air volume of 45 ft^3 of grade D air at a pressure of 2,216 psi is needed for a 30-minute supply. Cylinders are filled using a compressor or a cascade system of several large cylinders of breathing air. If the cylinder is overfilled, a rupture disc releases the pressure. The rupture disc is located at the cylinder valve, along with a cylinder pressure gauge accurate within ±5%. Because the gauge is exposed and subject to abuse, it should be used only for judging if the cylinder is full, not for monitoring air supply to the wearer.
- **High-pressure hose:** The high-pressure hose connects the cylinder and the regulator. The hose should be connected to the cylinder only by hand, never with a wrench. An O-ring inside the connector assures a good seal.

- **Alarm:** A low-pressure warning alarm is located near the connection to the cylinder. This alarm sounds to alert the wearer that only 20% to 25% of the full cylinder air supply is available for retreat, usually 5 to 8 minutes.
- **Regulator assembly:** Air travels from the cylinder through the high-pressure hose to the regulator. There it can travel one of two paths. If the bypass valve is opened, air travels directly through the breathing hose into the face piece. If the main line valve is opened, air passes through the regulator and is controlled by that mechanism. Also at the regulator (before air enters one of the valves) is another pressure gauge that also must be accurate to ±5%. Because it is visible and well-protected, this gauge should be used to monitor the air supply.

  Under the normal conditions, the bypass valve is closed and the main-line valve opened so air can enter the regulator. Once in the regulator, the air pressure in the regulator is reduced from the actual cylinder pressure to approximately 50 to 100 psi by a reducing mechanism. A pressure-release valve is located after the pressure reducer for safety should the pressure reducer malfunction. The airflow rate to meet NIOSH standards must meet or exceed 40 lpm. NFPA 1981 states the airflow rate must meet or exceed 100 lpm.
- **Breathing hose and facepiece:** The breathing hose connects the regulator to the face piece. Rubber gaskets at both ends provide tight seals. The hose is usually constructed of neoprene and is corrugated to allow stretching.

  Above the point in the mask where the hose is connected is a one-way check valve. This valve allows air to be drawn from the hose when the wearer inhales but prevents exhaled air from entering the breathing hose. If the check valve is not in place, the exhaled air may not be completely exhausted from the face piece.

  The face piece is normally constructed of neoprene, but sometimes of silicone rubber. Five- or six-point suspension is used to hold the mask to face. The visor lens is made of polycarbonate or other clear, shatterproof, and chemically resistant material. At the bottom of the facepiece is an exhalation valve.

  Some masks include an airtight speaking diaphragm, which facilitates communications while preventing contaminated air from entering.
- **Backpack and harness:** A backpack and harness support the cylinder and regulator, allowing the user to move freely. Weight should be supported on the hips, not the shoulders.

Inspection and Checkout

The SCBA must be inspected according to manufacturer's as well as 29 CFR recommendations. In addition, the SCBA should be checked immediately prior to use. Checkout and inspection procedures should be followed closely to assure safe operation of the unit.

DERMAL PROTECTION

Chemical protective clothing (CPC) is worn to prevent harmful chemicals from coming in contact with skin or eyes. It provides a barrier between the body and chemicals that have a detrimental effect on the skin or that can be absorbed through the skin and affect other organs. Used with respiratory protection, properly selected chemical protective clothing can protect personnel who must work in a chemical environment from injury.

Protecting workers against skin exposure requires using the most effective chemical protective clothing. Of primary importance is selecting clothing made from material most resistant to chemicals. The style and design of clothing also is important and depends on the nature of the hazard, i.e., whether a substance is a vapor/gas in the air or skin exposure will be from splash or direct contact with solids or semi-solids. Other selection criteria to consider include the probability of exposure, concentration of contaminant, ease of decontamination, mobility while wearing clothing, and to a lesser degree, cost.

Chemical protective clothing is made of a variety of manufactured materials. Each material provides skin protection against a range of chemicals, but no one material affords maximum protection against all chemicals. The chemical protective clothing selected must be made from a material that affords the greatest deterrent against the chemicals known or expected to be encountered.

Properly selected CPC can minimize risk of exposure to chemical substances, but may not protect against physical hazards. The use of other PPE also must be determined. Head protection is provided by hard hats; eye and face protection by goggles or impact-resistance lenses in spectacles; hearing protection by earmuffs or earplugs; and foot protection by impact-resistant and chemically resistant boots. Highly hazardous materials or special hazards may require specialized equipment for adequate protection.

CLASSIFICATIONS OF CHEMICAL PROTECTIVE CLOTHING

Chemical protective clothing is classified by style-design, protective material from which it is made, and whether the clothing is single-use (disposable).

Style-Design

Fully Encapsulating Suits (FES)

A fully encapsulating chemical protective suit is a one-piece garment that completely encloses the wearer. Boots, gloves, and facepiece are an integral part of the suit, but may be removed. If removable, they are connected to the suit by devices that provide a vapor- or gas-proof seal.

Respiratory protection and breathing air are provided to the wearer by a positive-pressure SCBA worn under the suit or by an air line respirator that maintains positive pressure inside the suit.

Fully encapsulating suits are primarily for protecting the wearer against toxic vapors, gases, mists, or particulates in air. Also, they protect against the splashes of liquids. The protection they provide against a specific chemical depends upon the material from which they are constructed.

Non Encapsulating Suits

A non-encapsulating chemical protective suit (frequently called a splash suit) does not have a face piece as an integral part of the suit. A positive-pressure SCBA or an air line respirator is worn outside the suit, or an APR is used. Splash suits are of two types: a one-piece coverall or a two-piece pants and coat. Either type may include a hood and other accessories.

Non-encapsulating suits are not designed to provide maximum protection against vapors, gases, or other airborne substances but against splashes. By taping at wrists, ankles, and neck, splash suits can be made to totally enclose the wearer such that no part of the body is exposed, but they still are not considered to be gas tight. They may be an acceptable substitute to a fully encapsulating suit if the concentration of airborne contaminant is low and the material is not extremely toxic to the skin.

Protective Materials

Chemical protective clothing is also classified based on the material from which it is made. All materials fall into two general categories, elastomers and non-elastomers.

Elastomers

Elastomers are polymeric (plastic-like) materials that, after being stretched, return to about their original shape. Most protective materials are elastomers. These include PVC, neoprene, polyethylene, Viton, Teflon, nitrile and butyl rubber, and others. Elastomers may be supported (layered onto cloth-like material) or unsupported.

Non-elastomers

Non-elastomers are materials that after being worn do not return to their original shape. Non-elastomers include Tyvek-coated fabrics.

Single-use Garments

A third classification is single-use or disposable garment. This classification is relative and based on cost and ease of decontamination. Disposable CPC is normally considered less than $25 per garment. In situations where decontamination is a problem, more expensive clothing may be considered disposable.

PERFORMANCE REQUIREMENTS

There are a number of performance requirements that must be considered in selecting the appropriate protective material. The relative importance is determined by the particular work activity and site-specific conditions.

Chemical Resistance: The ability of a material to withstand chemical and physical change. A material's chemical resistance is the most important performance requirement. The material must maintain its structural integrity and protective qualities upon contact with a hazardous substance.

Durability: The ability to withstand wear; the ability to withstand punctures, abrasions, and tears; the material's inherent strength.

Flexibility: The ability to bend or flex; pliability. It's extremely important both for glove and full body suit materials, for it directly affects the worker's mobility and agility.

Temperature Resistance: The ability of a material to maintain its chemical resistance during temperature extremes (especially heat) and to maintain flexibility in cold weather. The general tendency for most materials is that higher temperatures reduce their chemical resistance and lower temperatures reduce flexibility.

Service Life: The ability of a material to resist aging and deterioration. Factors such as chemicals, extreme temperatures, moisture, ultraviolet light, oxidizing agents, and others decrease a material's service life. Storage away from and proper care against these conditions can help prevent aging. Manufacturers should be consulted regarding any recommendations on a suit's shelf life.

Cleanability: The ability to effectively decontaminate protective materials. Cleanability is a relative measure of the ability of a material to release the contact substance. Some materials are nearly impossible to decontaminate, so it may be important to cover those materials with disposable garments to prevent gross contamination.

Design: The way a suit is constructed, which includes the general type and specific features. Features include:

- Fully encapsulating or non-encapsulating suits.
- One-, two-, or three-piece suits.
- Hoods, facepieces, gloves, and boots (attached or unattached).
- Location of zippers, buttons, storm flaps, and seams (front, side, back).

- Pockets, cloth collars, and velcro straps.
- Exhalation valves or ventilating ports.
- Ease of compatibility with wearing respiratory protection.

Sizing: the physical dimensions and proportions of clothing. Size is directly proportional to comfort and influences the number of unnecessary physical accidents. Ill-fitting clothing limits a wearer's mobility, dexterity and concentration. Manufacturers offer standard sizes in boots and gloves for both men and women; however, standard suit sizes for women are not available.

Color: Brightly colored suits make it easier to maintain visual contact between personnel. Suits of darker colors (black, green) absorb radiant heat from external sources and transfer it to the worker, increasing heat-related problems.

Cost: The cost of CPC varies considerably. Cost often will play a role in the selection and the frequency of use of CPC. In many situations, less expensive, single-use garments are more appropriate and as safe as more costly clothing. Other situations require high quality, costly clothing that may have to be discarded after limited use.

DONNING AND DOFFING OF FULLY ENCAPSULATED SUITS AND SCBAS

In responding to episodes involving hazardous substances, it may be necessary for response personnel to wear SCBA and fully encapsulating suits to protect against toxic environments. Donning and doffing of both is a relatively simple task, but a routine must be established and practiced regularly. Not only do correct procedures help instill confidence in the wearer, they reduce the risk of exposure and the possibility of damage to the suit. It is especially important to remove the equipment systematically to prevent or minimize the transfer of contaminants from suit to wearer.

The following procedures for donning/doffing apply to certain types of suits. They should be modified if a different suit or extra boots or gloves are worn. These procedures also assume that:

- The wearer has been trained to use SCBA, and the SCBA has been checked out.
- Appropriate decontamination steps have been taken prior to removal of the suits or other components.
- Sufficient air is available for routine decontamination and doffing of suit.

Donning and doffing an encapsulating suit is more difficult if the user has to do it alone because of the physical effort required. Also, the possibility of wearer exposure to contaminants or damaging the suit greatly increases. Therefore, assistance is needed in donning and doffing the equipment.

Donning

- Before donning a suit, thoroughly inspect for deficiencies that will decrease its effectiveness as the primary barrier for protecting the body. Do not use any suits with holes, rips, malfunctioning closures, cracked masks, etc. If a suit contains a hood or a hard hat is worn, adjust it to fit the user's head. If suit has a back enclosure for changing air bottles, open it.
- Use a moderate amount of talcum powder or cornstarch to prevent chafing and increase comfort. Both also reduce rubber binding.
- Use an antifog agent on suit and mask face pieces.
- While sitting, step into legs, place feet properly, and gather suit around waist.
- While sitting put on chemical resistant, steel-toe and-shank boots over feet of suit. Properly attach and affix suit leg over top of boot. For one-piece suits with heavy soled protective feet, wear leather or short rubber safety boots inside suit. Wear an additional pair of disposable boot protectors if necessary.
- Put on SCBA air tank and harness assembly. Don face piece and adjust it securely yet comfortably. Do not connect breathing hose. Open valve to air tank. (The air tank and harness assembly also could be put on before stepping into legs of suits).
- Put on inner gloves. For suits with detachable gloves, secure gloves to sleeves, if this has not been done prior to entering the suit. In some cases, extra gloves are worn over suit gloves.
- While standing, put arms into sleeves, and then head into hood of suit. The helper pulls suit up over SCBA, resting hood on top of SCBA, adjusting suit around SCBA backpack and users shoulders to assure unrestricted motion. To facilitate entry into the suit, bend at the knees as the hood is placed over wearer's head. Avoid bending at the waist, as this motion tends to use up room in the suit rather that provide slack. For a tall or stout person, it is easier to put on the hood of the suit before getting into the sleeves.
- Begin to secure suit by closing all fasteners until there is only room to connect the breathing hose. Also secure all belts and/or adjustable leg, head, and waist bands. Connect the breathing hose while opening main valve.
- When breathing properly with the SCBA, complete closing the suit.
- Helper should observe for a time to assure that wearer is comfortable and equipment is functioning properly.

Doffing

Exact procedures must be established and followed to remove the fully encapsulating suit and SCBA. Adherence to these procedures is necessary to minimize or prevent contamination of the wearer by contacting the outside surface of the suit.

The following procedures assume that before the suit is removed, it has been properly decontaminated, considering the type and extent of contamination, and that a suitably attired helper is available.

- Remove any extraneous or disposable clothing, boot covers, or gloves.
- If possible, wearer kicks off chemical resistant boots unassisted. To achieve this, oversize boots are often selected. Otherwise, helper loosens and removes chemical-resistant boots.
- Helper opens front of suit to allow access to SCBA regulator. As long as there is sufficient air pressure, hose is not disconnected.
- Helper lifts the hood of the suit over the wearer's head and rests hood on top of SCBA air tank. For a tall or stout person, it is easier to remove arms from sleeves of the suit prior to removing the hood.
- Remove external gloves.
- To minimize contact with contaminated clothing, helper touches only outside of the suit, and wearer touches the inside. Remove arms, one at a time, from suit. Helper lifts suit up and away from SCBA back pack, avoiding any contact between outside surface of suit and wearer's body. Helper lays suit out flat behind wearer.
- While sitting, remove both legs from suit.
- After suit is completely removed, roll internal gloves off hands, inside out.
- Walk to clean area and follow procedure for doffing SCBA.
- Remove inner clothing and clean body thoroughly.

ADDED CONSIDERATIONS

- If work is at a very dirty site or the potential for contamination is extremely high, wear disposable Tyvek or PVC coveralls over a fully encapsulating suit. Make a slit in back to fit around the bulge of the SCBA backpack.
- Wear clothing inside the suit appropriate to outside temperatures. Even in hot weather, wear cotton underwear, which absorbs perspiration and acts as a wick for evaporation, thus aiding body cooling. Long underwear also protects skin from contact with the hot surface of suit, reducing the possibility of burns in hot weather.
- Monitor wearer for heat stress.
- If a cooling device is used, modify donning and doffing procedures.
- If low-pressure warning alarm sounds signifying approximately 5 min of air remaining, follow these procedures.
  1. Quickly hose off suit and scrub especially around entrance/exit zipper. Remove any disposable clothing.
  2. Open zipper sufficiently to allow access to regulator and breathing hose.

3. Disconnect breathing hose from regulator as main valve is closed.
4. Immediately attach canister for vapor, acid gas, dust, mist, or fume to breathing hose. This provides protection against contaminants still present.
5. Continue doffing suit as described in the previous section. Take extra care to avoid contaminating helper and wearer.

CONCLUSIONS

For all practical purposes, there are four levels of protection criteria. Choosing a level of protection depends on hazardous incident of interest. Before entering the hazardous incident site, choose one of these levels of protection based on the anticipated hazard.

LEVEL A: A fully encapsulated suit with gloves, shoes, and hood attached, and SCBA. This is a maximum level of protection against toxic chemical vapors and particulates and should be used in situations where there is the least knowledge about the incident site.

LEVEL B: A splash suit with SCBA. Gloves, shoes, face piece, and hard hat are not an integral part of the suit. This level is used in situations where the chemicals are not toxic to skin and are low in concentration.

LEVEL C: A splash suit and an APR with gloves, shoes, and hard hat. An APR is good only when contaminants are known. And APR meets the requirements.

LEVEL D: A hard hat, shoes and goggles for eye protection. This is a minimum level of protection required to get into any hazardous incident site.

CHAPTER 14

AIR POLLUTION PERTAINING TO CONFINED SPACES

Peter Mizerek

Engineering & Environmental Services, Inc.
1259 Route 46, Bldg. 4, PO Box 5703
Parsippany, NJ 07054, USA

CONTENTS

INTRODUCTION

A "confined space" is generally defined by one, two or three of the following characteristics: It is a space that by design has limited openings for entry and exit; it is a space that has unfavorable natural ventilation and could contain or produce dangerous air contaminants; and it is a space that is not intended for continuous employee occupancy.

The worker who enters a confined space may be, or often is, exposed to multiple hazards because of primarily ignorance of the potential hazards or negligence in the enforcement of safety regulations. Ignorance and negligence have led to deaths by asphyxiation, fire, explosion, and exposure to toxic materials. Because injuries and deaths occurring from confined space are recorded across several different categories, the exact number of injuries and deaths specifically from confined space entry can only be estimated. However, the National Institute for Occupational Safety and Health (NIOSH) and the Occupational Safety and Health Administration (OSHA) are aware that a number of deaths occur each year directly because workers must enter and work in confined spaces.

The hazards encountered and associated with entering and working in confined spaces are capable of causing bodily injury, illness, and death to the worker. Accidents occur among workers because of failure to recognize that a confined space is a potential hazard. It should therefore be considered that the most unfavorable situation exists in every case and that the danger of explosion, poisoning, and asphyxiation will be present at the onset of entry.

Before forced ventilation is initiated, information such as restricted areas within the confined space, voids, the nature of the contaminants present, the size of the space, the type of work to be performed, and the number of people involved should be considered. The ventilation air should not create an additional hazard by recirculating contaminants, improperly arranging the inlet duct, or substituting anything other than fresh (normal) air (approximately 20.9% oxygen, 79.1% nitrogen by volume). The terms "air" and "oxygen" are sometimes considered synonymous; however, this is a danger assumption because the use of oxygen in place of fresh (normal) air for ventilation will expand the limits of flammability and increase the hazards of fire and explosion [1].

Hazardous conditions to be discussed in this chapter will primarily be hazardous atmospheres (flammable, toxic, irritant/corrosive, and asphyxiating), but physical and mechanical hazards also exist.

TYPES OF CONFINED SPACES

Confined spaces can be categorized generally as those with open tops and with a depth that will restrict the natural movement of air and enclosed spaces with very limited openings for entry.

In either of these cases, the space may contain mechanical equipment with moving parts. Any combination of these parameters will change the nature of the hazards encountered. Degreasers, pits, and certain types of storage tanks may be classified as open-topped confined spaces that usually contain no moving parts. However, gases that are heavier than air (butane, propane, and other hydrocarbons) remain in depressions and will flow to low points where they are difficult to remove. Therefore, these heavier-than-air gases are a primary concern when entry into such a confined space is planned. Other hazards may develop from the work performed in the confined space or because of corrosive residues that accelerate the decomposition of scaffolding supports and electrical components.

Confined spaces such as sewers, casings, tanks, silos, vaults, and compartments of ships usually have limited access. The problems arising in these areas are similar to those that occur in open-topped confined spaces. However, the limited access increases the risk of injury. Gases heavier than air, such as carbon dioxide and propane, may lie in a tank or vault for hours or even days after the containers have been opened. Because some gases are odorless, the hazard may be overlooked with fatal results. Gases that are lighter than air may also be trapped within an enclosed type of confined space, especially those with access from the bottom or side.

Hazards specific to a confined space are dictated by: the material stored or used in the confined space, for example, damp activated carbon in a filtration tank will absorb oxygen thus creating an oxygen-deficient atmosphere; the activity carried out, such as the fermentation of molasses, which creates ethyl alcohol vapors and decreases the oxygen content of the atmosphere; or the external environment, as in the case of sewer systems that may be affected by high tides, heavier-than-air gases, or flash floods [2].

The most hazardous kind of confined space is the type that combines limited access and mechanical devices. All the hazards of open-top and limited-access confined spaces may be present together with the additional hazard of moving parts. Digesters and boilers usually contain power-driven equipment that, unless properly isolated, may be inadvertently activated after entry. Such equipment also may contain physical hazards that further complicate the work environment and the entry and exit process.

REASONS FOR ENTERING CONFINED SPACES

Entering a confined space as part of the industrial activity may be done for various reasons. It is done usually to perform a necessary function, such as inspection, repair, maintenance (cleaning or painting), or similar operations that are an infrequent or irregular function of the total industrial activity [3].

Entry may also be made during new construction. Potential hazards should be easier to recognize during construction because the confined space has not been used. The types of hazards involved will be limited by the specific work practices. When the area meets the criteria for a confined space, all ventilation and other requirements should be enforced.

One of the most difficult entries to control is that of unauthorized entry, especially when there are large numbers of workers and trades involved, such as welders, painters, electricians, and safety monitors.

A final and most important reason for entry would be emergency rescue. This, and all other reasons for entry, must be well planned before initial entry, and the hazards must be thoroughly reviewed. The standby person and all rescue personnel should be aware of the structural design of the space, emergency exit procedures, and life support systems required.

HAZARDOUS ATMOSPHERES

"Atmosphere" is defined as the gases, vapors, mists, fumes, and dusts within a confined space. Hazardous atmospheres encountered in confined spaces can be divided into four distinct categories: flammable, toxic, irritant and/or corrosive, and asphyxiating.

Flammable Atmospheres

A flammable atmosphere generally arises from enriched oxygen atmospheres, vaporization of flammable liquids, by-products of work, chemical reactions, concentrations of combustible dusts, and desorption of chemicals from inner surfaces of the confined space.

Oxygen in place of air increases the flammability range of combustibles. Enrichment of the atmosphere with only a few percent of oxygen above 21% will cause an increase in the range of flammability. Hair and clothing will absorb the oxygen and burn violently. Enriched-oxygen atmospheres that expand the region of flammability could be the result of improper blanking off of oxygen lines, chemical reactions that liberate oxygen, or inadvertently purging a confined space with oxygen in place of air.

An atmosphere becomes flammable when the ratio of oxygen to combustible material in the air is neither too rich or too lean for combustion to occur. Combustible gases or vapors will accumulate when there is inadequate ventilation in areas such as a confined space. Flammable gases such as acetylene, butane, propane, hydrogen, methane, natural gas, or manufactured gases or vapors from liquid hydrocarbons can be trapped in confined spaces. Because many gases are heavier than air, they will seek lower levels as in pits, sewers, and various types of storage tanks and vessels. In a closed-top tank, it should also be noted that lighter-than-air gases may rise and develop a flammable concentration if trapped above an opening [4].

The by-products of work procedures can generate flammable or explosive conditions within a confined space. Specific kinds of work such as spray painting can result in the release of explosive gases or vapors [1]. Welding in a confined space is a major cause for explosions in areas that contain combustible gas [5].

Chemical reactions forming flammable atmospheres occur when surfaces are initially exposed to the atmosphere or when chemicals combine to form flammable gases. This condition arises when dilute sulfuric acid reacts with iron to form hydrogen or when calcium carbide makes contact with water to form acetylene. Other examples of spontaneous chemical reactions that may produce explosions from small amounts of unstable compounds are acetylene-metal compounds, peroxides, and nitrates. In a dry state, these compounds have the potential to explode upon percussion or exposure to increased temperature. Another class of chemical reactions that form flammable atmospheres arises from deposits of pyrophoric substances (carbon, ferrous oxide, ferrous sulfate, iron, etc.) that can be found in tanks used by the chemical and petroleum industry. These tanks containing flammable deposits will spontaneously ignite upon exposure to air [6].

Combustible dust concentrations usually are found during the process of loading, unloading, and conveying grain products, nitrated fertilizers, finely ground chemical products, and any other combustible material. It has been reported that high charges of static electricity, which rapidly accumulate during periods of relatively low humidity (below 50%), can cause certain substances to accumulate electrostatic charges of sufficient energy to produce sparks and ignite a flammable atmosphere [1]. These sparks also may cause explosions when the right air or oxygen-to-dust or -gas mixture is present.

Desorption of chemicals from the inner surfaces of a confined space is another process that may produce a flammable atmosphere. This is often a natural phenomenon in which the partial pressure at the interface between the surfaces and the stored chemical is radically reduced. For example, after liquid propane is removed from a storage tank, the walls of the vessel may desorb the remaining gas from the porous surface of the confined space.

Toxic Atmospheres

Substances regarded as toxic in a confined space can cover the entire spectrum of gases, vapors, and finely divided airborne dust in industry [7]. The sources of toxic atmospheres encountered may arise from the following:

- The manufacturing process—For example, production of polyvinyl chloride uses hydrogen chloride and vinyl chloride monomer, which is carcinogenic.
- A stored product—For example, removing decomposed organic material from a tank can liberate toxic substances such as hydrogen sulfide.
- The operation performed in the confined space—For example, welding or brazing with metals capable of producing toxic fumes.

Toxic gases may be evolved when acids are used for cleaning. Hydrochloric acid can react chemically with iron sulfide to produce hydrogen sulfide. Iron sulfide is formed on the walls of cooling jackets when only several parts per million sulfide are in the water used for the cooling process.

Another area where the hydrogen sulfide hazard exists is in the tanning industry. Lime pits used in the process of removing hair from the hides contain, in addition to lime, a 1% solution of sodium sulfate. Acid dichromate solution also is used in the tanning process. If these two solutions (sodium sulfate and acid dichromate) are combined, hydrogen sulfide will be produced.

During loading, unloading, formulation, and production, mechanical and/or human error also may produce toxic gases that are not part of the planned operation.

Toxic solvents, such as trichloroethylene, methyl chloroform, and dichloromethane, are used in industry for cleaning and degreasing. Acrylonitrile, infrequently used, has been encountered as an ingredient in a protective coating applied to tank interiors.

Trichloroethane and dichloroethane are widely used in industry as cleaning solvents because they are among the least toxic of the chlorinated aliphatic hydrocarbons. These solvents have been used as replacements for carbon tetrachloride and trichloroethylene [8].

The compatibility of materials must be considered when structural members and equipment are introduced in confined spaces. The previous history of the confined space must be carefully evaluated to avoid reactions with residual chemicals, wall scale, and sludge, which can be highly reactive. Other cases of incompatibility arise from the use of chemical cleaning agents. The initial step in chemical cleaning usually is the conversion of the scale or sludge into a liquid state, which may cause poisonous gases to be liberated.

Another hazardous gas that may build up in a confined space is carbon monoxide. This odorless, colorless gas that has approximately the same density of air is formed from incomplete combustion of organic materials such as wood, coal, gas, oil, and gasoline. It can be formed from microbial decomposition of organic matter in sewers, silos, and fermentation tanks. Carbon monoxide is an insidious toxic gas because of its poor warning properties. Early stages of carbon monoxide intoxication are nausea and headache. Carbon monoxide may be fatal at 1,000 ppm in air and is considered dangerous at 200 ppm because it forms carboxyhemoglobin in the blood, which prevents the distribution of oxygen in the body [9].

Carbon monoxide is a relatively abundant gas, therefore any untested atmosphere must be suspect. It must also be noted that a safe reading on a combustible gas indicator does not ensure that carbon monoxide is not present. Carbon monoxide must be tested for specifically. The formation of carbon monoxide may result from chemical reactions or work activities, so the fatalities from carbon monoxide poisoning are not confined to any particular industry. There have been fatal accidents in sewage treatment plants from decomposition products and lack of ventilation in confined spaces. Another area where carbon monoxide results as a product of decomposition is in the formation of silo gas in grain storage elevators. In still another area, the paint industry, varnish is manufactured by introducing the various ingredients into a kettle and heating them in an inert atmosphere such as carbon dioxide and nitrogen. The inert gas supply can contain a carbon monoxide content of more than 1% (10,000 ppm).

In welding operations, oxides of nitrogen and ozone of major toxicologic importance, and incomplete oxidation may occur forming carbon monoxide as a by-product [10].

Another poor work practice that has led to fatalities is the recirculation of diesel exhaust emissions [11]. Tests have shown that although the initial hazard from exhaust toxicants may be from increased carbon dioxide levels (or depleted oxygen levels), the most immediate hazard to life processes is carbon monoxide [12]. Increased carbon monoxide levels can be prevented only by strict control of the ventilation or the use of catalytic converters.

Irritant/Corrosive Atmospheres

Irritant or corrosive atmospheres can be divided into primary and secondary groups. The primary irritants exert no systemic toxic effects because the products formed by them on tissues of the respiratory tract are non irritant, and other irritant effects are so violent as to obscure any systemic toxic reaction. Examples of primary irritants are chlorine, ozone, hydrochloric acid, hydrofluoric acid, sulfuric acid, and sulfur dioxide. A secondary irritant is one that may produce systemic toxic effects in addition to surface irritation. Examples of secondary irritants include benzene, carbon tetrachloride, ethyl chloride, trichloroethane, trichloroethylene, and chloropropene [13].

Irritant gases vary widely among all areas of industrial activity. They can be found in plastic plants, chemical plants, the petroleum industry, tanneries, refrigeration industries, paint manufacturing, and mining operations.

Prolonged exposure at irritant or corrosive concentrations in a confined space may produce little or no evidence or irritation. This has been interpreted to mean that the worker has become adapted to the harmful agent involved. In reality, it means there has been a general weakening of the defense reflexes from changes in sensitivity, due to damage of the endings in the mucous membranes of the conjunctive and upper respiratory tract. The danger in this situation is that the worker is usually not aware of any increase in his exposure to toxic substances.

Asphyxiating Atmospheres

The normal atmosphere is composed approximately of 20.9% oxygen, 78.1% nitrogen, 1% argon and small amounts of various other gases. Reduction of oxygen in a confined space may be the result of either consumption of displacement.

The consumption of oxygen takes place during combustion of flammable substances, as in welding, heating, cutting, and brazing. A more subtle consumption of oxygen occurs during bacterial action, as in the fermentation process. Oxygen also may be consumed during chemical reactions as in the formation of rust (iron oxide) on the exposed surface of the confined space. The number of people working in a confined space and the amount of their physical activity also will influence the oxygen consumption rate.

A second factor in oxygen deficiency is displacement by another gas. Examples of gases that are used to displace air and therefore reduce oxygen level are helium, argon, and nitrogen. Carbon dioxide also may be used to displace air and can occur naturally in sewers, storage bins, wells, tunnels, wine vats, and grain elevators. Aside from the natural development of these gases or their use in the chemical process, certain gases also are used as inerting agents to displace flammable substances and retard pyrophoric reactions. Gases such as nitrogen, argon, helium, and carbon dioxide are frequently referred to as non-toxic inert gases but have claimed many lives. The use of nitrogen to inert a confined space has claimed more lives than carbon dioxide. The total displacement of oxygen by nitrogen will cause immediate collapse and death. Carbon dioxide and argon, with specific gravities of 1.53 and 1.38, respectively, may lie in a tank or manhole for hours or days after opening. Because these gases are colorless and odorless, measurements and ventilation should be adequately carried out [13].

Oxygen deprivation is one form of asphyxiation. While it is desirable to maintain the atmospheric oxygen level at 21% by volume, the body can tolerate deviation from this ideal. When the oxygen level falls to 17%, the first sign of hypoxia is a deterioration to night vision, which is not noticeable until a normal oxygen concentration is restored. Physiologic effects are increased breathing volume and accelerated heartbeat. Between 14% and 16%, physiologic effects are increased breathing volume, accelerated heartbeat, very poor muscular coordination, rapid fatigue, and intermittent respiration. Between 6% and 10%, the effects are nausea, vomiting, inability to perform, and unconsciousness. Less than 6%, causes spastic breathing, convulsive movements, and death in minutes.

In discussing oxygen and what constitutes an oxygen deficient atmosphere from a physiologic view, the concept of partial pressures must be addressed. At sea level, the normal atmospheric pressure for air (20.9% oxygen, 78.1% nitrogen, 1% argon, and trace amounts of various inert gases) is 14.7 psi or 760 mm Hg absolute. The partial pressure of oxygen at sea level will be approximately 160 mm Hg. The concept of partial pressures is that in any mixture of gases, the total gas pressure is the sum of the partial pressures of all the gases [14].

The partial pressure of oxygen in ambient air can be decreased by a reduction in the oxygen level at constant pressure or by maintaining the percentage of oxygen constant and decreasing the total atmospheric pressure, as in the case at high altitudes. It is important not only to know the oxygen percent by volume but also to understand the relationship of oxygen to the altitude and the concept of partial pressure. For example, 20.9% oxygen in air at sea level constitutes a greater partial pressure of oxygen than 20.9% oxygen at 5,000 feet because the total atmospheric pressure at 5,000 feet is less than at sea level. As the partial pressure of oxygen in the atmosphere drops, the volume of air required to maintain a partial pressure of oxygen of 60 mm Hg in the alveolar space of the lungs increases. A partial pressure of oxygen below 60 mm Hg in the alveolar space is considered oxygen deficient [14].

Absorption of oxygen by the vessel or the product stored therein is another mechanism by which the partial pressure of oxygen may be reduced and result in an oxygen-deficient atmosphere. For example, activated carbon, usually considered as an innocuous material free of occupational hazard and toxicity, can absorb oxygen in a closed vessel. Dry carbon would not reduce the oxygen level significantly. Damp activated carbon, however, has been known to decrease the oxygen level from 21% to 4% in a closed vessel. Figures 1, 2, and 3 show respiratory protective equipment that can be used for confined space entry.

STANDARDS FOR WORKING IN CONFINED SPACES

There are five general areas to be discussed when establishing standards for working in confined spaces: testing and monitoring, medical requirements, safety equipment and clothing, training, and work practices.

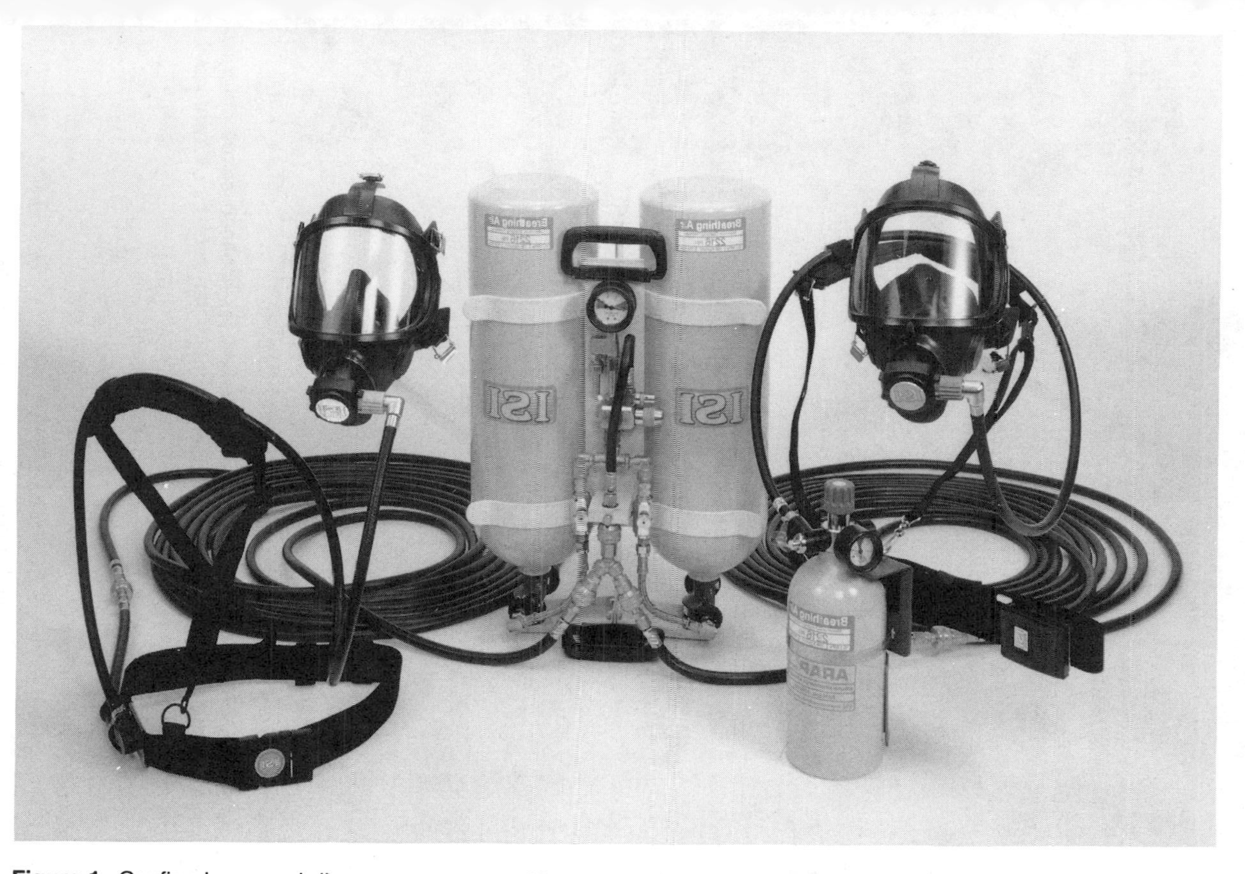

Figure 1. Confined-space air line rescue system. (Courtesy of International Safety Instruments, Inc., Lawrenceville, GA 30243.)

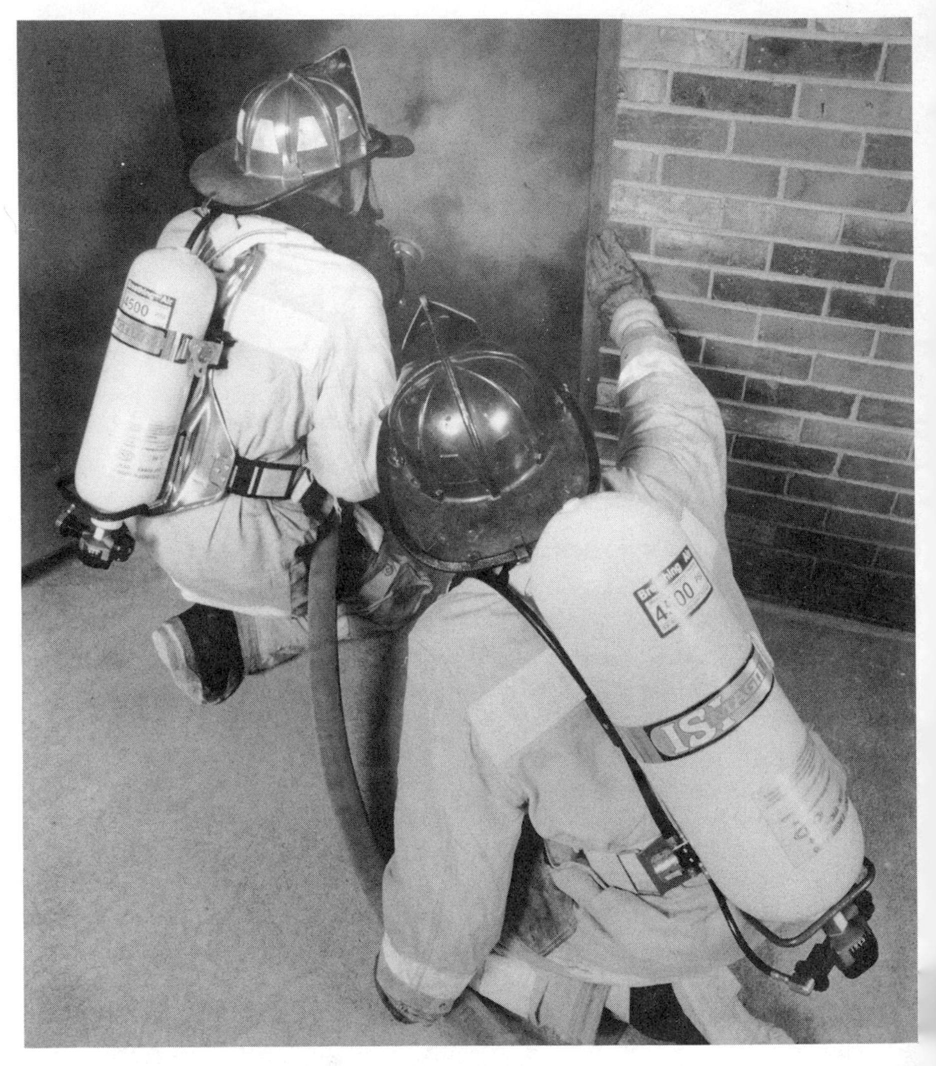

Figure 2. Self-contained breathing apparatus (SCBA) for confined space entry and emergency response. (Courtesy of International Safety Instruments, Inc., Lawrenceville, GA 30243.)

Testing and Monitoring

Prior to entering a confined space, workers should know the space's potential hazards. The various tests to be performed prior to entry should include tests for flammability, toxic agents, oxygen deficiency, and harmful physical agents.

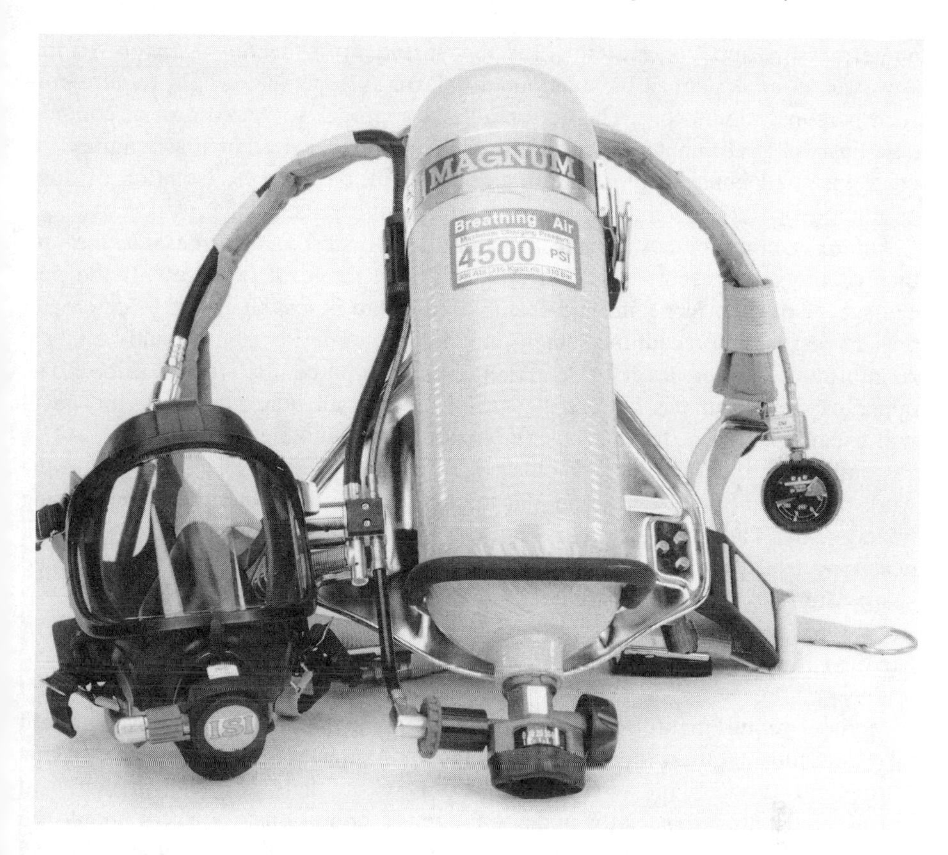

Figure 3. Self-contained breathing apparatus (SCBA). (Courtesy of International Safety Instruments, Inc., Lawrenceville, GA 30243.)

Specific instruments are required for testing the atmosphere for flammability, oxygen deficiency, carbon monoxide, and physical agents. For example, combustible gas indicators are designed to measure the concentration of flammable gases but will not measure or indicate the presence of carbon monoxide at toxic levels. Conversely, a carbon monoxide detector is designed for the measurement of carbon monoxide in ppm and not flammables in percent. It should be noted that combustible gas indicators respond differently to different flammable hydrocarbons and should be calibrated for the specific contaminant. The flammability measurement may be erroneous if the oxygen level is less or greater than at normal atmospheric concentrations. Therefore, it is recommended that the oxygen level be determined prior to flammability testing to make any necessary corrections in the flammability measurement.

When the materials may form a combustible dust mixture, special precautions must be taken to prevent an explosive atmosphere from developing. There are numerous instruments available for measuring airborne dust concentrations; however, none appear to have automatic alarm systems and would require constant personal monitoring. The only practical approach to the control of combustible dusts is to eliminate the hazard by preventive measures, such as: engineering controls, good housekeeping, elimination of ignition sources, isolation of dust-producing operations, and/or training of the employees.

The oxygen-deficiency measuring instrument is designed to measure the volume of oxygen present, usually scaled within a range of 0.0–25%. If the percentage of oxygen in a confined-space atmosphere is less than 19.5% or greater than 25%, special precautions, determined by a qualified person, should be taken. A minimum oxygen level of 19.5% has been adopted for worker safety. The upper oxygen limit has been set at 25% because an increase above this level will greatly increase the rate of combustion of flammable materials.

Continuous and/or frequent monitoring becomes necessary in cases where the work being performed within the confined space has the potential of generating toxic agents. The toxic gas or oxygen deficiency can be generated by the work occurring in the space or by gas being unexpectedly admitted into the confined space after the worker has entered.

Medical Requirements

Medical requirements for workers who might enter a confined space should take into consideration the increased hazard potential of confined spaces. In this setting, the workers must rely more heavily upon their physical, mental, and sensory attributes, especially under emergency conditions. Workers should be evaluated by competent medical personnel to ensure that they are physically and mentally able to wear respirators under simulated and actual working conditions. Because of the additional stress placed on the cardiopulmonary system, some pathologic conditions, such as cardiovascular diseases or those associated with hypoxemia, should preclude the use of respiratory protective devices.

In areas where the hazard potential is high, a person certified in CPR and first aid should be in attendance. Because irreversible brain damage can occur in approximately four minutes in an oxygen-deficient atmosphere, it is essential that resuscitation attempts occur within that time.

Safety Equipment and Clothing

Eye Protection

For persons who wear corrective spectacles, either prescription-ground safety glasses or plano-goggles should be provided. Additionally, if eye-irritating chemicals, vapors, or dusts are present, safety goggles should be required, and if both the face and eyes are exposed to a hazard, a full-coverage face shield with goggles shall be used.

Head Protection

Hard hats should be worn at all times.

Foot Protection

All foot protection should provide protection from falling objects and protection from any other hazard identified.

Body Protection

All personnel entering a confined space should wear full-coverage work clothing. Gloves and clothing made of impervious rubber or similar material should be worn to protect against toxic or irritating materials. If the hazards of heat or cold stress exist in the confined space, clothing tested to provide protection from over-exposure to these hazards should be worn. Other body protection required in specific operations such as welding (flame proofed), riveting (heat resistant), and abrasive blasting (abrasion resistance) should be provided to ensure worker safety.

Hearing Protection

Hearing protection should be required when engineering technology is insufficient to control the noise level. Emergency alarms still should be distinguishable when the hearing protection is worn.

Respiratory Protection

Respiratory protection should be determined by a qualified person based upon conditions and test results of the confined space and the work activity to be performed. Half-mask respirators are not recommended for use in any atmosphere greater than ten times the PEL because of the probability of accidentally breaking the face piece-to-face seal because of the work condition in a confined space. Also, gas masks designed for the same respiratory protection may be substituted for chemical cartridge respirators, but they are more cumbersome and restrictive to movement. The minimum service time of self-contained breathing apparatus should be calculated on the entry time, plus the maximum work period, plus twice the estimated escape time for safety margin.

Hand Protection

If hands are exposed to rough surfaces or sharp edges, the degree of protection can range from canvas to metal mesh gloves. Gloves made of impervious rubber or similar material should be worn to protect against toxic or irritating materials. Heat-protective gloves should be used when employees handle objects with temperatures greater than 140°F. Where a current flow through the body of more than 5 milliamperes can result from contact with energized electrical equipment, employees should wear insulating gloves that have been visually inspected before each use. Above 5,000 volts, rubber gloves should be used.

Additional safety equipment necessary to protect a worker in the environment of a confined space includes: a safety belt with "D" rings for attaching a lifeline; the combination of a body harness and/or safety belt with lifeline when an employee is required to enter to complete the gas analysis, when an employee is working in an area where entry for purposes of rescue would be contraindicated (special limitations or fire hazard); when any failure of ventilation would allow the buildup of toxic or explosive gases within the time necessary to evacuate the area, or when the atmosphere is immediately dangerous to life and health. Safety belts should be used as the primary means of suspension for the lifeline. If the exit opening is less than 18 inches in diameter, a wrist type harness should be used.

When it is determined by a qualified person that none of the special hazards associated with confined spaces pose an immediate threat to life, as in Class C (described later), then lifelines shall be readily available but not used during entry and work procedures.

Training

Training employees for entering and working in confined spaces is essential because of the potential hazards and the use of life-saving equipment. To ensure worker safety, the training program should be especially designed for the type of confined space involved and the problems associated with entry and exit. If different types of confined spaces are involved, this will require additional training. Areas that should be covered in an effective training program are:

- Emergency entry and exit procedures.
- Use of applicable respiratory equipment.
- First aid and cardiopulmonary resuscitation (CPR).
- Lockout procedures.
- Safety equipment use.
- Rescue drills.
- Fire protection.
- Communications.

Work Practices

Purging and Ventilation

Poor natural ventilation is one of the defining parameters of a confined space, therefore, purging and mechanical ventilation must be closely evaluated when safe work practices are developed for entering and working in confined spaces. Purging is the initial step in adjusting the atmosphere in a confined space to acceptable standards (PELs, LELs, and LFLs). This is accomplished either by displacing the atmosphere in the confined space with fluid or vapor (inert gas, water, steam, cleaning solution) or by forced air ventilation. The standard to bring the atmosphere in the confined space into equilibrium with the external environment is 20 air changes.

After purging, one should establish general and/or local exhaust ventilation to maintain a safe uncontaminated level. In addition, other information applicable to the special problems of confined spaces must be considered at this point. Entering into an inert atmosphere is one of the most hazardous activities associated with working in a confined space. Work in an inert atmosphere usually is performed by employees of companies who specialize in this because of the high degree of training and expertise needed to perform inert entry operations safely.

Isolation/Lockout/Tagging

The use of tags, while valuable for identification and/or information purposes, appears to have been inadequate in preventing accidents. Proper isolation and lockout procedures are more effective than tagging. Included in this category are warning signs to be posted at all entrances to any confined space.

Cleaning

Decontaminating a space by cleaning is necessary to provide for worker safety. However, it must be recognized that the cleaning process itself can generate additional hazards. Continuous and/or frequent monitoring is required during this process to determine that flammable mixtures and hazardous concentrations of contaminants are adequately diluted before safe entry can be made.

Equipment and Tools

Equipment and tools for use in a confined space should carefully inspected and should meet the following requirements:

- Hand tools should be kept clean and in good repair.
- Portable electrical tools, equipment, and lighting should be equipped with a ground fault interrupter.
- Electrical cords, tools, and equipment should be of heavy duty type with heavy insulation.
- Air-driven power tools should be used when flammable liquids are present.
- Lighting used in Class A or B confined spaces should be of explosion-proof design.

- Cylinders of compressed gases never should be taken into a confined space and should be turned off at the cylinder valve when not in use.
- Ladders should be adequately secured.
- Scaffolding should be properly designed to carry maximum expected load and should be equipped with traction-type planking.
- Electrical connections should be sealed.
- Hose lines and components should be designed for compressed gas work with a relief valve outside the confined space.
- Equipment used in a flammable atmosphere should be approved as explosion-proof or intrinsically safe for the atmosphere involved.

Permit System

The inherent dangers associated with a confined space indicate the need for strict control measures of employees and equipment. The use of a permit system provides written authorization for entering and working in confined spaces, clearly states all known or potential hazards, and identifies the safety equipment required to ensure the safety of the worker.

Entry and Rescue

The potential hazards associated with a confined space must be evaluated prior to entry. These hazards would include the following: oxygen level, flammability characteristics, toxic agents, and physical hazards such as limited openings or communications. To simplify entry and rescue, an entry classification has been set up. Three classes of severity have been defined.

Class "A" is a confined space that presents a situation that is immediately dangerous to life or health (IDLH). These include but are not limited to oxygen-deficient atmospheres, explosive or flammable atmospheres, and concentrations of toxic substances.

Class "B" is a confined space that has the potential for causing injury and illness, if preventive measures are not used, but not immediately dangerous to life and health.

Class "C" is a confined space in which the potential hazard would not require any special modification of the work procedure.

The classification shall be determined by the most hazardous condition of entering, working in, and exiting of the confined space. If the work practice involved in the confined space has the potential to increase existing hazards or generate additional ones, it shall be necessary to frequently evaluate the space to determine if a classification change is warranted.

Rescue procedures should be specifically designed for each entry. If a confined space has an A or B classification, there should be a trained standby person assigned to that confined space with a fully charged, positive pressure, self-contained breathing apparatus (SCBA) at hand. Provide additional communications to all workers within the confined space to summon rescue personnel if necessary. Under no circumstances should the standby person enter the confined space until he or she is relieved and is assured that adequate assistance is present. However, while waiting for rescue personnel, the standby person should make rescue attempts utilizing the lifelines from outside the confined space. Rescue teams entering a Class A or B confined space should be equipped with all the safety equipment of the standby person and required lifelines.

In the event of a Class C confined space rescue, a supplied-air respirator or a self-contained breathing apparatus should be used. A person summoned or one who recognizes the need for rescue should summon assistance and wait outside the confined space. Respirators and lifelines should be donned by rescue personnel with necessary equipment for removal of the victim or victims.

It is essential that well-planned rescue procedures and the proper use of personal protective equipment be followed. The literature reviewed has shown a very poor record in successful rescue efforts. Therefore, the standby and/or rescue team should be properly equipped and trained in all aspects of rescue.

Record Keeping

It is apparent that record-keeping systems should be kept to identify areas where accidents occur so underlying causes can be determined. The records to be kept by the employer should contain such information as employee name, age, training, job description, number of years on the job, accident location and severity, underlying causes, and action taken to ensure future worker safety.

REFERENCES

1. *Working in Confined Spaces,* (NIOSH Publication No. 80-106), San Diego, California, Safety Sciences, December 1980.
2. *Atmospheres in Sub-Surface Structures and Sewers,* Data sheet 550, Chicago, National Safety Council, 1970.
3. Kleinfeld, M., Feiner, B., "Health Hazards Associated with Work in Confined Spaces," *Occup. Med. 8*(7), July 1976, pp 358–364.
4. *Confined Spaces—An Overview of the Hazards and Recommended Control,* OSHA Regulations, Raleigh, North Carolina, North Carolina Department of Labor, 1977, pp. 95–126.
5. "Confined Spaces," *National Safety News,* Chicago, National Safety Council, October 1967, pp. 40–43.
6. *Safety Measures for Work in Tanks,* ENPI Technical Committee for the Chemical Industry, Securitas 2-3, 1978, pp 281–357.
7. *Encyclopedia of Occupational Health and Safety,* Volume 1, Geneva, International Labor Office, 1978, pp. 330–332.

8. Saunders, R.A., "A New Hazard in Closed Environmental Atmospheres," *Arch Environ Health* Vol. 14, March 1967, pp 380–84.
9. "Effects of Carbon Monoxide," *Fire Journal,* July 1967, pp 46–47.
10. Brief, R.S., Raymond, L.W., Meyer, W.H., Yoder J.D., *Better Ventilation for Close-Quarter Work Spaces, Air Conditioning, Heating, and Ventilation,* 1971, pp 74–88.
11. Marshall, W.F., Hurn, R.W., Hazard from Engines Rebreathing Exhaust in Confined Space, *US Bureau of Mines Report of Investigation 7757,* US Dpt of Interior, 1973.
12. "Preventing Confined Space Accidents," *MOH,* Volume 22(2), Lansing, Michigan, Michigan Dpt of Public Health, 1977, pp 1–8.
13. "Working Safely in Confined Spaces," *Safety Newsletter,* Chicago, National Safety Council, August 1975.
14. Allison, W.W., "Work in Confined Areas," *National Safety News,* Chicago, National Safety Council, February 1976, pp 45–50, April 1976, pp 61–67.

CHAPTER 15

NOISE MEASUREMENT

Paul N. Cheremisinoff

Department of Civil and Environmental Engineering
New Jersey Institute of Technology
Newark, NJ 07102, USA

CONTENTS

INTRODUCTION

Criterion level is the continuous equivalent A-weighted sound level that constitutes 100% of an allowable exposure. For OSHA purposes, this is 90 dBA for an 8-hour exposure.

Threshold level is the A-weighted sound level at which a personal noise dosimeter begins to integrate noise into a measure exposure. Sound levels below these thresholds are not included in the computation of a noise dose.

Paragraphs 29 CFR 1910.95 (a) and (b) of the OSHA Occupational Noise Exposure Standard date back to the 1969 Walsh-Healy Standard. Because the standard predates noise dosimetry and OSHA instructions for noise measurement were absent, dosimeters were developed that considered the threshold level to be the same as the 90 dBA criterion level.

Paragraph 1910.95(d) of the March 8, 1983, Hearing Conservation Amendment to the Occupational Noise Exposure Standard prescribes the use of an 80 dBA threshold level for monitoring situations (while the criterion level is still 90 dBA) where employees may be required to be included in a hearing conservation program.

Therefore, OSHA compliance officers must be aware of the differences from the use of dosimeters with either the 80 dBA or 90 dBA threshold levels. Typically, an 80 dBA threshold dosimeter will yield a noise dose greater than or equal to the dose from a 90 dBA threshold dosimeter when both are used simultaneously to evaluate the same noise exposure.

A dosimeter readout can be used to calculate both the continuous equivalent A-weighted sound level (L_A) and the 8-hour, time-weighted average (TWA) for the time period sampled, using the following formulas:

$$L_A = 90 + 16.16 \log \left[\frac{D}{12.5T}\right] \tag{1a}$$

$$TWA = 90 + 16.61 \log \left[\frac{D}{100}\right] \tag{1b}$$

where:

L_A = the continuous equivalent A-weighted sound level in decibels for the time period sampled

D = the dosimeter readout in percent noise dose

T = the sampling time in hours

TWA = the 8-hour, time-weighted average in decibels.

Equation 1b is used for enforcement purposes while equation 1a can be used to assist in evaluating hearing protectors and engineering controls.

Exchange Rate

The exchange rate is a trade-off between the A-weighted sound level in decibels and the duration of exposure in hours. The OSHA exchange rate is 5 dB, Table D-2 of the construction noise standard, 29 CFR 1926.52, and Tables G-16 and G-16a of the general industry noise standard, 29 CFR 1910.95.

Table 1
Exposure/Threshold Relationship

Exposure Conditions	90 dBA Threshold	80 dBA Threshold
90 dBA for 8 hours	100%	100%
89 dBA for 8 hours	0%	87%
85 dBA for 8 hours	0%	50%
80 dBA for 8 hours	0%	25%
79 dBA for 8 hours	0%	0%
90 dBA for 4 hours plus 80 dBA for 4 hours	50%	62.5%
90 dBA for 7 hours plus 89 dBA for 1 hour	87.5%	98.4%
100 dBA for 2 hours plus 89 dBA for 6 hours	100%	163.5%

Only 5 dB instruments may be used for OSHA compliance measurements. Certified Safety and Health Officers (CSHOs) should be aware that noise dosimeters used by the Department of Defense use a 4 dB exchange rate, and instruments used by the Environmental Protection Agency and most foreign governments use a 3 dB exchange rate.

The hypothetical situations listed in Table 1 illustrate the relationship between criterion level, threshold, and exchange rate. Given a 90 dBA criterion level and 5 dB exchange rate, assume ideal threshold activation and continuous sound levels.

EFFECTS

Auditory Effects

Chronic noise-induced hearing loss is a sensorineural permanent condition and cannot be treated medically. It is initially characterized by a declining sensitivity to high frequency sounds, usually to frequencies above 2,000 Hz.

Exposure of a person with normal hearing to intense noise will cause a temporary hearing loss (temporary threshold shift).

Extra-auditory Effects

- Speech interference
- Physiological effects:
  - Sleep reaction
  - Stress reaction
- Behavioral effects:
  - Morale
  - Reduced output efficiency
- Annoyance
- Fatigue
- Mental health effects

INSTRUMENT PERFORMANCE

Effects of the Environment

Temperature

Sound-measuring equipment should perform within design specifications for an ambient temperature range between –20°F and 140°F (–7°C to 66°C).

If the temperature at the measurement site is outside this range, refer to the manufacturer's specifications and determine if the sound level meter or dosimeter is capable of performing properly at the site.

Sound-measurement instruments should not be stored in automobiles during temperature extremes. Warm-up drift, moisture condensation, and weakened batteries may result and adversely affect performance.

Humidity

OSHA noise instruments will perform accurately as long as moisture does not condense or deposit on the microphone diaphragm. Where special problems concerning moisture condensation or rain are present, refer to the OSHA Assistant Regional Administrator (ARA) for Technical Support.

Atmospheric Pressure

Both atmospheric pressure and temperature affect the output of sound-level calibrators. The more significant of the two is atmospheric pressure. When checking an acoustical calibrator, **always** apply the corrections for atmospheric pressure as designated in the manufacturer's instruction manual. Both atmospheric pressure and temperature also affect the measured sound level.

In general, if the altitude of the measurement site is less than 10,000 feet above sea level, the correction is negligible and need not be considered further. If the measurement site is at an altitude higher than 10,000 feet or if the site is pressurized above ambient pressure, as is typical in submarine tunnel construction, use the following equation to correct the instrument reading:

$$C = 10 \log \left\{ \frac{460 + t}{528} \frac{0.530}{B} \right\} \tag{1.3}$$

where:
C = Correction, in decibels, to be added to or subtracted from the measured sound level.
t = Temperature in degrees Fahrenheit.
B = Barometric pressure in inches of mercury.
NOTE: For high altitude locations, C will be positive, while for hyperbaric conditions, C will be negative.

Wind or Dust

Wind or dust blowing across a microphone produces turbulence noise that may cause positive error in the measurement. Use a windscreen for all out-of-doors measurements and whenever there is significant air movement or dust inside a building (e.g., cooling fans or gusting wind through open windows). The ARA for Technical Support shall be consulted for special instrumentation when extreme air turbulence is encountered.

Magnetic Fields

Certain operations such as heat sealers, induction furnaces, generators, transformers, electromagnets, arc welding, and radio transmitting devices generate electromagnetic fields that may induce current in the electronic circuitry of sound-level meters and noise dosimeters, causing erratic reading. If sound-level meters or dosimeters must be used near such devices, attempt to determine if the effect of the magnetic field is negligible. Follow the manufacturer's instructions or consult with the ARA for Technical Support for assistance.

Effects of Sound

Microphone Placement

For sound-level meters and noise dosimeters equipped with omnidirectional microphones as used by OSHA, the effects of microphone placement and orientation of instrument reading in relation to the sound field are negligible in a typically reverberant environment.

To minimize body-shielding effects on microphones, consult the manufacturer's recommended placement. If the measurement site is nonreverberant and/or the noise source is highly directional, consult the manufacturer's literature to determine proper microphone placement and orientation.

Impulse Noise

For determining compliance with the impulse noise provision covered in 29 CFR 1910.95 (b)(1) or 29 CFR 1926.52(e), use the unweighted peak mode setting of the GenRad 1982 or, Quest 155, or an equivalent impulse-precision sound-level meter.

NOISE MEASUREMENTS

Instruments

Noise Dosimeter

The noise dosimeters used by OSHA meet the American National Standards Institute (ANSI) standard S1.25-1978, "Specifications for Personal Noise Dosimeters," which set performance and accuracy tolerances.

For OSHA use, the dosimeter must have a 5 dB exchange rate, 90 dBA criterion level, slow response, and either an 80 dBA or 90 dBA threshold gate for the appropriate standard to be evaluated.

Sound Level Meter

All sound-level meters used by OSHA meet ANSI Standard S1.4-1971 (R1976) or S1.4-1983, "Specifications for Sound Level Meters," which sets performance and accuracy tolerances. The sound level meter may be used for five purposes:

- To spot-check noise dosimeter performance.
- To determine an employee's noise dose by time-motion studies whenever a noise dosimeter is unavailable or inappropriate.
- To identify individual noise sources for abatement purposes.
- To aid in determining engineering feasibility of controls for individual noise sources for abatement purposes.
- To evaluate hearing protectors.

Octave-band Noise Analyzers

The Type 1 sound-level meters (such as the GenRad 1982 and 1933 and the Quest 155) used by OSHA feature built-in octave-band analysis capability. This can be of assistance in determining the feasibility of controls for individual noise sources for abatement purposes and also can be used to evaluate hearing protectors.

The analyzer segments noise into its component parts. Octave-band filter sets provide filters with the following geometric-mean frequencies: 31.5, 63, 125, 250, 500, 1,000, 2,000, 4,000, 8,000 and 16,000 Hz.

The spectral signature of a given noise situation can be obtained by taking sound-level meter readings at each of these settings, assuming the noise situation is fairly constant over time. The result usually indicates octave-bands that contain a majority of the total radiated sound power.

This can assist CSHO in determining the adequacy of various types of frequency-dependent noise controls. The result also can be used to determine the efficacy of various hearing protectors by examining the amounts of attenuation offered by the protectors in the octave bands corresponding to the octave bands where most of the sound energy is present.

The ARA for Technical Support can provide assistance in the use of octave-band analysis.

For compliance measurements, the A-weighted network and slow response shall be used exclusively. All continuous, intermittent, and impulse noise shall be included in computations of noise dose for compliance with any OSHA noise standard.

Sound-level reading in a nonreverberant environment should be taken by following the manufacturer's instructions. For practical purposes, this procedure should be followed for all sound-level measurements.

The microphone should be in an employee's hearing zone. OSHA defines the hearing zone as a sphere with a 2 ft diameter surrounding the head. Considerations of practicality and safety for each survey location will dictate the actual microphone placement.

When noise levels at an employee's two ears are different, the higher level shall be used for compliance determinations.

Accuracy

For compliance purposes, readings with an ANSI Type 2 sound-level meter and readings with a noise dosimeter are considered to have an accuracy of ±2 dBA. Readings with an ANSI Type 1 sound-level meter are considered to have an accuracy of ±1 dBA. For unusual measurement situations, refer to the instrument instruction manual and appropriate ANSI standards for guidance in interpreting instrument accuracy.

Calibration

Calibrate all noise measuring instruments according to the manufacturer's instructions before and after each day of use, and whenever the environment temperature and relative humidity change significantly.

Sampling Strategy

Results of the walk-through survey are used to assist the CSHO in planning the work shift sampling strategy.

Walk-through Survey

When screening for noise exposures, only sound-level meter measurements and estimates of exposure duration need to be collected. An integrating sound-level meter may be useful for this purpose. These spot readings are used to determine the need for a more complete evaluation.

Workshift Sampling

Whenever the results of the walk-through survey indicate that the provisions of the applicable OSHA noise standards may be violated, sample representative employees from each job classification who may be exposed above the permissible sound levels or action levels.

Use both a dosimeter with a threshold of 90 dBA and one with a threshold of 80 dBA on those employees identified during the walk-through whose exposure may exceed the levels of Table G-16 in 29 CFR 1910.95.

In general, for enforcement purposes, a 90 dB threshold dosimeter is necessary to establish noncompliance with Table G-16 and D-2 (see Table 2). However, in unusual situations the results of the 80 dB threshold dosimeter may be discussed with the *ARA for Technical Support* to determine whether they sufficiently demonstrate noncompliance with Tables G-16 and D-2. See Table 3 for exposure exceeding 230 percent.

Use one dosimeter with a threshold of 80 dBA on those employees who can be identified during the walk-through as having noise exposure that do not approach noncompliance with Tables G-16 of 29 CFR 1910.95 but whose exposure may exceed levels in Table G-16a.

Use one dosimeter with a threshold of 90 dBA on those employees who can be identified during the walk-through as having noise exposure that exceeds sound levels in Table G-16 of 1910.95 or Table D-2 of 1926.52.

29 CFR 1926 does not contain any of the monitoring requirements of the 1983 29 CFR 1910 hearing conservation amendment. Therefore, 29 CFR 1926 addresses only the 90 decibels threshold.

As a minimum strategy, conduct sampling for the time necessary to establish that an exposure above levels in Tables G-16, G-16a, or D-2 exists, taking into account instrument accuracy.

Table 2
Interpretation of Noise Dosimeter Readings Using a 90 dBA Threshold Level, 90 dBA Criterion Level, and a 5 dBA Exchange Rate

Dosimeter Reading	Tables G-16 and D-2	Table G-16a
Less than or equal to 50%	Within PEL[1]	Inclusive to
Greater than 50% but less than 66%[2]	Within PEL	Inconclusive
Greater than or equal to 66% but less than 100%	Within PEL	Over action
Greater than or equal to 100% but less than 132%[2]	Inconclusive	Over action
Greater than or equal to 132%	Over PEL	Over action

[1]Permissible Exposure Level (PEL)
[2]Uncertainty range with respect to the ±2dBA accuracy of a Type 2 noise dosimeter.
Data from 29 CFR 1910.95 Tables G-16 and G-16a and 29 CFR 1026.52 Table D-2

Table 3
Interpretation of Nose Dosimeter Readings Using an a 90 dBA Threshold Level, 90 dBA Criterion Level, and a 5 dBA Exchange Rate

Dosimeter Reading	Table G-16	Table G-16a
Less than or equal to 50%	Within PEL[1]	With Action Level
Greater than 50% but less than 66%[2]	Within PEL	Inconclusive
Greater than or equal to 66% but less than 100%	Within PEL	Over Action Level
Greater than or equal to 100%	Inconclusive	Over Action Level
Greater than 230%[3]	Over PEL	Over Action Level

[1]Permissible Exposure Level (PEL)
[2]A 50 percent reading corresponds to 85 dBA for eight hours or 80 dBA for 16 hours. The enforcement level is 66 percent with respect to the ±2dBA accuracy of a Type 2 noise dosimeter.
[3]The 230% value results from a possible 99% dosimeter reading below the 90 dBA threshold coupled with the 132% reading necessary for instrument uncertainty.
Data from 29 CFR 1910.95

Sampling Protocol

- Inform the employee that the dosimeter should not interfere with normal duties, and emphasize that the employee should continue to work in a routine manner.
- Explain to each employee being sampled the purpose of the dosimeter and that it is not a speech recording device.
- Instruct the employee being sampled not to remove the dosimeter unless absolutely necessary and not to cover the microphone with a coat or outer garment. Inform the employee when and where the dosimeter will be removed.
- When the dosimeter is positioned (normally in the shirt pocket or at the waist), clip the microphones to the employee's shirt collar at the shoulder close to the employee's ear. Care should be taken to ensure that the microphone is in the vertical position. Placement should be according to the manufacturer's instructions.
- Position and secure any excess microphone cable to avoid inconveniencing to the employee and snagging.
- Check the dosimeter periodically to ensure that the microphone is oriented properly. Take and note sound-level meter readings during the different phases of work performed by the employee during the shift. Take enough readings to identify work cycles. Widely fluctuating noise needs more sound-level meter readings for statistical purposes.
- Record the information required.
- Some dosimeters indicate when a 115 dBA sound level has been exceeded. This indication is not to be used for compliance determinations.

ULTRASONICS

Ultrasound is high frequency sound that is inaudible to the human ear. The human ear frequency audibility limit is approximately 20 kHz. This limit is not fixed in nature, and individuals can have higher or lower (usually lower) personal limits. The frequency limit normally declines with age.

Most of the audible noise associated with ultrasonic sources such as welders or cleaners is actually comprised of subharmonics of the machines' fundamental ultrasonic frequencies. For example, many ultrasonic welders have a fundamental operating frequency of 20 kHz, which is right at the human frequency audibility limit. However, a good deal of noise is also present at 10 kHz, the first subharmonic frequency of the 20 kHz operating frequency, and is audible to most people.

Applicability of 29 CFR 1910.95

29 CFR 1910.95 addresses airborne ultrasonic noise in relation to OSHA criterion parameters. This regulation sets employees' exposures based on the A-weighting filter. At 20 kHz, an A-weighted levels is down 10 dB from the unweighted

sound-pressure level. At 50 kHz, it will be −25 dB. Although these are large reductions, they still are notable and measurable if the unweighted sound-pressure levels are intense and no other interfering sources are present. This means that if one were to measure a 50 kHz, 110 dB tone using an A-weighting network, the instrument would indicate 85 dBA. However, ANSI S1.4(R1976) or S1.4-1983, which sets the performance and accuracy tolerances of all OSHA sound-level meters, requires a Type II microphone accuracy of only ±5 at frequencies above 10,000 Hz.

Health Effects

Research indicates that there is little effect on general health from ultrasonics except for direct body contact with a radiating source. Reported cases of headaches and nausea associated with airborne ultrasonic exposures appear to be from high levels of audible noise from source subharmonics. The American Conference of Governmental Industrial Hygienists (ACGIH) has adopted 1988–1989 permissible ultrasound exposure levels. However, these recommended limits—set at the middle frequencies of the one-third octave bands from 10 kHz to 50 kHz—are designed for the prevention of possible hearing loss from the subharmonics of the set frequencies.

Any attempted measurement of any source suspected of producing sound exceeding the ACGIH recommended limits requires the use of a precision sound-level meter equipped with a suitable microphone of adequate frequency response and a portable third-octave filter set.

Controls

High frequency noise is very directional and is relatively easily reflected or blocked by any type of barrier. The wavelength of a 16 kHz tone, for example, is about 3/4 inch, so a barrier of 1 to 2 inches should be sufficient to reflect noise of approximately the same frequency away from a nearby worker. Such barriers are inherent in some machine surfaces.

The ACGIH permissible levels are listed in Table 4.

High frequency audible noise also is easily absorbed by so-called acoustical materials such as glass fiber or foam. Moreover, higher frequency sound is absorbed readily by ambient air, which implies that any high frequency noise problem is almost certain to be a very local problem in a given workplace layout.

Information to be Collected

In addition to the information required for a health inspection, collect the following information where pertinent and where necessary to document a violation of 29 CFR 1910.95, 1926.52, or 1926.101.

Table 4
ACGIH Permissible Levels

Mid-frequency of Third-Octave Band kHz	One-Third Octave Band Level dB (unweighted)
10	80
12.5	80
16	80
20	106
31.5	115
40	115
50	115

Employee Data

- Distance from the employee to the primary noise source(s).
- Whether the employee's presence or proximity is required in the noise field.
- Employee exposure time pattern.
- Existence of any known employee auditory problems (e.g., ear infections, ringing in the ears, or trouble with hearing after the workshift).
- Employee's opinion of the practicality of potential noise controls (considering machine operation), where relevant.
- Hearing protection provided and any problems with its use or acceptance. Octave band levels and/or dBC and dBA data may aid in the evaluation of hearing protectors.
- Time since the last audiometric examination and frequency of the examinations.

Machine and/or Process Data

- Brief description of the type of machine and/or process, including identifying numbers, sketches, and photographs whenever possible.
- Condition of the machine, (e.g., age and maintenance).
- Machine operation (e.g., speed, cycle times, parts/minute, and materials used).
- Apparent existing noise and/or vibration controls.
- Source(s) and characteristics of the noise (e.g., fan noise, discrete and broadband components, continuous or noncontinuous). Use of octave-band analyzers, real-time analyzers, and narrow-band analyzers may be useful in determining sources of noise.
- Practical engineering and/or administrative controls and cost of controls.

Building Data

- Size and shape of the room.
- Layout of equipment, workstations, and break areas.
- Surface materials (e.g., ceiling/steel; walls/cinder block; floor/concrete).
- Existing acoustical treatment.
- Potential acoustical treatment.
- Noise from other sources (spill-over noise).
- Presence of barriers or enclosures.

Employer Data

- What has been done to control the noise? Are consultants used? Is plant noise monitored? Are controls implemented?
- What changes are planned?
- Are administrative controls used? How are they enforced?

Hearing Loss Data

Document the following when hearing loss is used to support a citation:

- The amount of the threshold shift and date it was recorded.
- Employee exposure level.
- Frequency and duration of employee exposure.
- Length of employment.
- Explanation of any followup measure taken.
- Duration of audiometric testing program.

Evaluation of Hearing Protection

The actual effectiveness of any individual hearing protector cannot be determined under work place conditions; however, OSHA noise standards 29 CFR 1910.95(j)(2) and 1926.52(B) require that personal hearing protection must attenuate the occupational noise received by the employee's ears to within the levels given in Tables G-16, G-16a, and D-2. Hearing protectors are evaluated under rigorous laboratory conditions specified by the American National Standards Institute in ANSI Z24.22-1957 (R1971) and ANSI S3.19-1974. OSHA experience and published scientific literature indicate that laboratory-obtained real ear attenuation data for hearing protectors are seldom achieved in the workplace.

29 CFR 1910.95(c)-(p)

When evaluating compliance with the hearing conservation amendment, use Appendix B of 29 CFR 1910.95 to determine the laboratory-based noise reduction for the use of a given hearing protector.

Field Attenuation of Hearing Protection

When analyzing the attenuation a personal hearing protector may afford a noise-exposed employee in an actual work environment, the hearing protector shall be evaluated as follows:

- To determine the laboratory-based noise reduction for a given hearing protector, use Appendix B of 1910.95.
- To adjust for the lack of attainment of the laboratory-based noise reduction calculated according to Appendix B estimating techniques, apply a safety factor of 50% (divide the calculated laboratory-based attenuation by two).
- For dual protection (i.e., ear plugs and muffs), add 5 dB to the Noise Reduction Rating of the higher-rated protector. Examples of calculating the hearing protector attenuation.
- NOTE: A different safety factor may seem appropriate in a particular instance. Where it appears that the attenuation of the hearing protector is not sufficient to reduce employee noise exposure below the levels listed in Tables G-16 and D-2, the employer should be advised that a greater degree of employee protection appears necessary. The degree of employee protection afforded the hearing protection in use is of particular significance when a determination is being made as to whether engineering and/or administrative controls are not feasible.

EVALUATION OF THE HEARING CONSERVATION PROGRAM

In all cases where employees are exposed to sound levels that exceed those in Tables D-2 or G-16a, a continuing effective hearing conservation program is required. The program must include all those employees whose work brings them either steadily or intermittently into areas in which the sound levels exceed those permitted by 29 CFR 1926.52 or 1910.95.

Agricultural Work Sites

Because paragraph 29 CFR 1928.21(a) does not reference the general industry noise standard (29 CFR 1910.95), a mandatory hearing conservation program does not apply to agricultural operations. However, if the CSHO inspects such operations likely to cause employees to be exposed to noise in excess of an 8-hour, time-weighted average of 85 dBA, then the employer should be advised that it is good practice to provide ear protection and to train employees in the proper use and fit of ear protection and in the hazards of noise exposure. Wherever it is practical, periodic audiometric testing should be encouraged to ensure the effectiveness of hearing protection.

Maritime Work Sites

Shipyard and longshoring operations come under the requirements of the general industry noise standard; therefore, employers in such operations must meet the elements of the general industry Hearing Conservation Amendments (29 CFR 1910.95(c) through (P)).

Construction Work Sites

Construction employees are not covered by the Hearing Conservation Amendment. However, certain aspects of hearing conservation are covered by the construction noise standards 29 CFR 1926.52 and 1926.101. In evaluating hearing conservation programs in construction employment, consider the following elements:

Monitoring

For the employer to know who must wear hearing protection and to be able to evaluate the adequacy of hearing protection used, the sound created by the different machines and operations should be known. This information can be obtained through a variety of sources, such as sound surveys, full-shift dosimetry, utilization of the OSHA-funded consultation program, insurance carriers, equipment manufacturers, or technical literature. The sophistication of monitoring will be limited by the employer's size and the practicalities of the job site. Although lack of monitoring does not constitute a violation, it is indicative of a less-than-fully effective hearing conservation program.

Training

Training and motivation are crucial to the effectiveness of hearing protection and the hearing conservation program in general. Training is much more difficult to provide to the transient workforce in construction than to general industry or maritime operation employees. However, training is required as outlined in 29 CFR 1926.21 (2).

Hearing Protection

20 CFR 1926.52(b) requires hearing protection to be provided and used to reduce the sound levels reaching the cochlea to within the levels of Table D-2. In addition, paragraph 1926.101(b) requires all insert hearing protectors to be

fitted individually to each overexposed employee by competent persons. A competent person is defined as a person trained in how to fit ear protection and one who is able to recognize the difference between a good fit and a poor fit.

Audiometric Testing

When employees are overexposed to noise, periodic audiometric testing is required if it is practical to do and if followup tests would track the effectiveness of the program on individual employees. The only way to ensure that a hearing conservation program is continuing and effective is to perform periodic audiograms.

General Industry Work Sites

Requirements

The Hearing Conservation Amendment (29 CFR 1910.95 (c)-(p)) establishes the minimum requirements for a continuing effective hearing conservation program. This program is mandated for all employees exposed at or above a dose of 50%.

Standard Threshold Shift (STS)

The standard threshold shift is an indicator of hearing loss that may be revealed by an annual audiogram of an employee. As defined in 19 CFR 1910.95 (g)(10), a standard threshold shift is "a change in hearing threshold relative to the baseline audiogram of an average of 10 dB or more at 2,000, 3,000, and 4,000 Hz in either ear." Table 5 gives examples of computing the STS.

CONTROL

Noise control should

- Control sources of noise;
- Preclude the propagation, amplification, and reverberation of noise; and
- Isolate the workers.

Engineering control could include antivibration machine mountings, acoustical enclosures, and so forth. Administrative practices may require shift rotation or exposure limitation. Personal protection equipment such as ear muffs and ear plugs can be used.

CALCULATING ATTENUATION

For a muff or plug, obtain the Noise Reduction Rating (NRR). Calculate the laboratory-based attenuation using this manual or NIOSH procedures when no

Table 5
Computing the Standard Threshold Shift

Example 1:

Frequency in Hertz	Baseline Audiogram Threshold in dB	Annual Audiogram Threshold in dB	Change
500	5	5	0
1000	5	5	0
2000	0	10	+10
3000	5	20	+15
4000	10	35	+25
6000	10	15	+5

Considering the values for 2000, 3000, and 4000 Hz, there are changes in hearing threshold of 10, 15, and 25 dB, respectively.

Thus:

$$STS = \frac{(10+15+25)}{3} = \frac{50}{3} = 16.7 \text{ dB}$$

Conclusion: The STS is +16.7 dB; hearing has deteriorated; the employee must be notified in writing within 21 days; and, depending on professional discretion, the employer may elect to revise the baseline.

Example 2:

Frequency in Hertz	Baseline Audiogram Threshold in dB	Annual Audiogram Threshold in dB	Change
500	5	5	0
1000	5	0	-5
2000	0	-10	-10
3000	5	-5	-10
4000	10	-5	-15
6000	10	5	-5

Again considering the values for 2000, 3000, and 4000 Hz, the hearing threshold is lowered by 10, 10, and 15 dB, respectively.

Thus:

$$STS = \frac{(-10-10-15)}{3} = \frac{-35}{3} = -11.6 \text{ dB}$$

Conclusion: The STS is –11.6 dB; hearing has improved significantly; the employee should be notified; and, depending on professional discretion, the baseline may be revised.

NRR is available for the protector. Because the noise dosimeter gives a TWA in dBA, subtract 7 dB from the NRR for spectral uncertainty pursuant to Appendix B in 1910.95. Consider the following examples:

Example 1: TWA_8 = 100 dBA and muff NRR = 19
Approximately Field Attenuation is: (19-7) × 50% = 6 dB
100 dBA TWA_8 – 6dB = 94 dBA

Conclusion: The protected TWA can be assumed to be 94 dBA. If feasible, engineering controls must be implemented. The protected TWA, without using the 50% safety factor, is 88 dBA. Therefore, better hearing protection is not required by the standard.

Example 2: Given TWA_8 = 98 dBA and plug NRR = 29
Approximately Field Attenuation is: (29-7) × 50% = 11 dB
98 dBA TWA_8 – 11 dB = 87 dBA

Conclusion: The protected TWA can be assumed to be 87 dBA. Engineering controls do not have to be implemented unless an employee exhibits an STS. Better hearing protection is not required because the protected TWA, without the safety factor, is 76 dBA. If an employee exhibits an STS, better hearing protection should be recommended.

Example 3: Given TWA = 110, plug NRR= 29 and muff NRR = 25
Employer requires dual protection.

a. Calculate field adjusted NRR for the higher-rated protector: (29-7) × 50% = 11 dB
b. Add 5 dB to this field adjusted NRR: 11 + 5 = 16 dB
c. Calculate the protected TWA_8:
110 dBA TWA_8 – 16 dB = 94 dBA.

Conclusion: The protected TWA_8 can be assumed to be 94 dBA. If feasible, engineering controls must be implemented. The protected TWA_8, without using the 50% safety factor, is 83 dBA. Therefore, better hearing protection is not required by the standard; however, it should be recommended.

Comments: In example 2, the conclusion states that engineering controls must be implemented if the exposed employee has an STS. However, because the noise exposure is less than 100 dBA, the *CSHO* must first consider if improvements can be made to the existing hearing conservation program to meet a protected TWA_8 of 85 dBA.

BIBLIOGRAPHY

Alberti, P.W., Ed. 1982. *Personal Hearing Protection in Industry.* New York: Raven Press.

Beranek, L.L., Ed. 1971. *Noise and Vibration Control.* New York: McGraw-Hill.

Berger, E.H., W.D. Ward, J.C. Morrill and L.H. Royster, Eds. 1986. *Noise and Hearing Conservation Manual.* Fourth Edition. Akron, OH: American Industrial Hygiene Association.

Cheremisinoff, P.N., 1993. *Industry Noise Control,* Englewood Cliffs, NJ 07632, Prentice Hall, Inc.

Code of Federal Regulations. 1989. Title 29, Parts 1900 to 1990. Washington, DC: U.S. Government Printing Office.

Ellerbusch, F. 1990. *Guide for Industrial Noise Control.* Matawan, NJ: SciTech Publishers.

Harris, C.M. 1957. *Handbook of Noise Control.* Second Edition. New York: McGraw-Hill.

Harris, C.M. Ed. 1988. *Shock and Vibration Handbook.* Third Edition. New York: McGraw-Hill.

Kryter, K.D. 1970. *The Effects of Noise on Man.* New York: Academic Press.

Kryter, K.D. 1985. *The Effects of Noise on Man.* Second Edition. New York: Academic Press.

Olishifski, P.E. and E.R. Harford, Eds. 1975. *Industrial Noise and Hearing Conservation.* Chicago: National Safety Council.

OSHA Field Operations Manual. 1989. OSHA Instruction CPL 2.45B. June 15, 1989. Washington,D.C.: U.S.Government Printing Office.

Sataloff, J. and P. Michael. 1973. *Hearing Conservation.* Springfield, IL: Charles C. Thomas.

Sataloff, R.T. and J.T. Sataloff. 1987. *Occupational Hearing Loss.* New York: Marcel Dekker.

CHAPTER 16

UNDERGROUND STORAGE TANKS

Paul N. Cheremisinoff

Department of Civil & Environmental Engineering
New Jersey Institute of Technology
Newark, NJ 07102, USA

CONTENTS

INTRODUCTION

A primary concern in groundwater contamination is leaking underground storage tanks. Such tanks store all types of liquids, including gasoline, oil and fuels, process chemicals, hazardous and toxic materials, and wastes. Of all potential sources of groundwater contamination, this source may pose the most serious risk to health and the environment.

The major federal legislation affecting underground storage tanks is the Resource and Recovery Act of 1976 (RCRA), amended in 1984. Subtitle I regulates underground tanks storing petroleum products, including gasoline and crude oil, and any substance defined as hazardous under Superfund. Subtitle I does not regulate tanks storing hazardous wastes as defined by RCRA. Such tanks are regulated under Subtitle C, which requires stringent design standards, including secondary containment for hazardous waste tanks. Subtitle I provides for the development and implementation of a comprehensive regulatory program for underground storage tanks containing "regulated substances" and the release of these substances into the environment.

LEAK DETECTION REQUIREMENTS

EPA's underground storage tank (UST) regulations require leak detection systems. Tanks and piping installed after December 22, 1988, must have leak detection when they are installed. For USTs installed before that date, the leak detection requirements were phased in over five years, depending on the age of the tank. By 1993, all USTs had to meet federal leak detection requirements. State and local regulations may be more stringent than federal requirements. Federal requirements phased in over five years for existing USTs are as follows:

If your UST was installed	It must have had leak detection by December of
before 1965 or unknown	1989
1965–1969	1990
1970–1974	1991
1975–1979	1992
1980–Dec. 1988	1993

Because of the threat to the environment, existing pressurized piping systems had to meet leak detection requirements by December 22, 1990. Other kinds of existing piping must have complied by the dates shown in the schedule above. Figure 1 shows leak detection methods for tanks and piping.

State or local regulations may differ from federal requirements, so it is necessary to check and see which requirements apply to your UST. Rather than requiring specific technologies, the EPA has identified a variety of general leak detection methods that owners and operators can use to meet the federal requirements. You can use:

- Groundwater monitoring,
- Vapor monitoring,
- Secondary containment with interstitial monitoring, or
- Automatic tank gauging.

These are all monthly monitoring methods, and eventually everyone must use at least one of them. However, as a temporary method (for ten years after new tank installation and for up to ten years for some older existing tanks), you can combine tank tightness testing and manual monthly inventory control (or manual tank gauging if you have a very small tank).

Not all of these leak detection methods can be used for both tanks and piping. Leak detection methods for piping include groundwater monitoring, vapor monitoring, secondary containment with interstitial monitoring, and tightness testing. Pressurized piping also must have an automatic line leak detector.

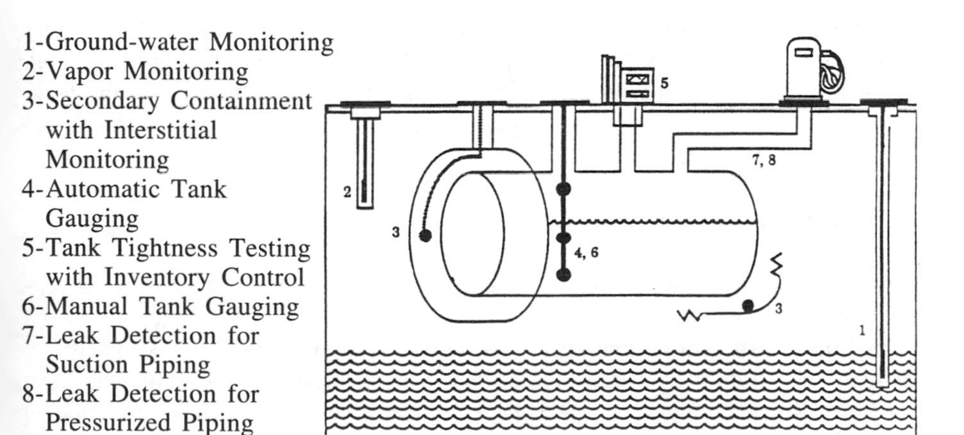

Figure 1. Leak detection methods for tanks and piping.

Groundwater Monitoring

Groundwater monitoring senses the presence of liquid product floating on the groundwater. This methods requires installation of monitoring wells at strategic locations in the ground near the tank and along the piping runs. To discover if leaked product has reached groundwater, these wells can be checked periodically by hand or continuously with permanently installed equipment. This method cannot be used at sites where groundwater is more than 20 feet below the surface.

Vapor Monitoring

Vapor monitoring senses and measures product "fumes" in the soil around the tank piping to determine the presence of a leak. This method requires installation of carefully placed monitoring wells. Vapor monitoring can be performed manually on a periodic basis or continuously using permanently installed equipment.

Secondary Containment with Interstitial Monitoring

Secondary containment consists of placing a barrier—a vault, liner, or double-walled structure—around the UST. Leaked product from the inner tank or piping is directed toward an interstitial monitor located between the inner tank or piping and the outer barrier. Interstitial monitoring methods range from a simple dip stick to a continuous automated vapor or liquid sensor permanently installed in the system.

Automatic Tank Gauging

Monitors permanently installed in the tank are linked electronically to a nearby control device to provide information on product level and temperature. During a test period of several hours when nothing is put into or taken from the tank, these monitors are used to automatically calculate the changes in product volume that can indicate a leaking tank.

Tank Tightness Testing and Inventory Control

This is a combination of periodic tank tightness testing and monthly inventory control.

Tightness tests require temporarily installing equipment in the tank. There are two types of tightness tests: volumetric and non-volumetric. A volumetric test involves filling the tank to a specified level and precisely measuring the change in level and temperature over several hours. Non-volumetric test methods include ultrasound techniques and tracer gas detectors. These are sophisticated tests and must be performed by trained, experienced professionals.

In addition to tightness testing, monthly inventory control should be used. Every month the product volume is balanced between what is delivered and sold from the tank with daily measurements of tank volume taken with a gauge stick, which measures what is actually in the tank. If the volumes don't balance, there may be a leak. This combined method can only be used during the first ten years after new tank installation or upgrade of the existing UST.

Manual Tank Gauging

To meet the federal leak detection requirements, this method can be used by itself only for tanks with capacities of up to 550 gal. It requires keeping the tank undisturbed for at least 36 hours, during which time no product can be added or removed. During that period, measure the contents of the tank twice, for example, at the beginning and at the end of a weekend. At the end of the month, average the weekly tests and compare the volume lost, if any, to the permissible standard in the federal regulations. For tanks with capacities between 550 gal and 2,000 gal, this method is allowed only in combination with tank tightness testing. This combined method, however, can be used only during the first ten years following tank installation or upgrade.

Choosing a Method

Choosing a leak detection method is not a cut and-dried process. There is no one leak detection system that is best for all sites, nor is there a particular type of leak detection that is consistently the least expensive.

Each of the leak detection methods has advantages and disadvantages. For example, vapor detection devices work rapidly and most effectively in dry soils, while liquid detectors are most appropriate for areas with a high water table. Identifying the correct option or combination of options depends on a number of factors including cost, tank type, groundwater depth, soil types, and other variables.

Table 1 is a listing of a few of the factors that could influence selection of the leak detection method considered best for a particular site.

GROUNDWATER MONITORING

Groundwater monitoring cannot be used at sites where ground water is more than 20 ft below the surface. When installed and operated according to manufacturer's instructions, a groundwater monitoring system meets federal leak detection requirements for new and existing underground storage tanks (USTs). Operation of a groundwater monitoring system at least once each month fulfills the requirements for the life of the tank. Groundwater monitoring also can detect leaks in piping. New USTs must have leak detection when they are installed.

Table 1
Factors to Consider in Selecting Leak Detection

Detection Option	Site-Specific Factors	Tank-Related Factors	Cost Factors
Groundwater Monitoring	Do not use if groundwater level is greater than 20 ft, if clay soil is present, or if existing product is already on the ground water.	Product must be able to float on water and not mix easily with water.	Well installation: $15–$70/ft depth Equipment: $200–$5,000 per tank.
Vapor Monitoring	Do not use at sites where soil is saturated with water, the backfill is clay, or soil vapor levels are too high.	Product must evaporate easily or substance that evaporates easily must be added to the tank.	$1,200–$6,000 per tank for equipment and installation.
Secondary Containment with Interstitial Monitoring	Site conditions (such as too much water) may require use of containment that completely surrounds tank or piping.	A double-walled system must be able to detect a release through the inner wall.	Total installed cost of $5,000–$12,000 per tank.

Automatic Tank Gauging (ATGS)	If water collects in excavation, ATGS must have water sensor.	To date, used primarily at sites with gasoline and diesel in tanks under 15,000 gallons.	Cost per tank: Equipment = $2,300–$3,900 Installation = $500–$3,000.
Tank Tightness Testing	Volumetric methods must account for presence of ground water and product temperature.	To date, used primarily at sites with gasoline and diesel tanks under 15,000 gallons.	$250–$1,000 per test per tank for problem-free test. If problems occur, costs may be much higher.
Inventory Control	None	None	Under $200, but *must be combined with tank tightness testing.*
Manual Tank Gauging	None	Limited to tanks under 550 gallons when used alone or under 2,000 gallons when combined with tightness testing.	Under $200, but *may also require tightness testing.*
Automatic Line Leak Detectors	None	Used only for pressurized lines.	Total installed cost of $400–$2,000 per line.
Line Tightness Testing	None	Used only for piping.	$50–$100 per test per line if conducted with tank test. May be more expensive if conducted alone. Must do test every 3 years.

Groundwater monitoring involves the use of one or more permanent monitoring wells placed close to the UST. The wells are checked at least monthly for the presence of product that has leaked from the UST and is floating on the groundwater surface. The two main components of a groundwater monitoring system are the monitoring well (typically a well of a 2 to 4 in. in diameter) and the monitoring device.

Groundwater monitoring can be used only if the stored substance does not easily mix with water and floats on top of water.

If groundwater monitoring is to be the sole method of leak detection, the groundwater must not be more than 20 ft below the surface, and the soil between the well and the UST must be sand, gravel, or other coarse materials.

Monitoring wells must be properly designed and sealed to prevent contamination from outside sources. The wells also must be clearly marked and locked.

Wells should be placed in or near the UST backfill so leaks can be detected as quickly as possible.

Product detection devices must be able to detect 1/8 in. or less of leaked product on top of the ground water.

In general, groundwater monitoring works best at UST sites where:

- The groundwater surface extends beneath the tank;
- Monitoring wells are installed in the tank backfill;
- Ground water is 2 to 10 ft from the surface; and
- There are no previous releases of product that would falsely indicate a current release.

A site assessment is critical for determining these site-specific conditions.

The proper design and construction of a monitoring system is crucial to effective detection of leaked product and should be performed by an experienced contractor. Before construction begins, any specific state or local requirements should be identified.

Purchasing a groundwater monitoring system is similar to other major purchase. Shop around, ask questions, get recommendations, and select a company that meets the needs of your UST site. Generally the capital costs for groundwater monitoring are much greater than the annual operating costs.

Installation

The number and placement of wells is very important. Many state and local agencies require one to four monitoring wells per UST, and additional wells may be required for piping. Before installation, a site assessment is necessary to determine the soil type, groundwater depth and flow direction, and the general geology of the site.

Variations

Detection devices may be permanently installed in the well for automatic, continuous measurements of leaked products.

Detection devices are also available in manual form. Manual devices range from a bailer used to collect a liquid sample for visual inspection to a device that can be inserted into the well to electronically indicate the presence of leaked product. Manual devices must be operated at least once a month.

VAPOR MONITORING

When installed and operated according to manufacturer's instructions, vapor monitoring meets the federal leak detection requirements for new and existing USTs. Operation of a vapor monitoring system at least once each month fulfills the requirements for the life of the tank. Vapor monitoring also can be installed to detect leaks from piping. State or local limitations on the use of vapor monitoring or requirements different from those presented herein should be followed. New USTs must have leak detection when they are installed.

Vapor monitoring measures "fumes" from leaked product in the soil around the tank to determine if the tank is leaking. Fully automated vapor monitoring systems have permanently installed equipment to continuously gather and analyze vapor samples and respond to a release with a visual or audible alarm. Manually operated vapor monitoring systems range from equipment that immediately analyzes a gathered vapor sample to devices that gather a sample that must be sent to a laboratory for analysis. Monitoring results from manual systems are generally less accurate than those from automated systems. Manual systems must be used at least once a month to monitor a site.

All vapor monitoring devices should be calibrated annually to ensure that they are properly responding to vapor. Maintenance items vary depending upon the system. Manual systems usually require more maintenance than automated systems.

Vapor monitoring requires the installation of monitoring wells within the tank backfill. Usually one well per 20 to 40 ft surrounding tanks and piping is sufficient, but the proper number depends upon the site conditions. The UST backfill must be sand, gravel, or another material that will allow the vapors to easily move to the monitor. The backfill should be clean enough that previous contamination does not interfere with the detection of a current leak. The substance stored in the UST must vaporize easily so that the vapor monitor can detect a release.

High groundwater, excessive rain, or other sources of moisture must not interfere with the operation of vapor monitoring systems for more that 30 consecutive days.

Monitoring wells must be locked and clearly marked.

Before installing a vapor monitoring system, a site assessment should determine whether vapor monitoring is appropriate at the site. A site assessment usually includes at least a determination of the groundwater level and background contamination. Some vapor monitoring systems can overcome site problems, such as clay backfill. Discuss any problems that may apply to your site with the equipment salesman and contractor to ensure they have considered the problems and will compensate for them, if necessary, when installing the vapor monitoring system.

The cost of a vapor monitoring system is influenced by the UST site condition, the required number and depth of monitoring wells, whether an automated or manual system is chosen, and the complexity of the chosen system and its maintenance. If a site needs to be cleaned up before a system can be installed, costs would increase. However, vapor monitoring has very low annual operating costs, unless a manual system requires laboratory analysis.

SECONDARY CONTAINMENT WITH INTERSTITIAL MONITORING

When installed and operated according to manufacturer's specifications, secondary containment with interstitial monitoring meets the federal leak detection requirements for new and existing USTs. Operation of the monitoring device at least once each month fulfills the requirements for the life of the tank. Secondary containment with interstitial monitoring also can be used to detect leaks from piping.

State or local requirements may differ from those described herein. In some jurisdictions, secondary containment is required for all USTs.

Correct installation is fairly difficult yet is crucial both for the barrier and the interstitial monitor. Therefore, trained and experienced installers are necessary. The purchase of secondary containment with interstitial monitoring is similar to any other major purchase. Shop around, ask question, get recommendations, and select a method and company that can meet the needs of the UST site.

Secondary Containment

Secondary containment provides a barrier between the tank and the environment. The barrier holds the leak between the tank and the barrier long enough for the leak to be detected. The barrier is shaped so that a leak will be directed towards the monitor. Barriers include:

- Double-walled tanks, in which an outer tank partially or completely surrounds the primary tank;
- Leakproof excavation liners that partially or completely surround the tank;
- Leakproof liners known as "jackets" that closely surround the tank; and

- Concrete vaults, with or without lining.
- Clay and other earth materials cannot be used as barriers.

Regulatory requirements include:

- The barrier must be immediately around or beneath the tank.
- The interstitial monitor must be checked at least once every 30 days.
- A double-walled system must be able to detect a release through the inner wall.

An excavation liner must:

- Direct a leak towards the monitor
- Not allow the specific product being stored to pass through it any faster than 10^{-6} cm/sec;
- Be compatible with the product stored in the tank;
- Not interfere with the USTs cathodic protection;
- Not be disabled by moisture;
- Always be above groundwater and the 25-year flood plain; and
- Have clearly marked and locked monitoring wells, if they are used.

In areas with high groundwater or a lot of rainfall, it may be necessary to select a secondary containment system that completely surrounds the tank to prevent moisture from interfering with the monitor.

Interstitial Monitors

Monitors are used to check the area between the tank and the barriers for leaks and alert the operator if a leak is suspected. Some monitors indicate the physical presence of the leaked product, either liquid of gaseous. Other monitors check for a change in condition that indicates a hole in the tank, such as a loss of pressure or a change, in the level of water between the walls of a double-walled tank.

Monitors can be as simple as a dipstick used at the lowest point of the containment to see if liquid product has leaked and pooled there. Monitors can also be sophisticated automated systems that continuously check for leaks.

AUTOMATIC TANK GAUGING SYSTEMS (ATGS)

When installed and operated according to manufacturer's specifications, automatic tank gauging systems (ATGS) meet federal leak detection requirements for new and existing USTs. A test performed each month fulfills the requirements for the life of the tank.

The product level and temperature in a tank are measured continuously and automatically analyzed and recorded by a computer. In the inventory mode, the ATGS replaces the use of the gauge stick to measure product level and perform inventory control. This mode records the activities of an in-service tank, including deliveries.

The ATGS must be able to detect a leak at least as small as 0.2 gal per hour. By December 1990, the ATGS also must meet federal regulatory requirements regarding probabilities of detection and false alarm.

ATGS have been used primarily on gasoline or diesel fuel tanks with a capacity of less than 15,000 gal. If considering using an ATGS for larger tanks or products other than gasoline or diesel, discuss the applicability with the manufacturer's representative.

Water around a tank may hide a leak by temporarily preventing the product from leaving the tank. To detect a leak in this situation, the ATGS should be capable of detecting water in the bottom of a tank.

The ATGS probe is permanently installed through a pipe (not the fill pipe) on the top of the tank. Each tank at a site must be equipped with a separate probe. The ATGS probe is connected to a monitor that displays ongoing product level information and the results of the monthly test. Printers can be connected to the monitor to record this information. For most ATGS, up to eight tanks can be connected to a single monitor.

ATGS usually are equipped with alarms for high and low product level, high water level, and theft.

ATGS can be linked with computers at other locations, from which the system can be programmed or read. No product should be delivered or withdrawn from the tank for a least six hours before the monthly test or during the test, which generally takes one to six hours.

An ATGS can be programmed to perform a test more often than once per month, if so desired.

Purchasing an ATGS is similar to any other major purchase. Shop around, ask questions, get recommendations, and select a method and company that can meet the needs of the site.

Costs

Equipment Costs

- **Monitor:** $1,700–2,700; varies with manufacturer and whether a printer is included.
- **Probe:** $500–1,100/probe; varies with manufacturer.
- **Cables:** $0.15–1.00/foot; varies with the contractor and the part of the country.

Installation Costs for a Typical 3-tank System

- For a site that already has conduits for cables: $500–1,500.
- For a site in which conduit must be laid and average distance: $2,500–3,000.

- For a very complex site with many conduits running long distances and where rewiring is necessary: up to $10,000.

Annual Operating Costs for a Typical 3-Tank System

- About $50–100 for electricity, printer paper, and maintenance.
- For a typical 3-tank system, the total equipment costs range from about $5,00 for a basic system to $10,000 for a top-of-the-line system. The installation cost averages about $2,000–3,000.

TANK TIGHTNESS TESTING AND INVENTORY CONTROL

When performed according to manufacturer's specifications, periodic tank tightness testing combined with monthly inventory control can temporarily meet the federal leak detection requirements for new and existing USTs. State and local requirements may differ from those presented herein. These two leak detection methods must be used together because neither method alone meets federal requirements for leak detection.

Tank Tightness Testing

Tightness tests include a wide variety of methods. Other terms used for these methods include precision testing and volumetric testing. There are a few methods that do not measure the level or volume of the product. Instead, these methods use a principle such as acoustics to determine the physical presence of a hole in the tank. With such methods, all of the factors in the following bullets may not apply.

- Most tightness test methods are volumetric methods in which the change in product level or volume in a tank over several hours is measured very precisely (in millimeters or thousandths of an inch).
- For most methods, change in product temperature also must be measured precisely (thousandths of a degree) with level measurements because temperature changes can cause volume changes that interfere with finding a leak.
- For most methods, a net decrease in product volume (subtracting out temperature-induced volume changes) over the time of the test indicates a leak.
- The testing equipment is temporarily installed in the tank, usually through the fill pipe.
- The tank must be taken out of service for the test, generally six to 12 hours, depending on the method.
- Many test methods require that the product in the tank be at a certain level before testing, which often requires adding product from another tank on site or purchasing additional product.

Some tightness test methods require all of the measurements and calculations to be made by hand by the tester. Other tightness test methods are highly automated. After the tester sets up the equipment, a computer controls the measurements and analysis.

There are several different acceptable ways to measure product temperature: mixing the product so it is all one temperature; using a sensor that calculates an average temperature throughout the depth of the product; and using at least three temperature sensors at different product levels to calculate an average temperature. A few methods measure properties of the product that are independent of temperature, such as the mass of the product, and so do not need to measure product temperature.

The tightness test method must be able to detect a leak at least as small as 0.1 gal per hour. By December 1990, the tightness test method also must be able to meet federal regulatory requirements regarding probabilities of detection and false alarm.

Testing Frequency

Tightness test must be performed periodically as shown in the following table:

New tanks	Every five years for ten years following installation.
Existing tanks	Every five years for ten years upgraded following upgrade.
Existing tanks	Every year until not upgraded.
1998	Upgraded tanks have corrosion protection and spill/overfill prevention devices.

After the applicable time period listed above, monitoring at least once per month.

Tank tightness testing has been used primarily on gasoline and diesel fuel tanks less than 15,000 gal in capacity. To use tightness testing for larger tanks or products other than gasoline or diesel fuel, discuss the method's applicability with the manufacturer's representative.

For most methods, the test is performed by a testing company. Manifolded tanks generally should be disconnected and tested separately. Depending on the method, up to four tanks can be tested at one time. Generally, an automated system is necessary to test three or four tanks at one time. Procedure and personnel, not equipment, are usually the most important factors in a successful tightness test. Therefore, well-trained and experienced testers are very important. Some states and local authorities may have tester certification programs.

Purchasing a tightness test is similar to any other major purchase. Shop around, ask questions, get recommendations, and select a method and a company that can meet the needs of the site.

There are no capital costs for test equipment. The total cost per test is highly variable. The prices quoted by testing companies range from about $250 to $1,000 per tank, with most between $500 and $800. These prices are for a simple test with no problems. The final cost for a tank tightness test can be significantly higher. Some factors that would add to the cost of a test one would ask are:

- Product to fill the tank to the minimum testing level, if it is product that you would not buy otherwise.
- Lost business from shutting down the tank during normal business hours.
- Replacing or repairing parts of the tank system before a test can be performed.
- Uncovering part of the tank system and then recovering it, to correct problems such as vapor pockets or piping that must be valved off.
- Travel charges.

Costs can be reduced if a large number of tanks are tested and if repairs and replacements are done before the test crew arrives.

Inventory Control

Inventory control is basically like balancing a checking account. Every month the product volume is balanced between what is delivered and what is sold from the tank with daily measurements of tank volume taken with a gauge stick. If the numbers don't balance, you may have a leak.

UST inventories are determined in the morning and in the evening or after each shift by using a gauge stick, and the data is recorded on a ledger form. The level on the gauge stick can be translated to a volume of product in the tank using a calibration chart, which is often furnished by the UST manufacturer. The amounts of product delivered to and withdrawn from the UST each day also are recorded.

At least once each month, the gauge stick data and the sales and delivery data are reconciled, and the month's overage or shortage is determined. If the overage or shortage is greater than or equal to 1.0% of the tank's flow-through volume plus 130 gallons of product, the UST may be leaking.

Inventory control must be used in conjunction with periodic tank tightness tests. The gauge stick should be long enough to reach the bottom of the tank and should be marked so the product level can be determined to the nearest 1/8 inch. A monthly measurement must be taken to identify any water at the bottom of the tank. Deliveries must be made through a drop tube that extends to within one foot of the tank bottom. Product dispensers must be calibrated to the local weights and measures standards.

MANUAL TANK GAUGING

Manual tank gauging can only be used for smaller tanks. Tanks up to 550 gal can use this method alone, but tanks from 550 to 2,000 gal can use manual tank gauging only when it is combined with tank tightness testing. Manual tank gauging cannot be used for tanks with capacities exceeding 2,000 gal. As described below, when performed according to recommended practices, manual tank gauging meets the federal leak detection requirements for USTs with a capacity of 550 gal or less for the life of the tank.

Four liquid level measurements must be taken weekly, two at the beginning and two at the end of at least a 36-hour period during which nothing is added or removed from the tank. The average of the two consecutive ending measurements are subtracted from the average of the two beginning measurements to indicate the change in product volume.

Every week, the calculated change in tank volume is compared to the standards shown in Table 2. If the calculated change exceeds the weekly standard, the UST may be leaking. Also, monthly averages of the four weekly test results must be compared to the monthly standard in the same way.

Liquid level measurements must be taken with a gauge stick that is marked to measure the liquid to the nearest 1/8 in.

Manual tank gauging may be used as the sole method of leak detection for tanks with a capacity of 550 gal or less for the life of the tank. For tanks with a capacity of 551 to 2,000 gal, manual tank gauging must be combined with a tightness testing shown in Table 3.

Table 2
Manual Tank Gauging Standards

Tank Capacity	Weekly Standard (one test)	Monthly Standard (avg. of 4 tests)
up to 550 gal	10 gal	5 gal
551–1,000 gal	13 gal	7 gal
1,001–2,000 gal	26 gal	13 gal

Table 3
Minimum Tightness Testing Frequency

New tanks	Every 5 years for 10 years following installation
Existing tanks, upgraded	Every 5 years for 10 years following upgrade
Existing tanks, not upgraded	Every year until 1998

Upgraded tanks have corrosion protection and spill/overfill prevention devices.

Note that this combined method will meet the federal requirements only temporarily. After the applicable time period listed, a monitoring method must be performed at least once a month.

Tanks greater than 2,000 gal in capacity may not use this method of leak detection to meet these regulatory requirements. Manual tank gauging is inexpensive and can be an effective leak detection method when used as described above with tanks of the appropriate size. For tanks less than 550 gal, the only costs are the price of a gauge stick and perhaps some product-finding paste. These costs are nominal (less than $200). For 550 to 2,000 gal tanks, there is the additional cost of periodic tank tightness testing. These costs are highly variable (from about $250 to well over $1,000 for each test).

LEAK DETECTION FOR PRESSURIZED UNDERGROUND PIPING

When installed and operated according to manufacturer's specifications, the leak detection methods discussed here meet the federal regulatory requirements for the life of new and existing pressurized underground piping systems. State and local requirements may differ from those described herein. New pressurized piping must leak detection when it is installed. Existing pressurized piping had to meet the leak detection regulatory requirements by December 22, 1990.

Each pressurized piping run must have one leak detection method from each of the following sets

An Automatic Line Leak Detector:

- Automatic flow restrictor,
- Automatic flow shutoff, or
- Continuous alarm system.

And One Other Method:

- Monthly groundwater monitoring,
- Monthly vapor monitoring,
- Monthly interstitial monitoring, or
- Annual tightness test.

The automatic line detector (LLD) must be able to detect a leak at least as small as three gallons per hour at a line pressure of ten pounds per square in within one hour by shutting off the product flow, restricting the product flow, or triggering an audible or visual alarm. The line tightness test must be able to detect a leak at least as small as 0.1 gal per hour when the line pressure is one and one-half times its normal operating pressure. By September 22, 1991, automatic LLDs and line tightness tests also must be able to meet the federal regulatory requirements regarding probabilities of detection and false alarm. Groundwater, vapor, and interstitial monitoring have the same regulatory requirements for piping as they do for tanks.

Automatic Line Leak Detectors (LLDs)

Flow restrictors and flow shutoffs can monitor the pressure within the line in a variety of ways: whether the pressure decreases over time; how long it takes for a line to reach operating pressure; and combinations of increases and decreases in pressure.

If a possible leak is detected, a flow restrictor keeps the product flow through the line at three gallons per hour, well below the usual flow rate. If a possible leak is detected, a flow shut-off completely cuts off product flow in the line or shuts down the pump. A continuous alarm system monitors line conditions continuously and immediately triggers an audible or visual alarm if a leak is suspected. Automated vapor or interstitial line monitoring systems also can be set up to operate continuously and sound an alarm, flash a signal on the console, or even ring a telephone in a manager's office when a leak is detected.

Both automatic flow restrictors and shutoffs are permanently installed directly into the pipe or the pump housing. Vapor and interstitial monitoring systems can be combined with automatic shutoff systems so that whenever the monitor detects a possible release, the piping system is shut down. This would qualify as a continuous alarm system. Such a setup would meet the monthly monitoring requirement as well as the LLD requirement.

Line Tightness Testing

The line is taken out of service and pressurized, usually above the normal operating pressure. A drop in pressure over time, preferably one hour, suggests a possible leak. Tightness tests must be conducted annually. Most line tightness tests are performed by a testing company. Some tank tightness test methods can be performed to include a tightness test of the connected piping. For most line tightness tests, no permanent equipment is installed. The line must be taken out of service for the test, ideally for several hours to allow the line to stabilize the test. In the event of trapped vapor pockets, it may not be possible to conduct a valid line tightness test. There is no way to tell definitely before the test begins if this will be a problem, but longer, complicated piping runs with a lot of risers and dead ends are more likely to have vapor pockets.

Secondary Containment with Interstitial Monitoring

A barrier is placed between the piping and the environment. Double-walled piping or a leakproof liner in the piping trench can be used. A monitor is placed between the piping and the barrier that senses a leak if it occurs. The simplest monitor is a stick that can be put into a sump to see if a liquid is present, such as those that monitor for the presence of evaporated product.

Proper installation of secondary containment is the most important and the most difficult aspect of this release detection method. Trained and experienced installers are necessary.

Groundwater or Vapor Monitoring

Groundwater monitoring checks for leaked product floating on the groundwater near the piping. Vapor monitoring detects product that leaks into the soil and evaporates there. A monitoring well should be installed every 20 to 40 ft. UST systems using groundwater or vapor monitoring for the tanks are well suited to use the same monitoring method for the piping.

Costs

Automatic LLDs

Automatic flow restrictors can be installed for about $300–$400 per line. Automatic shutdown devices cost about $2,000 for one line. There may be cost savings for multiple lines or when tank monitoring is included in the same system. Annual operating costs are negligible.

Line Tightness Tests

When performed at the same time as a tank tightness test, a typical line test costs about $50 to 100. The price varies with the length and complexity of the piping. If a testing company comes on site to perform only a line tightness test, the cost will probably be much higher unless you can negotiate a package deal for a large number of tests. Not all tightness testing companies will do independent line tests.

Secondary Containment with Interstitial Monitoring

The total installed cost for double-walled piping at a typical three-tank station is about $10,000, not including digging the trenches. Costs vary with the size of pipe, length of run, site conditions, and contractor. In general, double-walled piping systems cost about three times as much as single-walled systems.

For a typical station, trench liners cost about $25 to $40 per linear foot for 2-in. pipes, depending on the number of pipes. Installation is about $800 to $1,500, depending on site conditions.

Costs for interstitial monitoring devices range from essentially nothing for a dipstick to a total installed cost of about $1,000 per line for an electronic sensor, not including the control panel.

HEATING OIL TANKS GREATER THAN 5,000 GALLONS

Some states go further than the EPA and regulate as USTs all heating oil tanks with capacity greater than 5,000 gal. Such heating oil UST systems are required to meet the same requirements as those required by petroleum UST systems, except that those tanks installed before 1965 or of age unknown were allowed until December 22, 1990, to comply with leak detection.

Most heating oil UST systems are designed to have the product circulating through a pressurized underground delivery line from the tank to the boiler, with a suction return line from the boiler back to the tank. Modifications may be required to use some of the leak detection methods discussed in this chapter.

A large percentage of heating oil UST systems contain No. 2 fuel oil, which is virtually the same as diesel fuel and is easily leak-detected using some of the more common internal and external methods. Another smaller percentage of heating oil tanks contain the heavier No. 4 and No. 5 fuel oils, with the final group containing the very viscous No. 6 or Bunker C grades. The heavy grades slack large concentrations of volatile materials, so some external vapor monitoring devices may need modifications to be effective.

Groundwater Monitoring

To use groundwater monitoring, the regulated substance must be tested to verify that its specific gravity is less than 1.0 and will therefore float on the groundwater surface. Most heating grades of oil meet this requirement. In addition, the groundwater table must never fall below 20 ft from the land surface for more than 30 days.

Vapor Monitoring/Interstitial Monitoring

Because of low volatilization properties of Nos. 4, 5, and 6 fuel oils, vapor monitoring or secondary containment with interstitial monitoring may need modifications for leak detection to occur. One of these modifications is the addition of a more volatile substance to the fuel oil in the tank. The vapor monitoring system is then designed to test for the added substance.

Automatic Tank Gauging

Conversations with several manufacturers of automatic tank gauging equipment have indicated that most of these systems are not usable for leak detection for heating oil systems because of the cycling of the product through the boiler and inability to accurately determine the inventory within 1% plus 130 gal as with daily inventory control methods. In addition, those automatic tank gauging systems that tightness test the tank monthly require tank quiet periods that are not usually achievable in a boiler system, which need to run continuously.

Annual Tank Tightness Testing with Daily Inventory Control

To use annual tank tightness testing with daily inventory control, meters must be installed in the delivery and return lines to monitor the circulation of the product and achieve 1% plus 130 gal throughput accuracy required by the regulations. This method cannot be used past December 22, 1988, for USTs existing prior to December 22, 1988, and for only at a ten-year period for tanks installed on or after December 22, 1988. In addition, filters may be required to protect the meters from product debris.

Manual Tank Gauging

This method cannot be used because the tanks are greater than 5,000 gal in capacity.

BIBLIOGRAPHY

Cheremisinoff, P. N. (1992). *A Guide to Underground Storage Tanks Evaluation, Site Assessment and Remediation,* Prentice Hall, Englewood Cliffs, NJ.

CHAPTER 17

SAMPLING AIRBORNE ISOCYANATES

Giulio Sesana

Health and Safety Institute (PMIP)
USSL 69, Via Spagliardi 19, Parabiago (Mi) Italy

Giuseppe Nano

Dipartimento di Chimica Fisica Applicata
Politecnico di Milano, via Mancinelli 7
Milano, Italy

Arturo Baj

Health and Safety Institute (UOOML)
USSL 63, Piazza Benefattori
Desio (Mi) Italy

CONTENTS

INTRODUCTION

The isocyanates are a group of products derived from primary amines whose general formula is R-N=C=O [1]. They are often classified as esters of isocyanic acid. They are contained in many industrial products as basic compounds in the formulation of polyurethanic resins; this is why in the past they usually were called polyurethanes.

Polyfunctional isocyanates have been widely used in industry because they give rise to the formation of strongly reticulated macropolymers. Diisocyanates especially have had a wide application, and for this reason they have received greater attention in regard to the problems involved in sampling and analysis in industrial hygiene.

The isocyanates most used in industrial processes are at present toluene 2,4 diisocyanate (2,4 TDI), toluene 2,6 diisocyanate (2,6 TDI), 4,4' diphenylmethane

Table 1
Structure Formula, Name, Abbreviated Commercial Name and CAS Number of Main Isocyanates

Structure	Name	Synonym	CAS Number
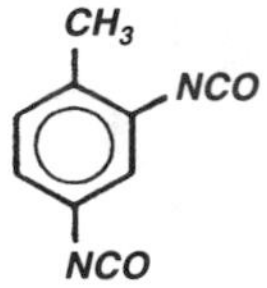 (benzene ring with CH_3, NCO, NCO)	2,4-Toluene diisocyanate	2,4 TDI	584-84-9
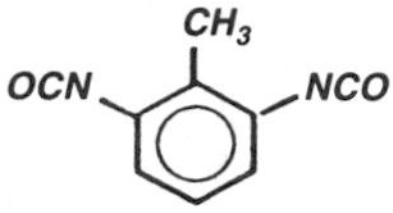 (benzene ring with CH_3, OCN, NCO)	2,6-Toluene diisocyanate	2,6 TDI	91-08-7
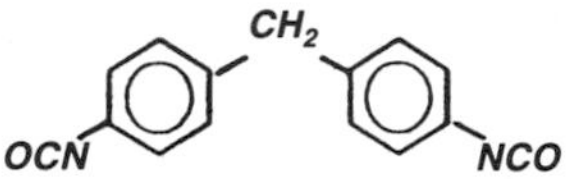 OCN–(phenyl)–CH_2–(phenyl)–NCO	4,4'-Diphenylmethane diisocyanate	MDI	101-68-8
$OCN - (CH_2)_6 - NCO$	1,6-Examethylene diisocyanate	HDI	822-06-0
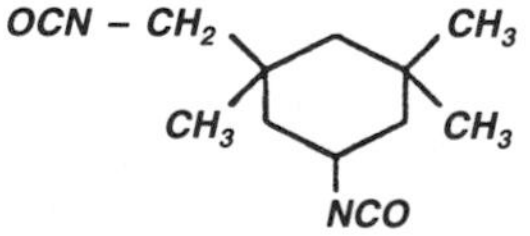 (cyclohexane ring with $OCN - CH_2$, CH_3, CH_3, CH_3, NCO)	Isophorone diisocyanate	IPDI	4098-71-9
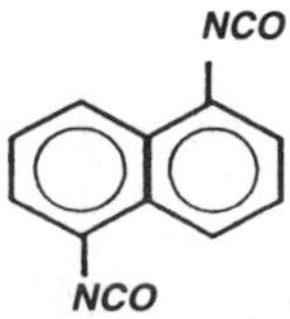 (naphthalene with NCO, NCO)	1,5-Naphthylene diisocyanate	NDI	3173-72-6
$CH_3 - NCO$	Methyl isocyanate	MIC	624-83-9

diisocyanate (MDI). Also used are 1,5 naphthalene diisocyanate (NDI), hexamethylene diisocyanate (HDI), and isophorone diisocyanate (IPDI).

Methylisocyanate (MIC), which was widely used in the past as an intermediate in the synthesis of pesticides (carbamates), special heterocyclic polymeric compounds and their derivatives, is at present little used, especially after the Bhophal catastrophe in India in 1984, where a breakdown in a part of the pesticide production plant caused a high number of deaths and widespread injuries among the population [2].

Table 1 shows the structure formula, the abbreviated commercial name and the CAS number of the main isocyanates, and Table 2 lists their chemical and physical properties.

(text continued on page 390)

Table 2
Chemical and Physical Properties

2,4-Toluene diisocyanate

Synonyms:
2,4-Diisocyanatotoluene
2,4-Diisocyanato-1-methylbenzene
Molecular Weight: 174.17
Physical description: Liquid
Melting Point: 19.5–21.5°C
Boiling Point: 250–251°C
Specific Gravity: 1.22 (4°/20°C)
Vapor Specific Gravity (air=1): 6.0
Flammability: Combustible
Flash Point: 132–135°C

Vapor Pressure:
0.13 kPa at 80–85°C
0.27 kPa at 94–95°C
0.67 kPa at 106–107°C
1.3 kPa at 120–121°C
1.5 kPa at 126°C

Solubility:
Water: Decomposes
Acetone: Miscible
Benzene: Miscible
Ether: Miscible
Ethanol: Miscible (Decomposes)
DMSO: Not available

Reactivity: Concentrated alkaline may cause polymerization. May react violently with compounds containing active hydrogen. Sensitive to sunlight and heat. Decomposes at 275°C.

2,6-Toluene diisocyanate

Synonyms:
2,6-Diisocyanato-1-methylbenzene
2,6-Diisocyanatotoluene
1,3-Diisocyanate-2-methylbenzene

Molecular Weight: 174.17
Physical description: Liquid
Melting Point: Not available

(table continues)

Table 2 (continued)

Boiling Point: Not available
Specific Gravity: Not available
Vapor Specific Gravity (air=1): Not available
Flammability: Combustible
Flash Point: >112°C

Vapor Pressure:
0.4–0.5 kPa at 97°C
0.8 kPa at 101–103°C
2.4 kPa at 129–132°C

Solubility:
Water: React
Acetone: <1 mg/ml at 20°C
Benzene: Miscible
Ether: Miscible
Ethanol: React
DMSO: >100 mg/ml at 20°C

Reactivity: Sensitivity to prolonged exposure to light.

4,4'-dphenylmethane diisocyanate

Synonyms:
1,1-Methylenebis (4-isocyanatobenzene)
Isocyanic acid, metylene di-p-phenylene ester
Molecular Weight: 250.26
Physical description: Solid
Melting Point: 37-46°C
Boiling Point: Not available
Specific Gravity: 1.225 (4°/20°C)
Vapor Specific Gravity (air=1): 8.5
Flammability: Combustible
Flash Point: 218°C

Vapor Pressure:
0.03 Pa at 55°C
1.73 kPa at 100°C
0.04 kPa at 150°C
0.13 kPa at 152–154°C
0.4 kPa at 184°C
0.67 kPa at 192–199°C

Solubility:
Water: <1 mg/ml
Acetone: Soluble
Benzene: Soluble
Ether: Not available
Ethanol: React violently
DMSO: >10 mg/ml at 20°C

Reactivity: React violently with alcohol's
Decomposes at 274°C

1,6-Hexamethylene diisocyanate

Synonyms:
1,6-Diisocyanatohexane
1,6-Exanediol diisocyanate
Isocyanic acid hexamethylene ester
Molecular Weight: 168,22
Physical description: Liquid
Melting Point: –67°C
Boiling Point: Not available
Specific Gravity: 1.05 (4°/20°C)
Vapor Specific Gravity (air=1): 6.0
Flammability: Combustible
Flash Point: 135°C

Vapor Pressure:
1.4 Pa at 20°C
0.13 kPa at 81.5–96°C
0.4 kPa at 102–106°C
0.67 kPa at 112°C

1.3 kPa at 120–128°C
2.7 kPa at 140–142°C

Solubility:
Water: React
Acetone: at 20°C
Benzene: Not available
Ether: Not available
Ethanol: React
DMSO: at 20°C

Reactivity: Concentrated alkaline may react violently. Decomposes at 255°C

Isophorone diisocyanate

Synonyms:
3-Isocyanatomethyl-3,5,5-trimethylcyclohexylisocyanate
Molecular Weight: 222.29
Physical description: Liquid
Melting Point: -60°C
Boiling Point: 158.5°C at 1.3 kPa
Specific Gravity: 1.06 (4°/20°C)
Vapor Specific Gravity (air=1): 7.67
Flammability: Combustible
Flash Point: 43°C

Vapor Pressure:
0.04 Pa at 20°C
0.9 Pa at 50°C
1.3 kPa at 158°C
2.0 kPa at 159°C
13.3 kPa at 217°C

Solubility:
Water: <1 mg/ml at 25°C (Decomposes)
Acetone: > 100 mg/ml at 18°C
Ether: Miscible
Ethanol: > 100 mg/ml at 18°C
DMSO: > 100 mg/ml at 18°C

Reactivity: React with substances containing active hydrogen. Decomposes at 260°C. Lower exposure limit 1%.

1,5-Naphthylene diisocyanate

Synonyms:
1,5-Diisocianato-Naphthalene
1,5-Diisocianato-Naphthalin
Molecular Weight: 210.2
Physical description: Solid
Melting Point: 129–132°C
Boiling Point: Not available
Specific Gravity: Not available
Vapor Specific Gravity (air=1): Not available
Flammability: Not available
Flash Point: Not available

Vapor Pressure:
5.3 kPa at 220–221°C

Solubility:
Water: Not available
Acetone: Not available
Benzene: Not available
Ether: Not available
Ethanol: Not available
DMSO: Not available

Reactivity: Not available

(table continues)

Table 2 (continued)

Methyl isocyanate

Synonyms:
Isocyanatomethane
Isocyanic acid methyl ester
Molecular Weight: 57.05
Physical description: Liquid
Melting Point: –45°C
Boiling Point: 39–59.6°C
Specific Gravity: 0.92 (4°/27°C)
Vapor Specific Gravity (air=1):
Flammability: Flammable
Flash Point: <–6.6°C

Vapor pressure:*
30 kPa at 8°C
50 kPa at 20°C
81 kPa at 32°C
263.7 kPa at 70°C
*Estimated by Ridel's method

Solubility:
Water: Soluble
Acetone: Not available
Benzene: Not available
Ether: Not available
Ethanol: Soluble
DMSO: Not available

Reactivity: React with water, iron, tin, cooper and catalyst. React rapidly in presence of acids, alkalis and amines. Lower exposure limit 5.3%

Notes:
Synonyms: *The most common synonyms are listed for each preferred name.*
Melting Point: *Melting points are drawn from literature sources and reported as a range when two or more melting points are recorded.*
Boiling Point: *Boiling points are drawn from literature sources and reported as a range when two or more melting points are recorded. Boiling point is reported at standard pressure (101.3 kPa).*
Specific Gravity: *Specific gravity is a ratio of the mass of the chemical at a given temperature (generally 20°C) to the mass of an equal volume of water at 4°C.*
DMSO *= Dimethyl sulfoxide*
After Jedrzejczak [19], Keith [20], INRS [21], Bellsteim [22], and DIPPR [23].

(text continued from page 387)

The commercial product usually consists of a prepolymer with low vapor pressure that, when required for use, is mixed with the isocyanate; therefore, occupational exposure depends on the way the product is used, which may involve the formation of mists and consequent exposure via inhalation (for example, by spraying) [3]. Isocyanates are mainly used for the formation of polyurethane foams and resins. The polymers are thermosets and are suitable for moulding and for producing synthetic fibers.

A feature of the -N=C=O functional group is its marked chemical reactivity with a large number of other functional groups, including the isocyanate group, forming in this case dimers, trimers, oligomers, and ureic and carbodiimide derivates (see Figure 1). The reaction of isocyanates is spontaneous with compounds containing atoms of active hydrogen: hydroxylated compounds spontaneously form esters (urethanes). If alcohol is bifunctional, a linear product is obtained.

If at least one of the base products is polyfunctional, the reaction leads to the formation of tridimensional macromolecules (see Figures 2 and 3).

Polyurethane foam usually is obtained from a mixture of an isocyanate with a polyalcohol (polyol) with immediate and spontaneous development of carbon dioxide (CO_2). The reaction of isocyanate with water develops carbon dioxide (see Figure 4).

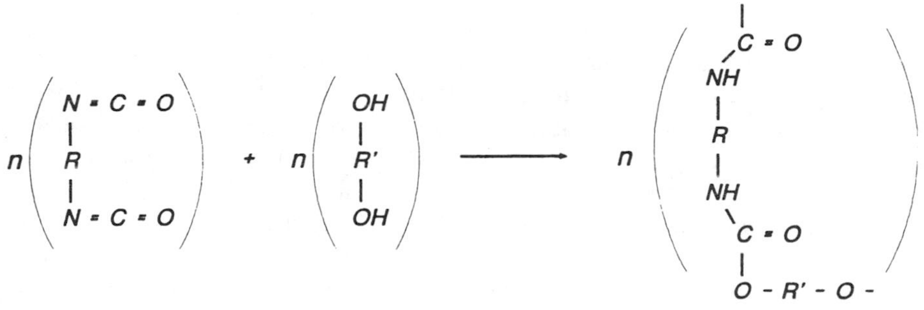

n (N = C = O | R | N = C = O) + n (OH | R' | OH) ⟶ n (| C = O / NH | R | NH \ C = O | O - R' - O -)

Figure 1. The -N=C=O functional group.

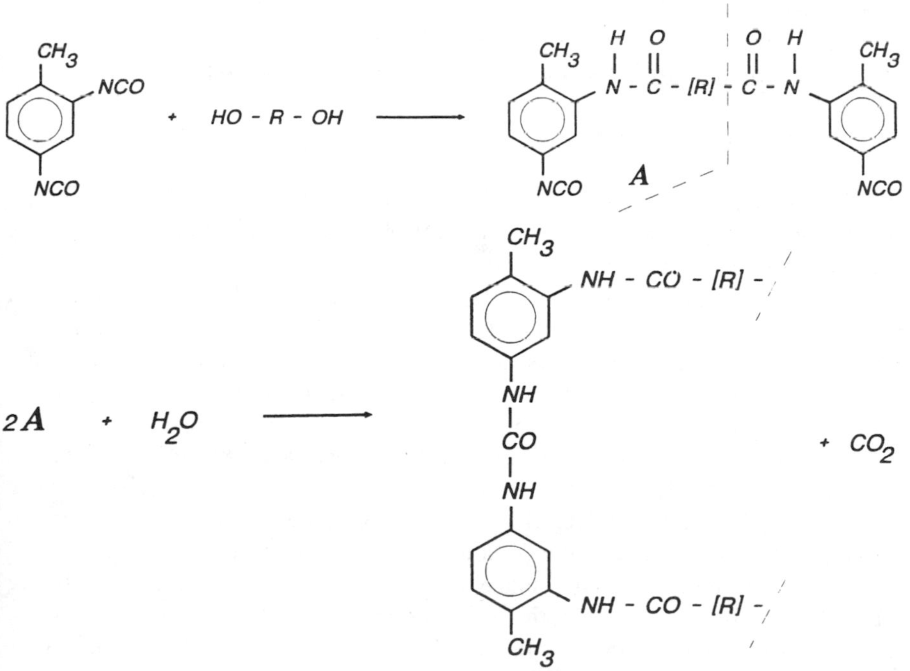

CH_3 NCO NCO + HO - R - OH ⟶ CH_3 NCO H O | ‖ N - C - [R] - C - N O H ‖ | CH_3 NCO

A

2A + H_2O ⟶ CH_3 NH - CO - [R] - NH | CO | NH NH - CO - [R] - CH_3 + CO_2

Figures 2 and 3. The reaction leading to the formation of tridimensional macromolecules.

Figure 4. The reaction of isocyanate with water forms CO_2.

If polymerization takes place in moulds, carbon dioxide (CO_2) can act as a foaming agent. Also, amines, usually tertiary amines, which act as catalyzing and cross-linking agents between the chains of carbon atoms, are generally added to the isocyanate-polyol mixture. Various other active or curing substances can be used to give the mixture the required consistency.

Isocyanates are used in the synthesis of commonly used elastomers: plastics, synthetic rubbers, adhesives, anti-corrosion agents, varnishes, paints, and lacquers. Isocyanates are used in the shoe manufacturing industry, in the automobile industry, in the furniture industry, in flexible polyurethane foam production, in the textile industry, in acoustic and heat insulation manufacture, and in the pesticide industry [4-16]. Exposure problems also may arise from the release of isocyanates form polymerized products even though free TDI usually can no longer be detected 24 hours after synthesis of the polymeric material [17-19].

TOXICOLOGICAL ASPECTS

The high reactivity of isocyanates is the reason for their toxic human effects. Isocyanates are irritants for the skin and mucous membranes. Eye contact causes lachrymation and sometimes conjunctivitis.

Inhalation exposure give rise to asthmatic syntoms, rhinopharingitis, and rhinitis. Above all, asthmatic symptoms can be acute and severe; cases can even be fatal. The asthmatic reactions are allergic in origin. Isocyanates have low volatility, so the concentration during the workshift is generally low. However, they can trigger asthmatic crises even at low concentrations.

Each isocyanate possesses a different toxicological activity [24]. Exposure to 2,4 TDI or to a 2,4-2,6 TDI mixture has been most extensively studied because of its widespread use. After brief periods of exposure via inhalation, irritation of the pharynx and of the conjunctiva appear, followed by breathing difficulties and dryness of the mouth, particularly at night. The symptoms are acute during the night but are not present in the morning, and return on resumption of work. Chronic intoxication can occur following long-term low-level occupational exposure, with typical asthma manifestations [25, 26].

Typical exposure to MDI and HDI, which are less volatile than TDI (2,4 or 2,6 TDI), occurs due to inhalation of mists. HDI is a potent irritant for the skin and mucous membranes.

Diagnosis of asthma can be made by completing the clinical data with laboratory tests; allergy tests are useful for epidemiological purposes. The most easy-to-use diagnostic tool is the challenge test, consisting of a brief exposure in a controlled atmosphere, and respiratory function tests. Recent data on the onset of asthma in workers exposed to isocyanates in Italian furniture varnishing have shown that asthma occurs in 2.7% of the exposed subjects [9, 27].

SAMPLING METHODS

The systems used for determination of isocyanates in air have undergone, starting from the Marcali method, several modifications both in the sampling procedure and in the analytical method. The most common sampling procedures involve either bubbling of air through an impinger or drawing samples on solid sorbent media; the most widespread reagents are N-4-nitrobenzyl-N-n-propylamine (nitro reagent) and 1-(2-methoxyphenyl) piperazine; the most common analytical methods are spectrophotometry and high-pressure liquid chromatography (HPLC).

The methods using bubbling through impinger certainly are more consolidated; also, the latest method published by NIOSH for isocyanates involves this sampling procedure, which, however, is of difficult application for personal sampling. More convenient procedures for personal sampling include the collection of the material using solid sorbent media, which only differs from the previous system in that the reagent is supported by an inert instead of being contained in an impinger; the same analytical methods can therefore be used.

The only limitations are represented by a low retaining capacity and by sampling inefficiency for some isocyanates, especially for hexamethylenediisocyanate (HDI) and for 4-4' diamminodiphenylmethanediisocyanate (MDI). Both limitations are connected to the support used.

For sampling of isocyanates, either glass wool, glass fiber filters, or sintered glass have been used, properly treated with the different reagents. The use of these impregnated supports requires much attention in the various phases of impregnation, sampling, storing, and handling before analysis. Direct reading methods also have been used for continuous control.

LIQUID SORBENT MEDIA

Marcali Method

With the Marcali method, air containing toluene diisocyanate is drawn through a glass bubbler at a collection rate of about 0.9 l/min; 15 ml of acid solution (aqueous solution with acetic acid 0.4N and hydrochloric acid 0.4N) is used as absorbent medium [28]. Under these conditions, for a concentration of 2,4-toluene diisocyanate less than 2 ppm, the collection efficiency is about 95% of diisocyanate in the air. Above 2 ppm, about 90% of diisocyanate is recovered.

Modified Marcali Method

The NIOSH method P&CAM 141 for TDI and P&CAM 142 for MDI are a modification of the Marcali method [29-32]. Using a midget impinger filled with 15 ml of acid solution (2.2% acetic acid, 3.5% concentrated hydrochloric acid in water), isocyanate is hydrolyzed to the corresponding diamine derivative. The diamine is then diazotized by sodium nitrite and sodium bromide solution. The sample is collected at a flow rate of 1 l/min for 20 minutes; the range covered by the method is 0.007 ppm to 0.140 ppm.

Nitro Reagent Method

The air is drawn through two midget impinger connected in series, containing each 10 ml of absorption solution (2×10^{-4}M nitro reagent in benzene or toluene), at a flow rate in between 1-2 l/min [33]. After sampling, the two solutions are combined and evaporated to 1 ml. For an air sample of about 10 liters, the lower detection limit is 0.08 mg/m^3 for HDI TDI and MDI; the upper limit is 33 mg/m^3 for HDI, 34 mg/m^3 for TDI and 50 mg/m^3 for MDI.

Methoxyphenyl Piperazine Methods

Air contaminated with isocyanates is drawn, at a flow rate of 1 l/min, through a midget impinger containing a solution of 1-(2-methoxyphenyl)-piperazine in toluene (≈ 50 μg/ml). The collection capacity is in the range 0.2 - 3.5 μmol NCO group using 15 ml sampling solution (34-36). An investigation of this method demonstrates a loss of reagent, which increased with the volume of air sampled [37].

Ethanol Absorption Method

An alkaline ethanolic absorption medium (38, 39) is used both for aromatic and aliphatic isocyanates [38, 39]. An impinger filled with 10 ml of ethanol potassium hydroxide (0.2%) solution is used to sample air at a rate of 1 l/min. The sampling efficiency for HDI and 2,6 TDI is between 93% and 98% (21, 40).

SOLID SORPTION MEDIA

Nitro Reagent Methods

A tube containing N-4-nitrobenzyl-N-n-propylamine on a suitable support is used to sample isocyanates in air.

J. Keller describes a tube packed with surgical cotton and glass powder [41]. NIOSH method P&CAM 326 uses a glass tube containing two sections of glass wool coated with nitro reagent to sample 2,4 TDI at a flow rate of 1 lpm; the precision for total sampling and analytical method (in the range 0.039 to 0.53 mg/m^3) is 0.060 (CV) [42]. NIOSH method P&CAM 347 uses a glass-fiber filter impregnated with nitro reagent to collect MDI [43]. At a flow rate of 1 lpm, MDI can be quantified at concentrations ranging from 4.4 to 800 µg/m^3 for 180 l air sample. A modification of NIOSH method uses impregnated glass-fiber filters at a flow rate of 1 lpm to sampling HDI [3, 44].

Methoxyphenyl Piperazine Methods

TDI vapors are collected on 1-(2- methoxyphenyl)-piperazine impregnated glass-fiber filters (45). For 15 liter samples at a concentration range from 0.03 to 0.14 mg/m^3, no breakthrough was observed. Glass tubes, with an internal tubular sintered glass support coated with the reagent, are used to collect 2,4 TDI; 2,6 TDI, HDI and MDI (see Figure 5) [46, 47]. At a flow rate of 1 l/min, a tube containing about 5 µmol of reagent can be used for eight hours at a concentration near the threshold limit value–time weighted average (TLV-TWA) without any significant breakthrough.

MAMA Method

A dual filter cassette, in which the first filter captures aerosol phase and the second captures vapor phase, is used to sampling TDI, HDI and MDI (see Figure 6) [48]. The first filter is a 5 µm Teflon®, the second is a 0.8 µm glass-fiber filter impregnated with 9-(N-methylaminomethyl) anthracene (MAMA).

Tryptamine Method

Several solid sorbent media coated with tryptamine were used to collect isocyanates [49-53]. Sampling tubes were prepared with Amberlite (XAD-2, XAD-4, XAD-7), molecular sieve 13X, silica gel, charcoal, and glass beds. Glass-fiber filters and nylon 6-6 filters also were used. A compared study indicates that Amberlite XAD-2 coated with tryptamine was the most efficient sorbent media.

DIRECT READING METHODS

A sampling unit that uses chemically treated paper tapes continuously drawn through a sampling head is used as a sequential monitor for toluene diisocyanate [12, 54]. TDI reacts with the chemicals in the tape to produce a red stain, and the instrument measures the intensity of the reflectance of any stain and converts the result in equivalent TDI concentration.

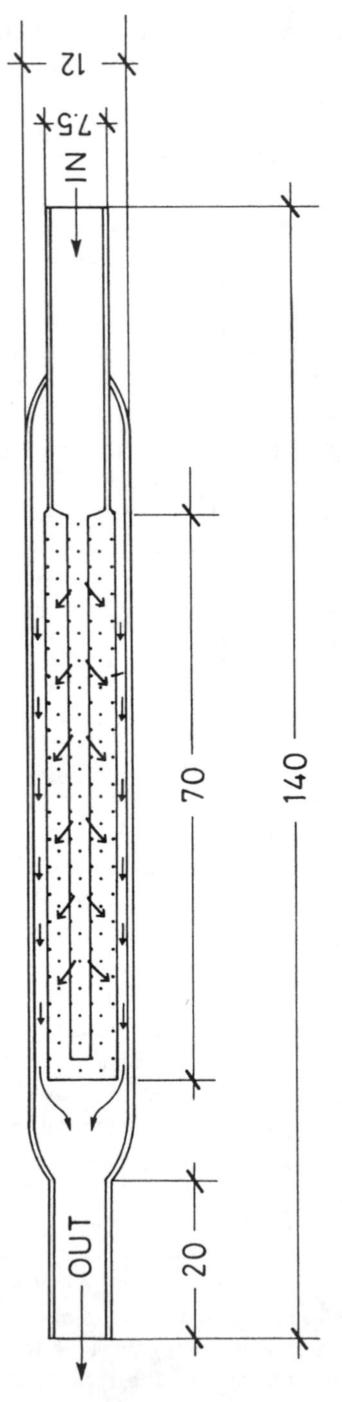

Figure 5. A sampling tube showing the pathway of sampled air. Dimensions are in millimeters.

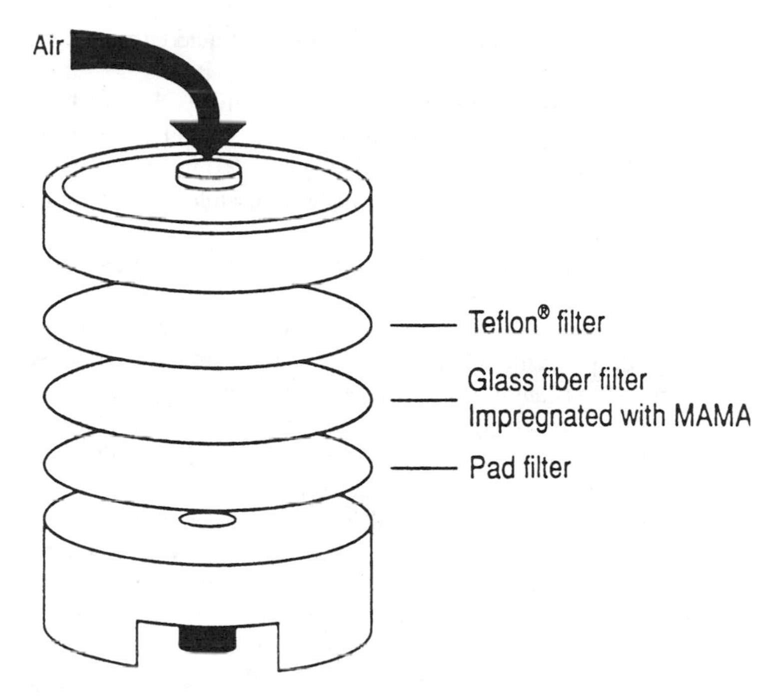

Figure 6. A sampling system for collecting gas and aerosol phase isocyanates [48].

ANALYTICAL METHODS

Traditionally, isocyanates were determined with the colorimetric method via development of color obtained from the reaction of coupling between the diazonium salt of the amine corresponding to the isocyanate, from which it is derived via acid hydrolysis and subsequent diazotation, and N-1-naphthylethylenediamine at 550 nm [28]. The method was later modified and adopted by NIOSH, especially to optimize the reaction mechanism of the 2,4 TDI-2,6 TDI mixture and eliminate the isomer effect [29, 31, 32]. A further modification recently has been proposed, which uses the amine as a coupling agent instead of a base for the formation of diazonium salt to avoid preferential influences of one of the isomers of TDI on the development of color [55]. The colorimetric method is simple to perform but has low sensitivity, is non-specific, and is not particularly suitable for the determination of aliphatic isocyanates.

At the end of 1970s, chromatographic analytical techniques began to be used, initially thin-layer chromatography (TLC), and later high pressure liquid chromatography (HPLC) [3, 33, 34, 41, 44, 45, 57, 58]. With the use of thin layer chromatography, starting in 1974, particular attention was given to determination of aliphatic as well as aromatic isocyanates, especially HDI, MIC, and prepolymers [33]. However, since 1975, high pressure liquid chromatography has been the most commonly used technique.

In 1980 NIOSH adopted a more sensitive and specific HPLC analytical technique suitable for isolating the isocyanate types present in the environmental sample [42, 43]. The identification of derivatives that are highly absorbent in ultraviolet and/or fluorescent light also has made it possible to successfully detect both aromatic and aliphatic isocyanates.

Table 3 summarizes the HPLC methods most frequently reported in the literature; UV detection and electrochemical detection have been used.

The most commonly used reagents are: nitroreagent (N-4-nitrobenzyl-N-n-propylamine), 1,2-pyridilpiperazine, 1-(2-methoxy) phenylpiperazine, fluorescamine, 9-(N-methylaminomethyl) anthracene (MAMA). 1-(2-methoxy) phenylpiperazine is at present the most widely used derivative because it is recommended by NIOSH and because it poses the least problems in use, especially in regard to maintaining the efficiency of the analytical columns over time.

The NIOSH method, which involves indirect determination of total isocyanates present in the sample by measuring the consumption of the piperazine reagent, has been the subject of numerous and justified criticism [35-37].

Figure 7 shows a typical analytical tracing obtained from a mixture of 2,4 TDI, 2,6 TDI, HDI and MDI in liquid chromatography with UV detection [46, 47].

There is a linear response in calibration graphs: The coefficients of the regression line are shown in Table 4 for the four main isocyanates. The detection limits in the solutions under analysis (0.2 ml) were 0.81 μg for 2,4 TDI; 0.08 μg for 2,6 TDI; 0.31 μg for HDI; and 0.53 μg for MDI. Table 5 gives data on accuracy and recovery offered by the analytical method [46].

Recently, 9-(N-methylaminomethyl) anthracene (MAMA) has been extensively used as a derivative of isocyanate, especially in the vapor phase in an attempt to determine the oligomers present in the air as well as the free monomers [48]. The oligomers react well with 1-(2-methoxy) phenylpiperazine on filter supports, whereas the monomers do not and are subsequently made to react with MAMA. Subsequent HPLC analysis is performed on a C18 ODS-1 column (5 μm, diameter 2,5 mm, length 150 mm) at a flow rate of 2 ml/min, using a UV detector (254 nm) or a fluorescence detector (ex. 254 nm, em. 412 nm) with acetonitrile-acetate buffer (pH 6) eluent, in acetonitrile gradient from 70% to 75%. Figure 8 shows a typical chromatographic analysis.

Even more recently, determination methods have been reported based on the reaction between the isocyanate group and tryptamine. Here, too, emphasis is laid on separating the monomer from the oligomers and on determination of both together, although it is not possible to detect the individual isocyanates separately if they are present as a mixture [49-53]. The methods use high pressure liquid chromatography and fluorescence detection.

In the past few years, to avoid derivative reactions that are difficult to effect and time-consuming, some authors have proposed direct methods of determination

Table 3
Analytical Conditions Most Frequently Used for the Analysis of Isocyanates

Reagent	Detector	Conditions nm	Sample (µl)	Analytical Conditions Column	Length	Gradient	Eluent	Ref.
Nitro reagent	UV-VIS	254 nm	90	HC Pellosil	2 foot	Linear	Ethanol/Hexane	58
Nitro reagent	UV-VIS	254 nm	50	Partisil 10	25 cm	No	i-PrOH/DCM	42
Nitro reagent	UV-VIS	254 nm	50	Partisil 10	25 cm	No	i-PrOH/DCM	43
Nitro reagent	UV-VIS	275 nm	100	Hypersyl ODS	12.5 cm	No	ACN/H_2O/TEA	3
Nitro reagent	UV-VIS	254 nm	50	Hypersyl ODS	12.5 cm	No	ACN/H_2O/TEA	45
Nitro reagent	UV-VIS	254 nm	50	Zorbax golden series silica	8 cm	Yes	Hexane/DCM/ MeOH	60
Methoxyphenyl piperazine	Electrochemical	0.8V(Ag/AgCl)	20	Radial pak C18	10 cm	No	ACN/NaOAc pH 6	34
Methoxyphenyl piperazine	UV-VIS	242 nm	20	Radial pak C18	10 cm	No	ACN/NaOAc pH 6	34
Methoxyphenyl piperazine	UV-VIS	254 nm	10	C8	25 cm	No	ACN/NaOAc pH 6	35
Methoxyphenyl piperazine	UV-VIS	254 nm	10	C8	25 cm	No	ACN/NaOAc pH 6	36
Methoxyphenyl piperazine	Electrochemical	0.8V(Ag/AgCl)	10	Spherisorb S5 ODS 2	15 cm	No	ACN/NaOAc pH 5	45
Methoxyphenyl piperazine	UV-VIS	254 nm	20	C8	25 cm	Yes	ACN/ACOH/ NaOAc pH 5	46
Methoxyphenyl piperazine	UV-VIS	242 nm	50	C 18 ODS 1	15 cm	No	ACN/NaOAc pH 6	48
Methoxyphenyl piperazine	UV-VIS	242 nm	50	C 18 ODS 1	15 cm	No	ACN/NaOAc pH 6	48

(table continues)

Table 3 (continued)

Reagent	Detector	Conditions nm	Sample (μl)	Analytical Conditions Column	Length	Gradient	Eluent	Ref.
Methoxyphenyl piperazine	UV-VIS	240-252 nm	10–50	Novo pak C18	30 cm	Yes	ACN/H_2O	61
Methoxyphenyl piperazine	Electrochemical	0.8V(Ag/AgCl)	10–50	Cyclobond I	25 cm	Yes	ACN/H_2O	61
Methoxyphenyl piperazine	UV-VIS	254 nm	10	Partisil 5	25 cm	No	i-Ottano/DCM/ MeOH	62
Hydroalcohol	UV-VIS	245 nm	50	Radial pak C 18	110 cm	No	THF/ACN/H_2O pH 5.5–7	39
MAMA	Fluorescence	254 EX/412 EM nm	50	C18 ODS 1	15 cm	No	ACN/MeOH/TEA pH 3	48
MAMA	UV-VIS	245 nm	15	C 18	10 cm	No	H_2O/TEA/H_3PO_4 pH 3	63
Pyridil piperazine	Fluorescence	240 EX/370 EM nm	10	Supelcosil LC-CN	25 cm	Linear	H_2O/$AcONH_4$/ AcOH	64 65
Fluorescamine	Fluorescence	240-390 EX/ 377 EM nm	10	Micropak CH 10	25 cm	No	MeOH/H_2O	16

Note: iPrOH = Isopropyl Alcohol; DCM = Dichloromethane; ACN = Acetonitrile; TEA = triethylamine; MeOH = Methyl Alcohol; AcOH = Acetic Acid; NaOAc = Sodium Acetate; $AcONH_4$ *= Ammonium Acetate, THF = Tetrahydrofurane*

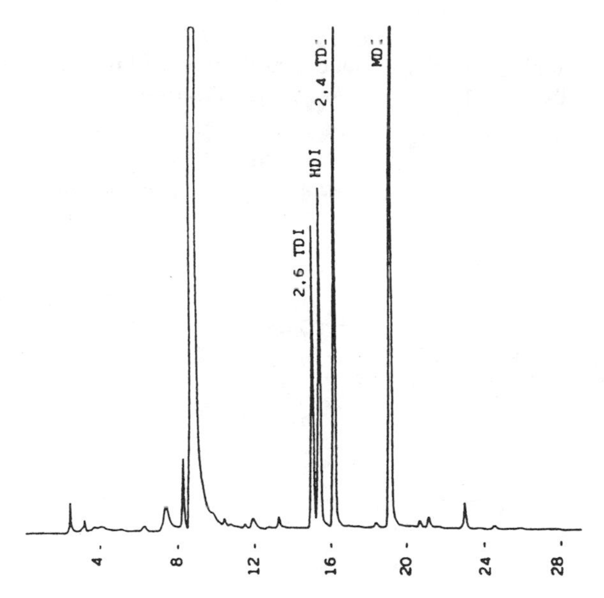

Figure 7. HPLC chromatogram of 2,4 TDI; 2,6 TDI; MDI; and HDI as 1-(2-methoxyphenyl) piperazine derivatives.

Table 4
Regression Line and Correlation Coefficient of the Analytical

Isocyanate	Method y = a + b ×	R
2,4 TDI	y = 0.55 + 2.72 ×	0.999
2,6 TDI	y = 0.16 + 5.21 ×	0.999
MDI	y = 1.17 + 31.6 ×	0.999
HDI	y = 0.55 + 1.76 ×	0.999

based on capillary gas chromatographic analysis (FID and NPD detectors) of the urethane derivatives obtained from isocyanates via hydro- alcohol hydrolysis [38, 40, 59]. Analysis is performed on a Duran 50 column in boron silicate glass, deactivated with 1,3 divinyltetramethyldisilazane, stationary phase PS 255 (Fluka, Buchs, CH) or OV 73 and NPD detector (sensitivity: TDI=10 fmol; IPDI=35 fmol; HDI=15 fmol) or FID detector (sensitivity higher than 200 fmol for each isocyanate) [38, 40, 59].

Table 5
Expected Quantity, Average Recovery, Precision and Recovery of the Analytical Method in Assessing Airborne Isocyanates

Isocyanate	Expected (μg)	Average Recovery (μg)	Precision (CV%)	Recovery (%)
2,4 TDI	1.96	1.83	5.1	93
	3.92	3.52	5.9	90
2,6 TDI	0.49	0.45	5.6	92
	0.98	0.89	6.4	91
HDI	2.09	2.25	5.3	107
	4.18	3.96	5.7	95
MDI	2.98	2.82	3.9	95
	5.96	5.57	6.4	93

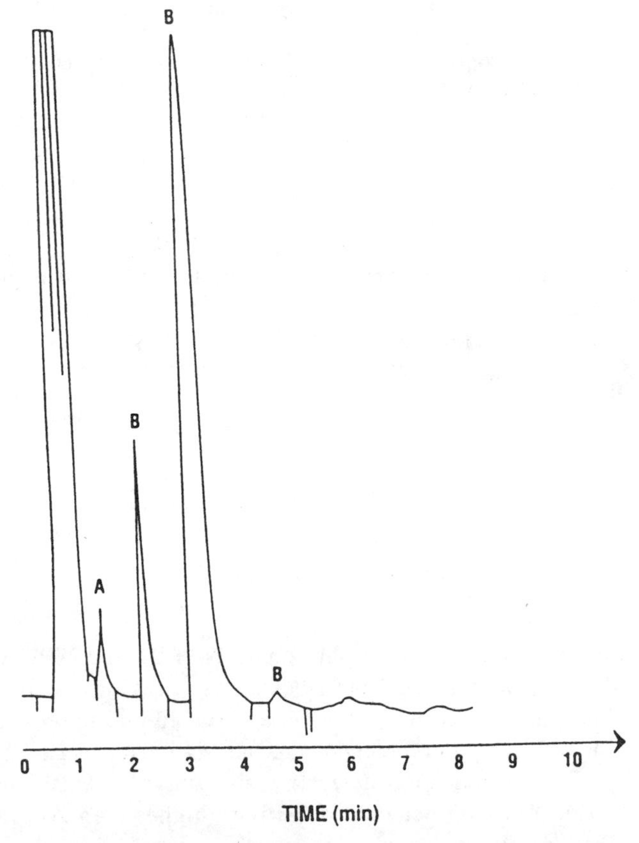

Figure 8. Desmodur N 3200 Chromatogram. A: HDI monomer. B: HDI oligomers [48].

REFERENCES

1. Kirk-Othmer, *Encyclopedia of Chemical Technology,* Vol. 13, New York (John Wiley and Sons ed.), 1981, p. 788–817.
2. Crowl, D.A., Louver, J.F., "Chemical Process Safety Fundamentals with Application" (Prentice Hall, ed.), Englewood-Cliffs, New York, 1991, p. 17–18.
3. Rosenberg, C., Tuomi, T., "Airborne Isocyanates in Polyurethane Spray Painting: Determination and Respiratory Efficiency," *Am. Ind. Hyg. Ass. J. 45*(2): 117–121 (1984).
4. Silant'ev, V.P., Blagodarnya, O.A., "Measures for prevention of the contamination of Air by Chemicals in Shoe Manufacturing," *Kozh.-Obuvn. Prom-st 11:*43–45 (1988).
5. Kuz'min, V.I., Bel'chenko, A.G., Zavgorodnii, A.S., Sinitsyna, E.L., "Working Conditions in the Manufacturing of Polyurethane Shoe Soles," *Kozh.-Obuvn. Prom-st. 11:*39–40 (1989).
6. Delfosse, M., Laureillard, J., "Paint spray Booths in Bodywork Shops in the Automobile Industry. Determination of Isocyanates and Solvents. Conformity to Booth Specifications," *Cah. Notes Doc. 138:*65–72 (1990).
7. Tornling, G., Alexandersson, R., Hedenstierna, G., Plato, N., "Decreased Lung Function and Exposure to Diisocyanates (HDI and HDI-BT) in Car Repair Painters; Observations on Re-examination of 6 Years After Initial Study," *Am. J. Ind. Med. 17*(3):299–310 (1990).
8. Sesana, G., Maggi, W., D'Angelo, R., Panzeri, R., Fattori, M., Cortona, G., "Indagine di Comparto nel settore Artigianale Finitura del Mobile in Brianza, nota 2: Esposizione a Polveri di carteggiatura Solventi e Isocianati," (Monduzzi) 1987, p. 567–572.
9. Cortona, G., Baldasseroni, A., Sesana, G., "Asma da Isocianati: Studio Epidemiologico Trasversale di un Campione di Verniciatori Del Legno," *Aspetti Epidemiologici dell'Asma Bronchiale* (Università di Padova), 1988, p. 175–181.
10. Taki, K., Kurachi, N., "Water-Based Polymer Isocyanate Adhesives," *Mokuzai Kogyo 43* (497):364–8 (1988).
11. Boeniger, M.F., "Air Concentration of TDI and Total Reactive Isocyanate Group in Three Flexible Polyurethane Manufacturing Facilities," *Appl. Occup. Environ. Hyg. 6*(10):853–858 (1991).
12. Rando, R.J., Kader, H.A., Hughes, J., Hammad, Y.Y., "Toluene Diisocyanate Exposure in the Flexible Polyurethane Foam Industry." *Am. Ind. Hyg. Ass. J. 48*(6):580–585 (1987).
13. Cirla, A.M., Pisati, G., Ratti, R., "Occurrence of asthma Induced by Toluene diisocyanate in the Manufacture of Synthetic Textiles," *Cah. Notes Doc. 140:* 686–689 (1990).
14. Sukhanov, V.V., Putilina, N., Shubina, Y., "Synthetic Materials for Sealing of Ventilation Structures in Mines," *Bezop. Promst. 11:*34–35 (1990).

15. Hirabayashi, Y., Goto, M., Sekamoto, A., "Pesticide-coated Granules as Sustained-Release Formulations," *Jpn. Kokai Tokkyo Koho J. P. 01 56,*601 (89 56,601) (1989); from *C. A.-Agrochemicals,* Vol. 112, 72343t (1990).
16. Dollberg, D.D., Verstuyft, A.W., "Analytical Techniques in Occupational Health Chemistry," ACS Symposyum Series 120, American Chemical Society 1980.
17. Rastogi, S.C., "Analysis of Diisocyanate Monomers in Chemical Products Containing Polyurethanes by High Pressure Liquid Chromatography," *Chromatographia 28*(1,2):15–18 (1989).
18. Phil, J.D., Whiton, R.S., "Determination of Organic Emissions from New Carpeting," *Appl. Occup. Environ. Hyg. 5*(10):693–699 (1990).
19. Jedrzejczak, K., Gaind, V.S., "Determination of Free Toluene Diisocyanates in Flexible Polyurethane Foams using Negative Chemical-ionization Mass Spectrometry," *Analyst 118*(2):149–152 (1993).
20. Keith, L.H., Walters, D.B., "Compendium of Safety Data Sheets for Research and Industrial Chemicals," VCH Publishers, Inc., Deerfield Beach, FL, 1985.
21. INRS, Supplement au Cahiers de Notes Documentaires 126, Fiches Toxicologique n° 46, 129, 164, 166, Institute National de Recherche et de Securite, Paris, 1987.
22. *Bellstein Handbook of Organic Chemistry File Data,* STN software, April 1993.
23. *DIPPR Data Compilation File,* STN software, April 1993.
24. International Labour Office (ILO) Geneva, "Encyclopaedia of Occupational Health and Safety," Vol. 1, 3ed, (L. Parmeggiani) (1983).
25. Sartorelli, E., "Trattato di Medicina del Lavoro," 1981, p. 711.
26. IARC (International Agency for Research on Cancer) IARC Monographs, "Some Monomers, Plastics and Synthetic Elastomers and Acrolein," Vol. 19, Lyon, 1979, pp. 303–320.
27. Cortona, G., Sesana, G., Ghezzi, I., "Indagine di Comparto nel Settore Finitura del Mobile in Brianza, Nota 3: Patologia Respiratoria, Atti 50 Congresso Societa," Italiana di Medicina del Lavoro ed Igiene Industriale (Monduzzi), 1987, pp. 573–578.
28. Marcali, K., "Microdetermination of Toluenediisocyanates in Atmosphere," *Anal. Chem. 29*(4):552–558 (1957).
29. Niosh, "Manual of Analytical Methods," Vol. 1, 2nd Ed., Method No. P&CAM 141, U.S. Department of Health, Education and Welfare, Washington DC, (1977).
30. Niosh, "Manual of Analytical Methods," Vol. 1, 2nd Ed., Method No. P&CAM 142, U.S. Department of Health, Education and Welfare, Washington, DC, (1977).
31. Larkin, R.L., Kupel, R.E., "Microdetermination of Toluenediisocyanate using Toluenediammine as the Primary Standard," *Am. Ind. Hyg. Ass. J. 30:* 640–642 (1969).

32. Rando, R.J., Hammad, Y.Y., "Modified Marcali Method for the Determination of Total Toluenediisocyanate in Air," *Am. Ind. Hyg. Ass. J. 46*(4): 206–210 (1985).
33. Keller, J., Dunlap, K.L., Sandridge, R.L., "Determination of Isocyanates in the working Atmosphere by Thin-Layer Chromatography," *Anal. Chem. 46:*1845–1846 (1974).
34. Bagon, D.A., Warwick, C.J., Brown, R.H., "Evaluation of Total Isocyanate-in Air Method using 1-(2-Methoxyphenyl)piperazine and HPLC," *Am. Ind. Hyg. Ass. J. 45*(3):39–43 (1984).
35. Niosh, "Manual of Analytical Methods," Vol. 1, 3nd Ed., Method No. 5505, U.S. Department of Health, Education and Welfare, Washington, DC, (1984).
36. Niosh, "Manual of Analytical Methods," Vol. 1, 3nd Ed., Method No. 5505 (revision #1), U.S. Department of Health, Education and Welfare, Washington DC, 1985.
37. Seymour, M.J., O'Connor, P.F., Teass, A.W., "Investigation of the inaccuracy of Niosh Method 5505 for Estimating the Concentration of Isocyanate in Air," *Appl. Occup. Environ. Hyg. 5*(2):115–122 (1990).
38. Skarping, G., Renman, L., Sango, C., Mathiasson, L., Dalene, M., "Capillary Gas Chromatographic Method for the Determination of Complex Mixture of Isocyanates and Amines," *J. Chromat. 346:*191–204 (1985).
39. Nieminen, E.H., Saarinen, L.H., Laakso, J.T., "Simultaneous Determination of Aromatic Isocyanates and some Carcinogenic Amines in the Work Atmosphere by Reversed-Phase High Performance Liquid Chromatography," *J. Liq. Chrom. 6* (3) 453–469 (1983).
40. Skarping, G., Smith, B.E.F., Dalene, M., "Trace Analysis of Amines and Isocyanates using Glass Capillary Gas Chromatography and Selective Detection. V. Direct Determination of Isocyanates using Nitrogen-Selective and Electron-Capture Detection," *J. Chromat. 331:*331–338 (1985).
41. Keller, J., Sandridge, R.L., "Sampling of Isocyanates in Air," *Anal. Chem. 51:*1868–1870 (1979).
42. Niosh, "Manual of Analytical Methods," Vol. 6, 2nd Ed., Method No. P&CAM 326, U.S. Department of Health, Education and Welfare, Washington, DC, (1980).
43. Niosh, "Manual of Analytical Methods," Vol. 7, 2nd Ed., Method No. P&CAM 347, U.S. Department of Health, Education and Welfare, Washington, DC, (1980).
44. Tucker, P.S., Arnold, J.E., "Sampling and Determination of 2,4-bis(carbonylamino)toluene and 4,4'-bis(carbonylamino)diphenylmethane in Air," *Anal. Chem. 54:*1137–1141 (1982).
45. Rosenberg, C., Savolainen, H., "Determination of Occupational Exposure to Toluene Diisocyanate by Biological Monitoring," *J. Chromat. 367:*385–392 (1986).

46. Sesana, G., Nano, G., Baj, A., "A new Tool for Sampling Airborne Isocyanates," *Am. Ind. Hyg. Ass. J. 52*(5):183–186 (1991).
47. Sesana, G., Nano, G., Baj, A., "Trace Aldehydes and Isocyanates Sampling," *G. Ig. Ind. 16*(2):83–92 (1989).
48. Lasage, J., Goyer, M., Desjardins, F., Vincent, J.Y., Perrault, G., "Workers' Exposure to Isocyanates," *Am. Ind. Hyg. Ass. J. 53*(2):146–153 (1992).
49. Wu, W.S., Nazar, M.A., Gaind, V.S., Calovini, L., "Application of Tryptamine as a Derivatising Agent for Airborne Isocyanates Determination. Part 1," *Analyst 112* (6) p 863–866 (1987).
50. Wu, W.S., Szklar, R.S., Gaind, V.S., "Application of Tryptamine as a Derivatising Agent for Airborne Isocyanates Determination. Part 2," *Analyst 113* (8) p. 1209–1212 (1988).
51. Wu, W.S., Stoyanoff, R.E., Szklar, R.S., Gaind, V.S., Rakanovic, M., "Application of Tryptamine as a Derivatizing Agent for Airborne Isocyanate Determination. Part 3," *Analyst 115*(6):801–807 (1990).
52. Wu, W.S., Stoyanoff, R.E., Gaind, V.S., Rakanovic, M., "Application of Tryptamine as a Derivatising Agent for Airborne Isocyanate Determination, Part 4," *Analyst 116* (1):21–25 (1991).
53. Wu, W.S., Gaind, V.S., "Application of Tryptamine as a Derivatising Agent for the Determination of Airborne Isocyanates, Part 5, Investigation of Triptamine coated XAD-2 Personal Sampler for Airborne Isocyanates in Workplaces," *Analyst 117:*9–12 (1992).
54. Rando, R.J., Duvoisin, P.F., Kader, H.A., Hammad, Y.Y., "A Sequential Tape Monitor For Toluene Diisocyanate," *Am. Ind. Hyg. Ass. J. 48*(6): 574–579 (1987).
55. Rando, R.J., Abdel-Kader, H.M., Hammad, Y.Y., "Isomeric Composition of Airborne TDI in the Polyurethane Foam Industry," *Am. Ind. Hyg. Ass. J. 45:* 199–203 (1984).
56. Keller, J., Vogel, J., "Determination of Toluenediisocyanate (TDI) in the Workplace Atmosphere by Isomer-Independent Photometric Method," Zentralbl. Arbeitsmed., Arbeitsschutz, *Prophyl. Ergon. 42*(1):26–31 (1992).
57. Dukhovnaya, I.S., Yun, E.M., "Determination of Diisocyanates by Thin-Layer Chromatography," *Zh. Anal. Khim. 44*(7):1296–1301 (1989).
58. Dunlap, K.L., Sandridge, R.L., Keller, J., "Determination of Isocyanates in Working Atmospheres by High Speed Liquid Chromatography," *Anal. Chem. 48:*497–499 (1976).
59. Skarping, G., Renman, L., Dalene, M., "Trace Analysis of Amine and Isocyanates Using Glass Capillary Gas Chromatography and Selective Detection, Determination of Aromatic Amines as Perfluorofatty Acid Amides Using Nitrogen Selective Detection," *J. Chromat. 270:*207–18 (1983).
60. Hakes, D.C., Johnson, G.D., Marhevka, J.S., "An Improved High Pressure Liquid Chromatographic Method for the Determination of Isocyanates using Nitroreagent," *Am. Ind. Hyg. Ass. J. 47*(3):181–184 (1986).

61. Schmidtke, F., Seifert, B., "A Highly Sensitive High-Performance Liquid Chromatographic Procedure for the Determination of Isocyanates in Air," *Fresenius J. Anal. Chem. 336*(8):647–654 (1990).
62. Simon, P., Moulut, O., "Separation of the Urea Piperazine Derivatives of Polyisocyanate Monomers and Prepolymers by Normal Phase Chromatography," *J. Liq. Chromatogr. 11*(9-10):2071–2089 (1988).
63. Andersson, K., Gudehn, A., Hallgren, C., Levin, J.O., Nilsson, C.A., "Monitoring 1,6-Hexamethylene Diisocyanate in Air by Chemosorption Sampling," *Scand. J. Work Environ. Health 9*:497–503 (1983).
64. OSHA Method No. 47 "Methylene bisphenyl isocyanate (MDI)" April, 1985.
65. Supelco, "Come Semplificare il Monitoraggio dei Diisocianati in Aria, Usando i Filtri ORBO™ 80" The SUPELCO Reporter XI (2):8–9 (1992).

CHAPTER 18

ORGANOPHOSPHATE POISONING: A CLINICAL PERSPECTIVE

Philip G. Bardin M.B.Ch.B., F. C. P. (SA)[1,2]
Stephan F. Van Eeden M.B.Ch.B., F.C.P.(SA); M. Med.[2,3]

Critical Care Unit A5
Department of Internal Medicine
University of Stellenbosch
Tygerberg Hospital
Cape Town

CONTENTS

INTRODUCTION

Organophosphate compounds are at present the most important group of insecticides used in industry, agriculture and in household gardens. Poisoning, especially by oral ingestion, can be rapidly fatal and should be aggressively managed

---

[1]Address for correspondence and current address:
Dr. Philip G. Bardin
Immunopharmacology Group
Medicine I
Level D, Centre Block
Southampton General Hospital
Tremona Rd.
Southampton SO9 4XY, U.K.
[2]Supported by a post-doctoral research fellowship from the Medical Research Council of South Africa.
[3]Current address:
UBC Pulmonary Research Laboratory
St. Pauls Hospital
Vancouver, Canada.

by resuscitation and adequate anticholinergic medication. Glycopyrrolate, a compound that does not cross the blood-brain barrier, may exhibit benefits on pulmonary secretions and subsequent infections that is superior to that of traditional atropine usage. Grading of the degree of intoxication can help to identify high-risk cases who merit admission to an intensive care unit to prevent or decrease pulmonary infections and ultimately decrease both morbidity and mortality.

Organophosphates can be lethal poisons, and their sale to the public should be rigorously and strictly controlled. Public education concerning safe storage, cautious use, symptoms of poisoning, and other sensible preventive measures should be encouraged to generate increased awareness of the dangerous nature of such poisons and to assist in the prevention of harmful insecticide exposure. Physicians must remain aware of the potential dangers related to these compounds and of the pathophysiology and management of acute organophosphate poisoning to make it possible for them to provide skilled management for a potentially lethal condition.

An alkaloid substance obtained from the Calabar bean found in West Africa (physostigmine) was brought to England in 1840 by a British medical officer. It was first tried as a therapeutic agent for glaucoma in 1877 by Laqueur, and basic research in the 1920s produced neostigmine for use in myasthenia gravis, followed by the development by Schrader in Germany of compounds for use as insecticides and as potential chemical warfare agents. One compound in this early group, parathion, later became the most widely employed insecticide of this class. In the 1950s, the carbamates were synthesized and shown to possess a high degree of selective toxicity against insects by means of potent but reversible inhibition of the enzyme acetylcholinesterase. In contrast, organophosphate compounds cause irreversible cholinesterase enzyme inhibition.

Today, organophosphates and carbamates are used worldwide in agriculture as well as in most household gardens [1]. This availability of the compounds has resulted in a gradual increase in accidental and suicidal poisoning mainly in Third World countries, where organophosphates and carbamates have replaced the chlorinated hydrocarbons such as DDT as insecticides because they are rapidly hydrolyzed after contact with soil and plants and because no accumulation occurs in the environment [2,3]. In Africa and the Far East, their widespread use has resulted in serious increases in poisoning because of easy availability and indiscriminate handling and storage [2,4,5]. Recent World Health Organization estimates suggest that more than three million cases of acute serious pesticide poisoning occur worldwide annually, the majority of which are caused by organophosphates used for agricultural purposes in Third World countries [3].

Although life-threatening insecticide poisoning now occurs infrequently in the United States and other developed countries, poisoning with these compounds remains an important public health priority [6,7]. In the United States, educational and regulatory changes have prevented severe pesticide poisoning from occurring with any frequency, except in the case of suicide [3]. In states such as Florida and California with prominent agricultural activity, poisoning also has decreased markedly and usually relates to the mild cholinergic illness that follows "picker" poisoning [6,8]. Attention recently has been focused on the possible health effects arising from chronic exposure to low levels of pesticides, and in California a proposal to ban at least 20 pesticides, known as Big Green, was recently the focus of state elections. Although the measure was defeated, it remains a highly politicized problem under continuing debate. Serious concern also exists in Europe and Britain over unsuspected acute and chronic effects of poisoning, and regulatory authorities are taking another look at public use and exposure to a group of potentially dangerous chemicals. The decreases in the incidence of acute insecticide poisoning is encouraging, but insecticides remain important agents involved in accidental ingestion or attempted suicide. Physicians must remain vigilant and able to recognize the clinical manifestations of a particularly treatable condition.

In the United States, this group of compounds accounts for as much as 80% of pesticide-related hospital admissions [9]. Many people are exposed to the compounds, often inadvertently, and subclinical poisoning may occur in as many as 20% to 40% of people who work with or come into contact with the compounds [5,6]. This can be expected to increase with the relatively easy availability of these agricultural pesticides of high toxicity to the amateur gardener. A study from Texas demonstrated that organophosphate poisoning often occurs in children, and the authors postulated that it may result from storage of pesticides in households throughout the year making them available to curious children [7]. A recent report emphasized the public importance of organophosphates in the USA as measured by reported yearly exposures (approx. 16,000) and treatment in a health-care facility (approx. 5,000 subjects) [9].

Ingestion of the organophosphates in an attempt to commit suicide has become a major problem in some communities and accounts for the majority of serious cases of poisoning [9]. This problem seems to be more frequent in the Third World, probably because of the wide availability of pesticides as a result of extensive use in agriculture. Many Third World countries also allow the sale of organophosphates directly over the counter and have inadequate regulations controlling the use and storage of the compounds [2,5,10]. Reliable statistics are not available in most of these countries, but evidence suggests that it is a significant problem and accounts for at least 40% to 60% of all cases of serious poisoning in some African countries [2,11].

Early and correct diagnosis and treatment may be life-saving in organophosphate poisoning (OPP), and a good understanding of the pathophysiology and clinical manifestations is essential for physicians. Although the clinical evaluation and treatment of OPP has not changed much during the past decade, a new grading system and treatment with alternative anti-cholinergic drugs have recently been proposed [12]. This chapter reviews the pathophysiology, diagnosis, grading, treatment, and complications of this potentially fatal disorder. Because the pharmacological actions of both organophosphate and carbamate compounds are qualitatively similar, they will be discussed as a group.

PATHOPHYSIOLOGY OF OPP

Acetylcholine is a neurotransmitter released by the terminal nerve endings of all postganglionic parasympathetic nerves and in both sympathetic and parasympathetic ganglia. It is also released at skeletal muscle myoneural junctions and functions as a neurotransmitter in the central nervous system. Normally the enzyme acetylcholinesterase hydrolyzes the neurotransmitter into two inactive fragments, choline and acetic acid. There are two types of this enzyme present in the body: red cell or true cholinesterase and serum or pseudo-cholinesterase [13]. Organophosphate compounds bind both types of cholinesterase covalently at their anionic binding site for acetylcholine, changing them into enzymatically inactive proteins. This binding of organophosphates with acetylcholinesterase is irreversible, except with early pharmacological intervention, and leads to the accumulation of acetylcholine at synapses and to overstimulation and subsequent disruption of nerve impulse transmission in both the peripheral and central nervous systems (see Figure 1). Pharmacological properties can be predicted by knowing the loci where acetylcholine is physiologically released by nerve impulses and the central and peripheral muscarinic and nicotinic responses of the corresponding effector organs (see Table 1).

The onset, intensity and the duration of the pharmacological effects that occur after poisoning are determined largely by the nature of the particular compound and whether it exhibits reversible or irreversible cholinesterase binding, as well as by the degree and route of exposure, lipid solubility, and rate of metabolic degradation. Reversible inhibition of cholinesterase by some compounds (neostigmine, physostigmine, edrophonium) is frequently employed for therapeutic purposes in clinical medicine. The pathophysiological derangements are identical to that caused by long-acting organophosphates with a speedier recovery as a result of the reversible nature of enzyme inhibition. Inhibition of cholinesterase by organophosphates is not immediately spontaneously reversible, and exposure to a substantial dose of these compounds causes a protracted end-organ effect. Some compounds (e.g., malathion) are metabolized to an inactive product at a more accelerated rate in higher animals and man than in insects, thereby exhibiting useful selective toxicity as pesticides [14]. Highly lipid-soluble agents such as chlorfenthion may produce symptoms and signs of cholinergic overactivity for an extended period of days to weeks, caused by subcutaneous lipid storage followed by subsequent chronic systemic release after redistribution [15,16]. These compounds also cause repeated relapses after apparently successful management [17].

Most organophosphates are highly lipid-soluble agents and are well absorbed from the skin, oral mucous membranes, conjunctiva, and gastrointestinal and respiratory routes (if in a vapor or aerosol). The effector organ dysfunction is at

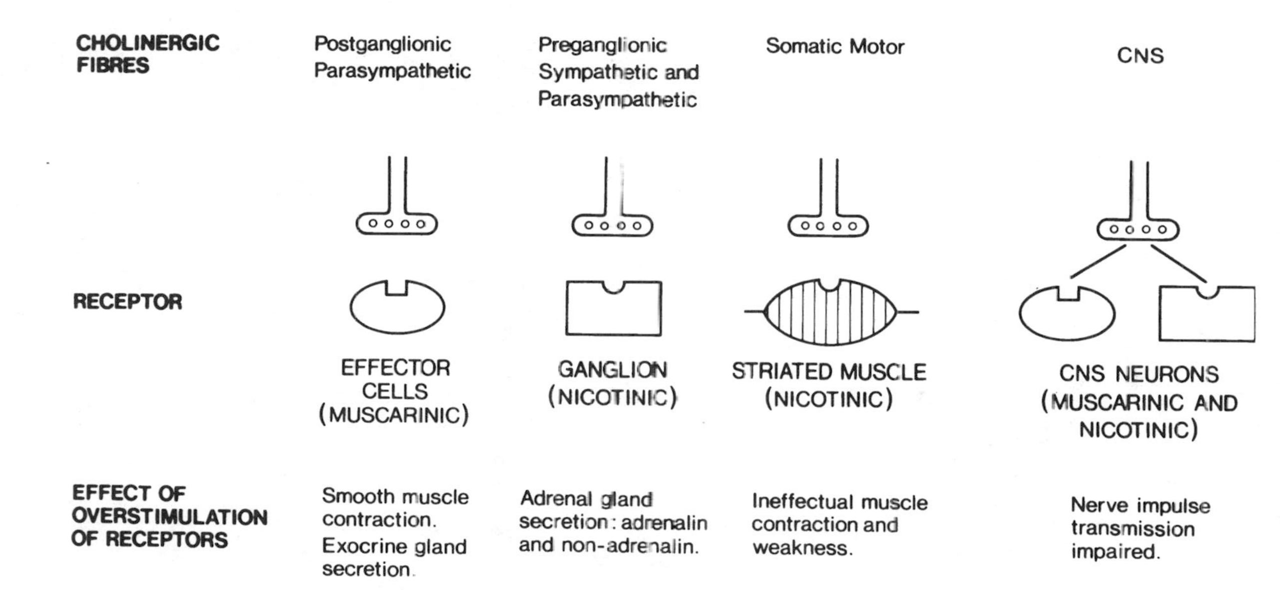

Figure 1. Acetylcholine (depicted as o) released by cholinergic nerve endings combines with four different groups of receptors to produce characteristic effects in different end organs. Organophosphate poisoning leads to decreased destruction of acetylcholine and overstimulation of the receptors (cholinergic crisis). CNS = Central nervous system.

Table 1
Symptoms and Signs of Cholinergic Receptor Overactivity in Acute OPP as Manifested in Different Organ Systems

Muscarinic receptors	Nicotinic receptors	Central receptors
CVS	MS	Altered consciousness*
Bradycardia	Fasciculations	Seizures
Hypotension	Weakness	Respiratory depression
	Paralysis	Cheyne-Stokes
GIT	Cramps	respiration
Salivation		Ataxia
Nausea	CVS	Dysarthria
Vomiting	Tachycardia	Tremor
Abdominal pain	Hypertension	
Diarrhoea		
Tenesmus		
Fecal incontinence		
RS		
Bronchorrhea		
Wheezing		
Cough		
EYE		
Miosis (may be unequal)		
Lacrimation		

**Wide spectrum of abnormalities: slight confusion to stupor and coma. CVS = Cardiovascular system, GIT = Gastrointestinal system, RS = Respiratory system, MS = Musculoskeletal system.*

first local and then rapidly spreads to become generalized. Following absorption, most organophosphates are hydrolyzed by enzymes known as the A-esterases or paroxonase, which are widely distributed in all tissues and hydrolyze a large number of these insecticides by splitting the anhydride, P-F, P-CN, or ester bonds [14]. The metabolic breakdown products are then excreted in the urine.

CLINICAL SYNDROME

The clinical manifestations of OPP, the result of cholinergic overactivity in the body, can be conveniently memorized by consideration of the effects of overstimulation of the muscarinic, nicotinic, and CNS receptors. Cholinergic overactivity at these sites results in a variable but reasonably diagnostic syndrome in acute OPP, which should lead to an early correct diagnosis, especially if the physician maintains a high index of suspicion (see Table 1). A history of possible intake of poison is extremely important, and it is always worthwhile contacting family or companions to obtain a thorough account of the patient's premorbid actions. Although many of the symptoms of OPP are non-specific, a few clinical signs are reasonably specific and should be a good clue to the diagnosis. They include excessive salivation and lacrimation, muscle fasciculations and weakness, constriction of the pupils, and finally the typical smell of the poison about the patient. Depressed levels of consciousness are common, and some patients may have seizures. Classical teaching has stressed the importance of a bradycardia for the diagnosis, although recent studies have shown a tachycardia to be a more frequent finding in acute OPP, probably resulting from preganglionic nicotinic receptor stimulation followed by release of adrenalin and nor-adrenalin from the adrenal gland, leading to a predominance of adrenergic effects on the heart [2,11]. Pulmonary insufficiency with hypoxia may be a contributory factor to the tachycardia. Hyperglycemia and glycosuria are at times transiently detected, although the pathogenesis and clinical importance of these findings remain unresolved [2].

The clinical manifestations of chronic OPP may be completely different from the acute disorder and frequently is not correctly diagnosed [1,10]. Patients may have vague symptoms that include anxiety, tiredness, depression, and other neuropsychiatric symptoms, diarrhea and abdominal pain, wheezing and cough, urinary problems, paresthesia, and muscular weakness [1,18]. This is often the clinical presentation after occupational exposure with chronic poisoning, and an occupational history is of great importance to permit diagnosis and prevention of further poisoning that may go undetected for a prolonged period.

DIAGNOSIS

Criteria for the diagnosis of OPP has been proposed using a combination of clinical manifestations, a decrease in serum pseudocholinesterase or red cell cholinesterase activity, and the response to treatment [19]. Combinations of these different criteria serve to differentiate other conditions such as infections, metabolic derangements (hypoglycemia, uremia), trauma, and acute neurological conditions, as well as other types of poisoning. The latter group may be difficult to distinguish, particularly opiates, nicotine poisoning, muscarine-containing mushrooms, and venomous snakebite; however, a careful history and physical examination usually will suggest the appropriate diagnosis (see Table 2).

Measurement of the serum pseudocholinesterase activity is the most valuable special investigation to assess suspected poisoning [20]. The esterase activity is invariably decreased in OPP, and the sensitivity is close to 100% for cases with significant poisoning [1]. Falsely lowered activity may be encountered in liver disease, malnutrition, chronic alcoholism, dermatomyositis, poisoning with carbon disulphide and organic mercury compounds, and as an autosomal recessive

Table 2
Differential Diagnosis of Organophosphate Poisoning

Acute poisoning	Chronic poisoning
Overdose	
Opiates	Alcohol
Phenothiazines	Opiates
Nicotine	
Venomous arthropod bites	
spider	
scorpion	
Venomous snake bite	
Mushrooms containing muscarine	
Infective/Other causes	
Pneumonia and aspiration	Gastroenteritis
Septicemia	Irritable bowel syndrome
Meningitis	Bronchitis/asthma
Encephalitis	Chronic fatigue syndrome
Leptospirosis, Shigellosis	
Botulism	
Neurological causes	
Epilepsy	Depression*
Subarachnoid bleed	Guillain-Barre syndrome
Subdural hematoma	Other polyneuropathies
Cerebral vasculitis	Motor neurone disease
Metabolic causes	
Uremia	Chronic renal failure
Hypo/hyperglycemia	Thyrotoxicosis
Myxedema coma	
Thyrotoxic crisis	
Reye's syndrome	

**Various other neuropsychiatric symptoms and signs have been described [10,18].*

condition in families who exhibit an increased sensitivity to succinylcholine because of decreased metabolic degradation by serum pseudocholinesterase. In the appropriate context, the finding of a lowered level of the enzyme is very suggestive of OPP, however, no association has been established between the magnitude of the decrease and the severity of poisoning [21,22]. Similarly, patients who have recovered fully from OPP still may have undetectable or very low serum pseudocholinesterase levels, and it has been suggested that red cell cholinesterase activity may better reflect tissue levels of the enzyme and correlate better with clinical recovery [18]. This possibility has not been verified, but it may help to give some useful indications of when to consider decreasing anticholinergic medication when other parameters are equivocal.

The response to treatment with atropine or other anticholinergic medication may be helpful in diagnosis. After therapeutic doses of atropine (bolus administration of 0,5-2.0 mg), there should be a noticeable decrease in the amount of upper airway secretions, although some patients will only become "dry" after relatively higher doses. Obviously this can only serve as a guide, and the definitive diagnosis must be made by a combination of clinical and biochemical findings. The response to treatment with oximes is seldom of value because of the multiplicity of different organophosphate compounds and the consequent varying efficacy and beneficial clinical effect of oximes after poisoning with these chemicals. Clinical improvements after oxime administration also may be difficult to assess and are prone to subjective misinterpretation.

Specific organophosphate compounds can be identified with the aid of high-profile liquid chromatography (HPLC) methods [1]. This is seldom of clinical or therapeutic value because treatment is primarily directed toward control of the cholinergic overactivity caused by poisoning and is not individualized for the specific type of compound. However, identification of the specific pesticide is needed in forensic and litigation situations. It has been suggested that measurement of some of the urinary metabolites might add specificity to organophosphate etiology in mixed exposures as well as provide some clues to the quantity of poison still present in body stores, thus giving some indication of how long antidotal therapy must continue.

GRADING SEVERITY OF OPP

Namba et al. have assessed the severity of acute OPP on the basis of clinical signs and symptoms, as well as the degree of inhibition of serum pseudocholinesterase activity, in an attempt to facilitate diagnosis and treatment and to give an indication of prognosis [19]. However, this proposed grading has proved unworkable in clinical practice because of the many and varied criteria in the different grades, as well as the difficulty of remembering and applying them in an acute clinical situation. Furthermore, it relies heavily on the level of activity of serum pseudocholinesterase, which has been shown to have an inconsistent association with the severity of clinical signs and degree of poisoning with these pesticides [11,17,21,22]. This grading was never validated.

A recent study in Taiwan found that a grading based on a severe decrease in pseudocholinerase was associated with an unfavorable outcome in OPP [4]. However, application of that grading was of limited value to identify high-risk subjects because 20% of cases with only mild to moderate initial decreases in pseudocholinesterase developed respiratory failure, 28% had sudden cardiovascular collapse, and 18% died. Although this study was retrospective, it suggested that the serum cholinesterase activity alone is not sufficient to identify serious OPP, probably because this particular grading does not incorporate any of the physiological derangements commonly found in this form of poisoning.

In a retrospective study of 61 cases with confirmed OPP, we have identified important prognostic indices and have devised a grading system based on specific clinical findings pertaining mainly to the respiratory and central nervous systems [11]. It also incorporated standard procedures such as a chest radiograph and arterial blood gas measurements but did not include serum pseudocholinesterase activity as one of the criteria for the reasons discussed above. The value of this grading was chiefly to facilitate the recognition of seriously poisoned subjects to permit their admission to an intensive-care unit (ICU) for meticulous bronchial toilet, optimal atropinization, and immediate resuscitation. It was anticipated that it may assist diagnosis of acute OPP by placing emphasis on the few relatively specific clinical signs incorporated into the grading. The proposed grading was then prospectively evaluated in a further study of 44 consecutive patients with acute OPP admitted to an ICU [12]. We graded patients immediately on admission, and this initial grading was correlated with three outcome criteria that reflect the prognosis. They were: the requirement for mechanical ventilation, the presence of any complications, and the duration of admission to an ICU. Analysis showed that a serious grading (grade 3) on admission was associated with an unfavorable outcome compared to lesser grades of severity (see Table 3). The most important factors related to outcome were the amount of poison ingested and involvement of the pulmonary system caused by bronchorrhea, often leading to impairment of gas exchange and the eventual development of respiratory failure. These aspects were incorporated into the grading system in an indirect

Table 3
Grading Versus Outcome in 39 Patients with OPP

	Grading		
	0 + 1 **n = 13**	**2** **n = 12**	**3** **n = 14**
Ventilation required	0	4	10[b]
Complications present[a]	1	9	13[b]
Duration of ICU stay (mean + SD)	3.5 + 0.5	4.0 + 1.5	12.5 + 8[c]
Death	0	0	3

[a]At least one complication during ICU stay.
[b]$p < 0.005$; [c] $p < 0.05$
Source: Bardin and Van Eeden [12].

Table 4
Revised Grading for Organophosphate Poisoning[a]

Grade[b]	Criteria
Mild poisoning	History of intake/exposure; Mild signs: normal consciousness secretions 1 + fasciculations 1 +
Severe poisoning	Severe signs: altered consciousness secretions 3 + fasciculations 3 +
Life-threatening poisoning	Suicide attempt Stupor PaO_2 < 75 torr (<10 kPa) Abnormal chest radiograph

[a]*May not be applicable <8 hours after poisoning.*
[b]*At least 2 criteria required for grade. If fewer, use previous grade.*
1+ = mild secretions; few fasciculations.
3+ = copious secretions; generalised fasciculations.
Source: Bardin and Van Eeden [12].

fashion (secretions, attempted suicide, level of consciousness) as well as in a direct manner (decreased PaO_2, abnormal chest X-ray), thus combining and reflecting the important variables that may be pivotal in the ultimate prognosis.

A revised simplified grading was devised that places subjects in one of three groups, on which basis further management can be determined (see Table 4). Based on this grading, all cases of life-threatening OPP merit immediate transfer and admission to an ICU. Although less of a priority, the same approach should be applied to severe poisoning if adequate facilities are available, whereas suspected mild poisoning warrants admission to a hospital for at least 72 hours for observation and treatment to ensure that the initial uncomplicated clinical syndrome does not deteriorate. The clinical manifestations of acute OPP may take at least six hours to develop fully. Caution should be exercised when applying this grading system to patients shortly after contact with or ingestion of the poison.

MANAGEMENT

Initial treatment should be directed toward ensuring an adequate upper airway by sectioning copious nasopharyngeal secretions or vomitus if present. Endotracheal intubation and mechanical ventilation is often required in unconscious patients to protect the lower airway from aspiration and to ensure adequate tissue oxygenation. Patients at risk of pulmonary aspiration of gastric or nasopharyngeal contents should be intubated at an early stage to prevent the subsequent pulmonary complications, factors that are the predominant cause of serious morbidity and death [4,11,12,19]. Hypoventilation as a result of muscle weakness may be difficult to diagnose, may develop insidiously, and should be identified by regularly monitoring maximal inspiratory pressure and vital capacity (if awake), breathing rate, and blood gases. It is preferable to identify and transfer patients with life-threatening poisoning to an ICU as soon as possible.

Treatment with anticholinergic medication is still the mainstay of treatment and should be started as soon as the airway has been secured. Atropine sulphate is best administered as a continuous infusion instead of as repeated boluses. We have successfully used an infusion of 30 mg of atropine sulphate in 200 ml of normal saline to titrate against the important parameters for adequate atropinization, and a recent uncontrolled study has suggested a reduction in mortality if a continuous infusion is employed [23]. This method can be supplemented by giving additional boluses of atropine (1-5 mg) to regain quick control of secretions or severe bradycardia when indicated. The most valuable indicators of sufficient anticholinergic treatment are control of secretions as manifested by dry mucous membranes and mydriasis of the pupillary sphincters [1,24]. It is essential to control and continuously monitor excessive secretions as effectively as possible to prevent the pulmonary sequelae of pneumonia and respiratory failure. The presence of a tachycardia can be misleading and should not be used as an important guide to treatment. High doses of atropine may be required, and in most hospitals the requisite expert nursing care as well as optimal atropinization can only be satisfactorily achieved in an ICU [17].

Atropine crosses the blood brain barrier and may cause severe toxic effects such as confusion, psychosis, coma, and seizures, all of which may be indistinguishable from acute OPP, relapse after treatment, cerebral hypoxia, and infection with septicemia [22,27]. Glycopyrrolate, a quaternary ammonium compound, has many theoretical advantages over atropine for the treatment of acute OPP. This includes better control of secretions, less tachycardia, and fewer CNS side effects because unlike atropine, glycopyrrolate does not cross the blood-brain barrier [28-30]. In a double-blind controlled study, we have recently shown that atropine and glycopyrrolate are of equal efficacy in acute OPP and that there may be a trend toward fewer respiratory infections in the glycopyrrolate group [12]. This may be a consequence of better control of endobronchial secretions with less mucus impaction in the smaller airways, followed by obstruction and infection, and should encourage more studies to confirm these observations and to define the role of glycopyrrolate in the management of acute OPP.

Other measures for decontamination of patients are of extreme importance. If the subject has ingested the poison and is awake, vomiting can be induced with ipecac followed by activated charcoal to absorb remaining insecticide from the gastrointestinal tract. In unconscious patients, a large-bore gastric tube should be placed (after endotracheal intubation) followed by lavage of the stomach and the administration of activated charcoal. Contaminated clothing must be removed and the skin washed with soap and water or a mildly alkaline detergent, particularly in cases of occupational exposure. Care should be taken to prevent accidental dermal contact by all personnel involved in the management of the patient [31].

There has been considerable controversy regarding the use of oximes in acute OPP [32-34]. The major pharmacologic action of oximes such as pralidoxime and obidozime is to reactivate acetylcholinesterase by removal of the phosphate group bound to the esteratic site [24]. This action only occurs shortly after human poisoning and inhibition of the enzyme, after which the enzyme "ages" and becomes more firmly bound to the esteratic site [24]. If the oximes are administered, they should be given as soon as possible following poisoning before aging takes place rendering the bond between cholinesterase and the organophosphate immutable by the antidote. Consequently, the oximes probably are most effective if given within six hours of poisoning, although some authors have reported beneficial responses after treatment had been delayed for as long as 24 hours [1]. The dose of pralidoxime is 1 gram IV over 45 to 60 minutes by infusion at a maximum infusion rate of 0.25 to 0.5 g per minute. It may be repeated two hours after the initial dose and every 12 hours as required [35]. The oximes never have been evaluated in a formal placebo controlled study, but individual experience and published anecdotal reports have suggested varying efficacy against a wide range of organophosphates [36]. The antidote appears to be effective against parathion, mevinphos, and Diazinon, while it appears to be less useful for the treatment of poisoning with other compounds such as malathion and methyl demeton [37-41]. Metabolic degradation of the oximes yields potentially harmful products such as hydrogen cyanide (19), and side effects such as nausea and vomiting, dizzyness, and disturbances of consciousness have been described (19,42), all of which may be confused with the effects of either organophosphates or toxic doses of atropine [19,42]. Administration of the oximes, especially in high doses, may not be entirely beneficial and has prompted a reassessment of this form of treatment in some institutions along with suspension of the use of oximes pending further definitive studies. Oxime reactivators are not useful and may be contraindicated in the treatment of poisoning with n-methyl carbamates [1].

The correct initial diagnosis, evaluation, and management of OPP by the primary care physician or other health professionals are of utmost importance. Once the diagnosis has been established and the airway is stabilized, atropine should be administered to the patient in adequate doses to ensure that mucous membranes are clearly dry, although toxic doses should be avoided. This is particularly important if the patient is to be transferred from a rural area to a hospital with ICU facilities. It has been noted in the past that patients with OPP are frequently undertreated initially with anticholinergic medication, to the detriment of the subsequent course of their illness [11].

COMPLICATIONS

Complications resulting from OPP occurred in up to 40% of cases of acute intoxication [2,7,11]. Death often occurs early (within 24 hours) in untreated cases and after variable periods of up to 10 days in those who reach a hospital and are given optimal management [43]. Early deaths are chiefly related to central nervous system depression, seizures, ventricular arrhythmias (for example, torsade des pointes) or respiratory failure from excessive bronchial secretions, bronchospasm, pulmonary edema, aspiration of gastric contents, paralysis of respiratory muscles, or apnea associated with depression of the medullary respiratory center [44]. Late mortality is caused by respiratory failure, often associated with infections (pneumonia, septicemia) or complications related to a protracted period of mechanical ventilation and intensive care management [4,11,12].

Respiratory failure is the most common complication following acute OPP and carbamate poisoning, occurs in up to 40% of patients, and is associated with a high morbidity and mortality [4,7,12]. The pathogenesis is multifactorial and related to aspiration of gastric contents, excessive secretions in the airways, pulmonary infection and pneumonia, septicemia, and the development of the adult respiratory distress syndrome (ARDS) [4]. Respiratory consequences of muscarinic overstimulation—including rhinorrhea, bronchorrhea, bronchoconstriction, and laryngeal spasm—may contribute to respiratory failure and often are combined with nicotinic effects such as respiratory muscle weakness and paralysis, including paralysis of the tongue and nasopharynx. Central depression of respiration occurs following cholinergic overstimulation of synapses in the brain stem and is a prominent cause of hypoxia, respiratory failure and death in the early period after acute OPP [44]. Peripheral neuromuscular block may produce respiratory muscle weakness and paralysis, and can contribute to the development of respiratory insufficiency at a later stage [45]. Respiratory failure during acute intoxication generally responds well to anticholinergic treatment, upper airway support and mechanical ventilation, if indicated. Weakness of the respiratory muscles does not improve on this form of treatment, and prolonged respiratory support may be necessary [12]. Sudden cardiovascular collapse often is the first indication of unsuspected or incipient respiratory failure, a presentation that is associated with a high mortality [5].

The development of pneumonia is the most important cause of delayed respiratory failure after OPP and occurs in up to 43% of patients [4,7,12]. Up to 80% of patients with pneumonia had respiratory failure, and the majority of these can be diagnosed within 96 hours of poisoning [4]. Inadequate or delayed atropinization and aspiration of gastric contents are the principal reasons for the development of pneumonia and underlines the importance of skilled medical assessment and treatment at an early stage after poisoning.

Central nervous system (CNS) manifestations of the acute cholinergic crisis after acute OPP are common (disturbances of consciousness and depression of vital centers in the medulla, seizures). Delayed CNS toxicity may occur and includes impaired memory, confusion, irritability, lethargy, and psychosis, all of which are usually reversible [2,19]. Peripheral neuropathy, usually with a latency of greater than ten days, can follow exposure to organophosphates, and there also may occur an intermediate neuropathic syndrome of uncertain pathogenesis that involves neural degeneration and signs of demyelination after acute poisoning with several compounds [46]. The latter syndrome is probably not from inhibition of acetylcholinesterase but from secondary inhibition of other enzymes in the myelin sheath.

Relapse complicating apparently successful management of OPP is well described [12,15,16]. Inadequate decontamination of the skin or gastrointestinal tract, premature termination of anticholinergic therapy, and redistribution of poison from subcutaneous fat stores are the major reasons. Patients with clinical manifestations following OPP should be adequately atropinized for at least 24 to 48 hours and carefully observed for 72 hours following termination of atropine treatment, before discharge from hospital.

Other complications such as liver function abnormalities, blood dyscrasias, coagulopathies, pancreatitis, and ulcerative stomatitis have been described, but a firm causative relationship has not been established [20,48,49].

REFERENCES AND BIBLIOGRAPHY

1. Hayes, W. J. "Organophosphate Insecticides." *Pesticides Studied in Man.* Baltimore: Williams and Wilkins, 1982: 285–315.
2. Hayes, M. M., Van der Westhuizen, N. G., Gelfand, M. "Organophosphate Poisoning in Rhodesia." *S Afr Med J* 1978; 54: 230–4.
3. Jeyaratnam, J. "Acute Pesticide Poisoning: A Major Health Problem. *Wld Hlth Statist Quart* 1990; 43: 139–145.
4. Tsao, T.C., Juang, Y., Lan, R., Shieh, W., Lee, C. "Respiratory Failure of Acute Organophosphate and Carbamate Poisoning." *Chest* 1990; 98: 631–636.
5. Innes, D. E., Fuller, B. H., Berger, G. M. B. "Low Serum Cholinesterase Levels in Rural Workers Exposed to Organophosphate Pesticide Sprays." *S Afr Med J* 1990; 78: 581–583.
6. Brown, S. K., Ames, R. G., Mengle, D. C., "Occupational Illness from Cholinesterase-Inhibiting Pesticides among Agricultural Applicators in California 1982-1985." *Arch Environ Health* 1989; 44: 35–39.
7. Zwiener, R. J., Ginsburg, C. M. "Organophosphate and Carbamate Poisoning in Infants and Children." *Paediatrics* 1988; 81: 121–126.
8. Reich, G. A., Davis, J. H., Davies, J. E. "Pesticide Poisoning in South Florida." *Arch Environ Health* (Chicago) 1968; 17: 768–772.
9. Litovitz, A. "1990 AAPCC Annual Report." *Am J Emerg Med* 1991; 9: 468–492.

10. Perold, J. G., Bezuidenhout, D. J. J. "Chronic Organophosphate Poisoning." *S Afr Med J* 1980; 57: 7–9.
11. Bardin, P. G., Van Eeden, S. F., Joubert, J. R. "Intensive Care Management of Acute Organophosphate Poisoning: A 7-Year Experience in the Western Cape." *S Afr Med J* 1987; 72: 593–597.
12. Bardin, P. G., Van Eeden, S. F. "Organophosphate Poisoning: Grading the Severity and Comparing Treatment between Atropine and Glycopyrrolate." *Crit Care Med* 1990; 18: 956–960.
13. Moss, D. W., Henderson, D. R., Kachmar, J. F. Tietz, N. W., ed. "Enzymes." *Textbook of Clinical Chemistry.* Philadelphia: W. B. Saunders, 1986: 619–774.
14. Taylor, P., Gilman, A. G., Goodman, I. S., eds. "Anticholinesterase agents." *The Pharmacological Basis of Therapeutics.* New York: Macmillan Publishers, 1985: 110–128.
15. Merrill, D., Mihu, P. "Prolonged Toxicity of Organophosphate Poisoning." *Crit Care Med* 1982; 10: 550–557.
16. Davies, J. E., Barquet, A., Freed, V. H., Haque, R., et al. "Human Pesticide Poisoning by a Fat-Soluble Organophosphate Insecticide." *Arch Environ Health* 1975; 30: 608–613.
17. Du Toit, P. W., Muller, F. O., Van Tonder, W. M. "Experience with the Intensive Care Management of Organophosphate Insecticide Poisoning." *S Afr Med J* 1981; 60:227–229.
18. Wadia, R. S., Sadagopan, C., Amin, R. B., Sarelesai, H. V. "Neurological Manifestations of Organophosphate Poisoning." *J Neurol, Neurosurg, Psychiatry* 1974; 37: 841–847.
19. Namba, T., Nolte, C. T., Jackrel, J., Grob, D. "Poisoning due to Organophosphate Insecticides." *Am J Med* 1971; 50: 475–492.
20. Fuller, B. H., Berger, G. M. B. "Automation of Serum Cholinesterase Assay - Pediatric and Adult Reference Ranges." *S Afr Med J* 1990; 78: 577–580.
21. Molphy, R., Ratthus, M. "Organic Phosphorus Poisoning and Therapy." *Med J Aust* 1964; 2: 337–340.
22. Wyckoff, D. W., Davies, J. E., Barquet, A., Davis, J. E. "Diagnostic and Therapeutic Problems of Parathion Poisonings." *Ann Intern Med* 1968; 68: 875–881.
23. Ye, C. Y., He, X. H., Chen, J. S. "Prognosis of Severe Organic Phosphorus Pesticide Intoxication and the Effect of Atropine Treatment." *Chung Hua Nei Ko Tsa Chih* 1990; 29: 76–78.
24. Durham, W. F., Hayes, W. J. "Organic Phosphorus Poisoning and its Therapy." *Arch Environ Health* 1962; 5: 27–53.
25. Ketchum, J. S., Sidell, F. R., Crowell, E. B., Hayes, A. H., et al. "Atropine, Scopolamine and Ditran: Comparative Pharmacology and Antagonists in Man." *Psychopharmacologia* 1973; 28: 121–145.
26. Rumack, B. H. "Anticholinergic Poisoning: Treatment with Physostigmine." *Paediatrics* 1973; 52: 449–551.

27. Birdsall, N. J. M., Burgen, A. S. V., Hulme, E. C. "The Binding of Agonists to Brain Muscarinic Receptors." *Mol Pharmacol* 1978; 14: 723–736.
28. Mirakhur, R. K., Dundee, J. W. "Glycopyrrolate: Pharmacology and Clinical Uses." *Anaesthesia* 1983; 38: 1195–1203.
29. McCubbin, T. D., Brown, J. H., Dewar, K. M. S., Jones, C. J., et al. "Glycopyrrolate as a Premedicant: Comparison with Atropine." *Br J Anaesth* 1979; 51: 885–889.
30. Proakis, A. G., Harris, G. B. "Comparative Penetration of Glycopyrrolate and Atropine across the Blood-Brain and Placental Barriers in Anaesthetized Dogs." *Anaesthesiology* 1978; 48: 339–344.
31. Dekker, M. "Organophosphate Insecticide Poisoning." *Clin Toxicol* 1979; 15: 189–191.
32. Lund, C., Monteagudo, F. S. E. "Early Management of Organophosphate Poisoning." *S. Afr Med J* 1986; 69: 34–35.
33. Jager, B. V., Stagg, G. N. "Toxicity of Diacetyl Monoxime and of Pyridine-2-Aldoxime Methiodide in Man." *Bull Johns Hopkins Hosp* 1958; 102; 203–206.
34. Tafuri, J., Roberts, J. "Organophosphate Poisoning." *Ann Emerg Med* 1987; 16: 193–202.
35. Namba, T. "Diagnosis and Treatment of Organophosphate Insecticide Poisoning." *Medical Times* 1972; 100. 100–126.
36. Namba, T., Hiraki, K. "PAM Therapy for Alkyl-Phosphate Poisoning." *JAMA* 1958; 166: 1834–36.
37. Namba, T., Taniguchy, Y., Okazaki, S., Homa, S., et al. "Treatment of Severe Organophosphate Poisoning with Large Doses of PAM." *Naika Ryoiki* 1959; 7: 709.
38. Warren, M. C., Conrad, J. P., Bocian, J. J., Hayes, M. "Clothing-Borne Epidemic: Organophosphate Poisoning in Children." *JAMA* 1963; 184: 266–267.
39. DePalma, A. E., Kwalick, D. S., Zukerberg, N. "Pesticide Poisoning in Children." *JAMA* 1970; 211: 1979.
40. Namba, T., Greenfield, M., Grob, D. "Malathion Poisoning: A Fatal Case with Cardiac Manifestations. *Arch Environ Health* 1970: 21: 533–536.
41. Barr, A. M. "Further Experience in the Treatment of Severe Organic Phosphate Poisoning." *Med J Aust* 1966; 1: 490–495.
42. Worrell, C. L. "The Management of Organophosphate Intoxication." *South Med J* 1975; 68: 335–339.
43. De Condole, C. A., Douglas, W. W., Evans, C. L., Holmes, R., et al. "The Failure of Respiration in Death by Anticholinesterase Poisoning." *Br J Pharmacol* 1953; 8: 446–475.
44. Steward, W. C., Anderson, E. A., "Effect of a Cholinesterase Inhibitor when Injected into the Medulla of the Rabbit." *J Pharmacol Exp Ther* 1968; 163: 309–317.

45. Vallilescu, C., Alexander, M., San, A. "Delayed Neuropathy after Organophosphate Insecticide Poisoning: A Clinical, Electrophysiological and Nerve Biopsy Study." *J Neurol Neurosurg Psychiatry* 1984; 47: 543–548.
46. Senanayake, N., Karalliedde, L. "Neurotoxic Effects of Organophosphate Insecticides." An Intermediate Syndrome." *N Engl J Med* 1987; 316: 761–763.
47. Dressel, T., Goodale, R., Arweson, M. "Pancreatitis as a Complication of Anticholinesterase Intoxication." *Ann Surg* 1979; 189: 199–204.
48. "Household Insecticides associated with Blood Dyscrasias in Children." *Am J Pediatr Hematol Oncol* 1982; 4: 438–439.
49. Milby, T. H., Epstein, W. L., "Allergic Contact Sensitivity to Malathion." *Arch Environ Health* (Chicago) 1964; 9: 434–435.

CHAPTER 19

HEAVY METAL POLLUTION IN THE BIOSPHERE AND BIOINDICATORS AND BIOMONITORING OF HEAVY METALS

G. N. Mhatre

H-3, Vishwakutir
Shankar Ghanekar Marg
Dadar, Bombay-400 028, India

CONTENTS

INTRODUCTION

Heavy metals as pollutants have attracted a great deal of attention after the much talked about Minamata and Niigata diseases caused by mercury and Itai-Itai disease caused by cadmium. Voluminous work has been done since then, especially on heavy metal occurrence, toxicity, tolerance, and bioscavenging of these pollutants with the help of plants. These facts are collated in this chapter, along with other data in a review form on the basis of studies carried out by various workers. A need for bioindicators of heavy metals and their possible use in biomonitoring is also discussed.

Pollution as a consequence of increased industrialization is a well-known phenomenon. Pollution and its adverse effects have become a household topic. Various types of pollutants have been identified, and their effects on living systems have been studied in detail by many workers all over the globe. Chaphekar and Mhatre have assessed and reviewed the effects of human activities on the Ganga river ecosystem, the oldest and holiest of India's rivers [1].

The presence of any pollutant in the environment influences the abiotic and biotic components of an ecosystem by changing the functions of normal life systems. In the process, organisms are either eliminated or they undergo changes, resulting in the species adapting to stresses and thus surviving the adverse situations. Accordingly, the species may be sensitive or tolerant. An organism is a product of its environment and hence has an indicator value.

Microbes, plants, and animals respond in a similar way to environmental changes. Floristic composition of a plant community and morphological alterations of a plant are commonly used to indicate changes in an environment. Vital life functions are affected by the pollutants to a degree that causes retardation in growth, reproduction, and life spans of the individuals and finally, of the whole population. Various species are differently sensitive to pollution: Some of them decrease in number and may totally disappear, while others become dominating. The resulting effect is a change of the community structure [2].

A close and continuous monitoring of any environmental variable should be based on periodic sampling because of the variation of pollutant stress with time and space. Thus plants and animals serve as man's early warning systems against environmental pollution. However, these monitors often are local and hence require intensive trials before their widespread use for this purpose.

This chapter is a collation of voluminous information and scientific data extensively reviewed by Nriagu and Mhatre on heavy metal pollution in general and the biosphere's relevance and significance especially in relation to bioindicating and biomonitoring of these heavy metals [3,4].

HISTORY OF ENVIRONMENTAL CONTAMINATION BY METALS

Environmental contamination with trace metals is as ancient as discovery of fire. Small amounts of metals are always released into the air by burning firewood, and the metal-enriched wood ash is then discarded in the environs. With the development of mining and metal-working techniques, the close links between metals, metal pollution, and human culture were established. During the Roman Empire, large quantities of metals, especially lead (aptly known as Roman Metal), were required to sustain the high standard of living [5]. The mines were operated on a small scale, but uncontrolled smelting in open fires often resulted in severe local contamination. Both Xenophon and Lucretius observed that noxious emissions from precious metal mines in Laurion, Greece, and other places were harmful to human health.

Vitruvius spoke of extensive water pollution in his time, and Pliny observed that emissions from mines and smelters were dangerous to animals, especially dogs. The interdiction of mining operations near ancient cities and the Roman edict forbidding any mining activities in Italy have been attributed to concern for environmental quality [3]. The well-documented, pandemic lead poisoning during the period of the Roman empire, however, had little to do with environmental pollution, but was a consumer disease stemming from extensive contamination of food, water, and wines with lead.

The development of large furnaces equipped with tall stacks during the 16th century drastically extended the sphere of influence of smelters and industrial installations. Trace metal profiles in peats, lake sediments, and ice sheets show that, by the end of the 17th century, metal pollutants released by industries in Britain and central Europe were reaching most regions of Scandinavia [6]. Accelerated accumulation of lead pollution in Greenland ice fields began around the turn of the 18th century [7]. Historical records in soils, peats, ice sheets, sediments, and tree rings in most parts of the Northern Hemisphere document pervasive trace metal pollution of increasing severity since the beginning of the Industrial Revolution. Between 1850 and 1900, worldwide industrial emissions of cadmium, copper, lead, nickel, and zinc to the atmosphere averaged about 380, 1,800, 22,000, 240 and 17,000 tonnes per year, respectively [8,9].

By the turn of this century, ever expanding technological development sharply increased the industrial consumption and discharge of toxic metals. Between 1900 and 1980, mine production of aluminum, nickel, chromium, copper, and zinc increased 114-, 35-, 18-, 5-, and 4-fold, respectively, while mine outputs of copper, lead, nickel, and zinc amounted to 250, 160, 17 and 185 tonnes, respectively [10]. Industrial emissions of cadmium, copper, lead, nickel, and zinc to the air increased 8-, 6-, 9-, 31-, and 8-fold, respectively [8,9]. These figures certainly indicated the large quantity of toxic metals being discharged into the environment. Sooner or later, traces of all the "new" metals that are mined will be discharged in the biosphere. Once dispersed in the biosphere, the metals cannot be degraded or recovered using current technology. Any environmental effects of trace metal pollution, therefore, tend to be permanent.

HEAVY METALS AS POLLUTANTS

Heavy metals and their salts occur naturally in the environment. Heavy metals are those with atomic numbers greater than 23 and densities more than 5. Until recently, metals were defined only on the basis of their densities. Passow *et al.* have listed 38 elements with densities greater than five; however, Nieboer and Richardson proposed that the term "heavy metals" be abandoned in favor of a classification of metal ions into CLASS A (oxygen-seeking), CLASS B (nitrogen/sulphur-seeking) and BORDERLINE (intermediate) [11,12]. The classification is related to atomic properties and the solution chemistry of metal ions.

Heavy metals are non-degradable, persist in nature for a longer period, and are toxic to living organisms at fairly low concentrations as stated in Table 1 [13].

The natural levels of heavy metals in the environment had never been a threat to health. But in the recent years, increased industrial activities leading to airborne emissions, auto exhausts, effluents from industries, as well as solid-waste dumpings have become the sources of large quantities of heavy metals in the

Table 1
Heavy Metals, Their Mode of Contamination, and Effects on Human Beings

Name	Mode of contamination	Physical effect on human beings	Maximum allowable concentration
1. Asbestos fibres	Airborne	Carcinogenic	—
2. Cadmium-Cd	Food borne	Nausea, griping pain, vomiting and diarrhea	None
3. Methyl mercury	Food borne	Affects central nervous system, brain damage	0.500 ppm
4. Arsenic-As	Water borne	100 mg causes severe poisoning	0.100 mg/l
5. Barium-Ba	Water borne	A muscle stimulation, affects heart	0.100 mg/l
6. Cadmium-Cd	Water borne	Cumulative amounts lead to heart disease	0.100 mg/l
7. Chromium-Cr^{6+}	Water borne	Carcinogenic when inhaled	0.050 mg/l
8. Lead-Pb	Water borne	Serious cumulative and acute body poison	0.050 mg/l
9. Mercury-Hg	Water borne	Damages brain Central nervous system	0.005 mg/l
10. Selenium-Se	Water borne	Cause loss of hair and dermal changes	0.010 mg/l
11. Silver-Ag	Water borne	Large amounts cause irreversible skin grayness	0.050 mg/l

Source: Berkson, et al. [13].

environment. The potential ecological effects of rising levels of heavy metal concentrations in the environment have evoked increasing concern as can be realized from the levels of some metals in an urban environment like Bombay as indicated in Tables 2 and 3 [14].

Chaphekar and Mhatre have reviewed the environment hazards posed by mercury [15]. Similar reviews also are available for other heavy metals.

Table 2
Airborne Concentrations of Pb, Cd, Zn and Cu in Bombay during 1980–81

Location	Pb	Cd	Zn	Cu
Deonar	62	3.5	185	118
Trombay	67	3.6	172	154
Parel	280	26.0	680	615
Byculla	320	32.0	1040	538
Worli	290	11.3	480	286
Phule Market	550	44.0	920	640
King's Circle	520	—	—	—
Fort	490	41.0	810	410
Matheran*	16	1.5	—	—

**A small hill station approximately 80 km east of Bombay.*
Airborne concentration geometric mean value in ng/m³
Source: Khandekar [14].

Table 3
Pb, Cd, Zn and Cu Concentrations in Surface Soils in Greater Bombay

	μg/g of soil			
Place	**Pb**	**Cd**	**Zn**	**Cu**
Haji Ali	192.0	1.6	308	216
Dadar	54.5	9.0	260	160
Santacruz	42.0	0.7	182	164
Deonar	8.6	0.6	141	101
Matheran*	4.3	—	78	63

**A small hill station, devoid of industries and automobiles. Values of metals in soils may be treated as background levels.*
Source: Khandekar [14].

INDUSTRIAL EMISSIONS OF METALS

Metals are discharged into the air, water, and soils from a wide variety of natural and industrial sources. Although the available information is very scanty, recently published information leaves no doubt as to the great impact of industrial activities on the regional and global cycling of trace metals.

Emissions of trace metals from primary natural sources to the atmosphere are shown in Table 4 [3,16]. Windblown soil particles account for more than 50% of the manganese, chromium, and vanadium, and for 30% to 50% of the antimony, nickel, molybdenum, and zinc. Volcanoes contribute more than 60% of the cadmium, 40% to 50% of the mercury and nickel, and 30% to 40% of the arsenic, chromium, and copper fluxes from natural sources. Biogenic sources are the leading contributors of arsenic, mercury, and selenium, while fires and sea salt sprays generally account for less than 15% of the total natural emissions of each element. Of course, some of the metals emitted by natural sources are derived anthropogenically; for instance, metals in wind-blown dusts often are of industrial origin. Thus, the data in Table 4 probably overstate the actual natural fluxes of metals into the pre-industrial-era atmosphere.

The three main industrial sources of atmospheric metal pollution are mining, smelting, and refining of metals; burning of fossil fuels; and production and use of metallic commercial products (see Table 5) [3,17]. The burning of fossil fuels accounts for more than 95% of vanadium and 80% of nickel. Hence, scientists use these two elements to identify air samples coming from power plants. Fossil fuel combustion also accounts for about 60% of anthropogenic emissions of mercury, selenium, and tin, and significant amounts of antimony, arsenic, and copper. Smelters represent the leading source of arsenic, copper, and zinc, while the

Table 4
Worldwide Emissions of Trace Metals from Natural Sources to the Atmosphere

Element	Windborne soil particles	Sea salt spray	Volcanoes	Forest fires	Biogenic Sources	Total[(a)]
	(thousand tonnes per year)					
Antimony	0.78	0.56	0.71	0.22	0.29	2.6
Arsenic	2.60	1.70	3.80	0.19	3.90	12.0
Cadmium	0.21	0.06	0.82	0.11	0.24	1.4
Chromium	27.00	0.07	15.00	0.09	1.10	43.0
Cobalt	4.10	0.07	0.96	0.31	0.66	6.1
Copper	8.00	3.60	9.40	3.80	3.30	28.0
Lead	3.90	1.40	3.30	1.90	1.70	12.0
Manganese	221.00	0.86	42.00	23.00	30.00	317.0
Mercury	0.05	0.02	1.00	0.02	1.40	2.5
Molybdenum	1.30	0.22	0.40	0.57	0.54	3.0
Nickel	11.00	1.30	14.00	2.30	0.73	29.0
Selenium	0.18	0.55	0.95	0.26	8.40	10.0
Vanadium	16.00	3.10	5.60	1.80	1.20	28.0
Zinc	19.00	0.44	9.60	7.60	8.10	45.0

(a) Totals are rounded.
SOURCE: J. O. Nriagu, "A Global Assessment of Natural Sources of Atmospheric Trace Metals," Nature 338 (1989): 47–49 (From Nriagu, 1990)
Source: Nriagu [3,16].

Table 5
Worldwide Atmospheric Emissions of Trace Metals from Anthropogenic Sources

Element	Energy Production	Mining	Smelting and Refining	Manufacturing Processes	Commercial Uses[a]	Waste Incineration	Transportation	Total[b]
				(thousand tonnes per year)				
Antimony	1.30	0.10	1.42	—	—	0.67	—	3.5
Arsenic	2.22	0.06	12.30	1.95	2.02	0.31	—	19.0
Cadmium	0.79	—	5.43	0.60	—	0.75	—	7.6
Chromium	12.70	—	—	17.00	—	0.84	—	31.0
Copper	8.04	0.42	23.20	2.01	—	1.58	—	35.0
Lead	12.70	2.55	46.50	15.70	4.50	2.37	248	332.0
Manganese	12.10	0.62	2.55	14.70	—	8.26	—	38.0
Mercury	2.26	—	0.13	—	—	1.16	—	3.6
Nickel	42.00	0.80	3.99	4.47	—	0.35	—	52.0
Selenium	3.85	0.16	2.18	—	—	0.11	—	6.3
Thalium	1.13	—	—	4.01	—	—	—	5.1
Tin	3.27	—	1.06	—	—	0.81	—	5.1
Vanadium	84.00	—	0.06	0.74	—	1.15	—	86.0
Zinc	16.80	0.46	72.00	33.40	3.25	5.90	—	132.0

(a) including agricultural uses.
(b) Totals are rounded.
SOURCES: J. O. Nriagu and J. M. Pacyna, "Quantitative Assessment of Worldwide Contamination of Air, Water and Soils with Trace Metals", Nature 333 (1988): 134–39 (quoted from Nriagu, 1990)
Source: Nriagu [3,17].

manufacture of steel is responsible for the largest fractions of manganese and chromium released to the atmosphere by human activities. Despite the reduced use of leaded gasoline in many countries, the automobile tailpipe still accounts for roughly two-thirds of lead emissions. Also, a large fraction of airborne arsenic still is derived from the spray of arsenical insecticides.

A comparison of Tables 4 and 5 leads to the conclusion that industrial emissions have become dominantly responsible for most of the trace elements in the air [3,16,17]. Anthropogenic emissions of lead, cadmium, vanadium, and zinc exceed the fluxes from natural sources by 28-, 6-, 3-, and 3-fold, respectively. Industrial contributions of arsenic, copper, mercury, nickel, and antimony amount to 100% to 200% of emissions from natural sources. In many urban areas and around some point sources, the natural emissions are insignificant in comparison with the anthropogenic metal burden.

The concentrations of trace metals in continental waters are controlled by atmospheric precipitation and the weathering processes on soils and bedrock. Because these pathways and processes have been significantly altered by humankind, the flux and distribution of trace metals in a large fraction of all freshwater resources has increased. The atmosphere, in particular, has become a key medium in the transfer of pollutant trace metals to remote aquatic ecosystems. On a global scale, this pathway annually supplies more than 70% of the lead and vanadium, about 30% of the mercury, and about 20% of the cadmium flux into aquatic ecosystems (see Table 6) [3,17]. In many rural and remote regions, the atmosphere actually supplies most of the trace metal budgets of aquatic ecosystems. For example, more than 50% of all the trace metal entering the Great Lakes is transported via the atmosphere [18].

Domestic and industrial wastewaters, sewage discharges, and urban runoff also contribute large quantities of metal pollution to the aquatic environment. These discharges often occur at point sources and can lead to excessive local metal burdens in water. There are many reports available on case histories of lakes, rivers, and marine coastal zones in every part of the world that are being polluted with trace metals. The figures in Table 6 suggest that the problem has become global in scope [3,17].

Modern societies generate large quantities of various metal-containing wastes that are discarded on land (see Table 7) [3,17]. The disposal of coal and wood ashes, industrial installations, and commercial products (which corrode and decompose) on land together account for about 55% to 80% of the metal pollution in soils. The large volumes of wastes associated with animal husbandry, logging, agriculture, and food production also often affect quantities of the trace metal of many soils. Agriculture soils generally receive most of their trace metal pollution from the atmosphere and from fertilizers, pesticides, and manure. Because of their very high trace metal contents, both municipal and industrial sewage

(text continued on page 438)

Table 6
Worldwide Inputs of Trace Metals and Aquatic Ecosystems

Sources	Trace Elements (thousand tonnes per year)												
	Antimony	Arsenic	Cadmium	Chromium	Copper	Lead	Manganese	Mercury	Molybdenum	Nickel	Selenium	Vanadium	Zinc
Domestic wastewaters	2.20	9.2	1.70	46.0	28	6.80	110	0.30	2.20	62.0	3.80	2.30	48
Electric power plants	0.18	8.2	0.12	5.7	13	0.72	11	1.80	0.65	11.0	18.00	0.30	18
Base metal mining and smelting	3.80	7.4	2.00	12.0	14	7.00	40	0.10	0.51	13.0	12.00	0.60	29
Manufacturing processes	9.30	7.0	2.40	51.0	34	14.00	21	2.10	4.20	7.4	4.30	0.55	85
Atmospheric fallout	1.10	5.6	2.20	9.1	11	100.00	12	2.00	0.95	10.0	0.82	26.00	10
Sewage discharges	1.50	4.1	0.69	19.0	12	9.40	69	0.16	2.90	11.0	2.00	3.50	17
Total Input[(a)]	18.00	42.0	9.10	143.0	112	138.00	263	6.50	11.00	114.0	41.00	33.00	237

(a) Totals are rounded.
SOURCES: J. O. Nriagu and J. M. Pacyna, "Quantitative Assessment of Worldwide Contamination of Air, Water and Soils with Trace Metals", Nature 333 (1988): 134–39 (quoted from Nriagu, 1990)
Source: Nriagu [3,17].

Table 7
Worldwide Inputs of Trace Metals into Soils

	Trace Elements (thousand tonnes per year)											
Source	Antimony	Arsenic	Cadmium	Chromium	Copper	Lead	Manganese	Mercury	Molybdenum	Nickel	Selenium	Vanadium	Zinc
Agricultural and animal wastes	4.90	5.80	2.20	82.00	67.0	26.0	158.0	0.85	34.00	45.00	4.60	19.00	316.0
Logging and wood wastes	2.80	1.70	1.10	10.00	28.0	7.4	61.0	1.10	1.60	13.0	1.60	5.50	39.0
Urban refuse	0.76	0.40	4.20	20.00	26.0	40.0	24.0	0.13	2.30	6.1	0.33	0.20	60.0
Municipal sewage & organic waste	0.18	0.25	0.18	6.50	13.0	7.1	8.1	0.44	0.43	15.0	0.11	1.3	39.0

Solid wastes from metal fabrication	0.08	0.11	0.04	1.50	4.3	7.6	2.6	0.04	0.08	1.7	0.10	0.12	11.0
Coal ashes	12.00	22.00	7.20	289.00	214.0	144.0	[illegible]076.0	2.60	44.00	168.0	32.00	39.00	298.0
Fertilizers & peat	0.25	0.28	0.20	0.32	1.4	2.9	12.0	0.01	0.46	2.2	0.27	0.97	2.5
Discarded manufactured products[a]	2.40	38.00	1.20	458.00	592.0	292.0	300.0	0.68	1.90	19.0	0.15	1.70	465.0
Atmospheric fallout	2.50	13.00	5.30	22.00	25.0	232.0	27.0	2.50	2.30	24.0	2.0	60.00	92.0
Total input[b]	26.00	82.00	22.00	898.00	971.0	759.0	1669.0	8.30	87.00	294.0	41.00	128.00	1322.0

(a) Metals used for industrial installation and "durable" goods are assumed to have a definite life span and to be released into the environment at a constant rate.
(b) Totals are rounded.
Note: These inputs exclude mine tailings and slags at the smelter sites.
SOURCES: J. O. Nriagu and J. M. Pacyna, "Quantitative Assessment of Worldwide Contamination of Air, Water and Soils with Trace Metals", Nature 333 (1988): 134–39 (quoted from Nriagu, 1990)
Source: Nriagu [3,17].

(text continued from page 434)

sludges sometimes are considered unsafe for disposal on land and clearly represent one of the most important sources of trace metal contamination in local soils.

The anthropogenic release of trace metals into the biosphere is compared to their mobilization during the weathering cycle in Table 8 [3,10,17,19]. The data show the dominating influence of industrial discharges on the trace metal economy of the biosphere. Manganese and vanadium are possible exceptions because the geochemistry and industrial discharges of these elements are largely unknown. For antimony, copper, lead, and zinc, the industrial discharges exceed the weathering flux by about threefold, for mercury the difference is tenfold. When mine outputs of metals are taken into account, there can be no doubt that humankind has become the key agent in the global redistribution of trace metals into the biosphere. Most of the industrial discharges occur into the

Table 8
Industrial and Natural Mobilizations of Trace Metals in the Biosphere

Element	Production from mines[(a)]	Total Industrial discharges[(b)]	Weathering mobilization[(c)]
		(thousand tonnes per year)	
Antimony	55.0	41	15.0
Arsenic	45.0	105	90.0
Cadmium	19.0	24	4.5
Chromium	6,800.0	1,010	810.0
Cobalt	36.0		120.0
Copper	8,114.0	1,048	375.0
Lead	3,077.0	565	180.0
Manganese	16,000.0	1,894	4,800.0
Mercury	6.8	11	0.9
Molybdenum	98.0	98	15.0
Nickel	778.0	356	255.0
Selenium	1.6	76	4.5
Vanadium	34.0	75	855.0
Zinc	6,040.0	1,427	540.0

(a) Only a fraction of each metal mined each year is released into the environment in the same year.
(b) Industrial discharges are calculated as discharges into soils (Table 7) & water (Table 6) minus the emissions to the atmosphere (Table 4).
(c) Weathering mobilization is calculated using average trace metal concentrations in soils and suspended sediment flux of 1.5×10^{16} grams per year in rivers. The dissolved trace metal flux is generally much lower than the particulate flux.
Sources: Nriagu [3,10,17] and Adriano [19].

Northern Hemisphere, where perturbations of the trace metal cycles are likely to be profound.

WORLDWIDE DISTRIBUTION OF METALS

The metals released from industrial sources are rapidly dispersed into different environmental compartments. The strong influence of anthropogenic inputs on the global and regional cycling of trace metals is evident from the causal relationships that have been observed between emission rates and ambient concentrations in many ecosystems. Because of the high ambient concentrations of trace metals, urban environments can be regarded as ecological "hot spots" of toxic metals. Average concentrations of airborne trace metals in urban areas generally exceed the levels in rural areas by five- to tenfold and the levels in remote locations by more than 100-fold for many elements (Table 9) [3,8,20-25]. If current loading rates continue, the contamination of urban air with toxic metals probably will become one of the major public health issues for future generations. Moreover, concentrations of certain toxic metals observed in urban rainfall now sometimes exceed the levels considered safe for drinking water, and in some developing countries, rainwater frequently is consumed by people living in urban areas. Furthermore, most of the toxic metals are associated with fine particles (less than one micrometer in size) that are readily absorbed in bronchial tracts.

An extensive survey of the global distribution of lead in the marine atmosphere found the highest concentrations near the urban-industrial areas of Western Europe and Eastern North America and the lowest levels in Antarctica [26]. Another study has revealed that lead concentrations in recently formed Antarctic ice layers are four times higher than lead concentrations in ice layers formed prior to the Industrial Revolution [27]. Scientists also have found a strong association in the distribution of airborne lead and cadmium over the remote North Pacific and have suggested that most of the cadmium is derived from anthropogenic sources [28]. In late spring every year, the Arctic is covered by a haze containing elevated levels of trace metals derived from industrial sources in Eurasia [29]. The Arctic haze affects nearly 9% of the Earth's surface and is the most extensive air pollution system known. These examples and many other regional surveys illustrate that few places on Earth are free of trace metal pollution.

Freshwater ecosystems are particularly sensitive to external trace metal inputs. The volumes of freshwater ecosystems often are small, and background concentrations of trace metals in such waters are generally low. Small additions from anthropogenic sources can alter the size of a reservoir's metals pool and, hence, the distribution and flow of metals in the ecosystem. The chemical forms of trace metals in water often are available to the biota, and seemingly small concentrations may accumulate to high levels in the food chain. Even if one assumes that only 10% to 30% of the total discharges go into lakes and rivers, the calculated

Table 9
Average Particle Size and Ambient Concentrations of Trace Metal in Urban Rural and Remote Locations

Element	Mass median diameter (micrometers)	Average concentration (& range)[a] (nanograms per cubic meter of air) Urban	Rural	Remote
Antimony	0.90	7.7 (15–0.4)	0.48 (0.90–0.05)	0.03 (0.06–0.0005)
Arsenic	1.30	20.0 (35–5.0)	3.20 (6.0–0.3)	0.29 (0.5–0.07)
Cadmium	1.10	5.2 (10–0.3)	0.26 (0.50–0.02)	0.05 (0.10–0.002)
Chromium	0.96	60.0 (115–5.0)	14 (25–2.0)	2.30 (4.5–0.04)
Cobalt	2.80	39.0 (70–8.0)	5.10 (10–0.2)	0.15 (0.30–0.006)
Copper	1.30	155.0 (300–10)	7.90 (15–0.8)	1.20 (2.4–0.06)
Lead	0.56	715.0 (1400–30)	16.00 (30–1.0)	2.50 (5.0–0.01)
Manganese	2.10	92.0 (180–5.0)	6.20 (12–0.5)	1.50 (3.00–0.05)
Mercury	0.78	4.0 (7.5–0.5)b	1.60 (3.0–0.02)	0.38 (0.75–0.004)b
Nickel	0.73	52.0 (100–5.0)	4.30 (8.0–0.6)	0.80 (1.5–0.08)
Selenium	0.92	13.0 (35–0.4)b	2.60 (5.0–0.08)	0.45 (0.9–0.007)b
Vanadium	1.60	36.0 (70–2.0)	1.80 (3.5–0.2)	0.50 (1.0–0.004)
Zinc	1.20	432.0 (850–15)	26.00 (50–3.0)	6.20 (12–0.4)

(a) Ranges are given from highest to lowest average concentration.
(b) Includes particulate and gaseous phases.
Sources: Nriagu [3,8,20-24] and Davidson [25].

pollution load would, in fact, dwarf the usual natural baseline concentrations of trace metals in most lakes and rivers [17]. Thus, freshwater resources must be considered to be at greatest risk in terms of toxic metal pollution.

The human influence on the oceanic cycle of trace metal is becoming increasingly apparent. For example, the atmospheric flux of anthropogenic lead to the ocean is large enough to significantly change the normal marine cycle of this element. Evidence for the perturbation is unequivocal. The typical profile of lead in the ocean water column is now determined by the intense atmospheric flux and is unlike the profiles of the other trace metals, which have less anthropogenic input. The isotopic composition of lead in surface ocean waters has shifted toward the isotropic ratios of industrial lead used in nearby terrestrial sources. The average lead content of the mixed surface layers of the North Atlantic, which is polluted, is about 26 times higher than the levels in similar layers of the South Pacific, which is less polluted. The 15-fold difference in the lead content of recently deposited coral shells compared with growth layers deposited a century ago gives another indication of the magnitude of oceanic lead pollution [3]. Even sediments in the deep Atlantic Ocean show a sharp increase in lead content and a change in the isotopic signature of the most recent deposits. These changes also have been attributed to the increased flux in industrial lead [3].

Soils represent the ultimate sink for trace metals in continental areas. Because metals are fairly immobile in soils (except under the influence of acid rain), metal pollution tends to accumulate primarily in the surface layers. The buildup of toxic metals in the most biologically active part of the soil, the organic topsoil, makes the metals readily accessible to some crops and vegetables. The chemistry of trace metals in soils thus plays an important role in the suspected transfer of contaminant metals to the food supply.

Human influence on trace metals in soils is demonstrated dramatically by the highly elevated levels of metals that now characterize the soils in urban areas and around major industries [19,30]. Most of these metals are delivered via the atmosphere. The median values (and typical ranges) of values reported for atmospheric fallout of trace metals in urban areas of North America are 160 (20 to 980) grams per hectare per year for copper, 910 (140 to 3,500) for lead, 18 (7 to 36) for cadmium, and 3,200 (80 to 4,800) grams per hectare per year for zinc [31]. The values for urban areas of Europe are 320 (78 to 500) for copper, 400 (180 to 600) for lead, 310 (33 to 530) for nickel, 15 (3 to 28) for cadmium, and 1,000 (220 to 5,850) grams per hectare per year for zinc [32]. At these deposits rates, the levels of most trace metals in surface soils will double in two to ten years, depending on the baseline metal contents. Atmospheric fallout of metals in rural areas also has risen sharply; the rates in rural Europe have been estimated to be 150 (14 to 320) grams per hectare per year for copper, 550 (38 to 3,900) for zinc, 220 (63 to 550) for lead, 32 (7 to 100) for nickel, and 4 (2 to 13 for cadmium [31].

Because the atmosphere is just one of the media for industrial emissions, the human perturbation of the distribution of trace metals in urban soils is obviously substantial. Indeed, numerous studies have demonstrated that soils in urban areas—whether from parks or domestic gardens, cities or small towns—have become so highly contaminated with a wide variety of trace elements that any baseline differences originating from the local geology is difficult to trace [19,30]. Typical concentrations of many toxic metals in both urban and rural soils are now from two to more than ten times higher than the levels in uncontaminated soils [30]. Moreover, the available levels, defined as the fraction extractable using mild reagents, also have increased markedly in urban soils [30]. In terms of the elevated levels and chemical forms of trace metals, soils in urban areas have become quite unique and are unlike anything produced by natural weathering processes.

Soils downwind of urban and industrialized centers also receive large doses of trace metal pollution. For example, the concentrations of lead, cadmium, arsenic, and antimony in surface soils of southern Norway are about ten times higher than the levels in the northern part of the country [33]. The regional difference has been attributed to the exposure of pollutants from Britain and central Europe. The lead contamination of forests in remote parts of the northeastern United States is primarily caused by long-range transport of industrial emissions from New England and the Midwest [34]. For arable soils, primary sources of metal pollution also include fertilizers, agricultural chemicals, and liquid and solid wastes (see Table 7) [3,17]. It has been estimated that the average cadmium input into agricultural lands in Europe is about 8 grams per hectare per year from the atmosphere and 5 grams per hectare per year from the application of phosphate fertilizer [35]. Metal contamination of agricultural soils in Belgium from fertilizers and the atmosphere has been estimated to average 16, 20, 260, and 3,800 grams per hectare per year for arsenic, cadmium, lead, and zinc, respectively [36]. At such loading rates, many arable soils in Europe may be close to exceeding their carrying capacity for trace metal pollution. In Japan, on the other hand, the problem is no loner hypothetical. About 9.5% of rice paddy soils have been rendered unsuitable for growing rice for human consumption because of excessive metal contamination [37].

GLOBAL EFFECTS

Environmental metal pollution may either result in direct assaults on human health or have indirect effects on human welfare by interfering with the integrity and vitality of the life-support systems. Expectedly, concern about the wanton fouling of the biosphere with toxic metals has centered around the effects on human health and, specifically, on acute rather than chronic effects of toxic metals. Reported case histories of acute poisoning in the general population by metals in the environment are few: familiar examples are the Itai-itai (cadmium poisoning) and Minamata (mercury poisoning) diseases of Japan. On the other hand, the chronic effects of toxic metals on human and animal populations and on ecosystem health have yet to receive adequate attention.

The exposure dose of a given element to any organism is a function of the element's concentration in the environment and the duration of exposure. Because some trace metals tend to accumulate progressively in the body or an ecosystem, long-term exposure to low concentrations can lead to adverse effects when the toxic dose is reached. In general, however, the long-term effects of exposing human populations to small doses of toxic metals in the environment are likely to be subclinical, ranging from early lesions to nascent clinical diseases with unrecognizable symptoms, and can lead to inordinate depletion of the functional integrity of cells, tissues, and organs. With the possible exception of lead testing, the current level of sensitivity of clinical tests is inadequate for the diagnosis of subtle biochemical changes, or the "no-effect" distress syndrome. Changes in vital signs become manifested only after the intoxication process has advanced to the stage where homeostatic mechanisms no longer can maintain body function within the normal range. Because of the wide spectrum of non-specific adverse health effects, there is a lot of debate over the risks of exposure to elevated levels of trace metals in the environment.

Lead is one of the most pervasive and pernicious toxic substances being released into the environment. Exposure to ambient environmental lead has been associated with a wide range of metabolic disorders and neuropsychological problems, especially in children [38]. These health effects apparently have no known threshold in terms of measurable lead concentrations in the body fluids or organs. Indeed, lead-induced biochemical changes have been detected down to picomolar lead concentrations [39]. Depending on the criteria used to define excessive exposure, from 9% to 25% of the preschool children in the United States may be at risk of having their health impaired by lead in the environment [38]. It has been suggested that about one-third of the world's urban population is being exposed to either marginal or unacceptable air lead concentrations, and the number of persons with elevated levels of lead in their blood is estimated to be between 130 million and 200 million [10]. Thus it is not surprising that chronic lead poisoning has been called one of the major public health issues of our time [40].

The discharge of other metals in the environment also constitutes a health hazard to many people. It has been estimated that, worldwide, about 250,000 to 500,000 people may have renal dysfunction because of cadmium poisoning, while 40,000 to 80,000 people, mostly in fishing villages, may suffer from mercury poisoning as a result of eating mercury-contaminated seafoods. About 250,000 people are believed to suffer from arsenic poisoning [10]. Furthermore, exposure to pollutant trace metals increasingly is implicated in the etiology of a large number of ailments, including cardiovascular diseases, reproductive failures, dermatitis allergies, and some cancers. That a large number of people are at risk of poisoning by persistent exposure to environmental doses of trace metals is a public health problem likely to remain unresolved for a long time.

Human beings are by no means the only organisms at risk: A large number of cats and dogs living in cities are known to have lead poisoning. High concentrations of lead have been detected in pigeons living in London, where the most contaminated birds show signs of acute lead poisoning [41]. About 2% to 3% (or one million individuals) of the fall and winter population of waterfowl in North America suffer fatal plumbism (chronic lead poisoning) caused by ingesting spent lead pellets [42]. In central Canada, advisories have been issued against eating moose, deer, and bear kidneys and livers because they contain dangerous levels of cadmium [43], but it is not known if the cadmium has impaired the health of the animals in any way. In fact, news of cattle deaths from lead poisoning in Talasari, situated on the borders of Maharashtra and Dadra Nagar Haveli in India, in January and February 1991 evoked public furor, and a detailed investigation carried out by Jadhav and Sawant indicated large-scale lead poisoning in the area caused by the release of industrial effluents [44]. These examples are enough to suggest that few living organisms are safe from undue trace metal inputs over an extended period of time.

Aquatic organisms, especially those at the top of the food chain such as fish-eating birds and mammals, are much more sensitive to toxic metals in their habitat than are terrestrial biota. Although dose-effect relationships have been documented in local aquatic environments, adverse effects on a regional scale attributable directly to trace metal pollution have not been demonstrated unequivocally. Aquatic toxicologists still are struggling to develop proper test protocols for determining ecological stress from toxic metal inputs. There is, however, some circumstantial evidence to suggest that the "no-effect" thresholds for some metals is fairly close to their current levels in natural waters [45]. Apparently, the margin of safety for the most sensitive organisms either has been exceeded or is eroding rapidly.

The propensity of metals to accumulate in living tissue can lead to metal concentrations in fish and other seafood that are above the levels considered safe for human consumption. The most familiar example is the contamination of aquatic habitats with mercury, which accumulates in invertebrates, fish, and other wildlife to levels above the guidelines issued to protect human health. It has been estimated that about 10,000 lakes in Sweden contain northern pike (*Esox lucins*) with mercury concentrations in excess of 1,000 μg/kg, and about 42,000 lakes have fish containing more than 500 μg/kg. Highly elevated mercury concentrations also have been reported in fish in remote acidic lakes of Ontario, Wisconsin, and the northeastern United States [46]. In these regions, most of the mercury probably is derived from human activities. Acid rain exacerbates the hazards associated with trace metal pollution. Increased acidification can increase the rates of metal mobilization from soil and sediments, the capacity of suspended sediments to retain metals, as well as the speciation—and hence the toxicity—of metals. Because of the close link between acidification and the behavior of metals, it has been suggested that the adverse biological effects of acid deposition may stem, at least in part, from the increased concentration of biologically available forms of metals [47]. Increased trace metal concentrations also can affect biological processes in soils, including the rates of litter decomposition, soil respiration, and nitrogen mineralization, as well as the activity of key soil enzymes [48]. Although trace metals may not be the primary causative factor, they may mediate forest death in some acid-sensitive terrains.

ECOTOXIC EFFECTS OF HEAVY METALS ON PLANTS

When any pollutant affects a sensitive plant species, some specific symptoms result whereas the pollutant itself often undergoes some chemical change soon after it has attacked the plant tissue. The type and extent of damage to susceptible plants serve as indicators of pollution. Hence, close observations of the injuries and symptoms produced on susceptible species provide us with a means of monitoring pollution and also help in determining the geographical distribution of a pollutant over a large area [49].

The most characteristic symptom of adverse effects of heavy metals on plants is inhibition of root growth followed by stunting or dwarfing of shoots [50,51]. Germination and early growth of seedlings also were found to be affected when tested with heavy metals such as mercury, lead, and arsenic [52]. Leaf necrosis appears specific to a particular metal, but there is also a general chlorosis of the younger leaves common to all metals [53]. A distinct relationship between the leaf injury index (LII) and the photosynthetic ability of plants and their primary production estimated as chlorophyll content and biomass, respectively, was established when treated with mercury [54,55]. The LII was thus suggested as a simple index for biomonitoring purposes. Specific symptoms for individual metals are reported in Table 10 [56].

On the other hand, some plants do not show any signs of damage despite high levels of heavy metals in their systems. Ernst has compiled a list of plants without symptoms of metal toxicity despite accumulation (see Table 11) [57].

A classic example of accumulation of heavy metals in the plant organs without any visible symptoms of injury was reported along the Kalu river estuary polluted with heavy metals discharged from the industries located on the banks of the river (see Table 12) [58]. On the contrary, the plant community was reduced to a monoculture having only one plant species growing luxuriantly [58].

Such plant materials are envisaged as prospective candidates as bioindicators as well as biomonitors.

BIOINDICATORS AND BIOACCUMULATORS OF HEAVY METALS

A bioindicator is defined as a plant or an animal that reveals the presence of a substance in its vicinity by showing some typical symptoms that can be distinguished from the effects of other natural or anthropogenic stresses. Salanki states that a good bioindicator is one that shows the earliest responses to the

Table 10
Morphological Variations Occurring under the Influence of Mineral Absorption by Plants

Aluminum	Shortening of roots and leaf scorch.
Boron	Stunting, prostrate forms, deformation, blotching, and browning of leaves.
Chromium	Chlorosis of leaves.
Cobalt	Chlorosis or decrease of chlorophyll
Copper	Chlorosis of leaves and dwarfism, reduction of size of seeds and corollas.
Iron	Darkening of leaves.
Manganese	Chlorosis of leaves with white blotching.
Molybdenum	Abnormally colored shoots, chlorosis, and vulnerability to insect attack.
Nickel	Chlorosis and necrosis of leaves.
Uranium	Variation in flower color, abnormal fruits, more chromosomes, stimulated growth.
Zinc	Chlorosis of leaves, symptoms of manganese deficiency.

Source: Tiagi [56].

Table 11
Metal Indicator Species

Plant - Species	Metal	Leaf*	Root*
Psychotria dowarrei	Ni	772.0	1580.0
Thlaspi alpestre	Zn	382.0	173.0
Minuartia verna	Pb	55.0	127.0
Trachypogon spicatus	Cu	0.1	43.0
Pearsonia metallifera	Cr	9.0	31.0
Acrocephalus robertii	Co	25.0	—
Minuartia verna	Cd	3.1	3.4
Cynodon plectostachyum	As	0.3	0.3

**Metal content µg/atom dry matter*
Source: Ernst [57].

Table 12
Heavy Metals in Sediments and in the Organs of *Pycreus macrostachyos* at Polluted (P) and Non-Polluted (NP) Spots along the Kalu River Estuary

	μg/g dry weight					
	Sediment		Leaf		Rhizom	
Metals	P	NP	P	NP	P	NP
H_g	1.5–140	39–52	3.3–110	1.4	6.9–53.3	2.9
P_b	5.4–10.6	3.8–4.4	100	3.0	100	2.8
C_d	0.62–12.6	0.42–6.8	2.5–10	2.5	0.98–10.0	2.1
C_u	91–864	78–89	9.0–26.4	23.0	2.03–307	89.0

Source: Mhatre [58].

pollutant enabling indication of the presence and prediction of the consequences of undesirable anthropogenic effects [2].

However, in some cases, plant species under natural conditions are capable of evolving tolerance by inheritance or through mutation. This aspect was postulated first by Prat, who stated that tolerance is evolved by natural selection [59]. Investigations carried out since then have been reviewed by Bradshaw and Ernst who have summarized different aspects of evolution of metal tolerances in various species [57,60].

PLANT COMMUNITIES AS INDICATORS

It is generally accepted that plant communities serve as better, effective, and more reliable indicators than a single species. Chaphekar et al. have reported a reduced diversity in the plant community because of water pollution along the Ulhas River in an industrial suburb near Bombay [61]. A similar phenomenon of highly reduced diversity to a single monoculture stand at a heavily metal-polluted spot along the Kalu River estuary also near Bombay was reported by Mhatre et al. [58]. Such reduced diversity stands indicate the quality of water and sediment in those areas.

NON-ANGIOSPERMIC PLANTS

The use of epiphytes (e.g., lichens and mosses) for monitoring metal pollution of air has been reported by many workers at different places. Lichens have been demonstrated to be useful as indicators of heavy metals fallout [62]. Little and Martin used *Sphagnum* moss for detection of zinc, lead, and cadmium at 47 sampling points over a large area around a zinc and lead smelting complex at Avenmouth, near Bristol, U.K. [63]. Parkarinen and Tolonen also used the peat moss *Sphagnum fuscum* for monitoring different heavy metals like lead, zinc, mercury, cadmium, nickel, etc., at 15 sites in Finland and two sites in western continental Canada [64]. The use of peat mosses, especially *S. fuscum,* was suggested for extensive mapping of heavy metal deposition within the boreal forest vegetation zone around mining areas, and specific metal indicator species have been identified. Chaphekar and Kulkarni also have reported differential sensitivity of some commonly found bryophytes in western India to mercury [65].

ANGIOSPERMIC PLANTS

It is widely agreed that plants can serve as a device for quick assessment of environmental quality. Numerous vegetational studies in and around mining areas have been undertaken, and specific metal indicator species have been identified. Plants that are largely restricted to or particularly abundant on metal-contaminated soils have been cited as indicators of metal-containing soils, but there is no clearcut evidence that they have been used for prospecting [66].

Another indication given by an organism is the accumulation of stress-causing substances like heavy metals in its body. Thus, plants growing on soils containing high concentrations of minerals tend to accumulate them in unusually high quantities. The occurrence of accumulator plants indicates the mineral contents of the soils on which they are growing, while chemical analysis of the plant confirms it. Antonovics et al., Bradshaw and Ernst have given various examples of accumulators of different heavy metals [57,60,66]. Under Indian conditions, work also has been carried out especially after detection of reduced plant diversity due to water pollution and detection of higher-than-permissible levels of mercury in body tissues of some fish in Thane creek [61,67]. A universally available plant from the tropics, *Alternanthera sessilis,* was identified as an accumulator of aluminum, and the roots were found to contain as high as 7,100 ppm of the metal whereas neighboring plants contained less than 700 ppm [68]. Mhatre et al. have reported a cyperaceae member *Pycreus macrostachyos* as an accumulator of Hg, Pb, Cd, and Cu [57]. Khalap has investigated *Eichhornia crassipes, Typha angustata,* and *Lemna minor,* which were found to accumulate large quantities of chromium in their organs [69]. Accumulation of heavy metals like Zn, Cd, Cu, Pb, and Cr in the plant body of *Ipomoea carnea* was reported by Puri [70]. All these investigations indicated the possibility of these plants being used as potential biomonitors for heavy metals.

Plants also may act as bioscavengers of these metals from a particular localized situation because of their metal-accumulating capabilities. The role of aquatic plants like water hyacinth, duckweeds, water rosettes, bulrushes, etc., in scavenging water pollutants by way of bioaccumulation and biomagnification is a well-known fact and has been implemented in various countries for effective control of water pollution as a secondary tertiary treatment. Chaphekar and Mhatre have reviewed various similar biological methods that have been implemented for water pollution control, highlighting their uses and limitations [71].

BIOMONITORING OF HEAVY METALS

Biomonitoring may be performed by using natural vegetation and crops present in an area under study. However, the differences in soil, water quality, and other conditions may influence the effects and diminish the comparability of results between sites. It is therefore advisable to use selected indicator plants cultivated under standardized conditions.

There are two factors closely tied to exotoxicophysiology regarding the morphophysiological bioindication of heavy metals:

1. Internal Factors (genetic composition) and
2. External Factors (nutrition).

The second one is further divided as:

- Growth of plants (morphometry, physiological changes, etc.),
- Macro and micro characters (related to leaf structure, leaf morphology, like cuticular structure, etc.),
- Biochemical characters, and
- Accumulations of heavy metals in plant organs.

All these factors contribute in deciding the phytomonitoring ability of an indicator. However, there are some limitations to bioindicators for use in biomonitoring:

- More qualitative or semi-quantitative in nature.
- May be site selective.
- Not always universal in application.
- Many species under natural conditions are capable of evolving tolerance by genetic recombinations over one or few generations or by mutation.
- Many of the damage symptoms are identical.
- Because the organism has the capacity to respond to other environmental factors such as nutrients, water, stress, temperature, etc., the indicator value of the plant species to a pollution stress needs to be checked properly.

Despite the limitations, there are certain definite advantages in using plants as indicators. These are:

- Inexpensive device to serve as indicators of environmental quality on a semiquantitative scale, as sensitive species can serve as warning signals for environmental hazards.
- Mineral accumulation by plants is used as a tool in geobotanical prospecting.

- Wastewater treatment with proven bioaccumulators serving as bioscavengers and wasteland restoration with plants of proven indicator-cum-accumulator value for respective metals.

REFERENCES

1. Chaphekar, S. B., and Mhatre, G. N. (1986) *Human Impact on Ganga River Ecosystem,* Concept Publishing Co., New Delhi, pp. 1–186.
2. Salanki, J. (1986) *Biological Monitoring of the State of the Environment: Bioindicators,* I.U.B.S. Monograph series No. 1 for I.C.S.U. by I.R.L.
3. Nriagu, J. O. (1990). "Global Metal Pollution: Poisoning the Biosphere," *Environment 32* (7).
4. Mhatre, G. N. (1991) "Bioindicators and Biomonitoring of Heavy Metals," *J. Environ. Biol.,* pp. 201–209.
5. Nriagu, J. O. (1983) *Lead and Lead Poisoning in Antiquity,* Wily Interscience, New York.
6. Livett, E. A. (1988) "Geochemical Monitoring of Atmospheric Heavy Metal Pollution: Theory and Application," *Advances in Ecol. Res. 18,* pp. 65–176.
7. Murozumi, M., Chow, T. J., and Patterson, C. C. (1969) "Chemical Concentrations of Pollutant Lead Aerosols, Terrestrial Dusts and Sea Salts in Greenland and Antarctic Snow Strata," *Geochimica et Cosmochimica Acta 33,* pp. 1,247–1,294.
8. Nriagu, J. O. (1979) *Copper in the Environment, Vol. 1,* John Wiley & Sons, New York.
9. Nriagu, J. O. (1979) "Global Inventory of Natural and Anthropogenic Emissions of Trace Metals to the Atmosphere," *Nature 279,* pp. 409–411.
10. Nriagu, J. O. (1988) "A Silent Epidemic of Environmental Metal Poisoning?" *Environ. Poll. 50,* pp. 139–161.
11. Passow, H., Rothstein, A., and Clarkson, T. W. (1961) "General Pharmacology of Heavy Metals," *Pharmac. Rev. 13,* p. 185.
12. Nieboer, E., and Richardson, D. H. S. (1980) "The Replacement of the Nondescript Term 'Heavy Metals' by a Biologically and Chemically Significant Classification of Metal Ions," *Environ. Pollut. (Ser. B),* p. 1,326.
13. Berkson, H., Bingham, E. C., Durgan, P. R., Mayer, F. L., Parker, F. L., and Taylor, F. B.; Liptak, B. G., ed. (1974) "Types of Water Pollutants and their Effect," *Environmental Engineer's Handbook 1,* Chilton Book Co., Radnor, Pennsylvania, pp. 354–449.
14. Khandekar, R. N. (1985) "Toxic Heavy Metal Pollution from Acute Exhaust," paper presented at the All India Seminar on Vehicular Emission, Bombay, Nov. 25–26.
15. Chaphekar, S. B., and Mhatre, G. N. (1981) "Mercury: An Environmental Hazard," *Current Trends in Life Science Vol. 9,* Prof. L. P. Mall Commemoration Volume, Today and Tomorrow Publishers, New Delhi, pp. 277–288.

16. Nriagu, J. O. (1989) "A Global Assessment of Natural Sources of Atmo spheric Trace Metals," *Nature 338,* pp. 47–49.
17. Nriagu, J. O., and Pacyna, J. M. (1988) "Quantitative Assessment of Worldwide Contamination of Air, Water and Soils with Trace Metals," *Nature 333,* pp. 134–139.
18. Nriagu, J. O., Kullenberg, G., ed. (1986) "Metal Pollution in the Great Lakes in Relation to their Carrying Capacity," *The Role of the Oceans as a Waste Disposal Option,* Reidel, Dordrecht, Netherlands, pp. 441–468.
19. Adriano, D. C. (1986) *Trace Elements in the Terrestrial Environment,* Springer-Verlag, New York.
20. Nriagu, J. O., ed. (1978) *Biogeochemistry of Lead in the Environment, Vol. 1,* Elsevier, Amsterdam.
21. Nriagu, J. O., ed. (1980) *Zinc in the Environment, Vol. 1,* John Wiley & Sons, New York.
22. Nriagu, J. O., ed. (1980) *Cadmium in the Environment, Vol. 1,* John Wiley & Sons, New York.
23. Nriagu, J. O., and Davidson, C. I., eds. (1986) *Toxic Metals in the Atmosphere,* John Wiley & Sons, New York.
24. Nriagu, J. O., ed. (1980) *Nickel in the Environment,* John Wiley & Sons, New York.
25. Davidson, C. I., and Wu, Y. L.; Pacyna, J. M., and Ottar, B., eds. (1989) "Dry Deposition of Trace Elements," *Control and Fate of Atmospheric Trace Metals,* Kluwer Academic Publishers, Dordrecht, Netherlands.
26. Volkening, J., Bauman, H., and Heumann, K. (1988) "Atmospheric Distribution of Particulate Lead over the Atlantic Ocean from Europe to Antarctica," *Geochimica et Cosmochimica Acta 22,* pp. 1,169–1,174.
27. Boutron, C., and Patterson, C. C. (1987) "Relative Levels of Natural and Anthropogenic Lead in Recent Antarctic Snow," *J. Geophys. Res. 92,* pp. 8,454–8,464.
28. Patterson, T. L., and Duce, R. A. (1987) "Atmospheric Geochemistry of Cadmium over the Remote North Pacific," Proceedings of the International Conference on Heavy Metals in the Environment, CEP Consultants Ltd., Edinburgh, pp. 137–139.
29. Maenhaut, W., Cornille, P., Pacyna, J. M., and Vitols, V. (1989) "Trace Element Composition and Origin of the Atmospheric Aerosol in the Norwegian Artic," *Atmos. Environ. 23,* pp. 2,551–2,569.
30. Purves, D. (1985) *Trace Element Contamination of the Environment,* Elsevier, Amsterdam.
31. Jeffries, D. D., and Schneider, W. R. (1981) "Atmospheric Deposition of Heavy Metals in Central Ontario," *Water, Air, & Soil Poll. 15,* pp. 127–152.
32. Bergkvist, B., Folkeson, L., and Berggren, D. (1989) "Fluxes of Cu, Zn, Pb, Cd, Cr, and Ni in Temperate Forest Ecosystems," *Water, Air & Soil Poll. 47,* pp. 217–286.

33. Steinnes, E.; Hutchinson, T. C., and Meema, K. M., eds. (1987) "Impact of Long-Range Atmospheric Transport of Heavy Metals of the Terrestrial Environment in Norway," *Lead, Mercury, Cadmium and Arsenic in the Environment,* John Wiley & Sons, New York, pp. 107–117.
34. Friedland, A. J., Johnson, A. H., Siccama, T. C., and Mader, D. L. (1984) "Trace Metal Profiles in the Forest Floor of New England," *Soil Sci. Soc. of Amer. J. 48,* pp. 422–425.
35. Hutton, M. (1980) "Metal Contamination of Feral Pigeons: Biological Effects of Lead Exposure," *Environ. Poll. (Ser. A) 22,* pp. 281–293.
36. Navarre, J. L., Ronneanu, C., and Priest, P. (1980) "Deposition of Heavy Elements on Belgian Agricultural Soils," *Water, Air & Soil Poll. 14,* pp. 207–213.
37. Asami, T.; Nriagu, J. O., ed. (1983) "Pollution of Soils by Cadmium," *Changing Metal Cycles and Human Health,* Springer-Verlag, Berlin, pp. 95–111.
38. U. S. Environmental Protection Agency (1986) *Air Quality Criteria Document for Lead,* Research Triangle Park, North Carolina.
39. Markovac, J., and Goldstein, G. W. (1988) "Picomolar Concentrations of Lead Stimulate Brain Protein Kinase C," *Nature 334,* pp. 71–73.
40. Lin-Fu, J. S.; Mahaffey, K. R., ed. (1985) "Historical Perspective on Health Effects of Lead," *Dietary and Environmental Lead: Human Health Effects,* Elsevier, Amsterdam, pp. 43–63.
41. Hutton, M. (1982) "Cadmium in the European Communities Report No. 26," University of London Monitoring and Assessment Research Center, London.
42. Bellrose, F. C. (1959) "Lead Poisoning as a Mortality Factor in Waterfowl Populations," *Bull. Ill. Nat. Hist. Survey 27,* pp. 235–288.
43. Glooschenko, V., Downes, C., Frank, R., Braun, H. E., Addison, E. M., and Hickie, J. (1988) "Cadmium Levels in Ontario Moose and Deer in Relation to Soil Sensitivity to Acid Precipitation," *Sci. Total Environ. 71,* pp. 173–186.
44. Jadhav, A. N., and Sawant, A. D. (1991) "Environmental Impact Assessment of a Lead Smelter at Talasari bordering Maharashtra and Union Territory of Dadra and Nagar Haveli," *Scavenger 22* (3,4), pp. 27–37.
45. Nriagu, J. O. (1989) "Effects of Atmospheric Trace Metal Deposition on Aquatic Ecosystems," background paper prepared for a workshop on The Effects of Atmospheric Contaminants on Aquatic and Terrestrial Ecosystems, Center for Clean Air Policy, Washington, D.C.
46. Hakanson, L., Nilsson, A., and Anderson, T. (1988) "Mercury in Swedish Fish," *Environ. Poll. 49,* pp. 145–162.
47. Scheider, W. A., Jeffries, D. S., and Dillon, P. J. (1979) "Effects of Acid Precipitation on Precambrian Fresh Water in Southern Ontario," *J. Great Lakes Res. 5,* pp. 45–51.
48. Tyler, G., Pahlsson, A. M. B., Bengtsson, G., Baath, E., and Tranvik, L. (1989) "Heavy Metal Ecology of Terrestrial Plants, Microorganisms and Invertebrates," *Water, Air & Soil Poll. 47,* pp. 189–215.

49. Brennan, E. C., Leone, I. A., and Daines, R. H. (1967) "Characterization of Plant Damage Problems by Air Pollution in New Jersey," *Plant Disease Reporter 51,* p. 10.
50. Foy, C. D., Chaney, R. L., and White, M. C. (1978) "The Physiology of Metal Toxicity in Plants," *Ann. Rev. Plant Physiol. 29,* pp. 511–566.
51. Wong, M. H., and Bradshaw, A. D. (1982) "A Comparison of the Toxicity of Heavy Metals using Elongation of Rye Grass *Lolium perenne,*" *New Phytol. 91,* pp. 255–261.
52. Mhatre, G. N., and Chaphekar, S. B. (1982) "Effect of Heavy Metals on Seed Germination and Early Growth," *J. Environ. Biol. 3* (2), pp. 53–64.
53. Gemmell, R. P. (1977) "Colonization of Industrial Wasteland," *Studies in Biol. 80,* Edward Arnold, London, p. 75.
54. Mhatre, G. N., and Chaphekar, S. B. (1984) "Response of Young Plants to Mercury," *Water, Air & Soil Poll. J. U.S.A. 21,* pp. 1–8.
55. Mhatre, G. N., and Chaphekar, S. B. (1985) "Effect of Mercury on Some Aquatic Plants," *Environ. Poll. 39* (3), pp. 207–216.
56. Tiagi, Y. D., and Aery, N. C. (1985) "Plant Indicators of Heavy Metals," *Biological Monitoring of the State of the Environment: Bioindicators,* for I.C.S.U. by I.R.L., pp. 207–222.
57. Ernst, W.; Mansfield, T. A., ed. (1976) "Physiological and Biochemical Aspects of Metal Tolerance," *Effects of Air Pollution on Plants,* Cambridge University Press, London, pp. 115–133.
58. Mhatre, G. N., Chaphekar, S. B., Ramani Rao, I. V., Patil, M. R., and Haldar, B. C. (1980) "Effect of Industrial Pollution on the Kalu River Ecosystem," *Environ. Poll. (Ser. A.) 23* (1), pp. 67–78.
59. Prat, S. (1934) "Die Erblichkeit der Resistenz Gegen Kupfer," *Ber dr. bot. Ges. 52,* pp. 65–67.
60. Bradshaw, A. D.; Mansfield, T. A., ed. (1976) "Pollution and Evolution," *Effects of Air Pollutants on Plants,* Cambridge University Press, London, pp. 143–159.
61. Chaphekar, S. B., Divekar, D. P., Entee, D. F., Jayaraman, K., Jere, M. G., Khan, S. M., and Tipnis, P. P. (1973) "Effect of Industrial Pollution on the Angiosperm Flora along the Ulhas River at Kalyan, near Bombay," *The Bontanique 4* (2), pp. 85–92.
62. Nieboer, E., Ahmed, H. M., Puckett, K. J., and Richardson, D. H. S. (1972) "Heavy Metal Content of Lichens in Relation to Distance from a Nickel Smelter in Suburry, Ontario," *Lichenologist 5,* pp. 292–304.
63. Little, P., and Martin, M. H. (1974) "Biological Monitoring of Heavy Metal Pollution," *Environ. Poll. 6,* pp. 1–19.
64. Parkarinen, P., and Tolonen, K. (1976) "Regional Survey of Heavy Metals in Peat Mosses (*Spagnum*), *Ambio 5* (1), pp. 38–40.
65. Chaphekar, S. B., and Kulkarni, M. S. (1979) "Effects of Mercury on Some Cryptograms," *Geobios 6,* pp. 218–220.

66. Antonovics, J., Bradshaw, A. D., and Turner, R. G. (1971) "Heavy Metal Tolerance in Plants," *Advances in Ecol. Res. 7,* pp. 1–85.
67. Tejam, B. M., and Haldar, B. C. (1975) "A Preliminary Survey of Mercury in Fish from Bombay and Thana Environment," *Indian J. Environ. Health 17,* pp. 9–16.
68. Nadkarni, R. A., and Chaphekar, S. B. (1974) "A Plant Species of Suspected Accumulator Behavior," *Experientia 33,* 34.
69. Khalap, K. P. (1986) "Plants and Waterborne Pollutants: Studies of Their Interactions," Ph.D. thesis submitted to the University of Bombay.
70. Puri, A. R. (1988) "Studies to Explore Bioindicator Potentials of Angiospermic Plants," Ph.D. thesis submitted to the University of Bombay.
71. Chaphekar, S. B., and Mhatre, G. N. (1981) "Role of Aquatic Plants in Water Pollution Control," *Indian Assoc. for Water Poll. Control. Tech. Annual Vol. 8,* pp. 108–115.

CHAPTER 20

AIR POLLUTION PROBLEMS AT REMEDIATION SITES

Nick Gavrushenko
Paul N. Cheremisinoff

Department of Civil & Environmental Engineering
New Jersey Institute of Technology
Newark, NJ 07102, USA

CONTENTS

INTRODUCTION

Air pollution problems at remediation sites involve defining the hazards the site may pose, developing a plan of action to deal with those hazards, and implementing that plan of action to remediate the site. These hazards include not only the adverse health effects that may result from inhalation, dermal contact, or ingestion of pollutants at the site, but also include potential hazards from explosion or flammability. Those affected include workers at the remediation site and the general public, who may be affected by migration of gaseous emissions and fugitive dusts beyond the site boundaries.

Laws and regulations have been enacted on the federal and state levels to protect both workers and the general public from the adverse effects of pollutants, including air pollutants, at remediation sites. These include laws to prevent the establishment of uncontrolled waste sites to avoid future problems (RCRA), laws regarding remedial actions at abandoned hazardous waste sites and at existing sites in response to current situations such as spills (CERCLA), and laws to protect workers performing remedial actions (OSHA). Standards and recommendations for threshold values of acceptable exposure have been developed. Workers are required to receive specific levels of training before working at hazardous waste sites. Employers must provide appropriate personal protective equipment for the hazards to be encountered and training in the proper use of the equipment. Health and safety programs that include medical monitoring of employees must be established. These are but a few of the requirements established to protect workers and the general public.

The reduction of air pollution problems at remediation sites is often a subset in the resolution of the total problem, i.e., remove, treat, or destroy the substances causing the air pollution problem, and the air pollution problem will disappear.

There are a number of significant differences between air pollution at remediation sites and air pollution at industrial facilities. First, air pollution products at industrial facilities can be readily determined; the raw materials and processes are known. At remediation sites the raw materials and processes causing the air pollution are not always known, and considerable effort usually is required to identify what they are. Second, air pollutants at industrial facilities usually can be readily collected for treatment from point sources. At remediation sites, it is rare for air pollutants to be emitted from a point source. Air pollutants can be emitted from anywhere on the site, including the complete site, and usually it is very difficult to readily collect it for treatment. Third, air pollutants and their control from stationary point sources have been studied extensively, in particular SO_x, NO_x, O_3, along with some of the other toxic and hazardous air pollutants from specific industrial processes. Emissions from abandoned, uncontrolled hazardous wastes sites often are contaminated. However, air contaminants (gaseous emissions or fugitive dusts) have not received anywhere near the attention as other hazardous materials in the leachate, groundwater, or surface water at remediation sites.

Historically, waste material usually was disposed of in the most expeditious way possible, either at the facility where it was generated or off-site. Site selection usually was based on criteria such as cost of site, ease of disposal, size of site, etc. Until the passage and implementation of RCRA, long-term environmental and human health effects were rarely considered as waste disposal site selection criteria. In addition, most industrial facilities did not analyze the chemical constituents in their wastes. Thus, there usually is no historical documentation of the substances disposed of and present in most disposal sites. This is true for both on-site disposal, where waste from any number of processes could have been disposed of, and off-site disposal sites.

In landfill sites, gaseous emissions and fugitive dusts may be hazardous as defined under RCRA as, toxic, corrosive, reactive, or ignitable. The generation of gases under a surface cover or cap can cause cracking or perforation of the cap, allowing rainwater to enter and possibly leach other hazardous material beyond the disposal site and into groundwater or other surface water.

In addition to landfills, remediation sites also include surface impoundments, waste storage piles, land treatment facilities, aboveground and underground storage tanks and surrounding areas that may have become contaminated as a result of leaks or spills during operations, and other sites contaminated by accidental or intentional spills. Each of these sites usually are associated with air pollution problems. Hazardous chemicals less dense than water often are readily volatilized from surface impoundments. Hazardous emissions from spills can collect and remain in trenches, enclosed buildings, basements, or other depressions. RCRA regulations require underground storage tanks to be emptied and cleaned before any changes in service are permitted or before the tank is put of service. The hazards associated with storage tanks are primarily related to the products contained within the tank. Many tanks are large enough to require that someone enter the tank for proper emptying and cleaning. In these instances, oxygen depletion and ventilation problems must be considered, in addition to those hazards directly related to the contained product.

As can be seen, air pollution problems are usually a secondary but significant problem at remediation sites; i.e., if not for the spill or presence of another material, there would not be an air pollution problem. Many times it is not readily apparent which materials are causing the air pollution problems. Therefore, to deal with the problems of air pollution at remediation sites, workers must be properly trained. They must be able to recognize the conditions conducive to explosion or fire. Workers must be advised of the adverse health effects that can result from inhalation, dermal contact, or ingestion of hazardous contaminants, and then be trained in the proper use of personal protective equipment for the appropriate hazard. They also must know the proper procedures for confined space entry.

To minimize adverse effects on the health of workers, the general public, and the environment from air pollution during remediation, it is necessary to identify the types and sources of emissions and the transport processes to select the remedy to properly control the air pollution emissions.

TYPES OF EMISSIONS

One of the biggest problems in determining the necessary remedial action required is the availability and reliability of information regarding the pollutants at remediation sites. Usually only fragmentary knowledge exists as to the chemicals disposed of at most sites. This has a tremendous effect on risk assessment and makes risk management tenuous and difficult. In addition, it is seldom possible to determine the extent to which surrounding populations have been exposed.

Wastes at remediation sites often are a mixture of hazardous and nonhazardous wastes (e.g., hazardous wastes mixed with industrial process wastes or small quantities of hazardous wastes dissolved in process water). Many chemicals have been detected in the mixtures of waste materials at disposal sites.

Based on the toxicity, ignitability, corrosivity, or reactivity, the U.S. EPA has identified nearly 500 chemicals as "hazardous wastes" when they are found as constituents of solid waste mixtures, as required by RCRA. At least 135 of these chemicals have been identified as "acute hazardous wastes" because they have been found to be fatal to humans in low doses or are otherwise capable of causing or significantly contributing to an increase in serious irreversible or incapacitating reversible illnesses. More than 300 other chemicals have been identified as "toxic wastes" because they have been shown to have toxic, carcinogenic, mutagenic, or teratogenic effects in humans or other life forms. Most if not all have been reported as being present in the environment in the vicinity of the sites identified on the National Priority List (NPL) and can be expected to be found at other sites not on the NPL that require remediation. At least 65 of the chemicals or groups of chemicals identified as hazardous wastes were found in air samples that have been taken [1].

The ten substances most commonly found at remediation sites are lead, trichloroethylene, toluene, benzene, polychlorinated biphenyls (PCBs), chloroform, phenol, arsenic, cadmium, and chromium [1]. While chloroform and trichloroethylene are highly volatile substances and benzene, toluene, and phenol are moderately volatile, each of these substances is easily released into the atmosphere. The metals are not volatile and PCBs have low volatility, but they often become attached or sorbed onto soil particles and may become airborne as a result of wind erosion.

SOURCES OF EMISSIONS

Gases from non-industrial processes may be emitted by the vaporization of liquids, venting of contained or entrained gases, or by chemical or biological reactions with solid or liquid waste materials. Various organic compounds may slowly but continually volatilize from landfills and from the exposed top surface of a surface impoundment. Low-boiling point organic materials, including contaminated solvent, if improperly contained will emit vapors that may be ignitable or toxic. Inorganic gases also can be emitted. Oxidizing gases such as chlorine may react with polymeric liner materials or other organic materials [2].

Waste sludges containing organic matter generally undergo decomposition or degradation from biological activity. Depending on the type of site, the biological degradation may either be aerobic or anaerobic. Under aerobic conditions, organic constituents are gradually oxidized to intermediate products such as organic acids and alcohols and then converted to organic residues and gases. Under anaerobic conditions, reduced sulfur, volatile intermediate products, and methane may be vented [3].

Wastes can change in physical and chemical character over the years because of degradation or as a result of reactions between the acids, bases, or other chemicals and hydrocarbons present in the waste that result in a mixture of complex organic compounds. Some components can harden or polymerize, causing a physical change in the waste material. The reactions between acids, bases, or other chemical reagents and hydrocarbons can produce gas emissions [4]. Degradation products usually are chemically simpler than the parent compound. But in some cases the degradation products may be more hazardous than the parent compound, as in the case of trichloroethylene, which results in the transient formation of intermediates such as vinyl chloride [1].

The rate and extent of chemical reaction or biological degradation are a function of physical and chemical reaction processes, diffusion, concentration of each chemical, soil or water temperature, pH, presence of oxygen and redox conditions, etc. The presence of high concentrations of some wastes may be toxic to microorganisms, and they would not be readily biodegraded. Some chemical reactions require energy or elevated temperatures, so even if reactive chemicals come into contact by mixing at the site, these reactions may not occur or may proceed at such a slow rate as to be immaterial. Other types of reactions may be explosive, and some waste mixtures may be easily ignitable [3].

Gaseous emissions may result during periods of remedial activity. Removal of drums may cause rupture or leakage of highly volatile or reactive materials. Excavation or grading operations may change the biological environment, causing action to shift from aerobic to anaerobic, or vice versa, producing new emissions. In surface impoundments, activities such as pumping, dredging, or excavating may lead to increased emissions from the removal or breakup of dried surface barriers or the mixing of the surface liquid layer with liquids of higher vaporization potential from subsurface regions. At sites with a cover (soil cap at landfills or a surface barrier at impoundments), gas emissions from the undisturbed site may be low in concentration. When the cover is disturbed and fresh waste is exposed, emissions may increase from 1,000 to 10,000 times that of the undisturbed site. These emissions are capable of resulting in significant downwind concentrations of contaminants [4].

Fugitive emissions of particulates containing hazardous constituents may be caused by wind erosion of exposed waste materials, wind erosion of cover soil, re-entrainment of particulates by vehicular traffic on haul roads and exposed surfaces, and during excavation of waste materials during remedial actions.

MAJOR ROUTES OF TRANSPORT AND EXPOSURE TO EMISSIONS

The extent and rate of chemical movement within or beyond a waste site or spill area are functions of the physical containment of the waste materials (e.g., in drums as liquids or semi-liquid sludges, bulk solids or liquids dumped onto the ground or buried, etc.), the chemical and physical properties of the substances, and the chemical and physical properties of the site. Substances released from disposal sites may move through various media in the environment, and humans are potentially exposed via a variety of routes of exposure. Figure 1 illustrates a generalized diagram for the transport of chemicals and potential exposure of humans to these chemicals.

In general, the higher the water solubility value, the more likely a compound will leach and migrate in groundwater. Low-density substances with low solubility float near the surface of the water, and high-density substances tend to sink. The more hydrophobic the compound, more likely that it has a high soil sorption coefficient and the less likely it will aqueously migrate through sorption to soil. The higher the vapor pressure and the higher the Henry's law constant of the compound, the more likely it will volatilize or evaporate from water or soil into the air [2].

Aqueous migration is an important route for the ultimate release of many air pollutants [1]. The leaching of chemicals depends on the water solubility of the substance, the sorption potential of the substance, soil type, and amount and intensity of water penetration. Leaching may occur in three directions: downward, laterally, and upward. The upward movement of leachate is the result of mass transfer and capillary action of the water under the influence of evaporation at the surface and may result in the accumulation of chemicals at the soil surface. A chemical with high water solubility, such as chloroform, and with a low potential for soil sorption will be leached rapidly and will tend to migrate upward, where it volatilizes from the water at the soil surface and results in an air emission. Other less-soluble chemicals may be carried into the groundwater away from the disposal site via groundwater transport to areas where upward movement to the soil surface is possible. They also may be carried to surface waters via runoff from the site or by groundwater movement. Once hazardous wastes enter the surface water, mixing and dilution occur much more rapidly than in groundwater, at which point certain hazardous wastes will volatilize rapidly. Biodegradation and chemical oxidation also may occur more rapidly in surface waters than in groundwater and produce hazardous products that readily volatilize.

The sorption of hazardous wastes onto soil particles provides another means for transport and exposure. Humans may be exposed to soil-bound hazardous wastes via direct contact at the disposal site, by inhalation or dermal contact of particles transported via wind erosion, or by exposure to suspended particles in surface water runoff. Soil-borne transport of hazardous waste may be especially important during remedial actions such as site excavation. The amount of wind erosion depends on the waste type, moisture content, wind velocity, and surface geometry. Contaminants sorbed to soil particles may not be detected by organic vapor detectors, which are used to determine atmospheric vapors, mists, and gases [2].

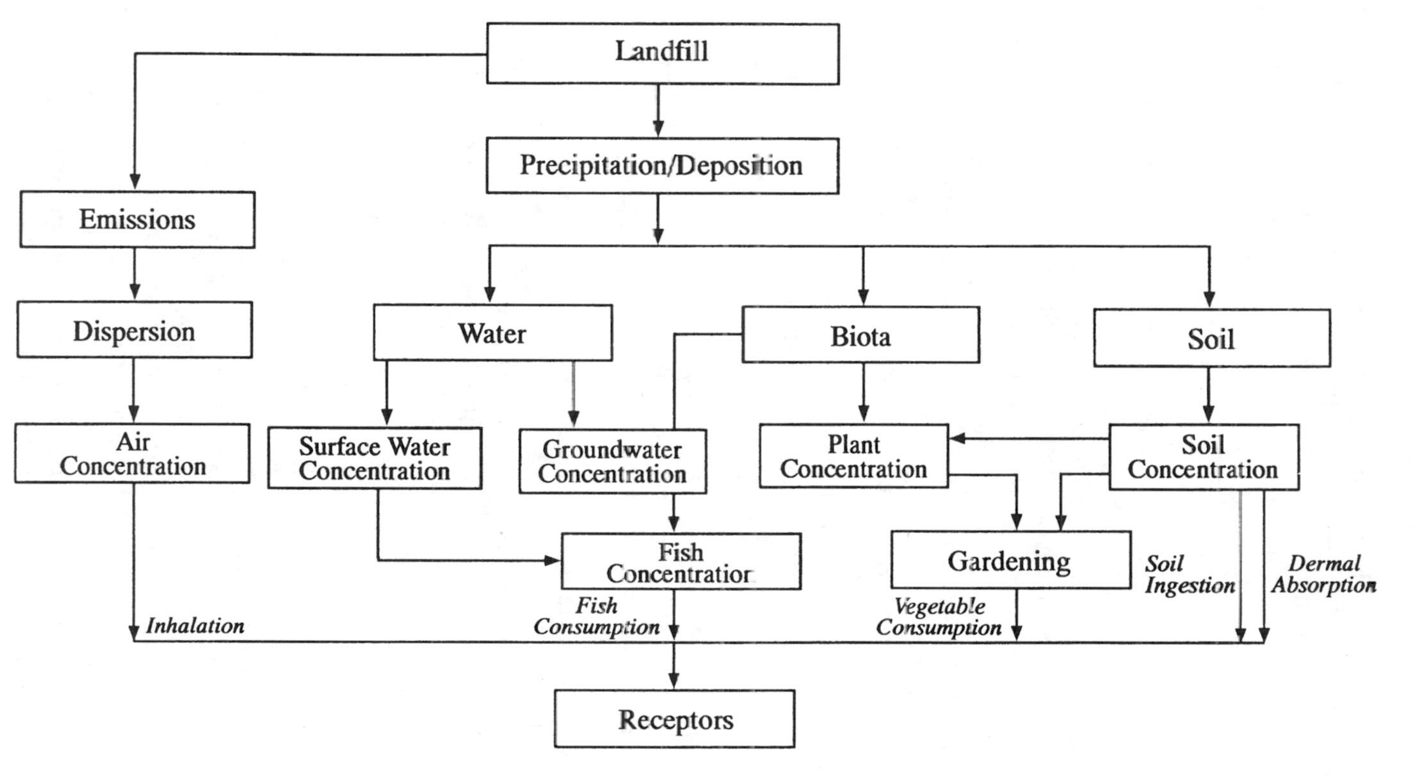

Figure 1. Transport pathway overview.

Compounds with low molecular weights, low water solubilities, and high vapor pressures evaporate more rapidly than compounds that do not have these characteristics. Some hazardous compounds have high vapor pressures and low solubility in water and thus exhibit high Henry's law constants. They will be readily transferred to the vapor phase and ultimately to the atmosphere. Evaporation from surface impoundments is one route of transport and usually occurs both by the formation of gas bubbles rising to the surface and by single molecules crossing the water-air interface. Compounds can also sublime from the surface of landfills or volatilize from within the ground [3].

Hazardous wastes buried in landfills also are transported upwards as gases through the soil and into the atmosphere. The gases diffuse through the air pores in the soil. At sites where biodegradation also occurs, additional gases will be generated and accelerate the movement of hazardous waste vapors. Although calculated rates of volatilization for covered landfills are usually low, the actual volumes of waste materials that evaporate into the atmosphere may be substantial [5].

As stated, gaseous emissions move from subsurface regions to the surface by diffusion and bulk gas flow in the soil. Gas flow in soil is dependent on free space diffusivity, porosity, and the degree of water saturation [5]. If there is a barrier to the flow at the surface (e.g., a man-made cover or cap or a natural barrier) then, unless physically constrained, gaseous emissions will move laterally through soil. This mechanism has shown to result in radon buildup in structures some distance from uranium tailing piles or in methane buildup near covered landfill sites. Gaseous emission also may be dissolved in groundwater and move in the direction of the water flow.

CONTROL OF EMISSIONS AND REMEDIAL ACTION

The problems with air emissions as identified above require careful assessment during site investigations. These problems will have a characteristic set of remedial actions quite different from those applicable to leachate, groundwater, or surface water contamination.

The emission control and remedy selection process for doing this generally consists of the following steps. Site conditions are evaluated to characterize and define the nature and extent of contamination. Field data, as necessary, is collected to identify waste types, concentrations, and distribution. Potential health problems are identified [6].

After the site is characterized, potential treatment technologies are identified. A range of alternatives attaining various cleanup levels are developed. The alternatives are screened to narrow the choices to be analyzed in detail. General performance criteria are developed. Each remaining alternative is analyzed in relation to the established performance criteria as well as to costs, long- and short-term effectiveness, and implementability. The protection of public health and environmental compliance is compared for each alternative as are reductions in the mobility or toxicity of the hazardous constituents during remediation. A remedy that will provide a permanent solution to protect human health and the environment is then selected and implemented.

Site Characterization

Site characterization is directly related to worker protection and the development of remedial actions. The more accurate, detailed, and comprehensive the information available about the site, the more the protective, control, and remedial measures can be tailored to the actual hazards encountered.

Site characterization generally proceeds in three phases: off-site surveys to gather all available information prior to site entry, on-site surveys by reconnaissance personnel, and monitoring to provide a continuous source of information about site conditions [6].

Off-site surveys focus on identifying all potential or suspected conditions that may be immediately dangerous to life or health (IDLH) so that they may be evaluated and preliminary controls can be instituted to protect initial entry personnel. Indicators of potential IDHL condition include: large containers or tanks that must be entered; enclosed spaces such as buildings or trenches that must be entered; the presence of potentially explosive or flammable situations; the presence of extremely toxic materials, such as cyanide or phosgene; visible vapor clouds; or areas where biological indicators such as dead animals or vegetation are found. Sources of information include historical records of site activities including hazardous substances used and their chemical and physical properties, company records, meteorologic data, aerial terrain photographs, land use maps, state and federal agency records, geologic and hydrologic data to determine potential pathways for dispersion, etc.

At sites in which the hazards are largely unknown or there is no need to go on-site immediately, off-site surveys can include: visual observations to develop a preliminary site map; review of historical and current aerial photographs to note the disappearance of depressions, pits, or other site changes; noting unusual conditions such as vapor clouds, discolored liquids, oil slicks, etc.; monitoring the ambient air at the site perimeter for toxic substances, combustible gases, organic and inorganic gases, oxygen deficiency, ionizing radiation, odor, etc.; collecting and analyzing off-site samples of soil, groundwater, site run-off, and surface water.

Based on the evaluation of preliminary off-site information, an initial safety plan for site entry is prepared, priorities for monitoring and investigation are identified, and protective equipment is selected for the initial site survey. Unless the off-site survey identifies the need for specific protective clothing or equipment, the minimum level recommended by EPA for protective clothing and equipment to be used for initial entry, until the site hazards have been further identified, is referred to as Level B.

The on-site survey is conducted to verify and supplement information from the off-site survey. The air is monitored for IDLH conditions and ionizing radiation; the physical condition of containers, storage systems, and materials are noted; land features and natural wind barriers are identified; potential pathways for pollution dispersion are determined; indicators of potential chemical exposure are noted; reactive, incompatible, or highly corrosive wastes are identified; air, soil, and water samples are taken; and, if necessary, remote sensing or subsurface investigative methods to locate buried wastes or contaminant plumes are used.

Once the presence and concentrations of specific chemicals or classes of chemicals has been established, the health hazards associated with these chemicals are determined. Using the data gathered off-site and on-site, a determination is made whether the site is or can be made safe for entry of cleanup personnel, or whether additional information is needed to define necessary protective measures. Once sufficient information is obtained, the site health and safety plan is updated, and cleanup begins.

Because site activities and weather conditions change, a continuous monitoring program must be implemented after site characterization has determined that the site is safe for commencement of other operations. Atmospheric and chemical monitoring is done using a combination of stationary sampling devices, personnel monitoring devices, and direct reading instruments. Monitoring also includes continual evaluation of any changes in site conditions or work activities that could affect worker safety.

Air Monitoring

Airborne contaminants can present a significant threat to worker health and safety. Reliable measurements of airborne contaminants are necessary to determine the level of personal protective equipment required, to define the areas where protection is needed, to assess the health hazards of exposure, and to determine whether continual exposure that may require the need for medical monitoring is occurring.

Remediation sites often poses a significant challenge to accurate and safe assessment of airborne contaminants because of uncontrollable variables. These include changes in: temperature—a 10°C increase can more than double the vapor pressure of PCBs; windspeed—a doubling of windspeed can double vapor concentrations near a liquid surface (dusts and particle-bound contaminants are also affected); rainfall—water from rainfall can plug vapor emission routes thereby reducing airborne emissions of certain substances; moisture—dusts, including hazardous wastes solids, are highly sensitive to moisture content; vapor emissions—the physical displacement of saturated vapors can produce short-term high vapor concentrations, and continuing diffusion may be an important long-term concentration phenomena; work activities—remedial action can lead to mechanical disturbances of contaminated materials thereby changing the concentration and composition of airborne contaminants [6].

Remediation Alternatives

Several remedial approaches can be used to treat gaseous emissions. Removal or deactivation of all sources of emission will control the problem permanently at the site. Removal may transfer the problem to another site, but it should be to a site where means are in place to handle the emissions under controlled conditions. The ultimate goal even in removal options is the permanent destruction or detoxification of hazardous wastes through the use of treatment technologies that reduce the toxicity, mobility, or volume of hazardous waste rather than prevention of exposure through isolation or pollution containment.

Surface Impoundments

Surface impoundments are used for a large variety of purposes: treatment by biodegradation, stabilization, equalization, oxidation, evaporation, or settling; and for holding sludges, mining tailings, and other substances.

Once a surface impoundment has been dewatered, it essentially becomes a landfill with the associated air pollution problems. However, at uncontrolled sites, even after additional liquid is not purposely added, all liquid may not evaporate from the impoundment. Precipitation and surface runoff will collect in the impoundment, or other natural sources of water may keep the impoundment filled. Each time water is added, soluble hazardous substances may leach into the water and then enter the atmosphere through vaporization [3].

Emissions from surface impoundments can be permanently controlled by stripping the entrained and dissolved gases from the liquid wastes. Chemicals may be added to the impoundment to increase the gases' solubility and prevent volatilization. For gaseous emissions that can be chemically or biologically degraded to nonhazardous substances, continual mixing of the impoundment to disperse the gases uniformly through the liquid layers will keep the gases within the liquid layers until chemical or biological degradation is complete. Emissions also can be reduced or eliminated by dewatering the lagoon by draining then removing the hazardous constituents for treatment or by adding bulking agents such as soil, cement, or crushed coral and capping the site to become a closed landfill [7].

Landfills and Waste Piles

There are several alternatives for remediation of contaminated landfill sites for *in situ* treatment and on-site or off-site aboveground treatment [7]. Each method has advantages and disadvantages.

Aboveground Technologies

The technologies for on-site and off-site aboveground treatment of soil are very similar. To provide on-site treatment, the systems used must be sized so that they are transportable. Most mobile treatment units are used for processing water for potable use, sewage, and industrial wastewaters for pollution control, and for emergency responses. The major difference between on-site and off-site systems is in the capacities of the facilities used. Commonly used treatment systems include high-temperature destruction systems such as rotary kilns or pyrolyzers, low-temperature stripping systems, and soil-washing systems.

High-temperature destruction systems treat soils, solids, and sludges contaminated with organic hazardous wastes. The pyrolysis reaction converts organics to carbon monoxide, hydrogen, and carbon. Sulfur and halogens in the original feed are released into the gas phase as corresponding hydrogen compounds. The gases are drawn off the reactor and into a cooling and cleanup system consisting of a cyclone, baghouse, and an acid gas scrubber prior to recycling or release into the atmosphere. Acidic wastewater is neutralized and either discharged to the sewer after sampling or subjected to additional treatment.

Low-temperature thermal stripping removes volatile contaminants in the soil by aeration using ambient or heated air. Soil is loaded into the unit, where it is shredded and aerated by churning and tossing. Aeration and shredding is a relatively slow process that merely transfers the VOCs from the soil to the air. Preheated air can be used to enhance volatilization and to remove VOCs. If the gases have commercial value, they may be passed through activated carbon to sorb the volatile constituents for reclamation. Otherwise, the offgases are passed through an afterburner, where the VOCs are destroyed at temperatures greater than 1,800°F with a residence time of more than two seconds.

Soil-washing systems extract hazardous materials from excavated soils. The soil is reduced to pieces less than 2.5 cm and then washed with water to consolidate the soil matrix and to strip any contaminant loosely absorbed on the solids or held in the void spaces of the soil. Additional extraction is then performed using solvents appropriate for the contaminants to be removed. The complete system for soil washing requires auxiliary equipment for processing wastewater for recycling and a system for the confinement, collection, and treatment of released gases and mists. Treatment residues consist of skimmings from froth flotation, fine particles discharged with used washing fluids, and spent carbon.

Major drawbacks to aboveground treatments include the additional health risks associated with excavation, transportation, storage, and handling of hazardous materials; excavation significantly increases the soil volume (by 20% to 30%) requiring additional disposal capacity; and additional land area for treatment systems such as land spreading for bioreclamation is required.

In Situ *Technologies*

In situ technologies generally are based on technologies developed for use in conventional water and wastewater treatment and in mining, oil and gas, and chemical process industries. The technologies use biological, chemical, and physical methods to degrade, detoxify, extract, or immobilize contaminants in place. Successful application of *in situ* technologies requires a good knowledge of subsurface characteristics, including the vertical and horizontal distribution of contaminants. Commonly suggested *in situ* technologies include bioreclamation, soil vapor extraction, air stripping, immobilization, and soil washing.

Bioreclamation involves the degradation of organic contaminants to inert or less-than-harmful products via the action of suitable microorganisms developed in or added to contaminated soil [8]. Bioreclamation has been attempted with some success in a number of situations. It is a simple, low-cost, and safe method, but it has several limitations. It generally is limited to applications involving localized organic groundwater contamination where contaminants are not present at toxic levels. Channeling and uneven treatment results in soils that are nonhomogeneous. It is also difficult to monitor or control the process.

Soil vapor extraction involves the application of a vacuum to the subsurface to remove volatile contaminants. The vacuum can be applied through vertical or horizontal extraction systems. The extracted air may require treatment before release into the atmosphere. It is intended to be a simple, low cost, and safe method that can treat significant soil depths in the unsaturated zone. Disadvantages include possible channeling and uneven treatment because of the nonhomogeneity of the subsurface soil. Treatment is limited to volatile compounds and can be very slow. Very few applications of this technology have been reported, therefore, its use as a viable alternative remains to be established.

Air stripping requires forcing clean air or steam or hot air into the subsurface to remove volatile contaminants. The contaminated air is collected at the surface or through extraction wells for aboveground processing before release into the atmosphere. Disadvantages include a limited treatment depth and the potential for groundwater pollution. Again, the treatment is limited primarily to volatile compounds.

The addition of physical or chemical agents that react with or encapsulate contaminants will immobilize them. This technology was originally developed for treating soil contaminated with radiation, but the same rationale can be applied for treatment of any contaminant. The treatment is applicable for both organic and inorganic wastes. Disadvantages are that it may not be feasible to use when complex wastes containing a range of contaminants with different characteristics are involved.

Soils also can be flushed or rinsed using suitable solutions to remove contaminants. Treatment solutions can be delivered and collected via gravity or forced systems. The treatment is applicable for both organic and inorganic wastes. Disadvantages are that is probably not feasible to use when complex wastes containing a range of contaminants with different solubility characteristics are involved. It also is difficult to limit reactions to target contaminants and to monitor and control treatment progress. Channeling and uneven treatment may result. Also, recovered solutions can be very dilute and large in volume and costly to treat and dispose of.

As can be seen, potential problems with *in situ* treatments include the nonhomogeneity and variable properties of the soil and waste media the treatment agents must move through, which can result in channeling and uneven or incomplete treatment; the slow treatment rates due to the slow rate of fluid flow within the soil; the potential spread of contamination during treatment; and the potential risk of generating or volatilizing other toxic constituents by the action of added reagents.

Control of Emissions

It also is important to control emissions while remedial activity is taking place or for control of emissions prior to remediation. This generally involves taking measures for the temporary control of emissions.

Surface Impoundments

Methods for controlling volatile emissions from surface impoundments involve providing some kind of barrier at the water-air interface. This includes the use of covers, immiscible liquids, and floating spheres [9].

Floating covers generally consist of a synthetic lining placed in one piece over an impoundment with proper anchoring at the edges and floats to prevent the lining from submerging. The cover functions as both a surface water control mechanism and as a mechanism for controlling gaseous emissions. Floating covers are often used when some period of time will elapse before the surface impoundment is remediated, but are effective usually for periods of less than one year where the primary mechanism for gas emission is surface volatilization. Surface covers also are less effective in situations where large volumes of gas are generated within the surface impoundments. In these cases, the cover and gas formation must be monitored and trapped gas collected and treated prior to release to the atmosphere. The removal of surface covers is required to perform remediation activities.

Floating immiscible liquids are suitable for controlling emissions of water-soluble organics only. Their effectiveness depends on the ability to form a continuous monolayer on the surface. A floating organic layer usually prevents the formation of the monolayer. Wave action also destroys the continuity of the monolayer. Immiscible liquids are suitable for use during remediation activity because the discontinuity of the monolayer can be localized to the site of remedial activity. The monolayer will not interfere with remedial activity.

Floating spheres, commonly made of polyethylene, are very effective in reducing volatile emissions by forming a closely packed configuration that reduces exposed surface area. They are compatible with a broad range of compounds, both organic and inorganic, with no major effect on remediation activities.

Waste Piles

The primary emission problem with waste piles is the control of fugitive dusts. Dusts can be controlled using water spraying, chemical dust suppressants, covers, and wind screens. Dust suppressants temporarily bind soil particles and reduce fugitive dust emissions. During active remediation, frequent reapplication of dust suppressants is required. Wind screens and fences are porous screens that take up or deflect a sufficient amount of wind lowering the wind velocity below the threshold required for initiation of soil movement. They are less effective for control of particles less than 10 microns. Effectiveness also depends on wind velocities at the site. Covers are very effective in controlling dust emissions but must be removed for remediation [9].

Landfills

Active interior gas collection and recovery systems are used to collect gases from beneath a landfill surface before they are vented to the atmosphere, if they are generated in sufficiently large volumes [9].

Pressure gradients and paths of gas migration are altered by mechanical means such as drilling extraction wells and using vacuum blowers or compressors to create the pressure gradient to induce gas flow from the landfill to the blower. The gas is then collected. If large volumes of methane are generated, the gas is passed through a unit that will remove carbon dioxide and trace contaminants prior to feeding the recovered methane into a natural gas pipeline. Volatile hazardous waste constituents have little commercial value and must be treated. Common treatments include the use of afterburners to incinerate the gases being emitted, activated carbon to adsorb emissions, or condensation of collected emissions.

TRAINING

In accordance with OSHA requirements of 29CFR1910.120, all employees, exposed to hazardous substances, health hazards, or safety hazards and their supervisors and managers must receive training before they are permitted to engage in operations that could expose them to such hazards [6,10].

Initial training required for general site workers includes a minimum of 40 hours instruction off-site and a minimum of three days actual field experience under the direct supervision of a trained, experienced supervisor. Workers on-site only occasionally for a specific limited task shall receive a minimum of 24 hours of off-site instruction and a minimum of one day actual field experience under the direct supervision of a trained, experienced supervisor. Any worker required to wear a respiratory must have the minimum of 40 hours instruction off-site, and a minimum of three days actual field experience under the direct supervision of a trained experienced supervisor. Employees and supervisors that have successfully completed the training and field experience required shall receive a written certificate. Workers must receive eight hours of refresher training annually to remain certified [10].

Items included in the training program include use of personal protective equipment; work practices to minimize risks from hazards; safe use of engineering controls and equipment on the site; safety, health, and other hazards present; and medical surveillance requirements including recognition of symptoms and signs that may indicate overexposure. Additional training is required for employees engaged in responding to hazardous emergency situations at hazardous waste cleanup sites or for any employees who will be on-site to provide technical advice or assistance in a specialized area. These individuals will be required to receive training or demonstrate competency in that area of specialization annually. Site-specific training also is required for each employee as part of the health and safety plan that must be developed for each site [10].

Individual states or municipalities may require training and certifications in addition to that specified by OSHA. For example, New Jersey requires additional training and certification for workers performing remedial work in confined spaces such as tanks.

PERSONAL PROTECTIVE EQUIPMENT

Anyone entering an uncontrolled hazardous waste site must be protected against any potential hazard. Personal protective clothing and equipment protects individuals from the chemical, physical, and biological hazards at a waste site. Personal protective equipment (PPE) should protect the respiratory system, skin and body, face and eyes, feet and hands, head and hearing. The primary health hazards from air pollutants are through inhalation, dermal contact, and ingestion, so particular care must be made in the selection of PPE to provide the proper protection [1].

Use of PPE is required by Occupational Safety and Health Administration (OSHA) regulations in 29CFR1910 and by EPA regulations in 40CFR300.71. A written PPE program must be part of an employer's safety and health program. No one piece of PPE, nor any single combination of PPE, is capable of protecting against all threats. Therefore, a site-specific PPE program also must be developed for each site. The selection of PPE must be based upon site hazards, limitations on use of particular PPE, and work duration. The PPE program also must specify maintenance and storage requirements, decontamination and disposal requirements, donning and doffing procedures, as well as limitations during temperature extremes and other appropriate medical considerations.

The use of PPE creates significant worker hazards, such as heat stress, physiological stress, and impaired visibility, mobility, and communication. The greater the level of PPE protection, the greater the associated risks. For any given situation, equipment and clothing should be selected that provide an adequate level of protection. Too much protection can be, and usually is, a safety hazard and should be avoided.

Key factors involved in selecting PPE are identification of the hazards, their routes of potential entry (inhalation, skin absorption, eye or skin contact, etc.), the performance of the PPE materials, and seams in providing a barrier to these hazards. Other factors include work requirements and conditions and durability of PPE materials in relation to worker tasks. OSHA provides general guidance in the selection of different protection categories based on the level of protection provided and the conditions for which each category is appropriate [10].

Level A PPE provides the greatest level of skin, respiratory, and eye protection and should be used when the hazardous substances have been identified that require the highest level of protection. The use of a self-contained breathing apparatus is required, as is a totally encapsulating chemical protective suit. Level B PPE provides for the highest level of respiratory protection but a lesser level of skin protection; it is used when air contaminants have not been completely identified but are not suspected to be at levels high enough to be harmful to the skin or be absorbed through the skin. Level C PPE is used when the concentrations and types of airborne contaminants are known and the criteria for using air-purifying respirators are met. A self-contained breathing apparatus is not required. Level D PPE involving minimal protection against nuisance contaminants is used when the atmosphere contains no known hazard [10].

Respiratory protection is of primary importance, as the lungs present the body's greatest exposed surface area. There are three major categories of respirators that differ with respect to the air or oxygen source.

- Self-Contained Breathing Apparatus (SCBA) supply air from a source carried by the user.
- Air-Line Respirators (ALR) supply air from a remote source connected to the user by a hose.
- Air-Purifying Respirators (APR) enable the user to inhale "purified" ambient air.

Other protective clothing includes face masks, fully encapsulating suits, nonencapsulating suits, aprons, leggings, sleeve protectors, gloves, proximity garments, blast and fragmentation suits, cooling garments, and anti-radiation suits. Each type of protective clothing has a specific purpose. The appropriate selection and use of PPE must be part of the training each worker receives before working at a hazardous site and should be part of the annual refresher training. Figure 2 shows a worker wearing protective gear.

Figure 2. Worker wearing personal protective gear.

CONFINED SPACE ENTRY

The hazards associated with confined spaces are primarily related to the products contained within the confined space. The most common confined spaces are storage tanks. In instances where a worker must enter a tank to properly empty and clean it, oxygen depletion and ventilation problems must be specifically considered in addition to those hazards directly related to the contained product. The possibility of explosion is much greater in confined spaces because combustible or explosive vapors can accumulate. If the vapors are not dispersed mechanically, protection must be provided against the possibility of explosion [11].

Remediation of sites with leaking underground storage tanks (USTs) is a particularly important problem. It has been estimated that 10% to 25% of the several million USTs containing petroleum or other hazardous substances in the United States are leaking. Subtitle I to RCRA was added as part of the Hazardous Solid Waste Amendments of 1984 to require the remediation of sites containing leaking USTs. Under the regulations in 40CFR280, not only must the contaminated soil or water be remediated, the leaking UST must be closed or removed. As part of the closure and removal process, USTs must be emptied and cleaned of liquid and accumulated sludges. In most cases, entry into the UST is required [12]. Because of the large number of UST sites requiring remediation, it is essential that all workers receive the proper training and supervision to perform this work [1].

Because of increased hazards of oxygen depletion and explosion not normally found at other hazardous waste remediation sites, some states and municipalities require additional training and certification of employees working in confined spaces in addition to the training required for working at hazardous sites. Good work practices require the use of safety lines and a "buddy" system when entering a confined space; i.e., one enters the confined space and a "buddy" remains outside to pull the worker out of the confined space with the safety line should something happen. The selection of PPE for working in confined spaces is given the same considerations as for any other hazardous site.

CONCLUSIONS

Air pollution problems at remediation sites are different from those at industrial facilities for a number of reasons. First, many times the air pollutants are not identified, and considerable effort and care is required to identify them to properly protect workers and the general public against any health hazards posed by the pollutants. Second, pollutants from industrial facilities are usually easily collected for treatment. Remediation sites are usually exposed and can cover large areas, which makes gas collection very difficult. A median sized landfill has a surface area of more than 22 acres (more than 950,000 square feet) [13]. The surface area of an average impoundment is about 1/2 an acre (more than 20,000 square feet), but there often are several impoundments at one site [13]. Third, emission processes at remediation sites differ from those at industrial facilities. The primary process for hazardous gas emission is through the volatilization of the hazardous compound. These compounds may volatilize slowly but continually for long periods of time. While the rate of volatilization may be slow, the volume of gas generated may be quite large. There also are many variables that influence the rate of pollutant formation and extent of pollution, many of which cannot be readily controlled.

Air pollution emissions from remediation sites may appear beyond the site boundaries. Hazardous gases may be soluble in groundwater and may travel long distances with the groundwater before volatilizing into the atmosphere. The presence of caps or covers may force gases to migrate laterally through soil utilities and other pathways to outlying areas.

Each remediation site is unique, but all have in common the potential for contaminating the environment by releasing toxic substances. The chemicals in remediation sites exhibit an enormous diversity of physical and chemical characteristics, and the sites themselves are diverse. At many of these sites, especially old sites, few precautions were taken to prevent environmental releases of hazardous substances, and the lack of documentation regarding quantities and types of chemicals disposed at the sites makes it difficult to define a "typical" remediation site.

It is therefore essential that anyone who enters a hazardous waste site recognize and understand the potential health and safety hazards associated with the cleanup of that site. Site workers must be trained to work safely in contaminated areas. The level of training should be consistent with the worker's job function and responsibilities. Workers must be thoroughly familiar with work practices and procedures contained in the site safety plan that has been developed. This includes the proper selection and use of personal protection equipment.

The general public also must be protected from the health hazards posed by the site. This includes protection from potentially increased concentrations of emissions that may result during remediation activities. This can include anything from spraying water to minimize fugitive dust emissions; to using foams, plastic sheets, or other substances to suppress vapor releases; to the use of large pressurized tents to cover active excavation areas.

The traditional approaches, consisting of site isolation and/or excavation and redisposal in land-based containment systems, do not provide the desired long-term environmental protection. Waste treatment and destruction can provide a more permanent solution. On-site treatments eliminate the risks associated with waste transportation to off-site locations. Treatments requiring excavation generally have been based on existing technologies and have been proven to be effective treatment methods. However, the major problem with treatments requiring excavation is the increased volume of soil that must be disposed of. Excavation and crushing of soil increases the soil volume by at least 20% to 30%, and possibly more [14]. *In situ* treatments avoid this problem, but there remains a substantial amount of uncertainty regarding the effectiveness of currently available technology. If concerns regarding effectiveness can be resolved, *in situ* treatments can many times be much more cost-effective than treatments requiring excavation.

The passage and enforcement of RCRA, with its requirements for cradle-to-grave tracking of hazardous substances, should limit the creation of new sites that require remediation because of improper disposal of hazardous wastes. Once existing uncontrolled disposal sites and leaking USTs have been remediated, it is hoped that only remediation of accidental spills and their associated air pollution problems will be required in the future.

REFERENCES

1. Grisham, Joe W., ed., *Health Aspects of the Disposal of Waste Chemicals,* Pergamon Press, New York, 1986.
2. Ehrenfeld, John R., and Jefferey Bass, *Evaluation of Remedial Action Unit Operations at Hazardous Waste Disposal Sites,* Noyes Publications, Park Ridge, 1984.
3. Ehrenfeld, John R., et al., *Controlling Volatile Emissions at Hazardous Waste Sites,* Noyes Publications, Park Ridge, 1986.
4. St. Clair, Ann E., and Kishore T. Ajmera, "Remedial Action at Uncontrolled Hazardous Waste Sites," *Environmental Progress,* v. 3, August 1984, pp. 188–193.
5. Hwang, Seong T., "Toxic Emissions from Land Disposal Facilities," *Environmental Progress,* v. 1, February 1982, pp. 46–52.
6. Martin, William F., John M. Lippit, and Timothy G. Prothero, *Hazardous Waste Handbook for Health and Safety,* Butterworths, Boston, 1987.

7. Fawcett, Howard H., ed., *Hazardous and Toxic Materials: Safe Handling and Disposal,* Second Edition, John Wiley & Sons, New York, 1988.
8. Kamnikar, Brian, "Bioremediaton of Contaminated Soil," *Pollution Engineering,* November 1, 1992, pp. 50–52.
9. USEPA, "Remedial Action at Waste Disposal Sites," EPA/625/6-85/006, October 1985.
10. Code of Federal Regulations, Occupational Safety and Health Administration, Labor, 29CFR1910.120.
11. Halvorsen, Fred, "Safety Procedures for Testing and/or Removing Underground Storage Tanks," *Storage Tank Compliance,* Hazardous Waste Technology Monograph Series, 1986, pp. 21–25.
12. Peterson, Jack M., "Safe Procedures for Storage Tank Investigation and Removal," *Storage Tank Compliance,* Hazardous Waste Technology Monograph Series, 1986, pp. 37–40.
13. Vogel, Gregory A., "Air Emission Control at Hazardous Waste Management Facilities," *Journal of the Air Pollution Control Association,* v. 35, May 1985, pp. 558–566.
14. Andrachek, Richard G., and Kevin M. Sullivan, "Cutting the Cost of Remediation," *Pollution Engineering,* November 1, 1992, pp. 45–47.

CHAPTER 21

HEAT STRESS

Seogi Ibrahim

SciTech Technical Services, Inc.
Morganville, NJ 07751

CONTENTS

INTRODUCTION

Operations where high air temperatures, radiant heat sources, high humidity, direct physical contact with hot objects, or strenuous physical activities are present have a high potential for heat stress. Such places include iron and steel foundries, non-ferrous foundries, brick-firing and ceramics plants, glass products facilities, rubber products factories, electrical utilities (particularly boiler maintenance jobs), bakeries, confectioneries, commercial kitchens, laundries, food canneries, chemical plants, mining sites, smelters, and steam tunnels.

Outdoor operations conducted in hot weather—such as construction, refining, asbestos removal, and hazardous waste-site activities (especially those that require semi-permeable or impermeable protective clothing)—also are likely to cause heat stress among exposed workers.

Age, weight, degree of physical fitness, degree of acclimatization, metabolism, use of alcohol or drugs, and a variety of medical conditions such as hypertension all affect a person's sensitivity to heat. However, even the type of clothing worn must be considered. Prior heat injury predisposes an individual to additional injury.

It is difficult to predict just who will be affected and when, as individual susceptibility varies.

In addition, environmental factors include more than the ambient air temperature; radiant heat, air movement, conduction, and relative humidity all are factors.

Definitions

- The American Conference of Governmental Industrial Hygienists (ACGIH) defines heat stress as "the total net heat load on the body" that results from exposure to external sources and from internal metabolic heat production [1].
- **Heat** is a measure of energy in terms of quantity.
- A **calorie** is the amount of heat required to raise 1 gram of water 1°C (based on standard temperature of 16.5 to 17.5°C).
- The **Btu** (British thermal unit) is the quantity of heat necessary to raise the temperature of one pound of water 1°F.
- **Conduction** is the transfer of heat from particles that are touching each other in a stationary situation such as in the transfer of heat from the skin to air. The air temperature must be cooler than skin temperature for this to occur.
- **Convection** is the circulation of that air next to the skin, which results in an increased cooling action.

- **Evaporation** is the cooling of the body that takes place when sweat evaporates on the skin surface.
- **Radiation** is the transfer of heat energy through space. For example, the heat from a boiler or from the sun will transfer to (or heat) objects in their surrounding areas.
- **Globe temperature** is the temperature inside a blackened, hollow, thin, copper globe.
- **Metabolic heat** is a by-product of the body's activity.
- A **natural wet-bulb temperature** (NWB) is usually used to measure humidity. It is obtained by a wetted sensor, such as wet wick over a mercury-in-glass thermometer that is exposed to natural air movement and unshielded from radiation.
- A **dry-bulb temperature** (DB) is essentially used to measure temperature, the temperature of air as registered by a thermal sensor, such as an ordinary mercury-in-glass thermometer shielded from direct radiant energy sources.

HEAT DISORDERS AND HEALTH EFFECTS

Heat Stroke

Heat stroke occurs when the body's system of temperature regulation fails and the body's temperature rises to critical levels. It is caused by a combination of highly variable factors and is difficult to predict.

The primary symptoms are confusion; irrational behavior; loss of consciousness; convulsions; a lack of sweating (usually); hot, dry skin; and an abnormally high body temperature (a rectal temperature of 41°C (105.8°F). If the body temperature rises too high, death will follow. The elevated metabolic temperatures caused by a combination of workload and environmental heat load, both of which contribute to heat stroke, are highly variable and difficult to predict.

Place the victim in a shady area, remove his or her outer clothing, wet the skin, and increase air movement to improve evaporative cooling until professional methods of cooling are initiated and the seriousness of the condition can be assessed. Fluids should be replaced as soon as possible. The medical outcome of an episode of heat stroke depends on the victim's physical fitness and the extent of first aid treatment.

Regardless of the worker's protestations, no employee suspected of being ill from heat stroke should be sent home or left unattended unless a physician has specifically approved such an order.

Heat Exhaustion

The symptoms are headache, nausea, vertigo, weakness, thirst, and giddiness. Fortunately, this condition responds readily to prompt treatment.

Heat exhaustion should not be dismissed lightly, for several reasons. One is that the fainting associated with heat exhaustion can be dangerous because the victim may be operating machinery or controlling an operation that should not be left unattended; moreover, the victim may be injured when he or she faints. Also, the symptoms seen in heat exhaustion are similar to heat stroke, which is a medical emergency.

Workers suffering from heat exhaustion should be removed from the source of heat and provided fluid and salt replacement along with adequate rest.

Heat Cramps

Heat cramps commonly result from performing hard physical work in a hot environment. These cramps are attributable to the continued loss of salt that occurs in sweating. Cramps often occur in the muscles used during work and can be alleviated readily be resting and drinking water.

Salt tablets should **not** be used for this purpose because they tend to cause retention of both salt and water in the digestive system, which deprives the rest of the body of water and electrolytes.

A commercial replenishing fluid, e.g., Gatorade, may be taken frequently, if medically recommended. Addition of salt in food should be done with care for those suffering from cardiovascular disorders.

Heat Collapse Fainting

Because of excessive pooling of the blood in the extremities, the brain does not receive enough oxygen. Consequently, the exposed individual loses consciousness.

The reaction is similar to that of heat exhaustion and does not affect the body heat balance. It is rapid and unpredictable.

For prevention, the worker can become acclimatized or encouraged to remain somewhat active.

Heat Rashes

Heat rashes are the most common problem in the work environment. Prickly heat appears as red papules, usually in areas where the clothing is restrictive and gives rise to a prickling sensation, particularly as sweating increases. Prickly heat occurs in skin that is persistently wetted by unevaporated sweat, and heat rash papules may become infected if they are not treated. In most cases, heat rashes disappear when the affected individual returns to a cool environment.

Heat Fatigue

A predisposing factor of heat fatigue includes the lack of acclimatization.

Symptoms include impaired performance of skilled sensorimotor, mental, or vigilance jobs. There are no specific recommendations for treatment unless accompanied by other heat illness. The use of a program of acclimatization and training in a hot environment is advisable.

INVESTIGATION GUIDELINES

Investigational guidelines are as follows:

Employer and Employee Interviews

Review for indications of any heat-stress problems.
Following are some questions for the employer interviews:

- What type of action, if any, has the employer taken to prevent heat stress problems?
- What are the potential sources of heat?
- What is the magnitude and distribution of employee complaints?

Following are some questions for the employee interviews:

- What heat-stress problems do they experience?
- What type of action has the employee taken to minimize heat stress?
- What is the employer involvement, i.e., employee training including the hazard communication of heat stress to employees?

Walk-through Inspection

General

- During the walk-through inspection, the investigator will determine building and operation characteristics, and whether any heat load per employee should be noted.

Workload Assessment

- Under conditions of high temperatures and heavy workload, one should establish the workload category of each job.
- The workload category is established by ranking each job into light, medium, heavy, and very heavy categories.

Work/Rest Periods

- For a more comprehensive investigation, a detailed "time and motion" study may be necessary to determine work/rest periods.

- In some cases, a videotape may be helpful in evaluating work practices, metabolic load, etc.

Activity Examples

- Light hand work: Writing, knitting
- Heavy hand work: Typewriting
- Heavy work with one arm: Hammering nails (shoemaker)
- Light work with two arms: Filing metal, planing wood, raking a garden
- Moderate work with the body: Cleaning a floor, beating a carpet
- Heavy work with the body: Railroad track laying, digging

Sample Calculation

Assembly line work using a heavy hand tool [1].

A. Walking along	2.0 kcal/min
B. Intermediate value between heavy work with two arms and light work with the body	3.0 kcal/min
C. Add for basal metabolism	1.0 kcal/min
Total:	6.0 kcal/min

SAMPLING METHODS

Body Temperature Measurements

Instruments are available that measure body temperature directly, e.g., ear canal probe or chest surface measurement.

Environmental Measurements

Environmental heat measurements should be made at or as close as possible to the specific work area where the worker is exposed.

When a worker is not continuously exposed in a single hot area but moves between two or more areas having different levels of environmental heat, or when the environmental heat varies substantially within a single hot area, environmental heat exposures should be measured for each area and for each level of environmental heat to which employees are exposed.

Wet-Bulb Globe Temperature Index

Hourly wet-bulb globe temperatures (WBGT) should be calculated for the combination of jobs, including all scheduled or unscheduled rest periods. These can be compared to the ACGIH values.

For indoor and outdoor conditions with no solar load, WBGT is calculated as:

$$WBGT = 0.7\ NWB + 0.3\ GT$$

For outdoors with a solar load, WBGT is calculated as:

$$WBGT = 0.7\ NWB + 0.2\ GT + 0.1\ DB$$

where NWB = natural wet-bulb temperature
DB = dry-bulb temperature
GT = globe temperature

In the case of workers wearing semipermeable or impermeable clothing or encapsulating ensembles, workers should be monitored when the temperature in the work area is above 70°F (21°C). Monitoring may include heat-stress measurements, observations, body temperatures, and verbal communication with the worker.

Use of the Table 1 requires the WBGT and the approximate type of workload. The workload type can be estimated using the data. Knowing these, one can recommend either controls to lower the WBGT or more rest breaks.

Table 1
Permissible Heat Exposure Threshold Limit Values

	Work Load*		
Work/Rest Regimen	**Light**	**Moderate**	**Heavy**
Continuous Work	30.0°C	26.7°C(80.0°F)	25.0°C(77.0°F)
75% Work, 25% Rest, Each Hour	30.6°C	28.0°C(82.4°F)	25.9°C(78.6°F)
50% Work, 50% Rest, Each Hour	31.4°C	29.4°C(84.9°F)	27.9°C(82.2°F)
25% Work, 75% Rest	32.2°C	31.1°C(88.0°F)	30.0°C(86.0°F)

**Values are in °C and °F, WBGT*
Source: ACGIH, 1989 [1].
These TLVs are based on the assumption that nearly all acclimatized, fully clothed workers with adequate water and salt intake should be able to function effectively under the given working conditions without exceeding a deep body temperature of 38°C. They also are based on the assumption that the WBGT value of the resting place is the same or very close to that of the workplace. Where the WBGT of the work area is different from that of the rest area, a time-weighted average value should be used.
These TLVs apply to physically fit acclimatized individuals wearing light summer clothing.

Measurement

Usually, a portable heat stress meter or monitors may be used. The meters and monitors may integrate this information and calculate both the indoor and outdoor WBGT index according to the established ACGIH threshold limit value equations. With this information and information on the type of work being performed, the heat-stress meter can determine how long a person can safely work or remain in a particular hot environment.

Other Thermal Stress Indices

The effective temperature index (ET) combines the temperature, the humidity of the air, and air velocity. The index has been used extensively in the field of comfort ventilation and air-conditioning work. ET remains a useful measurement technique in mines and other places where humidity is high and radiant heat is low.

Heat-Stress Index (HSI)

The heat-stress index (HSI), developed in 1965, considers all the environmental factors and the work rate, but it is not totally satisfactory for determining the heat stress on an individual worker and can be complicated to use.

CONTROL

General

Ventilation, air cooling, fans, shielding, and insulation are the five major types of engineering control used to reduce heat stress in hot work environments.

The same goal also may be achieved by using power assists and tools that reduce the physical demands placed on a worker. However, for this approach to be successful, the metabolic effort required for the worker to put these devices into use or to operate them must be less than the effort required without them.

Another method is to reduce the effort necessary to operate power assists, and to giveworkers frequent rest breaks in a cooler environment.

Acclimatization

The human body can adapt to heat exposure up to a point. This physiological adaptation is called acclimatization. After acclimatization, the same activity will produce lower-level cardiovascular demands. The worker will sweat more efficiently (causing better evaporative cooling), may lose less salt, and thus will more easily maintain normal body temperatures.

A properly designed and applied acclimatization program decreases the risk of heat-related illnesses and unsafe acts. Such a program basically involves exposing employees to work in a hot environment for progressively longer periods. NIOSH says that, for workers who have had previous experience in jobs where heat levels are high enough to produce heat stress, the regimen should be 50% exposure on day 1, 60% on day 2, 80% on day 3, and 100% on day 4. For new workers who will be similarly exposed, the regimen should be 20% on day 1, with a 20% increase in exposure each additional day.

Fluid Replacement

Cool (50° to 60°F) water or any cool liquid (alcoholic beverages excluded) should be made available to workers in such a way that they are stimulated to frequently drink small amounts, e.g., one cup every 20 minutes. Ample supplies of liquid should be placed close to the work area. Although some commercial replacement drinks contain salt, it is not necessary for acclimatized individuals because such individuals generally add enough salt in their summer diets. Unacclimatized individuals may need salted drinking water in a concentration of 0.1%.

Engineering Controls

Ventilation

General ventilation is used to dilute hot air with cooler air (generally cooler air that is brought in from the outside). This technique clearly works better in cooler climates than in hot ones. A permanently installed ventilation system usually handles large areas or entire buildings. Portable or local exhaust systems may be more effective or practical in smaller areas.

Air treatment and air cooling differ from ventilation because they reduce the temperature of the air by removing heat and sometimes humidity from the air. Air conditioning is a method of air cooling, but it is expensive to install and operate. An alternative is the use of chillers to circulate cool water through heat exchangers over which air from the ventilation system is then passed. Chillers, like general dilution ventilation, are more efficient in cooler climates than in warmer ones.

Local air cooling can be effective in reducing air temperature in specific areas. Two methods have been used successfully in industrial settings. One type, cool rooms, can be used to enclose a specific workplace or to offer a recovery area near hot jobs. The second type is a portable blower with built-in air chiller. The main advantage of a blower, aside from portability, is its minimal set-up time.

Convection

Another way to reduce heat stress is to increase the air flow or convection using fans in the work area, as long as the air temperature is less than skin temperature. Changes in air speed can help workers stay cooler by increasing both the convective heat exchange (the exchange between the skin surface and the surrounding air) and the rate of evaporation. Because this method does not actually cool the air, any increases in air speed must impact the worker directly to be effective.

If the temperature is higher than 95°F, the hot air passing over the skin may actually make the worker hotter and offset any increase gained in evaporative cooling. Increases in air speed have no effect on workers wearing vapor-barrier clothing.

Heat conduction solutions include insulating hot surfaces that generate the heat and changing the surfaces.

Simple engineering controls such as shields can be used to reduce the problem of radiant heat (heat coming from hot surfaces within the worker's line of sight). A matte black surface, through absorption, can reduce radiant heat more than a perfectly smooth, polished one. With some sources of radiation such as heating pipes, it is possible to use both insulation and surface modifications to achieve a substantial reduction in radiant heat. Instead of reducing radiation from the source, shielding can be used to interrupt the flow between the source and the worker. Polished surfaces make the best barriers, although special glass or metal mesh surfaces can be used if visibility is a problem.

Any shield, whether temporary or permanent, should be situated so it will not interfere with the air flow. The reflective surface of the shied should be kept clean to maintain its effectiveness.

ADMINISTRATIVE AND WORK PRACTICES

Training

Training is the key to good work practices. If all employees do not understand the reasons for using new or changing old work practices, the chances of such a program succeeding are greatly reduced.

NIOSH (1986) states that a good heat-stress training program should cover at least the following components:

- Knowledge of the hazards of heat stress.
- Recognition of predisposing factors, danger signs, and symptoms,
- Awareness of first-aid procedures for and potential health effects of heat stroke.
- Employee responsibilities in avoiding heat stress.
- Dangers of the use of drugs, including therapeutic ones, and alcohol in hot work environment.

- Use of protective clothing and equipment.
- Purpose and coverage of environmental and medical surveillance programs and the advantage of worker participation in such a program.

Schedule hot jobs for the cooler part of the day, and schedule routine maintenance and repair work in hot areas for the cooler seasons of the year.

Other Administrative Controls

- Reduce the physical demands of work such as excessive lifting or digging with heavy objects.
- Recovery areas should be provided, such as air-conditioned enclosures and rooms.
- Use shifts, e.g., early morning, cool part of the day, or night work.
- Use intermittent rest periods with water breaks.
- Use relief workers.
- Use worker pacing.
- Assign extra workers.
- Limit worker occupancy, or numbers of workers present especially in confined or enclosed spaces.

PERSONAL PROTECTIVE EQUIPMENT

Reflective Clothing

Reflective clothing, which can vary from aprons and jackets to suits that completely enclose the worker from neck to feet, can stop the skin from absorbing radiant heat. However, because most reflective clothing does not allow air exchange through the garment, the reduction of radiant heat exposure must more than offset the corresponding loss in evaporative cooling. For this reason, reflective clothing should be worn as loosely as possible.

Auxiliary Body Cooling

In extreme situations where radiant heat is high, auxiliary cooling systems can be used under the reflective clothing.

Ice Vests

Commercially available ice vests, though heavy, may accommodate as many as 72 ice packets, which are usually filled with water. Carbon dioxide (dry ice) also can be used as a coolant.

The cooling offered by ice packets lasts only two to four hours or less at moderate to heavy heat loads, making frequent replacement necessary. However, ice vests do not encumber the worker with air supply or power cords, and thus permit maximum mobility. Clothing with ice packets is also relatively inexpensive.

Wetted Clothing

Another simple and inexpensive personal cooling technique that is effective when reflective or other impermeable protective clothing is worn is the use of wetted terry cloth coveralls or wetted two-piece, whole-body cotton suits.

This approach to auxiliary cooling can be quite effective under conditions of high temperature and low humidity where evaporation from the wetted garment is not restricted.

Water-Cooled Garments

These garments range from a hood, which cools only the head, to vests and "long johns," which offer partial or complete body cooling. This equipment requires a battery-driven circulating pump, liquid-ice coolant, and a container.

Although this system has the advantage of allowing wearer mobility, the weight of the other components limits the amount of ice that can be carried and thus reduces the effective use of time. The heat transfer rate in liquid cooling systems may limit their use to low-activity jobs; even in such jobs, service time is only about 20 minutes per pound of cooling ice. An outer insulating jacket should be an integral part of these systems to keep outside heat from melting the ice.

The most highly effective, as well as the most complicated, personal cooling system is one that uses circulating air. By directing compressed air around the body from a supplied air system, both evaporative and convective cooling are improved. The greatest advantage occurs when circulating air is used with impermeable garments or double cotton overalls. One type, used when respiratory protection is also necessary, forces exhaust air from a supplied-air hood ("bubble hood") around the neck and down inside an impermeable suit. The air then escapes through openings in the suit.

Air also can be supplied directly to the suit without using a hood. This can be done in three ways:

- by a single inlet.
- by a distribution tree, or
- by a perforated vest.

Also, a vortex tube can be used to reduce the temperature of circulating air. The cooled air from this tube can be introduced either under the clothing or into a bubble hood. The use of a vortex tube acts as a heat pump that separates the airstream into a hot and a cold stream and also can be used to heat in cold climates. It is, however, noisy and requires a constant source of compressed air supplied through an attached air hose.

One problem with this system is the limited mobility of workers whose suits are attached to an air hose. Another is that of getting air to the work area itself. These systems should therefore be used in work areas where there is not much moving around or climbing.

Respirator Usage

The use of the self-contained breathing apparatus (SCBA) and its additional weight adds stress to the worker, and this stress will contribute to the overall heat stress of the worker. Chemical-protective clothing such as the totally encapsulated chemical protection suits also will add to the heat stress problem. Frequent shifting of workers may be necessary.

HEAT STRESS: GENERAL WORKPLACE REVIEW

Following is a list of sample questions that the compliance officer may wish to consider when investigating heat stress.

Workplace Description

- Type of Business
- Heat-producing equipment or processes used
- Previous history (if any) of heat-related problems
- "Hot" spots
- Steady or intermittent?
- Number of employees exposed? For how many hours per day?
- Is potable water available?
- Are supervisors trained to detect/evaluate heat stress symptoms?
- Are Exposures Typical for a Workplace in this Industry?

Weather at Time of Review

- Temperature
- Humidity (wet-bulb globe temperature)
- Air velocity

Is Day Typical of Recent Weather Conditions?

- Get information from the weather bureau.

Heat-Reducing Engineering Controls

- Ventilation in place?
- Ventilation Operating?
- Air conditioning in place?
- Air conditioning operating?
- Fans in place?
- Fans operating?
- Shields or insulation between sources and employees?
- Are reflective faces of shields clean?

Work Practices To Detect, Evaluate, and Prevent or Reduce Heat Stress

- Training program?
- Content?
- Where given?
- For whom?
- Liquid replacement program?
- Acclimatization program?
- Work/rest schedule?
- Scheduling of work (during cooler parts of shift, cleaning and maintenance during shut-downs, etc.).
- Cool rest areas, including shelter at outdoor work sites?
- Heat monitoring program?

Personal Protective Equipment

- Reflective clothing in use?
- Ice and/or water-cooled garments in use?
- Wetted undergarments (used with reflective or impermeable clothing)
- Circulating air systems in use?

First Aid Program

- Trained personnel?
- provision for rapid cool-down?
- Procedures for getting medical attention?
- Transportation to medical facilities readily available for heat stroke victims?

Medical Screening and Surveillance Program

- Content?
- Who manages program?

HEAT STRESS-RELATED ILLNESS OR ACCIDENT FOLLOW-UP

- Describe events leading up to episode.
- Evaluation/comments by other workers at the scene.
- Work at time of episode (heavy, medium, light)?
- How long was affected employee working at site prior to episode?
- Medical history of affected, if known.
- Appropriate engineering controls in place?
- Appropriate engineering controls operating?
- Appropriate work practices used by affected employee(s)?
- Appropriate personal protective equipment available?
- Appropriate personal protective equipment in use?
- Medical screening for heat stress and continued surveillance for signs of heat stress given other employees?
- Additional comments regarding specific episode(s).

MEASUREMENT OF WET-BULB GLOBE TEMPERATURE

Because measurement of deep temperature is impractical for monitoring a worker's heat load, measurement is required of those environmental factors that mostly nearly correlate with deep body temperature and other physiological responses to heat. At the present time, the wet-bulb globe temperature (WBGT) index is the simplest and most suitable technique to measure these environmental factors. WBGT values are calculated by the following equations:

1. Indoors or outdoors with no solar load:
   WBGT = 0.7 NWB + 0.3 GT
2. Outdoors with solar load:
   WBGT = 0.7 NWBb + 0.2 GT + 0.1 DB

   Where:
   WBGT = Web-bulb globe temperature
   NWB = Natural wet-bulb temperature
   DB = Dry-bulb (air) temperature
   GT = Globe thermometer temperature

The determination of WBGT requires the use of a black globe thermometer, a natural (static) wet-bulb thermometer, and a dry-bulb thermometer. The proper technique to measure environmental factors follows.

The range of the dry and the natural wet-bulb thermometers should be –5°C to +50°C, with an accuracy of ±0.5°C. The dry bulb thermometer must be shielded from the sun and the other radiant surfaces of the environment without restricting the airflow around the bulb. The wick of the natural wet-bulb thermometer should be kept wet with distilled water for at least 1/2 hour before the temperature reading is made. It is not enough to immerse the other end of the wick into a reservoir of distilled water and wait until the whole wick becomes wet by capillarity. The wick must be wetted by direct application of water from a syringe 1/2 hour before each reading. The wick must cover the bulb of the thermometer, and an equal length of additional wick must cover the stem above the bulb. The wick always should be clean, and new wicks should be washed before using.

A globe thermometer, consisting of a 15 cm (6-inch) diameter hollow copper sphere painted on the outside with a matte black finish or equivalent, must be used. The bulb or sensor of a thermometer (range –5°C to +100°C with an accuracy of ±0.5°C) must be fixed in the center of the sphere. The globe thermometer should be exposed at least 25 minutes before it is read.

A stand should be used to suspend the three thermometers so that they do not restrict free airflow around the bulbs, and the wet-bulb and globe thermometer are not shaded. It is permissible to use any other type of temperature sensor that gives a reading similar to that of a mercury thermometer under the same conditions. The thermometers must be so placed that the readings are representative of the employee's work or rest areas, as appropriate.

Once the WBGT has been estimated, employers can estimate workers' metabolic heat load and use the ACGIH method to determine the appropriate work/rest regimen, clothing, and equipment to control the heat exposures of workers in their facilities.

BIBLIOGRAPHY

American Conference of Governmental Industrial Hygienists. 1986. *Documentation of the Threshold Limit Values and Biological Exposure Indices* (5th ed.). Cincinnati: American Conference of Governmental Industrial Hygienists.

American Conference of Governmental Industrial Hygienists. 1988. *Threshold Limit Values and Biological Exposure Indices for 1988–1989.* Cincinnati: American Conference of Governmental Industrial Hygienists.

American Industrial Hygiene Association. 1975. AIHA, *Heating and Cooling for Man in Industry.* 2nd Edition, Akron, OH.

Electric Power Research Institute (EPRI). 1987. *Heat-Stress Management Program for Nuclear Power Plants.*

The Human Factors Section, Health, Safety and Human Factors Laboratory, Eastman Kodak Company. 1983. *Ergonomic Design at Work for People,* Vol. II, Lifetime Learning Publications, Belmont, CA.

National Institute for Occupational Safety and Health. *Criteria for a Recommended Standard-Occupational Exposure to Hot Environments,* DHHS (NIOSH) Publication No. 86-113, April 1986.

National Institute for Occupational Safety and Health. *Occupational Safety and Health Guidance Manual for Hazardous Waste Site Activities,* DHHS(NIOSH) Publication No. 85-115, 1985.

National Institute for Occupational Safety and Health. *Standards for Occupation Exposure to Hot Environments: Proceedings of a Symposium,* DHHS (NIOSH) Publication No. 76-100, January 1976.

National Institute for Occupational Safety and Health. *Working in Hot Environments,* DHHS (NIOSH) Publication No. 86-112 Revised 1986.

National Safety Council, 1985. NSC, *Pocket Guide to Heat Stress,* Chicago, IL.

Ramsey, J. D., Burford, C. L., Beshir, M. Y., and Jensen, R. C. *Effects of Workplace Thermal Conditions On Safe Work Behavior.* Journal of Safety Research 14: 105-114, 1983.

CHAPTER 22

TRANSPORTATION OF HAZARDOUS MATERIALS

Jose Abraham

South East Water Pollution Control Plant
Philadelphia Water Department
Philadelphia, PA 19116, USA

Paul N. Cheremisinoff

Department of Civil and Environmental Engineering
New Jersey Institute of Technology
Newark, NJ 07102, USA

CONTENTS

INTRODUCTION

Hazardous materials (hazmat) transportation is a large and growing segment of the transportation industry. Special concern is addressed to safety in the transportation of hazardous materials because of the potential for fires, explosions, groundwater contamination, and toxic effects on human health if hazardous materials are inadvertently released.

This chapter will review and identify issues and problems associated with transporting hazardous materials.

Reliable and comprehensive information on the movements of hazardous materials is lacking. About 5% to 15% of all inland freight transported is hazardous materials. Petroleum and gasoline shipments account for almost half of all truck transport of hazardous materials. More than 85% of the hazmat releases reported to federal agencies occur on the highways.

Federal regulations affecting the movement of hazardous materials have been promulgated since 1886. The authority to develop and enforce hazardous materials transportation regulations rests with United States Department of Transportation (DOT). Hazardous Materials Regulations (HMR) prescribe requirements for shippers, carriers, and manufacturers of packaging to hazardous materials. The HMR covers all aspects of the transportation of hazardous materials, from the classification of the material and its packaging to the delivery to its final destination.

The transportation of hazardous materials is regulated by both the Hazardous Materials Transportation Act (HMTA) and the Resource Conservation Recovery Act (RCRA). The EPA's regulations for the transportation of hazardous waste generally follow the manifest requirement and adopt the DOT's regulations for labeling, placarding, and packaging hazardous substances.

Under the Comprehensive Environmental Response Compensation and Liability Act of 1980 (CERCLA), the EPA can sue the potentially responsible parties involved in a toxic material incident, including transportation, for cleanup cost. A critical review of all these hazardous materials regulations are presented in this chapter.

Federal authority to enforce hazardous materials transportation regulations and to achieve safe transportation of hazardous materials is distributed among numerous federal agencies. The DOT is responsible for issuing and enforcing regulations applicable to both interstate and intrastate transportation of hazardous materials. The RCRA of 1976 requires the EPA to set standards for transportation of hazardous wastes. EPA regulations must be consistent with DOT regulations concerning any materials subject to the Transportation Safety Act. State and local governments and a number of industry-related agencies such as CHEMTREC also are involved in the safe transportation of hazardous material. The responsibility and activities of all these agencies is identified and presented in this chapter.

The New Jersey Department of Transportation adopted regulations governing the transportation of hazardous materials by truck and rail, as mandated by the state legislature. These rules and regulations are in substantial conformance with the federal requirements. This chapter outlines the conformance of New Jersey hazmat regulations with respect to federal requirements.

Application of different advanced technology in the transportation of hazardous material is helpful in reducing the risk involved in it. This chapter identifies the various applicable technologies. Also, some critical research and development (R&D) efforts are lacking. This chapter identifies and presents the future R&D efforts needed in the field of hazardous materials transportation.

Risk management in the transportation of hazardous wastes is the impartial adjustment of conflicting claims in selecting means to reduce the possibility of loss, injury, or death. Public concern is greatest about risks that are involuntary, uncontrolled, unfamiliar, immediate, man-made, and catastrophic. Hazmat transportation possess many and sometimes all of these attributes. Risk assessment involves estimating the frequencies and consequences of undesirable events, then evaluating the associated risk in quantitative terms. Various aspects of the transport of hazardous materials are highly interrelated, but unfortunately are often only seen and analyzed in virtual isolation. This chapter discusses the pathway approach as a tool to risk assessment.

As the transportation of hazardous materials is inherent in any advanced and technologically complex society, the problems and risks involved with it are likely to remain. But we can prevent or acceptably minimize the risks by simultaneously using advanced technology and by enforcing various regulations.

A number of industrial processes of vital economic importance are dependent on the uninterrupted flow of hazardous materials shipments. The production of hazardous materials associated with technological growth and economic development creates a potential danger to the population and the environment in the event of a release. The risks associated with the transportation of hazardous materials have drawn considerable attention at local, national, and international levels, resulting in the development of a regulatory framework to enhance the safety of hazardous materials movement.

All the hazardous materials regulations are made with one basic aim: to acceptably minimize the risk inherent in transportation of hazardous materials or wastes, oil, or other potentially harmful substances. These laws include the Hazardous Materials Transportation Act and the formidable regulations thereunder, the Oil Pollution Act of 1990, tank vessel statutes, Coast Guard regulations, some international agreements and treaties, Section 311 and 312 of the Clean Water Act, the RCRA manifest requirements, and other parts of the system of laws and regulations applicable to transportation.

According to data from Hazardous Materials Incident Report System (HMIRS) maintained by the Research and Special Programs Administration of the DOT, 7,481 hazardous waste incidents occurred during the year 1989. More than 5,000 incidents have occurred annually since 1973. The incident statistics also show that most of the incidents occur on the highways, accounting for more than 85% of the hazmat releases. About 5% to 15% of all inland freight transported is hazardous substances. The trends over the years in the transport of hazardous materials goes hand in hand with economic developments, but the increasing quantities of hazardous materials transportation is a threat to the general public and the environment.

The aim of this chapter is to:

- Identify the hazardous materials transportation issues and problems.
- Review the various hazardous material transportation laws and regulations.
- Explain the functions and responsibilities of various agencies with respect to hazardous materials transportation.
- Explain how advanced technology can help in achieving safe transportation goals.
- Identify states' concerns relative to federal hazardous waste transportation regulation.
- Identify critical R&D needed for the safe transportation of hazardous material.
- Explain how risk-management techniques help in reducing the potential risks associated with hazmat transportation.

REVIEW OF ISSUES AND PROBLEMS

The Challenge

Freight transport, including the transport of hazardous materials, is an essential activity upon which many sectors of industry and economy are dependent. The trend over the years in the transport of hazardous goods and materials goes hand in hand with economic developments. The hazardous materials transported pose risks to the general public and the environment. These interrelated factors can be better explained by Figure 1.

Reliable and comprehensive information on the movement of hazardous goods and materials in United States is lacking and needs to be developed. In the United States, a total of 1.5 billion tons of hazardous materials was transported in 1982. Trucks accounted for the largest mode of transport, constituting 60% (see Table 1). Petroleum and gasoline shipments account for almost half of all truck transport of hazardous materials. Chemicals represent the second-largest category of hazardous materials transport by truck. In 1986, almost 1.5 billion

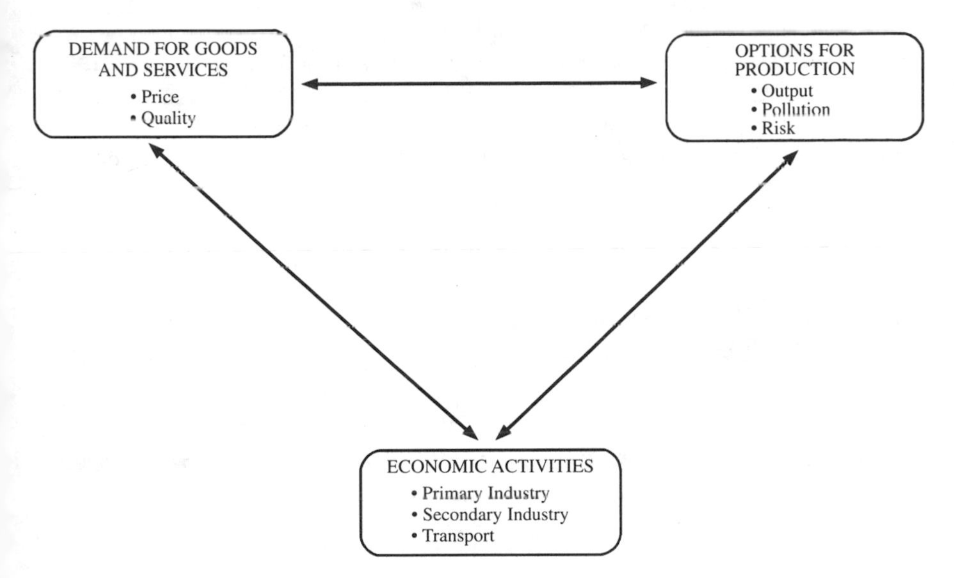

Figure 1. The interrelated factors of hazardous materials transport.

tons of freight moved by rail in the United States, of this 62 million tons or 4% was hazardous materials. In 1989, a total of 7,481 hazardous materials incidents were reported by the Hazardous Materials Incident Report of System (HMIRS) which is maintained by the Research and Special Programs Administration (RSPA) of the DOT (see Figure 2). Highway-mode transport accounts for most of the hazardous materials incident and damages (see Table 2). The types of vehicles carrying hazardous materials on the nation's highways range from cargo tank trucks to conventional tractor-trailers and flat beds that carry large portable tank containers or non-bulk packages, such as cylinders, drums, and other small containers. Rail shipments are usually bulk commodities, such as liquid or gaseous chemicals and fuels, carried in tank cars. Most hazardous materials transported by water are moved in bulk containers, such as tank ships or barges, while air shipments are typically small packages, often high-value or time-critical materials. Pipelines usually carry petroleum products.

Historical Development of Regulations

Federal regulations affecting the movement of hazardous materials toxic substances and wastes have been promulgated since 1886.

Beginning with the act to regulate the transportation of nitroglycerine, or glynoin oil, some 64 such laws have been identified. It was during the late 1880s that rail carriers initiated the first organized efforts to improve the safety of

Table 1
Estimated Transportation of Hazardous Materials in U.S.A. by Mode in 1982

Mode	Number of Vehicles/Vessels Used	Tons Transported	Miles
Truck	337,000 dry freight or flat bed 13,000 cargo tanks	927 million	93.6 billion
Rail	115,600 tank cars	73 million	53 billion
Waterborne	4909 tanker barges	549 million	636.5 billion
Air	3,772 commercial planes	285 thousand	459 million
Total		1.5 billion	784 billion

Estimated % of Hazardous Materials by Mode in Terms of Tonnage Moved

Country	Year	Road	Rail	Water
Canada	1983	57.3	30	12.7
Finland	1982	67.7	31.5	0.8
France	1983	75.8	17.4	6.8
Germany	1984	60.3	11.1	28.6
U.S.A.	1982	59.8	4.7	35.4
Norway	1984	73.4	3.3	23.3

increasing movements of goods considered to represent a danger to their employees, equipment, or the cargo they were transporting.

One of the pioneering steps in their effort was the creation of the Bureau for the Safe Transportation of Explosives and Other Dangerous Articles by the American Railway Association (ARA). Continuing problems in rail transportation prompted the 1908 Congress to regulate the transportation of explosives and other dangerous articles in interstate commerce.

This act adopted the existing framework of classifications of dangerous articles upon which subsequent regulations were structured. This commodity-oriented regulatory framework still prevails. The act also designated shippers, carriers, and the Interstate Commerce Commission as the parties vested with the responsibility for assuring transportation of these goods with no appreciable danger to people or property. This requirement secured safety in transit and established the levels of safety attained by the first regulations adopted.

Significantly, this legislation also provided for the procedural handling of the safety matters by the Interstate Commerce Commission, which prior to that time had been charged principally with economic regulation. Under these procedures,

(text continued on page 504)

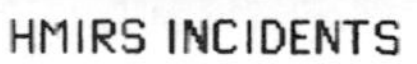

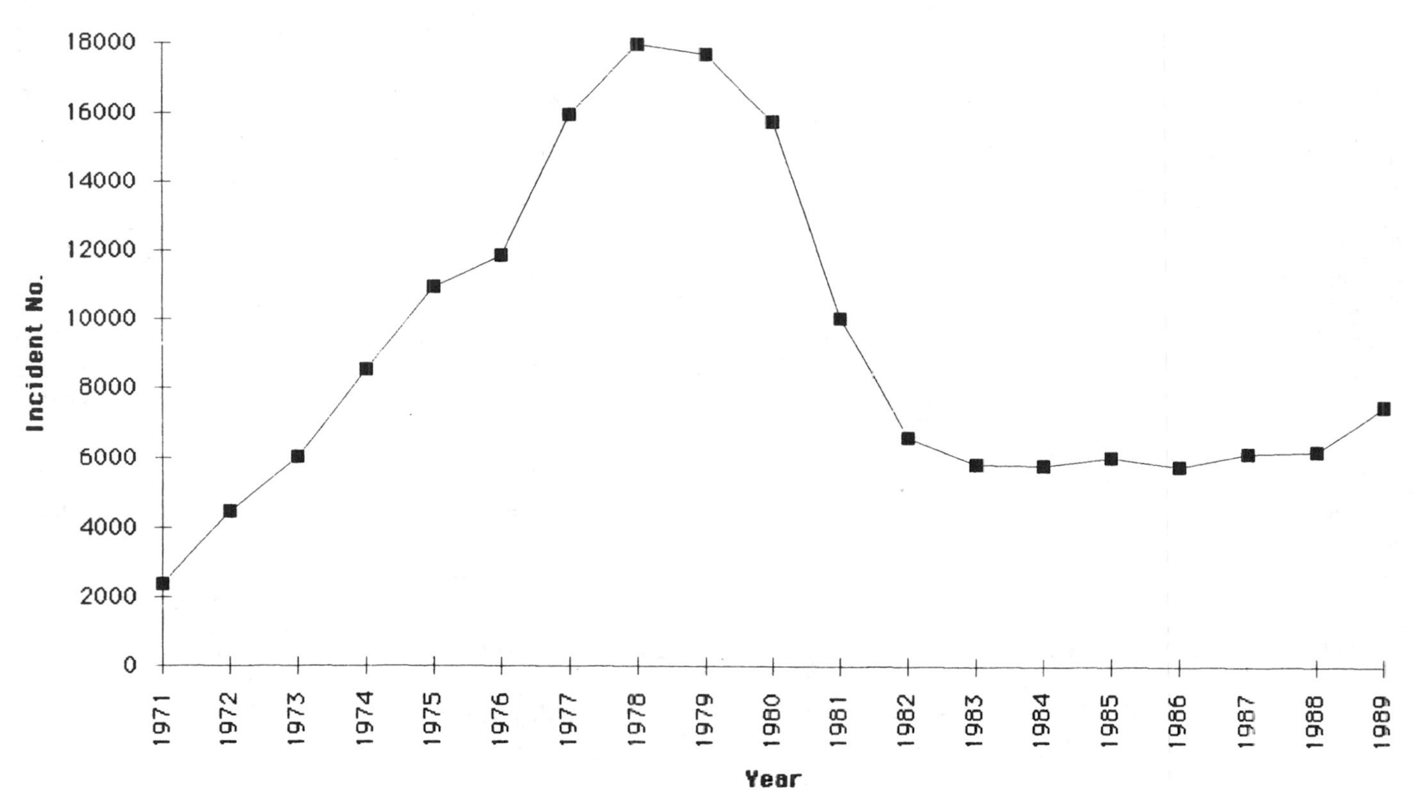

Figure 2. Number of hazardous materials incidents by year, 1971–1989.

Table 2
Incident Statistics by Mode and Reporting Year

Mode	1982	1983	1984	1985	1986	1987	1988	1989	Total
Incidents by Mode									
Air	95	66	102	114	120	172	172	187	1,019
Highway	5,662	4,872	4,508	4,752	4,614	4,952	4,904	5,977	40,241
Railway	830	868	996	842	855	886	1,019	1,178	7,474
Water	8	12	8	7	7	15	16	10	83
Freight Forwarder	6	1	145	298	150	118	78	127	923
Other	1	1	6	6	1	1	1	2	30
Total	6,602	5,820	5,765	6,019	5,758	6,135	6,190	7,481	49,770
Injuries by Mode									
Air	0	3	15	4	12	26	6	54	120
Highway	88	118	147	195	229	247	127	205	1,356
Railway	36	68	76	53	59	25	36	36	389
Water	1	0	18	0	2	8	0	7	36
Freight Forwarder	0	0	3	1	12	25	0	15	56
Other	0	0	0	0	2	0	0	0	2
Total	125	189	256	253	316	331	169	317	1,959

Deaths by Mode									
Air	0	0	0	0	0	0	0	0	0
Highway	13	8	6	8	16	10	19	8	83
Railway	0	0	0	0	0	0	0	0	0
Water	0	0	1	0	0	0	0	0	1
Freight Forwarder	0	0	0	0	0	0	0	0	0
Other	0	0	0	0	0	0	0	0	0
Total	13	8	7	8	16	10	19	8	89

Damages by Mode—$									
Air	26,826	52,525	770,956	12,299	62,813	13,779	562,176	105,011	1,606,385
Highway	11,381,564	9,253,755	11,118,351	12,689,492	13,106,727	15,648,693	18,551,864	15,320,205	107,070,651
Railway	4,331,465	2,559,130	3,353,339	10,273,671	3,077,825	7,554,815	2,432,476	10,265,205	43,847,927
Water	30,000	76,088	509,029	3,242	53,500	99,930	74,262	39,900	885,951
Freight Forwarder	35	300	14,011	13,918	102,117	51,126	15,009	37,655	234,171
Other	200	16,500	975	515	3,385	50	2,700	2,600	26,925
Total	15,770,090	11,958,298	15,766,661	22,993,137	15,406,367	23,368,393	21,638,487	25,770,577	153,672,010

(text continued from page 500)

the initiative for changing or improving regulations lay largely with parties having an interest in a single product or a single container or package. The record contains relatively few instances of changes made upon the commission's own motion. Thus, the body of regulations as it exists today grew item by item utilizing the basic concepts and framework first given cognizance in the 1908 act. As the new products, containers, and modes emerged, each was accommodated by the then-existing scheme of regulations. While the regulations were changed from time to time over the years, the underlying approaches and framework of the regulations remained intact.

An important aspect of past development of safety measures was the relatively low level of research activity required to sustain the unsophisticated level of the safety control activities, both voluntary and regulatory, needed to satisfy the demands of the interested parties. Until 1968, essentially no research was budgeted to the regulators of hazmat transportation. By considering matters at issue on a ad hoc basis for largely single-mode use, there was little need for more than arriving at a consensus of expert judgments for the three parties with direct interest in the problem—namely the shippers, the carriers, and their suppliers.

During World War II, two aspects of these approaches underwent a change in one of the modes: The regulatory jurisdiction over marine bulk flammable goods transportation was assigned to the Coast Guard, which then began to exercise the primary initiative for regulatory changes and to institute meaningful enforcement efforts in this safety area. The nature of the resulting regulations suggests that the most ethical approach and the consensus of experts continued to provide the basis for regulatory changes.

The creation of Department of Transportation (DOT) in 1967 resulted in significant changes in the administration of these regulations. Regulatory authorities for each mode began meeting regularly to exchange views, and important new rule-making procedures were developed. Regulatory staffs grew in numbers and expertise, and initiatives for regulatory changes began shifting from the shippers, carriers, and suppliers to the regulators. Regulatory changes, once published quarterly, came with increasing frequency. Pertinent research was initiated by different agencies within the department. Increasing emphasis has been directed at the consequences of accidents, by considering not only the nature but the degree of hazard of the dangerous commodity being analyzed by the regulators. Changes based on a subjective hazard rating scheme were developed and initiated by one of the modes.

The principal responsibility for regulating the transportation of hazardous materials rests with the federal government through the Commerce Clause of Section 8 of the U.S. Constitution. This has been broadly interpreted to include the movement of hazardous materials in both intrastate and interstate commerce.

The Transportation Safety Act, which is also known as HMTA, authorized the secretary of transportation to regulate the safe shipment of hazardous materials. The basic regulations are found in 49 CFR 100-109. Additional statutory authority in Title 49 includes the Natural Gas Pipeline Safety Act of 1968, the Hazardous Liquid Pipeline Safety Act of 1968, the Hazardous Liquid Pipeline Safety Act of 1979, the Dangerous Cargo Act of 1940, the Federal Aviation Act of 1979, the Parts and Waterways Safety Act of 1979, and the Federal Railroad Safety Act of 1970.

The U.S. Coast Guard (USCG) and the EPA were authorized in the Clean Water Act of 1977 to act directly to eliminate a release or a substantial threat of release of a hazardous substance into the surface waters and to mitigate any resulting damage. Title 40 of the Code of Federal Regulations (CFR) includes the National Contingency Plan based on this act, which gives broad powers to the EPA to respond to a hazardous materials incident. The Resource Conservation and Recovery Act (RCRA) of 1976 also provided the EPA with regulatory authority in hazardous waste and new chemicals, respectively. Related regulations based on these acts also are found in Title 40.

The Comprehensive Environmental Response, Compensation, and Liability Act (CERCLA) requires notification to the Coast Guard's National Response Center in the event of a release of a "reportable quantity" of a hazardous substances or a substantial threat of a release to the environment. Notification requirements of multimedia release—i.e., to air, water, and ground modes—are given for different substances. The purpose of this notification is to allow the EPA or the Coast Guard to decide whether or not government response to the release is necessary.

Title 46 of the CFR codifies regulations for waterborne bulk cargoes, and Title 33 codifies regulations for the safe handling of hazardous materials on commercial vessels and waterfront facilities. Title 23 regulates state programs for cleaning up hazardous debris after accidents.

LAWS AND REGULATIONS

Overview

Different laws and regulations are in effect to acceptably minimize risk inherent in the transportation of hazardous waste or materials, oil, and other potentially harmful substances. These laws and regulations intended to protect the environment from spills and other incidents occurring during the transportation process are an important part of the environmental law system. These laws include the Hazardous Materials Transportation Act and the formidable regulations thereunder, the Oil Pollution Act of 1990, the tank vessel statutes, Coast Guard regulations, some international agreements and treaties, Sections 311 and 312 of the Clean Water Act, the RCRA manifest requirements, and other part of the system of law and regulations applicable to transportation.

They require, in essence, that covered substances be identified; packed properly in proper containers; labeled; loaded onto a placarded, properly designed, built, and maintained truck, tanker, pipeline or other transportation mode; and delivered to an appropriate site. Adequate financial responsibility is to be maintained and demonstrated throughout the process and careful records are to be kept of everything picked up and everything delivered to its ultimate destination. In some instances, the ability to respond to a spill must be maintained. In every instance, appropriate notifications in the event of a spill must be given.

Hazardous Materials Transportation Act (HMTA)

Regulations governing the transportation of hazardous materials by highway are contained in Title 49 of the Code of Federal Regulations (49 CFR). The Hazardous Materials Regulations (HMR) prescribe requirements for shippers, carriers, and manufacturers of packaging for hazardous materials. The HMR require that hazardous materials are properly classified, marked, labeled, packaged, and described on shipping documentation. In addition, carriers are required to properly load and transport hazardous materials, properly placard their vehicles, and to otherwise transport these materials in compliance with the HMR.

Carriers also must comply with the Federal Motor Carrier Safety Regulations (FMCSR) regarding vehicle standards, driver qualifications, hours of service, financial responsibility requirements, driving and parking rules for hazardous materials shipments, and other such regulations.

Authority to develop and enforce hazardous materials transportation regulations rests with the DOT. The authority of the DOT extends to all shipments of hazardous materials transported within the United States, imported to or exported from the United States, or passing through the United States in the course of being shipped between places outside the United States. Responsibility for highway transportation within the DOT is divided between the Research and Special Programs Administration (RSPA) and the Federal Highway Administration (FHWA).

The RSPA develops and issues regulations applicable to all modes governing hazardous materials definition and classification, shipping and carrier operations, and packaging and container specifications.

The FHWA develops and issues the FMCSR and is responsible for enforcing all hazardous materials regulations applicable to the highway mode.

The HMR cover all aspects of the transportation of hazardous materials, from the classification of the material and its packaging to its delivery to its final destination. The shipper must first classify a material according to its hazard class. A proper shipping document containing information about the hazardous material must accompany all shipments. In addition, each package containing hazardous materials must be marked and labeled as prescribed in the HMR, and transport vehicles must be placarded if they contain certain classes and quantities of hazardous materials. Figure 3 lists requirements for the transportation of hazardous wastes.

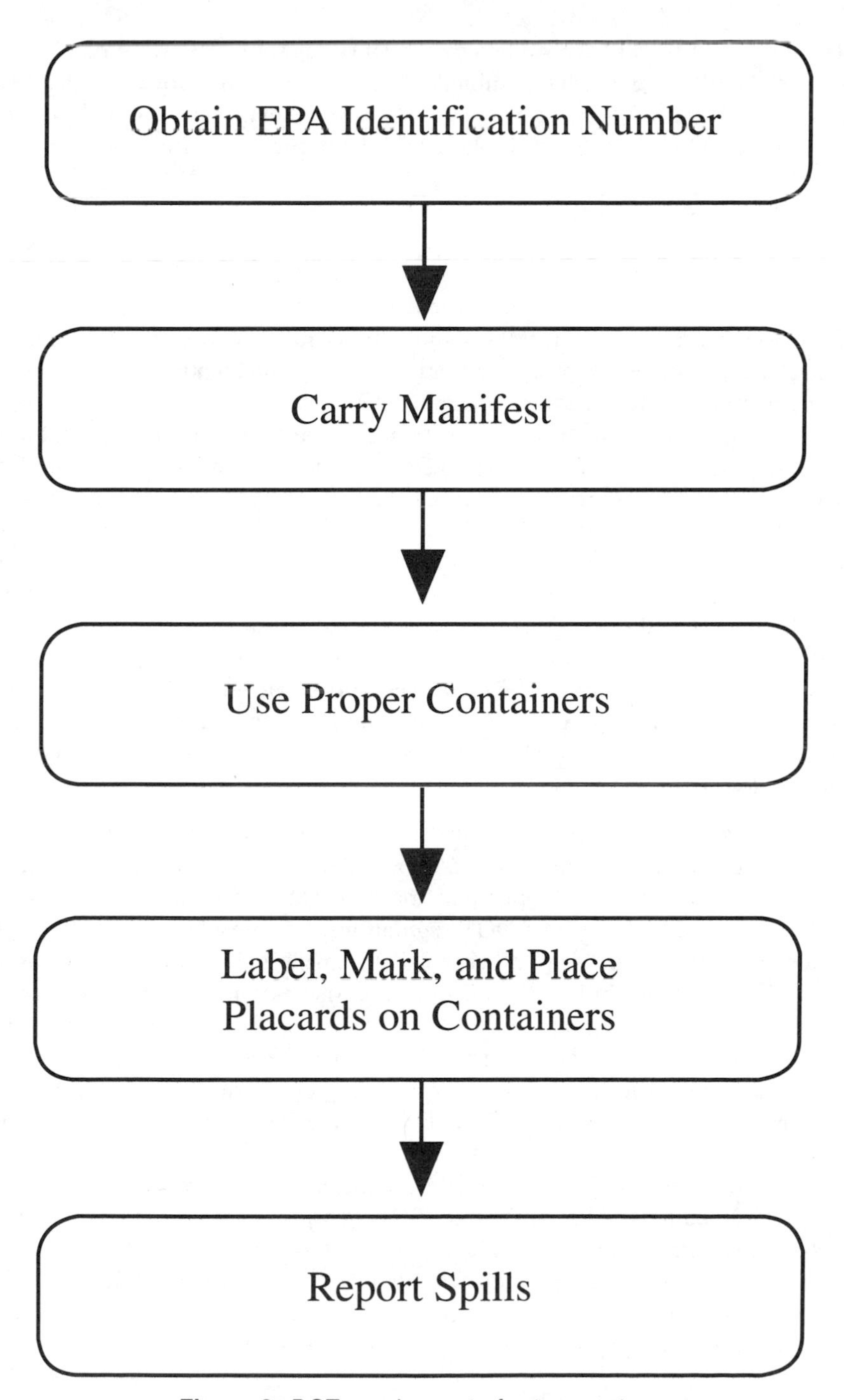

Figure 3. DOT requirements for transporters.

The labels and placards prescribed by the DOT for shipments within the United States are similar to most international labels and placards; however, the DOT labels and placards must be used for domestic shipments. International placards may be used in addition to DOT placards for international shipments.

Review of Title 49 of CFR

Part 106 prescribes general rule-making procedures for adopting Office of Hazardous Materials Transportation regulations.

Part 107 contains procedures for the submission and review of packaging exemption applications, inconsistency ratings, and nonpreemption determinations. Enforcement authorities also are described.

Part 171 is a general introduction to the hazardous materials regulations. Special requirements for hazardous wastes are included, as well as definitions of terms and list of technical documents incorporated by reference into the regulations. Reporting requirements for hazardous materials also are specified.

Part 172 contains the Hazardous Materials Table. The table lists the hazardous materials and hazard classes subject to regulation, and appropriate requirements for labels and packaging for air and water shipments (see Table 3). In addition, Part 172 includes detailed regulations for shipping papers, marking, labels, and placards. This part in essence forms the hazard communication requirements. These requirements are important because they are intended to furnish essential information about the cargo to emergency response personnel if an accident occurs.

- **Shipping papers.** Most shipments of hazardous materials must be accompanied by shipping papers that describe the hazardous materials and contain a certification by the shipper that the material is offered for transport in accordance with applicable DOT regulations. For most shipments, the DOT does not specify the use of a particular document and the information can be provided on a waybill or similar document is a sample shipping document. The exceptions are hazardous waste shipments, which must be accompanied by a specific document called the Hazardous Waste Manifest. A manifest lists EPA identification numbers of the shipper, carrier, and the designated treatment, storage, or disposal facility, in addition to the standard information required by the DOT.
- **Marking.** The DOT has established marking requirements for packages, freight containers, and transport vehicles. Shippers are required to mark all packages with a capacity of 110 gallons or less with proper shipping name of the hazardous material, including its UN/NA identification number. This is done so that the contents of a package can be identified if it is separated from its papers. Requirements for intermodal portable tanks, highway cargo

(text continued on page 512)

Table 3
Department of Transportation Hazard Classes

Hazard Class	Definition	Examples
Flammable liquid	Any liquid having a flash point below 100°F as determined by tests listed in 49 CFR 173.115(d). Exceptions are listed in 49 CFR 173.115(a).	Ethyl alcohol, gasoline, acetone, benzene, dimethyl sulfide.
Combustible liquid	Any liquid having a flash point at or above 100° and below 200°F as determined by tests listed in 49 CFR 173.115(d). Exceptions are listed in 49 CFR 173.115(b).	Ink, methyl amyl ketone, fuel oil
Flammable solid	Any solid material, other than an explosive, liable to cause fires through friction or retained heat from manufacturing or processing, or which can be ignited readily creating a serious transportation hazard because it burns vigorously and persistently (49 CFR 173.150).	Nitrocellulose (film), phosphorus, charcoal
Oxidizer	A substance such as chlorate, permanganate, inorganic peroxide, or a nitrate, that yield oxygen readily to stimulate the combustion of organic matter (49 CFR 173.151).	Potassium bromate, hydrogen peroxide solution, chromic acid
Organic peroxide	An organic compound containing the bivalent -O-O- structure and which may be considered a derivative of hydrogen peroxide where one or more of the hydrogen atoms have been replaced by organic radicals. Exceptions are listed in 49 CFR 173.151(a).	Urea peroxide, benzoyl peroxide
Corrosive	Liquid or solid that causes visible destruction or irreversible alterations in human skin tissue at the site of contact. Liquids that severely corrode steel are included (49 CFR 173.240(a)).	Bromine, soda lime, hydrochloric acid, sodium hydroxide solution
Flammable gas	A compressed gas, as defined in 49 CFR 173.300(a), that meets certain flammability requirements (49 CFR 173.300(b)).	Butadiene, engine starting fluid, hydrogen, liquefied petroleum gas
Nonflammable gas	A compressed gas other than a flammable gas.	Chlorine, xenon, neon, anhydrous ammonia

(table continues)

Table 3 (continued)

Hazard Class	Definition	Examples
Irritating material	A liquid or solid substance which on contact with fire or when exposed to air gives off dangerous or intensely irritating fumes. Poison A materials excluded (49 CFR 173.381).	Tear gas, monochloroacetone
Poison A	Extremely dangerous poison gases or liquids belong to this class. Very small amounts of these gases or vapors of these liquids, mixed with air, are dangerous to life (49 CFR 173.326).	Hydrocyanic acid, bromoacetone, nitric oxide, phosgene
Poison B	Substances, liquids, or solids (including pastes and semi-solids), other than Poison A or irritating materials, that are known to be toxic to humans. In the absence of adequate data on human toxicity, materials are presumed to be toxic to humans if they are toxic to laboratory animals exposed under specified conditions (49 CFR 173.343).	Phenol, nitroaniline, parathion, cyanide, mercury-based pesticides, disinfectants
Etiologic agents	A viable micro-organism, or its toxin, which causes or may cause human disease. These materials are limited to agents listed by the Department of Health and Human Services (49 CFR 173.386, 42 CFR 72.3).	Vibrio cholerae, clostridium botulinum, polio virus, salmonella, all serotypes
Radioactive material	A material that spontaneously emits ionizing radiation having a specific activity greater than 0.002 microcuries per gram (μCi/g). Further classifications are made within this category according to levels of radioactivity (49 CFR 173, subpart I).	Thorium nitrate, uranium hexafluoride
Explosive	Any chemical compound, mixture, or device, the primary or common purpose of which is to function by explosion, unless such compound, mixture, or device is otherwise classified (49 CFR 173.50).	
	Explosives are divided into three subclasses:	
	Class A explosives are detonating explosives (49 CFR 173.53).	Jet thrust unit, explosive booster
	Class B explosives generally function by rapid combustion rather than detonation (49 CFR 173.88); and	Torpedo, propellant explosive

	Class C explosives are manufactured articles, such as small arms ammunition, that contain restricted quantities of Class A and/or Class B explosives, and certain types of fireworks (49 CFR 173.100).	Toy caps, trick matches, signal flare, fireworks
Blasting agent	A material designed for blasting, but so insensitive that there is very little probability of ignition during transport (49 CFR 173.114(a)).	Blasting cap
ORM (Other Regulated Materials)	Any material that does not meet the definition of the other hazard classes. ORMs are divided into five substances:	
	ORM-A is a material which has an anesthetic, irritating, noxious, toxic, or other similar property and can cause extreme annoyance or discomfort to passengers and crew in the event of leakage during transportation (49 CFR 173.500(a)(1)).	Trichloroethylene, carbon tetrachloride, ethylene dibromide, chloroform
	ORM-B is a material capable of causing significant damage to a transport vehicle or vessel if leaked. This class includes materials that may be corrosive to aluminum (49 CFR 173.500(a)(2)).	Calcium oxide, ferric chloride, potassium fluoride
	ORM-C is a material which has other inherent characteristics not described as an ORM-A or ORM-B, but which make it unsuitable for shipment unless properly identified and prepared for transportation. Each ORM-C material is specifically named in the Hazardous Materials Table in 49 CFR 172.101 (49 CFR 173.500(a)(3)).	Castor beans, cotton, inflatable life rafts
	ORM-D is a material such as a consumer commodity which, although otherwise subject to regulation, presents a limited hazard during transportation due to its form, quantity, and packaging (49 CFR 173.500(a)(4)).	Consumer commodity not otherwise specified, such as nail polish; small arms ammunition
	ORM-E is a material that is not included in any other hazard class, but is subject to the requirements of this subchapter. Materials in this class include hazardous wastes and hazardous substances (49 CFR 173.500(a)(5)).	Kepone, lead iodide, heptachlor, polychlorinated biphenyls

SOURCE: 49 CFR 172.101 and 173.

(text continued from page 508)

tanks, and rail tank cars specify that the UN/NA identification number be displayed on a placard or orange rectangular panel. Additional requirements are also specified in 49 CFR 172. 300-338.

- **Labels:** Labels are symbolic representations of the hazard associated with a particular material. Figure 4 contains some examples of DOT labels. They are required on most packages and must be printed on or affixed near the marked shipping name. The Hazardous Materials Table indicates which materials require labels. Shipments of limited quantities of certain hazardous materials may not require labeling. Additionally, some hazardous materials are exempt from labeling requirements according to 49 CFR 172.400.
- **Placards** are symbols that are placed on the ends and sides of motor vehicles, rail cars, and freight containers indicating the hazards of the cargo. Figure 5 shows examples of Hazardous Materials placards. Placarding is the joint responsibility of shippers and carriers. Placard designs and rules for providing and affixing are specified by DOT in 49 CFR 172.506-.514. DOT has developed placards required for each hazard class (see Table 4).

Part 173 indicates the types of packaging that may be used by shippers of hazardous materials. General shipping and packaging regulations are followed by more specific requirements for certain hazard classes. Hazard class definitions also are contained in Part 173, which indicates the types of packages authorized for each hazard class as well as regulations governing the reuse and reconditioning of packaging, and qualification, maintenance, and use requirements for rail tank cars, highway cargo tanks, intermodal portable tanks, and cylinders.

Part 174 prescribes regulations for rail transport. General operating, handling, and loading requirements are specified, as well as detailed requirements for certain hazard classes. Carriers are required to forward shipments of hazardous materials within 48 hours after acceptance of the material. The Federal Railroad Administration enforces these regulations.

Part 175 applies to passenger and cargo aircraft shipments of hazardous materials. The regulations include quantity limitation, loading, and handling requirements, and special requirements for certain hazard classes. The regulations also require that pilots be informed of any hazardous materials carried in an aircraft. Responsibility for the enforcement of hazardous materials regulations for the air mode lies with Federal Aviation Administration (FAA).

Part 176 addresses nonbulk transportation of hazardous materials by waterborne vessels. Requirements for accepting freight, handling, loading, and storage are prescribed. Coast Guard regulations for bulk shipments of hazardous materials are contained in Title 46 of the Code of Federal Regulations. The Coast Guard regulates bulk transport by water.

(text continued on page 516)

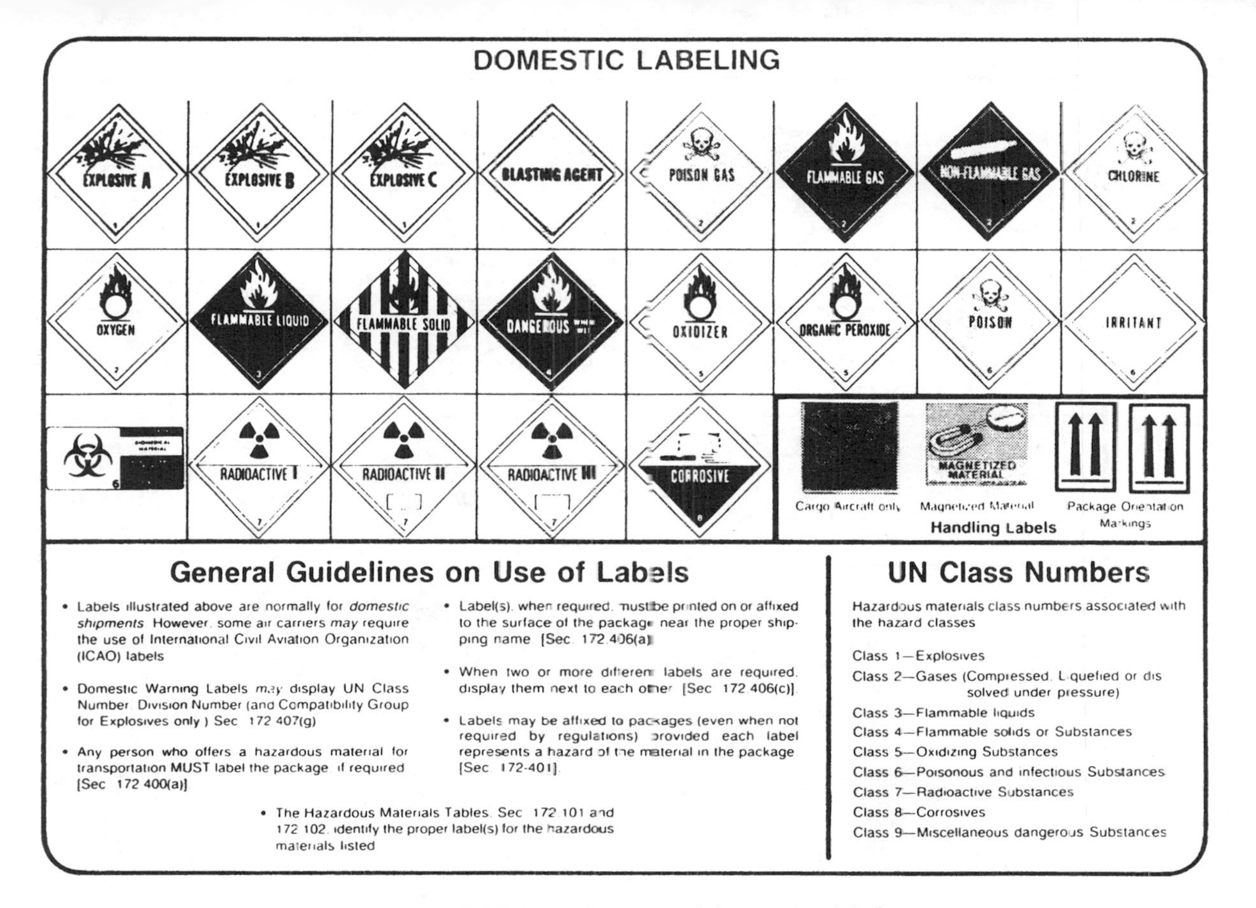

Figure 4. DOT hazardous materials warning labels.

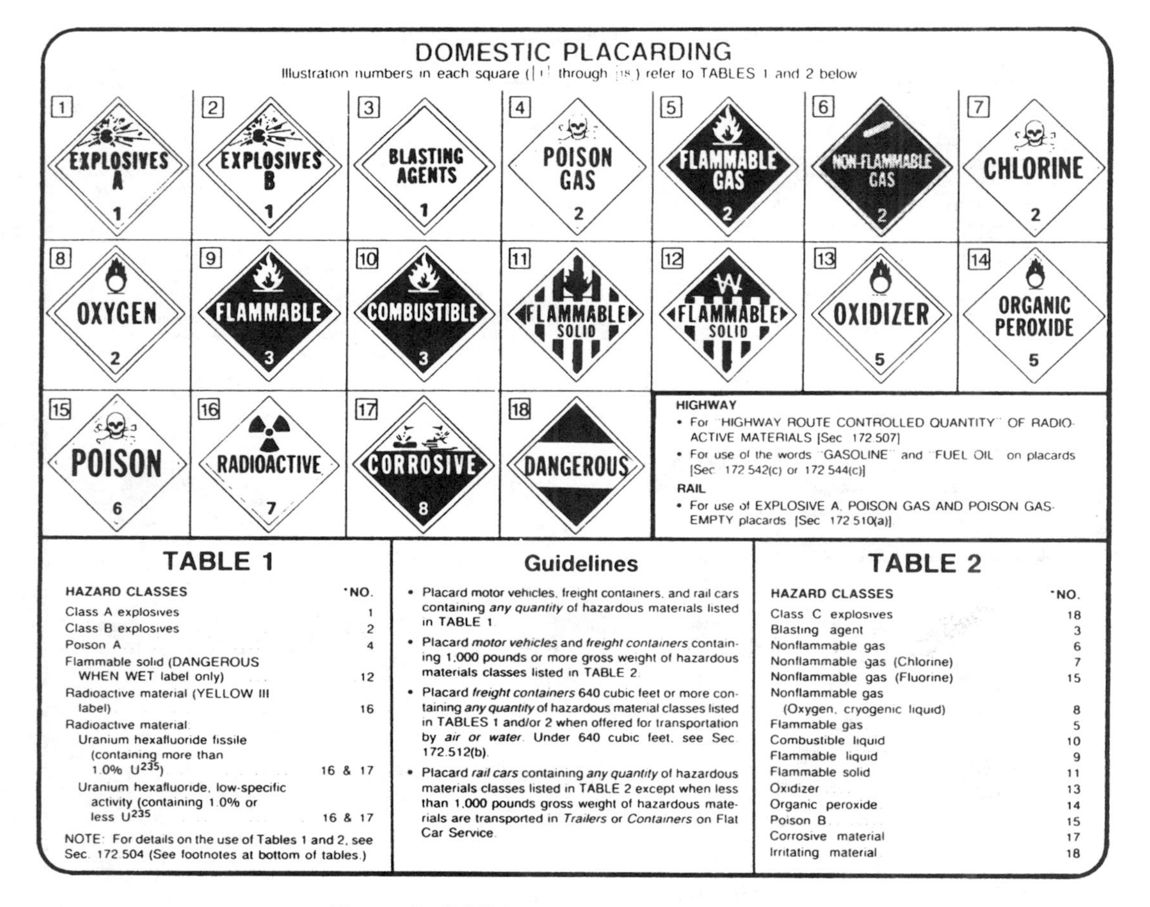

DOMESTIC PLACARDING

Illustration numbers in each square (1 through 18) refer to TABLES 1 and 2 below

HIGHWAY

- For "HIGHWAY ROUTE CONTROLLED QUANTITY" OF RADIOACTIVE MATERIALS [Sec 172.507]
- For use of the words "GASOLINE" and "FUEL OIL" on placards [Sec 172.542(c) or 172.544(c)]

RAIL

- For use of EXPLOSIVE A, POISON GAS AND POISON GAS-EMPTY placards [Sec 172.510(a)]

TABLE 1

HAZARD CLASSES	*NO.
Class A explosives	1
Class B explosives	2
Poison A	4
Flammable solid (DANGEROUS WHEN WET label only)	12
Radioactive material (YELLOW III label)	16
Radioactive material:	
Uranium hexafluoride fissile (containing more than 1.0% U^{235})	16 & 17
Uranium hexafluoride, low-specific activity (containing 1.0% or less U^{235}	16 & 17

NOTE: For details on the use of Tables 1 and 2, see Sec. 172.504 (See footnotes at bottom of tables.)

Guidelines

- Placard motor vehicles, freight containers, and rail cars containing *any quantity* of hazardous materials listed in TABLE 1.
- Placard *motor vehicles* and *freight containers* containing 1,000 pounds or more gross weight of hazardous materials classes listed in TABLE 2.
- Placard *freight containers* 640 cubic feet or more containing *any quantity* of hazardous material classes listed in TABLES 1 and/or 2 when offered for transportation by *air or water*. Under 640 cubic feet, see Sec. 172.512(b).
- Placard *rail cars* containing *any quantity* of hazardous materials classes listed in TABLE 2 except when less than 1,000 pounds gross weight of hazardous materials are transported in *Trailers* or *Containers* on Flat Car Service.

TABLE 2

HAZARD CLASSES	*NO.
Class C explosives	18
Blasting agent	3
Nonflammable gas	6
Nonflammable gas (Chlorine)	7
Nonflammable gas (Fluorine)	15
Nonflammable gas (Oxygen, cryogenic liquid)	8
Flammable gas	5
Combustible liquid	10
Flammable liquid	9
Flammable solid	11
Oxidizer	13
Organic peroxide	14
Poison B	15
Corrosive material	17
Irritating material	18

Figure 5. DOT hazardous materials warning placards.

Table 4

If the motor vehicle, rail car, or freight containter contains a material classed (described) as—	The motor vehicle, rail car, or freight container must be placarded on each side and each end—
Class A explosives	EXPLOSIVES A.
Class B explosives	EXPLOSIVES B.
Poison A	POISON GAS.
Flammable solid (DANGEROUS WHEN WET label only)	FLAMMABLE SOLID W.
Radioactive material	RADIOACTIVE.
Radioactive material:	
Uranium hexafluoride, fissile (containing more than 1.0 pct U^{239})	RADIOACTIVE AND CORROSIVE.
Uranium hexafluoride, low specific activity (containing 1.0 pct or less U^{239})	RADIOACTIVE AND CORROSIVE.
Class C explosives	DANGEROUS.
Blasting agents	BLASTING AGENTS.
Nonflammable gas	NONFLAMMABLE GAS.
Nonflammable gas (chlorine)	CHLORINE.
Nonflammable gas (fluorine)	POISON.
Nonflammable gas (oxygen, cryogenic liquid)	OXYGEN.
Flammable gas	FLAMMABLE GAS.
Combustible liquid	COMBUSTIBLE.
Flammable liquid	FLAMMABLE.
Flammable solid	FLAMMABLE SOLID.
Oxidizer	OXIDIZER.
Organic peroxide	ORGANIC PEROXIDE.
Poison B	POISON.
Corrosive material	CORROSIVE.
Irritating material	DANGEROUS.

(text continued from page 512)

Part 177 contains regulations for the highway mode that apply to common, contract, and private carriers. In addition to regulations for handling, loading, and storage, routing rules for high-level radioactive materials and other requirements are specified.

Part 178 presents detailed specifications for the fabrication and testing of packaging described in Part 173.

Part 179 prescribes detailed specifications for rail tank cars. Procedures for obtaining Association of American Railroad approval of new tank car designs or changes to existing ones are provided.

DOT Regulations for Asbestos

The DOT regulates the transport of asbestos in accordance with the provisions of the Hazardous Materials Transportation Act of 1975 (HMTA). Pursuant to its authority under HMTA, the DOT has designated asbestos as a hazardous material for purposes of transportation and has prescribed requirements for shipping papers, packaging, marking, labeling, and transport vehicle placarding applicable to the shipment and transportation of asbestos materials. In accordance with these requirements, commercial asbestos must be transported in

- rigid, leak-tight packaging,
- bags or other non-rigid packaging in closed freight containers, motor vehicles, or rail cars that are loaded by and for the exclusive use of the consignor and unloaded by the consignor, or
- bags or other nonrigid packaging that are dust- and sift-proof in strong outside fiberboard or wooden boxes.

Specific regulations have been promulgated for the transport of asbestos material by highway. Under these regulations, asbestos must be loaded, handled, and unloaded in a manner that will minimize occupational exposure to airborne asbestos particles released incident to transportation and any asbestos contamination of transport vehicles removed.

The Sanitary Food Transportation Act, passed by Congress in 1990, prohibits the transportation of asbestos in motor vehicles or rail vehicles that were used to transport food and other consumer products. Under this new law, the DOT is required to issue regulations prohibiting the use of motor or rail vehicles for other purposes if these vehicles provide transportation for asbestos wastes, extremely hazardous substances, or refuse.

Oil Pollution Act of 1990

In 1990, Congress enacted the Oil Pollution Act in response to the massive Exxon Valdex oil spill in Prince William Sound in March 1989. The new act mandates a comprehensive federal oil spill response system and establishes the Oil Spill Liability Trust fund to provide much greater federal resources for oil spill response and cleanup, prompt and adequate compensation for those harmed by oil spill, and an effective and consistent system of assigning liability.

The act also significantly strengthens requirements for the proper handling, storage, and transportation of oil, and for a full and prompt response in the event discharge occurs. The act made many of these regulatory changes, which have potentially significant effects far beyond the oil industry, by amending Section 311 of the Clean Water Act; many provisions of Title 46, which governs navigation safety; the Deepwater Ports Act; the Outer Continental Shelf Land Act (OCSLA); and the Trans-Alaska Pipeline Authorization Act (TAPAA).

Under Section 311 of the Clean Water Act, owners and operators of vessels and onshore and offshore facilities from which oil is discharged to navigable waters, shorelines, or into the exclusive economic zone are strictly responsible for the cost of conducting the cleanup of that spill and for natural resource damages suffered as a result of it. Section 311 prohibits discharges of oil in harmful quantities, which are defined as those which cause a sheen on the water. Section 311 of the Clean Water Act also governs the discharge of hazardous substances.

Hazardous Waste Regulations

Introduction

The transportation of hazardous wastes is regulated by both the Hazardous Materials Transportation Act (HMTA) and the Resource Conservation and Recovery Act (RCRA) of 1976 (see Table 5). Subtitle C of RCRA, which is administered by the EPA is the primary federal statute governing hazardous wastes. Although the regulatory program developed by the EPA is chiefly concerned with the disposal of hazardous wastes, Section 3003 of the RCRA directed the EPA to establish certain standards for transporters and to coordinate regulatory activities with the DOT.

Under Comprehensive Environmental Response Compensation and Liability Act of 1980 (CERCLA), the EPA can sue the potentially responsible parties, including transporters, for recovery of cleanup costs.

Resource Conservation and Recovery Act of 1976

The most significant and comprehensive hazardous waste legislation passed in the 1970s was the Resource Conservation and Recovery Act of 1976 (RCRA). RCRA was designed to provide a "cradle-to-grave" regulatory scheme to monitor hazardous wastes (see Figure 6). Under the act, the administrator of the EPA is empowered to promulgate regulations applicable to hazardous waste generators and transporters as well as to owners and operators of treatment storage and disposal facilities. RCRA recently has been amended to cover medical wastes

Table 5
EPA & DOT Hazardous Waste Transportation Regulations

Requirements	Agency	Code of Federal Regulations
Generator/Shipper		
1. Determine if waste is hazardous according to EPA listing criteria.	EPA	40 CFR 261 and 262.11
2. Notify EPA and obtain I.D. number; determine that transporter and designated treatment, storage or disposal facility have I.D. numbers	EPA	40 CFR 262.12
3. Identify and classify waste according to DOT Hazardous Materials Table and determine if waste is prohibited from certain modes of transport.	DOT	49 CFR 172.101 (also 49 CFR 173-177)
4. Comply with all packaging, marking, and labelling requirements.	EPA	40 CFR 262.32(b)
	DOT	49 CFR 173
		49 CFR 172, subpart D, and 49 CFR 172, subpart E
5. Determine whether additional shipping requirements must be met for the mode of transport used.	DOT	49 CFR 174-177
6. Complete a hazardous waste manifest.	EPA	40 CFR 262, subpart B
7. Provide appropriate placards to transporter.	DOT	49 CFR 172, subpart F
8. Comply with recordkeeping and reporting requirements	EPA	40 CFR 262, subpart D
Transporter/Carrier		
1. Notify EPA and obtain I.D. number	EPA	40 CFR 263.11
2. Verify that shipment is properly identified, packaged, marked, and labelled and is not leaking or damaged.	DOT	49 CFR 174-177
3. Apply appropriate placards.	DOT	49 CFR 172.506
4. Comply with all manifest requirements (eg., sign the manifest, carry the manifest, and obtain signature from next transporter or owner/operator of designated facility).	EPA	40 CFR 263.20
5. Comply with recordkeeping and reporting requirements.	EPA	40 CFR 263.22
6. Take appropriate action (including cleanup) in the event of a discharge and comply with DOT incident reporting requirements.	EPA	40 CFR 263.30-31
	DOT	49 CFR 171.15-17

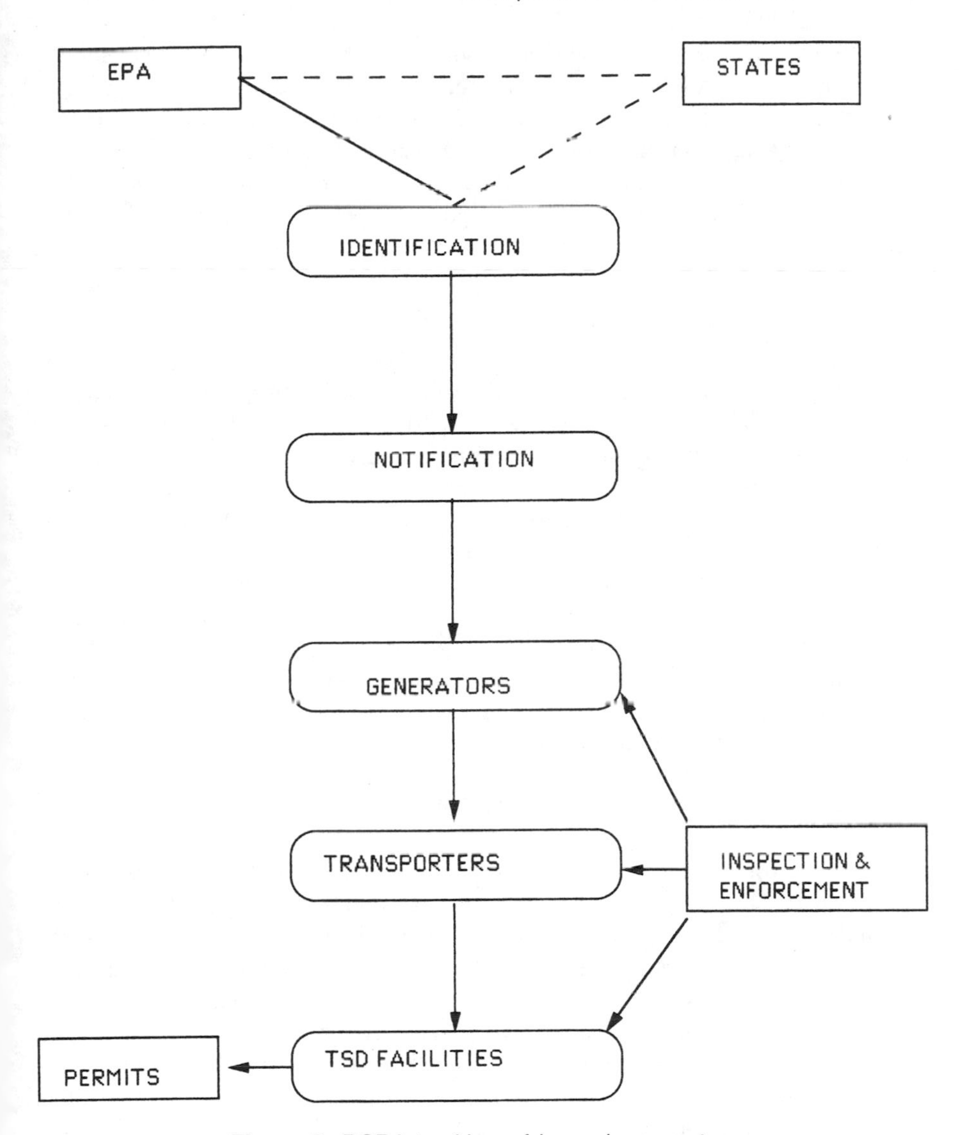

Figure 6. RCRA tracking of hazardous wastes.

generated in the diagnosis, treatment, or immunization of human beings or animals. The act now tracks medical wastes in all stages of generation, transportation, and disposal.

The standards applied under RCRA to transporters of hazardous wastes were established by the DOT under the Hazardous Materials Transportation Act. The EPA serves in a coordinating capacity.

The EPA's regulations for the transportation of hazardous waste generally follow on the manifest requirement and adopt the DOT's regulations for labeling, placarding, and packaging hazardous substances. Under these regulations, the transporter must be in possession of a manifest prepared by the generator and immediately respond to any spill by taking the appropriate measures, including reporting and cleaning up the discharge. Transporters also must comply with DOT regulations under the Motor Carrier Safety Act of 1984.

Transportation of Hazardous Wastes

A transporter is any person engaged in the off-site transportation of hazardous waste by air, rail, highway, or water. Off-site transportation includes both interstate and intrastate commerce, so the reach of RCRA includes not only shippers and common carriers of hazardous wastes, but also the company that occasionally transports hazardous wastes on its trucks solely within its home state.

Anyone who moves a hazardous waste that is required to be manifested off the site where it is being treated, stored, and disposed of will be subject to the transporter standards. The only persons not covered are generators or operators of transport, storage, and disposal (TSD) facilities who engage in on-site transportation of their hazardous wastes. Once a generator or a TSD facility operator moves its hazardous waste off-site—which can be any distance along a public road, he is then considered a transporter and must comply with these regulations. Thus, more than 500,000 companies and individuals in the United States who generate more than 172 million tons of hazardous waste must comply with this RCRA regulatory program.

EPA has promulgated standards for all transporters of hazardous wastes as 40 CFR part 263. These standards are closely coordinated with the standards issued by the DOT under the Hazardous Materials Transportation Act for the shipment of hazardous materials. For the most part, the EPA's regulations incorporate and require compliance with the DOT provisions on labeling, marking, placarding, using proper containers, and reporting spills. All transporters must obtain an EPA identification number prior to transporting any hazardous waste, and they may only accept hazardous waste that is accompanied by a manifest signed by the generator.

The transporter himself must sign and date the manifest acknowledging acceptance of the waste and return one copy to the generator before leaving the generator's property.

At all times, the transporter must keep the manifest with the hazardous waste. When the transporter delivers the waste to another transporter or to the designated TSD facility, he must

1. Date the manifest and obtain the signature of the next transporter or the TSD facility operator,

2. Retain one copy of the manifest for his own records, and
3. Give the remaining copies to the person receiving the waste (see Figure 7).

If the transporter is unable to deliver the waste in accordance with the manifest, he must contact the generator for further instructions and revise the manifest

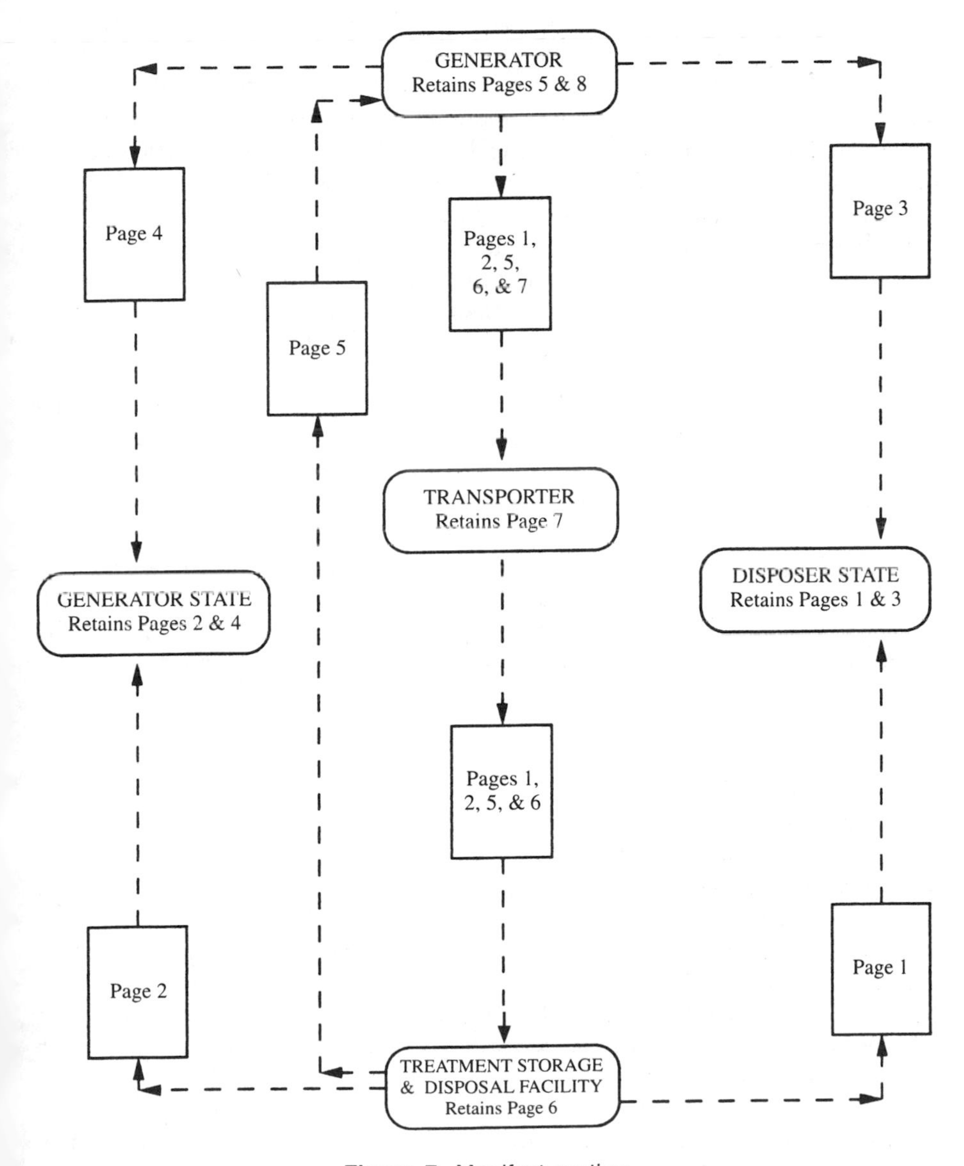

Figure 7. Manifest routing.

accordingly. The transporter must keep the executed copy of the manifest for three years.

Special requirements apply to rail or water transporters of hazardous waste and those who transport hazardous waste outside of the United States.

The transporter may hold a hazardous waste for up to ten days at a transfer facility without being required to obtain an RCRA storage permit. A transfer facility generally includes a loading dock, storage area, and similar areas where shipments of hazardous waste are held during the normal course of transportation.

Transporters of hazardous wastes may become subject to the Part 262 requirements for generators if, for example, the transporter mixes hazardous wastes of different DOT descriptions by placing them into a single container, or if he imports hazardous waste from a foreign country. Also, a hazardous waste that accumulates in a transport vehicle or vessel will trigger the generator standards when the waste is removed. If an accidental spill or other discharge of a hazardous waste occurs during transportation, the transporter is responsible for its cleanup. The transporter must take immediate response action to protect human health and the environment. Such action includes treatment or containment of the spill and notification of local police and fire departments. The DOT's discharge reporting requirements are incorporated into the RCRA regulations. They identify the situations in which telephone reporting of the discharge to the National Response Center and filing of a written report are required. Transporters are subject to both DOT and EPA enforcement.

The 1984 Hazardous Solid Waste Act (HSWA) affected transporters in minor respects. For example, the EPA has established requirements for the transportation of hazardous waste-derived fuels. In addition, railroads are shielded from the RCRA "Citizen Suit" and "imminent hazard" enforcement provisions if the railroad merely transports the hazardous waste under a sole contractual agreement and exercises due care.

Inspection and Enforcement

The RCRA provides that any officer, employee or representative of the EPA or a state with an authorized hazardous waste program may inspect the premises and records of any person who generators, stores, treats, transports, disposes of, or otherwise handles hazardous waste.

A manifest must identify the transporter. Any person knowingly omitting material information from a manifest, application, label, record, report, permit, or other compliance document to be filed with the EPA or a state agency can be criminally prosecuted by the federal government. Transporting hazardous waste without a manifest is federal crime.

The RCRA authorizes imposition of civil penalties of up to $25,000 per day, as well as criminal penalties of up to $50,000 per day. Criminal penalties may include imprisonment for as long as two years. Any "organization" knowingly placing any person in imminent danger of death or serious bodily injury will face penalties of up to $1 million.

The RCRA authorizes injunctive relief against potentially responsible parties whose wastes are no longer deposited at a particular waste site but who may be ordered to abate a hazard created by wastes deposited in the past. Such an abatement could cost a responsible party millions of dollars. The RCRA provides citizens' suits against responsible parties; it doesn't create a private cause of action for damages. But the RCRA does permit citizen enforcement of any standard, regulation, condition, or order promulgated under the authority of the RCRA.

Comprehensive Environmental Response Compensation and Liability Act (CERCLA)

In 1980, Congress enacted the Comprehensive Environmental Response Compensation and Liability Act of 1980 (CERCLA). CERCLA provides for the "establishment of a national inventory of inactive hazardous waste sites and a program for appropriate environmental response to protect the public health and the environment." This program, termed the National Contingency Plan, consists of "procedures and standards for responding to release of hazardous substances, pollutants, and contaminants."

Under CERCLA, Congress established the Superfund to facilitate the cleanup of inactive and uncontrolled waste sites. The president is authorized to use the Superfund to take whatever actions are necessary and consistent with the National Contingency Plan to protect the public health, welfare, or environment from a release or threatened release of hazardous substances. Costs incurred in taking such action may be recovered from the responsible parties.

Whenever there is release of hazardous substance in a reportable quantity, the owner or operator of a generation facility or transportation vessel must promptly report it to the National Response Center. "Release" is very broadly defined under CERCLA to include "any spilling, leaking, emitting, discharging, escaping, leaching, dumping, or disgorging into the environment." A transporter will not be liable under CERCLA for the cleanup of a site unless the transporter selected for site for delivery of the hazardous waste. The party selecting the site to receive the shipment will be a potentially responsible party.

The EPA has established reportable quantities by level of toxicity. A reportable quantity ranges from between 1 pound and 5,000 pounds, depending upon the toxicity of the substance. The regulation adopts an accumulation period of 24 hours to measure reportable quantities.

CERCLA targets its liability upon four categories of persons or entities.

1. Current owners or operators of hazardous waste facilities.
2. Past owners or operators of hazardous waste facilities.
3. Persons who arranged for treatment or disposal of waste at a facility.

4. Persons who transported hazardous waste substances for treatment or disposal at the facility.

As under the RCRA, liability under CERCLA does not require proof of intent or negligence. This standard is imposed without regard to whether the plaintiff is a governmental body or a private party seeking to recoup response costs from another potentially responsible party.

CERCLA is given retroactive effect in its application to generators, transporters, and any other potentially responsible party. However, courts may consider the potential harshness of the retroactive impact in fashioning relief under their broad equitable powers.

RESPONSIBILITIES OF DIFFERENT AGENCIES

Overview

Federal authority to enforce hazmat regulations is distributed among numerous federal agencies. Five of the agencies are within DOT—the Research and Special Programs Administration (RSPA) and four modal administrations: The U.S. Coast Guard, the Federal Aviation Administration (FAA), the Railroad Administration (FRA), and the Federal Highway Administration (FHWA). The other federal agencies are the Environmental Protection Agency (EPA), the Nuclear Regulatory Commission (NRC), and, peripherally, the Occupational Safety and Health Administration (OSHA), (see Tables 6 and 7).

The HMTA provides the DOT with the authority to impose both civil and criminal penalties against people who violate regulations. While RSPA is responsible for issuing the hazardous materials regulations under the act, it shares enforcement responsibilities with each of the DOT's modal administrations. RSPA's inspection and enforcement efforts are focused primarily on container manufacturers, reconditioners, retesters, and packaging exemption holders. The Coast Guard, with assistance from the National Cargo Bureau and the American Bureau of Shipping, conducts waterfront facility and vessel inspections. The FAA inspects freight at air carrier facilities, which serve as collection points for packages coming from freight forwarders and shippers. FRA has responsibility for rail shippers, carriers, and freight forwarder facilities. FRA also inspects railroad tank and freight cars as well as bulk container manufacturers. FHWA inspects motor carrier and shipper facilities in addition to roadside or terminal checks of motor vehicles.

Under a 1980 memorandum of understanding between the EPA and DOT, the EPA may bring an enforcement action involving a waste transporter if the transportation is ancillary to other activities normally under the EPA's jurisdiction, such as the storage or disposal of hazardous wastes. Additionally, EPA has agreed to make available to the DOT any information regarding possible Hazardous Materials Transportation Act violations.

Table 6
Federal Activities in Hazardous Materials Transportation

	DOT									
Regulation of	**RSPA**	**FHWA**	**FRA**	**FAA**	**USCG**	**FEMA**	**EPA**	**NRC**	**DOE**	**DOD**
Hazardous Materials	*				*			*	*	*
Containers	*	*	*	*	*			*	*	
Vehicles and Vessels		*	*	*	*					
Operators		*			*				*	*
Planning	*	*	*	*	*	*		*	*	*
Recordkeeping	*	*	*	*	*			*	*	*
Inspection	*	*	*	*	*			*	*	*
Enforcement	*	*	*	*	*			*	*	*
Training	*	*	*	*	*	*		*	*	*
Emergency Response					*	*		*	*	*

KEY: RSPA—Research and Special Programs Administrations
FHWA—Federal Highway Administration
FRA—Federal Railway Administration
FAA—Federal Aviation Administration
USCG—US Coast Guard
FEMA—Federal Emergency Management Agency
EPA—Environmental Protection Agency
NRC—Nuclear Regulatory Commission
DOE—Dept. of Energy
DOD—Dept. of Defense
**Represents activity*

Table 7
Federal and International Regulatory Framework for Transportation of Hazardous Materials

Standard-setting Body	Highway	Rail	Air	Water
Department of Transportation Administration				
Research and Special Programs Administration	*	*	*	*
Federal Highway Administration—Bureau of Motor Carrier Safety	*			
Federal Railroad Administration		*		
Federal Aviation Administration			*	
United States Coast Guard				*
Other federal agencies				
Environmental Protection Agency	*	*	*	*
Nuclear Regulatory Commission	*	*	*	*
Occupational Safety and Health Administration	*	*	*	*
International Organizations				
United Nations—Committee of Experts on the Transport of Dangerous Goods	*	*	*	*
International Atomic Energy Agency	*	*	*	*
International Civil Aviation Organization			*	
International Air Transport Association			*	
International Maritime Organization				*

Responsibility for regulating the transportation of radioactive materials is divided between the Nuclear Regulatory Commission (NRC) and the DOT. Under a memorandum of understanding, the NRC is responsible for the design and performance of packages used to transport high-level radioactive materials, and the DOT has regulatory authority over packages used to ship low-level radioactive materials. Inspection and enforcement authority is similarly divided, but the agencies have agreed to consult each other on the results of inspections when they are related to each other's requirements. Hazardous materials inspection authority in many states is divided among several agencies. Usually, the state police or highway patrol is charged with roadside inspections, and another agency, such as the DOT, has authority to conduct inspection of terminals. In addition, a special agency may be empowered to inspect carriers of radioactive materials.

Federal Agencies

Environmental Protection Agency

The Resource Conservation and Recovery Act of 1976 requires the EPA to set standards for transporters of hazardous wastes. These standards cover record keeping, labeling, and compliance with a manifest system for origination to disposal management of hazardous wastes. EPA regulations must be coordinated with the DOT and be consistent with DOT regulations concerning any materials subject to the Transportation Safety Act.

The Toxic Substances Control Act of 1976 authorized the EPA to regulate the manufacture, processing, and distribution in commerce of chemical mixtures and substances, the use or disposal of which may present an unreasonable risk of injury to health or to the environment. It doesn't authorize the EPA to regulate the transportation of such substances, but it does intend that there be coordination with agencies such as the DOT that could be used in achieving the act's goals.

The Clean Air Act of 1977 gives the EPA authority to respond to a hazardous substances incident if:

- The pollutant is identified as a hazardous substance,
- There is a discharge of or substantial threat of a discharge of a substance into U.S. waters, or
- The discharge or potential discharge involves such a quantity as to create an unacceptable threat to the environment or to the general public.

Many incidents that meet these criteria will occur during hazmat transportation. Once a discharge or a potential discharge to the environment exists, the EPA is authorized to respond. CERCLA extends response authorization to multimedia environmental releases of reportable quantities. In each regional EPA office, there are Regional Response Teams and an Environmental Response Team to mitigate spills.

Department of Transportation

The Transportation Safety Act authorized DOT to regulate the transportation of hazardous materials in commerce. The secretary of transportation may issue regulations applicable to both interstate and intrastate transportation of hazardous materials.

The Transportation Safety Act (HMTA) authorizes the secretary to:

- Establish and revise criteria for handling hazardous materials,
- Require hazardous materials carriers, shippers, and package container manufacturers to submit biannual registration statements,
- Grant exemptions to issued regulations,
- Inspect records and properties relative to packaging, containing, and transporting hazardous materials,
- Provide both civil and criminal penalties for violations of the HMTA or its regulations,
- Establish facilities and staff to evaluate risks and set up a central reporting and data system to facilitate hazardous materials emergency response, and
- Conduct a continuing review of all aspects of the transportation of hazardous materials to recommend steps to ensure safety and to prepare an annual report to Congress.

The DOT also is empowered, under various other pieces of legislation, to regulate the transport of hazardous materials by individual mode. For example, DOT regulation of rail transport of hazardous materials is provided by the Federal Railway Safety Act [45 usc 331 (1976)], and highway transport by the Department of Transportation Act [49 usc 1655 (e)(6)(c)(1976)] and under the Motor Carrier Act of 1980 (94 stat. 793), which requires motor carriers to carry liability coverage and to compensate injured parties for personal injuries and to replace and/or restore proper or natural resources destroyed in hazmat accidents. The minimum levels of financial responsibility for carriers were superseded by Section 406 of the Surface Transportation Assistance Act of 1982 (96 stat. 2097), which

- expanded the authority of the secretary of transportation to require minimum levels of financial responsibility for motor vehicles transporting hazardous materials, hazardous substances, or hazardous wastes by foreign carriers in the United States engaged in foreign carriers,
- required motor carriers domiciled in any contiguous foreign country to carry on board each vehicle it operates in the United States evidence of financial responsibility, and
- expanded the applicability of the minimum levels of financial responsibility requirements to include motor vehicles that have a gross vehicle weight rating (GVWR) of less than 10,000 lb when transporting certain hazardous materials.

In implementing the act, the secretary of transportation delegated authority to the Materials Transportation Bureau (MTB) in DOT's Research and Special Projects Administration (RSPA).

The MTB is primarily responsible for coordinating the efforts of the four modal administrations to provide comprehensive oversight to the hazardous materials industry. Its responsibilities include the development and issuance of regulations and exemptions that govern the transport of hazardous materials by all modes, with the exception of bulk transport by water. These regulations cover shipping and carrier operations, packaging and container specifications, hazardous materials definitions, and communication of hazards.

The MTB also is responsible for enforcement of regulations other than those applicable to a single mode of transport. Enforcement regulations applies to manufacturers, reconditioners, and retesters of DOT specification containers and to multimodal shippers of hazardous materials. The other modal administrations—U.S. Coast Guard (USCG), Federal Aviation Administration (FAA), Federal Highway Administration (FHWA) through its Bureau of Motor Carrier Safety (BMCS), and Federal Railroad Administration (FRA)—originate, issue, and enforce dedicated hazardous materials regulations for each of their respective modes of transportation. The MTB is responsible for pipeline commodity transport.

The MTB sponsors a number of training and education programs designed to foster compliance with hazardous materials transportation regulations. These programs cover training for rail, motor vehicle, and maritime operations relevant to the movement of hazardous materials, and develop hazmat training materials for both public and private sector parties. The DOT's Transportation Safety Institute in Oklahoma City routinely conducts in-depth training courses for government and industry personnel in all phases of hazmat transportation enforcement.

Coast Guard

The U.S. Coast Guard issues and enforces regulations related to the maritime transport of bulk shipments. It is responsible for federal activities required to mitigate oil and hazardous materials threats to the shoreline of the Great Lakes, navigable rivers, and coastal areas, and operates the National Response Center, located in Washington, D.C.

Established in 1974, the center receives initial reports of spills of hazardous materials and oil, as required under the provisions of the Clean Water Act, and provides facilities, communication, information, storage, and other needs for coordinating emergency response. The center's database contains information on the characteristics of some 900 chemicals and can be used to develop predictions of the behavior, movement, and human and environmental effects of chemicals. The center operates 24 hours a day and works as a cooperatively structured private-sector emergency center.

The USCG and EPA jointly share lead responsibility for on-site response. This responsibility is exercised locally by on-scene coordinators supplied by both agencies. These coordinators are empowered to redirect the response, if necessary—a rare occurrence.

Department of Energy

The U.S. Department of Energy (DOE) is the coordinating agency for the Interagency Radiological Assistance Plan (IRAP), which provides rapid and effective assistance in handing radiological incidents. The DOE is a participating agency along with the U.S. Nuclear Regulatory Commission (NRC) and about ten other agencies. To implement IRAP, the DOE has divided the nation into eight regions, each with a Regional Coordinating Office. DOE also operates the Joint Nuclear Coordinating Center in Albuquerque, New Mexico, which has special expertise of nuclear weapons.

Federal Emergency Management Agency

The Federal Emergency Management Agency (FEMA) is an independent federal agency. It was established in 1978 and has overall responsibility for coordinating federal response to disasters. But its role in the transportation of hazardous materials has not been fully clarified. FEMA operates the National Fire Academy, which provides training in fire service technology and in disaster planning for hazmat incidents.

Nuclear Regulatory Commission

The NRC is the designated agency to enforce nuclear materials regulations. The Nuclear Waste Transportation Safety Act of 1979 amended the Transportation Safety Act to:

- Define DOT and NRC responsibilities for nuclear materials shipment regulation, with the DOT primarily responsible,
- Require the DOT to develop a national plan for responding to radioactive materials transportation emergencies,
- Provide grants to states to study safety and logistics for nuclear materials shipments to disposal facilities, and
- Require the DOT to study and establish shipping standards for foreign nuclear materials received by the United States.

The NRC has divided the United States into five regions. An office for each region is available for assistance in an emergency.

National Transportation Safety Board

The National Transportation Safety Board (NTSB) investigates selected hazardous materials accidents and issues reports on them. It also makes recommendations to federal transportation agencies regarding technological and institutional measures that would enhance safety in all modes of hazmat transportation.

State and Local Government

State and local government responsibility for the safe movement of hazardous material is fragmented and widely varied. The following types of agencies within a state are responsible in some way for hazardous materials transport. They include departments of transportation, public service commissions, motor vehicle departments, state police or highway patrols, environmental agencies, civil defense agencies, health departments, and state fire marshals. At the local government level, involved agencies are police, fire, health, civil defense, and traffic engineering agencies.

International Agencies

International agencies such as the International Atomic Energy Agency, the International Civil Aviation Organization, the International Air Transport Association, and the International Maritime Organization are responsible for safe transportation of hazardous materials (see Table 8).

Industry

Association of American Railroads, Bureau of Explosives

The Bureau of Explosives (BOE) tests and classifies for transportation purposes those materials considered hazardous under government regulation and makes recommendations covering packaging, labeling, and handling. BOE works with member industrial groups as well as federal and military agencies, conducts safety audits of rail facilities and production and shipping sites, and assists railroad personnel in the handing of derailments where hazardous materials are involved.

Chemical Transportation Emergency Center

The Chemical Transportation Emergency Center (CHEMTREC) was established in 1971 by the Chemical Manufacturers Association. It is a toll-free telephone information hot-line that provides immediate and sustained communication with individuals at the site of a hazmat transportation emergency. The CHEMTREC database contains information on the hazardous properties of thousands of chemicals or chemical classes. It also has extensive information on hundreds of chemical companies, including their emergency response teams, and on ownership or leasing of tank cars, so it is likely to know whose product may be in which car.

On receipt of a call, usually from a carrier or emergency services personnel, the CHEMTREC communicator ascertains the location of the incident; the products involved; identity of manufacturer, shipper, carrier, and consignee; number of trucks or railcars; and conditions prevailing at the scene. The communicator provides emergency information about the specific hazards involved. CHEMTREC then notifies the shipper, who in turn assumes responsibility for the response. By using a teleconferencing bridge, CHEMTREC can simultaneously link all affected parties, including the National Response Center.

Since 1980, CHEMTREC officially has been recognized by the DOT as fulfilling the requirement of HMTA that such an emergency response center be created. CHEMTREC is required to notify U.S. Coast Guard National Response Center of significant hazardous materials transportation incidents. "Significant" is defined as those incidents that have caused or have potential for causing considerable harm to the public or the environment.

Chlorine Emergency Plan

CHLOREP, operated by the Chlorine Institute and composed of chlorine-producing companies, operates a number of regional emergency response teams in the United States and Canada. The member company closest to the scene of the accident will send a task force, regardless of whose product is involved.

REQUIREMENTS OF NEW JERSEY

Overview

The New Jersey Department of Transportation adopted regulations governing the transportation of hazardous materials by truck and rail, as mandated by the state legislature, that took effect on March 18, 1985. These rules and regulations are in substantial conformance with the federal requirements contained in Title 49 of the Code of Federal Regulations (CFR), Parts 100 to 199. Several additions and modifications to the original legislation were identified by NJDOT and the New Jersey State Police. These modifications were approved by state legislation effective January 1986.

New Jersey's Hazardous Materials Regulations were developed using a three-phase strategy. First, because hazardous material transportation involved a new area of expertise within the NJDOT, advanced technical training was undertaken to develop a working knowledge of the federal regulations. Second, a broad frame of reference was defined to set priorities and identify the problems that most likely would need to be addressed by a hazardous material safety program. Finally, a forum was established within the state that included agencies involved with or affected by the proposed action. All three phases occurred concurrently and still are active and available for response to specific needs.

Regulations

The state adopted the federal regulations, specifically 49 CFR Transportation, Parts 100 to 199, revised November 1, 1983.

Those sections include:

- Part 171, General Information, Regulations, and Definitions. Sections 171.15 and 171.16 were modified, and Sections 171.1, 171.4, 171.5, 171.10, and 171.20 were not adopted.
- Part 172, Hazardous Materials Table and Hazardous Materials Communications Regulations.
- Part 173, Shippers - General Requirements for Shipments and Packaging. Section 173.118(a) and Section 173.24 were modified; Section 173.32 was not adopted.
- Part 174, Carriage by Rail. Section 174.8 was omitted.
- Part 177, Carriage by Public Highway. Appendix A and Section 177.825(a), (b), (c), and (e) were excluded.
- Part 178, Shipping Container Specifications.
- Part 179, Specifications for Tank Cars. Sections 197.3, 197.4, and 199.5 were omitted.

These NJDOT regulations define the commodities that constitute a hazard to the general public and prescribe conditions under which they may be transported. When a commodity is defined within a particular hazard class, such as a Poison B or an Explosive A, shippers originating the material and carriers transporting the material must comply with specific requirements pertaining to shipping papers, packaging, labeling, marking, placarding, signed certification, loading and storage specifications, blocking and bracing requirements, etc. The regulations define standards that ensure that the materials remain safely contained while in transit. Also, the regulations demand that strict packaging and container requirements be met so that the public is adequately safeguarded.

New Jersey's Hazardous Materials Regulations are used in conjunction with the federal regulations. The intent is to create state requirements that are in substantial conformance with federal rules. State authorities are preempted by the federal government in regulating the transportation of hazardous materials. Uniformity is a key objective in national efforts aimed at cooperative enforcement and reciprocity. Uniformity also is desirable to achieve compliance from the motor carrier industry. Without endorsing nationally accepted standards, New Jersey or any other state would experience difficulty in achieving high levels of compliance.

Enforcement

The Office of Hazardous Materials Transportation Compliance and Enforcement of the New Jersey's State Police enforces New Jersey's Hazardous Materials Regulations. The state's current legislation authorizes the Port Authority of New York and New Jersey to enforce New Jersey's Hazardous Materials Regulations, and also authorizes NJDOT to inspect rail equipment used in the transportation of hazardous materials.

NJDOT also is involved with the creation of a database to monitor hazardous materials enforcement activities. The database will include violations records and will communicate with other information systems. These data will be used for impact analyses, and ultimately enforcement efforts will focus on problem locations such as high-incident sites and terminals with frequent or flagrant violations.

SAFETY AND ADVANCED TECHNOLOGY

The "Information Age" promises to have a dramatic impact on the transportation industry.

Applicable Technology

The risks involved in the transportation of hazardous materials arise primarily from release of such materials caused by accidents. A whole chain of events takes place before exposure to the effects of hazardous materials, i.e., the accident takes place, material is released, it spreads or explodes, and people are caught in the way. To combat this risk, the following can be done:

- Reduce the likelihood of the accident,
- Reduce the likelihood of a release in a given accident,
- Reduce the likelihood of severe consequences given a release, or
- Reduce the likely impact on people given a consequence.

The advanced technologies assist in each of these strategies to varying degrees.

Reducing the Likelihood of an Accident

There are various ways in which advanced technologies can help reduce the likelihood of an accident.

Vehicle Control

The technologies in this area assist the driver in controlling the vehicle. One of the most promising developments is a collision avoidance system. This involves tracking a vehicle in front using a radar or a laser. The distance between the two vehicles and their relative speeds are measured. If and imminent collision is expected, the system can alert the driver or apply the brakes automatically. Although these types of devices have been available for some time, they have not been perfected to the point that both false positive and false negative indications are completely eliminated. Also, there is the issue of liability, in which the manufacturer could face a product liability suit if the driver relies on such a system and gets into an accident.

The application of the device can be extended so that the vehicle is able to maintain a constant distance between itself and the vehicle in front through judicious use of the brakes and the cruise control system. This system can do wonders to reduce driver fatigue, which is the cause of accidents, when perfected.

Driver Information Systems

A variety of technologies are being developed to improve the quality of information being provided to the driver. One major group of technologies in this area informs the driver where he is in relationship to his destination. This requires finding the vehicle's location in an absolute term and superimposing his location on an electronic map to locate the vehicle with respect to the destination. There are five technologies that fit into category of automatic vehicle location.

1. **Dead-reckoning system:** The orientation and distance traveled are monitored, and given an initial position, the system can find vehicle's current location. An example of such a system is one made by ETAK.
2. **Ground-based-radio-determination system:** This system uses the principle of triangulation. An example is LORAN-C.
3. **Low-earth orbit satellite based system:** This system also employs triangulation. The only system in this category is Global Positioning System developed by Department of Defense. The satellites locate the position, and the calculations are performed on a vehicle-based device that is fairly expensive. The service not available everywhere.
4. **Radio determination satellite systems:** These systems use the distance from two satellites in geosynchronous orbit to determine location. The calculations are performed at a central facility, and the results are transmitted back to the truck.
5. **Proximity system:** In this system, fixed devices en route are used to determine the location of truck carrying transponders.

Also, databases are now available to display city maps on a video screen. These electronic maps are generally provided in the form of a compact disc. The present location of the truck can be displayed on such an electronic map to assist in driver navigation.

Driver/Vehicle Performance

Trucks now can be equipped with on-board computer and vehicle management systems to monitor the performance of trucks and drivers. Parameters such as hours on the road, vehicle speed, and time since last break can be measured and recorded. The vehicle maintenance parameters also can be measured and recorded. This information can be conveyed in real time to a dispatch center that monitors the performance of the driver and vehicle continuously, resulting in a safer operation.

Heavy Truck Operation

One potential cause of unsafe operation is a truck carrying more load than it should. Such overweight trucks can be detected using a weigh-in-motion system, along with an automatic vehicle classification system. If the trucks also are installed with transponders or the radio tags, the identity of the offending truck can be recorded and action taken to remove it from the road.

Reducing the Likelihood of Spills and Consequences

This includes design improvements in the tank car.

The consequences of some types of hazmat spills can be kept to a minimum if immediate and correct actions are taken. For this, the driver needs to communicate immediately with an emergency response center or with his dispatcher. Technologies that provide instant two-way communication can help under these circumstances.

Advanced technologies can reduce impact on people; the key to this is communication. The information on the spill of a toxic material and the need to evacuate people can be transferred rapidly from the truck to an emergency response center to the local police or fire department. A prompt evacuation then can be ordered before harmful vapors reach the population.

STATE LEGISLATIVE CONCERNS

A recent report to the U.S. Senate Committee on Commerce, Science and Transportation states, "There is overwhelming evidence that the Federal Highway Administration (FHWA), an arm of DOT, had not devoted the level of attention to hazardous materials transportation that many would judge necessary." For example, even though there are more than 1,200 motor carriers currently transporting hazardous materials that have received an unsatisfactory safety rating from the FHWA, FHWA has not used its authority to shut down any of these motor carriers.

Generally, states are concerned with the following four interrelated hazardous material transportation issues.

1. The complexity of existing hazardous material transportation regulations including the identification and definition of hazardous materials administered by varies federal agencies, the DOT, the EPA, and the NRC.
2. The lack of sufficient financial and manpower resources at the state and federal levels to ensure necessary levels of cargo inspection and enforcement of laws and regulations relating to hazmat transportation.

3. The lack of comprehensive and coordinated training programs and emergency response planning.
4. The preemption of state and local laws and regulations designed to ensure adequate protection of public health and safety, in light of the federal government's lack of financial manpower resources needed to effectively enforce federal regulations.

RESEARCH AND DEVELOPMENT EFFORTS

Research and development efforts should be focused toward the following aspects of hazardous materials transportation.

- **Performance Standards-Based Regulations:** Engineered design standards form the basis for many federal regulations and may inhibit innovative hardware design and development of new technology contributory to the safe transportation of hazardous materials. In addition, design specifications are quickly outmoded. The use of performance standards in government regulations would encourage industry to develop better technology to meet the standards.
- **Technological Data Base:** There is a need to pool all known technical data from both national and international sources to assess extent of data in the development of safety standards. Such assessment is necessary to performance standards.
- **Modification of Hazardous Materials:** Lowering the hazards posed by materials in transport by modification of the physical and chemical forms of the materials and, through packaging, to achieve safety represent an area requiring more research. The use of inhibitors, suppressants, neutralization, gelation, expanded metal mesh containment, and improved package design need to be studied from the perspective of preparing materials for transport, transportation itself, and in the event of release during an accident.
- **Evaluation of On-Scene Hazards:** Information available at the scene of a hazardous materials accident is critical to decision making. Information needed includes the state of integrity of the containment and rapid and clear identification of the materials involved.
- **Cargo Tank Safety Devices:** The safety devices on cargo tanks pose regulatory and design problems. Design improvements are needed for cargo tank piping, remote controls, fusible links, control cables, and other emergency systems. Greater reliability of these devices also are critical.
- **Hazard Classification Criteria:** Definitions, tests, and classification of hazardous materials are usually related to the in-plant use of the materials. Present safety regulatory criteria and schemes for classifying hazardous materials and wastes need to be examined for the adequacy of their relationship to the transportation environment. In addition, there is inconsistency

in the definitions and classification protocols used by the different federal regulatory agencies.

- **Protective Clothing and Personnel Equipment:** Emergency-response and cleanup teams need protective, full-body suits that are lightweight, comfortable, and that permit long working hours, allow full mobility and dexterity, and assure communications. Customized protective clothing to be worn in hazardous materials accident response and cleanup needs to be more effectively designed. Breathing apparatus that fits properly and allows work over a sustained period also is necessary.
- **Cargo Tank Truck Stability:** The on-road stability of cargo tank trucks is an issue in the truck transport of hazardous materials. Vehicle rollover is a major source of leakage. Regulatory vehicle performance standards that address the problem of the total vehicle design, as opposed to just tank design, are necessary. The effects of such vehicles on highway design and regulations also should be examined, as should the cost-risk benefit of design changes.
- **Pipelines:** Pipelines, currently used primarily for the transport of petroleum products and some anhydrous ammonia, may have some potential for the transportation of other hazardous materials and wastes. The technological, safety, and economic issues surrounding the potential use of pipelines for the transport of hazardous materials and wastes require more study.

RISK MANAGEMENT

All transportation of hazardous materials involves a certain degree of risk. The probability of death is the simplest definition of risk and is a measure closely correlated to injuries, property damage, and environmental degradation.

Transportation and management of hazardous materials can be conceptualized as shown in Figure 8. The probable consequences of hazardous materials release is shown in Figure 9. In this concept, it is assumed that the commodity has not been prohibited and is to be transported.

The means of controlling risk are evaluated and compared to the costs involved. The quality of the information on expected levels of compliance, extra costs incurred, cost of implementation, future accident rates, etc., makes this a difficult task. The result is a safety plan for risks. This plan is implemented, usually only partially, and the transportation takes place. There are two outcomes, either a safe trip or an accident. Given an accident, there may or may not be a release of the contents. Given a release of the contents and the resulting potential damage areas, there may or may not be population within the areas. Also, in the event of an accident, emergency response forces will be on the scene and modify the possible damage areas and the population within these areas. Again, there will be varying levels of readiness and performance of the response forces. There are many barriers to achieving a good safety plan, in particular implementing and maintaining the integrity of the plan. Lack of information, competitive

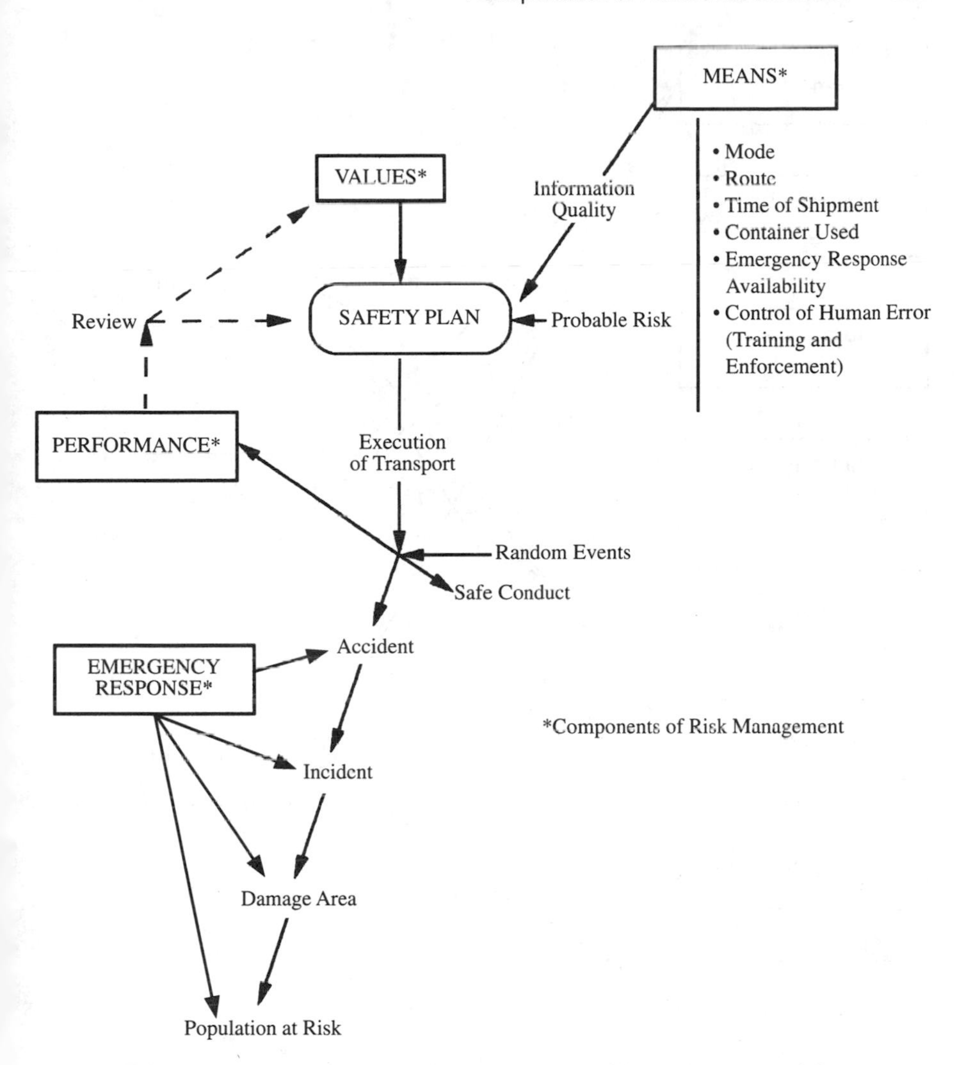

Figure 8. Transportation and management of hazardous materials.

forces within the transport sector, possibilities to expert risk to another country, and human native to relax precautions, if nothing happens for a while, are some examples of these barriers.

Approaches to Management

Figures 10 and 11 illustrate two approaches to managing the risk of dangerous good movements. In the traditional approach as shown in Figure 10, specific

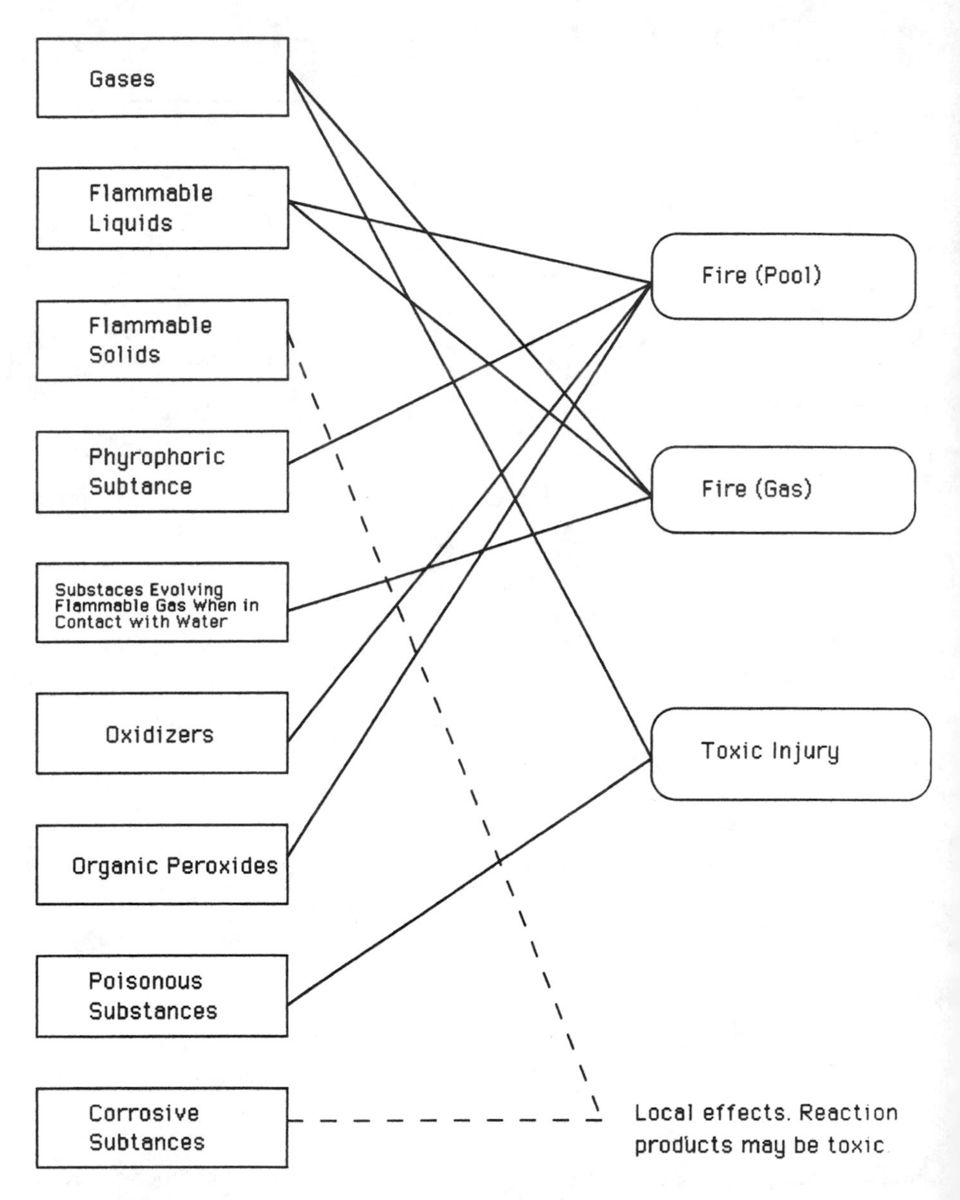

Figure 9. Most likely consequences of hazardous materials release.

regulation, enforcement, or training requirements are introduced usually in response to the disaster. Then, with the passage of time, these measures are evaluated by considering the improvements in the rate of incidents or the number of infractions, etc.

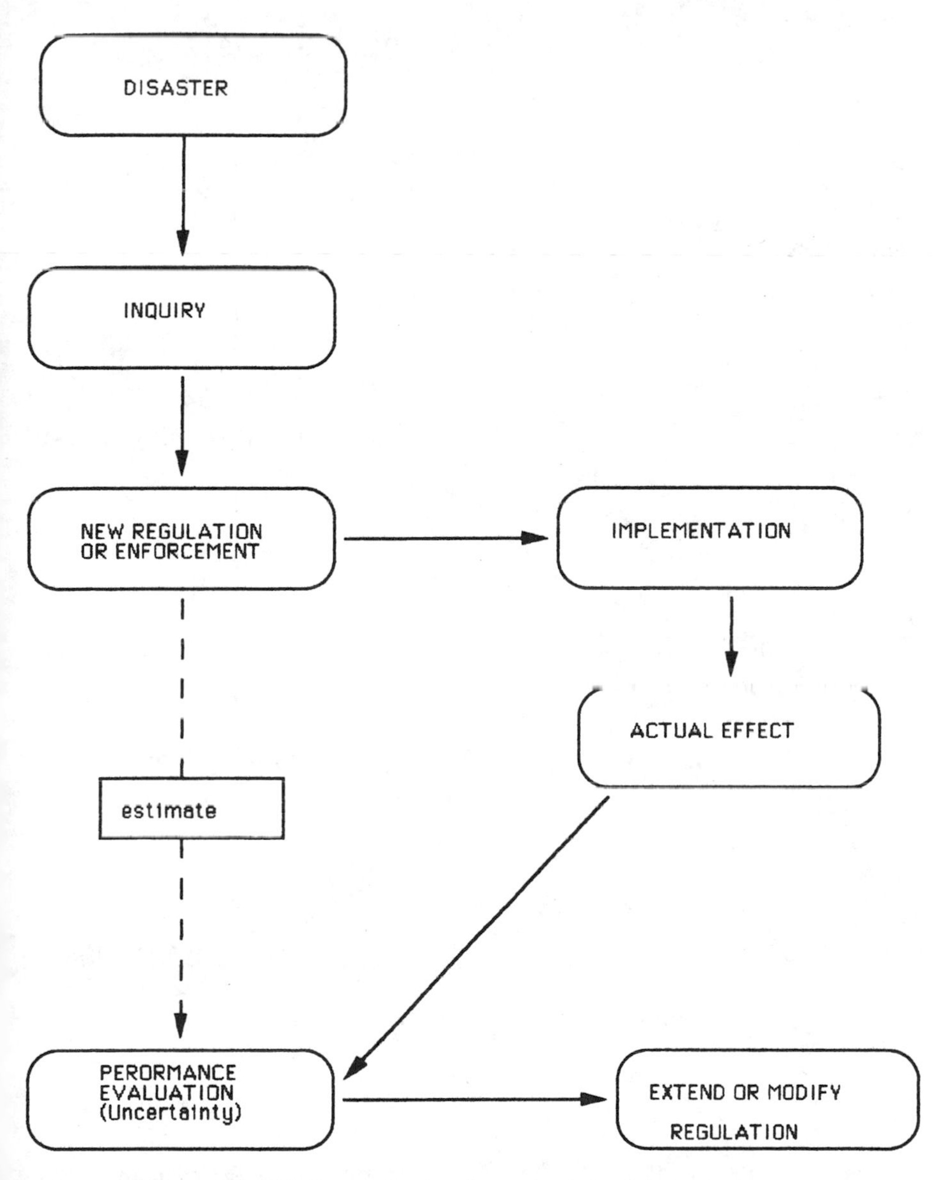

Figure 10. Existing methods of approach to risk management.

In the second scheme, as shown in Figure 11, involves government audit and approval for a total risk-management process and safety plan produced by the industry or transporter. The plan is evaluated on its rationality, conformance to man-machine capabilities, and likely achievement of acceptable risk levels. The

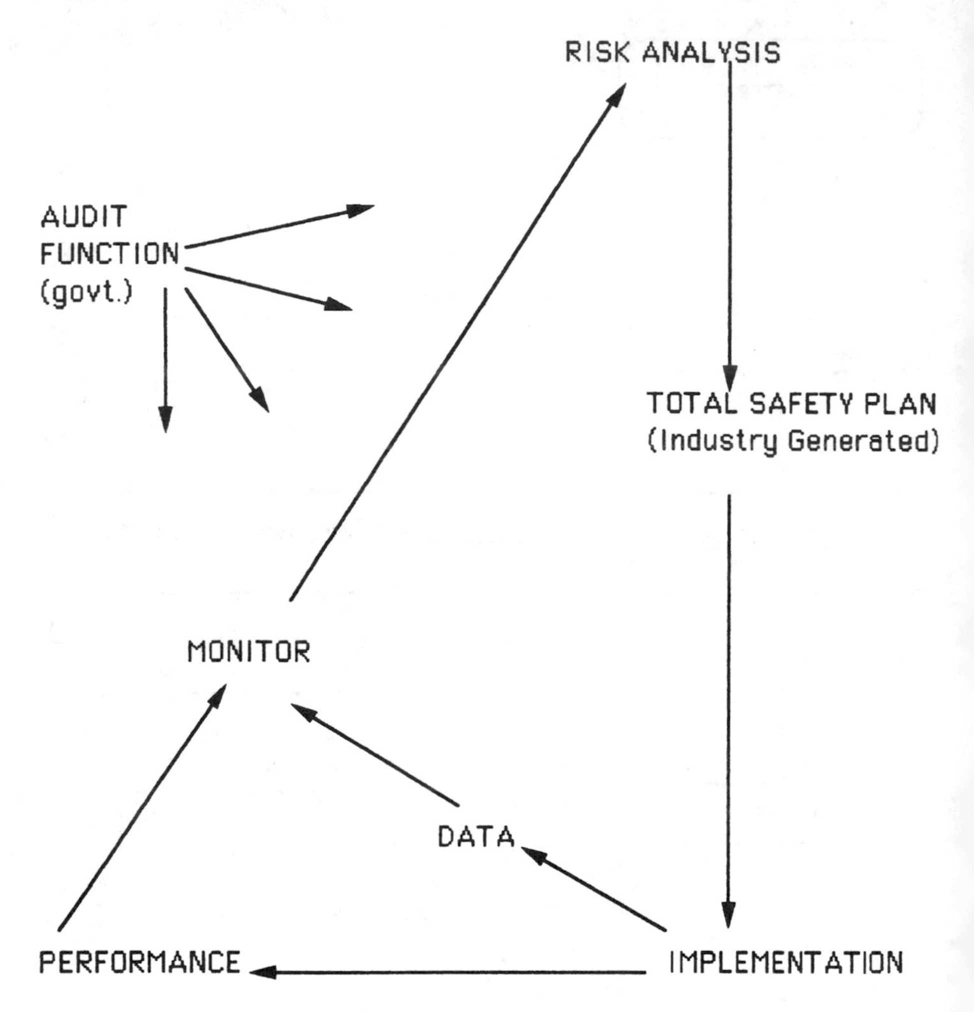

Figure 11. Industry safety plan approach.

plan also is judged in terms of the actual performance of the plan in action. Difficulties include how to deal with major events with very long return periods in measuring performance and public acceptance of the procedure given an incident.

The selection of a management approach should be evaluated in terms of the time required to change the system in response to new information or new techniques, the dislocations required in the existing decision process, and cost effectiveness of the risk reduction.

Risk Assessment

Much of the complexity of the dangerous goods problem arises from its tremendous scope, scale, and complexity. Numerous efforts have been addressed to isolate aspects of this larger problem in great detail, but these independent findings are often of limited collective benefits unless they are considered in terms of their system-wide role and significance. A tool for considering such system-wide significance is the pathway approach that can represent and enumerate the entire sequence of events.

The pathway approach is a simple but systematic procedure to establish a framework for analyzing important interrelationship and relative significance. The approach forces a comprehensive enumeration of the factors involved and points out any omissions. The method allows for the expertise of those involved on a daily basis to be input into the risk analysis. It also provides a basis for making balanced comparisons between the relevant factors, which in turn allow the relative strengths and weakness of the system to be identified. Finally, the pathway also serves as a means for evaluating past progress and identifying future research priorities. Two important pathway approaches to risk assessment are considered with examples.

Spill Pathway

This approach takes into consideration the consequence of a spill in a transportation accident.

The generalized scenarios following a spill is given below:

- Poisonous, toxic, flammable or explosive material endangers large numbers of trapped motorists, e.g., between interchanges, in a cut section, or in a traffic jam downwind of a poisonous or toxic gas release.
- A chemical spill of poisonous or explosive materials enters underground urban mass transit stations or transit tunnels through sidewalk vents, stairwells, etc.
- Hazardous materials accidents release toxic, flammable, or explosive materials in a tunnel.
- Gasoline, LPG, propane (flammable, explosive gases), etc., accidents cause releases on elevated facilities, including ramps, with people at risk below or in adjacent buildings.
- A release of poisonous, toxic, or explosive gases in populated areas in general or in locations and situations where special populations or institutions such as schools, hospitals, hostels, nursing homes, apartment complexes, etc., are at risk.
- Release from accidents between hazardous materials containers on highways and passenger trains or trains carrying hazardous cargo either at rail-highway

crossings at grade or in situations with shared right of way, such as freeways with transit in the median.

- A release of explosive materials from facilities in populated areas, and particularly in situations and areas where catastrophic consequences could occur to highway structures, adjacent housing, or air rights. Situations involving petrochemical plants could result in conflagration.
- Sufficient quantities of poisonous materials—such as herbicides or dangerous biological agents, or any material causing long-term or permanent damage—are released into a portable water supply, particularly reservoirs, aquifers, or watersheds.
- Spills of unknown wastes or nuclear materials, particularly in populated areas or in areas particularly difficult to respond to and clean up.
- Carriers of toxic, flammable, or explosive materials leak material during transit in heavily populated or congested areas.

Figure 12 shows the spill consequence pathway following ignition of a propane cloud after an accident. Pathway development is helpful in identifying the thermodynamic and physical relationships involved and in establishing the criteria to run an analytical model of this sequence of events.

As shown, the pathway considers what happens when a shipment is involved in an accident and a fire starts that heats the tank car. Tank car venting prior to rupture needs is examined, and the potential for ignition of the resulting vapor cloud is considered. Finally, the actual extent of shock wave is analyzed, the consequent damages are estimated, and cleanup or disruption costs are determined. The second step is to assist in each step in the pathway.

Following is an example of the criteria used for risk analysis.

I. Understanding of the process
   A. Analytical models
   B. Empirical data
II. Risk-Management Potential
   A. Pre-incident measure
   B. Emergency response
   C. Equipment
III. Developments
   A. Recent
   B. Expected in the future
IV. Research priority
   A. Potential benefit
   B. Cost effective

The model is developed using pathway analysis. To more clearly identify the data and analytical needs of the modeling approach, this pathway of events was analyzed based on the existing literature.

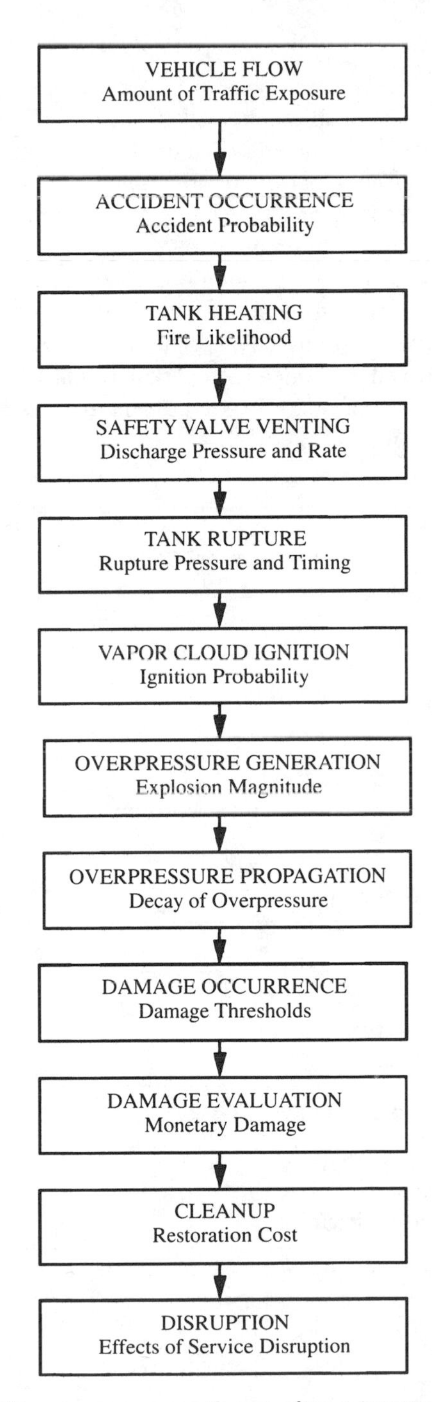

Figure 12. Spill consequence pathway after a transporation accident.

Shipment Pathway

The second pathway is the shipment path from the material source, the producer or bulk storage, to its ultimate destination, the consumer or final distributor. In between, the pathway attempts to identify the loading, control and transfer operations involved in transporting the dangerous commodity.

The descriptors in Table 8 attempt to capture those attributes that qualitatively describe the current operating conditions of each of the pathway elements. They were classified in terms of personnel, equipment, regulations, information exchange, and emergency response. Descriptors in Table 9 attempt to capture the potential for improved risk management in terms of a description of the frequency and magnitude of risks involved and the scope of possible risk-management improvements.

Table 8
Current Operating Conditions

Personnel
• training level • attitude • human error potential
Equipment
• design quality • maintenance/inspection • future probability
Regulations
• details/coverage • clarity and practicality • compliance
Information Exchange
• placarding/labelling • documents • advance notification
Emergency Response
• availability • response time • quality

Using the description of each of the pathway elements can be rated to evaluate the pathway as a whole as well as by its elements. Considering the pathway elements as one dimension and the element descriptors as another, a two-dimensional matrix can be constructed. The cells of this matrix can then indicate how a given pathway element rates in terms of a specific descriptor. Such rating systems are useful when expressed in terms of the qualitative rating system given in Table 10.

Table 9
Improvements in Risk Management

Risks
• Incident frequency • accident frequency • hazard/damage extent
Risk management
• potential for improvements • estimated cost-effectiveness • likely implementation success

Table 10
Qualitative Rating System of Pathway Elements

(1) Measures of quality.
(+) Acceptable (++) Very good/High (+++) Outstanding/Very High
(2) Measure of deficiency
(-) Marginal (--) Poor/low (---) Unacceptable/Very low
(3) Measure of Unknown.
(?) Unknown or Undecided (0) Irrelevant or not applicable.

An example of an application of the matrix and its rating is illustrated in Figure 13. It indicates the pathway scoring for each of the steps involved in chlorine shipments.

Hazardous Materials Routing

Routing is a tool available to reduce the risks associated with transportation of hazardous materials. Estimates of accident and release rates are essential in conducting risk assessments in routing studies for highway transportation of hazardous materials. The most widely accepted risk assessment model for identifying preferred routes for hazardous materials transportation is that presented in the DOT guidelines. This model was first presented in the 1980 FHWA publication *Guidelines for Applying Criteria to Designate Routes for Transporting Hazardous Materials* (Barber and Hindebrond, 1980). This document updated in 1989 and republished by the DOT Research and Special Programs Administration.

The DOT guidelines are based on selection of minimum risk routes, where risk is determined for individual route segments as:

$$\text{Risk} = \text{Accident Probability} \times \text{Accident Consequence}$$

The DOT guidelines contain procedures for determining accident risk on the basis of accident rates and route segment length, the number of people potentially exposed, and the value of property potentially exposed to hazardous materials releases. The FHWA route selection model is illustrated in Figure 14.

Evaluation of Hazardous Material Transportation by Rail

The Rail Hazardous Materials Routing System developed by ALK Associates, Inc., of Princeton, NJ, provides historical accident rate and population exposure statistics for rail routes under study for the purpose of transporting hazardous materials. This system is used by shippers, receivers and manufacturers of hazardous materials to evaluate current and proposed rail routings. In addition, the system suggests alternative routes that will minimize routing criteria such as accident rate, hazardous accident rate, route service ability, and population exposure.

The four major components of the Rail Hazardous Materials Routing System are the Princeton Transportation Network Model (PTNM), a computer-based transportation network and geographic information and mapping system; the yearly Federal Rail Administration Accident/Incident Files; aggregated traffic statistics from the Interstate Commerce Commission Carload waybill sample; and census information from the Department of Commerce. This system is used by shippers, receivers, and generators of hazardous wastes to evaluate current and proposed rail routes.

SHIPMENT PATHWAY	PERSONNEL			EQUIPMENT		
	Training Level	Attitude	Human Error Potential	Design Quality	Maintenance/ Inspection	Failure Probability
Source Storage Facility	+ + +	+ + +	+ +	+ +	+ +	+ + +
Loading Operation	+ + +	+ + +	+ +	+ +	+ +	+ +
Exit Control	+ +	+ + +	+ +	0	0	0
Transport Segment 1 (Rail)	0	0	0	0	0	0
Intermediate Transfer (Rail/Truck)	0	0	0	0	0	0
Exit Control	0	0	0	0	0	0
Transport Segment 2 (Truck)	0	0	0	0	0	0
Entrance Control	+ +	+ +	+	0	0	0
Unloading Operations	+	+ +	+	+	+ +	+
Destination Storage Facility	+	+ +	+	+ +	+ +	+ +

Figure 13. Example of a qualitative rating system matrix.

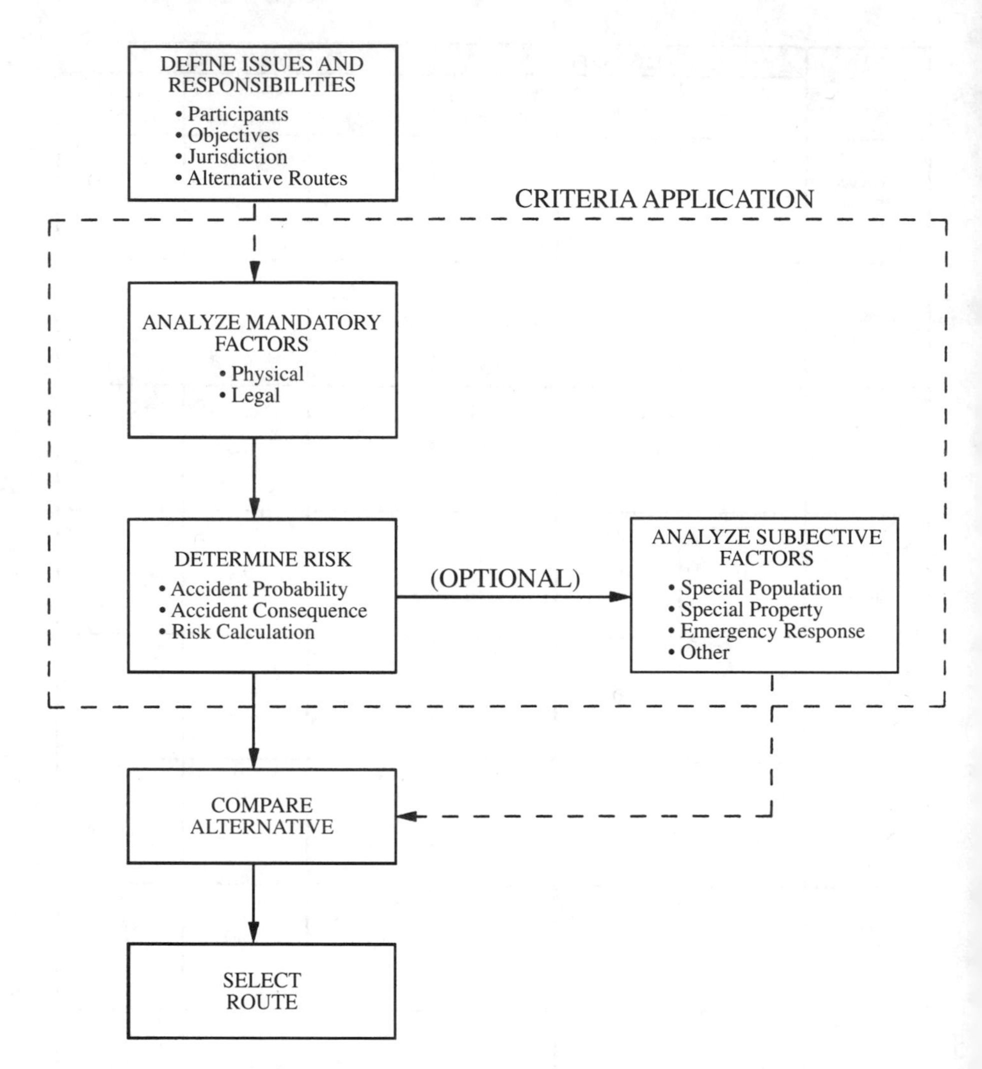

Figure 14. FHWA model of hazardous materials routing method.

CONCLUSION

Hazardous materials transportation is a large and growing segment of transportation industry. More than 1.5 billion tons of hazardous materials are transported annually in the United States. Special concern is addressed to safety in the transportation of hazardous materials because of the potential for fires, explosions, ground water contamination, and toxic effects on human health if hazardous materials are inadvertently released.

Most of the hazardous materials reach their destination safely because

- Manufacturers, shippers, and carriers are for the most part, aware of the dangers of the personal, property, and environmental damage and expense that an accident could cause, and they take appropriate precautions; and
- Hazardous materials transportation is heavily regulated by several government bodies.

The Hazardous Materials Transportation Act, passed in 1975, is the primary federal law regulating this transportation. The basic regulatory structure has been developed, largely by industry, during the past 100 years, and mostly before public awareness of the dangers of toxic substances and understanding of the complex measures necessary to protect public health and environment reached their present levels. There have been no far-reaching regulatory reforms and no strategic changes to help the system cope with late 20th century technologies and public awareness.

Highway transportation is a predominant part of the hazmat transportation safety problem, accounting far more than 85% of the hazmat releases reported to federal agencies. Hazardous material accidents usually are caused by inadequately trained personnel, poor coordination and communication, or lack of information and advance planning. This shows the need for training all people involved in the transportation of hazardous materials and setting national standards for truck drivers who transport hazardous materials.

A harmonious international, federal, state, and local regulatory control system is essential for the safe transportation of hazardous materials. Some of the regulatory policies of the federal agencies—such as DOT, EPA, DOE, FEMA, and NRC—overlap and conflict. As first step, federal agencies should consolidate their lists of hazardous materials waste and substances into a single comprehensive listing and consolidate federal programs.

Assessing hazmat transport risk involves selecting an appropriate risk estimation methodology. Risk associated with hazardous materials transportation can be reduced by stricter standards for containers, selecting highway and rail routes in which materials can be safely shipped, and by enforcing federal manufacturing, testing, repair, and reconditioning standards for hazmat containers and packaging. Enforcing standards for labeling, marking, placarding, designating, and classifying hazardous materials is necessary for safe transport. The responsibility of emergency response lies with state and local government. Prenotification of hazardous material transportation is helpful in emergency response.

Future technological innovations and the advent of the information age definitely will help in attaining the goal of safe transportation.

BIBLIOGRAPHY

Abkowitz, M., and List, G. (1987). "Hazardous Materials Transportation Incident-Accident Information Systems." *Transportation Research Record 1148*, 1–8.

Barkan, C., Glickman, T., Harvey, A. "Benefit-Cost Evaluation of Different Specification Tank Cars to Reduce the Risk of Transporting Environmentally Sensitive Chemicals." *Transportation Research Record 1313*, 33–43.

Code of Federal regulations (CFR), Title 49-Transportation, United States Government Printing Office, Washington, D.C., 1991.

Environmental Law Handbook, Government Institutes, Inc., Rockville, MD, 1991.

Harwood, D., Russell, E., and Viner, J. "Characteristics of Accidents and Incidents in Highway Transportation of Hazardous Material." *Transportation Research Record 1245*, 23–33.

Mintz, M., Bertram, K., Saricks, C., and Rowland, R. "Hazardous Materials Emergencies in Railyards: Preparedness Guidance for Railroads and Adjacent Communities." *Transportation Research Record 1313*, 44–47.

Risk Concepts in Dangerous Goods Transportation Regulations, National Transportation Safety Board, Washington, D.C., 1971.

Risk Management for Dangerous Goods, University of Waterloo Press, Ontario, Canada, 1986.

Sharp, J., Novack, R., and Anderson, M. "Purchasing Hazardous Waste Transportation Service: Federal Legal Considerations." *Transportational Journal Vol. 31/no. 2*, 4–14.

Soeteber, P. (1986), "Incidence, Regulation, and Movement of Hazardous Materials in New Jersey." *Transportation Research Record. 1063*, 8–14.

State and Local Issues in Transportation of Hazardous Waste Materials: Toward a National Strategy, American Society of Civil Engineers, New York, N.Y; 1990.

Transporting Hazardous Goods by Road, Organization for Economic Co-operation and Development, Paris, France, 1988.

Transportation of Hazardous Materials: Toward a National Strategy (Vols. 1 & 2), Transportation Research Board (TRB) Special Report 197, Washington, D.C., 1983.

U.S. Congress, Office of Technology Assessment, *Transportation of Hazardous Materials*, OTA-SET-304, US GPO, Washington, D.C., July 1986.

CHAPTER 23

RADON IN HOMES—HEALTH EFFECTS, MEASUREMENT AND CONTROL

Victor E. Archer

Rocky Mountain Center for Occupational and Env. Health
University of Utah Medical Center
Salt Lake City, UT 84112, USA

CONTENTS

INTRODUCTION

Our knowledge of the health effects of radon and its short-lived progeny has come primarily from the tragic experience of underground miners. Central European miners of silver, cobalt, and radium ores were found to have high rates of lung cancer in the 1870–1890 period [1,2]. As many as 70% of them died of the disease, which was later traced to high levels of radon and radon progeny in the air of the mines [3]. Little attention was paid to this observation until the United States began mining uranium to make nuclear bombs in the 1940s. At that time, the United States had no regulatory system to control exposures. The U.S. miners, as well as other miners around the world, have been well studied [4,5]. These studies have yielded good data on the risk of lung cancer from radon progeny exposure, based on mine measurements and long-term mortality studies of miners.

Figure 1 summarizes the exposure-response curves from two of the uranium miner studies and from a large French study of rats [6]. The exposure unit is in working level months (WLM), the unit used to express cumulative exposure to the short-lived radon progeny. One WLM is the exposure received during a

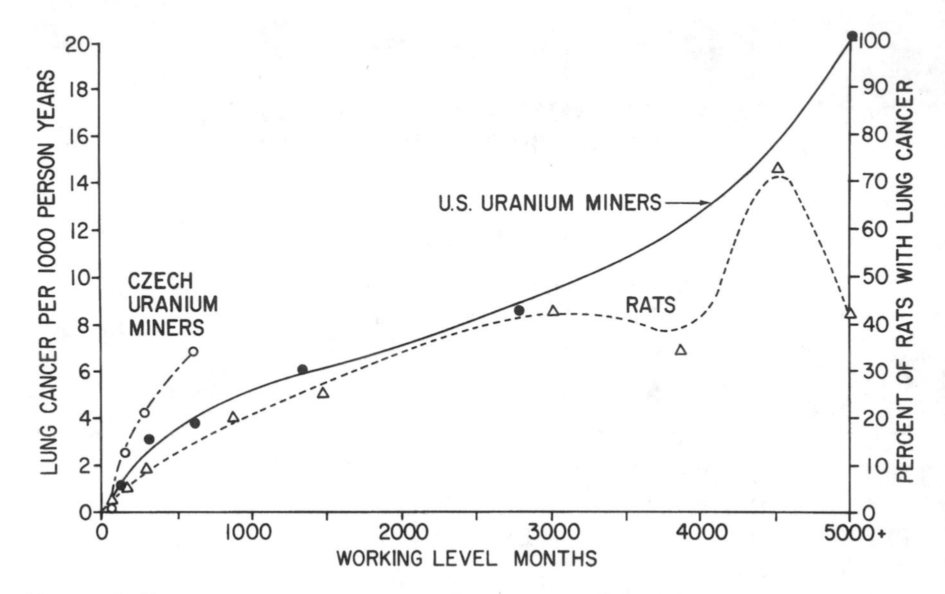

Figure 1. Exposure - response curves from two studies of uranium miners and for the large French study of rats. "Working Level Months" represents lifetime cumulative exposure of bronchi to alpha radiation from radon progeny.

month's work in a mine that has an average of 1.3×10^5 MEV/liter of potential alpha energy from short-lived radon progeny (1WL). Note the similarity of exposure-response curves for man and rats. The Czechoslovakian and other studies indicate a somewhat higher risk per WLM from radon progeny than that found in the first U.S. study. The human data extends down to about 40 WLM, and the rat data extends down to 20 WLM, where a cancer excess still is found [7,8]. The average exposure to radon progeny in homes is between 10 and 20 WLM per lifetime, so little extrapolation is needed from the miner and animal data.

Radon (^{222}Rn) is one of the decay products in the middle of the radioactive decay chain of uranium (^{238}U), the most abundant uranium isotope. All of the newly created elements other than radon in this decay chain are heavy metals, so they tend to stay in the place where they are created. But radon, from radium decay, is a noble gas. Because it is a gas, it tends to diffuse out of the place where it is created to collect in empty spaces such as mine drifts, caves, and basements. The next four elements in this radioactive decay chain are collectively known as short-lived radon progeny or daughters because they act as a unit, having a combined half-life of about 30 minutes. They are isotopes of metals (lead, bismuth, and polonium), so they tend to attach themselves to particles in the air. They are then inhaled as free ions or attached to particles. The radiation from radon progeny is delivered where these ions and particles are deposited in the lung. This is mostly alpha radiation, which can penetrate only about 70 μm of tissue. This distance, however, is sufficient so that most of the nuclei of cells of the bronchial epithelium are within range [9]. Most of the free ions and a large portion of the dust carrying radon progeny is deposited in the larger bronchi. (The electric charge on particles resulting from radioactive decay tends to collect clusters of polarized water molecules and aids in the deposition.) It then decays where deposited before the natural lung clearance mechanisms can move it very far. The rest of the body, however, receives only minimal exposures. There are other radon isotopes in other radioactive decay chains, but they are rarely of concern in homes.

Most of the radiation dose comes from alpha particles. When an alpha particle strikes the nucleus of a cell, the intense ionization usually breaks chromosomes or deletes genes. Such breaks and deletions are difficult for DNA mechanisms to properly repair, so they sometimes result in missing genes, cross-linking or other rearrangements. Such changes can represent a malignant transformation and can result in cancer manifested many years later (5 to 70 years). At the low levels of radon encountered in homes, no effect other than lung cancer is anticipated, although other effects have been suggested [10].

LUNG CANCER RISK

Because lifetime risk data from the studies of miners cannot be applied directly to the home radon situation, models have been created to extrapolate and calculate the lifetime risk to humans from radon progeny in their homes. The National Council of Radiation Protection and Measurement, the International Committee for Radiation Protection, and the National Research Council have created the most highly regarded models [1,2,11]. The Environmental Protection Administration (EPA) has used these models to determine the radon risk in homes [12,13]. It has leaned most heavily on the Committee on Biological Effects of Ionizing Radiation (BEIR IV) of the NAS-NRC [1]. Although different models used somewhat different assumptions, their results are surprisingly close. A recent reconsideration of the BEIR IV model resulted in a downward revision by 20% for children and 30% for adults, on the grounds that dust conditions are different and that people in homes are less active than miners, spending most of their time sleeping or sitting [14]. There is less air exchange and less radon progeny inhalation at such times. This adjustment, however, has been challenged [15]. All of these scientific bodies consider the radon level in homes a matter for concern and recommend that action be taken to reduce the level in those homes with the highest concentration.

Table 1 summarizes the radon risks calculated by EPA along with other comparable risks [16]. This table gives the radon level in homes in pCi/L of radon, an old term, as well as in Becquerels per cubic meter, the new SI term. The reason it is given in radon rather than radon progeny is that measurement of radon is much easier, less costly, and more constant than is the measurement of

Table 1
Relative Risks for Radon and Lung Cancer*

Radon Level			Lung Cancer		
	pCi/L	Bqm^{-3} [b]	Excess deaths/1,000 (Lifetime Exposure Risk)	Comparable Lung Cancer Risk	Other Comparable Death Risk
	More than 100	More than 3,700	300+	Uranium Miner of 1960s	
Max. permissible in mines now[a]	65	2,400	360	3–4 Packs/day of cigarettes for 50 years	
	20	740	135	1.5 Packs/day of cigarettes for 50 years	about 100 times the risk of drowning in U.S.
	10	370	71	Light Smoker	about 90 times the risk of dying in a home fire
EPA Action Level	4	150	29	150 Chest X-rays/year	about 100 times the risk of dying in an airplane crash
Average Indoor Level	1.3	48	9	Passive smoking	about the same risk as dying in a car crash
Average Outdoor	0.4	15	3	15 Chest X-Rays/year	

**Adapted from "A Citizens Guide to Radon," USEPA (reference 16).*
[a]The maximum permissible value for miners appears high because they are exposed only 40 hours per week.
[b]Bqm^{-3} is the SI unit of radioactivity which replaces the older unit "Curie per liter".

radon progeny. The relationship between the two is sufficiently constant to make the measurement of radon more practical. The advantage of measuring radon is greatest for long-term measurements, such as a year, which are needed to fully characterize the exposure in homes. Table 1 also gives some equivalents to make it easier to understand the risk involved. The EPA estimates that radon in homes causes about 14,000 deaths per year in the United States [16,17]. This is comparable to the risk from passive smoking, but far less than the risk from smoking a pack of cigarettes a day for 50 years.

HOME RADON HEALTH STUDIES

There have been a number of exploratory epidemiology studies done in an attempt to verify or refute the applicability of studies of miners and of animals to the home situation. These studies have all been plagued by the confounding effects of cigarette smoking and chemical air pollution. Most have been ecologic studies; some of these have given negative results, but others are consistent with the previously stated estimate of the radon effect [14,18]. The best studies have been of the case-control type, which measured radon in homes and controlled for cigarette smoking. The first of these was a Swedish study on an island where radon levels in homes were measured and found to be much higher on one side than on the other [19]. The higher radon levels were strongly associated with lung cancer, but the numbers were too small to be conclusive. A recent Swedish study is much stronger: The authors measured radon levels in 85% of residences between 1945 and five years prior to death for 210 women with lung cancer and in the similar period for residences of two sets of controls—one from hospital patients and one from the general population [20]. Geometric mean radon levels for homes of cases was 127.7 Bqm^{-3} vs. 96.0 Bqm^{-3} for controls. Lung cancer risks increased with estimated radon exposure reaching a relative risk of 5.2 ($p < .01$). The risk was 1.7 in women having an average radon level above 150 Bqm^{-3} (4 pCi/L). Smoking was considered by adjustment in the regression analysis. These risks were similar to the risks for radon in Table 1.

Another study was of lung cancer among women in New Jersey in which radon in residences of cases and controls was measured for about the latter third of their lives [21]. A consistent statistically significant relationship with radon was found among light smokers who developed lung cancer, but not among heavy smokers or among nonsmokers. The nonsmoking group would have been expected to give the best correlation, but confounding by passive smoking, by community air pollution, and by the inclusion of cases with metastatic disease to the lungs (rather than lung primary) may have led to the ambiguous result. A third case-control study was done in a Chinese city with very high lung cancer rates [22]. Lung cancer was found to be strongly associated with community air pollution, with pollution by open fires in homes, and with cigarette smoking, but not with radon in homes. This was a particularly bad choice of location to identify a radon effect because the small effect of radon could easily be obscured by the much larger effects of the other agents which were poorly quantitated.

Two of the ecological studies deserve comment: One is a Chinese study which was interpreted as negative [23]. It compared the lung cancer rates between two counties whose average radon levels differed by a factor of three. When the sexes were combined, the lung cancer rates for the two counties were about the same. But when examined by sex, the rate for women was doubled in the high-radon county whereas the male rate was higher in the low-radon county. Chinese women in this area have traditionally avoided tobacco smoking, but smoking by men is quite variable from one area to another, so the data from women appears to be more meaningful. Another ecological study that has gained some attention is one in which the average radon values were reported to vary inversely with the lung cancer rates [24]. The highest lung cancer rates are in urbanized areas where people are exposed to more community air pollution and who smoke more cigarettes. Average house radon levels are lower in urbanized areas [21]. The relationship found in this study may thus represent an association with population density rather than with radon. The author attempted to compensate for this difference, but it is unlikely that he was successful.

Because the case-control studies are not yet conclusive, there are some scientists who oppose government intervention in the control of radon for a variety of reasons. Their arguments, however, are not persuasive [25]. Most medical organizations that have reviewed the radon problem, such as the American Medical Association and the American Lung Association as well as the scientific groups noted above, have concluded that radon in some homes carries a sufficiently high risk that public action is warranted.

PREDICTION OF RADON LEVELS IN HOMES

Much effort has been put into the prediction of what the radon exposure in any given house might be. Geological approaches have found that radon is likely to be high in homes built on soil derived from rock that had higher-than-average concentrations of uranium and radium (detectible by aerial radiometric survey), if the water table is 20 or more feet below the surface, if the soil contains much coarse sand or gravel, if the house is over an old earthquake fault, if the house is built on a steep slope, or if the house is on a ridge [26,27]. Radon levels are likely to be low if soil conditions impede the flow of soil gas, e.g., if the water table is only a few feet from the surface near the bottom of a valley. Clay soils are tight and offer more resistance to the movement of radon and water than other soils, but clay has a tendency to crack when dry, such as under a house. Soil cracks are a ready conduit for radon. So the radon levels in homes built on clay soil are the hardest to predict. If no cracks exist, radon levels will likely be low; if there are deep cracks, the radon levels will likely be high.

Construction features are also of importance. Small houses with open crawl spaces beneath have the lowest radon levels. Mobile homes and apartments above the second floor also have low values because they have little ground contact. Large houses with full basements and a fireplace tend to have higher radon levels. Radon tends to be higher if basement walls are of cinderblock or stone rather than poured concrete. Basement rooms have higher radon levels than do upper-story rooms. Houses built on concrete slabs tend to have radon levels in their first floor similar to what is found in basements. Radon in homes is usually higher in winter than summer. Houses in the northern part of the United States tend to have higher radon levels than homes in southern states. The reason is that doors and windows are closed more days per year, and houses are tighter and tend to be tied more closely to the ground than in the south (e.g., more basements and fewer ventilated crawl spaces). Electrically heated homes tend to have higher and less-uniform radon levels than houses with central heating systems.

The lifestyle of the occupants also influences radon exposure. Sleeping with an open window or using a swamp cooler greatly reduces radon exposures. Daily exercise in a closed basement or closing crawl space vents during cold weather may greatly increase exposure. The location of a bedroom also may be of importance, i.e., radon will likely be higher in a basement bedroom than an upstairs bedroom. Living in a warm climate and using refrigerated air conditioning, keeping the windows closed during all seasons, may reverse the north-south trend in radon levels.

Although there is a tendency for a house to have a radon level similar to its neighbors, as in "radon-prone areas," this tendency is not strong enough to be reliable. Many houses with high and low radon levels are adjacent to each other. The only reliable way to know is to test for radon in each house.

RADON ENTRY

Radon, of course, can enter houses with outside air. Yet radon inside houses is generally higher than in outside air. Some indoor radon comes from building materials; from water obtained from drilled wells that had not been aerated; from natural gas when it is burned without a chimney, as in cooking; and from soil gas [28,29]. Soil gas moving upward from the soil beneath and around a house is by far the most important source of home radon [2,28,30]. Soil gas varies in its radon content, mainly because of soils' varying content of uranium and radium [31]. Radon is constantly being exhaled from the earth to the atmosphere. Enclosed spaces in contact with the soil always have higher radon levels than the ambient air. Because heated air tends to rise and contributes to a chimney effect for the interiors of houses, most houses have a slight negative pressure gradient near their contact with the soil. This chimney effect is enhanced by

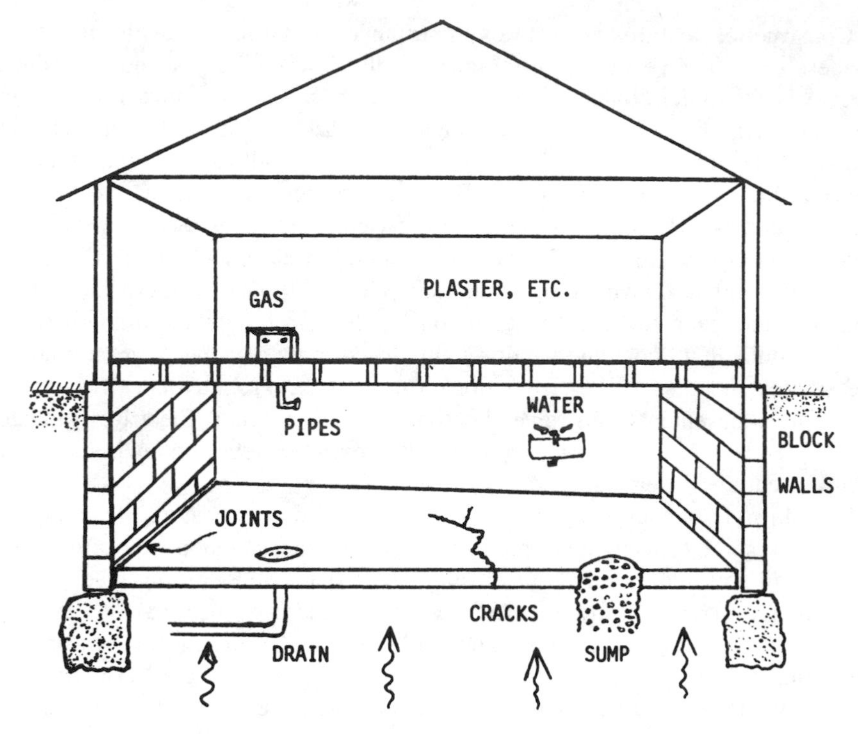

Figure 2. Sources and routes of radon entry into a house.

cracks or other openings, such as around an attic door or a fireplace damper in the upper parts of a house, exhaust fans, etc. Cracks to outside air near the basement floor, such as under an outside basement door, may minimize the negative pressure effect. This negative pressure pulls radon-carrying soil gas into the house through cracks and holes in basement walls and floors. A small amount of radon may diffuse through solid concrete. Changes in barometric pressure and outdoor winds also influence the movement of radon from soil into houses. When concrete sets, it always shrinks a little, so that even the highest quality concrete will have shrinkage cracks around its edges. Figure 2 illustrates radon sources and entry points.

MEASUREMENT

The EPA recommends that radon measurements in homes be done in three stages [16,32,33]. The first is a screening or short-term test of 1 to 30 days. It is usually done with a charcoal canister, but also may be done with short-term track-etch or electret detectors, with a continuous monitor, or with a charcoal-liquid scintillation detector [16,28,32,33]. This test is best taken under simulated "worst" conditions. That is, it is taken in the lowest living area, and the house is kept closed for the duration of the test. The dosimeter is placed in a relatively safe place where it will be undisturbed, like on a high shelf, or high on a wall, but away from drafts or heat sources such as furnace vents, washing machines, dishwashers, stoves, or direct sunlight from a window.

If this first test is near or above the EPA action level of 150 Bq m^3 (4 pCi/L) of radon, further testing is required. If below this level, the house has low radon levels, and no action is necessary, although some action may be prudent. A repeat of the initial test can be done to form the basis for action, but it is usually better to have the followup test be for a long term, 3 to 12 months, to average out the daily, weekly, and seasonal fluctuations in the radon level which make short-term measurements unreliable [30,34]. The long-term test is done with a track-etch or a long-term electret detector. It is placed in a prime living area such as the living room or bedroom—bathrooms and kitchens are not good places—in the spots previously suggested. The full 12 months is better than shorter time periods, but 3 to 9 months often will give acceptable precision. After the designated period, the detector is retrieved and returned to the manufacturer, who processes it and obtains a value translated by use of calibration curves into average concentration over the time exposed.

After remediation is accomplished, the long-term test should be repeated to be sure that the remediation was effective. This is called third-stage testing.

RADON CONTROL

There currently are no legal requirements regarding radon levels in U.S. homes, just the objective set by Congress in the Indoor Radon Abatement Act that targets a long-term goal of achieving indoor radon levels similar to ambient outdoor levels. However, there is a widespread move to incorporate radon-resistant techniques into building codes. The EPA "action level" of 150 Bqm^{-3} (4 pCi/L) is usually used for guidance [16]. Although this level is not a "safe level," it is an arbitrary level above which the lung cancer risk is high enough to warrant action by most prudent persons. The ALARA principle (as low as reasonably achievable), however, when applied to radon in homes implies that prudent action should sometimes be taken to reduce radon levels as far below 150 Bqm^{-3} as is possible without undue expense.

Radon in homes is best controlled during construction by using radon-resistant building techniques [35]. This consists of using poured concrete instead of cinderblocks or stones for all parts that are in contact with the ground and making sure that all cracks or gaps at joints, pipe penetrations, etc., are sealed with flexible caulking, such as catalyzing urethane rubber. In addition, outside walls in contact with soil and the underside of basement floors should be covered with a plastic vapor barrier with extensive overlap of all loose edges. In radon-prone areas [36] it also may be wise to install a subfloor ventilation system during construction. Crawl spaces must be well ventilated.

Radon Reduction in Existing Houses [28,30,32,33,37]

There are four fundamental mitigation methods. The first is source-material removal. If tailings from uranium or phosphate mills or other materials contaminated with uranium and radium are located beneath the house or were used as fill material around basement walls, then removal of these materials may be the first choice for mitigation. The second is the use of techniques or materials to increase the resistance the house offers to radon entry, generally referred to as sealing. The third tries to reduce the pressure gradient between the building and soil, generally referred to as soil ventilation. The fourth increases the removal rate of radon from the building, usually referred to as increasing ventilation.

The first step in any radon-mitigation effort is to identify radon entry points. This may be easy or difficult, depending on the house structure and condition. Some entry points such as French drains or sump holes may be obvious, but others, such as cracks or porous material like cinderblock, may be more difficult to identify. Few houses are built with airtight foundations, even though solid concrete is nearly impermeable to soil gas. There are usually many holes and cracks in the walls and floors of the basement. Gaps around pipes through which utilities (water, sewer, gas, oil, and electricity) come into the house are common entry points. Crawl spaces with a dirt floor are common in houses that have no basement or a partial basement. All such spaces must be ventilated with outdoor air, either with large openings or with fans.

If simple inspection does not reveal the likely routes of entry, then some diagnostic tests can be done. A "smoke gun" can be used to shoot a small amount of smoke into holes, drains, sumps, or along cracks, or even holes drilled for the purpose, and then watching the resultant air flow to show possible radon entry routes. Inserting a vacuum cleaner hose into a hole that goes through the basement wall or floor and sucking, along with use of the smoke placed strategically about the basement, may identify routes of entry.

When the basement walls and floor are of concrete (this also applies to slab-on-grade), and entry routes can be identified, sealing all openings and cracks may provide adequate remediation. Joints and other openings must be chiseled wide enough to allow a channel-type sealant to be inserted.

Preparation for the sealant involves removal of the surface layer of concrete on both sides of the joint or around an opening with a powered chipping gun, vacuum removal of the dust, followed by application of a concrete sealer. A catalyzing urethane rubber caulk is then applied liberally. This is a flexible sealant that allows some movement while maintaining the seal, but it does not produce new stresses that might produce new cracks. Other sealants such as epoxy will seal temporarily but are so rigid and strong that new cracks often appear at the sealant edges. Other sealants can be used but they must be flexible and have no acid components that react with concrete and may break the seal.

The elaborate preparation is needed because concrete surfaces are usually covered by a coating of laitance (a thin layer of cement with poor adhesion), and this is often painted. All paint and loose material must be removed to expose a fresh concrete surface before a reliable bond and effective closure of the opening can be obtained. The sealer penetrates the concrete pores and forms a thick skin of material on the surface that is firmly bonded to the concrete. The sealant can be used around pipes and drains that penetrate the basement walls and floor. Toilets can be lifted to expose the openings to be sealed.

Floor drains or sumps may not be connected to sanitary sewers but to the soil beneath the floor. These provide direct access for soil gas to enter. Such entries can be closed by replacing the drain or sump with one that incorporates a water trap and by connecting a priming line to a regularly used fixture (laundry machine, tub, or toilet) to maintain the water seal. It is not usually wise to put in a water trap with the expectation that a householder will regularly refill the trap. The junction between drain pipes and the edge of the sump crock or floor is often not airtight, and sealing is required.

A five-fold reduction of radon is often achievable by the careful sealing of identifiable entry routes. However, the occupant must remain alert to the possible development of new cracks or of cracks developing around the sealant. If the simple methods of ventilating crawl spaces and sealing of cracks and openings does not provide sufficient reduction in radon, then it probably means that radon is coming through concealed cracks, through mortar joints, or through the walls if they are of cinderblock or stone.

In this case, one must consider use of the third approach—reducing the pressure gradients. If the pressure is lower outside the basement walls and under the house than inside, then little radon can enter. The idea is to use a fan to draw radon-rich air away from the region around and under the house, thereby lowering the pressure immediately outside the house. There are three techniques for doing this.

1. Drain Tile Ventilation

Some houses have "weeping tile" placed all or partway around the house at a depth below the basement floor. It is designed to collect water and drain it away from the foundation. The collected water is either taken to a sump in the basement from which it is pumped away, or taken to an open low point outside the house. The perforated pipe itself might serve as a collection system and route of entry for radon, but it can be used to reduce radon. If the system empties into a sump, an airtight cover can be placed over the sump and a pipe can be run from beneath the cover to a fan located outside the house. This provides a double benefit, for not only is the suction transmitted to the weeping tile via the drainpipe, but any openings between the edge of the sump and the floor slab allow the fan to draw soil gas from beneath the floor slab. This widespread pressure reduction often is effective in reversing the soil gas flow over most of the basement area.

On sloping sites or if the weeping tiles are connected to a floor drain, it may be best to lay pipes so that collected water discharges to a lower level on the site, and, fan situated to exhaust air, with water traps located strategically (see Figure 3). In areas with cold winters, the fan must be mounted on the end of a long vertical riser to minimize condensation and freezing. Cost for this system

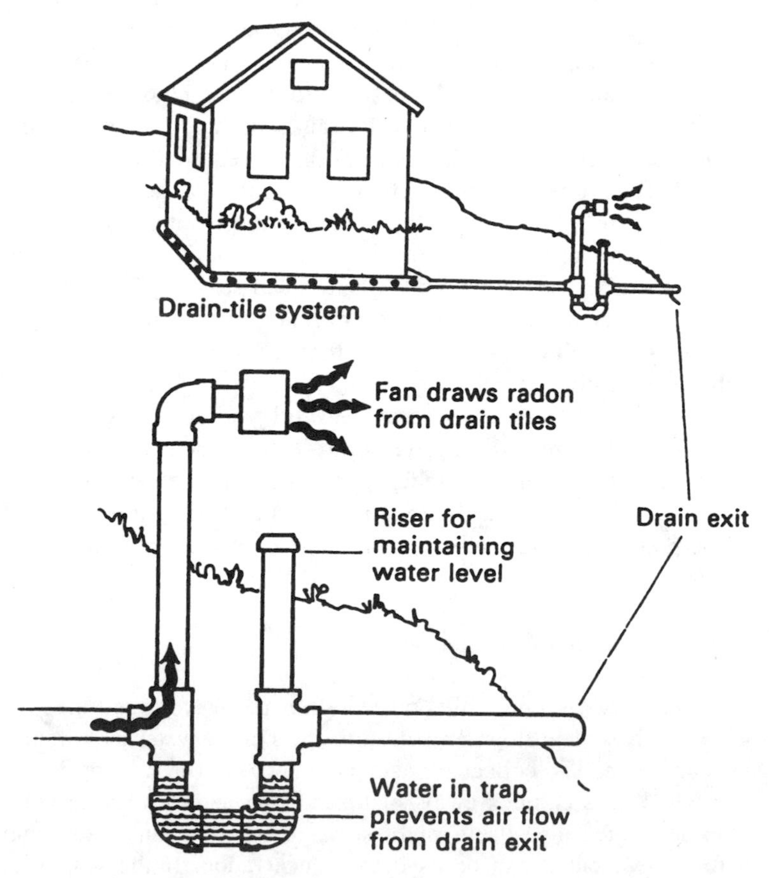

Figure 3. Radon removal via drain tile system. From Ref. 37.

can range from $250 to $1,000. Running costs would be about $100 per year. Radon reduction factors with this system have ranged from 4 up to 80.

2. Subfloor Suction

If a building has a concrete floor slab, subfloor exhaust ventilation can be used. One or more pipes, usually plastic, are inserted through the floor into the soil. The opening around the pipe is sealed, the pipe is extended outside the building, and a small exhaust fan is placed at the elevated end of the pipe (see Figure 4). The fan draws the soil gas to the cavity in the soil through the layer of coarse material present under most floor slabs, and, in turn, draws air from the building atmosphere through every unsealed crack and opening in the floor slab or at its edges, thus preventing soil gas and radon from moving from the soil into the building through these openings. This can be very effective if the basement walls are of poured concrete. If the walls are stone or concrete block, the system may

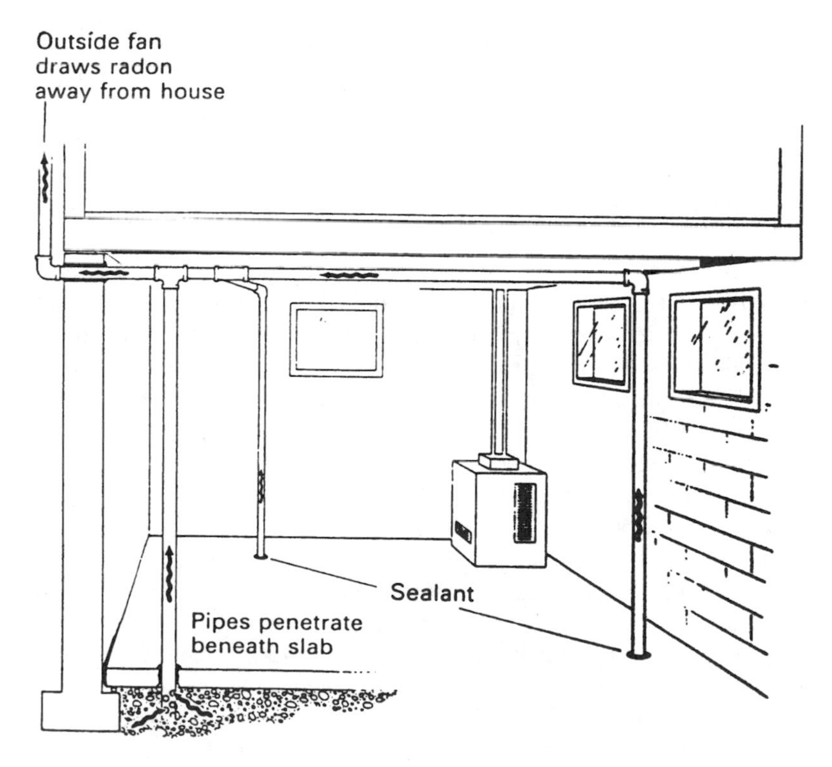

Figure 4. Radon subfloor suction system. From Ref. 37.

be less effective because it may not reduce the flow of soil gas into the building through the blocks or between stones.

The radius over which the exhaust system will be effective is difficult to predict, as it depends on the permeability of the subfloor fill and the size and distribution of openings. Usually, however, it will extend up to 4 to 5 meters with a fan that develops a 50 to 100 pascal pressure differential. Small houses may require only a central exhaust location for satisfactory performance, but larger houses generally require two or more locations so that no portion of the floor is farther than 4 or 5 meters from a point of exhaust. All the exhaust pipes may be connected so they can be served by a single fan.

The cost of a good subfloor system varies depending on the number and length of pipes needed, but would range from about $500 to $2,000, with running costs of about $100 per year. Radon reduction factors have varied from 4 to 850, with an average of about 700.

For houses that have dirt, tile, or other porous material as a basement floor rather than poured concrete as assumed above, the above system can be adapted as follows: A network of perforated pipes, like those used as drainage tiles, can be placed over the floor material and connected to a fan that exhausts to the outside. The pipes must then be covered with a nearly airtight cover, such as polyethylene or vinyl sheets. Loose edges of the plastic sheets should have a large overlap or be sealed with a sealant as previously described.

3. Block-Wall Exhaust

The major radon entry routes in buildings with hollow concrete block foundation walls may be the walls themselves and the wall/floor joint. Both of these flow paths can be controlled by using the wall cavities as the collection system, exhausted either by a small fan inserted into each wall, or by a header system tapped into each wall. Suction is needed at each wall because the blocks at the corners are often filled with concrete for reinforcement, and therefore there is no connection between walls. If the pressure in the block cavities is significantly lower than in the building, then air will flow into the wall at all openings—including the wall/floor joint (see Figure 5).

A modification of this system involves the installation of a sheet metal baseboard duct around the perimeter of the basement. Holes are drilled behind the ducts into the hollow spaces within the blocks (see Figure 5). This approach produces more uniform ventilation and may be more pleasing in appearance. It is also more costly. This system has the advantage of also covering the joint between the wall and the floor, which may be a major route of radon entry.

Exhausting the walls is appropriate only if the leakage areas to the house can be controlled such that there is no danger of back-drafting combustion appliances. If this is a problem, blowing air into the walls is an alternative. The goal then

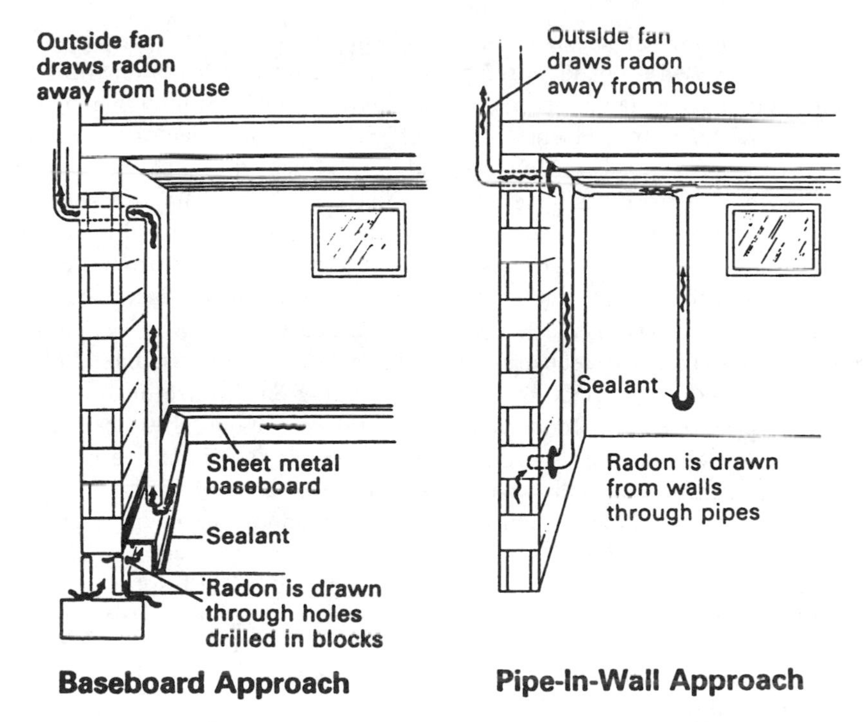

Figure 5. Block wall ventilation, two approaches. From Ref. 37.

is to displace soil gas from the walls with fresh outside air, so that leakage into the house consists only of air with low radon concentration. This effectively prevents back-drafting, but the walls tend to be colder than before.

For these wall ventilation techniques to work well, all major holes must be sealed. The tops of the walls must be sealed, and it may be necessary to coat the inside block wall with something like an epoxy paint. Openings between the blocks and an exterior brick veneer or openings concealed by masonry fireplaces and chimneys may defeat the radon reduction effort.

The cost of this system would range from about $1,000 to $5,000 with running costs of $100 to $500 per year, depending on the amount of heated air that is exhausted. Radon reduction factors for block-wall exhaust systems have varied from 2 to 800.

In some cases, the fan used in these radon reduction methods can be replaced by a pipe that goes to the top of the house roof, providing a stack effect that may be sufficient to exhaust the radon-rich soil gas. It provides less suction and may, therefore, be less effective than a fan, but it has the advantage of requiring little maintenance.

The fourth approach to radon control is ventilation—replacing radon-rich air within the house with outside air. This works fine during the summertime, but in colder climates it may be impractical during the winter. Simply leaving basement windows open may be surprisingly effective. A fan at one of the windows increases the effectiveness. However, this limits use of the basement during winter, so this approach is generally considered for temporary use only. This approach can be used on a permanent basis when it is supplemented with a heat-exchange unit (see Figure 6). This is a unit that transfers the heat from the heated interior air to outdoor air coming in through a special intake duct. The cooled interior air can then be exhausted, and the warmed exterior air is added to the room air. These units can reduce radon by factors of 5 to 30. Installation of such a system may cost from $500 to $2,000, depending on the size of the unit that is needed. Electricity to operate the unit would cost about $100 per year, but there would be an added cost in the loss of heated air as the unit's efficiency in transferring heat is usually between 60% and 75%.

Any radon remediation system that uses a fan should be equipped with an emergency warning system (lights, bell, etc.) to warn the occupants of a malfunction such as a fan failure. In addition, occupants should inspect the system at regular intervals to see that the system is intact and functioning as intended.

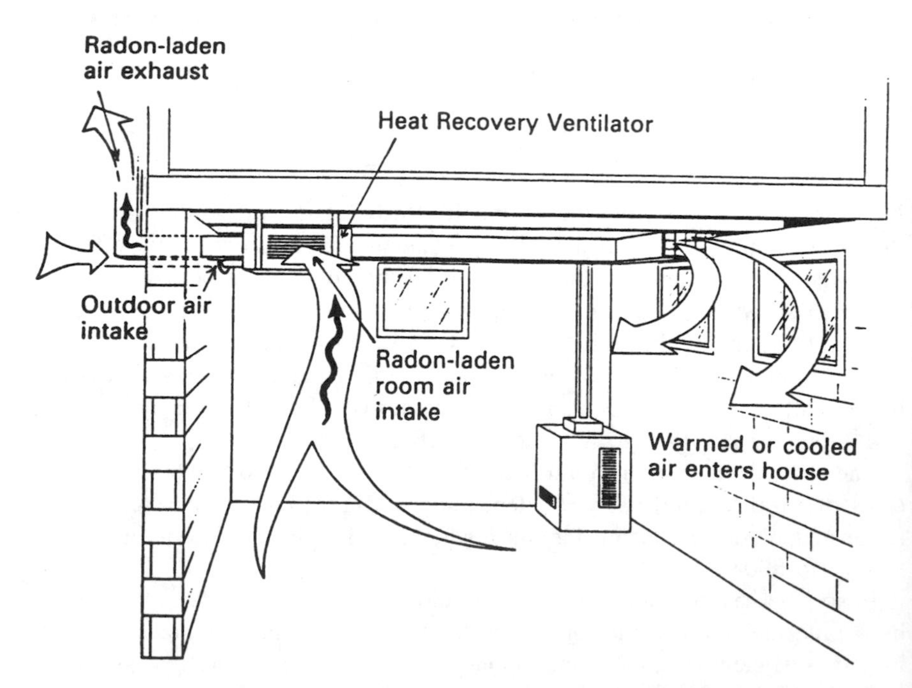

Figure 6. Radon removal from basement, using ventilation with heat exchanger system. From Ref. 37.

A fifth approach, not recommended here, is air cleaning. Radon progeny, when inhaled, are usually attached to airborne particles. By removing these particles, it is possible to remove some of the radon progeny from the air. However, because the radon is not removed, radon progeny rapidly builds up again, so the cleaning must be continuous. Cleaning systems are of three types: electrostatic precipitators, unipolar space charging, and air filters. Standard air filters do not remove the very small particles to which radon progeny is mostly attached, but there are high-efficiency filters available that will remove a substantial fraction of it. Electrostatic precipitators also will remove some of the very small particles. With unipolar space charging, positive ions appeared to be more effective than negative ions, but neither appeared to be of much value [38]. At best, these methods can reduce radon progeny by a factor of 3 to 8 without removing any radon, so they will be inadequate if much reduction is needed. However, by removing the larger particles, more of the radon progeny will be inhaled as free ions or attached to extremely fine particles. These particles are most hazardous, so the reduction in lung cancer risk may be less than gross measurements of radon progeny would indicate.

Post-Mitigation Testing

After a mitigation system is installed, it must be tested to be certain that it is working. One way is to use a charcoal canister detector and repeat the initial screening test. As far as possible, the conditions should be the same as when the first measurement was taken. That is, the detector should be in the same place as the original test, and the doors and windows kept closed as much as possible. However, a long-term (6–12 months) test would be better.

RADON IN THE WATER

Although it is rare, the household water supply may contribute as much or more radon to the house air than the soil gas from under a house. Most public water systems have extremely low radon content in their water. There are two reasons for this. The first is that much of the water supply comes from streams and lakes, which always have negligible radon levels because the water is well-aerated. Even when the public water supply comes from springs or wells, it is stored for a few days and treated [29]. During this storage, most contained radon can escape to the air or will decay, as it has a half-life of 3.7 days.

About one person in five in the United States gets water from an individual well or a small community system that does not store or treat the water. When the water comes from a dug well similar to what our ancestors used, it will be aerated and have little radon. When it comes directly from a drilled well, it will contain some radon which has been picked up from the rocks and sand through which it has flowed. The radon content of such water is extremely variable and in most areas will be negligible [39]. About 370,000 Bqm^{-3} (10,000 pCi/L) in culinary water will typically contribute about 37 Bqm^{-3} (1pCi/L) of radon to the air of the house, so levels below this are of little concern. The drinking of water containing levels of radon that occur naturally is of little concern, but when the radon is released into the atmosphere of a home, it will add to the radon level in the house and may be significant. After water leaves the faucet, the radon quickly escapes from water and goes into the air. The activities that release most radon from water into the house are, in order of importance: laundry, showers, dishwashing, flushing toilets.

Most of the radon testing companies that supply dosimeters for measuring radon in air also can test for radon in water. They will supply a plastic vial with instructions for filling it. It should be returned to the laboratory as quickly as possible.

If it is determined that there is sufficient radon in the well water to be of concern, there are two ways in which it can be controlled. The first is by aerating the water before use, which allows the radon to escape. The second method is by using activated charcoal filters that are used in about the same manner as water softeners. They must be "backwashed" annually, but otherwise require little attention. If radon in the water is very high, the tank should be checked occasionally with a gamma ray detector to see if radioactivity is building up in the tank to significant levels.

RADON AND REAL ESTATE TRANSACTIONS

When buying or selling a home, decisions about radon must be made [32,40]. Until the 1960s, there was a universal doctrine called "caveat emptor," or "let the buyer beware." However, since then, most states have recognized the fact that most buyers of homes are unable to determine for themselves all the inapparent hazards related to a home. The Minnesota Court of Appeals ruled in 1993 that a partnership that sold a building whose structural steel was coated with asbestos breached their "duty to disclose" by not disclosing the presence of the asbestos coating, and was liable for the cost of removing it. As a result of this and similar cases, in most states when a house is sold, there is an "implied warranty," which means that the seller is responsible for disclosing any hidden defects of which he is aware that might be injurious to the purchaser. This includes such items as asbestos, termites, flood danger, hazardous chemical contamination, earthquake faults, and high radon levels.

The law is not entirely clear as to how much of this responsibility is assumed by the real estate agent who is a "middle man" in most house purchases. But if the agent knows of dangerous defects and conceals them, it is fraud. Some states go further and state that the agent "has an affirmative obligation to discover adverse factors that a reasonably competent and diligent investigation would disclose."

Sellers who make a full disclosure about hazards or the results of any radon tests that have been done in a house are clear of legal liability. This has led to the use of hazard-disclosure forms that are attached to home-sales contracts. When applied to radon, they disclose the results of any radon tests that have been taken in the house; they give the buyer the right to have the house tested for radon; and they specify options if the test result is high. That is, it might specify that if the result is high, the buyer can either negate the contract without penalty or that the seller must take remedial action to reduce the radon level below 150 Bqm^{-3}.

If no radon measurements have been made in a house prior to its being offered for sale, it is unwise for the seller to obtain a short-term measurement just before the sale. The opportunities for obtaining a misleading measurement at such times are so great that such measurements should be disregarded by the buyer. There are two ways to handle the radon problem in this situation. One is by negotiation: The seller can agree to reduce the cost of the house by an amount equal to the average cost of radon remediation in the area (perhaps $1,500), and the buyer will then agree to free the seller of any further responsibility. The second approach is to set up an escrow account: Part of the purchase price, perhaps $2,500, is placed by the seller in an escrow account. After the buyer moves into the house, he obtains a long-term radon test. If the results require it, radon mitigation would be done, using money from the escrow account. Any money in the account not spent would then be divided between the buyer and seller, to reduce the incentive for the buyer to spend more of the money than necessary. An important advantage of these two strategies is that the radon testing and remediation can be conducted under optimal conditions, without the pressure of time when a contract is to be signed [32].

Radon measurements should be made promptly after purchase of a building to avoid problems with the statute of limitations. There is an experimental procedure that may make the escrow deposits unnecessary. It involves counting the radiation from lead and polonium (^{210}Po) embedded in glass surfaces, such as windows [41,42]. This count reflects the cumulative exposure to radon progeny since the glass was first put into the home. It involves short-term tests that are probably not subject to interference. It would not, however, be applicable to newly constructed houses.

Each of the ten EPA regional offices (Boston, New York, Philadelphia, Atlanta, Chicago, Dallas, Kansas City, Denver, San Francisco, and Seattle) and each of the states in the United States have radiation offices that can provide help with radon problems. Some are much better staffed than others, but all can advise homeowners as to where radon test kits can be obtained. Some also can recommend contractors for remediation or can give advice on remediation.

REFERENCES

1. National Research Council: BEIR IV. *Health Risks of Radon and Other Internally Deposited Alpha-emitters.* Nat'l Acad Press, Washington, DC, 1988.
2. NCRP Report No. 78. "Evaluation of Occupational and Environmental Exposure to Radon and Radon Daughters in the United States." National Council on Radiation Protection and Measurements, Bethesda, MD, 1984.
3. Hueper, W., *Occupational Tumors and Allied Diseases.* CC Thomas, Springfield, IL (1942).
4. Archer, V.E., "Lung Cancer Risks of Underground Miners: Cohort and Case-Control Studies," *Yale J. Biol. Med. 61*:183 (1988).
5. Samet, J.M., "Radon and Lung Cancer," *J. Natl Cancer Inst. 81*:745 (1989).
6. Chameaud, J., Perraud, R., Masse, R., Lafuma, J., "Contribution of Animal Experimentation to the Interpretation of Human Epidemiological Data," *Radiation Hazards in Mining: Control, Measurement and Medical Aspects.* (M. Gomez, ed.), American Institute of Mining, Metallurgical and Petroleum Engineers, New York, NY, 1981, pp. 222.
7. Svec, J., Kunz, E., Tomasek, L., "Cancer in Man after Exposure to Radon Daughters," *Health Phys. 54*:27 (1988).
8. Department of Energy, "Radon Inhalation Studies in Animals." DOE/ER 0396. National Technical Service, Springfield, VA 1988.
9. Baldwin, F., Hovey, A., McEwen, T., et al., "Surface to Nuclear Distances in Human Bronchial Epithelium: Relationships to Penetration by Radon Daughters," *Health Phys. 60*:155 (1991).
10. Henshaw, D.L., Eatough, J.P., Richardson, R.B., "Radon as a Causative Factor in Induction of Myeloid Leukemia and other Cancers," *Lancet 335*: 1008 (1990).
11. ICRP Publication 50, "Lung Cancer Risk from Indoor Exposures to Radon Daughters." International Commission on Radiological Protection, Pergamon, Oxford, England, 1987.
12. Puskin, S., Nelson, C.B., "EPA's Perspective on Risks from Residential Radon Exposure," *J. Air Poll. Cont. Assoc. 39*:915 (1989).
13. Puskin, J.S., Yang, Y., "A Retrospective Look at Radon-induced Lung Cancer Mortality from the Viewpoint of a Relative Risk Model," *Health Phys. 54*:635 (1988).
14. National Research Council: "Comparative Dosimetry in Mines and Homes," National Academy Press. Washington, DC, 1991.
15. Harley, N., "Comparative Dosimetry of Radon Mines and Homes," *Health Phys. 63*:238, (1992).
16. Environmental Protection Agency: "A Citizen's Guide to Radon (2nd Edition). The Guide to Protecting Yourself and Your Family. Air and Radiation." ANR 464. USEPA and USPHS. Superintendent of Documents, (ISBN 0-116-036222-9) Washington, DC, 1992.

17. Guimond, R.J., "Radon Risk and EPA," *Science 251*:724 (1991).
18. Neuberger, J.S., "Residential Radon Exposure and Lung Cancer. An Overview of Published Studies," *Cancer Detect. Prev. 15*:435 (1991).
19. Edling, C., Kling, H., Axelson, O., "Radon in Homes—A Possible Cause of Lung Cancer," *Scand. J. Work Environ. Health 10*:25 (1984).
20. Pershiagen, G., Liang, Z-H, Hrubec, Z., et al., "Residential Radon Exposure and Lung Cancer in Swedish Women," *Health Phys. 63*:179 (1992).
21. Schoenberg, J.B., Klotz, J.B., Wilcox, H.B., et al., "Case-control Study of Residential Radon and Lung Cancer among New Jersey Women," *Cancer Res. 50*:6520 (1990).
22. Blot, W.J., Xu, Z-Y, Boice, J.D., et al., "Indoor Radon and Lung Cancer in China," *J. Natl. Cancer Inst. 182*:1025 (1990).
23. Hofmann, W., Katz, R., Zhang, C., "Lung Cancer Risk at Low Doses of Alpha Particles," *Health Phys. 51*:457 (1986).
24. Cohen, B.L., "Expected Indoor ^{222}Rn Levels in Counties with Very High and Very Low Lung Cancer Rates," *Health Phys. 57*:897 (1989).
25. Archer, V.E., "A Review of Radon in Homes: Health Effects, Measurement, Control and Public Policy," *Applied Occup. Environ. Hyg. 6*:665 (1991).
26. Jackson, S.A., "Estimating Radon Potential from an Aerial Radiometric Survey," *Health Phys. 62*:450 (1992).
27. Janssen, I., Stebbings, J.H., Essling, M.A., et al., "Prediction of ^{222}Rn from Topography in Pennsylvania," *Health Phys. 61*:775 (1991).
28. NCRP Report, No. 103. "Control of Radon in Homes," National Council on Radiation Protection and Measurements, Bethesda, MD, 1989.
29. Hess, C.T., Michel, J., Houten, T.R., et al., "The Occurrence of Radioactivity in Public Water Supplies in the U.S.," *Health Phys. 48*:553 (1985).
30. Department of Energy, "Indoor Radon and Decay Products: Concentrations, Causes, and Control Strategies." DOE/ER-0480P. Technical Report Series. NTS, US Department of Commerce, Springfield, VA, (1990).
31. Reimer, G.M., Gunderson, L.C.S., "A Direct Correlation Among Indoor Rn, Soil Gas Rn and Geology in the Reading Prong Near Beyertown, PA," *Health Phys. 57*:155 (1989).
32. Brenner, D.J., *Radon, Risk, and Remedy,* W.H. Freeman Co., New York, 1989.
33. Environmental Protection Agency, "Consumers Guide to Radon Reduction. How to Reduce Radon Levels in Your Home, Air and Radiation," 66045; 402-K92-003, USEPA, Washington, DC, 1992.
34. Harley, N.H., Terilli, T.B., "Predicting Annual Average Indoor ^{222}Rn Concentration," *Health Phys. 59*:205 (1990).
35. Environmental Protection Agency, "Radon Reduction in New Construction. An Interim Guide." OPA-87-009. USEPA, Washington, DC 1987, and Proposed Model Standards and Techniques for Control of Radon in New Buildings. Federal Register, Vol. 58, No. 68, Apr. 12, 1993, pp. 19097–19106.

36. Scott, A.G., "Comparison of Criteria to Define Radon Prone Areas," *Health Phys. 64*:435 (1993).
37. Environmental Protection Agency, "Radon Reduction Methods, A Homeowners Guide." OPA-86-005. USEPA, Washington, DC, 1986.
38. Maher, E.F., Rudnick, S.N., Moeller, D.W., "Effective Removal of Airborne ^{222}Rn Decay Products Inside Buildings," *Health Phys. 53*:351 (1987).
39. Nazaroff, W.W., Doyle, S.M., Nero, A.V., Sextro, R.G., "Potable Water as a Source of Airborne ^{222}Rn in U.S. Dwellings: A Review and Assessment," *Health Phys. 52*:281 (1987).
40. Environmental Protection Agency, "Home Buyer's and Sellers Guide to Radon." 402-R-93-003, Superintendent of Documents, Washington, DC, 1993.
41. Lively, R.S., Steck, D.J., "Long-Term Radon Concentrations Estimated from Polonium-210 Embedded in Glass," *Health Phys. 64*:485 (1993).
42. Mahaffey, J.A., Parkhurst, M.A., "Estimating Past Exposures to Indoor Radon from Household Glass," *Health Phys. 64*:381 (1993).

CHAPTER 24

INDUSTRIAL ODORS—PROBLEMS AND CONTROL

Paul N. Cheremisinoff

Department of Civil & Environmental Engineering
New Jersey Institute of Technology
Newark, NJ 07102, USA

CONTENTS

INTRODUCTION

Odor is probably one of the most complex of all air pollution problem. The only good measuring device is the human nose. Odorous materials can range from inorganic gases and vapors—such as ammonia, hydrogen sulfide, and hydrogen chloride, across the full spectrum of organic compounds—such as mercaptans, amines, acids, aldehydes, and ketones. No instrument has been devised with the capability and the sensitivity of human olfactory system. The human nose can distinguish thousands of different materials that may exist in a wide range of concentrations. Methods of odor control include odor modification and masking, combustion, absorption, and adsorption. Selection of a control technique depends on the type and nature sources of the odor problem and desired end results.

Odor can be caused by fumes and gaseous emissions from many diversified industrial processes and can be the result of organic components in the effluent waste gas. Unlike some other forms of air pollution, odors are difficult to quantify and inventory. There are few if any scientific measuring devices or instruments available for accurately determining the concentration of a particular odor, such as there are for measuring particulate levels or sulfur dioxide concentrations. Odors differ from typical pollutants in that they may be mixtures of elements or compounds that can be quantified, and are a characteristic of a particular group of elements or compounds. The best odor detection and measurement is the human nose. Compared to sophisticated instrumentation used to measure other air pollutants, the nose makes odor surveillance somewhat arbitrary and subjective.

Odor has two basic characteristics that determine its effect on human: intensity and quality. The intensity of an odor is a quantification, either through subjective words and numbers of the strength of an odor. The units for the reporting of odors are intensity and description. The intensity describes an odor's strength; whereas the description designates the nature or character of the odor.

CHARACTERISTICS OF ODOR

Intensity of an odor varies logarithmically with the concentration of the odorous compound, according to the Weber-Fechner Psychophysical Law:

$$I = K \ln C$$

where

I = the intensity of the odor
K = a constant related to the pervasiveness
C = the concentration of the odorous compound

Odors are classified by either the intensity or the description. The typical odor intensity scale ranges from zero to five as shown in Table 1.

Odor descriptions designate the nature or character of the odor including flowery, fruity, burnt, decomposition, and chemical.

The threshold limited value (TLV) is given in parts per million (ppm) and represents the lowest level at which an odor is detected or recognized in ambient air by the participants of an odor panel. The threshold limited value may vary widely for a particular odor depending on the panel making the determination.

Table 1
Odor Intensity

Intensity	Strength
0	No odor
1	Very faint
2	Faint
3	Moderate
4	Strong
5	Very strong

Because all people do not have the same sensitivity for the detection of odors, the threshold concentration may vary from person to person. A concentration at or greater than this represents a danger to workers and exposure at this level should be avoided. The threshold limit value and the hazardous exposure level may differ by several magnitudes as shown in Table 2.

The level at which a compound is present determines the control necessary. The hazardous exposure level presents the actual danger, whereas the TLV indicate a nuisance concentration. Levels at or above the hazardous exposure level require control at the source, while concentrations only representing a nuisance can be controlled in the other ways.

Table 3 is a partial listing of suggested descriptions used during an odor verification survey. Such descriptions can be very helpful in determining the chemical nature of the odorants.

The quality of an odor may be changed by dilution method. In mixture of odorous compounds, this may be due to differences in the pervasiveness of the compounds. The pervasiveness refers to the ability of an odor to maintain a detectable intensity despite dilution with large volume of air.

Table 2
TLVs vs. Hazardous Exposure Levels

Compound	Threshold Limit Value (TLV)	Hazardous Exposure Level (ppm)
Hydrogen Sulfide	0.00047	10
Ethyl Mercaptan	0.001	0.5
Pyridine	0.021	5

Table 3
Suggested Descriptive Terms and Associated Chemical Compounds

Description	Compounds
sweet/aromatic/oily/solvent	lower ketones, alcohols, ester, vanilla, etc.
fishy/musty	amines
sharp/purgent/rancid	acrylates, aldehydes, higher ketones
burnt/smokey	asphalt, tar
sulfur	mercaptans, sulfides
acidic	vinegars, resins

Table 4
Odorous Industrial Operations

Industry	Odorous Material
Chemical manufacture	Hydrogen sulfide, ammonia, amines, alcohols, aldehydes, phenols, mercaptans, esters, chlorine and chlorinated organics, etc.
Coke ovens	Sulfurous, ammoniacal, and phenolic compounds
Fertilizer	Bone meal, organic nitrogen compounds, ammonia
Food and kindred products	Dairy wastes, cannery wastes, fish, baking bread, chocolate, flavors, packinghouse wastes, meat products for rendering, coffee roaster effluents, cooking odors, etc.
Foundries	Core-oven odors, quenching oils
General industrial	Burning rubber, forming and molding plastics, incinerator smoke, solvents and lacquers, asphalt
Petroleum	Sulfur compounds from crude oil, cresols, asphalt
Pharmaceuticals	Biological extracts and wastes, spent fermentation liquors
Pulp and paper	Sulfurous compounds
Soap and toiletries	Perfumes, anima
Tanneries	Hair, flesh, hides

SOURCES OF ODOR

Many common chemical sources of odors are organic compounds that contain nitrogen or sulfur in their molecular configuration. Amines, mercaptans, and complex decomposition products of proteins, which contain nitrogenous and sulfurous compounds, are typical odorants.

Table 4 lists the different various industrial sources.

MEASUREMENT

Odor is a physiological response of individuals. Some instruments that measure the physical and chemical characteristics of an odor-laden gas as an indirect means of means of measuring odor are of questionable value in describing the odor potential of an unknown mixture. A more reliable and accurate method for measuring odor is the nose.

There are two methods for odor measurement — the organoleptic methods and the chemical/instrumental methods. The organoleptic methods that use the human olfactory system are completely subjective. Chemical/instrumental methods generally suffer from two major shortcomings: sensitivity and flexibility.

To compare the instruments with human nose, we can find the human olfactory system is capable of detecting and identifying a wide variety of chemical structures and giving different responses to different materials. Generally speaking, the instrumental methods or chemical methods are restricted to particular structures and give a similar response to all compounds with that structure.

The most successful instrumental methods for the measurement of odor are those using gas chromatography (GC). By using GC, we can determine the concentrations of sulfur dioxide, hydrogen sulfide, and other odorous gases (to concentration about 0.01 ppm). Using a colormetric cell with platinum electrodes as the agent to detect sulfur-containing compounds has been reported (to concentrations about 0.1 ppm). The threshold levels detected by the human nose are as low as 0.0021 ppm (trimethylamine).

CONTROL TECHNOLOGY

There are four important control methods for solving the odor problems: absorption, adsorption, incineration, and dispersion.

Absorption

Scrubbers used for odor control by absorption are usually the spray tower, packed tower, or wetted porous media types. They operate by dissolving, condensing, or chemically reacting the odorant with the scrubbing liquid. The odorant in the air exhaust can be present as vapor or as fine droplets or can be absorbed on the surface of particulate contaminants. It can usually be removed by liquid washing of the air. The same equipment can be used to control the temperature and humidity of the air or to recover vapor such as solvents from the air at the same time.

The odorant vapor may condense from the air when it contacts the cold scrubbing liquid. The scrubbers using this principle are also called contact condensers. This equipment can be effectively used for the removal of odorous contaminants from air exhausts that are rich in water vapors, such as are the rendering plant cooker exhausts. Contact condensers also are used to dehumidify the air. On the other hand, the odorant vapors may have sufficient solubility in the scrubbing liquid to be removed without chemical action. Inorganic gases such as HCl and NH_3 usually are removed by water scrubbing. The choice of scrubbing liquid is dependent on the odorant compound.

The addition of chemicals to the scrubbing solution often is effective in odor removal. Aqueous acid solutions effectively remove basic odorants such as amines or pyridine, and aqueous base solutions remove acidic odorants such as hydrogen sulfide. The addition of oxidizing agents such as potassium permanganate, hydrogen peroxide, and chlorine can destroy numerous organic odorants by oxidation.

The permanganate oxidation of simple sulfur compounds, such as hydrogen sulfide and sulfur dioxide, is quantitative and virtually instantaneous. Other odorants also can be oxidized to odorless compounds under a wide variety of reaction conditions; however, pH value, concentration, contact time, and reaction temperature are important variables.

When potassium permanganate ($KMnO_4$) is used in air scrubbing, most of the time in alkali, neutral, or slightly acid media, permanganate is reduced to insoluble manganese dioxide. In a strong acid media, permanganate is reduced to bivalent manganese. In both cases the nascent oxygen formed will immediately react with the oxidizable odorants to form odorless or less offensive compounds.

Chlorine has been successfully used to disinfect and control odors of water supply. Chlorine also has been used to deodorize gases from rendering plants, fish meal processing, and pharmaceutical manufacturing. Chlorine will react with many hydrocarbons, replacing one or more hydrogen atoms and forming hydrogen chloride as a by-product. With unsaturated hydrocarbons, chlorine produces saturated chloride. Aqueous solutions of chlorine (maximum solubility is 1% at 9.6°C) are strongly acidic and contain some elemental chlorine, which may chlorinate organic compounds. In weakly acidic solutions, chlorine is converted to chloride ion and hypochlorous acid. In natural to alkaline solutions, chlorine is converted to chloride ion and hypochlorite ion:

$$Cl_2 + H_2O \rightarrow HOCl + HCl$$
$$Cl_2 + 2\ NaOH \rightarrow NaOCl + NaCl + H_2O$$

Because of its toxic and odorous nature, chlorine, when it is used in conjunction with wet scrubbing units, is usually added to the odorous gases before entering the scrubber; sufficient retention should be given for maximum effect. Chlorine concentrations of approximately 20 ppm for a rendering plant and 20 to 50 ppm for pharmaceutical manufacturing have given satisfactory results.

The scrubbing liquid must be maintained in relatively odor-free condition to avoid recontamination of air. Therefore, either fresh water should be added continuously or the organic solvent should be regenerated continuously to assure this requirement. The discarded water is likely to be sufficiently contaminated to require further treatment.

Adsorption

Adsorption is a method for removing odorous substances from air streams. This process depends on the attractive forces between a solid surface and a gaseous molecule. These forces can be either chemical or physical in nature. If they are chemical, the process is usually not reversible. If physical, the adsorbent may be used over and over again after going through a regenerative cycle. Activated carbon is the most commonly used adsorbent.

If the odor is due to a solvent or some other compound with a recoverable value and is not contaminated with other substances, it may be recovered through the regenerative process and be reused. If this is the case, this system may pay for itself in a short period of time.

Adsorption may be an effective and economical control method for emission with low concentrations of odorous compounds. Adsorption is a physical process in which molecules form the gas phase are captured by and retained on the surface of a solid (adsorbent). The process is diffusion-limited and is dependent on the active surface area available. The mechanism is principally dependent on Van Del Waal's forces of attraction between the adsorbent and the contaminant to be removed from gas stream (adsorbate). The efficiency of operation is governed by the thermodynamic equilibrium that exists between the adsorbate and the adsorbent. This method is most commonly used in dealing with odors contained in large volumes of fairly dry air. The adsorption equipment is often used in conjunction with spot ventilating or area exhaust systems. While the initial capital costs for equipment and adsorbent can be high, the cost for operating such as a system is usually moderate or low.

The phenomenon of carbon adsorption has been applied in gas-mask filters, air conditioning systems, and sewage treatment ventilation. For example, toxic vapors, odorous compounds, or other undesirable contaminants adhere to the large surface area provided by the intricate pore structure of carbon granules.

To maximize adsorption efficiency, the greatest possible surface area should be contained in the smallest practical volume. This principle is first applied in the manufacture of granular carbon. The objective is to assure that each granule is provided with maximum surface area and controlled pore size while maintaining a desired density and hardness. The surface is created by permeating a carbonaceous material, such as coal or coconut char, with minute holes or pores. The number and diameter of these pores determine the total available surface area. Surface areas for commercial activated carbons vary from 500 to 1,400 square meter per gram, or about 125 acres in as little as one pound.

Due to the nonpolar nature of its surface, activated carbon has the ability to adsorb organic and some inorganic materials in preference to water vapor. The amount of materials adsorbed is partially dependent on the physical and chemical characteristics of the specific compound or compounds. In general, organics having molecular weights greater than 45 and boiling points over 0°C will be readily adsorbed.

Once saturated, the carbon can be reused after a portion of the adsorbed material has been removed by regeneration. This regeneration can be accomplished thermally in a furnace or by passing hot gas or steam through the carbon bed until sufficient material has been desorbed. The amount desorbed will be the design working capacity of the bed. It is usually preferable, for economic or safety considerations, to allow some adsorbed material to remain in the pores of the carbon.

Economic considerations are related to the increased regeneration costs resulting from the effort to strip the carbon clean of all adsorbed organics. Besides, safety considerations are necessary when easily oxidized organics at high concentration are being adsorbed. Allowing a residue of the volatile material to remain in the pores of the carbon minimizes the temperature rise associated with heat of adsorption and reduces the possibility of ignition in the carbon bed.

The actual adsorption of odor causing fumes and gases takes place on the internal surface area of the adsorbent. Three processes occur in the overall adsorption process:

- Transfer of the odor-causing molecules from the waste gas to the outer surface of the adsorbent,
- Diffusion by capillary action into the porous adsorbent, and
- Physical attraction to the inner surface area of the adsorbent.

Combustion and Incineration

Many odor problems are caused by organic pollutants in the waste fumes and gases from a variety of processes. Combustion and incineration are used for odor control where undesirable organic vapors are converted into carbon dioxide and water vapor through the oxidation of the pollutants at high temperatures. Assuming combustion is the solution to a particular odor problem, a determination must be made as to which combustion system would be most applicable.

There are four major types of incineration systems that may be applicable:

- Direct flame,
- Catalytic,
- Thermal, and
- Flare.

The direct flame and catalytic systems are applicable when the concentration of contaminants is below the lower explosive limit with the organic concentration not exceeding 25% of the lower explosive limit for the substance. Table 5 lists the lower explosive limit (LEL) in percent by volume for some common organic pollutants.

Thermal combustion is used when the organic, odor-causing constituent of the waste gas is very small. This the most commonly used system for odor control. The direct-flare combustion system is the least expensive because the waste gas

Table 5
Lower Explosive Limit for 80me Common Organics

Compound	LEL, % by volume
Acetone	2.5
Benzene	1.2
Carbon disulfide	1.1
n-Hexane	1.0
Isopropanol	1.8
Methane	6.1
Methyl ethyl ketone	1.8
Toluene	1.1
Xylene	0.9

supplies the fuel necessary for incineration and is not considered primarily as an odor-control technology.

Good combustion is dependent on the time, temperature, and turbulence. The temperature must be high enough to thoroughly oxidize the organic pollutant. This temperature usually must be several hundred degrees above the autoignition temperature. The time the waste gas remains in the incinerator (residence time) will vary depending on the temperature and turbulence. The turbulence determines the amount of mixing achieved between the pollutant and air such that the oxygen molecules have contacted with each organic molecule within the combustion chamber at the required temperature. Any one of the three variables can be independent and the other two dependent. Excellent turbulence will result in a reduction in temperature and residence time. Maintenance of good turbulence is paramount in incinerator design because it is the cheapest of the three factors to supply and will result in keeping the incinerator as small as possible (reduced residence time) and the use of auxiliary fuel at a minimum (reduced temperature). The primary cost associated with turbulence is for the energy required to ensure complete mixing. The desired goal is the optimization of the three factors to achieve the best results at the lowest overall cost.

Direct-Flame Incineration

Direct-flame incineration can be used when the waste gas is a combustible mixture not needing the addition of air as well as with materials below their lower explosive limit.

A well-designed incinerator can burn gases with heating values as low as 100 Btu per cubic foot without auxiliary fuel. Gases with even lower heating values also can sustain combustion without additional fuel as long as they are preheated to 600° to 700°F. Organic pollutants in concentrations below the lower explosive limit are incinerated at temperature of 900° to 1,400°F in the presence of a flame, providing proper conditions are present.

Small quantities of auxiliary fuel are needed to sustain combustion when the amount of combustible material is below the lower flammable limit. The pollutant serves as a significant contributor to the fuel required, thus keeping the amount of auxiliary fuel needed at a minimum. Many installations employ heat recovery systems to reduce fuel costs to a minimum.

Direct flame incineration should be used only where the pollutant supplies a minimum of 50% of the fuel value of the waste gas mixture. The direct-flame method of incineration is highly effective, obtaining continuous efficiencies of 90% to greater than 99% removal of organic odor-causing pollutants. Most direct flame installations are operated at temperatures of 850° to 1,500°F with a velocity of 15 to 20 feet per second and a residence time of 0.3 to 0.5 seconds.

Catalytic Incineration

Catalytic incineration, as its name implies, employs a catalyst to allow the direct flame process to occur at lower temperature and in the absence of a flame. Noble metals such as platinum or palladium are typical catalysts. The catalyst is present in the waste gas stream such that the maximum surface area is contacted by the waste gas. The most common methods of presenting the catalyst efficiently to the waste gas are by: a nichrome wire screen coated with the catalyst; airfoil-shaped aluminum rods with the precious metal deposited on it; a bed of spheres coated with the catalyst; and an aluminum honeycomb on which the catalyst has been deposited. Maintain the optimum surface area and present the buildup of catalyst poisons. the catalyst also will need reactivation after approximately three to five years of use.

The major advantage of a catalytic incinerator over other types of incineration is that the combustion reactions occur at lower temperatures. Catalytic reactions can be achieved at preheat temperatures of 600° to 1,000°F with maximum operating temperatures of 1,500° to 1,600°F, compared to 2,500°F for a direct-flame system. Operation at lower temperatures results in lower fuel costs; however, the capital investment is significantly more. Catalytic incinerators operate at or below 25% of the lower explosive level of the organic pollutant in the waste gas. A high concentration of the organic pollutant in waste gases may provided excessive internal heat, resulting in catalyst burnout. Thus, catalytic systems are most applicable to situations where there is a relatively small temperature rise across the catalyst.

The catalytic incinerator may be very similar in configuration to a direct-flame incinerator with exception of the catalyst section. Catalytic systems are widely used to control the odors associated with paint solvents, chemical manufacture, food preparation, wire enameling ovens, and lithographing ovens.

Efficiency is effected by the preheat temperature and the space velocity through the catalyst bed. Efficiency capabilities of approximately 85% to 92% are lower than that for the direct flame system.

Thermal Incineration

Thermal incineration can effectively destroy odor-causing gases and fumes at temperatures from 1,000° to 1,500°F, well below the 2,000°F or greater required by most industrial burners. The effluent gas streams treated by thermal incinerators contain weak mixtures of organics, primarily hydrocarbons, and air and have very low heating values ranging from 1 to 20 Btu per cubic foot. Direct-flame incinerators that were previously discussed are a form of thermal incineration using a gas flame in a direct process where the flame actually occurs in the waste gas effluent stream and the products of combustion enter the effluent stream. Thermal incineration also includes afterburning and the direct-flare processes. Afterburners are not generally thought of as odor-control devices, and their primary function is to provide secondary burning to remove unburned products of combustion from the exhaust gas after primary incineration. Direct-flare systems will be discussed very briefly in next section. Thermal incineration as discussed here includes only those systems designed for weak organic waste concentration where auxiliary fuel sources are required.

Perhaps the major application of thermal incinerators is to control the low-level emission of hydrocarbons and associated odors. These systems are commonly used for incinerating the waste gases from drying petroleum processes releasing light concentrations of hydrocarbons; baking and curing operations that result in the formation of oils or solvents in the waste gas; food processing where hydrocarbons present an odor problem; and, generally, any chemical process that produces odors associated with hydrocarbons.

Thermal incinerators require an auxiliary fuel source because of the weak mixtures present and their low Btu output. The process involves the heating of a combustion chamber through the use of a conventional fuel and then injecting the waste gas stream into the chamber. The waste gas can enter the chamber just downstream of the burner flame or, preferably, should pass through the flame to insure complete hydrocarbon conversion.

As mentioned, the waste gas is weak in organics and contains predominantly air. Therefore, an outside source of oxygen usually is not needed to insure the complete combustion of the odor-causing organic contaminants. In the situation where oxygen must be added, a fan or blower can be used to premix outside air with the waste gas or inject the air directly into the combustion chamber. The importance of time, temperature, and turbulence is significant in thermal incineration because outside fuel and, occasionally, air sources are used in the process. Thermal incinerators should maintain combustion chamber temperatures ranging from 1,200° to 1,500°F with a waste gas residence time of 0.5 to 0.6 seconds. Turbulence can be created by strategic baffling to insure thorough mixing and complete combustion, thus obtaining maximum efficiency from the system. The operating temperature is determined by identifying all the undesirable organic pollutants in the effluent gas and designing for a temperature several hundred degrees higher than the highest auto-ignition temperature for those specific pollutants.

The efficiency of thermal incinerators relative to the removal of organic odor-type compounds is very high, at near 100%. Complete conversion of the hydrocarbons to carbon dioxide and water vapor would require a combustion chamber temperature of greater than 1,800°F. Thermal incinerators do not achieve this complete conversion, and some of the hydrocarbons are exhausted as carbon monoxide rather than carbon dioxide. This usually presents no problem because the initial hydrocarbon concentration is very low.

Direct Flare Combustion

As with direct-flame incineration, direct flare combustion also is included in the broad category of thermal incineration. Direct-flare combustion is used when the organic concentrations in the waste gas stream are between the lower and upper explosive limits, in the flammable range. Auxiliary fuels can be used in the event the pollutant concentration is above the upper explosive limit. Should the concentration be above the upper explosive limit, air can be added to the waste gas to maintain a flammable status. Should the waste gas be in the flammable range, care must be taken to prevent flashback through the system.

Direct-flare systems are not designed specifically for odor control, even though this is achieved through the burning of combustible waste gases. The system is the least expensive of the various incineration processes and is essentially a pipe with a flame-holding device at its upper end and an ignitor to provide the initial spark.

Elevated direct-flare systems are most commonly found in many refineries and chemical plants. They are poor mixing devices and release combustible hydrocarbons to the air, depending on atmospheric turbulence for mixing and dilution. Ground flare systems also are used in refineries, chemical plants, and gas fields, and operate similarly to elevated-flare systems.

Direct-flare systems are typically inefficient and are used primarily to release combustible hydrocarbons and other organics present in waste gases in the flammable range into the atmosphere for diffusion. They provide some odor control: however, due to their inefficiency, odor detection is dependent more on the atmospheric and meteorological conditions than on the direct-flare process.

Dispersion

Dispersion can, under certain circumstances, do an adequate job. The technology makes maximum use of local meteorological factors — wind speed, topography, elevation, temperature, volume, and velocity of the gases containing the odorous substance. The use of tall stacks has been found effective in reducing the ground-level concentrations of air contaminations.

Many odors produced in the chemical industry, while objectionable, may be harmless and can only be detected in high concentrations. Odors of this type need only be exhausted from the roof of the manufacturing operation to achieve the desired dispersion. However, in an area subject to inversions, it is a method that can deliver the odor, completely undispersed, back to ground level, with embarrassing results. This is a method that is usually not recommended.

NEUTRALIZATION OF ODOR

Odor concentration or odor intensity, however measured, can be reduced by adding a nonchemical reactive controlling agent to a malodor. The concept of odor counteraction pairs different odors in proportions that render the mixture odorless or nearly odorless. Some of the more interesting pairs of odors compensating each other are, for example, ethyl mercaptan and eucalyptol; skatole and coumarin; butyric acid and oil of juniper.

Masking is based on the premise that people perceive a mixture of smell as a single odor. Masking consists of superimposing a pleasant odor over an unpleasant one. This effort raises the total odor level and creates an overpowering sensation, hopefully more pleasant but usually objectionable.

Another method of odor abatement exists based on the principle of modification or compensation. This involves lowering the malodor intensity and mixing it with selected chemical gases or vapors. It has been shown that when two substances of given concentration are mixed in a given ratio, the resulting odor may be different or less intense than that of the separate components. Eventually it may not be perceptible at all. This modification phenomenon is essentially a decrease of olfactory intensity. The results show that modification occurs in the majority of cases where two substances are mixed in uneven proportions. A pleasant smell and an unpleasant smell will sometimes neutralize or modify each other so that neither can be recognized.

The implementation of the modification concept has led to positive solutions for problems of nuisance malodors. There can be no objectionable odor unless there is someone to smell it and complain. This has brought about bathing malodorous air with gaseous odor-modifying chemicals before it reaches the individual. The procedure involves setting up a screen of modifying gas between the sources of malodor and the source of complaints. Odoriferous air passes through the screen and mixes with the proper modifier gas. Many of these chemical screen installations have operated successfully in varied applicants. Waste disposal plants, industrial facilities, feedlots, and lagoons have eliminated nuisance odor complaints. It has been established that odor modification materials have their proper place in the abatement of malodor nuisances.

There are several important considerations before selecting this route:

- No attempt should be made to modify or otherwise disguise any gas or vapor that may possibly be toxic or harmful.
- Odor modification should not be used as a substitute for good housekeeping or pollution control.

- Odor control technology is sophisticated, and only trained and experienced personnel should be involved, quite often a perfumer.
- Where it is feasible and economical, the malodor should be eliminated or destroyed at the source.
- Odor-modifying chemicals must conform to existing specifications and regulations.
- Odor-modifying materials should affect a lowering of olfactory intensity.

Ozone (O_3) is a triatomic allotrope of oxygen that oxidizes most oxidizable organic and inorganic compounds. The sulfur- and nitrogen-containing odorants of sewage treatment plants, food, and chemical processing plants are converted to sulfoxide, sulfones, and amine oxides that have a substantially higher odor threshold.

Ozone—like chlorine, fluorine, potassium permanganate, hydrogen peroxide, etc.—is an oxidant, but it differs from other oxidants in some very important ways, such as its potency and the facility and economy with which it can be generated on-site from air, electricity, and cooling water. The most practical and economical method of generating ozone employs a glass tube dielectric, around which is produced a corona discharge. The corona discharge converts a portion of the oxygen in the air feed gas to ozone. Ozone can be generated directly from ambient air, or preferably from ambient air that has been filtered, compressed, cooled, and dried.

Most odors are associated with molecules that have centers of high electron density, such as amines, sulfides, and unsaturated hydrocarbons. Ozone most often reacts chemically, as if it were an electron deficient molecule. Thus, a molecule having a site of excess electrons (i.e., the odoriferous molecule) is attracted to a molecule deficient in electrons, i.e., ozone. The two moleculars intract chemically to produce compounds, usually oxides and oxygen, that do not possess any odor.

The products of oxidation always should be less toxic than the original odoriferous product.

CONCLUSIONS

Odors are difficult to quantify and inventory because they may not be specific compounds, but rather a characteristic of a group or mixture. The majority of the odor problems in industry are caused by waste fumes and gases containing organic contaminants, particularly hydrocarbons. There is a wide array of control options.

BIBLIOGRAPHY

Cheremisinoff, P. N. (1992). *Industrial Odour Control,* Butterworth-Heinemann Ltd., Oxford, UK.

Cheremisinoff, P. N. (1993). *Air Pollution Control and Design for Industry,* Marcel Dekker, Inc., New York, NY.

Cheremisinoff, N. P. and Cheremisinoff, P. N. (1993). *Carbon Adsorption for Pollution Control,* PTR Prentice Hall, Englewood Cliffs, NJ.

Cheremisinoff, P. N. (1992). *Waste Incineration Handbook,* Butterworth-Heinemann Ltd., Oxford, UK.

CHAPTER 25

EFFECTS OF ACID DEPOSITION ON VEGETATION

Todd G. Grant
Richard B. Trattner

Department of Chemistry, Chemical Engineering & Environmental Science
New Jersey Institute of Technology
Newark, NJ 07102, USA

CONTENTS

INTRODUCTION

The title of this paper, "The Effects of Acidic Deposition on Vegetation," would lead the layman to conclude that this paper is an overview of acid rain. This is untrue; the paper will focus on how acidic deposition effects vegetation. This is a more accurate term than acid rain because the acidity that comes from atmospheric pollution has other components besides rain (wet deposition). Acidic deposition is commonly defined as the total hydrogen ion loading over a given period of time. The total hydrogen ion deposition to vegetation may come from several types of deposition. In addition to acid rain, vegetation may have to contend with acidic snow, acidic fog, acidic aerosols, and acidic gases. Therefore, acidic deposition is a better technical term to cover all sources of acidity to which vegetation is exposed.

The definition of acid deposition and particularly wet deposition (rain) must be identified. For most of the United States, pre-industrial rain was more acidic than pure water. The general definition of acidic precipitation is rain with an average annual pH less than 5.0. This definition calls for the precipitation to be more acidic than the best estimate of natural (pre-1500 A.D.) rain that was remote from arid regions, in large forested regions of the United States (See Figure 1).

This paper will focus mainly on the damage caused to vegetation by acidic deposition as a result of sulfur dioxide and nitrogen oxides. However, other oxidants—such as ozone, hydrogen peroxide, and organic free radicals—also cause acid deposition to some degree, and are discussed.

During the past several years, eight cases of apparent decline of regional forests in the United States have been suggested to have been caused by air pollution. Multiple stresses are probably to blame. However, in a few cases air pollution is known to have been the cause of the decline of vegetation. In these cases of damage, the air pollutant was ozone, possibly acting together with one or more associated oxidants. For most other cases, definitive conclusions about the relative importance of natural stresses and air pollution cannot yet be drawn.

Acidic deposition also affects foliage. Controlled exposure studies simulating acidic deposition on seedlings have not detected injury or other adverse effects on the foliage of conifer and hardwood species that were tested down to pH 3.5. The average pH of rainfall in the United States rarely falls below 4.1. Therefore, at current levels of acidic deposition, short-term direct foliar damage on healthy forests is unlikely. Surveys indicate that the majority of North American forests are healthy. In addition, no regional patterns of adverse effects on crop growth and production have been identified related to acidic deposition.

EMISSIONS OF ACIDIC DEPOSITION PRECURSORS

Sulfur Compounds

Historical Emissions of Sulfur Dioxide

Figure 1 shows the historical trends in man-made sulfur dioxide (SO_2) emissions. In 1900, annual national emissions of sulfur dioxide are estimated to have been about 9 million metric tons, and about 81% of those emissions resulted from coal combustion in industry and locomotives [1]. These estimates were constructed using historical emissions estimated by various researchers. Estimates

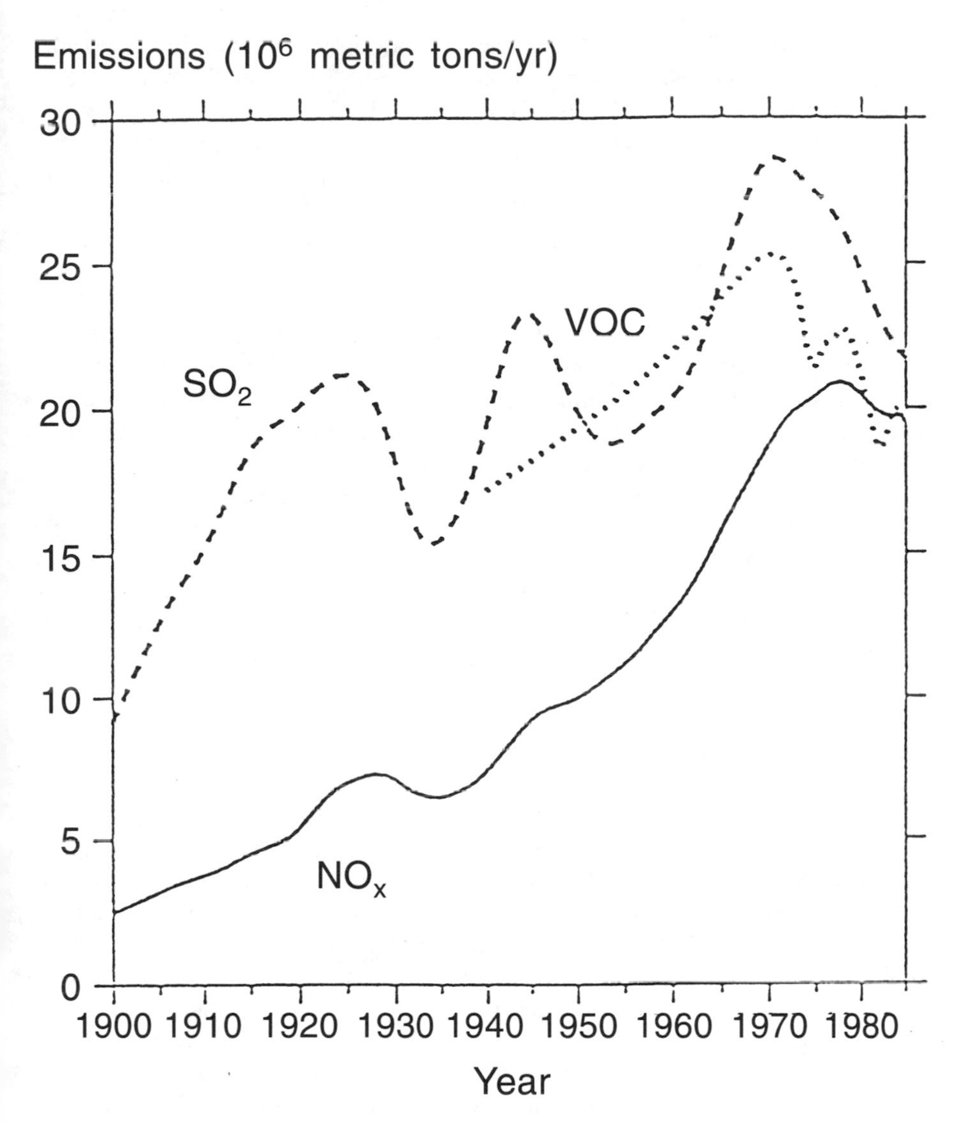

Figure 1. Generalized, interpretive view of historical trends in man-made SO_2, NO_x, and VOC emissions in the United States from 1900 to 1985.

from any one year can vary among researchers because of slight differences in methodology and the emission factors that a particular researcher may have used. Best estimates for sulfur compounds were delivered for 1980. Trends estimated for the time period before and after 1980 were normalized for the best estimates for 1980, and continuous smooth curves for the period between 1980 and 1985 were created.

As coal use and industrial activity such as non-ferrous smelting increased after 1900, annual emissions of sulfur dioxide rose, peaking at about 21 million metric tons in the late 1920s [1]. After 1930 emissions declined through the early part of the depression. They peaked again in the 1940s during the rise of manufacturing, commercial activity, and rail transportation associated with World War II. They peaked once again in the early 1970s before the implementation of air pollution control initiatives. As a result of regulations contained within the Clean Air Act, particularly those directed at power plants, annual emissions had decreased to about 22 million metric tons by 1983 [1].

Figures 2 and 3 show the historical emission of sulfur on a per capita and per dollar of Gross National Product basis, respectively. The graphs show that the trends in sulfur dioxide emissions per capita reflect changes in per capita use of coal and, secondarily, petroleum products (residential fuels in particular). The decrease in per capita sulfur dioxide emissions between 1920 and 1960 is primarily related to the increased use of natural gas and other fuels with low sulfur content to supply the growing demand for energy. The increase in per capita sulfur dioxide emissions between 1960 and 1970 is related to a per capita increase in energy consumption and, in particular, to a rise in per capita use of high-sulfur fuels (coals and residential oils) primarily in electrical power plants per capita consumption of sulfur dioxide declined after 1970 as a result of the implementation of air pollution regulations [1].

Major Sources of Sulfur Dioxide Emissions

Sulfur is found naturally in coal, oil, iron, and copper ores and several other substances used as fuels or processed into salable materials. It is during the combustion process that most of the sulfur is released. Some of the sulfur is trapped by pollution-control equipment and the remainder is emitted into the atmosphere. Sulfur compounds also are emitted naturally be terrestrial and marine biological processes, by chemical and physical activities in the earth's crust (e.g., volcanism and other geothermal activities), and in oceanic surface waters [1].

Most sulfur-compounded emissions from man-made sources are in the form of sulfur dioxide. A small portion is emitted as primary sulfates.

The combustion of coal and oil in electric utilities and industrial facilities are the largest sources of man-made sulfur dioxide emissions in the United States, accounting for 70% of the SO2 emissions in 1980. Emissions from non-ferrous smelters contributed about 4% or 5% of the emissions. Significant reductions in smelter emissions have occurred since 1980, primarily because of plant closures. Other industrial sources including petroleum refining, cement plants, pulp and paper manufacturing, and iron and steel production, which contributed about 9% or 10% of the total. Residential and commercial fuel use and transportation each contributed less than 5%.

Various estimates of sulfur dioxide emissions in 1980 are shown in Table 1. The estimates of national emissions vary only slightly, and there is fairly close

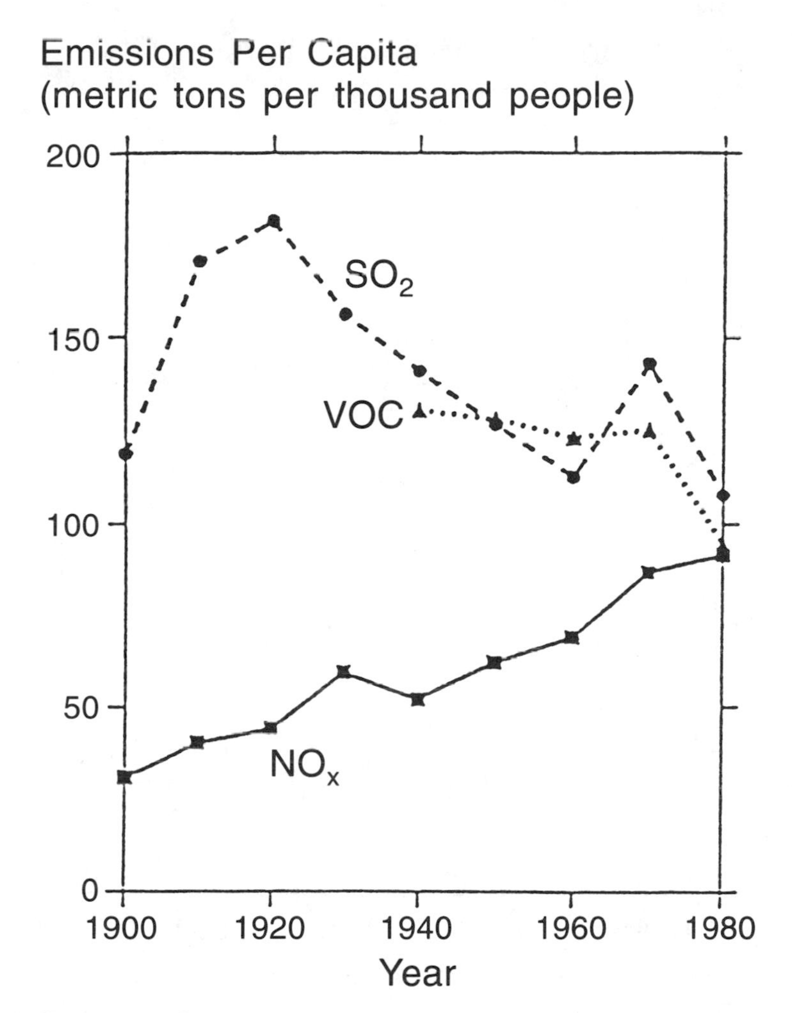

Figure 2. Trends in U.S. emissions per capita (in metric tons per thousand people). Values were computed for years between 1900 and 1980 in ten-year increments; trends shown do not reflect interannual trends within those ten-year periods.

agreement for individual source categories. It should be noted that a 1982 inventory estimates sulfur dioxide emissions of 15.8 million metric tons/year from electric utilities and 24.1 million tons/year from all sources.

Trends in Sulfur Dioxide Emissions

The combustion of coal has been the single largest source of sulfur dioxide emissions in the United States since 1900. Coal-fired power plants only recently have become a contributor of sulfur dioxide emissions because until about 1960

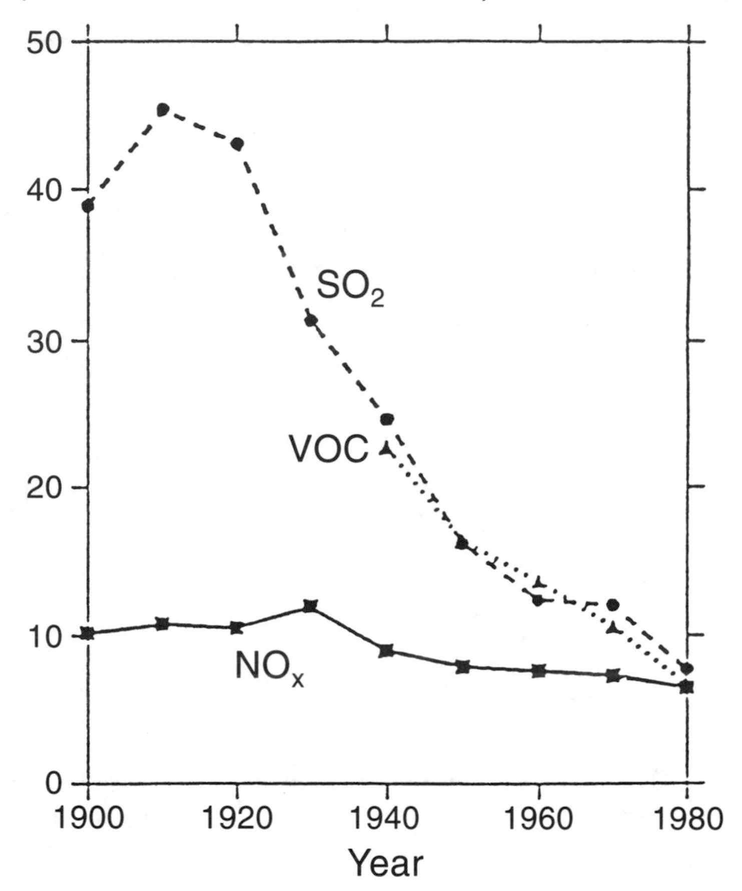

Figure 3. Trends in U.S. emissions per million dollars of GNP (metric tons/1982 dollars). Values were computed for years between 1900 and 1980 in ten-year increments; trends shown do not reflect interannual trends within those ten-year periods.

most coal combustion occurred in manufacturing facilities, residences, commercial establishments, and locomotives. The shift of coal combustion from small dispersed sources to large power plants may be significant to acid deposition studies [1].

The trend in sulfur dioxide emissions during the past 15 years has been strongly influenced by pollution control requirements as a result of the Clean Air Act. It was the Clean Air Act that required states to produce state implementation plans (SIPs) that outlined how standards for air pollutants would be set and achieved. It was under these SIPs that emission limits were set for major

Table 1
Estimates of Man-made SO_2 Emissions in the United States in 1980 (in million metric tons)

Emission Source	1980 NAPAP Emissions Inventory	NAPAP Monthly/State Data Base	EPA Trends Report	NAPAP Historical Data Base
Coal-fired power plants	14.6	14.7	14.2	15.1
Other power plants	1.3	1.2	1.3	1.2
Industrial combustion	3.3	2.4	2.4	2.9
Nonferrous smelters*	1.1	1.1	1.2	1.1
Other industrial processes	2.1	2.4	2.3	N/A
Residential/commercial fuel use	1.0	0.9	0.9	0.8
Transportation and misc.	1.1	0.9	0.9	N/A
Total	24.5	23.6	23.2	24.7

**Values shown include copper, lead, and zinc smelters.*

sources of existing air pollution sources. These regulations also are important because under these rules, the older not-as-stringently controlled plants are being phased out and are being replaced by more-controlled plants.

Economic factors also contributed to the decline of sulfur emissions in recent years. The price-competitiveness of low-sulfur coal compared to high-sulfur coal has resulted in overcompliance by some of the utility companies. Also, the increased use of some lower-emitting plants, along with the decreased use of some higher emitting plants, accounts for some of the recent reductions in sulfur dioxide emissions.

According to most of the available data, emissions of sulfur dioxide peaked in about 1970 at about 30 million metric tons. A steady decrease in sulfur dioxide emissions is estimated to have occurred between 1973 and 1983. Emissions of sulfur dioxide from coal-fired plants in 1983 are estimated to be about 13% lower than emissions in 1977 despite the fact that consumption of coal during these same years increased by about 28%.

Table 2 lists sulfur dioxide trends from 1975 to 1985. For the most part, sulfur dioxide emissions have been on the decline despite the increase in production of coal-fed power [1].

It is the ratio of sulfur dioxide emissions to energy produced that reflects the fact that even as the demand for and supply of energy increased, the emissions of sulfur dioxide decreased.

Nitrogen Compounds

Historical Emissions of Nitrogen Oxides

Figure 4 shows the historical trends in man-made nitrogen oxide emissions. Emissions of nitrogen oxides increased from about 2 million metric tons in 1900 to about 7 million metric tons in 1925. Between 1940 to the late 1970s, annual nitrogen oxide emissions increased three-fold—from 7 million to 21 million metric tons. Nitrogen oxides are emitted from the combustion of all fuels including coal, oil, natural gas. The historical trend in nitrogen oxide emissions reflects the increased use of fuels that emit nitrogen oxides during combustion, including coal, oil, and natural gas. Unfortunately, emission regulations promulgated by the Clean Air Act required smaller reduction in nitrogen dioxide than sulfur dioxide emissions. However, due to the required installation of automobile emission controls—a result of the act, nitrogen oxide emissions have remained fairly level during the past decade. They rarely fluctuated by more than about 6%. Annual emissions of nitrogen oxides are estimated at about 20 million metric tons [1].

Historical emissions in the United States on a per capita and per dollar basis can be seen in Figures 2 and 3. Emissions in nitrogen oxides per capita grew between 1900 and 1980, primarily because of the rise in per capita consumption

Table 2
Estimates of U.S. Coal Consumption in Electric Utilities and U.S. Man-made SO_2 Emissions: 1975 to 1985.

	Electric Utility Coal Consumption[a]		Coal-Fired Electric Utility SO_2 Emissions (Million Metric Tons)	Ratio, Million Metric Tons of SO_2 per 10^{18} J (Coal-Fired Utilities)	Total U.S. SO_2 Emissions From All Sources (Million Metric Tons)[c]
	Million Metric Tons	10^{18} J[b]			
1975	369	9.2	15.4	1.67	25.8
1976	407	10.2	15.9	1.56	26.6
1977	434	10.8	16.0	1.48	26.7
1978	437	10.7	14.5	1.34	24.9
1979	479	11.9	14.8	1.24	24.9
1980	517	12.7	14.7	1.16	23.6
1981	543	13.2	14.5	1.10	23.3
1982	540	13.2	13.8	1.05	21.6
1983	568	13.8	13.9	1.01	21.3
1984	603	14.7	14.3	0.97	21.7
1985	631	15.2	14.2	0.93	21.2[d]

[a] *Energy Information Administration, 1986.*
[b] *10^{18} Joules = 0.95×10^{15} Btu = 0.95 quad.*
[c] *Total U.S. SO_2 emissions presented here are apparently low by about 1 million metric tons due to possible omission of some emissions from nonpurchased fuels in the industrial sector.*
[d] *Total U.S. emissions were estimated by adding the 1985 value for electric utilities to 1985 values for other categories of sources from the Environmental Protection Agency, 1987.*

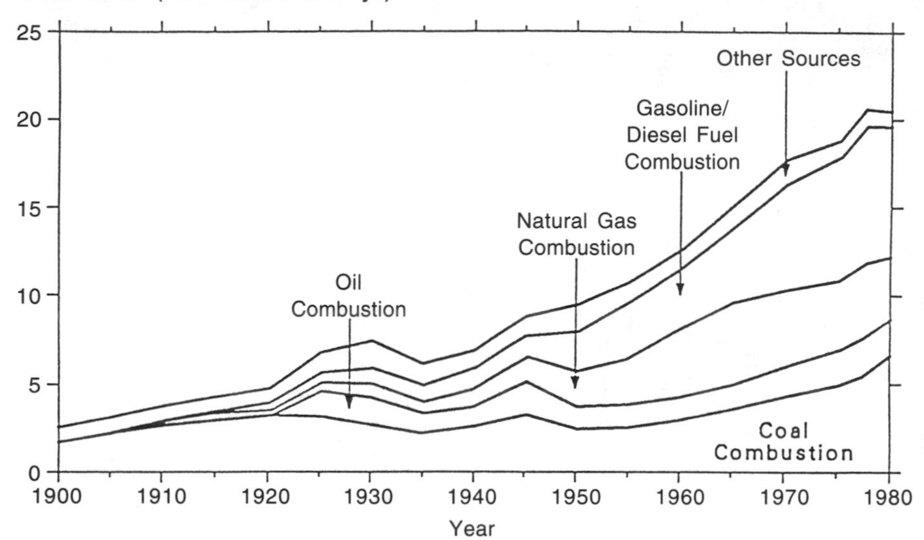

Figure 4. Historical trends in U.S. man-made NO_x emissions by fuel type from 1900 to 1980.

of fossil fuels and secondarily because of the increased use of fuel in high-temperature combustors such as power plant boilers [1].

Major Sources of Nitrogen Oxide Emissions

Although nitrogen found in fuels such as coal and oil is the source of some man-made nitrogen oxides emitted to the atmosphere, most of the emissions come from the combination at high temperatures of nitrogen (N_2) and oxygen (O_2) found naturally in the air. Therefore, combustion of a fuel like coal that has a lower nitrogen content than a fuel like natural gas will result in only slightly reduced emissions of nitrogen oxides. It is primarily the temperature at which combustion occurs, not the nitrogen content of the fuel, that controls nitrogen oxide emission rate. For example, the same amount of fuel burned in a high-temperature power plant boiler will produce more nitrogen oxides than if the same amount of fuel were burned in a home furnace at a lower temperature [1].

Fuel combustion in vehicles, power plants, industrial boilers, process heaters, and residential and commercial heaters in the principal source of nitrogen oxide (NO). However, as discussed later in this chapter, this is usually quickly converted.

Table 3 gives the various estimates of man-made sources of nitrogen oxides in the United States. Burner and combustion design strongly influence emissions,

Table 3
Estimates of Man-made NO_x Emissions in the United States in 1980 (in million metric tons)

Emission Source	Best Estimate	1980 NAPAP Emissions Inventory	NAPAP Monthly/State Data Base	EPA Trends Report	NAPAP Historical Data Base
Power plants	5.8	5.9	5.8	6.4	7.2
Industrial combustion	4.0	4.0	3.0	3.0	1.4
Highway vehicles	7.2	6.5	7.2	7.2	7.0
Other transportation activity	2.0	1.8	3.3[a]	2.0	1.3
Other sources	1.7	2.2		1.7	3.3
Total[b]	20.7	20.4	19.5	20.3	20.2

[a]*Off-highway vehicles and other miscellaneous sources were combined in this publication; the value does not include emissions from forest fires and other miscellaneous burning.*
[b]*Values may not sum to totals due to independent rounding.*

and the level of detail at which boiler characteristics are considered in the calculations affect emission estimates [1].

Trends in Nitrogen Dioxide Emissions

According to estimates, the amount of man-made nitrogen emissions during this century have increased proportionately with the increased use of fossil fuel.

National total nitrogen oxide emissions are estimated in Table 4. This table shows fluctuations during the past ten years. Since peaking in 1978, annual oxide emissions are estimated to have declined by about 10% by 1983, somewhat less than the sulfur dioxide decline during the same period. This decline resulted primarily from increased emission controls on cars and a decline in industrial fuel use. Between 1983 and 1985, estimated nitrogen dioxide emissions increased slightly.

In 1975, about 7.1 million metric tons of nitrogen oxides were emitted by highway vehicles, and an additional 1.8 million metric tons came from other

Table 4
Estimates of Man-made NO_x Emissions from Various Source Categories in the United States: 1975 to 1985 (in million metric tons/yr)

	Source Category				
Year	Highway Vehicles	Power Plants	Industrial Fuel Combustion	Other[a]	Total
1975	7.1	4.8	3.5	3.2	18.6
1976	7.4	5.1	3.7	3.3	19.5
1977	7.6	5.4	3.7	3.5	20.2
1978	7.6	5.4	3.7	3.6	20.3
1979	7.3	5.5	3.6	3.6	20.0
1980	7.2	5.8	3.0	3.3	19.3
1981	7.3	5.9	3.0	3.2	19.4
1982	7.0	5.6	3.1	3.0	18.7
1983	6.8	5.8	2.7	2.9	18.2
1984	6.8	5.8	2.9	3.1	18.6
1985[b]	7.1	6.3	2.9	3.0	19.3

[a]*The category "other" includes residential and commercial fuel use, industrial processes, aircraft, railroads, vessels, and off-highway vehicles. It does not include from forest fires and other burning.*

[b]*The 1985 value for total emissions was obtained by adding the 1985 power-plant estimate to 1985 values for other categories of sources from the U.S. Environmental Protection Agency 1987.*

transportation sources such as railroads. Gasoline-powered autos contributed 4.4 million metric tons, heavy-duty diesel vehicles contributed 1.4 million metric tons, and light trucks and other miscellaneous vehicles contributed the remainder (1.4 million metric tons) for 1975. By 1984, emissions from gasoline-powered autos had dropped to 3.3 million metric tons—a 25% decrease; however, emission from heavy-duty diesels increased by 38% to almost 2 million metric tons, and emissions from light trucks and other miscellaneous sources increased by 12% to 1.6 million metric tons. Therefore, although large reductions in nitrogen oxide emissions from gasoline-powered autos are estimated to have occurred due to fairly stringent controls resulting from the Clean Air Act, high growth in nitrogen oxide emissions from other highway categories generally has reduced the effect of those controls on total emissions. Additionally, from 1984 to 1985, nitrogen oxide emissions from gasoline powered autos are estimated to have increased by about 5%—the first increase since 1975.

Increases in emissions from power plants also offset the decreases from automobiles. Power plant emissions of nitrogen oxides increased by 33% in the 1975 to 1985 period shown in Table 4 because of increased use of fossil fuels in power plants. Emissions of nitrogen oxides from power plants are not as stringently controlled as those of sulfur dioxide. Newer emission control regulations on power plants are not much more stringent than older regulations [1]

Other Species

Volatile Organic Compounds

Emissions of volatile organic compounds (VOCs) are more difficult to characterize and estimate than those of sulfur dioxide or nitrogen oxide because of the wide diversity of sources. Most man-made VOC emissions come from small dispersed sources rather than large point sources (see Table 5).

Transportation activity is the largest source of VOC emissions, contributing about 38% in 1980. It is estimated that more than 08% of transportation emissions come from gasoline combustion. Use of organic solvents and paints in industry and commercial operations accounted for about 35% of VOC emissions in 1980 [1].

Table 6 shows the estimated contribution of man-made and natural sources of volatile organic compounds. It should be noted that estimates of man-made emissions in this table are subject to an uncertainty of plus or minus 50%. Natural emission estimates are subject to an uncertainty factor of 3%.

Volatile organic compounds are important in the formation of hydrogen peroxide and, hence, in the production of acidic deposition. Therefore, the uncertainties associated with the absolute levels of VOC emissions and natural sources should be reduced [1].

Table 5
Estimates of Man-made VOC Emissions in the United States in 1980 (in million metric tons)[a]

Emission Source	Best Estimate	1980 NAPAP Emissions Inventory	EPA Trends Report
Transportation activity	8.2	9.4	8.2
Fuel combustion	2.5	2.5	2.2
Industrial surface coating[c]	1.9	1.9	2.3
Other industrial processes	3.9	2.0	4.5
Misc. organic solvent use	1.9	1.9	1.9
Petroleum product storage and marketing	1.4	1.4	2.1
Forest fires[d]	0.8	0.8	0.9
Other[e]	0.8	0.8	0.7
Total[f]	21.4	21.0	22.8

[a]*Values exclude methane and ethane emissions. Emissions are expressed in terms of tons of the individual hydrocarbon compounds emitted.*
[c]*Application of paint.*
[d]*Forest fires can be considered either a man-made or natural source.*
[e]*Includes solid waste disposal, incineration, open burning, and fires (except forest fires).*
[f]*Values may not sum to totals due to independent rounding.*

Ammonia

Most fixed nitrogen emitted from biogenic sources into the atmosphere is in the form of ammonia. A majority of the ammonia that enters the atmosphere is produced by the biological decomposition of nitrates in organic matter in soils, plant residues, and wastes from animals both wild and domestic. Also, ammonia can be directly volatilized by burning biomass and from agricultural fertilizers. Annual global emissions of ammonia from biogenic sources, not including the effects of ammonia exchange with surface vegetation, amount to approximately 87 million metric tons of ammonia. Other estimates of annual global ammonia emissions form biogenic sources range from 22 million to 300 million metric tons.

Industrial sources of ammonia emissions include the manufacture of ammonia, ammonium nitrate, urea, fertilizer, petroleum refining, cooking, fertilizer application, and fuel combustion. Estimates of the man-made sources of ammonia emissions are listed in Table 7.

Ammonia is unique among atmospheric trace species as the only gaseous basic compound present in significant concentrations. Most of the emitted ammonia

Table 6
Estimated Emissions of VOCs from Man-made and Natural Sources by Region (in million metric tons)

	Season/Source														
	Winter			Spring			Summer			Fall			Total[a]		
Region	Man-Made	Natural	Total[a]	Man-Made	Natural	Total[a]	Man-Made	Natural	Total[a]	Man-Made	Natural	Total[a]	Man-Made	Natural	Total
New England	0.3	0	0.3	0.3	0.1	0.4	0.2	0.4	0.6	0.3	0.2	0.5	1.1	0.7	1.8
New York/New Jersey	0.5	0	0.5	0.4	0.1	0.5	0.4	0.3	0.7	0.4	0.1	0.5	1.7	0.5	2.2
Middle Atlantic	0.6	0	0.6	0.5	0.2	0.7	0.5	0.7	1.2	0.5	0.3	0.8	2.0	1.3	3.3
Southeast	1.0	0.5	1.5	1.0	1.4	2.4	0.9	3.3	4.2	0.9	1.6	2.5	3.7	6.8	10.5
Great Lakes	1.1	0	1.1	1.0	0.4	1.4	1.0	1.5	2.5	1.0	0.6	1.6	4.0	2.5	6.5
Total eastern U.S.[a]	3.4	0.5	3.7	3.2	2.2	5.4	3.0	6.2	9.2	3.1	2.8	5.9	12.5	11.8	24.3
South Central	0.8	0.5	1.3	0.8	1.6	2.4	0.9	4.0	4.9	0.8	1.8	2.6	3.2	7.8	11.0
Central	0.3	0	0.3	0.3	0.3	0.6	0.3	0.9	1.2	0.3	0.3	0.6	1.2	1.6	2.8
Mountain	0.2	0	0.2	0.2	0.8	1.0	0.3	2.5	2.8	0.3	1.0	1.3	0.9	4.3	5.2
West	0.6	0.4	1.0	0.6	1.0	1.6	0.6	2.7	3.3	0.6	1.4	2.0	2.3	5.4	7.7
Northwest	0.2	0.2	0.4	0.3	0.8	1.1	0.2	1.8	2.0	0.2	1.0	1.2	0.9	3.8	4.7
Total western U.S.[a]	2.0	1.1	3.1	2.2	4.5	6.7	2.3	11.9	14.2	2.2	5.5	7.7	8.5	22.9	31.4
Total U.S.[a]	5.4	1.7	7.1	5.2	6.8	12.0	5.2	18.1	23.3	5.3	8.3	13.6	21.0	34.8	55.8

[a] *Values may not sum to totals due to independent rounding.*

Table 7
Estimates of Man-made Emissions of Ammonia in 1980 in the United States (in million metric tons)

Emission Source	Emissions
Livestock/Waste Management	0.54
Fertilizer Production	0.11
Anhydrous NH_3 Application	0.05
NH_3 Synthesis	0.04
Petroleum Refining	0.04
Motor Vehicles	0.03
Fossil Fuel Combustion	0.02
Coke Manufacture	0.01
Total	0.83

is removed by processes such as wet and dry deposition; therefore ammonia plays an important role in aerosol formation and precipitation chemistry. About 20% of the ammonia in the atmosphere is photochemically oxidized in the atmosphere. It has been estimated that ammonia in the atmosphere acts as a small nitrogen oxide sink over North America.

Because large amounts of ammonia are volatilized into the atmosphere from biogenic sources and most are returned to the surface in wet or dry deposition, ammonia represents an important intermediary for the redistribution of fixed nitrogen through the biosphere. It also is important to note that when ammonia is redeposited on soils, nitrification of ammonia can form acids that lower soil pH.

Alkaline Aerosols

The most important thing to note about alkaline aerosol emissions is that they can significantly influence precipitation acidity by neutralizing some fraction of the acids. It is thought that soil sources probably supply most of the alkali and alkaline content in precipitation. Alkaline aerosols are introduced into the atmosphere from open sources, industrial sources, and other miscellaneous sources. Open sources include traffic or unpaved roads and wind erosion. Industrial sources include iron and steel manufacture, magnesium production, cement and concrete production, fuel combustion, and solid-waste disposal. According to the National Acid Precipitation Assessment Program (NAPAP), about 0.16 million metric tons of alkaline particulate matter was emitted from these various industrial sources in the United states in 1980 [1]. It is thought that of all alkaline

Table 8
Estimated U.S. Emissions of Alkaline Aerosols from Open Sources (in million metric tons/yr)

Region	Calcium	Magnesium	Sodium	Potassium
	1.0	0.1	0.1	0.1
Northwest	2.8	0.4	0.3	0.3
Southwest	2.2	0.6	0.8	0.7
Central	2.5	0.3	0.2	0.2
Northeast	1.6	0.2	0.8	0.1
Gulf Coast				
Total U.S.	1.0	1.7	1.4	1.3

emissions about, 90% are from open sources. Recent estimates of these emissions are presented in Table 8.

Emissions presented in this table include only the fraction of aerosols that influence precipitation chemistry. Alkaline aerosols from open sources are suspended in the atmosphere by the wind or other mechanical disturbances of the land surface. The potential effect of this resuspended material on acidic precipitation depends on several factors. The total flux of material into the atmosphere depends on the composition, condition, and extent of a surface coupled with the nature of the physical and mechanical forces acting to displace the surface material. Although large amounts of material are raised from the surface, most of the material falls back to the surface within several meters of the source. However, some of the material can remain suspended in the atmosphere. The amount of material that stays suspended depends largely on the particle size. Finally, the influence of this material depends on the elemental and chemical composition of the particles and their solubility and chemical stability in the atmosphere. All of these factors were taken into consideration when the table was prepared [1].

Hydrogen Chloride

While sulfuric and nitric acids are thought to be the major constituents of acidic deposition, contribution of hydrogen chloride and hydrogen fluoride could be locally significant. Research conducted by NAPAP estimated emissions of hydrogen chloride and hydrogen fluoride by identifying significant man-made emission sources and estimating emission rates through a literature search. The estimates are shown in Table 9.

Table 9
Estimated U.S. Emissions of HCL and HF
(in million metric tons/yr)

	Emissions	
Emission Source	**HCL**	**HF**
Coal Combustion		
Utilities	0.42	0.05
Industrial Boilers	0.11	0.01
Residential Boilers	–0.	–0.
Industrial Processes		
Propylene Oxide Manufacture	0	—
Hydrogen Fluoride Manufacture	—	–0.
Primary Aluminum Production	—	0.01
Phosphate Fertilizer Production	—	0.01
Incineration	0.07	—
Total	0.06	0.08

ATMOSPHERIC CHEMICAL PROCESSES UPON PRECURSORS

Nitric Acid

Most of the anthropogenic nitrogen dioxide emission is in the form of nitric oxide. The oxidation of nitric oxide to nitrogen dioxide is accomplished by ozone and periaxial radicals. The transformation of nitrogen dioxide to nitric acid in the daytime is primarily due to the oxidation of nitrogen dioxide by hydroxyl radicals:

$$NO_2 + OH \rightarrow HNO_3 \quad (1)$$

The uncertainty of the conversion rate of nitrogen dioxide to nitric acid is due to not knowing the abundance of the hydroxyl radicals in the atmosphere, but the reaction rate has been determined accurately by laboratory studies. It is very difficult to gauge the amount of hydroxyl radicals in the atmosphere. Through different types of modeling, it has been shown that the concentration of the hydroxyl radicals is a function of many parameters, including solar UV radiation, water, vapor, nitrogen oxides, volatile organic compounds, and carbon monoxide.

A particularly effective sink for hydroxyl radicals that has not been considered in most models is the heterogenous removal of the hydroperoxyl radicals by aerosols. Recently, laboratory measurements of the mass accommodation coefficient of hydroperoxyl radicals with aerosols suggests that the removal of hydroperoxyl radicals, and thus hydroxyl radicals, by chemically reactive aerosols is highly efficient. If the atmospheric aerosols can remove hydroperoxyl radicals as efficiently as the laboratory particles, the rate of reaction could be reduced by more than a factor of two [1].

Another factor that must be considered in the above conversions is nighttime chemistry. In summer, the daytime conversion can be carried out at about the same rate in the nighttime, whereas in winter the nighttime conversion rate can be 10 times greater than the daytime rate [1].

The nighttime conversion process may play an important role in the seasonal behavior of nitrate ion deposition. Observed nitrate ion concentration in precipitation show little seasonal variation, whereas the sulfate concentration in the winter is two to three times lower than in the summer in the United States. Because neither nitrogen oxides nor sulfur dioxides has any large seasonal variation, it is argued that the nighttime conversion of nitrogen oxides compensates for part of the seasonal asymmetry in the daytime conversion rate. The ratio of summer to winter conversion rate is about 12 to 1 without the nighttime conversion and about 3 to 1 with it. Apparently a more stable boundary layer in winter is needed to compensate for the rest of the seasonal asymmetry [2].

Sulfuric Acid

The gas-phase formation of sulfuric acid occurs primarily through reactions of sulfur dioxide with hydroxyl radicals:

$$SO_2 + OH \rightarrow HSO_3 \qquad (2)$$

The hydrogen sulfite radical (HSO_3) is responsible for the formation of sulfuric acid through the following reactions:

$$HSO_3 + O_2 \rightarrow HO_2 + SO_3 \qquad (3)$$

$$SO_3 + H_2O \rightarrow H_2SO_4 \qquad (4)$$

The formation of the hydroperoxyl radical is important because it is converted back the hydroxyl radical via reactions (3), (4), and nitric oxide. This allows reactions (2), (3), and (4) to proceed catalytically without consuming hydrogen radicals; otherwise reaction (2) would consume hydrogen radicals and the equation would then be non-linear, especially when sulfur dioxide concentrations are high. To date, there have been indirect measurements supporting reaction (3). Based on this and other recent information, it is estimated that the gas-phase formation of sulfuric acid is probably a linear process, i.e., it is not limited by oxidant availability, thus the more sulfur dioxide going into the atmosphere the more sulfuric acid is produced [2].

The conversion rate for reaction (2) depends on the concentration of hydroxyl radicals. As mentioned previously, this is a function of ambient conditions. A typical summer diurnal average conversion rate is about 0.5% per hour in the boundary layer of the rural atmosphere [2].

It is important to remember that the major oxidants, hydrogen peroxide and ozone, for sulfur dioxide for aqueous-phase reactions are produced in gas-phase reactions. A wide range of hydrogen peroxide concentrations has been predicted from photochemical simulations. Consideration of the various sources and sinks of the free radicals in the atmosphere leads to predicted hydrogen peroxide levels of about 3 to 5 ppb for summertime in the Northeast. This is consistent with recent observations [1].

Model calculations predict a gas-phase production of hydroxyl radicals and hydroperoxyl radicals in winter that is ten times less than that in summer. As a result, the gas-phase formation of sulfuric acid and the gas-phase abundance of hydrogen peroxide in winter will decrease accordingly. This reduction in photochemistry contributes to the observed lower sulfate concentrations in precipitations in winter than in summer [1].

Other Species

Recent measurements show that relatively large amounts of formic and acetic acids are present in precipitation in both remote and polluted areas. These acids contribute significantly to the acidity of precipitation, although their full impact nationwide has not been determined.

The photochemical processes that lead to the formation of organic acids are complex, and knowledge of them is quite uncertain. Both gaseous- and aqueous-phase reactions contribute to their formation. A number of gas-phase processes lead to the formation of organic acids. A laboratory study of the products of alkenes in the presence of nitric oxide found that 5% to 20% is converted to organic acids in the gas phase. Their measurements can be explained if the gas-phase reactions are responsible for most of the organic acids [1].

The sources of organic acids have not been quantitatively identified. Natural hydrocarbons must be a significant source, as indicated by the presence of organic acids in significant amounts in remote areas. On the other hand, anthropogenic impacts also should be large either through direct VOC emissions or via influences on the chemistry because of the close correlations found between organic acids and nitrogen oxides at rural as well as urban stations.

PRECIPITATION PROCESSES

Air Motion Within Cloud Systems

Air motion has considerable effect on precipitation scavenging and should be noted for the following reasons:

- Often, particularly on a local scale, air motions are driven primarily by the energetics of the storm/cloud system that are inducing the scavenging process;
- Air motion in and around a storm determines the amount of air processed by the storm/cloud system, and thus the amount of pollutant exposed to the possibility of wet removal;
- Mixing processes within the storm/cloud system control the interaction of relative pollutants, hence influencing the amount of reaction that can take place and affecting the efficiency of the scavenging process; and
- Air motions within a storm/cloud system are responsible for the movements of all pollutants, regardless of whether they are redeposited. Pollutants that are not deposited are transported vertically in storm/cloud systems updrafts, and such motions strongly influence subsequent long-range transport.

Attachment Processes

Once a pollutant particle or molecule is transported into a region of a storm/cloud system where condensed water is present, it may attach to cloud water, ice, and precipitation by a variety of mechanisms. As can be seen in Figure 5, these

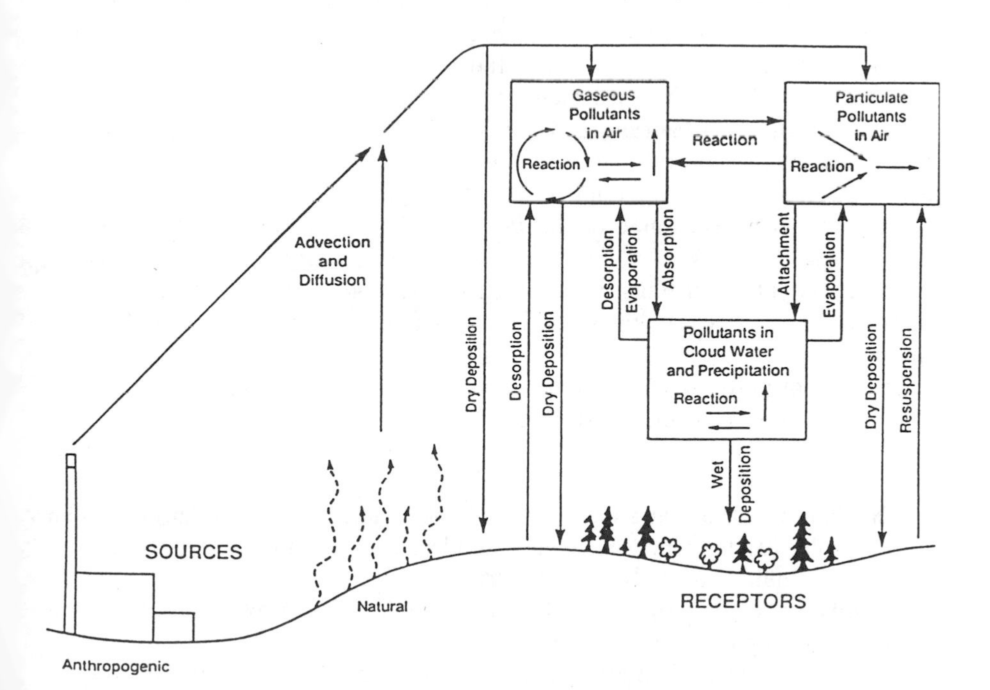

Figure 5. A schematic diagram of the interrelationships of processes that modify air pollutants as they are transported from sources to deposition.

processes can be reversible in the sense that both attachment and detachment can occur. The rates at which these processes take place depend on the amount of condensed water that is present, the conversion processes occurring locally between various classes of water (e.g. ice, vapor, cloud water, etc.), and the affinity of the water for a particular pollutant.

Physical attachment of dry pollutants to condensed water is absolutely necessary to the scavenging process, and the rate at which such attachment occurs has a large bearing on the overall wet removal rate. There is a substantial mechanistic difference between the attachment, of aerosol particles and gas molecules.

Both particle size and chemical composition are important factors determining the rate at which aerosol particles are attached to droplets and thus removed from the atmosphere. Aerosol particle size is described in terms of modes on a size-distribution plot (Figure 6).

These modes are associated with individual formation and removal processes. Within this scheme, the so-called fine-particle mode consists of particles 0.2 μm in diameter or less, which are formed by condensation and reaction of gaseous precursors. Owing to their small size, these particles have high mobility and rapidly agglomerate into larger particles to make up the accumulation mode, with particle sizes in the range between 0.2 μm and 2 μm. The preponderance of material (sulfate, nitrate particles, etc.) of interest to acidic precipitation formation usually exists in the accumulation range. Above about 2 μm, typical aerosol size distribution enters the so-called mechanical mode, made up largely of wind-blown dust particles, fly ash, and products of incomplete combustion. Gravitational settling and inertial effects become important elements in this size range and limit the abundance of coarse material in the atmosphere at any one time. It is important to note that typical cloud-droplet spectra are distributed in the size range of the mechanical aerosol mode, and thus cloud droplets exhibit many of the physical properties usually considered to be in this category. Physical transformations, such as agglomeration, of aerosol particles can be described by kinetic equations that are similar to those of chemical rate processes [1].

Dry aerosol particles can become attached to condensed water by a number of individual mechanisms. These include:

- **Nucleation:** The migration of water vapor molecules to the surface of an aerosol particles with subsequent condensation in sufficient quantities to form a liquid drop.
- **Brownian diffusion:** The diffusional transport of aerosol particles to an existing droplet.
- **Impact and interception:** The initial collection of aerosol particles by falling droplets.

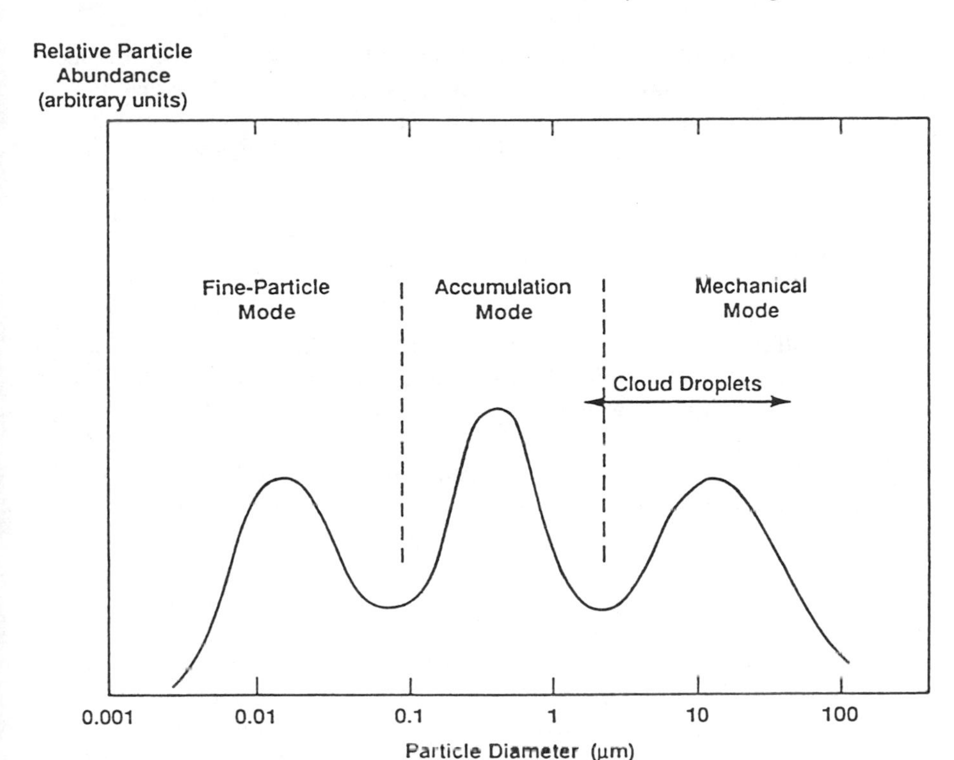

Figure 6. Hypothetical categories of particle sizes in clean air, showing the sub-micrometer "fine particle" class of major importance to long-range transport, and the "coarse particle" mode of more local interest. The vertical scale is intended to be indicative of particle concentrations in air, by number, surface area, or mass.

- **Diffusionphoresis:** The transport of aerosol particles to the surface of a droplet induced by the flux of water vapor molecules migrating to the surface.
- **Thermophoresis:** The thermally induced transfer of aerosol particles to the surface of a droplet that is cooler than its surroundings.
- **Electrostatic transport:** The migration of particles to a droplet surface caused by a difference in electrical charge.
- **Pollutant gas attachment:** Owning to the high diffusive motility of gases, advective/diffusive attachment dominates other mechanisms for pollutant gas uptake by all classes of *hydrometers.* Well-established mathematical formulations exist to describe gas-phase transport under such circumstances. Several additional features become active in these situations, however, lending substantial complexity to the overall attachment process.

The relative importance of these mechanisms differs depending upon the local environment. Most acid-forming aerosol particles are so hygroscopic that nucleation mode of attachment dominates whenever cloud formation processes are active. Under such conditions other mechanisms usually are important in special circumstances and are typically ignored in acid rain models.

The question of gas solubility does not pose any large practical problem for acid precipitation assessment as long as numerical values for pertinent solubility parameters are known. Fortunately, solubility parameters for most common gases that act as acid precursors have been measured. For example, the water solubility sulfur dioxide, nitric acid, peroxyacetyl nitrate, ozone, hydrogen peroxide, hydrogen chloride, ammonia, formaldehyde, and carbon dioxide all have been quantified to a point where values can be employed reliably in scavenging calculations [1].

A second set of features complicating the question of pollutant gas uptake involves interfacial and aqueous-phase transport. Molecular exchange across a gas-liquid interphase is determined partly by gas-phase transport to the aqueous surface, but transport across any surface—such as an organic monolayer—and mass transfer within the liquid also potentially are important. In addition, there is usually a finite chance that pollutant molecules approaching the liquid surface may be reflected on contact rather than being accommodated by the aqueous medium. This later feature generally has been treated mathematically using the concept of an accommodation coefficient whose value ranges between zero (total reflection) and one (total accommodation). In aggregate, each of these features can be visualized as a series of "resistance" to pollution uptake; comparatively high uptake resistance is considered to be "rate influencing" while particularly low resistances are unimportant as controlling features of the uptake of pollutants [1].

While the first three transport mechanisms listed can be predicted with generally acceptable accuracy, the accommodation coefficient is not well understood at the present time. Most work has assumed that the impedance to mass transfer is negligibly small compared to other factors and thus does not constitute a rate-influencing step. More recently, several authors have questioned this assumption. The accommodation coefficiences for both ozone and sulfur dioxide have been measured, and the results have been incorporated with mass transfer chemical reaction equations to determine the associated rate-influencing effect. Although the measured value of the ozone accommodation coefficient is rather loose, this recent work indicates that other limiting effects render accommodation considerations, at least for sulfur dioxide and ozone, negligible as barriers to the sulfur dioxide scavenging and oxidation process.

In summary, many of these microscopic phenomena are sufficiently well understood to allow reliable estimates of pollution attachment, but additional work is required in specific areas before a total picture is available [1].

Transformations

Sulfur Chemistry

In cloud and precipitation systems, sulfur usually exists in the water in its (IV) and (VI) oxidation states. Aqueous phase sulfur in the (IV) oxidation state appears as a consequence of absorption of gaseous sulfur dioxide. This absorption process generally is limited by solubility equilibria, which are pH dependent. In addition, the acid-base equilibria, sequestered by organic complexing agents, is possible; dissolved formaldehyde, for example, is known to din into S(VI), stabilizing it in a form that is immune to oxidation and enhances its total solubility as well.

Aqueous-phase sulfur in the (VI) oxidation state (sulfate ion) is incorporated both by scavenging of preexisting sulfate particles and by aqueous phase oxidation of S(IV). The relative importance of the direct and reactive mechanisms appears to depend strongly on storm conditions and chemical mix and has been the subject of continued conjecture. It has been predicted that in-cloud oxidation mechanisms account for 55% to 70% of precipitation sulfate in summertime convective storms. The fraction of sulfur dioxide converted to sulfate in clouds in the wintertime is estimated to be similar, but the total sulfur deposition is only about one-third of that produced in the summertime. High regional conversion amounts also have been estimated based on the calculated abundance of aerosol and precipitation sulfate. Measurements made upwind and downwind of individual clouds, in most cases have indicated aqueous-phase formation of S(VI), and measurements upwind and downwind of urban areas appear to indicate substantial in-cloud production of S(VI) as well. The acidity of cloud water has been observed to be greater than can be explained by composition of aerosols and gases in nearby clear air [1].

Measurements of oxidants and the distribution of sulfur pollutants between the gaseous and aqueous phases are consistent with a large fraction of sulfur dioxide being converted to aqueous phase S(IV) in the summer. A different situation, in which preexisting aerosol clouds account for all sulfate in precipitation, also has been observed in winter places such as Michigan. These measurements, combined with enhanced understanding of oxidant behavior acquired during recent years, suggest a substantial attenuation of reactive sulfur scavenging during winter months. Typically the percentage of sulfate aerosol that is neutralized, mostly by ammonia, is approximately 75%, but neutralization varies significantly between storms. In extreme cases, very little neutralization takes place.

Reaction rate coefficients for S(IV) oxidation by different chemical pathways have been measured in a number of laboratories. Pursuant to assumptions on atmospheric composition, reaction rates with hydrogen peroxide, ozone, HNO_2, nitrogen dioxide, and oxygen as a function of pH, are summarized as follows.

- Reaction with oxygen in the absence of a catalyst is negligibly low.
- Reaction with hydrogen peroxide is nearly independent of pH over the pH range of cloud water and precipitation at a hydrogen peroxide level equivalent to a gas-phase partial pressure of 1 ppb. Conversion rates of sulfur dioxide of several hundred percent per hour can occur. This means that most of the sulfur dioxide is converted in a very short time. Hydrogen peroxide concentrations of a few parts per billion are common in the eastern United States during the summer.
- Reaction with ozone is expected to be important above pH 5. The inverse relation between sulfur dioxide solubility and hydrogen concentration causes this reaction to strongly decrease in importance at a lower pH level.
- At the high end of the plausible range for transition metal ion concentrations, canalized reactions with oxygen may be significant.

In-cloud oxidation of sulfur dioxide by hydrogen peroxide can lead to a nonlinear relation between ambient sulfur dioxide and precipitation S(VI). The nonlinearity arises because there will be circumstances in which ambient levels of sulfur dioxide exceed ambient levels of hydrogen peroxide, locally in summer, probably generally in winter. There is observational evidence that under these circumstances the hydrogen peroxide oxidizes an equivalent amount of sulfur dioxide. The remaining sulfur dioxide is not readily oxidized because of the low pH and consequent low sulfur dioxide solubility. Because clouds generally are more acidic in the summertime and sulfur dioxide solubility also decreases with increasing temperature, dissolved sulfur dioxide concentrations in summertime are considerably lower than in winter. Accordingly, if sulfur dioxide concentrations are reduced by a value equal to the concentration of hydrogen peroxide, it is expected that there would be little change in the amount of S(IV) formed. The prevalence and regional scale implications of the local imbalance between oxidant and sulfur dioxide, however, have not been quantified. Nonlinear relations between sulfate ion formation and sulfur dioxide are expected also for other oxidation mechanisms in conjunction with the pH dependence of sulfur dioxide solubility [1].

Nitrogen Chemistry

Nitrate in precipitate can result in the scavenging of gas-phase nitric acid, specific gas-phase nitrogen oxides, and aerosol nitrate ions. In contrast to sulfur chemistry, there is no evidence that significant amounts of acid are formed from aqueous-phase oxidation of the emitted pollutants, nitric oxide, and nitrogen dioxide. According to recent measurements of the solubility and kinetics of peroxyacetyl nitrate and nitrous acid, these compounds are likewise not expected to be significant sources of nitrate ions in precipitation [1].

Measurements of cloud water and precipitation composition during a study of the frontal storms suggests the formation of nitric acid consistent with an in-cloud aqueous-phase source. In the absence of sunlight, it is expected that nitric acid can be produced from the gas-phase reaction sequence:

$$NO_2 + O_3 \rightarrow NO_3 + O_2$$
$$NO_3 + NO_2 \rightarrow N_2O_5$$

followed by:

$$N_2O_5 + H_2O \rightarrow 2HNO_3 \text{ (aqueous)}$$

with conversion rates of nitrogen dioxide on the order of 10% per hour. During the day, nitrate is rapidly destroyed via photolysis or reaction with nitric oxide. However, even in the presence of sunlight or nitric oxide, this reaction sequence could be driven toward formation of nitric acid if mass transfer of nitrogen pentoxide to water, currently unknown, can compete effectively with gas-phase back reactions. The solubility and kinetics of nitrate also are not well understood. Continued production of gas-phase nitric acid in air within clouds also can occur in daytime via the usual gas-phase photochemical route:

$$NO_2 + OH \rightarrow HNO_3$$

The rate of conversion of nitrogen dioxide from this process can be several tens of percent per hour but will differ in the cloud due to a changed chemical environment and solar intensity. The removal of all soluble chemicals by the cloud water causes the change in the chemical environment. The corresponding sulfur reactions:

$$SO_2 + \tfrac{1}{2}O_2 \rightarrow SO_3$$
$$SO_3 + H_2O \rightarrow H_2SO_4$$

occurs with a rate constant of an order of magnitude lower than is the case for nitrogen dioxide. As a consequence, an unrealistically long cloud lifetime would be required for this gas-phase reaction to yield an appreciable apparent in-cloud sulfuric acid source.

Other sources of aqueous-phase nitrate ions may exist. Gas-phase measurements at Niwot Ridge, Colorado, indicate that the nitrogen pollutants detected by a nonselective chemiluminescent nitrogen oxide detector is often twice that which could be identified on the basis of measurements of individual compounds. More recent work suggests another organic nitrate compound in addition to peroxyacetyl nitrate [1].

Aqueous Phase Chemistry

Recent calculations have explored the possibilities of an aqueous-phase photochemistry involving interactions with radicals produced in sites or transferred from the gaseous to the aqueous phase. These studies have a much shorter history than their gas-phase counterparts. Many interesting mechanisms have been suggested, but the quantitative evaluations of rates depends upon further direct measurements in laboratory investigations.

As in the gas phase, the hydroxyl radical can oxidize hydrocarbons, and recombination reactions of hydro peroxy radical (or its reaction products with hydrogen oxide) can form hydrogen peroxide. Hydroxyl radical and hydroperoxy radicals are rapidly interconverted, as in the gas phase, via reactions with ozone and hydrocarbons. Formation of ozone via reactions with nitric acid, which is a prominent feature of gas-phase photochemistry, is not expected to occur to any significant extent in the aqueous phase because of the very low solubility of nitrogen oxide.

Both primary and secondary photochemical phenomena occurring in cloud environments depend strongly on local actinic flux. Although model estimates of such flux levels have been made, the question of exact values and spatial variabilities of these values within clouds remain major factors that limit progress in this area [1].

Delivery by Precipitation

As noted previously, pollutants attached to various condensed aqueous elements within a storm may be advected from the storm's domain and reinjected to the gaseous atmosphere, or they may be delivered to the surface by precipitation. The fraction of water vapor entering the storm that is removed by precipitation is often referred to as the water extraction efficiency. Within a cloud, the process leading to the formation of precipitation—such as cloud droplet coagulation, accretion, and freezing—also acts as the mechanism for creating chemical loadings in precipitation, thus the water extraction efficiency is strongly linked to the storm's efficiency for pollutant removal.

Both water extraction and pollutant removal efficiencies can range from zero to nearly unity, with the cloud physics of precipitation formation representing a strong determinant of the actual magnitude of any given storm. This final stage represents an important rate-influencing step of the scavenging sequence. Because of this, it is essential that these processes by well understood and simulated adequately in models.

Most scavenging modeling efforts within NAPAP have approached this problem by treating the cloud physics in a consolidated form, similar to that for nucleation processes. Thus, all cloud droplet sizes are lumped together into a single "cloud" category, and so forth. Interconversions of water and pollutants between classes are described in terms of rate equations whose coefficients are estimated on the basis of both theoretical considerations and experience.

Such lumping of cloud physics in applied scavenging models is appropriate because of computer time constraints and because present theoretical understanding does not warrant more elaborate alternatives for widespread practical use. One the other hand, predictive output of these models depends critically upon just how well these parameterizations are chosen. Significant improvements are being made to the lumped-cloud physics parameter presently in use. Tests have demonstrated that currently accepted rainfall speeds and evaporation parameters perform reasonably well for most frontal storms, but grossly underpredict rain and pollutant delivery for many types of summertime convective storms. Progressive improvements of these parameterizations, based on both theoretical and observational factors, are continuing under NAPAP and are intended to provide the most efficient and credible parameters for scavenging modules existing within regional models.

A second concern with presently applied parameter lumping techniques relates to their implicit assumption that pollutant concentrations in the various classes of hydrometers are essentially invariant with hydrometer size. Early field information has suggested that this assumption may be unwarranted in the case of raindrops, and recent Swedish measurements have indicated a considerable size dependence of concentration in cloud droplets. These studies should be extended and examined carefully.

Cloud and Fog Depositions

This section will describe how depositions from the atmosphere take place in the absence of precipitation, largely by mechanisms associated with turbulence, interception, impaction, and gravitational settling. There are occasions in which the mechanisms associated with dry deposition apply to the small droplets present in clouds and fog. In some context, the deposition of pollutants in cloud and fog water droplets are of prime importance. These processes are like those of dry deposition and therefore are discussed in this section [1].

The deposition of cloud and fog droplets usually is not effectively measured by wet-collection buckets; even if a wet-only device were to open in the presence of fog, it is likely that most of the deposition would take place on the outside of the bucket, not the inside. It is now known that the deposition of small water droplets generated in clouds and fogs can serve as a highly efficient mechanism for the transfer of pollutants from the air to the surface, a process not easily measured by any collection device.

Various mechanisms of cloud and fog formation exist, but it is adequate to point out that fogs occur in humid environments, on cool nights, and in low-lying areas. In contrast, clouds are formed by dynamic processes and can be intercepted by mountains. Furthermore, the presence of a mountain changes airflow, forcing air to flow upwards along mountain slopes, which can result in the formation of clouds around the peak.

This mechanisms of cloud droplet deposition are interception, impaction, and gravitational setting. As in the case of dry deposition of submicrometer aerosol particles, the lack of devices routinely measuring the flux from the atmosphere imposes the need to use indirect methods. In most studies, deposition rates have been computed from measurements of the concentration of some chemical species in liquid cloud water, from which the flux to the surface is computed using a deposition velocity model. In this case, multiple process models are again appropriate. Table 10 lists some predictions of cloud droplet deposition velocity produced by multiple resistance models.

Effective total deposition velocities are estimated to be about 7% of the wind speed. The total deposition velocity is made up of contributions by both impact and sedimentation. At low wind speeds sedimentation is comparable to impaction, but with very light winds sedimentation is more important than impaction. The model predicts that impaction is the dominant mechanisms throughout the depth of the canopy but that sedimentation becomes increasingly more important towards the floor of the forest. Field verification of this model is lacking but is a major goal of ongoing research [1].

Transport and Dispersion

The transport and dispersion of gases and particles over distances of tens to thousands of kilometers has been an active area of scientific study for many years. In this context, transport refers to the movement of a substance with the mean wind, and dispersion is the lateral and along-wind spreading of cloud material by both turbulent diffusion and vertical wind shear. The later process is the dominant component of dispersion associated with regional transport. Even before the problem of acidic deposition had been recognized, research on how

Table 10
Predictions of Cloud Droplet Deposition Velocity

Wind Speed At Canopy Top (M/S)	Deposition Velocity (CM/S)	Relative Importance of Gravitational Settling
2	14	39%
4	29	17%
6	44	10%
8	60	7%
10	75	6%

materials are carried by the winds was already prolific. This earlier body of work together with more recent studies forms the basis for present efforts to understand the transport of acid-forming substances. Achieving this understanding is complicated by complex physical and chemical interactions transforming and removing these materials, sometimes in a nonlinear fashion.

The major factors controlling transport and dispersion are the character and movement of a meteorological storm. Two extremes are recognized in meteorological storms with a whole section of variations in between:

- Strong fast-moving cyclones that have the potential to transport large amounts of pollutants rapidly across North America. Generally occurring in winter and spring, these storms bring with them strong horizontal winds, vertical wind shear, and widespread precipitation.
- Weak, slow-moving high-pressure systems that allow sufficient time for slower-acting processes such as thorough mixing to take place. These systems occur more frequently in the summer and fall and are characterized by local convective storms.

In the case of the former, pollutants may move rapidly over hundreds to thousands of kilometers downwind from their sources with limited removal, while in the latter case, pollutants move more slowly and are mixed vertically causing a buildup of concentrations. Figure 7 shows the major storm tracks across North America. The paths in Figure 7 represent the direction the predominantly winter storms follow, promoting strong flow conditions.

Four different scales of transportation/dispersion must be considered: local, mesoscale, regional, and global. Meteorologists define these scales as follows: local, less than 10 km; mesoscale, 10 to 500 km; regional, 500 to thousands of kilometers; and global, thousands of kilometers. The differentiation is important for acidic deposition and air quality, and the resultant effects differ for each of these scales [1].

Local and Mesoscale

Transport and dispersion on the local scale and the mesoscale are determined by the type of pollutant sources, the height of the release, and the structure and dynamics of the atmospheric boundary layer. There are two major configurations of man-made sources: point emissions from stacks, and pollutant emissions. In 1980 in the United States, more than 60% of the man-made sulfur dioxide was attributed to coal-fired electrical generation plants many from tall stacks. Regulatory pressure to reduce local pollution at ground level has encouraged construction of these tall stacks in the past few decades. The percentage of stacks taller than 7 meters has risen from 25% to 67% of the total from 1950 to 1980 [1]. In recent years, the height of new stacks has leveled off. Tall stacks are suspected of increasing atmospheric residence times for pollutant emissions and

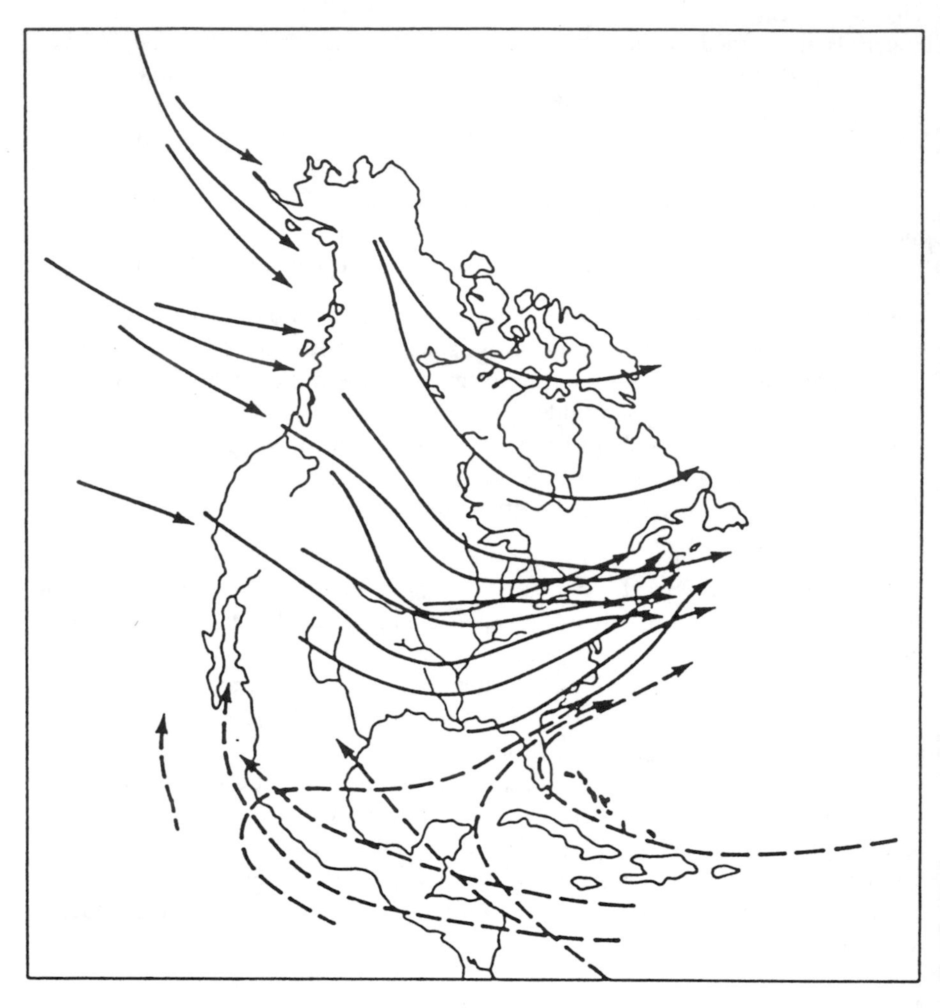

Figure 7. Major storm tracks for North America. Dashed lines denote tropical cyclone paths, and solid lines denote extratropical cyclone paths.

substantially reducing nearby dry deposition. The increased residence times of the primary pollutants, before wet or dry deposition, have increased the total ratio of secondary products, such as sulfates and nitrates, to the primary precursors, sulfur dioxide and nitrogen oxides, at long distances from the source. On the other hand, more of the emissions from urban industrial complexes tend to be near sulfur sources. As expected, urban industrial emissions are quite important on the local and mesoscale because substantial decreases ground-level pollutants are experienced, particularly from dry deposition [3].

The fundamental differences in the behavior of emissions from these two types of sources are from characteristics in the mixing layer. Under clear skies during warmer seasons, the daytime mixing layer can be extremely efficient in quickly mixing both low-level and elevated pollutant releases throughout its entire vertical extent. During the same conditions at nighttime, near-ground releases are trapped within the shallow mixing layer and produce relatively high pollutant levels on the local and the mesoscale. Within strong storm systems, especially in the winter, this diurnal variation of atmospheric stability is not as great.

Under certain conditions, convective non-precipitating clouds may play an important role in mixing pollutants into higher levels of the atmosphere. This phenomenon is called venting and was investigated with a field experiment during the summers of 1983, 1984, and 1985. This study also concentrated on the chemical transformations of sulfur and nitrogen pollutants in clouds. The findings suggest that even nonprecipitating clouds may vent significant amounts of mixed-layer pollutants into the free troposphere while at the same time accelerating the conversion of primary pollutants to their oxidized acidic form.

To help quantify transport and dispersion on the local and mesoscale, meteorological transport and plume models have been developed to estimate pollutant concentrations at ground level. Models incorporating chemical processes have been combined with air transport to deal with urban and mesoscale photochemical problems. Mesoscale field programs have been conducted in Philadelphia to characterize the near-source effects of urban and industrial complexes on the chemistry of rainwater. It was found that these urban areas affect wet deposition out to distances of 150 km and greater during certain seasons of the year. Increases of sulfur and nitrogen deposition by as much as a factor of two were observed in some cases downwind of the urban areas as compared to upwind. These data are being used to evaluate mesoscale models of acidic deposition.

The ability to specify the movement of air over local and mesoscale distances is limited. Observations of the winds are available at the surface from National Weather Services sites every hour, but these winds do not necessarily represent the wind flow at the altitude of the plume. Because non-surface measurements are available only every 12 hours from sources 400 km to 500 km apart, interpolation to these scales may be misleading. Trajectories using these wind observations to estimate the position of the mean mass of individual plumes are limited by the misrepresentation of the observed winds. Few studies have been conducted with artificial tracers to identify the motions in and around precipitating clouds on the local and mesoscale [4].

Regional and Global

The most significant determinant of long-range transport of air pollutants is the wind field. Day-to-day movement of pollutant material in the atmosphere on regional and global scales is controlled by fronts and cyclonic and anticyclonic systems. On this scale, the spreading of pollutants is dominated by the action of vertical wind shear and wind direction changes acting in combination with the diurnal cycle of vertical mixing and nighttime stabilization of the atmosphere.

Emissions of acid precursors from tall stacks represent a large fraction of the total sulfur dioxide and are prime sources of long-range transport particles. Volatile organic compounds, which are essential in production of oxidants for acid formation, are emitted by area low-level sources. For example, as a result of significantly lower mixing heights in winter than in summer, a large percentage of the elevated emissions from tall stacks in the northeastern United States may remain elevated and relatively intact for more than a day and for distances greater than 500 km in the winter. It should be noted, however, that the distances over which emissions from tall stacks maintain their identity is highly variable. In a deep daytime missing layer, typical of weak horizontal flow in the summertime, these emissions may be rapidly brought down to ground level within a few kilometers of the sources as they are vigorously mixed throughout the layer. Under these conditions, tall stack emissions are little different from ground-level emissions in contributing to the pollutant concentrations within the mixed layer a short distance from the stack [1].

Because of the long-range transport of acid and acid precursors, it is desirable to estimate pollutant fluxes out of the United States. A long standing experiment of this type is the Western Atlantic Ocean Experiment, which attempts to identify the advection and deposition of atmospheric emissions of sulfur dioxide, nitrogen oxides, and synthetic organic compounds from North America to the western Atlantic Ocean. Measurements are made at ground sites, on ships, and by aircraft. The measurements include precipitation chemistry and air quality with the goal of determining the eastward fluxes during different meteorological conditions. Early results based on average air quality data and climatological wind data indicated that of the man-made sulfur dioxide emitted to the atmosphere of eastern North America, about 30% to 35% is transported off the continent. The movement of nitrogen oxide is less certain. It would be expected that transport to the Atlantic Ocean in summer would be less than in the winter, due to greater oxidant levels, larger dry deposition on the foliage crops and deciduous trees, and lighter winds.

When the chemistry of the precipitation of the northeast United States is compared with the value of the western Atlantic, Atlantic precipitation is found to have a higher ratio of sulfuric acid to nitric acid contributing to the total acidity. This implies that sulfur has an important long-range transport component. Nitrogen oxide emissions are transformed to nitric acid efficiently and are more likely to be deposited on the local and mesoscale downwind of large urban centers of the East Coast and therefore affect the precipitation quality closer to the sources. On Bermuda, for example, concentrations of hydrogen ions, sulfur, nitrogen compounds, and organic compounds in precipitation in storms originating in the eastern United States and Canada are about three times greater than concentrations in storms from other parts of the Atlantic Ocean. Supported by precipitation chemistry, measurements taken aboard ships suggest that of the approximately 3 million metric tons of sulfur exported from North American to the Atlantic annually, about half deposits to the Western Atlantic while the other half is deposited further to the east. Of the about 1 million metric tons of exported nitrogen, most is deposited to the Western Atlantic. Aircraft measurements confirm that a significant amount of sulfur transport takes place in the upper atmosphere [1].

Conclusions

Transport and dispersion play an important role in determining how pollutants are physically moved from source to receptor. Horizontal winds and vertical mixing are the main mechanisms that determine the paths that trace materials take in the atmosphere. These mechanisms act on all scales from a few kilometers to thousands of kilometers. Vertical transport by storm systems and venting by cumulus clouds are now recognized as important components of transport and dispersion. In general, three-dimensional characterization of the flow fields is necessary to model these systems. Simulation of these complex interacting flows is being accomplished through a combination of predictive models and observing wind fields [1].

THE EFFECTS OF ACID PRECIPITATION ON VEGETATION

Exposure Characteristics

Effects on Agricultural Soils

When assessing the potential effects of acidic deposition to agricultural soils, the increased benefits from sulfur and nitrogen fertilization also must be considered as well as the possible need for increased liming to counteract soil acidity. Crop sulfur requirements for maximum yield range from 10 to 40 kg ha^{-1} yr^{-1}, while nitrogen requirements are 298 and 153 kg/yr for corn and wheat, respectively. According to calculations using 1983 data, the annual wet deposition of sulfur was 7 to 10 kg/yr east of the Mississippi and about 3 kg/yr in the Great Plains. More than half the nitrogen and sulfur depositions occur during the growing season. The addition of dry deposition to these calculations may add 10% to 100% to these amounts depending on proximity to strong point sources or urban plumes. Thus, atmospheric deposition can play a beneficial role in sulfur and nitrogen fertilization of crops [1].

In the Northeast and Midwest, the atmospheric deposition of sulfur in combination with soil mineralization may meet most of the requirements for corn, wheat, and soybeans in many regions of the United States. Although sulfur is not a normal component of fertilizer applications, its use recently has increased. About 800,000 tons of sulfur was applied directly to cropland in 1986. Atmospheric nitrogen, even in the Northeast, meets only 5% to 10% of the annual requirements for corn and wheat. However, repeated intermittent foliar applications from rain provides a more rapid and efficient supply of nutrients to the foliage than manual addition of nutrients to the soil. The application of nitrogen fertilizers through agronomic management practices and the fixation of atmospheric nitrogen by leguminous crops can enhance soil by an order of magnitude of ten times above the contributions of wet deposition. Liming needs as a result of wet deposition are small in comparison with other factors.

Research in this area, although limited, supports this information. For example, a two-year study of the effects of acid rain on poorly managed plant soil systems typical of pasture land reported that soil biota appear to function optimally at ambient levels of acid rain (pH 4.2) compared with higher acidity (pH 3.5 to 3.2) and lower acidity (pH 4.5 to 5.4) levels. The results also indicated that vegetation may benefit from the sulfur and nitrogen present in the rain that is more acidic than ambient. The study also demonstrated that when land is not cultivated, surface soil (top 5 cm) may become acidified within a two-year period of acid rain with average pH < 3.5 at ambient rainfall amounts. Results may be interpreted to indicate that soil organisms at equilibrium with ambient conditions may be affected negatively by both increased soil acidity from rain more acidic than ambient (pH < 4.2) and also by decreased sulfur and nitrogen input from rain less acidic than ambient (pH > 4.2) over the growing season. Continuation of such a study over a longer term may reveal whether or not microorganisms would achieve a new equilibrium at rain pH values different from ambient levels [1].

Effects of Sulfur Dioxide

Plant metabolism is affected by sulfur dioxide in a variety of ways. In some cases at low concentrations, sulfur dioxide may act as a fertilizer by providing sulfur needed. At high concentrations, sulfur dioxide can cause visible damages and mortality. Carbohydrate levels have been increased by low levels of sulfur dioxide and decreased by high concentrations.

The mechanism of action of sulfur dioxide on plants has been studied by comparing susceptible versus resistant cultivars of the same species. Differences in sensitivity of soybean cultivars were related to there capacity to metabolize toxic sulfite to the less toxic sulfate. Differences also often are related to sulfur dioxide uptake and, thus, to stomatal activity.

Plant species are known to adapt to sulfur dioxide stress. For example, geraniums reproducing for a period of more than 31 years near a point source of sulfur dioxide were more resistant to the sulfur dioxide than were plants further from the source.

Effects of relatively low sulfur dioxide concentrations on carbon transloca-tion and partitioning and on plant growth and yield have been determined by the US EPA. The results support the contention that plants are more sensitive to lower sulfur dioxide concentrations (less than 260 μg m^{-3}) when exposed continuously than when intermittently exposed. For example, continuous expo-sure of winter wheat to 180 μg M^{-3} sulfur dioxide for 22 days straight produced leaf injury in young plants. Such injury would not be expected with intermit-tent exposures. Periods without exposure may be critical in recovery potential following their exposure to higher levels of sulfur dioxide.

Extensive dose-response studies with an open-air exposure system have been used at Argonne National Laboratory to simulate exposures of crops to sulfur dioxide near point sources. The yield of five soybean cultivars was consistently decreased by periodic exposure of plants after flowering to total sulfur dioxide doses greater than 10 ppm/hour. Doses were products of mean sulfur dioxide concentrations ranging from 80 ppb to 400 ppb, exposures durations of 2.5 to 4.2 hours per event. Maximum peak to mean sulfur dioxide concentrations ratios were about 2.5. On the average, doses in the 5 ppm/hour range essentially had no effect. However, concentrations in as low as 19 ppb during 18 four-hour fumigation episodes resulted in a particular yield suppression for a particular year. Similar sulfur dioxide exposure regimes have been measured in ambient air in non-urban areas of the Ohio River Valley. There were also some apparent cultivar differences in response. For example, studies in 1980 using Williams and Corsoy cultivars exposed to 19 sulfur dioxide episodes averaging 2.5 hours per episode suggested a possible stimulation in yield at low sulfur dioxide doses for Corsoy and a reduction for Williams. The reason for the difference in response could be due to differences in stage of development at the time of the sulfur dioxide treatments. In contrast to soybeans, two field corn cultivars examined in similar studies were reported to be resistant even to acute sulfur dioxide exposure (24 ppm) for 12 three-hour fumigation periods [1].

Sulfur dioxide concentrations in most agricultural areas of the United States rarely exceed 10 ppb for extended periods of time. However, short-term exposures (one hour or less) of 100 ppb were measured 40 to 60 times in rural areas near the source during the growing season in the Ohio Valley area. Other areas affected have sig-nificantly fewer pollutants than the Ohio River Valley. For example, in rural areas affected by refineries near Chicago, the one-hour mean sulfur dioxide concentra-tion was 31 ppb, with a one-hour maximum of 175 ppb during the growing season of 1982. Daily four- or 7-hour exposures of cotton, tomatoes, or soybean in open-top chambers throughout the growing season have demonstrated that sulfur dioxide concentrations likely to occur in much of the United States do not suppress yield. Other studies of corn, wheat, alfalfa, barley, and kidney beans support this conclu-sion. However on a local rather than regional basis, concentrations of sulfur dioxide near point sources can potentially cause a decrease yield in some crop species such as soybeans, wheat, and potatoes.

Effects of Nitrogen Oxides

Of the various nitrogen compounds only PAN (peroxyacetyl nitrate), nitric oxide and nitrogen dioxide reach concentrations locally that could possibly be phytotoxic. Reports on the direct effect of PAN and nitrogen dioxides (NO_x) on vegetation usually are associated with areas of specific industrial sources or where vehicular traffic is frequently congested. For example, vegetation injury has been noted near nitric acid factories and arsenals and in the Los Angeles basin. No reports of injury have been published about PAN or other species of nitrogen.

The response of vegetation to high concentrations of nitrogen dioxide shows considerable variation ranging from leaf chlorosis to necrosis to subtle alterations of leaf metabolism and premature senescence. These processes can be explained by the physiological process affecting nitrogen dioxide uptake into the leaf, pollutant toxicity at target sites, and cellular repair capacity.

In a recent controlled exposure study, the yield of two greenhouse grown potatoes cultivars was significantly reduced by twice-weekly five-hour exposures to 200 ppb nitrogen dioxide throughout the growing season. The yield effects were contributed to accelerated defoliation in exposed plants, which presumably decreased photosynthesis carbon accumulation. Tomato yield was reduced by 22% in plants continuously exposed to nitrogen dioxide for 128 days. Experiments using low nitrogen dioxide doses (150 ppb) for four hours three times a week for four weeks reported having no effect on leaf and stem dry weights of tomato and snap beans. Ten three-hour exposures of nitrogen dioxide in eight treatment groups ranging in concentration from 70 to 400 ppb over the growing season had no effect on soybean yield. Episodes of nitrogen dioxide, even extremely close to a power plant, are only occasionally above several ppb [1].

According to the Environmental Protection Agency's Air Quality Criteria for Nitrogen Oxides, true concentration and exposure frequencies of nitrogen dioxide that produce measurable injury to crop plants are higher than those normally found in the eastern United States even with plants considered to be sensitive to nitrogen dioxide. A 30 minute acute exposure of 6,000 to 10,000 ppb or an eight-hour exposure of 2,000 to 5,000 ppb to nitrogen dioxide was required to produce foliar injury levels of 5%. Ambient levels of nitrogen oxides, however, may be deleterious in combination with sulfur dioxide in limited regions of the United States, even though independent nitrogen dioxide effects are not noted until concentrations repeatedly reach 200 ppb, nearly two orders of magnitude greater than levels for average growing season concentrations over broad agricultural regions. It has been concluded that at current ambient conditions, nitrogen dioxide alone does not measurably affect agricultural production in the United States.

Impact of Acid Deposition on Crop Deposition

Early assessment of crop loss because of pests, disease, and climatic factors generally were based on field surveys in which visible injuries in the field were correlated with loss in yield. A particular stress could be identified through a particular form of injury symptom. Research on air pollutants was begun because certain visible injury symptoms appeared to be correlated with high levels of air pollution. Sulfur dioxide was the first pollutant to be recognized for its toxicity to plants as early as the late 1800s. In the first half of this century, sulfur dioxide injury to vegetation was documented in the vicinity of large smelting operations. Power production became an important source of sulfur dioxide in the middle of the century. Most of these sources have reduced their emissions through control devices or have built taller stacks so that pollution is carried away farther from the source, thus reaching concentrations that cause damage less frequently.

Specific symptomology has been described for sulfur dioxide, ozone, and other pollutants on a number of crops. When the existence of acid rain gained worldwide attention in the early 1970s, research was initiated on agricultural crops to determine if acidic rain could reduce yield. Appearance of visible injury symptoms associated with acidic rain has not been reported from ambient levels of acidity; however, other pollutants are known to cause physiological and yield effects without visible injury. In addition to possible direct effects of acidic precipitation on plants, impacts on the soil, particularly in pasture and unmanaged grasslands also were postulated [1].

Crop yield is a measure of the economically marketable portion of the plant. An air pollutant can concernably injure a plant but not produce economic damage. For example, in certain instances soybean plants exhibit visible injury to leaves from ozone but without any observable reduction of the mass of seeds produced. Conversely, visible injury to spinach leaves lowers the crop's market value.

To estimate crop loss, a standard reference yield is needed to compare each crop type. When crop cultivars are grown under optimal conditions of water, nutrition, solar energy, etc., using available modern technology to eliminate the presence of weeds, pests, and disease, production is considered the maximum yield. Attainable yield results from the use of the best adapted cultivar for a given geographic location, assuming average climate, air quality, and soil, and with the application of the best available cost-effective control technology for pest and disease. The actual yield is what farmers obtain under ambient environmental conditions, including air pollution, using average technological and management practices in a particular year. Crop loss is generally defined as a difference between attainable and actual yield. Some of the losses can be attributed to inherently unfavorable environmental conditions and some lack of use of known management practices, often for economic reasons. The research evaluated here was performed to determine how actual yield is influenced by the presence of acidic rain and associated atmospheric pollutants. In this assessment, regional crop loss from air pollutants is defined as the difference between the actual regional yield at current ambient air pollutant levels and the yield at the preindustrial (background) level. It has been estimated that in the United States, average preharvest crop losses of about 37% are due to biological agents (i.e. insects, weeds, disease) even though pesticides and herbicides are typically used. This translates to a loss of about 1.5 billion tons of production in the United States. Unfavorable environmental conditions such as drought, excessive moisture, and frost have been estimated to reduce maximum yield for a particular farm by as much as 69% in a given year.

Modern technology has played an important role in crop production. Anthropogenic input of chemical substances such as fertilizers and pesticides and genetic selection resulting from crop breeding programs have been responsible for dramatic increases in crop yield, taking into account the variations of environmental conditions, from 1925 to 1984. Nitrogen and sulfur, which are components of acidic deposition, are required plant nutrients and are present in most fertilizer regimes to achieve maximum yield.

Damage to vegetative growth from gaseous byproducts of fossil fuel combustion, smelters, and chemical plants has been observed near local sources. In laboratory experiments, visible damage has been observed on many species from short-term exposures of simulated acidic rain and associated gaseous pollutants at concentrations much higher than ambient levels, which demonstrates what can happen in extreme conditions [1].

Impacts on Vegetation as a Result of Soil Acidification Due to Acid Deposition

Indirect effects of air pollutants on forests involve the relationship among acidic deposition, soil characteristics, and the physiology of forest organisms. Acidic deposition falls directly on the soil, or is mediated through the canopy where its chemical composition is modified by reaction with the constituents on the leaves, leaching of dry deposited material on the surface, and leaching materials from the leaf surface. In general, the acidity and the nitrate/ammonia concentration falling on the soil through a forest canopy will be less than the direct wet precipitation due to neutralization and foliar uptake of the nitrogen for plant metabolism. Of concern is either acidic deposition has changed the chemical characteristics of the soil on soil solution, thereby contributing to any of the current declines, or will alter the chemistry of the soil or soil solution over time, leading to future stress or decline.

The indirect effects of air pollutants on forests as mediated by soils are much more difficult to establish than direct effects. To relate a change in soil properties to a particular decline, it must be established that the decline occurred and that the present conditions are deleterious to forest growth.

Possible adverse impacts to the health of trees from acidic deposition to soil may be grouped as follows:

- Reduced nutrition due to leaching of potassium, calcium, or magnesium after replacement by hydrogen ions and aluminum, i.e., base cation exchange.

- Aluminum metal toxicity to fine roots.
- Aluminum toxicity to roots from release of monomeric aluminum.
- Excess nitrogen that may destabilize nutritional balance.
- Damage to microbial systems by excess hydrogen ions.

Several general concepts may assist in the analysis of each of the potential effects:

- Most forest soils have an adequate buffering system.
- The buffering process may in some cases lead to the acidification of the soil. Cumulative effects of base exchange and leaching, if in excess of weathering, may reduce available nutrients, but rates of change and threshold levels are highly uncertain. There are reports of long-term soil acidification in some soils.
- Many forest soils have such high changeable acidity compared with the addition from acidic deposition that little change in acidity can be expected to occur.
- Soil solutions can be acidified by the leaching of a mobile anion (sulfate) through the acidic soil even if the acidity of the soil itself is not measurably affected.
- Nearly all the forest soils are deficient in available nitrogen. Hence, the effect of nitrogen deposition may be a positive nutritional one, and nitrate does not act as a mobile anion in most systems.
- The microbiological systems that support tree nutrition are naturally adapted to acidic soil solutions. Studies of how they respond to changes in soil solution pH as additions of potentially toxic substances are currently underway.

The possible effects of nutrient leaching form the foliage was discussed previously; of equal importance when discussing soil acidification is the possibility of the leaching of nutrients from soils.

Several irrigation field tests have been conducted in forests with a range of simulated acidic rain. The tests, however, were under extreme acidic deposition conditions and did not simulate the additional dry deposition component of calcium, magnesium, etc. Attempts also have been made to measure long-term changes in soil nutrients of change in pH under field conditions. A study found no evidence of pH changes in podzol in Sweden in over a 39-year period. A Swedish survey reported a strong correlation between increasing age of the forest and decreasing pH in the humus layer. His results showed that from 1961–63 to 1971–73, exchangeable potassium, calcium, and magnesium decreased significantly, while hydrogen ions and aluminum increased. Troedsson concluded that the acid production of the coniferous trees was more important than the atmospheric deposition.

Other studies have shown significant long-term increases in soil acidity but are difficult to intercalibrate because of different methods of pH measurements. For example, forested soil with chemical characteristics are potentially subject to reduced nutrient availability due to leaching by acidic deposition. Substantial areas in the Appalachians and Southeast fit this criteria. The leaching hypothesis requires that replacement of nutrients by weathering and by atmospheric deposition is insufficient to offset the leaching. No field sites have been found in the United States where this process has been demonstrated to produce a detrimental effect to trees at ambient levels of deposition. Search for such soils should be focused in areas of highest acidic deposition loading, low to moderate base saturation soils, and low cation exchange capacity [1].

Another factor that must be taken into consideration with soil acidification is the possibility of aluminum toxicity. At high enough acidity, low enough calcium concentrations, and low enough levels of organic complexing agents, monomeric aluminum concentrations may become high enough in the soil solution to damage the fine root systems the trees. The question is whether there are forest soil conditions under which these are triggered by ambient acidic deposition, and if so whether the impact on the health of forests may be observed. The hypothesis explains the die back of silver fir and other species in the Sulling beach forest area of West Germany. The only field evidence was epidemiological, i.e., that soil aluminum concentration rose in 1969, the same time the fine roots declined. Close inspection of the data show that root decline was almost over before the rise in aluminum concentration started [1].

There is strong opposing evidence. Experiments were conducted over seven years in which Norway spruce was watered with simulated acid rain pH 5.7 to 2.0 Even trees watered at pH 2.0 (over 100 times more than ambient in the United States) and exposed to the abnormally high aluminum concentration in soil of 30-125 mg/L showed no symptoms of aluminum toxicity. The tree growth was normal. The aluminum concentrations were high, but the fine roots were healthy and the tree growth was normal [1].

Although aluminum toxicity to plants has been demonstrated in hydroponic environments and calcium/aluminum ratios estimated for acidified soils, there are other factors, including the degree of organic complexing and the influence of mycorrhizal, that could modify any direct relationship between acidic deposition and injury due to aluminum. In any event, the large number of non-sensitive soil types in the eastern United States precludes the aluminum toxicity from having a general role in forest damage throughout the east from acidic deposition. It has been estimated that 4% of the forest soils in the eastern United States might be subjected to pulses of high aluminum concentrations in soil solutions, located primarily in the Adirondacks, northern New England, and the southeastern outercoastal plain, but no such cases have been established [1].

Another factor that must be considered with soil acidification is the effects of trace metals. It is hypothesized that when trace metals accumulate on the surface layer of the soil in sufficient quantities and concentrations, when mobilized by acidic deposition into the soil solution, they may injure tree roots and other tissue uptake.

Five species of tree seedlings were grown from seeds in a greenhouse for 17 weeks in a sandy soil that had been pretreated with cadmium, lead, copper, and zinc at 0.6, 11.4, 2.0, and 20.6 ppm, respectively. Treatment consisted of amending the soil with an additional 0, 15, or 100 ppm of cadmium. The treatment was applied in distilled water at the surface; leachate was resprayed on the surface. Most of the cadmium added remained on the upper one-third of the pot soil, which confirmed that cadmium is rather immobile in soil. The 15 ppm treatment produced 25–180 ppm in the roots and 2–40 ppm in the shoots after 17 weeks but had no effects on height, growth, or total dry weight of the seedling. For the 100 ppb treatment, however, height, weight, and dry weight were significantly effected, i.e., reduced from the control. These results put an upper and lower level (600–1,200 ppm root or 28–117 ppm shoot) for toxic levels of cadmium in young tree tissue.

One other consideration for trace metal effects is the suggestion that at certain concentrations acidic precipitation or trace metals may reduce the activity of organisms that decompose the organic layer, thereby restricting the process of cycling nutrients back to the soil for use by the tree. No evidence of trace metal effects at ambient rural concentration into the upper soil layer has been produced.

Excess nitrogen is another consideration with soil acidification. A fertilization effect from elevated atmospheric nitrate (NO_3^-), ammonium (NH_4^+) or ammonia (NH_3) deposited on forests in the fall season might upset development of cold-hardiness in trees.

Generally, nitrate deposition in dry form from nitric acid and nitrogen dioxide is incorporated into the plant tissue. Wet nitrate falling on the foliage is also partially absorbed into the foliage in young shoots. Evidence on absorption comes from studies that show that through fall from a forrest canopy contains less nitrate than the bulk precipitation. It has been suggested that severe damage occurs when excess nitrogen delays the physiological changes that normally occur in late summer or early fall to "harden" the tree to winter cold. If first harding is delayed, the tree might be damaged by a very severe winter cold period or a cold snap after an early thaw. The authors were able to correlate missing red spruce needles with 1981 and 1984 warm spells (e.g., 1984 affecting new 1983 needles). The question is whether or not atmospheric nitrogen played a significant role.

At lower levels, almost all forest soils are deficient in nitrogen, so the deposition of nitrate may have a fertilization effect. Should nitrate deposition cause the soil to become saturated in a given area, the excess nitrate will pass through the soil without any buildup, although some acidification of the soil will occur. The saturation has happened in the soils of a red adler forest that became saturated due to the natural fixation of nitrogen by this species. Red adler forests soils that receive more than 100 kg/yr through symbiotic N_2 fixation have been found. Excess nitrate leached the soil down to the 40 cm level, at which point the nitrate concentration exceeded the drinking water standard of 45 mg NO_3^-L on occasion. The hydrogen ion and nitrate releases occurred in the top 10 cm of the soil underlying the red adler stand so that percolate and the soil solution were acidified to pH 4.6 in the A-1 horizon compared with a pH of 5.0 in the horizon of an adjacent douglas fir stand. The soil underlying both stands had a high buffering capacity so that by the time the soil solutions reached 30 cm depth under either species, the pH had risen to above 5.2

Ultimately detrimental effects of deposited nitrogen could occur when a tree becomes nutritionally unbalanced because some other element such as magnesium or potassium is available in insufficient quantities in the soil to match the nitrogen component. High elevation red spruce forests may present such a situation, as deposition is estimated to be 20 kg/ha/yr as determined by the NAPAP Mountain Cloud Program for total cloud deposition. In addition, this nitrate is deposited rather uniformally throughout the year; whenever the tree is metabolically active, it will be uptaken and reduced with the expenditure of energy supplied by the oxidation of stored carbohydrates. During the autumn, the tree is still metabolically active enough to restore the carbohydrates that are used to supply the energy for nitrate reductase. This, in turn, may lead to impairment of the process for winter hardening and needle cuticular development. An experiment in Finland with much higher levels of nitrogen was shown to cause injury to Norway spruce.

CONCLUSIONS

Research on the effects of gaseous pollutants suggests that a number of factors can influence the effects of pollutants on crop yield and vegetative damage. These factors include dose rate, recovery interval, temperature, moisture, and true presence of pests, disease, or other pollutants. Additional information on the interactions between abiotic and biotic stresses with acid rain and gaseous pollutants, as well as information on the relative sensitivity of a wider range of species and cultivars, will reduce uncertainties. Such study also would provide more accurate estimates of pollution damage.

As for the future, because most agricultural crops are harvested on an annual basis and soils are managed, no long-term cumulative effects from acidic deposition or associated gaseous pollutants are expected if present levels of wet and dry deposition remain constant over time. If the best available emission control technology were to be implemented over time, damage from gaseous pollutants and possible second-order effects from acidic deposition would be reduced.

REFERENCES

1. The National Acid Precipitation Assessment Program (NAPAP), "Interim Assessment on the Causes and Effects of Acid Deposition," *Emissions and Control,* Vol. 2, U.S. Government Printing Office, Washington, DC, 1986.

2. White, James C. *Global Climate Change Linkages: Acid Rain, Air Quality, and Stratospheric Ozone,* Elsevier Science Publishing Co., Inc., New York, NY, 1989.
3. Battye, R., Demmy, D., Smith, M., Tax, W., Zimmerman, D., "Anthropogenic Emissions Data for the 1985 NAPAP Inventory," Alliance Technologies Corp., NC.
4. Mackenzie, J., and El-Ashvy, M., "Ill Winds: Air Pollution Toll on Trees and Crops," *Tech. Rev.,* Vol. 92, April 1989.

CHAPTER 26

OCCUPATIONAL SAFETY AND HEALTH IN THAILAND

Tsuyoshi Kawakami

Division of International Cooperation
The Institute for Science of Labour, Japan

Chalermchai Chaikittiporn and Pramuk Osiri

Department of Occupational Health, Faculty of Public Health
Mahidol University, Thailand

CONTENTS

INTRODUCTION

Thailand, located in southeast Asia, has a population of 56,923,000 (1991). Thailand is a rich agricultural nation and is one of the world's largest rice exporters. Among the labor force of 29,866,000 in 1991, 17,997,000 (60.26%) was in the agricultural sector [1]. However, as with many other southeast Asian countries, Thailand is rapidly industrializing, resulting in more people involved in industrial sectors. Foreign investment has been promoted, resulting in an increase of manufacturing industries such as textiles, electronics, and food processing. Recently, large-scale petrochemical industries have been developed from the rich natural gas resources in the Bay of Thailand.

SAFETY AND HEALTH PROBLEMS

As shown in Table 1, the number of registered cases of occupational injuries has increased sharply year by year [1]. The increased number of the injuries reflects the seriousness of occupational safety and health problems in the nation. By improving reporting systems and widening coverage of occupational injury cases by the Department of Welfare and Labour Protection, the Ministry of Interior is also a factor in fully understanding injury cases. As the country continues to industrialize, workers in mushrooming small-sized enterprises and seasonal migration groups, which move from the rural regions to the industrial

Table 1
Occupational Injuries Reported to the Workmen's Compensation Fund of the Department of Labour

Year	No. of Reported Injuries	No. of Employees Who Are Entitled to Compensation	Fatalities	Permanent Total Disability*
1986	39,416	37,445	363	36
1987	43,359	42,811	396	43
1988	55,966	49,874	352	37
1989	67,912	63,857	442	30
1990	82,280	79,787	634	32

From Ref. 1.
**Does not include fatalities.*

regions, have become more difficult to track and evaluate. Recently, several cases of large-scale accident such as severe fire and chemical explosions, have raised concern about controlling hazards.

Occupational disease statistics, obtained through an epidemiological survey implemented by the Ministry of Public Health (see Table 2), show a conspicuous increase in pesticide poisoning in the agricultural sector [2]. In contrast, the number of reported cases of occupational diseases in industrial sectors is relatively small. Cases of occupational diseases seem to be overlooked due to the constraints of diagnostic skills and the current reporting system. To clarify the real magnitude of the occupational disease problem, the Ministry of Public Health recently has reviewed current research and information on occupational diseases [3]. The review showed that ameliorating conditions causing occupational diseases, such as lung diseases and musculoskeletal disorders, and hazards, such as exposure to hazardous chemicals like lead and manganese, is a priority. For example, a lead poisoning surveillance program covering 2,143 workers in 38 industries revealed that 809 cases had a blood lead level about 40 μg% and 43 cases above 80 μg%.

In addition to governmental statistics, several studies have been conducted in Thailand providing valuable information. Chaikittiporn et al. conducted an epidemiological survey on byssinosis and found a high prevalence among cotton textile workers [4]. Chavalitsakulchai et al. carried out a survey on noise-induced hearing loss among textile workers and found that many young workers in their 20s and 30s already had developed noise-induced hearing loss and that hearing loss increased with the number of years on the job [5].

Table 2
Occupational Diseases Reported to the Ministry of Public Health

Diseases/year	**1986**	**1987**	**1988**	**1989**	**1990**	**1991**
Insecticide poisoning	3,107	4,633	4,234	5,348	4,827	3,828
Lead poisoning	51	51	32	18	4	25
Manganese, mercury and Arsenic poisoning	13	22	6	11	6	10
Petroleum products poisoning	6	13	16	23	7	25
Gas, vapor poisoning	28	32	51	43	6	51
Caisson's disease	9	8	9	0	0	0
Total	3,214	4,759	4,348	5,443	4,850	3,939

From Ref. 2.

Attention also has been paid to work organization and work scheduling aspects. Osiri et al. revealed through their questionnaire of 97 enterprises that among the 58,833 employees surveyed, 32,383 were shift and night workers, mainly in three-shift system in three teams. Of those, 71.7% of the shift workers were female [6]. Ton That in his survey of a motorcycle factory found night workers complained more of social and family problems compared to day workers [7]. Wongphanich et al. stressed that female night workers, who are commonly seen in Thailand, need improved working conditions, especially working hours, work load, and wages [8]. High priority has been placed on organizing work better and improving scheduling.

MEASURES FOR IMPROVEMENT

All the related sectors are becoming more aware of the seriousness of occupational health and safety problems in the nation. The Department of Welfare and Labour Protection in the Ministry of Interior has been active in increasing the effectiveness of labour inspection and establishing a legal framework for occupational health and safety. While establishing the mechanism for solving occupational health problems, the agency is emphasizing tripartism and encouraging the active participation of employers, employees, and government officials. The National Institute for the Improvement of Working Conditions and Environment in the Department of Welfare and Labour Protection has been playing a tremendous role in improving working conditions and environment. In 1993, the Ministry of Labour and Social Welfare was newly established and started to strengthen the above policies. The Ministry of Public Health is promoting greater working knowledge and better diagnostic skills about occupational diseases to medical professionals. The Ministry of Industry has been playing an active role, especially in regard to machine safety at various workplaces. The Ministry of Science, Technology, and Energy has paid particular attention to industrial waste control. To promote the exchange of information and experience, coordination among the governmental agencies is essential.

Several large-scale enterprises have succeeded in establishing well-organized safety control measures and have exhibited their experience during the National Safety Week organized by the Department of Welfare and Labour Protection. Some labor unions have paid greater attention to health and safety and have been conducting participatory training to access the necessary information and increase problem-solving skills [9].

In the academic and technical field, Mahidol University has long been playing a leading role in this field since its establishment in 1969. It educates occupational health, safety, and industrial hygiene specialists every year up to the graduate level and it is attempting to establish a doctorate degree course in occupational health. The Faculty of Engineering at Chulalongkorn University also has been playing a vital role, especially in its unique ergonomic research and machine redesigns appropriate to the body size of Thai workers [10]. The Faculty of Medicine at Thammasat University started a new medical education curriculum, stressing that all medical students have a firm background in practical occupational health.

A successful model of activity based on local improvement efforts recently has been formulated in small-scale factories by the Occupational Health Training and Demonstration Center of the Ministry of Public Health [11]. In that, setting up a supportive framework to facilitate self-generated occupational safety and health action at various workplaces has been noted as a key factor for practical improvement. The intervention team, consisting of the staff of the center and external researchers, played advisory and facilitative roles to act on instead of pointing out the weak points of the workplaces. In Figures 1 and 2 and Table 3, some examples of improvement implemented by the project are shown. Taking a positive approach was the key to success, stressing locally made low-cost solutions, encouraging self-generated efforts of employers and workers, and emphasizing sustainable development of the action.

Figure 1. In a small-scale pesticide factory, the bottle-filling process had no ventilation and workers had been exposed to the chemicals.

Figure 2. They have newly introduced a closed system and installed the local exhaust ventilation. It is noted that the improvement was planned by workers and carried out by using locally available materials (Courtesy of Occupational Health Training and Demonstration Center, the Ministry of Public Health, Thailand).

FUTURE PERSPECTIVES

Thailand, facing rapidly increasing occupational safety and health problems, is working to improve its situation. A stronger commitment from all related sectors is essential. For further progress, coordination in sharing the occupational health experience of all the concerned bodies is important because there is overlap. Furthermore, development of occupational health experts who can facilitate improvement and provide ready-to-use information directly to workers and managers is desired at the workplace. Establishment of the tripartite participatory mechanism and the appropriate legal framework for it and further practical support to workplaces are recommended. This will increase self-generated improvement actions at the local level. Appropriate international cooperation schemes focusing on improvement can play an effective role in accelerating the efforts in Thailand.

Table 3
Some Examples of Low-Cost Improvements Implemented by Small-Scale Factories.

Before Innovation	After Innovation
Control panels had no sign for switches.	Put labels indicating the function of the switches using the local language.
No guarding for electric fans, which were dirty.	Installed guard covers and cleaned them.
Chemical storage tanks had no signs and labels.	Indicated the chemical storage tanks and information label for safety.
Emergency bathrooms for exposure to chemicals were out of order.	Repaired the emergency bathrooms for workers.
In some sections, working chairs were not available.	Provided suitable ergonomic chairs.
Floor surface was rough, and workers had difficulty in carrying products.	Adjusted the rough surface by iron plates, which made the floor smooth and safe.
Drinking water for workers was unsuitable. It could be contaminated by chemical dust.	Made small cabinets for keeping water-drinking cups inside.
No canteen was available. Workers had to eat lunch in their workplaces.	Established a canteen for the workers.

Courtesy of the Occupational Health Training and Demonstration Center, the Ministry of Public Health, Thailand.

REFERENCES

1. Department of Labour, Ministry of Interior, Thailand. *Yearbook of Labour Statistics in Thailand.* 1991.
2. Division of Epidemiology, Ministry of Public Health, Thailand. *Epidemiological Survey Yearbook.* 1991.
3. Division of Occupational Health, Ministry of Public Health, Thailand. "Preliminary Review of Occupational Health in Thailand." 1990.
4. Chaikittiporn, C. "A Six-Year Followup Study of Byssinosis," *Research Perspectives on Occupational Health and Ergonomics in Asia and Other Countries* (M. Wongphanich et al., eds.), Department of Occupational Health, Faculty of Public Health, Mahidol University, Bangkok, 1992, 462–467.

5. Chavalitsakulchai, P., Kawakami, T., Kongmuang, U., Vivatjesadawut, P., Leongsrisook, W., "Noise-induced hearing loss of textile workers in Thailand," *Ind. Hlth. 27:* 165–173 (1989).
6. Osiri, P., Kawakami, T., Sakai, K., Chaikittiporn, C., "A survey on Shift Work System in Thailand," *J. Human Ergology 23* (1994).
7. Ton That, K., "Factors Affecting the Health and Living of Shift Workers in a Motorcycle Factory in Thailand," *Research Perspectives on Occupational Health and Ergonomics in Asia and Other Countries* (M. Wongphanich et al., eds.), Department of Occupational Health, Faculty of Public Health, Mahidol University, Bangkok, 1992, 576–582.
8. Wongphanich, M., Saito, H., Kogi, K., Temmyo, Y., "Conditions of Working Life of Women Textile Workers in Thailand on Day and Shift Work Systems," *J. Human Ergology 11*(Suppl): 165–175 (1982).
9. Intaranont, K., "Anthropometry and Physical Work Capacity of Agricultural and Industrial Populations in Northeast Thailand," (A research monograph), Chulalongkorn University Printing House, Bangkok, 1991.
10. Kawakami, T., Wongphanich, M., "Experience of Participatory Training in Thailand," *Research Perspectives on Occupational Health and Ergonomics in Asia and Other Countries,* (M. Wongphanich et al. eds.), Department of Occupational Health, Faculty of Public Health, Mahidol University, Bangkok, 1991, 562–565.
11. Tandhankul, N., Yuangsaard, S., Pongpanich, C., Punpeng, T., Punngok, T., Kawakami, T., "Action-Oriented Occupational Health Promotion of Small Enterprises in Samutprakarn Province in Thailand," *Proceedings of 13th World Congress on Occupational Safety and Health,* New Delhi, 1993, 511–514.

CHAPTER 27

PULMONARY FUNCTION IN WORKERS EXPOSED TO DIESEL EXHAUST

Ulf Ulfvarson and Monica Dahlqvist

Department of Environmental Technology and Work Science
Royal Institute of Technology
S-10044 Stockholm, Sweden

CONTENTS

INTRODUCTION

Occupational exposure to diesel exhaust occurs in several occupations, e.g., stevedores, work with loading and unloading of vehicles in ferry traffic or when automobiles are loaded or unloaded from freight vessels, work in garages and stores, work in city traffic with diesel vehicles, and in under-earth work (construction of tunnels and in mines).

Diesel exhaust contains a number of substances that will affect the health of exposed human beings if the substances are present in concentrations high enough. Therefore, many investigations have been performed to study the health effects of exposure to diesel exhaust, e.g., effects on mucous membranes, effects on the lung function, effects on the heart and blood vessels, fatigue, etc., and genetic effects including cancer [1-5].

Diesel exhausts are detrimental to the general environment, especially depending on the concentrations of carbon dioxide and nitrogen oxides.

Up to the present time, the negative properties of the diesel exhaust are balanced by the positive properties of the diesel engine. First of all, the good operational economy of the diesel engine should be mentioned. The diesel engine has the highest efficiency of all combustion engines and therefore will emit the smallest possible amount of carbon dioxide per unit of output energy [6]. The diesel engine therefore has become the dominating energy converter in heavy vehicles. In this perspective, it is important to continuously work with techniques that will decrease the emissions harmful to the environment and the health of exposed workers.

In this chapter we will discuss effects of diesel exhaust on pulmonary function with special regard to occupational exposure and physiological measurements of the lungs that may be useful in evaluating the effects of diesel exhaust on lung function. We also will discuss the importance of the composition of diesel exhaust to these effects and the possibilities of developing a generally applicable exposure indicator.

LUNG FUNCTION

Definitions

Exposure to diesel exhaust might result in histological, morphological, and physiological deteriorations in lung function. We want to emphasize that the descriptions of the different methods are meant to serve as a brief overview. The different indices of lung function may be measured in humans as well as in animals, although there sometimes are differences in methodology.

Measurement of Lung Function

Measurements concerning ventilatory capacity are obtained by dynamic spirometry. The breathing maneuver is usually performed in the form of a maximal inspiration followed by a maximal expiration, vital capacity, (VC). The volume may be directly measured by a water-sealed or a low-resistive bell or indirectly measured by pneumotachography (i.e., integration of a flow signal over time).

Forced expired vital capacity (FVC) is defined as a VC measurement performed with a maximally forced expiratory effort. Due to the simplicity of the test and the relatively inexpensive equipment, the forced expirogram has become a useful test in epidemiological studies. Associated variables that can be calculated from the forced expirogram are forced expired volume in one second,

(FEV_1), FEV_1 in per cent of FVC (FEV%), peak flow (PEF), maximal flows at 50% and 75% of forced vital capacity (MEF_{50} and MEF_{25}, respectively). FVC, FEV_1, and PEF reflect the conditions in the upper airways, and MEF_{50} and MEF_{25} are considered to reflect the conditions in lower airways.

The proportion of residual volume (RV), i.e., the volume of air still in the lungs after a maximal expiration, can be determined by gas dilution or by body plethysmography. The gas dilution technique requires less-complicated equipment and is therefore more convenient to use in studies carried out at the workplace. The technique is based on dilution of an indicator gas in a rebreathing circuit. The subject and the apparatus form a closed unit, and the initial concentration of the gas is thus reduced when it is diluted into the gas volume in the lungs. At equilibration, the concentration of indicator gas is the same in the lungs as in the apparatus, and functional residual capacity (FRC), can be calculated by means of a simple dilution equation. Two to three VC-maneuvers are carried out to provide a reliable base for the calculation of total lung capacity (TLC).

Values for airway closure and gas distribution in the lungs can be obtained by the same maneuver as the single breath wash-out technique. The maneuver is carried out by means of a maximal inspiration of pure oxygen. When the expiration from the basal parts of the lungs grows, the nitrogen concentration will rise abruptly in case of airway closure in dependent lung regions. This volume above RV, at which airways close during an expiration, is usually expressed as closing volume (CV) in percentage of VC (CV%). Distribution of the inspired air in the lungs is expressed as the slope of the alveolar plateau, phase III. The variables CV% and phase III are considered to reflect the conditions in the peripheral small airways with an internal diameter about 2 mm.

Transfer factor for carbon monoxide (TL_{CO}) is an expression of the diffusion capacity of oxygen transport into the lung capillaries. Determination of TL_{CO} is carried out by means of a breathing maneuver including a maximal expiration, followed by a maximal inspiration of a gas mixture containing carbon monoxide, helium, oxygen, and nitrogen. After a breath-holding period, a maximal exhalation is done, reflecting the content in the alveolar air. TL_{CO} is then calculated, assuming an exponential uptake of carbon monoxide.

The resistance in the airways essentially depends on the radius and length in the airways but also on air viscosity. The airway resistance (R_L) can be determined in humans by use of body plethysmography, where the changes in flow and pressure during breathing maneuvers are recorded. In anesthetized animals, R_L may be determined by measuring the changes in flow and pressure in the airways during controlled respiration as aided by a ventilator.

Compliance is a function of, inter alia, the elastic properties in the lungs. The compliance of the lung is defined as the difference in the transpulmonary pressure divided by the change in volumes during a breathing maneuver. The transpulmonary pressure is the difference between the pressures in the mouth (atmospheric) and in the lung. During inspiratory and expiratory maneuvers, the changes in volume and pressure are recorded by means of a spirometer and pressure transducer, respectively. When the measurements are performed during tidal breathing, dynamic compliance (C_{dyn}) is measured. Static compliance (C_{st}) is obtained when a slow VC maneuver is carried out. In the latter case, the measurements are carried out in a body plethysmograph, and the expiration is intermittently interrupted by means of a shutter. However, measurements of C_{dyn} and C_{st} often are cumbersome to perform when examining exposure effects on lung function in studies on humans at the work site. This measurement is therefore considered to be more appropriate in animal studies. A useful approach in this context is to measure C_{st} of the lung and chest wall, in anesthetized animals, by recording the pressure change in trachea after administration of a small volume of air.

By applying Hooke's law and assuming a simplified geometry of the lung and chest wall, it can be shown that C_{st} is proportional to the expression $D^4 \times E^{-1} \times t^{-1}$. In this expression, D is the average diameter of the lung and chest, E is the elasticity module, and t is the average thickness of the lung and chest wall. The elasticity module is a property measured in strength tests of materials and is, for instance, high for steel and low for lead.

Acute and Chronic Effects of Occupational Exposure on the Pulmonary Function

Work-related, temporary decrease of the lung function over a work shift has been recorded in cotton workers [7]. Later, several authors reported work-related temporary changes of the lung function in hemp and textile workers, in coal miners, in workers exposed to toluene diisocyanate, firemen, in rubber processing workers, in moulders and coremakers, in welders, in ski waxers, in workers exposed to organic dust, and workers exposed to biocides in water-borne paints [8-24].

One special question is if repeated, temporary effects can result in chronic effects. An acute, temporary lung function decrease may not only be a biologic exposure indicator but a predictor for a future chronic lung function decrement. The results of a study on cotton textile workers with characteristic work-related respiratory symptoms suggest that across-shift change in FEV_1 was predictive for pulmonary function impairment [25]. Similar relationship between acute, temporary lung function decrease during a working week and an accelerated decline in lung function also was found in asymptomatic wood trimmers [26].

Chronic effects on the lungs in occupationally exposed subjects may be detected by different investigation designs. Several study designs have helped detect harmful effects on lung function in exposed subjects.

A common study design is to compare the actual values in exposed subjects with lung function values obtained in a reference population without occupational exposure to the specific agent. The reference subjects may be recruited from the same or nearby workplaces or from the same geographical area. Multivariate analysis has been used in some studies to assess differences between exposed subjects and their referents, e.g., subjects exposed to oil-mist, in workers exposed to paper dust, and in vehicle mechanics with an exposure to asbestos [27,28]. Lung function values in exposed subjects also may be standardized by means of reference equations based on lung function values in unexposed subjects, e.g., in subjects exposed to cotton dust and in asbestos cement plant workers [29,30].

Another approach is to study the differences between the lung function values and external reference values, calculated by means of prediction equations based on healthy subjects, in exposed and unexposed workers, respectively. This design has been used, e.g., in workers exposed to saw fumes containing terpenes and in wood trimmers exposed to organic dust [23,31].

The reference population also may be matched to the exposed subjects according to sex, age, height, and smoking habits to control for confounding factors. This procedure has been useful in detecting effects of exposure to diisocyanates in car painters, in subjects exposed to phthalic anhydride, and in bookbinders with exposure to paper dust [32-34].

Finally, chronic effects on lung function also may be studied by examining the individual changes in lung function in exposed and unexposed subjects over a period of several years. One advantage with the longitudinal study design is that the intersubject variability is reduced. The design is considered to be very time-consuming and therefore expensive, however. Nevertheless, this approach has been used successfully in wood workers exposed to formaldehyde, in car painters exposed to diisocyanates, in wood trimmers exposed to organic dust, and in saw-mill workers exposed to saw fumes [35-38].

Lung Function Decrease Associated with Occupational Exposure to Diesel Exhaust

Acute, Temporary Effects

It is generally agreed that diesel exhaust causes an unpleasant smell, which is easy to recognize. This smell is associated with the existence of particles in the exhaust. The effects of diesel exposure was investigated during loading by means of diesel-propelled trucks aboard a roll on-roll off ship. The employees experienced that the exhaust smell vanished when specially constructed microfilters were mounted on the exhaust tubes on all trucks in use [39].

An increased prevalence of respiratory symptoms over a work shift also was noted in diesel bus garage workers. Symptoms of eye irritation, labored breathing, chest tightness, and wheeze were strongly correlated with exposure [40].

Exposure to diesel exhaust also gives measurable irritative effects on the lungs, shown as an acute, temporary lung function decrease [2]. Mechanical filtration of diesel exhaust from trucks used in loading by stevedores relieves subjective disorders and reduces the acute, temporary lung function decrease [39,41]. When filters were used, the average decline in lung function (FVC) over a work shift was reduced by 50% to 60%. In the nonexposed control subjects, a tendency toward an increase in FVC was found over a work shift (see Figure 1). The particles therefore play a role in the irritative effect.

In situations where mixed exposure occurs, e.g., exposure to diesel exhaust and other dusts in mines, acute lung function changes may not be detected. Lung function during work shift was examined in coal miners with diesel exhaust exposure. The results were compared to the lung function of coal miners with no diesel exposure. No significant difference in the ventilatory function changes during a workshift was found [42]. The authors state that the potential for an interactive effect of coal mine dust aerosols and diesel emissions could not be excluded [42].

There is reason to assume a threshold effect in diesel exhaust exposure. Exposures below a certain level will not be sufficient to induce an acute, temporary decrease in lung function. This was indicated in workers on car ferries and in bus garages [1,2]. In addition, subjects exposed to diesel exhaust in an exposure chamber displayed an increase in FVC [1] (see Figure 2). In the experiment, the concentration of nitrogen dioxide was ten times as high as was found for the studied stevedores. The lack of a generally applicable exposure indicator is a problem in comparing different exposure situations. The measured value of particle concentration in the exposure chamber, after a dilution 1:40, was only 55% of the value expected due to the dilution effect only. The particles consisting of condensed liquids probably precipitated in the dilution tunnel [43].

Diesel exhaust exposure has been shown to cause acute inflammatory changes in the lungs, shown as neutrophilia and decreased phagocytosis. Eight nonsmokers underwent bronchoalveolar lavage (BAL) before and after exposure to

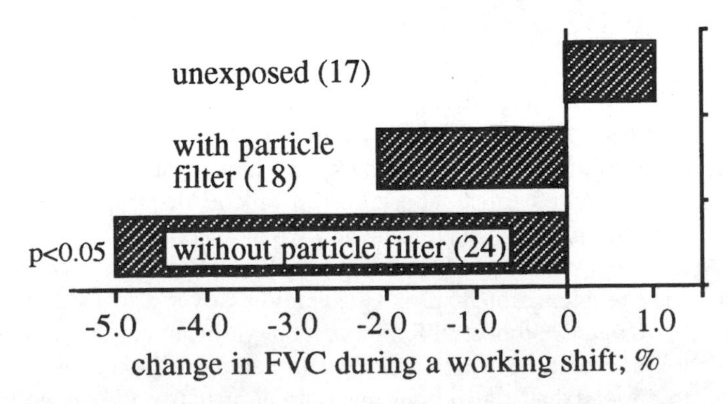

Figure 1. Change in lung function (FVC) during a working shift in nonexposed subjects and stevedores exposed to filtered and unfiltered diesel exhaust, respectively. Daily average concentrations of nitrogen oxides and respirable dust during a work shift without filtering, were 1.1 ppm and 0.2 mg/m^3, respectively. (Based on reference 39.)

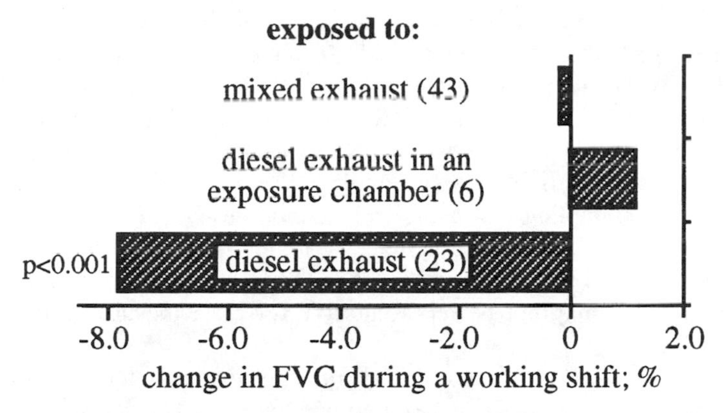

Figure 2. Change in lung function (FVC) during a working shift in subjects working in bus garage and on car ferries exposed to mixed exhaust, subjects exposed to diesel exhaust in an exposure chamber, and stevedores working on roll-on-roll-off ships exposed diesel exhaust. The concentrations of NO_2 were 0.2, 2.1, and 0.2 ppm, respectively, and of dust 0.2, 0.6, and 0.4 mg/m^3. (Based on reference 1.)

diesel exhaust in an exposure chamber [44]. The diesel exhaust was generated in an idling motor and the diesel exhaust exposure was adjusted to a NO_2 level of 1.5 parts per million (ppm), i.e., similar to a level that may be found in occupationally exposed workers.

The inflammatory lung function effects in humans or some irritating gases included in motor exhaust, i.e., nitrogen dioxide and sulfur dioxide, have been studied as well [45]. The subjects experienced only very mild symptoms, and lung function was clinically unaffected, as measured with dynamic spirometry. Despite the absence of clinical symptoms, inflammatory changes were found in the lung as reflected in BAL fluid [45].

Due to the complexity of diesel exhaust, it is probable that the irritating effect on mucous membranes, airways, and lungs depends on a complicated interaction among several components in the exhaust. The particles may transport irritating substances into the lungs where they can exert their effects.

Chronic Effects

Several studies to investigate if diesel exhaust exposure results in lung function impairment have been performed. The results of the studies have been contradictory, however. Relationships between exposure to diesel exhaust and a chronic lung function deterioration have been found in diesel bus garage workers and in stevedores [46-47].

Other studies on locomotive repairmen and salt miners have failed to show a relationship between the exposure to diesel exhaust and lung function deterioration

[48-51]. In some of the studies, where no lung function effects were found, the investigated workers have been exposed not only to diesel exhaust but to other pollutants as well, including coal dust [51-54]. This may have caused difficulties in interpreting the results. In many studies of chronic effects, the exposure assessment is insufficient, which also will contribute to the difficulties in interpretation. Measurements of exposure indicators are affected by several factors, for instance the sampling and analytical methods used, the investigation strategy, the distribution of exposure periods over time, rotation of workers between different tasks, and seasonal variations [55]. Exposure assessment during one or a few work shifts may not be representative for the exposure during a longer period of time.

A careful selection of a reference group may decrease the variation in the examined spirometric variables in exposed and unexposed subjects. This approach may to some extent explain why it was possible to show a statistically significant impairment of the lung function (FVC, FEV_1, MEF_{50}, and MEF_{25}) in coal miners when compared with their carefully matched referents [55].

Despite the contradictory results on the lung function of diesel exposure, there are many reports on symptoms from the mucous membranes, like cough and phlegm, associated with exposure to diesel exhaust [46,49,52,54].

Chronic effects of exposure to diesel exhaust on lung function also has been studied in different animal species. In drawing conclusions from animal experiments, several circumstances have to be considered. There are differences between species and also between animals and human beings. Differences in methodology also may be of importance when examining the results. The exposure levels sometimes have been much higher than will occur in humans with occupational exposure to diesel exhaust.

Increased R_L^v and deposition of black pigment-laden macrophages in the lungs were demonstrated in a study on guinea pigs, where the concentration of nitrogen dioxide and particulate matter were 2.5 ppm and 7 mg/m^3, respectively. After recovery for 90 days in clean air, the lungs of the animals still retained pigmented macrophages [56]. Decreased lung function (VC, TLC, and TL_{CO}) was demonstrated in cats exposed to diluted diesel exhaust corresponding to >6 mg/m^3 for 124 weeks [57]. On the other hand, no statistically significant differences in lung function, except a decrease in CV, were noted with an exposure level of 6 mg/m^3 for one year. The authors conclude that this decrease was not to be interpreted as an adverse effect on lung function because decreased CV reflects a decreased airway closure, and thus an improvement in lung function [58]. Cats exposed to diesel exhaust, diluted 1:14 for 28 days, showed a significant decrease in MEF_{10}, but no statistically significant differences were noted in C_{dyn}, R_L, or TL_{CO} [59]. The same exposure level and duration of the study demonstrated a mild effect on lung function in rats [60].

Rats exposed to diluted diesel exhaust six months, with a particle concentration of 3.5 mg/m^3, showed chronic inflammation, decreased lung volumes, and decreased alveolar/capillary gas exchange efficiency [61]. After analyzing the BAL fluid in the rats, the authors suggest that the inflammatory responses were qualitatively related to the concentration of diesel soot. They also noted a no-effect level of 0.35 mg/m^3 [62].

Long-term exposure to diesel exhaust over a two to three year period in rats and monkeys resulted in deposition of particles in pulmonary tissues and inflammatory alterations, shown as decreased phagocytes. The results also indicate that diesel exhaust exposure at airborne concentrations of 2 mg/m^3 is more hazardous to the health of the animals than the same exposure concentration of respirable coal dust [63]. Exposure of rats to diesel exhaust with a particulate concentration of 1.5 mg/m^3 for 267 days did not result in any harmful lung function effects [64]. However, the rats had a statistically significant increase in FRC after 600 days with the same exposure level [65].

A comparison can be made between the results of occupationally exposed workers and experimentally exposed animals. The effects on lung function in animals are similar to the effects in workers. This is found although the exposure in animal experiments, as judged by the concentrations of common exposure indicators, is much higher than in occupational exposure of human beings. The main findings are decreased lung volumes, sometimes in combination with a decreased TL_{CO}, suggesting restrictive impairment. In addition, inflammatory reaction in the lungs has been demonstrated. The chronic lung function impairment caused by long-term exposure to diesel exhaust is consistent with the acute, temporary effects, e.g., decreased FVC and FEV_1, in occupationally exposed workers [2,39].

Although there are several reports of chronic lung function impairment of workers occupationally exposed to diesel exhaust, there are several other studies where diesel exhaust exposure was not associated with lung function effects. It is possible that this is due to differences in exposure levels. The exposure to diesel exhaust may in some cases have been too low to induce effects on lung function. In studies where the workers have been exposed to high diesel exhaust exposure, i.e., of bus garage workers and of stevedores, there are clear effects on the lung function, however. This is true for acute as well as chronic effects on the lungs. Further research is needed to attribute the effects to certain components in the diesel exhaust and to investigate the importance of all factors that influence the composition of diesel exhaust, e.g., different fuels and after treatment of the exhaust by particle traps. This knowledge is necessary to minimize the effects on the lungs of diesel exhaust-exposed workers.

THE COMPOSITION OF EXHAUST

When fuel is burnt in gasoline or diesel engines, mostly water and carbon dioxide is formed—the same substances found in exhalation air. Other substances are nitrogen and some remaining oxygen from the air, soot particles (especially in diesel exhaust), a great number of hydrocarbons and their derivatives, and carbon monoxide, formaldehyde, nitrogen oxides, and sulfur dioxide. The concentration of the substances included varies depending on the composition of the fuel, the content of the lubricant, the motor construction, the operating temperature, the load, the condition of the motor, the supply of fuel, etc.

The composition of diluted diesel exhaust generated under controlled conditions is exemplified in Table 1 [1]. The test vehicle was a 1980 Volvo 244D automobile with a manual, four-speed transmission. The vehicle was powered by a six-cylinder, precombustion chamber diesel engine (2,383 cm^3). The compression was 23:1, and the engine output 60 kW. During the experiment, the vehicle was run at a constant speed, equivalent to 60 km/h in third gear. This rate produced an engine speed of about 2,580 revolutions/min. The engine load was calculated as 18 kW (maximum engine output about 35 kW at the indicated engine speed).

Employees occupationally exposed to exhaust generally are in contact with exhaust from many engines. The exhaust usually has been diluted with a factor about 50 times or more. Table 2 gives an overview of concentrations of respirable dust and some gaseous air pollutants observed in workplaces with diesel emissions.

THE IMPORTANCE OF FUEL TO EMISSIONS

Fuel for diesel engines usually contains heavy hydrocarbons, 12 to 16 carbon atoms per molecule, with boiling points from 180° to 360 °C. Table 3 presents data on standard requirements on characteristic properties of three qualities of diesel fuels. Urban diesel 1 and 2 are intended for densely populated areas and therefore contain a minimum of sulfur. There is also a certain supply

Table 1
Composition of Diesel Exhaust after Dilution 1:40

Substance	Concentration, mg/m^3
Nitrogen dioxide	4,0
Nitrogen monoxide	6,7
Carbon monoxide	5,3
Benzene	0,3
Toluene	1,3
Formaldehyde	0,05
Acetaldehyde	0,3
Sum of hydrocarbons, ppm	35
Total dust	0,6

Based on Ref. 1.

Table 2
Overview of Observed Concentrations of Some Pollutants found in Workplaces with Diesel Engine Exhaust

Type of Work Place	Acetic-aldehyde mg/m^3	Respirable dust mg/m^3	Form-aldehyde mg/m^3	Carbon monoxide mg/m^3	Nitrogen dioxide mg/m^3	Nitrogen monoxide mg/m^3
Bus garages[a)]	0,28–1,5	0,46	0,04–0,8	1,7–24	0,2–1,1	0,3–1,0
Car ferry (2-h run)[a)]	0,49–1,5	..[d)]	0,03–0,31	13–100	<0,6	0,06
Car ferry (20-min run)[a)]	1,02–2,1	0,1–0,3	0,1–0,3	5–190	0,2–0,8	0,2–1,0
roll-on roll-off ship[a)]	<1,6	0,13–0,59	≤0,03	1,4–2,7	0,15–1,0	0,1–0,8
roll-on roll-off ship, second study occasion[a)]	..[d)]	0,3–1,0	0,1–0,5	1,1–5,1	0,06–2,3	0,02–0,7
Truck drivers in tunnel-construction[b,c)]	..[d)]	1,11±0,37	..[d)]	7,07±3,08	1,27±0,71	3,54±0,94

a) Ref. 2.
b) Ref. 41.
c) ±standard deviation.
d) Data not available.

of diesel fuel that contains only paraffin hydrocarbons, but this quality probably still will be difficult to obtain everywhere. In practice, the manufacturers often supply fuels with lower concentrations of sulfur and aromatics than specified in the standard.

Different fuels affect the emission of gases and particles [66,67]. The emission of particles is strongly dependent on the composition of the diesel fuel. This is especially evident at low loading at a high number of revolutions per minute [66]. Heavier grades of diesel fuel—with high molecular weight, high boiling point interval, and with a high content of aromatic hydrocarbons or sulfur—yield a great deal of particles. A linear relationship between the sulfur content of the fuel and the particle emission has been observed [68]. A high concentration of sulfur in the fuel will result in a high concentration of sulfate in the exhaust. Due to the hygroscopy of sulfates and the water content, the related particle mass concentration increases [66]. When a catalytic particle trap is used, the high concentration of sulfur in the diesel fuel may contribute to the emission of particles. This is due to the formation of sulfur trioxide by oxidation of sulfur dioxide in contact with the catalyst and the subsequent formation of sulfuric acid particles from water and sulfur trioxide.

A diesel engine fuel of current interest is ethanol. The lower combustion temperature for ethanol will give less nitrogen oxides than when hydrocarbons are used as fuel. An important advantage is that ethanol as a fuel for diesel engines will not emit soot, yet leakage of lubrication oil into the cylinder may produce a certain amount of particles. At low load, there is an oxidation to acetic acid, which is a nuisance when diesel buses operating on ethanol are waiting at bus stations.

Table 3
Requirement on Diesel Qualities for High-Speed Diesel Engines

Property	Urban diesel 1	Urban diesel 2	Standard summer diesel 3
Sulfur, weight-%	≤0,001	≤0,02	≤0,3
Aromatics, volume-%	≤5	≤20	no requirement
Boiling point interval, °C	180–300[a)]	180–300[a)]	250–370[b]
Density[c)], kg/m^3	0.8087	0.8143	0.8487

(Swedish Standard SS 155435) (Urban diesel 1 and 2 are intended for densely built-up areas.)

a) Concerns initial and final boiling points.

b) Concerns minimum temperature at 65% recovery and maximum temperature at 95% recovery respectively.

c) At 15°C in air.

EXPOSURE INDICATORS

Due to complex composition of diesel exhaust, quantification is very difficult. Quantification of diesel exhaust exposure has to be done measuring one or several determinants or exposure indicators, i.e. substances that use to be present in the exhaust. The indicators cannot be assumed to be well-correlated to other components of the exhaust. As a consequence, a relationship has to be investigated between the biologic effect and the exposure indicator, which may or may not be the biologically active component of the diesel exhaust. There is also a possibility to use transient changes in biologic effects as a biologic exposure indicator. The problems involved are discussed below.

Low Molecular Inorganic Gases

Diesel exhaust contains many substances each one of which may be injurious to health—carbon monoxide, nitrogen dioxide, formaldehyde. They are all relatively easy to analyze, and therefore the exposure judgment traditionally has been founded on the concentrations of such substances. The understanding has grown that it is unsatisfactory to describe the health effects of diesel exhaust in such simple terms. It is often an open question which substance or combination of substances will cause known health effects. The composition of the exhaust is very complex and varies from one occasion to another. If many exposure situations are observed, it is therefore very unlikely that the concentration of an exposure indicator, e.g., carbon monoxide, is correlated with other substances of possible interest to the health of exposed persons.

In an investigation of possible gaseous exposure indicators the following substances were discussed as exposure indicators: carbon monoxide, formaldehyde, and nitrogen dioxide [69]. The authors concluded that none of these three gases meets the demands that have to be put on an indicator for all kinds of exhaust. The investigation recommended special "dimension values" for these substances that should be used when exposure to motor exhaust is present. The occupational health standards issued in 1990 by the Swedish National Board of Occupational Safety and Health laid down a special condition for the standards for carbon monoxide and nitrogen dioxide: When the source of these two gases is motor exhaust, the standards are much lower than otherwise.

Exposure Indicators Based on Characterization of Particles

The use of "dimension values" no doubt is only a temporary solution to the problem. Occupational health standards for exhaust should not ignore the particles. Both the chronic and acute effects on the lungs depend at least partly on the presence of particles. If the occupational health standard does not reflect the concentration of particles, control measures that concern a decrease of the concentrations of particles are not rewarded. Such measures may concern the construction and maintenance of the motor, the operation conditions, or filtration of exhaust. Some of these measures may affect the composition of gases as well, but filtration will not. Control measures directed to the concentration of gaseous substances are much more difficult to achieve, and their usefulness in reducing the health risks of exhaust are debatable.

A coarse method to measure soot particles is the filter-blackening method. In this method, the exhaust-polluted air is pumped through an analytical filter and the blackening of the filter is measured photometrically. The blackening is usually recorded in Bosch units [70,71]. A personal diesel aerosol sampler for underground coal mines has been developed [72]. The method is based on the idea that the predominant fraction of diesel exhaust particles is less than 0.8 μm while mineral dust is greater than 0.8 μm [72,73]

It is possible that these methods are useful in certain work environments. One drawback with both the size-selective sampler and the blackening method is the lack of specificity. Not only soot particles but other types of particles as well may have a typical size less than 0.8 μm or have the property to blacken a filter, e.g., coal dust in a coal mine. The methods therefore cannot be used in all possible work environments.

A method to measure the workplace concentration of particles from an engine should be able to differentiate diesel soot particles from other particles in the workplace. Identification by means of substances that can be extracted from particles with various solvents has been tried [4,74-76]. Not only soot particles from diesel engines but also particles from other sources may contain extractable substances. The problem is to find unique components to correctly determine the origin of the particles. Investigations to use the complex composition of the exhaust to identify the origin of the exhaust have been performed. Another idea is to analyze many substances in the extract and study if there is a certain pattern that can be recognized [77]. So far methods of this kind can be used only in certain work places.

Different experiments to use added labelling substances or radioactive labelling to indicate exhaust have been carried out [78-81]. This is not possible in routine monitoring, however.

Experiments also have been performed to measure the ash residue after combustion of a dust sample from workplaces with diesel exhaust exposure [41,82] Not only soot particles but also other dust particles that may be present in a workplace will burn, and therefore this method is not generally applicable.

Diesel particles have been characterized by means of transmission electron microscopy [83,84]. Results from the latter study, where particles were sampled on 0.2 μm gold-covered filters in a truck garage, suggested that the distribution of particle size, based on the particle volume, constituted of two modes, where particles with an aerodynamic diameter of more than 1.0 μm were found to be aggregated particles [84].

Biological Indicators

Although there is no generally applicable chemical or physical exposure indicator for diesel exhaust exposure, biological monitoring is a possibility. Usually, biological monitoring consists of an assessment of overall exposure to chemicals present in the workplace through measurement of the appropriate determinants in biological specimens collected from exposed persons. The determinant can be the chemical itself or the metabolites, or a characteristic reversible biochemical change induced by the chemical [85].

To indicate the presence of irritating substances, it also has been suggested to use an acute, temporary medical effect in an exposed person or in a biologic model system [29]. We found FVC, FEV_1, and possibly TL_{CO} to be particularly useful lung function variables for this purpose in humans. In most situations, the acute, temporary changes in the lung function variables are comparatively small. To minimize the effect of the biologic variability and the diurnal variations, each person or test animal should be used as his own control. The lung function after exposure is then compared to the person's unexposed lung function value at the same time of the day.

This approach may be used in animals as well. An animal model for noninvasive lung function measurements before and after short-term exposure, i.e., 5 hours, has been modified for this purpose [86]. In animal studies we have found the compliance of the respiratory system to be a useful biologic exposure indicator for lung irritation.

NATURE AND IMPORTANCE OF PARTICLES IN DIESEL EXHAUST

A typical regulation requires cooling of the exhausts to 52°C before sampling particles on a filter. This means that substances in the exhaust that will condense at that temperature—water, lubricant oil, and fuel—will be found in the mass of particles [68]. Less than half of the mass fraction of particles will be soot particles. This fraction increases with the load. An important part of the particles will be formed by water and sulfates [68]. Table 4 presents the content of the particle fraction according to a compilation [87]. The particles were sampled 10 cm from the end of the exhaust tube.

The carbon particles consist of basically spherical primary particles or spherules that agglomerate into aggregates. A particle may contain several thousands of spherules. Carbon particles also will contain adsorbed hydrocarbons [88]. Information about the size of the particles varies in different sources of literature. According to one publication, the volume-mean diameter is found to range from 50 to 220 μm with no consistent trend for the variation with either speed (i.e., of a vehicle with a diesel engine) or load [89].

Table 4
Compilation of Content of the Particles Sampled 10 cm from the End of the Exhaust Tube

Fraction	Substance/substance group	mg/m³ in sample about 10 cm from the exhaust tube
Particles	total amount	30–75
Carbon		20–45 (about 60 % of the weight of the particles
Soluble in diethyl ether	total amount	10–13 (about 30 % of the weight of the particles)
	alkanes	10–12
	alkane acids	0,1–1
Water soluble extract	total amount	0,3–2 (about 3 % of the weight of the particles)
	alkane acids	0,1–0,6
	nitro alkanes	0,03–0,1
	aldehydes	0,03–0,1
	nitrite	0,03–01
	sulfate	0,2–0,3

Based on Ref. 87.

Sulfur content in diesel fuel will generate less-soluble particles in the exhaust. This content of sulfur in the less-soluble fraction of the particles will increase with the dilution of the exhaust with air which indicates that air oxidation is a contribution factor [90].

According to model calculation, the number of particles in dilute exhaust will decrease due to aggregation. Within a minute, the number will decrease to about 10%. The fate of a particle after leaving the exhaust pipe and until it is inhaled by an exposed person is due inter alia to the efficiency and swiftness of dilution, the weather condition, and the transport time.

Besides aggregation of the particles, rapid chemical changes of adsorbed substances may be expected even at room temperature [91]. Heterogeneous reactions are difficult to investigate. Investigation methods based on extraction and analysis of extracted material are difficult to implement due to the changes in the conditions at the surface. It is therefore a very challenging task to study the fate of adsorbed substances during transport from where the particles originate in the engine to the point where the particle settles in the body in contact with the cells.

It is possible that very small particles are especially toxic in the lungs. Experiments with rats were performed by intratracheal inhalation of particles of titanium dioxide and aluminium oxide in two sizes, 0.015–0.050 μm and 0.2–0.5 μm [92,93]. The authors concluded that the ultrafine particles were very toxic to the lungs. They penetrated the lung tissue more easy than the coarser particles. In the lungs, particle aggregates were dispersed into the primary particles [93]. It is possible that particles in diesel exhaust act as carriers for adsorbed substances [94]. In the exhaust there seems to be a partition equilibrium of volatile substances, especially hydrocarbons between the gas phase and the solid phase, the soot particles. Droplets concentrated around agglomerates of soot particles have been observed in transmission electron microscopes [83]. These droplets were strongly bound hydrocarbon compounds of high molecular weight because lighter hydrocarbons would vaporize in the microscope's vacuum.

On the surface of the soot particles, substantial amounts of hydrocarbons and derivatives of hydrocarbons, sulfur compounds, and other substances may be adsorbed. Five to 50% of the total particle mass may be organic substances that can be extracted by organic solvents [95]. The chemical properties of substances extracted from the particle surface by ether or water may change by the process of extraction. Substances on the particle surface have been studied by X-ray methods (ESCA) showing that nitrogen is present in the form of nitrate and ammonia ions [96,97]. A very small amount of the total amount of inorganic nitrogen compounds in the exhaust are bound to the particles. This small proportion may not be unimportant when it is considered that the small particles are able to reach the alveoli where they may precipitate.

Irritation of the lungs could depend on substances adsorbed on particles. A water soluble, irritating gas like sulfur dioxide is dissolved in the mucous membrane of the nose and the airways. If gas molecules are adsorbed on the surface of small particles, they can be taken deep into the lungs. In laboratory experiments with guinea pigs, the irritating effect of sulfur dioxide on the lung function was increased if the guinea pigs inhaled zinc oxide particles simultaneously [98]. Zinc oxide alone had no effect on the lung function [99].

Contradictorily to this finding, there was no evidence of acute temporary lung function deterioration in normal male human subjects when they were exposed for 4 hr to 0.5 ppm nitrogen dioxide or 0.75 ppm sulfur dioxide together with the following aerosols: ammonium bisulfate, ammonium sulfate, or ammonium nitrate [100]. One explanation for the difference in results is that these authors made the comparisons on a group basis with 10 to 15 subjects in each group and did not consider the diurnal variations in lung function variables. We have found that it is important to minimize individual variability by standardizing each individual's exposed value by the corresponding unexposed variable value of each subject recorded at the same time of the day as the exposed value.

In a recent investigation we found evidence that only the soot particle fraction of diesel exhaust reached and had an effect on the airways and lungs [43].

In an animal experiment with diesel exhaust exposure, the compliance of the respiratory system was used as a biologic exposure indicator. The concentration of all particles sampled immediately after dilution contained a mixture of soot particles and condensed particles. The concentration of all particles was not correlated to either the concentration of soot particles or a change in compliance Table 5.

There was a clear association between the change in compliance and the exposure concentrations of the soot particles and formaldehyde but no association between the change in compliance and the concentration of nitrogen dioxide. The lack of association between the effects on lung function and the concentration of nitrogen dioxide is corroborated by the outcome of human studies [2,44].

The blackening of a filter paper immediately after dilution in Bosch units measures the soot particle concentration. These particles were small enough to be transported through the comparatively long tubes between the motor and the exposure chambers without precipitation. The particle sample immediately after the dilution, on the other hand, contained an aerosol of condensed liquid (lubricating oil, fuel, and perhaps water). These particles later on may have been precipitated in the tubes before the exposure chambers.

Table 5
Correlations between Bosch units, Concentration of Particle Samples Immediately after Dilution, Particles in the Exposure Chambers, and Change in Compliance

	Particles immediately after dilution	Particles in the exposure chambers	Compliance
Bosch units	r = –0.059 n = 33 p = 0.7	r = 0.875 n = 33 p < 0.0001	r = 0.43 n = 33 p = 0.01
Particles immediately after dilution	1	r = –0.183 n = 33 p = 0.3	r = –0.0082 n = 33 p = 1
Particles in the exposure chambers		1	r = 0.47 n = 33 p = 0.005

r = product moment correlation coefficient, n = number of observations, p = significance level as probability of no correlation. Based on Ref. 43.

CONCLUSIONS

There are conflicting views in the literature regarding chronic effects on the pulmonary function by diesel exhaust. It seems reasonable that this is a question of exposure level. In a number of reports on chronic effects on the pulmonary function, the exposure can be assumed to involve high exposures. Investigations reporting negative findings, on the other hand, can be assumed to concern low-level exposures.

It is very difficult to assess the exposure to diesel exhaust due to the complex composition of this pollutant. This is especially true for chronic exposure. Relevant exposure indicators have to be developed. So far there is no generally applicable exposure indicator available, and before this problem is solved—if it can be solved—it will be very difficult to determine the effects of diesel exhaust exposure.

Acute effects on the pulmonary function become important in this context. In acute exposure situations, it is much easier to measure the individual exposure to one or several exposure indicators. The exposure is also more homogenous during a period of a working day or a working week. It is likely that one or several exposure indicators can be associated with acute, temporary effects on the pulmonary function. It is therefore also possible to attribute the effects to certain components in the exhaust.

So far there is no conclusive evidence that acute, temporary effects on the respiratory system due to diesel exhaust exposure can be used as a predictor for chronic changes of the respiratory system. There are, however, some indications that this is the case, and the possibility should be investigated. The literature survey shows that chronic and acute, temporary effects regard the same type of changes of the pulmonary function.

REFERENCES

1. Ulfvarson, U., Alexandersson, R., Aringer, L., Anshelm-Olson, B., Ekholm, U., Hedenstierna, G., Hogstedt, C., Holmberg, B., Lindstedt, G., Randma, E., Rosén, G., Sorsa, M., Svensson, E. "Health Effects after Exposure to Motor Exhaust." [Hälsoeffekter vid exponering för motoravgaser]. National Institute of Occupational Health. Sweden, Solna, Arbete och Hälsa 5:1–83 (1985).
2. Ulfvarson, U., Alexandersson, R., Aringer, L., Svensson, E., Hedenstierna, G., Hogstedt, C., Holmberg, B., Rosén, G., Sorsa, M. "Effects of Exposure to Vehicle Exhaust on Health." *Scand J Work Environ Health 13*:505–512 (1987).
3. Gustavsson, P., Plato, N., Lidström, E-B., Hogstedt, C., "Lung Cancer and Exposure to Diesel Exhaust among Bus Garage Workers." *Scand J Work Environ Health 16*:348–354 (1990).

4. Harkov, R., Shiboski, S. "Can Extractable Organic Matter (EOM) Be Used as a Motor Vehicle Tracer in Multivariate Air Quality Receptor Models?" *J Environ Sci Health Part A21*(2): 177–190 (1986).
5. Hedberg, G., Jacobsson, K-A., Langendoen, S. "Factors Influencing the Labor Turnover among Drivers: A Retrospective Cohort Study." [Faktorer som påverkar omsättningen bland yrkesförare. En retrospektiv kohortstudie]. National Institute of Occupational Health. Sweden, Solna, Arbete och Hälsa 11:1–58 (1988).
6. Egnell, R. "The Diesel Motor and its Potential for Development." [Dieselmotorn och dess utvecklingspotential]. *NUTEK B 1992:*4, Stockholm (in Swedish).
7. McKerrow, C.B.; McDermott, M., Gilson, J.C., Schilling, R.S.F. "Respiratory Function during the Day in Cotton Workers: A Study in Byssinosis." *Br J Ind Med 15:*75–83 (1958).
8. Bouhuys, A., Barbero, A., Lindell, S.E., Roach, S.A., Schilling, R.S.F. "Byssinosis in Hemp Workers." *Arch Environ Health 14:*533–544 (1967).
9. Valic, F., Zuskin E. "Respiratory Function Changes in Textile Workers Exposed to Synthetic Fibers." *Arch Environ Health 32:*283–287 (1977).
10. Reger, R. "Ventilatory Function Changes over a Work Shift for Coal Miners Exposed to Diesel Emissions." In: Bridbord, K., French, J., eds. *Toxicological and Carcinogenic Health Hazards in the Workplace.* Proceedings of the first annual National Institute for Occupational Safety and Health scientific symposium. Washington: Department of Health, Education, and Welfare. Pathotox Publ., Inc. 1978:346–347.
11. Love, R.G. "Lung Function Studies Before and After a Work Shift." *Br J Ind Med 40:*153–159 (1983).
12. Peters, J.M., Murphy, R.L.H., Pagnotto, L.K., Van Ganse, W.F. "Acute Respiratory Effects in Workers Exposed to Low Levels of Toluene Diisocyanate (TDI)." *Arch Environ Health 16:*642–647 (1968).
13. Loke, J., Farmer, W., Matthay, R.A., Putnam, C.E., Smith, G.J.W. "Acute and Chronic Effects of Fire Fighting on Pulmonary Function." *Chest 77:*369–373 (1980).
14. Unger, K.M., Snow, R.M., Mestas, J.M., Miller, W.C. "Smoke Inhalation in Firemen." *Thorax 35:*838–842 (1980).
15. Sheppard, D., Distefano, S., Morse, L., Becker, C. "Acute Effects of Routine Firefighting on Lung Function." *Am J Ind Med 9:*333–340 (1986).
16. Governa, M., Comai, M., Valentino, M., Antonicelli, L., Rinaldi, F., Pisani, E. "Ventilatory Function in Rubber Processing Workers: Acute Changes over the Workshift." *Br J Ind Med 44:*83–89 (1987).
17. Åhman, M., Alexandersson, R., Ekholm, U., Bergström, B., Dahlqvist, M., Ulfvarson, U. "Impeded Lung Function in Moulders and Coremakers Handling Furan Resin Sand." *Int Arch Occup Environ Health 63:*175–180 (1991).

18. McMillan, G.H.G., Heath, J. "The Health of Welders in Naval Dockyards: Acute Changes in Respiratory Function during Standardized Welding." *Ann Occup Hyg 22:*19–32 (1979).
19. Kilburn, K.H.R., Warshaw, C.T., Boylen, J.C., Thornton, S.M., Hopfer, F.W., Sunderman, J., Finklea, J. "Cross-Shift and Chronic Effects of Stainless-Steel Welding Related to Internal Dosimetry of Chromium and Nickel." *Am J of Ind Med 17:*607–615. (1990).
20. Oleru, G., Ademiluyi, S.A. "Some Acute and Long-Term Effects of Exposure in Welding and Thermal-Cutting Operations in Nigeria." *Int Arch of Occup Environ Health 6:*605–612 (1987).
21. Dahlqvist, M., Ulfvarson, U., Bergström, B., Ekholm, U. "Temporary Effects on Lung Function in Welders Engaged in Cored Wire Welding." *Occup Hyg.* In press (1993).
22. Dahlqvist, M., Alexandersson, R., Andersson, B., Andersson, K., Kolmodin-Hedman, B., Malker, H. "Exposure to Ski-Wax Smoke and Health Effects in Ski Waxers." *Appl Occup Environ Hyg 7:*689–693 (1992).
23. Dahlqvist, M., Johard, U., Alexandersson, R., Bergström, B., Ekholm, U., Eklund, A., Milosevich, B., Tornling, G., Ulfvarson, U. "Lung Function and Precipitating Antibodies in Low Exposed Wood Trimmers in Sweden." *Am J Ind Med 21:*549–559 (1992).
24. Ulfvarson, U., Alexandersson, R., Dahlqvist, M., Ekholm, U., Bergström, B., Scullman, J. "Temporary Health Effects from Exposure to Water-borne Paints." *Scand J Work Environ Health 18:*376–387 (1992).
25. Christiani, C., Wegman, D.H., Eisen, E.A., Ting-ting, Y., Pei-Iian, L. "Cotton Dust Exposure and Longitudinal Change in Lung Function." *Am Rev Respir Dis 141;*A:589 (1990).
26. Dahlqvist, M., Ulfvarson, U. "Acute Effects on Forced Expiratory Volume in One Second and Longitudinal Change in Pulmonary Function among Wood Trimmers." *Am J Ind Med. 25:*551–558 (1993).
27. Järvholm, B., Thorén, K., Brolin, I., Ericsson, J., Morgan, U., Tylen, U., Bake, B. "Lung Function in Workers Exposed to Soft Paper Dust." *Am J Ind Med 14:*457–464 (1988).
28. Dahlqvist, M., Alexandersson, R., Hedenstierna, G. "Lung Function and Exposure to Asbestos among Vehicle Mechanics." *Am J Ind Med 22:*59–68 (1992).
29. Berry, G., McKerrow, C.B., Molyneux, M.K.B., Rossiter, C.E., Tombleson, J.B.L. "A Study of the Acute and Chronic Changes in Ventilatory Capacity of Workers in Lancashire Cotton Mills." *Br J Ind Med 30:*25–36 (1973).
30. Ohlson, C.G., Rydman, T., Sundell, L., Bodin, L., Hogstedt, C. "Decreased Lung Function in Long-Term Asbestos Cement Workers: A Cross-Sectional Study." *Am J Ind Med 5:*359–366 (1984).
31. Hedenstierna, G., Alexandersson, R., Wimander, K., Rosén, G. "Exposure to Terpenes: Effects on Pulmonary Function." *Int Arch Occup Health 51:*191–198 (1983).

32. Alexandersson, R., Plato, N., Kolmodin-Hedman, B., Hedenstierna, G. "Exposure, Lung Function, and Symptoms in Car Painters Exposed to Hexamethylene Diisocyanate and Biuret Modified Hexamethylene Diisocyanate." *Arch Environ Health 42:*367–373 (1987).
33. Nielsen, J., Bensryd, I., Almquist, H., Dahlqvist, M., Welinder, H., Alexandersson, R., Skerfving, S. "Serum IgE and lung function in workers exposed to phtalic anhydride." *Int Arch Occup Environ Health 63:*199–204 (1991).
34. Dahlqvist, M. "Lung Function and Exposure to Paper Dust in Bookbinders: A Pilot Study." *Upsala J Med Sci 97:*49–54 (1992).
35. Alexandersson, R., Hedenstierna, G. "Pulmonary Function in Wood Workers Exposed to Formaldehyde: A Prospective Study." *Arch Environ Health 44:*5–11 (1989).
36. Tornling, G., Alexandersson, R., Hedenstierna, G., Plato, N. "Decreased Lung Function and Exposure to Diisocyanates (HDI and HDI-BT) in Car Repair Painters: Observations on Re-Examination 6 Years after Initial Study." *Am J Ind Med 17:*299–310 (1990).
37. Hedenstierna, G., Alexandersson, R., Belin, L., Wimander, K., Rosén, G. "Lung Function and Rhizopus Antibodies in Wood Trimmers: A Cross-Sectional and Longitudinal Study." Int Arch *Occup Health 58:*167–177 (1986).
38. Dahlqvist, M., Alexandersson, R., Ulfvarson, U. "Pulmonary Function Changes in Saw-Mill Workers: A Prospective Study of Occupational Exposure to Saw Fumes." *Occup Hyg. 1:*17–26 (1992).
39. Ulfvarson, U., Alexandersson, R. "Reduction in Adverse Effect on Pulmonary Function after Exposure to Filtered Diesel Exhaust." *Am J Ind Med 17:*341–347 (1990).
40. Gamble, J., Jones, W., Minshall, S. "Epidemiological-Environmental Study of Bus Garage Workers: Acute Effects of NO_2 and Respirable Particulate on the Respiratory System." *Environ Res 42:*201–214 (1987).
41. Ulfvarson, U., Alexandersson, R., Dahlqvist, M., Ekholm, U., Bergström, B. "Pulmonary Function in Workers Exposed to Diesel Exhausts: The Effect of Control Measures." *Am J Ind Med 19:*283–289 (1991).
42. Ames, R.G., Attfield, M.D., Hankinson, J.L., Hearl, F.J., Reger, R.B. "Acute Respiratory Effects of Exposure to Diesel Emissions in Coal Miners." *Am Rev Respir Dis 125:*39–42 (1982).
43. Ulfvarson, U., Dahlqvist, M., Sandström, T., Bergström, B., Ekholm, U., Lagerstrand, L., Figler, B., Nilsen, A., Bjermer, L., Trønnes, T., Nilsen, O. "Experimental Evaluation of the Effect of Filtration of Diesel Exhaust by Biologic Exposure Indicators." *Am J Ind Med.* In press. (1994).
44. Rudell, B., Sandström, T., Stjernberg, N., Kolmodin-Hedman, B. "Controlled Diesel Exhaust Exposure in an Exposure Chamber: Pulmonary

Effects Investigated with Bronchoalveolar Lavage." *J Aerosol Sci 21*(suppl 1):411–414 (1990).
45. Sandström, T. Pulmonary effects of air pollutants. "Bronchoalveolar Lavage Studies on the Effects on NO_2 and SO_2 Exposure in Healthy Humans." National Institute of Occupational Health. Solna, Sweden. Arbete och Hälsa 1:1–58 (1989).
46. Gamble, J., Jones, W., Minshall, S. "Epidemiological-Environmental Study of Bus Garage Workers: Chronic Effects of Diesel Exhaust on the Respiratory System." *Environ Res 44:*6–17 (1987).
47. Purdham, J.T., Holness, D.L., Pilger, C.W. "Environmental and Medical Assessment of Stevedores Employed in Ferry Operations." *Appl Ind Hyg 2:*133–139 (1987).
48. Battigelli, M.C., Mannella, R.J., Hatch, T.F. "Environmental and Clinical Investigation of Workmen Exposed to Diesel Exhaust in Railroad Engine Houses." *Ind Med and Surgery 33:*121–124 (1964).
49. Gamble, J., Jones, W. "Respiratory Effects of Diesel Exhaust in Salt Miners." *Am Rev Respir Dis 128:*389–394 (1983).
50. Gamble, J., Jones, W., Hudak, J. "An Epidemiological Study of Salt Miners in Diesel and Nondiesel Mines." *Am J Ind Med 4:*435–458 (1983).
51. Ames, R.G., Hall, D.S., Reger, R.B. "Chronic Respiratory Effects of Exposure to Diesel Emissions in Coal Mines." *Arch Environ Health 39:* 389–394 (1984).
52. Reger, R., Hancock, J., Hankinson, J., Hearl, F., Merchant, J. "Coal Miners Exposed to Diesel Exhaust Emissions." *Ann Occup Hyg 26:*799–815 (1982).
53. Reger, R., Attfield, M.D. "Diesel Emissions and Associated Respiratory Health Effects in Mining." *Health Issues Related to Metal and Nonmetallic Mining,* Wagner W.L., Rom, W.N., Merchant, J.A., eds. Boston, Massachusetts, Butterworth Publishers, pages 393–412 (1983).
54. Reger, B., Hancock, J. "Coal Miners Exposed to Diesel-Exhaust Emissions," *Health Implications of New Energy Technologies,* W.N. Rom, and V.E. Archer, eds. Ann Arbor Science Publishers, Inc., Michigan, pages 213–231 (1980).
55. Ulfvarson, U. "Limitations to the Use of Employee Exposure Data on Air Contaminants in Epidemiologic Studies." *Int Arch Occup Environ Health 52:*285–300 (1983).
56. Wiester, M.J., Iltis, R., Moore, W. "Altered Function and Histology in Guinea Pigs after Inhalation of Diesel Exhaust." *Environ Res 22:*285–297 (1980).
57. Moorman, W.J., Clark, J.C., Pepelko, W.E., Mattox, J. "Pulmonary Function Responses in Cats Following Long-Term Exposure to Diesel Exhaust." *J Appl Toxicol 5:*301–305 (1985).
58. Pepelko, W.E., Mattox, J., Moorman, W.J., Clark, J.C. "Pulmonary Function Evaluation of Cats after One Year of Exposure to Diesel Exhaust."

Health Effects of Diesel Engine Emissions: Proceedings of an international symposium, Vol. 2, Health Effects Research Laboratory, U.S. EPA, pages 757–765 (1980).

59. Pepelko, W.E., Mattox, J.K., Yang, Y.Y., Moore, W. Jr. "Pulmonary Function and Pathology in Cats Exposed 28 Days to Diesel Exhaust." *J Environ Pathol Toxicol 4:*449–458 (1980).
60. Pepelko, W.E. "Effects of 28 Days Exposure to Diesel Engine Emissions in Rats." *Environ Res 27:*16–23 (1982).
61. Mauderly, J.L., Bart, E.B., Bice, D.E., Eidson, A.F., Henderson, R.F., Jones, R.K., Pickrell, J.A., Wolff, R.K. "Inhalation Exposure of Rats to Oil Shade Dust and Diesel Exhaust." *Annual Report of the Inhalation Toxicology Research Institute,* October 1, 1985, through September 30, 1986, Muggenburg, B.A., Sun, J.D., eds; Lovelace Biomedical and Environmental Research Institute, Albuquerque, New Mexico, pages 273–278 (1985).
62. Henderson, R.F., Pickrell, J.A., Jones, R.K., Sun, J.D., Benson, J.M., Mauderly, J.L., McClellan, R.O. "Response of Rodents to Inhalated Diluted Diesel Exhaust: Biochemical and Cytological Changes in Bronchoalveolar Lavage Fluid and in Lung Tissue." *Fundam Appl Toxicol 11:*546–567 (1988).
63. Lewis, T.R., Green, F.H.Y., Mooorman, W.J., Burg, J.R., Lynch, D.W. "A Chronic Inhalation Toxicity Study of Diesel Engine Emissions and Coal Dust, Alone and Combined," *J Am College Toxicol 8:*345–375 (1989).
64. Gross, K.B. "Pulmonary Function Testing of Animals Chronically Exposed to Diluted Diesel Exhaust." *Health Effects of Diesel Engine Emissions,* Proceedings of an International Symposium, December 3–5, 1979, Pepelko, W.E., Danner, R.M., Clarke, N.A., eds; Cincinnati, Ohio, U.S. Environmental Protection Agency, pages 606–624 (1979).
65. Gross, K.B. "Pulmonary Function Testing Chronically Exposed to Diluted Diesel Exhaust." *J Appl Toxicol 1:*116–123 (1981).
66. Egebäck, K.E., Grägg, K. "Impact of Fuels on Diesel Exhaust Emissions." MTC 9010A. AB Svensk Bilprovning, Motortestcenter, Stockholm (1990).
67. Wall, J.C., Hoekman, S.K. "Fuel Composition Effects on Heavy Diesel Particulate Emissions." Chevron Research Company Fuels Division. SAE Paper 841364, Oct. 8–11 (1984).
68. Cartelliere, W. "Die Partikelproblematik aus der Sicht des Motorenentwicklers." Motor und Umwelt, Tagung Motor und Umwelt 89, AVL List GmbH, Graz, 1./2.8.89 127–153 (1989).
69. Lundberg, P., Camner, P., Gustavsson, P. "Discussion about Indicator Substances for Acute Effects of Occupational Exposure to Motor Exhaust." [Överväganden rörande indikatorsubstans för akuta effekter av yrkesmässig exponering för motoravgaser]. National Institute of Occupational Health. Solna, Sweden. Arbete och Hälsa 5:1–58 (1986).

70. Anonymous. "Test Procedure for Verifying the Smoke Density of the Exhaust." In: Regulation with Prescriptions about Exhaust Purification for Diesel Engines in Certain Automobiles. A 30 Regulation (Kungörelse med föreskrifter om avgasrening för dieseldrivna motorer till vissa bilar). Swedish National Environment Protection Board, SNFS 1991:11, MS 40, Annex 3:139–144.
71. Christian, R., Knopf, F., Jaschek, A., Schindler, W. Eine neue Messmethodik der Bosch- Zahl mit erhöhter Empfindlichkeit. MTZ Motortechnische Zeitschrift 54:16–22 (1993).
72. Rubow, K.L., Marple, V.A., Tao, Y., Liu, D. "Design and Evaluation of a Personal Diesel Aerosol Sampler for Underground Coal Mines." SME Annual Meeting, Salt Lake City, Utah, preprint number 90–132. Society for Mining, Metallurgy, and Exploration, Inc., Littleton, Colorado (1990).
73. Cantrell, B.K., Rubow, K.L. "Development of Personal Diesel Aerosol Sampler Design and Performance Criteria." SME Annual Meeting, Salt Lake City, Utah, preprint number 90–74. Society for Mining, Metallurgy, and Exploration, Inc., Littleton, Colorado, (1990).
74. Brorstöm-Lundén, E., Lindskog, A. "Characterization of Organic Compounds on Airborne Particles." *Environ Int 11:*183–188 (1985).
75. Brorström Lundén, E., Lövblad, G. "Deposition of Soot-Related Hydrocarbons." IVL-report L90/209, B990 Göteborg, July 1990.
76. Hammond, S.K., Smith, T. J., Woskie, S., Schender, M B., Speizer, F. E. "Characterization of Diesel Exhaust Exposures for a Mortality Study." *Polynuclear aromatic hydrocarbons: mechanisms, methods and metabolism.* Cooke, M., Denis, A.J. , eds; Battelle Press, Columbus, Ohio, pages 533–541 (1985).
77. Maenhaut, W., Thiessen, L., Verduyn, G. "Study of the Respirable Immission Levels of a Cyclist in Brussels' Traffic Using PIXE as an Analytical Technique." *Nucl Instrum Methods Phys. Res.,* Sect B B49 (1–4), 406–413 59–2 (1990).
78. Wells, A.C., Venn, J.B., Heard, M.J. "Deposition in the Lung and Uptake to Blood of Motor Exhaust Labeled with Lead-203." Inhaled Part 4, In Symp, 4th Meeting Date 1975, Vol 1 175–189 Pergamon Oxford, England, (1977).
79. Wolff, R.K., Henderson, R.F., Snipes, M.B., Griffith, W.C., Mauderly, J.L., Cuddihy, R.G.; McClellan, R.O. "Alterations in Particle Accumulation and Clearance in Lungs of Rats Chronically Exposed to Diesel Exhaust." *Fundam Appl Toxicol 9:*154–166 (1987).
80. Chan Tai, L., Lee, P.S., Hering, W.E. "Pulmonary Retention of Inhaled Diesel Particles after Prolonged Exposures to Diesel Exhaust." *Fundam Appl Toxicol 4* 624–631 (1984).
81. Sun, J.D., McClellan, R.O. "Respiratory Tract Clearance of Carbon-14-Labeled Diesel Exhaust Compounds Associated with Diesel Particles or as a Particle-Free Extract." *Fundam Appl Toxicol 4* 388–393 (1984).

82. Lehman, E., Rentel, K.H., Allesher, W., Hohmann, R. "Measurement of Workplace Exposure to Diesel Exhaust." Zentralbl. Arbeitsmed., Arbeitsschutz, *Prophyl. Ergon. 40*(1), 2–11, 59–5 (1989).
83. Carpenter, K., Johnson, J.H. "Analysis of the Physical Characteristics of Diesel Particulate Matter Using Transmission Electron Microscope Techniques." SAE technical paper series 790815. Off-Highway Vehicle Meeting and Exposition. MECCA, Milwaukee, Sept 10–13, 1979. Society of Automotive Engineers, Inc. 400 Commonwealth Drive, Warrendale, Pennsylvania, 15096.
84. Figler, B., Sahle, W., Ulfvarson, U., Krantz, S. "Electron Microscopic Investigation of Diesel Smoke." [Elektronmikroskopisk undersökning av dieselrök.] 41 Nordiska arbetsmiljömötet i Reykjavik 2–4 September 1992:116–117 (1992).
85. American Conference of Governmental Industrial Hygienists. 1992–1992 "Threshold Limit Values for Chemical Substances and Physical Agents and Biological Exposure Indices." Cincinnati, Ohio (1992).
86. Dahlqvist, M., Lagerstrand, L., Nilsen, A. "Repeated Measurements of Transfer Factor in Rabbits: An Animal Model Suitable for Evaluation of Short-Term Exposure." *Clin Physiol. 14:*53–61 (1993).
87. Stokinger, H.E. "Toxicology of Diesel Emissions." In Proceedings of the Symposium on the Use of Diesel-Powered Equipment in Underground Mining, Pittsburgh, Pa., January 30–31, 1973; 147–157. Bureau of Mines Information Circular 8666/1975 (PB 243463). United States Department of the Interior.
88. Ahmann, C.A., Siegla, D.C. "Diesel Particulates—What They Are and Why." *Aerosol Sci and Techn 1:*73–101 (1982).
89. Groblicki, P.J., Begeman, C.R. "Particle Size Variation in Diesel Car Exhaust," SAE paper 790421 (1979).
90. Engeljehringer, K., Schindler, W. "The Organic Insoluble Diesel Exhaust Particulates: Differences between Diluted and Undiluted Measurements." *J Aeorsol Sci 20:*1377–1380 (1989).
91. Chang, S.G., Novakov, T. "Formation of Pollution Particulate Nitrogen Compounds by NO-Soot and NO_2-Soot Gas-Particulate Surface Reactions." *Atmospheric Environment 9:*495–504 (1975).
92. Oberdorster, G., Ferin, J. Finkelstein, G., Wade, P. Corson, N. "Increased Pulmonary Toxicity of Ultrafine Particles?" II. Lung Lavage Studies. *J Aerosol Sci 21:*384–387 (1990).
93. Ferin, J., Oberdorster, G. Penney, D.P., Soderholm, S.C., Gelein, R., Piper, H.C. "Increased Pulmonary Toxicity of Ultrafine Particles?" I. Particle Clearance, Translocation, Morphology. *J Aerosol Sci 21:*381–384 (1990).
94. Vergeer, H.C., Lawson, A., Mitchell, E.W., Eng P. "Developments in Diesel Exhaust Emission Control Technology for Underground Mining." Annals of the American Conference of Governmental Industrial Hygienists 14:249–255 (1986).

95. Kittelson, D.B., Dolan, D.F. "Diesel Exhaust Aerosols." In: Willeke, K, ed. *Generation of Aerosols and Facilities for Exposure Experiments.* p. 337–359. Ann Arbor Science Publishers, Inc., Ann Arbor Mich, p. 597 (1980).
96. Novakov, T., Mueller, P.K., Alcocer, A.E., Otwos, J.W. "Chemical Composition of Pasadena Aerosol by Particle Size and Time of Day. Chemical States of Nitrogen and Sulfur by Photoelectron Spectroscopy." *J Colloid Interface Sci 39:*225–234 (1972).
97. Novakov, T., Chang, S.G., Harker, A.B. "Sulfates in Pollution Particulates: Catalytic Oxidation of SO_2 on Carbon Particles." *Science 186:*259–261 (1974).
98. Amdur, M.O., McCarthy, J.F., Gill, M.W. "Respiratory Response of Guinea Pigs to Zinc Oxide Fume." *Am Ind Hyg Assoc J 43:*887–889 (1982).
99. Amdur, M.O. "When One Plus Zero is More Than One." *Am Ind Hyg Assoc J 46:*467–475 (1985).
100. Stacy, R.W., Seal, E., House, D.E., Green, J., Roger, L.J., Raggio, L. "A Survey of Effects of Gaseous and Aerosol Pollutant on Pulmonary Function of Normal Males." *Arch Environ Health 38:*104–115 (1983).

CHAPTER 28

BIOMONITORING OF HUMAN POPULATIONS EXPOSED TO PESTICIDES

Claudia Bolognesi

Unit of Toxicologic Evaluation
Istituto Nazionale per la Ricerca sul Cancro
Viale Benedetto XV, 10, 16132
Genova, Italy

Franco Merlo

Department of Environmental Epidemiology and Biostatistics
Instituto Nazionale per la Ricerca sul Cancro
Viale Benedetto XV, 10, 16132
Genova, Italy

CONTENTS

INTRODUCTION

A pesticide is a substance capable of selectively killing a pest. Pesticides are used to control disease vectors, to improve agricultural production, and to protect stored crops. They are effective against undesirable plants and animals that destroy or cause disease to plants, animals, and humans, such as fungi, insects, mites, nematodes, rodents, and weeds. This large group of chemicals is classified in terms of the organisms they are intended to kill: insecticides, herbicides, nematocides, fungicides, insect- and plant-growth regulators, fumigants, and rodenticides. The demand of modern agriculture and the need to control disease vectors in developing countries have encouraged the rapid growth of the pesticide industry.

More than 15,000 individual compounds and 35,000 formulations have come into use since 1945 [1]. The Pesticide Manual lists about 680 individual compounds in current world use as pesticides [2]. They are formulated in different ways, such as liquids, dusts, granules, impregnated pellet-tablets, resin strips, concentrates, etc. The fundamental peculiarity property of pesticides should be the selectivity. These chemicals should act specifically against certain organisms without adversely affecting others [3]. Theoretically, this might be achieved by several means. A pesticide might act by modifying a functional biosystem that is peculiar to pest organisms, or reacting with specific target molecules in pest species. Pesticides also could be metabolically activated more rapidly or to a large extent in pest species than in non-pest species, or alternatively they could be detoxified by the non-pest species more rapidly than by the pest species. However, absolute selectivity is difficult to achieve because of the similarities in the biological macromolecules of all organisms. Undesirable toxic effects frequently occur in non-target species. Other effects could be induced by an improper use of these compounds.

PESTICIDE MECHANISMS OF ACTION

Use of pesticides in agriculture is not a new technology: They have been employed since ancient time mainly to control insects [4]. Inorganic compounds, particularly arsenical insecticides, were probably the first to be used. These were followed by plant extracts, like pyrethrum or nicotine, and since 1900 by petroleum products. Later, synthetic insecticides were developed. The first was DDT, which gave rise to the industry of manufacturing chlorinated hydrocarbons. Organophosphates became widely used following the banning of DDT [4,5]. Shortly afterwards, insecticidal carbamates were introduced. During the 1970s and 1980s, many new pesticides were developed. The most important group of recently discovered insecticides effective at lower doses than previous groups is that of synthetic pyrethroids. These compounds are derived from molecules originally isolated in pyrethrum flowers [6]. The main classes of pesticides, depending on their particular use, are: insecticides, fungicides, herbicides, fumigants, and rodenticides. The study of the mechanisms of action of pesticides is very important in the evaluation of the health hazards of these compounds to human and non-target organisms.

Insecticides

Chlorinated compounds were widely used in agriculture, mainly as insecticides, and in the malaria control program from mid-1940 to mid-1960 [7]. These pesticides have fallen into disuse because they are very persistent in the environment and tend to accumulate in biologic and non-biologic media. These chemicals demonstrate an excitatory effect on the central nervous system. The two major classes of insecticides actually used are organophosphates and carbamates. The mechanisms of action of these compounds are similar: They are potent irreversible or reversible cholinesterase inhibitors [8]. Pyrethroids modify the ionic permeability of nervous membranes and produce destructive poisoning in insects and mammals [9].

Fungicides

Fungicides exert toxic action on biochemical processes mainly by acting on vulnerable sites in metabolic pathways. Fungi differ greatly in physiology than other life forms, and pesticides must by specifically selective for these organisms. The chloroalkyl thio compounds, such as captan, captafol, and folpet, exert a specific inhibition of enzyme activity in mitochondria. Dithiocarbamates are toxic to fungi, inhibiting enzyme activity involved in the production of adenosine triphosphate. Another important chemical class of fungicides, benzimidazoles (benomyl or carbendazim), are effective antimicrotubule agents that generally disturb cellular division [10].

Herbicides

Herbicides are used to control unwanted plants. These compounds are classified by their selectivity, type of distribution, mode of action, time of application, and chemical structure [11]. They tend to be more toxic to plants than to other life forms. Herbicides may act as growth regulators, which modify the metabolism of the plant; defoliants that strip growing plants of their leaves; or desiccants, which dry up the plant [12,13]. These pesticides belong to different chemical classes such as inorganic chemicals, chlorophenoxy acids, dipyridyls, and substituted phenols. Herbicides usually take advantage of specific plant metabolic pathways, such as photosynthesis, plant hormone action, regulation of cell division, and synthesis of amino acids.

Fumigants

A fumigant is a substance or a mixture of chemicals that produces gas, vapor, fume, or smoke intended to destroy insects, bacteria, and nematodes. Fumigants are used predominantly in control of nematodes in soil. A derivative of dithiocarbamic acid, vapam, is an active soil sterilant with a wide spectrum of action, producing methyl isothiocyanate in soil. Haloalkane fumigant, methyl bromide, ethylene dibromide, and related fumigants act primarily by inhibiting the conversion of succinate to fumarate through binding with sulphydrylic groups in different enzymes [14].

TOXICOLOGY OF PESTICIDES

Occupational exposure may occur during the manufacture and processing of insecticides as well as during their use. Farmers are not the only people exposed to agrochemicals. A high amount of pesticides are used outside the agricultural practice. Home gardening and household pest control imply the use of insecticides, fertilizers, and weed killers. Pesticide residues on fruits or vegetables and contamination of drinking water may cause exposure to the general population [15,16]. In addition, the persistence of different insecticides and herbicides in the ecosystem imply further exposure.

Exposure to pesticides presents a peculiar hazard: It can lead to problems ranging from acute illnesses, such as acute toxicity; chronic disorders, such as reproductive effects; degenerative diseases; and cancer [17]. Assessing the extent and consequences of pesticide use is difficult due to the complexity of the subject. Worldwide, half a million cases of pesticide poisonings occur annually, and one tenth of them are fatal. The adverse reactions to pesticides, especially fatal reactions, are more common in Third World countries [1,18]. These effects are mainly caused by inappropriate handling of pesticides, by accidents involving lack of training, or insufficient information on the properties of pesticides. The consequences of human exposure to pesticides vary according to numerous factors such as type of exposure—inhalation or dermal; concentration and physical state of the compound; exposure time; single, repeated, or chronic exposure; potential chemical synergism; and environmental characteristics such as ambient temperature, and relative humidity. In addition, there is a variety of technical variables such as protective measures, hygienic habits, and use of drugs or other chemicals that could modify the extent of exposure.

CANCER HAZARD AND HUMAN EXPOSURE TO PESTICIDES

The carcinogenicity of specific classes of pesticides reported in experimental and animal systems (see, Carcinogenicity Studies), the dramatic increase of the number and amount of pesticides introduced in the market since the mid-1940s and the persistence and ubiquity of human exposure to complex mixtures of chemicals have stimulated the conduction of human studies on the carcinogenicity of pesticides to humans [1,2,5,15,16]. Several case reports have been published on the possible link between exposure to specific pesticides and aplastic anemia or blood dyscrasias showing potential associations between these diseases and exposures to DDT, lindane, pyrethrum, chlordane, dieldrin, toxaphene, 2,4-dichlorophenoxyacid (2,4-D), arsenical compounds, pentachlorophenol, zineb, ziram, malathion, carbaryl, dichlorvos, and propoxur [19-44]. However, no evidence of such an association was observed in a case-control study of fatal aplastic anemia and in population-based surveys conducted among subjects potentially exposed to pesticides [45-50]. The discrepancy between the hypotheses generated by case series and findings from health surveys may be explained by methodological problems frequently encountered in case-reports as well as in health surveys that make the interpretation of the results difficult and increase the chance of failure in detecting real causal relationships [51]. Exposure assessment in health surveys frequently is based on broad occupational categories (i.e., farmers, agricultural workers, sprayers, etc) or job titles so that relevant exposures and their health effects tend to be diluted resulting in an underestimate of the epidemiologic measures of the true strength of the association between exposure and health effect, i.e., the relative risk (RR) or the odds ratio (OR). It has been shown that nondifferential exposure misclassification results in risk estimates that are biased toward the null hypothesis of the study (52-54) of no association between exposure and adverse health effects (RR or OR = 1), and that even a small nondifferential misclassification may drastically reduce the estimated measure of effects [52-55].

The possibility that exposure to pesticides may increase the probability of developing site-specific cancers has been widely investigated through epidemiological studies of occupationally exposed populations (e.g., farmers, manufacturers, forest land and railroad workers, pesticides sprayers) and of accidentally exposed industrial workers or residents. Blair et al., in reviewing the published occupational health surveys independently conducted in the United States of America, Canada, Finland, New Zealand, Sweden, The Netherlands, Denmark, and Britain, summed the observed and expected numbers of specific cancers reported in each survey to compute meta-relative risks (MRR) through a procedure—i.e., meta-analysis—that should have "minimized the influence of unusual chance findings from individual studies" [56-79]. The computed MRR was slightly but significantly elevated among farmers when compared to the general population for Hodgkin's disease (MRR = 1.16), melanoma (MRR = 1.15), multiple myeloma (MRR = 1.12), leukemias (MRR = 1.07), and lip (MRR = 2.08), stomach (MRR = 1.12), and prostate cancers (MRR = 1.08). Nonsignificant increases were found for non-Hodgkin's lymphoma (MRR = 1.05) and for brain (MRR = 1.05) and connective tissue cancers (MRR = 1.06). The authors concluded that despite the fact that the health surveys examined suffer from the bias toward the null, which basically dilutes the magnitude of the effect measured by the study, "... they provide important clues for understanding the etiology of specific cancers frequently occurring among subjects exposed to pesticides, to address future research, and to promote public health and policy measures aimed at containing the impact of pesticides exposure in human health" [56]. Because exposure to pesticides is not limited to agricultural or industrial workers but occurs in the general population as well through indoor and outdoor household treatments and potentially, to a lesser extent, through pesticide residues that are found in food and water, it may be a reason of public health concern, particularly in the light of the fact that non-Hodgkin's lymphoma, multiple myeloma, leukemias, and brain and prostate cancers, which are found in excess in farmers and in other populations exposed to agricultural chemicals, are increasing in the general population as well [15,80-83]. With the exception of lip cancer and skin melanoma, which are clearly linked to ultraviolet light exposure, leukemias, multiple myeloma, brain cancer, non-Hodgink's lymphoma, and soft-tissue sarcomas have been associated with exposures to specific pesticides. These association have been detected by cohort and case-control studies summarized in Tables 1 and 2 [89-162].

Epidemiologic Evidence from Cohort Studies

Cohort studies estimate the frequency of adverse health outcomes (e.g., mortality or incidence rates) in subjects exposed to agents that are suspected of being related to the development of specific health effects (e.g., a type of cancer) and in unexposed individuals (i.e., referent population). Cohort studies imply selection of subjects based upon the presence or absence of a specific exposure. The exposed cohort is then followed for a sufficient period of time to assess incidence or death rates for the disease of interest as well as for other diseases. Comparison of rates in exposed and referent subjects allow to estimate whether an association exists between exposure and health outcome. Cohort studies (see Table 1) have been carried out in pesticide production workers, licensed pesticides sprayers, wood industry workers, grain millers, gardeners, farmers, and in the general population accidentally exposed to 2,3,7,8-tetrachlorodibenzo-para-dioxin (2,3,7,8-TCDD), 2,4,5-trichlorophenol (2,4,5-T), and other chemicals after a runaway reaction during the production of 2,4,5-T at the ICMESA plant in the border of the town Seveso, Italy. Excess risk for lymphoemopoietic cancers, soft-tissue sarcomas, and brain cancers have been reported in several cohort studies (see Table 1). Soft-tissue sarcomas and non-Hodgkin's lymphoma have been found to be associated with exposures to phenoxy acids herbicides, particularly 2,4-D, 2,4,5-T, and 2,3,7,8-TCDD, a highly toxic contaminant of 2,4,5-T (see Table 1). The fact that several studies of manufacturing workers have a low power of detecting an increased risk for rare cancers such as soft-tissue sarcomas

(text continued on page 692)

Table 1
Selected Findings in Epidemiologic Cohort Studies Conducted among Subjects Exposed to Pesticides

Reference (year,#)	Occupation/*Exposure*	Cancer site	Relative Risk	95%CI	Comments
Axelson, et al. (1980, *85*)	Railroad workers *Phenoxyl acids, Amitrol, Diuron*	Stomach	6.1	0.07–21.8	Risk confined among phenoxy acids sprayers, 10 years of latency
Zack and Suskind (1980, *86*)	Production workers accidentally exposed to *2,3,7,8, - TCDD*	Lymphatic & Hematopoietic	3.4	0.7–9.9	Small sample size (121)
		Skin	6.7	0.17–37	Based on 1 case of fibrous histiocytoma of soft-tissue origin
Barthel (1981,*87*)	Plant protection workers and agronomists *DDT, Hexachlorocylcoexanes, Methylparathion, 2,4-D, MCPA, Zineb, Maneb, Calcium Arsenate*	Lung	1.8	1.40–2.4	Trend with years (y) of exposure: <10 y, RR=1.2; 10–19 y, RR=1.7; >19 y, RR=3
Thiess and Frentzel-Beyme (1982,*88*)	Production workers accidentally exposed to *2,3,7,8-TCDD*	All Cancers	1.7	0.7–3.5	Small sample size, (74)
		Stomach	4.3	0.9–13	
Abate et al. (1982, *112*)	General population accidentally exposed to *2,3,7,8-TCDD*	Liver	1.8	1.02–3	Residents of the town of Seveso (54.5% of the population lived in polluted areas)
		Lymphomas and Leukemias	n.a.	n.a.	Clusters were observed in three Municipalities with high proportions of residents living in polluted areas

(table continues)

Table 1 (continued)

Reference (year,#)	Occupation/*Exposure*	Cancer site	Relative Risk	95%CI	Comments
Riihimaki et al. (1982, *89*)	Applicators *2,4,5-T; 2,4,-D*	Multiple Myeloma	5.0	0.00–18.4	Relative risk based on 1 case
Blair et al. (1983, *90*)	Licensed pest control workers *Chlorinated Hydrocarbons, Organophosphates, Arsenic, Phenoxy Acids, Carbamates*	Skin	1.3	0.20–4.8	Potential exposure to arsenical insecticides.
		Leukemia	1.3	0.40–3.4	
		Brain	2.0	0.60–4.7	Trend for lung cancer with years (y) of licensure: <=10 y, RR=1.01; 10–19 y, RR=1.55; >=20 y, RR=2.89;
Merlo (1985,*92*)	Production workers accidentally exposed to *2,3,7,8-TCDD*	Lymphatic and Hematopoietic	n.a.	n.a.	3 cases
		Liver & Intrahepatic Bile Ducts	n.a.	n.a.	3 cases
		Brain & Skin	n.a.	n.a.	3 and 2 cases, respectively 746 workers employed between 1947–76 expected deaths not provided
Lynge (1985, *91*)	Manufacturers *2,4,5-T, 2,4-D MCPA, 2,3,7,8-TCDD*	Soft-Tissue Sarcomas	3.67	1.00–9.31	Soft-tissue sarcomas confined among shippers, packagers and pigment millers after 10 years since first exposure
		Hodgkin's Disease and non-Hodgkin's Lymphoma	1.32	0.35–3.37	

Coggon et al. (1986, *94*)	Manufacturers, sprayers *MCPA, Organophosphorous Insecticides, Chlortriazine Pesticides, 2,4,5-T*	Multiple Myeloma Leukaemia Testis Nasal	1.62 1.75 2.27 4.93	0.53–3.8 0.97–2.9 0.61–5.7 1.02–14	High risk for soft-tissue sarcomas associated with potential exposure to phenoxy acids: RR=2.8
Kauppinen et al. (1986, *95*)	Wood industry workers *Chlorophenols, Arsenic, Lindane, Copper Sulphate, Parathion, Chromates*	Respiratory tract Prostate	13.1 1.9	2.22–71 0.80–3.8	Trend for respiratory cancers with length of exposure to pesticides in wood dust Cohort based case-control study
Puntoni et al. (1986, *96*)	General population accidentally exposed to *2,3,7,8-TCDD*	Soft-Tissue Sarcomas	2.36	n.a.	Ratio of the 1981 mortality rate relative to 1975
Steineck & Wiklund (1986, *93*)	Agricultural workers	Multiple Myeloma	1.20	1.09–1.3	Men employed in horticulture had a RR=1.27
Wang and MacMahon Mac Mahon et al. (1979–88, *84, 100*)	Pest control workers *Fumigants, Carbamates, Clorinated Hydrocarbons, Organophosphates*	Skin Lymphatic & Hematopoietic	1.3 1.0	0.65–2.2 0.67–1.4	Potential exposure to arsenical compounds
Bond et al. (1988–89, *99, 104*)	Manufacturers *2,4-D; 2,4,5-T, MCPA, Silvex, 2,3,7,8-TCDD*	Stomach Soft-Tissue Sarcomas non-Hodgkin's Lymphoma Lymphohemopoietic cancers	1.9 5.0 2.0 3.1	0.80–3.8 0.60–18 0.75–6.2 1.01–7.3	Two cases of soft-tissue sarcomas were observed in workers diagnosed with chloracne vs 0.005 expected

(table continues)

Table 1 (continued)

Reference (year,#)	Occupation/*Exposure*	Cancer site	Relative Risk	95%CI	Comments
Corrao et al. (1989, *102*)	Farmers licensed for use of pesticides *Dithiocarbamates, Organochlorine, Phenoxy Acids*	Skin	1.4	1.00–1.8	High risk for lymphomas in grain production areas
		Lymphomas	1.4	1.00–1.9	
		Haematopoietic	1.1	0.80–1.5	
Bertazzi et al. (1989, *103*)	General population accidentally exposed to *2,3,7,8-TCDD*	Gallbladder and Biliary Tract	12.1	1.6–88	Women living in the highly polluted area
		Gallbladder and Biliary Tract	21.9	2.9–166	Women living in the highly polluted area, 1st quinquennium following the contamination
		Hodgkin's Disease	9.1	1.1–78	Males living in the intermediate pollution area, 2nd quinquennium following the contamination
		Leukemia	4.9	1.1–2.21	
		Liver	1.96	0.6–6.2	
		Melanoma	11.4	1.3–102	
		Gallbladder and Biliary Tract	4.2	0.6–32	Women living the intermediate pollution area, 2nd quinquennium following the contamination
		Soft-Tissue Sarcomas	23.9	2.2–265	
		Thyroid Gland	24.8	2.2–274	

		Multiple Myeloma	4.9	1.1–22	Males living in the low pollution area, 2nd quinquennium following the contamination
		Myeloid Leukemia	4.1	1.3–12	
		Brain	6.7	1.4–34	Females living in the low pollution area, 2nd quinquennium following the contamination
Wiklund et al. (1987, 88–89, *97, 98, 101*)	Licensed pesticide applicators *DDT, Fenitrothion, Phenoxy Acids*	non-Hodgkin's Lymphoma	1.0	0.63–1.5	Trend for testicular cancer with years (y) since licensure: 0–4 y, RR=0.94; 5–9 y, RR=1.4; >=10 y, RR=2.5
		Hodgkin's Disease	1.2	0.60–2.2	
		Lip	1.8	0.96–2.9	
		Testis	1.6	0.92–2.5	
		Soft-Tissue Sarcomas	0.9	0.40–1.9	
Zober et al. (1990, *105*)	Production workers accidentally exposed to *2,3,7,8-TCDD*	All cancers	1.2	0.8–17	All workers
		All cancers	1.4	0.87–21	Workers diagnosed with chloracne
		All cancers	2.0	1.22–32	Workers diagnosed with chloracne, >=2 years since first exposure

(table continues)

Table 1 (continued)

Reference (year,#)	Occupation/*Exposure*	Cancer site	Relative Risk	95%CI	Comments
Alavanja et al. (1990, *106*)	Flour millers *Carbontetrachloride, Pyrethrum, Ethylene Dibromide, Malathion, Phosphine, Methylbromide*	Leukemia non-Hodgkin's Lymphoma Pancreas	1.4 1.5 1.3	0.90–2.0 0.90–2.3 0.90–1.9	Trend for non-Hodgkin's lymphomas (and pancreatic cancer) with years (y) since first employment: <5 y, RR=0.64; 5–9 y, RR=0.49; 10–19 y, RR=1.3; >20 y, RR=2.3
Wigle et al. (1990, *77*)	Male farm operators *2,4-D, 2,4,5-T, Propanil, Sodium Arsenite, Sulfuric Acid, Dinitrophenols, Bromoxynil, Cyanazine, Linaron*	non-Hodgkin's Lymphoma Brain Multiple Myeloma	0.9 1.03 1.02	0.75–1.1 0.83–1.3 0.80–1.3	Trend for non-Hodgkin's lymphomas with acres sprayed with herbicides as well as with fuel and oil for farm purposes. Exposure to solvents
Manz et al. (1991, *108*)	Production workers *2,3,7,8-TCDD, 2,4,5-TCP, Trichlorophenol*	All Cancers Hematopoietic System	1.39 2.65	1.10–1.75 1.21–5.03	Cancer mortality increased in men with >=20 years of employment; high risk in men highly exposed to 2,3,7,8-TCDD: RR=2.77 (1.6–4.5);
		Breast	2.15	0.98–4.09	in women

Saracci et al. (1991, *107*)	Sprayers	non-Hodgkin's Lymphoma	0.5	0.01–1.4	IARC-NIEHS
	Production workers	non-Hodgkin's Lymphoma	1.5	0.64–2.9	International Register
	Sprayers	Soft-Tissue Sarcomas	1.01	0.61–8.7	10–19 y since first
	Production workers	Soft-Tissue Sarcomas	0.97	0.03–5.4	exposure: RR=8.82 for
	2,4,5-T, 2,4-D, MCPA, MCPP, 2,3,7,8-TCDD	Thyroid Gland	3.67	1.00–9.4	sprayers RR=6.06 for the entire cohort
Coggon et al. (1991, *110*)	Manufacturers *Phenoxy Acids, Chlorophenol 2,3,7,8-TCDD*	non-Hodgkin's Lymphoma	2.29	0.28–8.3	Deaths occurred >10 years since first exposure to phenoxy compounds
Fingerhut et al. (1991, *109*)	Production workers *2,3,7,8-TCDD, 2,4,5-T, 2,4-D, Chlorophenols*	All Cancer	1.46	1.21–1.86	NIOSH, USA Registry in workers with >=1 year of exposure, >=20 years of latency
		Soft-Tissue Sarcomas	9.2	1.90–27	
Bertazzi et al. (1992, *113*)	General population accidentally exposed to *2,3,7,8-TCDD*	Leukemias	2.1	0.7–6.9	Males, 1–19 years old
		Leukemias	2.5	0.2–27	Females, 1–19 years old
		Lymphatic Leukemia	9.6	0.9–106	Males, 1–19 years old
Hansen et al. (1992, *111*)	Gardeners	Soft-Tissue Sarcomas	5.3	1.1–15	Males
		Chronic Lymphatic Leukemia	2.8	1.0–5.9	Males
		non-Hodgkin's Lymphoma	2.0	0.9–3.9	Males and females
Semenciw et al. (1993, *114*)	Farmers *Herbicides, Insecticides*	Multiple Myeloma	0.82	0.69–0.95	RR=1.69, >$900 spent in fuel/oil for farming purposes; trend with acres sprayed with insecticides

n.a. = not available

Table 2
Selected Findings in Case-control Studies on Cancer and Exposure to Pesticides

Reference (year,#)	Occupation/*Exposure*	Cancer site	Odds Ratio	95%CI	Comments
Hardell & Sandstrom (1979, *124*)	*Phenoxy acids*	Soft-Tissue Sarcomas	5.3	2.4–11.5	Possible effect of chlorinated dibenzodioxins and dibenzofurans
	Chlorophenols	Soft-Tissue Sarcomas	1.2	0.4–3.7	
	DDT	Soft-Tissue Sarcomas	6.6	1.76–25	
Hardell et al. (1981, *125*)	Railroad, agriculture and forestry workers *2,4,5-T, 2,4-D, MCPA, Picloram, Amitrol*	Malignant Lymphomas	4.8	2.9–8.1	OR=7 for men exposed to phenoxy acids >=90 days
Eriksson et al. (1981, *126*)	*Phenoxy acids*	Soft-Tissue Sarcomas	6.8	2.6–17.3	OR=8.5 for men exposed to phenoxy acids >30 days
	Chlorophenols	Soft-Tissue Sarcomas	3.3	1.3–8.1	
Cantor (1982, *128*)	Farmers, *Phenoxy acids*	non-Hodgkin's Lymphoma	1.22	0.98–1.51	Trend with acres treated with insecticides and wheat acreage
Hardell et al. (1982, *127*)	*Phenoxy acids*	Nasal and Nasopharyngeal	2.1	0.9–4.7	Most nasal cancers were of the squamous cell type
	Chlorophenols	Nasal and Nasopharyngeal	6.7	2.8–16	
Smith et al. (1984, *130*)	*2,4,5-T*	Soft-Tissue Sarcomas	1.3	0.6–2.5	OR=3.2 (1.3–7.8) in railways workers)
	Chlorophenols	Soft-Tissue Sarcomas	1.5	0.5–4.5	OR=2.8 (1.3–6.3) in meat workers
Donna et al. (1984, *143*)	Agricultural workers *Herbicides*	Ovary	4.38	1.0–16.0	Women
Cantor & Blair (1984, *129*)	Farmers, *Phenoxy acids*	Multiple Myeloma	1.4	1.0–1.8	High risk for use of fertilizers and acres treated with insecticides

Pearce et al. (1985, *131*)	Agriculture and forestry workers	Multiple Myeloma	2.22	1.3–3.8	Increased risk among workers aged 20–64 years
		Malignant Lymphoma and Multiple Myeloma	5.5	1.6–20.1	In orchard farmers
Hoar et al. (1986, *133*)	*2,4-D, Triazines*	non-Hodgkin's Lymphoma	2.6	1.4–5.0	Significant trend with years and days/year (d/y) of 2,4-D use: >=21 d/y OR=7.6 (1.8–32)
	Amides	non-Hodgkin's Lymphoma	2.5	1.2–5.4	
	Insecticides	non-Hodgkin's Lymphoma	2.9	1.1–7.6	
Morris et al. (1986, *132*)	*Organochlorine, Organophosphorous, DDT, Arsenical Pesticides, 2,4-D*	Multiple Myeloma	2.6	1.5–4.6	OR=1.3 (1.0–1.6) for those who lived on a farm
Vineis et al (1987, *135*)	Agricultural workers *2,4,5-T, 2,4-D, MCPA*	Soft-Tissue Sarcomas	2.7	0.59–12.4	Women, high prevalence of rice growing area; OR=0.9 in men
Woods et al. (1987, *134*)	*Phenoxy Acids, 2,4-D*	Soft-Tissue Sarcomas	0.9	0.4–1.9	Risk for non-Hodgkin's Lymphoma elevated among forestry herbicide applicators: OR=4.8
	Chlorophenols	non-Hogkin's Lymphoma	1.6	0.7–3.8	
	DDT, Chlordane	non-Hodgkin's Lymphoma	1.8	1.0–3.2	
Lowengart et al. (1987, *136*)	*Pesticides, Insecticides*	Childhood Leukemia	3.8	1.37–13	Parents use of pesticides in the home
		Childhoold Leukemia	6.5	1.47–59	Parents use of pesticides in the garden
Shu et al. (1988, *138*)	Maternal exposure to pesticides	Childhood Acute Lymphatic Leukemia	3.5	1.1–11.2	Exposure during pregnancy
Hardell & Eriksson (1988, *139*)	*Phenoxy Acids Chlorophenols*	Soft-Tissue Sarcomas	3.3	1.4–8.1	No association with exposure to chlorophenols

(table continues)

Table 2 (continued)

Reference (year,#)	Occupation/*Exposure*	Cancer site	Odds Ratio	95%CI	Comments
Musicco et al. (1988, *137*)	Farmers *Insecticides, Herbicides, Fertilizers*	Brain Gliomas	2.1	1.27–3.58	No association with herbicides use
Hoar Zahm et al. (1989, *147*)	Wood workers	Soft-Tissue Sarcomas	1.7	0.9–3.2	Increased risk with years (y) of employment: >6 y, OR=2.1
Pearce et al. (1989, *140*)	Farmers *Phenoxy Herbicides* 2,4,5-T	non-Hodgkin's Lymphoma non-Hodgkin's Lymphoma non-Hodgkin's Lymphoma	1.0 1.0 1.4	0.8–1.7 0.7–1.5 0.8–2.3	The highest risk was detected among orchard workers: OR=3.7; increased risk with days/year use of phenoxy herbicides: 10–19 d/y, OR=2.2
Brownson et al. (1989, *148*)	Farmers	Lymphatic-hematopoietic Prostate non-Hodgkin's Lymphomas	1.28 1.33 1.40	1.06–1.56 1.18–1.51 1.04–1.85	High risk for Rectal Cancer (OR=1.2) and Nasal Cavity/Sinuses (OR=1.7)
Boffetta et al. (1989, *141*)	*Pesticides, Herbicides*	Multiple Myeloma	4.3	1.7–10.9	Exposed farmers only
Buckley et al. (1989, *149*)	*Pesticides*	Childhood Acute Leukemia	2.7	1.0–7.0	Paternal exposure to pesticides
		Childhood Acute Leukemia	3.5	0.9–14	Child exposure to pesticides in the home
Donna et al. (1989, *146*)	Agricultural workers *Triazine Herbicides*	Ovary	2.7 3.0	1.0–6.9 1.1–8.5	Women, trend with duration and probability of agricultural exposure

Reif et al. (1989,89, *142, 144*)	Farmers	Malignant Melanoma	1.25	1.05–1.50	Lip cancer among dairy farmers: OR=4.88 (2.8–8.6)
		Lip	2.43	1.81–3.27	
		Prostate	1.26	1.13–1.41	
		Brain	1.34	1.04–1.74	Brain cancer among livestock farmers: OR=2.96 (1.5–5)
		Lymphatic-haematopoietic	1.44	1.08–1.42	
		Leukemia	1.24	0.99–1.55	
		non-Hodgkin's Lymphomas	1.24	0.99–1.56	
La Vecchia et al. (1989, *145*)	Farmers and food processing	Hodgkin's Disease	2.1	1.0–3.8	Trend with years (y) of exposure to herbicides for Hodgkin's disease: >10 y, OR=8.7 and non-Hodgkin's lymphoma: >10 y, OR=5.2
		non-Hodgkin's Lymphoma	1.9	1.2–3.0	
		Multiple Myeloma	2.0	1.1–3.5	
Eriksson et al. (1990, *151*)	*Phenoxy Acids*	Soft-Tissue Sarcomas	1.8	1.02–3.18	
	2,4,5-T	Soft-Tissue Sarcomas	3.85	1.15–13	2,4,5-T contaminated by TCDDs
	Chlorophenols	Soft-Tissue Sarcomas	3.25	1.69–16	
Morris Brown et al. (1990, *153*)	Farmers	All Leukemias	1.2	0.9–1.3	Increased risk for users of organophosphate insecticides on animals: OR=11.6. Trend with days/ year use of chlordane, dichlorvos, DDT, malathion, lindane
	Herbicides, Pesticides, Organophosphates, Triazines Phenoxy Acids	Chronic Lymphocitic Leukemia	1.4	1.1–1.9	
Pasqualetti et al. (1990, *152*)	Farmers *Pesticides*	Multiple Myeloma	2.83	1.87–4.8	Farmers, no details
Hoar Zahm et al. (1990, *154*)	2,4-D	non-Hodgkin's Lymphoma	1.5	0.9–2.5	Among mixers or sprayers mixing or applying 2,4-D >21 days/year
		non-Hodgkin's Lymphoma	3.3	0.5–22	

(table continues)

Table 2 (continued)

Reference (year,#)	Occupation/*Exposure*	Cancer site	Odds Ratio	95%CI	Comments
Wingren et al. (1990, *150*)	*Phenoxy Acids,*	Soft-Tissue Sarcomas	4.4	1.0–18	Gardeners, adjusted for age and smoking
	Chlorophenols	Soft-Tissue Sarcomas	4.8	0.8–24	Railroad workers, adjusted for age and smoking
Smith & Christophers (1992, *156*)	*Phenoxy Herbicides, Chlorophenols*	Soft-Tissue Sarcomas	1.0	1.3–3.1	OR=3.0, 5 years latency
		Malignant Lymphoma	1.5	0.6–3.7	OR=3.0, 5 years latency Increased risk with days (d) of use: >3 d, OR = 2 and OR = 2.7 for soft-tissue sarcomas and malignant lymphoma
Scherr et al. (1992, *157*)	Agriculture, forestry and fishing industry	non-Hodgkin's Lymphoma	3.01	1.1–10.8	Increased risk for plant farmers and gardeners
Richardson et al. (1992, *159*)	*Herbicides, Insecticides*	Acute Leukemia	3.5	1.1–10.8	Adjusted for exposure to benzene Increased risk with years (y) of exposure: OR=6 for >10 y herbicides use OR=4 for >10 y insecticides use

Holly et al. (1992, *158*)	Agricultural Occupations *Herbicide, Pesticides, Fertilizers*	Childhood Acute Leukemia	8.8	1.8–43	Based on father's agricultural occupations
		Childhood Acute Leukemia	6.1	1.7–22	Based on father's use of pesticides
Cantor et al. (1992, *155*)	Farmers *Insecticides, Fungicides, Fertilizers*	non-Hodgkin's Lymphoma	1.2	1.0–1.5	Significantly increased risks for handling animal insecticides: chlordane, OR=2.2; malathion, OR=1.2; nicotine, OR=1.8 and crop insecticides: carbaryl, OR=3.8; DDT, OR=1.8; diazinon, OR=2.6; lindane, OR=2.2; malathion, OR=2.9; OR generally increased when protective measures were not used.
Hoar Zahm et al. (1993, *161*)	*Atrazine Organophosphate Insecticides, 2,4-D*	non-Hodgkin's Lymphoma	1.2	0.9–1.7	Adjusted for use of 2,4-D and organophosphates
Franceschi et al. (1993, *160*)	Farmers	Larynx	5.2	1.7–16	Increased risk confined among farmers born >=1930 adjusted for smoking
		Oral cavity & Pharinx	4.1	1.7–9.6	
Davis et al. (1993, *162*)	Family pesticide use *Lindane, Chlordane, Carbaryl, Dichlorvos*	Childhood Brain Cancer	2.9	1.3–7.1	Increased risk for termite treatment, insecticides used in garden, orchard, yard

(text continued from page 678)

and lymphomas, that exposure definition is not comparable from study to study, and that most studies include workers whose historical occupational exposure is to complex chemical mixtures, including 2,3,7,8-TCDD, prevent from making any definitive judgment of causality [115—122,167].

Examination of the findings generated by the cohort studies published so far reveals that several occupational cohorts have not been followed up for a sufficient time that would allow for detecting excess cancer risk in exposed workers. However, despite these limitations, recently published epidemiologic studies of large cohorts reported excess risk for soft-tissue sarcomas, haematopoietic malignancies, thyroid gland, and other endocrine tumors, cancers of the respiratory tract, breast cancer in women, and for all cancers in subjects exposed to phenoxy herbicides and/or 2,3,7,8-TCDD [91,103,105,107-109]. These recent findings, along with the evidence that humans show interindividual variation in enzyme induction by TCDDs, may explain the discrepancies among cancer risks detected by epidemiologic studies and suggest the need for future studies that identify subgroups of exposed workers genetically susceptible to the health effects induced by toxic chemicals [123].

Epidemiologic Evidence from Case-Control Studies

The relationship between occupational exposures to phenoxy herbicides and chlorophenols has been investigated widely in case-control studies (see Table 2). This type of epidemiologic design investigates the frequency of exposure to specific agents suspected to be linked with the adverse health outcome of interest in individuals with a specific disease, e.g., a type of cancer, and in subjects without that disease (i.e., controls or referents) and without other diseases that are known to be associated with the candidate exposure. Case-control studies imply selection of individuals based upon the presence or absence of a specific disease. Cases and referent subjects are interviewed by trained personnel, who should be unaware of the case-control status to avoid information bias inflating the exposure-effect relationship estimated by the study, to collect individual information on past exposure to environmental agents as well as other general characteristics.

The available epidemiologic evidence from case-control studies (see Table 2) suggests that a causal relationship may exist between exposure to phenoxy herbicides and chlorophenols and the development of non-Hodgkin's lymphoma and soft tissue sarcomas. Non-Hodgkin's lymphoma was found to be associated with frequent use of phenoxy acetic acid herbicides, particularly 2,4-D, and, with less consistency, with exposure to organophosphate insecticides, fumigants, fungicides, and triazine herbicides. Studies conducted independently in the USA and in New Zealand have reported dose-response patterns between non-Hodgkin's lymphoma and use of 2,4-D and 2,4,5-T that could not be explained by other environmental risk factors such as exposure to solvents, other agrochemicals, zoonotic viruses, and by the potential contamination of these herbicides with 2,3,7,8-TCDD [133,140,154]. The association between soft-tissue sarcomas and phenoxy herbicides and chlorophenols reported by case-control studies (see Table 2) may be because in most instances cases were exposed to 2,3,7,8-TCDD-contaminated herbicides. The evidence of a causal relationship between exposure to phenoxy herbicides, chlorophenols and 2,3,7,8-TCDD remains subject of scientific controversy [115-122,163,164].

A limited number of case-control studies have investigated the associations between other types of cancer and exposure to pesticides (see Table 2). Significant relationships have been reported between ovarian cancer and exposure to triazine herbicides, brain cancers and insecticides use, and non-Hodgkin's lymphoma and leukemia and several insecticides, suggesting that exposures to specific pesticides may increase cancer risk in humans [137,142,146,153,155]. Increased risks have been detected also for brain cancer, leukemias, and Ewing's bone sarcomas in children who were exposed to pesticides in their home or whose parents were occupationally exposed to pesticides, suggesting that children may be particularly susceptible to pesticide-related delayed adverse health effects [136,149,158,162].

CARCINOGENICITY STUDIES IN EXPERIMENTAL SYSTEMS

Pesticides, because of their high biological activities and for regulatory purpose, have been extensively studied for their toxicological properties, but only a fractions of these studies have been published in public literature [165,166]. The carcinogenicity of a substance in animals is established when administration in adequately designed and conducted experiments results in an increased incidence of one or more types of malignant neoplasm in treated animals in respect to untreated animals maintained under identical conditions. In the absence of adequate data on humans, it is biologically plausible to regard agents for which there is sufficient evidence of carcinogenicity in experimental animals as though they present a carcinogenic risk to humans. The chemicals are classified for carcinogenicity by International Agencies into different categories on the basis of the evidence of the experimental data available in the scientific literature [167-169]. The evidence that a chemical causes tumors in experimental animals is of two degrees:

- **Sufficient evidence for carcinogenicity** is provided by experimental studies that show an increased incidence of malignant tumors in multiple species and following multiple routes and doses, and
- **Limited evidence of carcinogenicity** in animals because of inconclusive results.

The concepts of "sufficient evidence" and "limited evidence" indicate varying degrees of experimental evidence. This categorization has to be related to the knowledge available and presupposes a continuous change with the acquisition of new experimental data. Forty-four pesticides have been evaluated by the International Agency for Research on Cancer (IARC), Lyon, France, as adequately studied in laboratory animals [170,219]. Twenty-five agrochemical active ingredients demonstrate with sufficient evidence a carcinogenic effects in laboratory animals (see Table 3). For 19 compounds the evidence is limited (see Table 4). A high number of pesticides have been extensively studied in laboratory animals, but the available studies are not adequate to make an evaluation of the carcinogenic potential.

Some pesticides with sufficient or limited evidence of carcinogenicity such as chlordecone, DDT, pentachlorophenol, captafol, aldicarb, and others have been banned or now have restricted use in some countries. These compounds have been considered as widespread environmental pollutants due to their bioaccumulation and persistence in the ecosystem. Residues of these pesticides have been detected in the food chain and in different biological media in humans [15,220,221]. Many types of organochlorine pesticides cause malignant tumors in lower mammals, but the potential risk of these carcinogens in human population is different. HCB, for example, may be considered more hazardous. A high number of long-term experiments have been conducted with DDT. The experimental evidence for carcinogenicity of this insecticide is principally based on the induction of liver tumor cells in mice. The studies on rats have provided contradictory results. These discrepancies and the negative results in hamsters suggest different metabolic pathways in the different species. The present knowledge on the carcinogenicity of DDT in animals does not allow a carcinogenic risk assessment of this pesticide in humans.

In contrast, in the case of HCB, there is a wide spectrum of carcinogenic activity for different organs on different species. Chlorophenols, such as trichlorophenol and pentachlorophenol isomers could be considered as hazardous chemical carcinogens, the first one demonstrating limited evidence in humans. Few pesticides belonging to the chemical families of carbamates or dithiocarbamates have been demonstrated as animal carcinogens. Sulfallate induces malignant tumors in rats and mice. Diallate and ziram have shown limited evidence of carcinogenicity. The principal hazard of these compounds is the presence of impurities and the metabolic and degradation products of thiocarbamates. In addition, recent reports have described the production of carcinogenic N-nitrosocompounds by the reaction of many thio- and alkylcarbamates with nitrite. As an example, N-nitrosocarbaryl, a derivative of carbaryl, is a potent carcinogen in rats. The ethylene thiourea, a degradation product of ethylene bisdithiocarbamate fungicides, such as mancozeb, maneb, metiram, zineb, nabam and others, is a potent thyroid carcinogen in rats. Chlordecone, chlordane, heptachlor, and mirex are also other animal carcinogens of this chemicals class.

(text continued on page 700)

Table 3
Pesticides with Sufficient Evidence of Carcinogenicity in Experimental Animals

Compound	C.A.S. No.	Target Organs		References
		Mouse	Rat	
Insecticides				
Aramite	140.57.8	Liver	Liver	IARC, 1974. 170
Chlordane	57.74.9	Liver	Liver, thyroid	IARC, 1991. 171 NCI, 1977. 172
Chlordecone	143.50.0	Liver	Liver	IARC, 1979. 173 NCI, 1976. 174
DDT	50.29.3	Liver, blood, lung	Liver	IARC, 1991. 175
Dichlorvos	62.73.7	Forestomach, esophagus	Pancreas, blood	IARC, 1991. 176
Dimethylcarbamoyl chloride	79.44.7	Skin	Nasal tract	IARC, 1987. 177
Hexachlorocyclohexanes (HCH)				
technical-grade HCH	608.73.1	Liver, blood		IARC, 1987. 178
alpha-HCH	319.84.6	Liver	Liver	IARC, 1987. 178
Heptachlor	76.44.8	Liver	Thyroid	IARC, 1991. 171 NCI, 1977. 179
Mirex	2385.85.5	Liver	Liver	IARC, 1979. 180
Toxaphene (Polychlorinated camphenes)	8001.35.2	Liver	Thyroid	IARC, 1979. 181

(table continues)

Table 3 (continued)

Compound	C.A.S. No.	Target Organs		References
		Mouse	Rat	
Fungicides				
Captafol	2425.06.1	Small intestine, heart, spleen, liver	Kidney, liver	IARC, 1991. 182
Chlorophenols				
Pentachlorophenol	87.86.5	Liver, spleen		IARC, 1991. 183
2,4,6-Trichlorophenol	88.06.2	Liver	Blood	IARC, 1979. 184 NCI, 1979. 185
Ethylene Thiourea	96.45.4	Liver	Thyroid	IARC, 1987. 186
Hexachlorobenzene	118.74.1	Liver	Liver, kidney	IARC, 1987. 187
Sodium ortho-phenylphenate	132.27.4	Liver	Urinary tract	IARC, 1987. 188
Herbicides				
Amitrole	61.82.5	Thyroid, liver	Thyroid	IARC, 1987. 189
Nitrofen (technical-grade)	1836.75.5	Liver, spleen	Pancreas	IARC, 1983. 190
Sulfallate (technical)	95.06.7	Mammary gland, lung	Mammary gland, forestomach	IARC, 1983. 191 NCI, 1978. 192

Fumigants				
1,2-Dibromo-3-chloropropane	96.12.8	Forestomach. nasal cavity, lung	Forestomach, mammary gland, nasal cavity, tongue, adrenal cortex	IARC, 1987. 193 NTP, 1982. 194
1,2-Dichloroethane	107.06.2	Lung, blood, mammary gland, liver, uterus	Blood, forestomach, mammary gland	NCI, 1978. 196 IARC, 1979. 195
1,3-Dichloropropene	542.75.6	Bladder, lung, forestomach	Liver, forestomach	NTP, 1985. 198 IARC, 1986. 197
Ethylene dibromide	106.93.4	Forestomach, lung, nasal cavity, spleen, esophaugs, mammary gland, subcutaneous	Forestomach, nasal cavity, liver, spleen, mammary gland, lung	IARC, 1987. 199
Tetrachloroethylene	127.18.4	Liver	Blood	IARC, 1987. 200

Table 4
Pesticides with Limited Evidence of Carcinogenicity in Experimental Animals

Compound	C.A.S. No.	Target Organs		References
		Mouse	Rat	
Insecticides				
Aldrin	309.00.2	liver	Inconclusive	IARC, 1987. 201
Arsenic and comp.	7440.38.2	lung	lung, stomach	IARC, 1987. 202
Chlorobenzilate	510.15.6	liver	Inconclusive	IARC, 1983. 203
Dicofol	115.32.2	liver(M)	Inconclusive	IARC, 1983. 204
Dieldrin	60.57.1	liver	—	IARC, 1987. 205
beta-HCH	319.85.7	liver	Inconclusive	IARC, 1987. 206
gamma-HCH (lindane)	58.89.9	liver	thyroid	IARC, 1987. 206
Terpene polychlorinates (strobane)	8001.50.1	liver(M)	—	IARC, 1974. 207
Tetrachlorvinphos	22248.79.9	liver	thyroid, adrenal gland(F)	IARC, 1983. 208

Fungicides				
Captan	133.06.3	Duodenum	—	IARC, 1983. 209
Chlorothalonil	1897.45.6	—	kidney	IARC, 1983. 210
Pentachloronitrobenzene	82.68.8	liver(M)	Inconclusive	IARC, 1974. 211
Ziram	137.30.4	lung(F)	thyroid(M)	IARC, 1991. 212
Herbicides				
Atrazine	1912.24.9	NT	mammary gland, uterus, blood	IARC, 1991. 213
Diallate	2303.16.4	Liver, lung(M)	inconclusive	IARC, 1983. 214
Monuron	150.68.5	—	kidney, liver(M)	IARC, 1991. 215
Picloram	1918.02.1	—	liver, thyroid(F)	IARC, 1991. 216
Trifluralin	1528.09.8	lung, forestomach(F)	thyroid	IARC, 1991. 217
Fumigants				
Trichloroethylene	79.01.6	lung, liver	kidney	NCI, 1976. 218 NTP, 1988. 219

(text continued from page 694)

The number of carcinogens in the organophosphate chemical class of pesticides is limited. Dichlorvos, widely used as an insecticide, has been demonstrated as an animal carcinogen inducing a rare aesophageal squamous-cell tumor in mice treated with the compound in the diet. Captafol and captan, chloroalkyl compounds widely used as fungicides, demonstrate a carcinogenic effect with sufficient and limited evidence. Captafol acts on multiple target organs in two rodent species.

Published carcinogenicity studies on pesticides in experimental animals are generally carried out using the active agrochemical products rather than the commercial formulations. Humans are generally exposed to a complex mixture of different types of pesticides. Chemical agents and by-products in technical or formulated preparations also can induce biological consequences. For example, formulations of the herbicide 2,4,5-T may contain highly toxic dioxins. Nitro-derivatives of pesticides also may occur as impurities in technical products such as N-nitrosodiethylamine and N-nitrosodiisopropylamine in trifluralin. Although the solvents, carriers, emulsifiers, and surfactants generally have been considered to be inert ingredients in pesticide formulations, the presence of aromatic hydrocarbon solvents or surfactants in a complex mixture may have a biological implication in the induction of the neoplastic process. Finally, biologically active products also may be formed during the storage under inappropriate conditions of the pesticides. The extrapolation from experimental data based on pure compounds to humans appears an uncertain process.

GENETIC TOXICOLOGY

The carcinogenic process should not be considered as an inherent property of a chemical, but rather a result of the interaction of a chemical with a complex biological system influenced by many factors. The development of cancer has been divided in three major stages: initiation, promotion, and progression. Genotoxic events are prevalent in the first stage, while nongenotoxic effects prevail in the second phase. Carcinogenic substances could be divided into genotoxic and nongenotoxic agents. The genotoxic compounds are those that act through direct interaction with and modification of DNA or by a clastogenic event. Nongenotoxic carcinogens act through nongenotoxic mechanisms such as promotion, modulation of a normal tissue differentiation, disturbance of hormonal balance, impairment of the immune system. Different short-term tests have been widely used to evaluate the genotoxic potential of a chemical compound. The evaluation of the results from short-term genetic toxicity studies about the capacity of these assays to predict carcinogens show that mutagenic potential is a primary risk factor for carcinogen identification [222]. Pesticides long have been considered potential chemical mutagens. Experimental data revealed that various agrochemical ingredients possess mutagenic properties, inducing gene mutation, chromosomal alteration, or DNA damage. A large part of organophosphate, chlorinated, carbamate, and pyrethroid pesticides were reported to be positive for genetic and cytogenetic effects in mammalian systems in vivo or in vitro. An analysis of genotoxic profiles of 115 pesticides revealed that about 57% of the chemicals analyzed were active in some short-term tests [223]. Forty-nine agrochemical ingredients gave negative results in all tests. Pesticides are not potent genotoxic agents: They generally give positive results in few genetic tests. Despite some discordance in the results of short-term tests, pesticides with similar chemical structure produce similar profiles of genotoxic activity. Some nongenotoxic agrochemical ingredients that could be considered as potential human carcinogens may act through a nongenotoxic mechanism, such as microsomal enzyme induction or disturbance of hormone balance.

BIOMONITORING OF HUMAN POPULATION

The process of risk assessment associated with environmental exposure involves:

- hazard identification,
- exposure assessment,
- dose-response assessment, and
- risk characterization.

Because of the already-mentioned difficulties in conducting sound epidemiological studies, the hazard identification is primarily based on animal toxicology data. Acute and chronic human toxicity usually are determined by extrapolation from animal studies. Estimation of health risk associated with environmental exposures is composed of two primary steps: individual exposure evaluation and biological effects assessment. This process has been greatly improved by development of analytical methods and biomonitoring techniques to measure chemical compounds or early biological events that provide individual markers of exposure, effect, or susceptibility in humans. The U.S. Environmental Protection Agency has adopted a general definition of the biological marker: "a measurement of environmental pollutants or their biological consequences after the contaminants have crossed one of the body's boundaries and entered human tissues or fluids and which serve as an indicator of exposure, effect, and/or susceptibility" [224]. A series of events has been described from the release of pollutants in the environment, to human exposure, to early biological events and finally to health effects (see Figure 1).

Biological Markers of Exposure

Information on human exposure to environmental agents is a crucial component in the risk-assessment process. Many human exposures occur through a wide spectrum of environmental pathways and by different routes. Biological markers

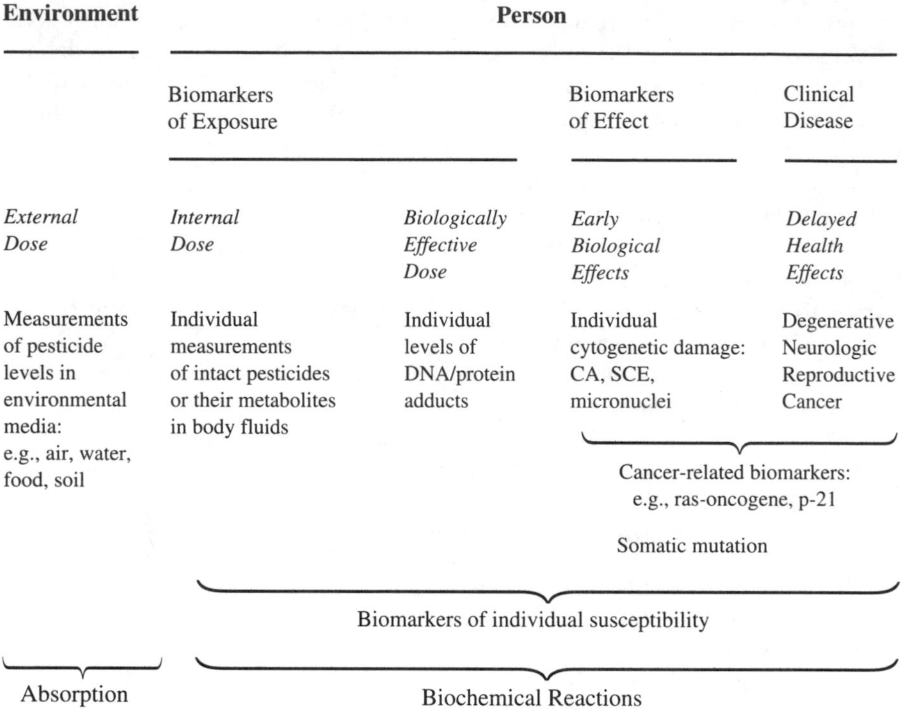

Environment	**Person**			
	Biomarkers of Exposure		Biomarkers of Effect	Clinical Disease
External Dose	*Internal Dose*	*Biologically Effective Dose*	*Early Biological Effects*	*Delayed Health Effects*
Measurements of pesticide levels in environmental media: e.g., air, water, food, soil	Individual measurements of intact pesticides or their metabolites in body fluids	Individual levels of DNA/protein adducts	Individual cytogenetic damage: CA, SCE, micronuclei	Degenerative Neurologic Reproductive Cancer

Cancer-related biomarkers: e.g., ras-oncogene, p-21

Somatic mutation

Biomarkers of individual susceptibility

Absorption — Biochemical Reactions

Figure 1. Biological markers in the continuum of events from exposure to health effects.

of exposure are indicators of internal or biologically effective doses. Markers measure the amount of a compound or its metabolites in biological specimens, or interaction products between that chemical or its metabolites with proteins or nucleic acids.

Monitoring of Pesticides and Their Metabolites (Body Burden)

Historically, chemical exposure in the workplace has been performed through environmental monitoring. The principal problem of this approach is that the airborne concentration of a chemical is not necessarily linearly correlated with the amount adsorbed, given the great variation in physicochemical and biological properties between chemicals, including different relationships between body burden and excretion of the compound. Biological monitoring, by measuring the concentration of the chemical or metabolite in biological media, such as blood, urine, saliva, sweat, exhaled breath, hair, and adipose tissues could be a valid complement to environmental monitoring. Residues of specific chemicals or their metabolites in various human fluids or tissues are indicative of the total body burden of the products and of past and present exposure. These methods yield data directly useful in risk-assessment models. Analytical procedures for the detection of intact pesticides and their metabolites in biological fluids have been developed to study patterns of absorption, metabolism, and excretion of these compounds; to establish exposure limits and field re-entry intervals; and to evaluate the efficacy of work practices and protective clothing. Available techniques for detecting a variety of pesticides in urine and blood have been reviewed with particular emphasis on technical information, costs, and validation in terms of dose-response relationship [225,226].

Organophosphates

Exposure to organophosphate insecticides is measure by determination of alkyl phosphate or phenolic metabolites excreted in the urine [227]. Organophosphates are hydrolyzed very rapidly (in terms of hours), but their urinary metabolites can be detected for several days after exposure. This determination is proposed as a sensitive and specific indicator enabling the detection of a low level of exposure. The specific metabolic residues are known for many organophosphates, but monitoring for a single compound is very difficult because of simultaneous exposure to many agrochemical ingredients [228]. The measurement of urinary metabolites has been considered very useful, but the complexity of exposure in field studies, including variation of kinetics among compounds, have limited the applicability of these methods.

Carbamates

Few analytical methods have been developed to detect carbamates or their metabolites in biological samples. The determination of phenolic metabolites appears to be the most sensitive indicator for this class of chemicals [226]. The detection of 1-naphthol in urine and in blood is one of the methods applied in field studies to evaluate the exposure to carbaryl [229,230].

Dithiocarbamates

Dithiocarbamates are metabolized and excreted too rapidly to be detected in biological samples. A group of fungicides belonging to this family of compound, ethylenebisdithiocarbamates (EBCD) such as nabam, maneb, zineb, mancozeb, are metabolized to ethylene thiourea (ETU), which can be measured in urine.

Organochlorines

The analytical methods used to detect organochlorine pesticides have been extensively developed and widely used. Most of these compounds are lipid-soluble and accumulate in adipose tissues over long periods of time. Serum determinations have been widely applied to evaluate cumulative exposure and also to monitor a large fraction of nonoccupationally exposed populations. Residues of pesticides and their metabolites in human fluids collected from the general U.S. population are indicative of the total body burden of these pesticides and of past and present exposure. A significant proportion of serum samples tested had detectable levels of residues originating from DDT and its analogs, BHC, aldrin, dieldrin, chlordane, heptachlor, and other compounds belonging to the chemical class of chlorinated benzenes [220,221]. The detection of chlorophenols and nitrophenols in the blood or urine can be used assess exposure to intact nitrophenolic herbicides, pentachlorophenol, other chlorinated phenols, metabolites of lindane hexachlorobenzene, and organophosphorus insecticides [231].

Phenoxy Acids

Common analytical methods are available to detect intact phenoxy acids in blood and urine. Most phenoxy herbicides are rapidly excreted in hours or days. Urinary metabolites of all the phenoxy acids compounds have been monitored in human exposure studies. The most significant concern in occupational exposure to 2,4,5-T has been its contamination with the 2,3,7,8-TCDD, a highly toxic compound that accumulates in adipose tissues. Exposure to 2,4-D can be adequately assessed by urinary determination because approximately 95% of the compound is excreted in the urine with a pH-dependent rate. A recent GC multiresidue screen for acid herbicide residues in urine has been developed to detect different 2,4-D, 2,4,5-T, and other relevant phenoxy herbicides [232]. Biological monitoring is generally used to evaluate exposure levels in occupational settings. However, the interpretation of the toxicological significance of the observed metabolite levels is often difficult because the relationship between these levels and toxic dose are generally unknown.

For most agricultural chemicals, international agencies have not established biological thresholds for health effects other than acute illness. Monitoring programs could allow the correlation of indexes of exposure to chronic effects such as cancer, neurologic diseases, and reproductive effects. A recent National Health and Nutrition Examination Survey revealed the presence of different intact pesticides or their metabolites in urine samples collected from the U.S. general population [221]. The most frequently occurring residue in urine was pentachlorophenol (PCP) found in detectable concentrations in 71.6% of the general population. Other pesticides such as 3,5,6-trichloro-2-pyridimol, 2,4,5-trichlorophenol, paranitrophenol, dicamol, and malathion were quantified, but at lower frequencies [221]. Although no correlation can be made between health effects and residue levels, the extent of exposure to pesticides in the general population can be used in risk-management strategies.

Immunoassays for Pesticides

Analytical chemical procedures to detect intact pesticides and their metabolites have limited use in monitoring programs because they are expensive, time consuming, and often require sophisticated equipment and trained personnel. Immunoassays constitute low-cost, rapid, automated residue detection methods that have become common analytical methods in the clinical laboratory today. Within the past 15 years, there has been a heightened interest in generating antibodies to environmentally significant chemicals for trace residue analysis [233]. These assays, when applied to human blood, urine, and tissue samples, can provide a quantitative assessment of dose. Radioimmunological (RIA) and enzyme-linked immunoadsorbent assays (ELISA) have been developed for numerous pesticides [234]. Most reported techniques use polyclonal antibodies, but in some case monoclonal antibodies have been developed. These methods have sensitivities of detecting nanograms and, in some cases, picograms of chemicals.

Determination of Cholinesterase Activity

Nonspecific markers of exposure, such as conjugate complexes or elevated/depressed enzyme levels, provide indirect evidence about exposure to a specific chemical or a class of chemicals. Occupational exposure to organophosphate and carbamate insecticides can be monitored indirectly by measuring cholinesterase activity in plasma, red blood cells, and whole blood. The main mechanism in the toxicity of organophosphate and carbamate insecticides is the covalent binding of phosphate radicals to the active sites of the cholinesterase. The inhibition of cholinesterase activity leads to the accumulation of acetylcholinesterase at synapses, causing overstimulation and subsequent disruption of transmission in both the central and peripheral nervous systems.

A diagnosis of organophosphate insecticide poisoning could involve measurement of cholinesterase activity in blood [235]. The serum cholinesterase level is a sensitive but not specific indicator of organophosphate poisoning. The normal range of cholinesterase levels is quite broad, and individuals may manifest organophosphate toxicity with normal RBC cholinesterase levels. In addition, whole blood and plasma enzyme activities were found to be unrelated to the urinary excretion of OP pesticides metabolites. Different absorption of OP pesticides in several groups of workers generally was not reflected by equivalent differences in their plasma and whole blood cholinesterase activities. An accelerated rate of enzyme regeneration in the body has been described due to successive multiple exposure of workers to small amounts of OP pesticides. In addition, the validity of plasma cholinesterase levels monitoring in risk assessment studies is questionable because the long term effects associated with elevated levels of this enzyme are not well known.

A review of the scientific literature on biological monitoring of cholinesterase revealed that elevated levels can be found in 1% to 12.9% of individuals, depending on the kind of testing and population surveyed [236]. Low levels of plasma cholinesterase may be from causes apart from insecticide exposure such as liver disease, including hepatitis and cirrhosis; malnutrition; infections; anemia; and myocardial infarction. High levels of activity also have been observed in association with hyperlipoproteinemia, nephrosis, insuline-dependent diabetes, overweight, alcohol consumption in females, and skin-fold thickness in males. The biological functions of this enzyme were almost unknown, as well as its natural substrates and inducers. However, for practical purposes, in biomonitoring studies the exposure to pesticides, carbamate, or organophosphate insecticides, has been considered the only cause of significant cholinesterase depression among healthy workers [237].

Adducts of Carcinogens with Cellular Targets

Measurement of internal dose is crucial for relating exposure to effects. Measurement of DNA and protein adducts has been recognized as a marker of exposure and biologically effective dose. Carcinogen-DNA adduct determination as a means of monitoring human exposure to environmental and occupational carcinogens has become feasible with the development of sensitive techniques that do not require radiolabeled carcinogens. These methods have been applied to the detection of DNA adducts in white blood cells or tissues of individuals with exposure to chemical, occupational, or environmental carcinogens [238,239]. They include immunoassays with policlonal and monoclonal antisera, 32P-postlabeling of modified nucleotides, fluorescence spectroscopy, and gas-chromatography/mass spectroscopy [240-246]. These techniques quantitate adducts at a level of detection of one adduct / $10^{8 \text{ to } 10}$ nucleotides.

Large interindividual variability in adduct levels has been observed in people with similar levels of exposure. These differences probably are due to differences in absorption, metabolism, and DNA repair capability and demonstrate the advantage of biomonitoring over environmental monitoring in determining the individual exposure. The use of DNA adducts as a measure of exposure and cancer risk is complicated by different factors, such as interindividual metabolism and DNA repair that influence DNA adduct kinetics. In addition, adduct determination in surrogate cells, such as peripheral blood lymphocytes, may not reflect adduct levels in target tissues. Protein adducts are primarily considered as a measure of exposure. The quantification of hemoglobin adducts has technical advantages because the protein can be readily obtained in abundant quantities. In addition, hemoglobin adducts have a long biological lifetime, approximately 120 days for humans [247]. Hemoglobin adducts of aromatic amines released from pesticides have been studied: Adducts were detected for linuron, diuron, monuron, monolinuron, chlorpropham, chlordimeform, and propham [248].

Biological Markers of Effect

Biological markers have the ability to detect early biologic changes predictive of a disease. The primary goal of the research on biomarkers is improved interpretation of these methods as early indicators of adverse health effects, with primary emphasis on cancer risk, neurotoxicity, and reproductive developmental toxicity. Biological markers of carcinogenic exposure have been developed to study the relationship between exposure and early events in the oncogenic process. Genetic markers have been considered of particular relevance in epidemiological studies of human populations exposed to potential carcinogenic agents [249]. These include in addition to traditional cytogenetic markers, other alterations in chromosomal structure such as restriction fragment length polymorphisms, loss of heterozygosity, and translocation markers.

Specific genetic changes recently have been identified as critical molecular events in the initiation and development of many cancers. Among these, oncogenes activation, especially those of the ras family, and the inactivation of tumor-suppressor genes, such as p53, by point mutations, chromosomal deletion, and other structural changes are considered relevant in the process of carcinogenesis [250]. Some of these changes have been observed in chemically induced tumors in experimental animals, but the role of the chemical induction of these genetic abnormalities in human cancers remains unclear. Nowadays, cytogenetic tests are the most frequently used methods to establish human exposure to mutagens and/or carcinogens.

Three major types of cytogenetic changes have been identified as biological indicators of genotoxic exposure: chromosomal aberrations, micronuclei, and SCEs [251-254]. Conventional techniques for measuring chromosomal changes require that the cells be proliferating so that chromosomes in interphase can be seen at mitosis or as micronuclei. Sister chromatid exchanges methodology has been widely applied to monitor human exposure to potentially genotoxic agents. The micronucleus assay is increasingly being used as an alternative cytogenetic end point. More than 100 cytogenetic studies in occupational exposed populations have been published. The studies of micronuclei frequency are very limited. Increased chromosomal damage as a function of exposure has been reported for different chemical carcinogens such as vinyl chloride, benzene, ethylene oxide, and styrene [255-257]. The results of these studies are often contradictory. The sources of these variations are mainly the exposure conditions.

There are 34 published studies on human populations exposed to pesticides measuring alterations in mutagenic parameters [258-293]. Their findings are reported in Table 5. Twenty-six studies gave positive results. Cytogenetic analysis of chromosomes from peripheral lymphocytes of patients suffering from acute intoxications by pesticides were positive in all cases [259,266,267]. Positive findings have been reported in plant workers acutely exposed to tetrachlorobenzene, chlorinal, buvinol, DDT, 1,2-dibromo-3-chloropropene, and organophosphate compounds such as dimethoate, dichlorvos, mevinphos, methyl parathion, trichlorphon, and malathion [259,261,267]. A chromosomal aberration study in patients suffering acute organophosphate insecticide intoxication evidenced a temporary but significant increase in the frequency of chromatid breaks and stable chromosome-type aberrations in acutely intoxicated persons [259]. A large frequency of chromosomal aberrations has been observed also in mildly intoxicated persons. The frequency of chromosomal damage was significantly elevated during the high intoxication stage, but the normal frequency was restored in six months after the acute exposure. A possible dose-effect relationship has been hypothesized, but the limited number of cases and the uncertainty of chemical exposure data did not allow a clear conclusion.

A study in populations of exposed people revealed that SCE frequencies in floriculturists with intoxication symptoms are higher than those of asymptomatic persons [275]. Although acute intoxications with pesticides are relatively rare events, exposure during occupational practices may result in cytogenetic damage. Different studies have reported high incidences of chromosomal aberrations in agricultural workers, floriculturists, vineyard cultivators, and cotton agriculturists. A correlation between occupational exposure to pesticides and high frequencies of SCE also has been established in a few studies.

Only one study is available on micronuclei frequency in humans occupationally exposed to pesticides (293). This study reported an increase of this parameter in one group of floriculturists. A clear association has been found between micronuclei frequency, duration and condition of exposure. The floriculturists considered in this study were exposed to a mixture of pesticides, mainly fumigants, fungicides, insecticides, and only a few herbicides. Frequent changes in the agrochemical formulations were also verified in this area. In a recent study conducted in our laboratory, a population of floriculturists working in the same area revealed lower levels of micronuclei, probably associated with the withdrawal from the market of some genotoxic compounds, but still showed a positively graded response with the amounts of genotoxic pesticides used (Kg/year).

Because there is evidence that the increased levels of a cytogenetic damage may result from high exposure to pesticides, biological monitoring of accidentally exposed victims could provide useful information for future application of biomarkers in the genetic hazard of pesticides in humans. The negative results obtained in a limited number of studies could be attributable to the fact that the people were exposed to low levels of pesticides. A cytogenetic study on papaya workers exposed to low levels of ethylene dibromide failed to show any significant difference in SCE levels and CA frequencies. A recent paper failed to detect cytogenetic damage, measured as SCE frequency, in a rural population exposed to low levels of insecticides, fungicides, and herbicides [290]. In all these studies, the cytogenetic effects have not been associated with the exposure to any particular compound or class of chemicals because exposure generally involves complex mixtures of compounds.

Today, the somatic mutation theory of carcinogenesis is substantiated by increased experimental evidence [257]. Genetic changes, such as gene mutations,

(text continued on page 715)

Table 5
Cytogenetic Studies on Populations Exposed to Pesticides

Population	Exposure	Duration of Exposure	Assay	Result	References
42 Applicators a)26 herbicide sprayers b)16 insecticide sprayers	a) 2,4-D; 2,4,5-T; others b) multiple insecticides	1–25 y	CA (48)	pos (peak spray season)	Yoder et al., 1973. 258
31 Poisoning patients or applicators	Dichlorvos, mevinphos, trichlorphon, methyl-parathion, diazinon, malathion, dimethoate	Acute intoxications	CA	pos	Van Bao et al., 1974. 259
25	DDT	2 mo–10 y	CA	neg	Rabello et al., 1975. 260
8		20 d–2 y	CA	neg	
				pos	
Factory production workers					
14	tetrachlorobenzene	—	CA	pos	Czeizel et al. 1975, 1976. 261,262
36	Klorinal		CA	pos	
26	Buvinol		CA	pos	
Factory production workers					
ND	ziram	1–10 mo	CA	pos	Pilinskaya, 1974, 1982. 263,264
15	zineb		CA	pos	
ND	pirimor		CA	pos	
Applicators					

(table continues)

Table 5 (continued)

Population	Exposure	Duration of Exposure	Assay	Result	References
57 Pesticide sprayers	herbicides insecticides fungicides fumigants	1 y–>1 y	SCE	pos	Crossen et al., 1978. 265
18 Chemical plant workers	1,2-dibromo-3-chloropropene	3–18 mo	Y-chromosomal non-disjunction in sperm	pos	Kapp et al., 1979. 266
25	1,2,4,5-tetrachlorobenzene	at least 6 mo	CA	pos	Kiraly et al., 1979. 267
17 Chemical plant	trichlorphon		CA	pos	
34 workers	diazinon		CA and Chrom.	pos	
25	safidon 40 WP		CA and Chrom.	pos	
ND	lindane		CA	neg	
109 Agriculturists	Not specified	1–20 y	CA	pos	Nehez et al., 1981. 268
Agriculturists	—	—	CA	pos	Volnjanskaya and Vesilos 1981. 269
15 Chemical plant workers	methyl parathion	1 w–7 y	CA	neg	de Cassia Stocco et al., 1982. 270
35 Applicators (Sprayers in forestry)	2,4-D, MCPA	July–October spraying period	SCE	neg	Linnainma, 1983. 271

20 Accidental exposure in chemical plant workers	dimethoate	Acute exposure	SCE	pos	Larripa et al , 1983. 272
36 14 Floriculturists	organophosphorous carbamates organochlorine	at least 10 y	CA and SCE	pos	Dulout et al., 1985. 273
80 Agriculturists	pesticides (ca. 80)	1–>15 y	CA	pos	Paldy et al., 1987. 274
40 Plant breeders	organophosphorous carbamates organochlorines	at least 10 y	SCE	neg	Dulout et al , 1987. 275
15 Vineyard cultivators	DDT, lindane, quinalphos, diethane, metasystox, parathion, copper sulfate, dichlorvos, dieldrin.	5–12 y	CA	pos	Rita et al., 1987. 276
60 Papaya workers	ethylene dibromide (fumigation)	average 5 y	SCE CA	neg	Steenland et al., 1986. 277
12 Sprayers in greenhouses	pesticides (20)	1–4 y	Mutagenicity of urine	pos	Shane et al., 1988. 278

(table continues)

Table 5 (continued)

Population	Exposure	Duration of Exposure	Assay	Result	References
11 Pesticide sprayers	organophosphates carbamates pyrethroids	/	CA	neg	Desi et al., 1986. 279
55 Applicators	insecticides fungicides acaricides	2–15 y	CA	pos in openfields neg in greenhouses	Nehez et al., 1988. 280
25 Cotton agriculturists in India	DDT, malathion, BHC, parathion, dimethoate, fenitrothion, urea.	2–18 y	CA SCE	pos pos	Rupa et al., 1989a. 281
50 Cotton agriculturists in India	DDT, BHC, endosulfan, malathion, methyl parathion, phosphamidon, monocrotophos, quinolphos, dimethoate, fenvalerate, cypermethrin	1–25 y	CA	pos	Rupa et al., 1989b. 282
50 Cotton agriculturists in India	DDT, BHC, endosulphan, malathion, methyl parathion, phosphamidon, monocrotophos, quinolphos, dimethoate, fenvalerate, cypermethrin	1–25 y	SCE	pos	Rupa et al., 1989c. 283

44 Factory production workers	fungicide Novozir Mn80 macozeb-containing	up to 2 y	CA SCE	pos pos	Jablonicka et al., 1989. 284
22 Applicators	organophosphates, organochlorine, carbamates	spraying period	urinary clastogenicity	pos	See et al., 1990. 285
24	Phosphine (grain fumigant)	6 w to 3 mo	CA	pos	Garry et al., 1989. 286
9	phosphine alone				
11	phosphine and other pesticides				
4	other pesticides and fumigants				
15 Fumigant applicators	state grain workers				
17 Agriculturists	organophosphates carbamates		SCE	neg	Carbonell et al., 1990. 287
26 Applicators	organophosphates organochlorine pyrethroids	2–18 y	CA	pos	Rupa et al., 1991. 288
64 (32) Applicators (healthy people and cancer patients)	ca. 60 pesticides	ND	SCE CA	pos pos	De Ferrari et al., 1991. 289

(table continues)

Table 5 (continued)

Population	Exposure	Duration of Exposure	Assay	Result	References
94 Agriculturists	many pesticides used in agriculture	1–35 y	SCE	neg	Gomez-Arroyo et al., 1992. 290
27 Floriculturists	many pesticides	/	SCE	neg	Dulout et al., 1992. 291
29 Applicators in greenhouses	organophosphates, organochlorines, carbamates dithiocarbamates	4–30 y	CA	pos	Kourakis et al., 1992. 292
71	organophosphates, dithiocarbamates	2–55 y	MN	pos	Bolognesi et al., 1993. 293

(text continued from page 708)

chromosomal alterations or rearrangements, gene amplification, and aneuploidy, have been observed in different tumors. Point mutations have been described to activate proto-oncogenes and to inactivate tumor-suppressor genes in certain cancers. Chromosomal rearrangements, gene amplification, and numerical chromosomal changes also have been documented. Therefore, chemicals that induce genetic events could heritably alter a critical target gene necessary for neoplastic development. These observations support the use of mutagenicity tests in biomonitoring studies to evaluate the carcinogenic risk of chemicals to humans.

One of the most important health risks of human populations exposed to pesticides is the potential genotoxic hazard. There is evidence of a chromosomal damage in occupationally exposed workers. The increases of the cytogenetic parameters generally have been correlated with the intensity of exposure and/or the lack of protective measures. Although on the basis of the available knowledge, the relevance of cytogenetic damage in predicting individual health risk is still subject of study, positive results in biomonitoring studies could allow identification of hazardous exposures and implementation of protective measures aimed at preventing irreversible health effects.

Immunosuppressive Activity of Pesticides.

Various environmental pollutants have demonstrated a capacity to modify normal immune processes. The types of effects that occur are often chemical-specific as well as species-specific and include immunosuppression, targeting either systemic or local immunity (lung or skin), hypersensitivity diseases manifested as respiratory allergies or contact dermatitis, and in certain instances autoimmune diseases [294]. There is an increasing number of reports describing various immune changes in individuals occupationally exposed to chemical agents. The immune system of laboratory animals, including rodents, is remarkably similar to that of humans with respect to organization, function, and responsiveness. *In vitro* and *in vivo* experimental studies in chemical-induced immunosuppression suggest that many environmental chemicals can inhibit the immune system and alter host resistance to infectious agents or tumor cells. Pesticides, as such as trimethyl phosphorothioate, carbofuran, and chlordane have been identified as immunotoxic in these studies [295]. Organophosphorous insecticides have been known or presumed to cause allergic problems in humans, producing contact hypersensitivity [296,297].

Markers of Individual Susceptibility

Human subjects are not equally susceptible to the toxic and carcinogenic effects of chemicals. A variety of factors such as age, sex, metabolism, DNA repair capability, and genetic makeup have been described as affecting the human susceptibility to cancer. Most variability may be due to diet, coexisting exposure, medical conditions, dietary consumptions, protective measures and lifestyle behavioral factors [298-300]. Susceptibility markers may be used for predicting human response to carcinogenic exposures such as production of active metabolites, excretion of detoxified products, DNA and protein adduct formation in cellular targets, and DNA repair efficiency [301-303]. Individual differences in susceptibility to chemical carcinogens are one of the most important issues in the risk assessment of human cancers. Although different types of host factors may be involved in the differential susceptibility, genetic differences in the metabolism of carcinogens have been suggested to be associated with different predispositions to cancers. Comparative studies show significant individual and species-to specific variation in the excretion of metabolites of different chemical compounds. Genetic polymorphisms have been established to occur in phase-I as well as in phase-II enzymes of drug metabolism. Genetic variation in P450 metabolism has emerged for pharmaceutical agents, but etiological association between cancer incidence and genetic variations of P450 is controversial. A number of human microsomal P450 species have been characterized and some of their properties have been determined, including chromosomal localization, main organ of P450 expression, and class of carcinogens activated [304,305].

An example of the interindividual variability in pesticide metabolism is the difference in the detoxification of parathion. Paraoxon (diethyl-p-nitrophenyl phosphate) is the active metabolite of parathion, a widely used organophosphorous insecticide. After activation of parathion by the cytochrome P450 system, the paraoxon is hydrolyzed by a serum enzyme, i.e., paraoxonase, to generate p-nitrophenol and diethylphosphoric acid. On the basis of measurement of paraoxonase activity, humans can be divided into three serum paraoxonase phenotypes: A (low activity), AB (intermediate activity), and B (high activity). There is a 10 to 15-fold difference in serum paraoxonase activity between the extremes of low and high activity groups. If such serum paraoxonase is important in the detoxification of paraoxon, individuals with low activity would be expected to have a diminished ability to metabolize paraoxon and therefore to be more sensitive to paraoxon toxicity [306].

A genetic polymorphism recently has been reported for the glutathione-S-transferase (GST) present in human blood [307]. GSTs are able to conjugate glutathione with different chemical compounds, including small molecules such as monohalogenated compounds. The existence of two human subpopulations has been demonstrated: "conjugators" and "nonconjugators." A high enzyme activity was found in conjugators toward the pesticide methyl bromide. The erythrocytes of non-conjugators lack enzyme activity for the substrate. This enzymatic activity could constitute a marker of human susceptibility to different chemical compounds, including pesticides. Pseudocholinesterase variants have been identified that confirm an increased susceptibility to exposure to organophosphate and carbamate insecticides. A significant interindividual variation in mixed function oxidase activation by compounds belonging to the chemical class of dioxins has been found [287]. Differences in human susceptibility to chronic effects induced by TCDD could be attributed to increased enzymatic capabilities. Progress in human genetics is leading to a rapid proliferation in the number of tests that have been introduced for genetic screening in the workplace. Genetic tests may be used to detect those individuals who carry a genetic marker that makes them more susceptible to toxic agents of the workplace. A variety of legal and regulatory measure in the USA have been proposed to prevent genetic discrimination in the workplace [308].

CONCLUSIONS

Biomonitoring studies incorporating individual biomarkers of exposure, effect, and susceptibility may be particularly useful in understanding the continuum of biological events that occur between human exposures to toxic pesticides and related adverse health outcomes, particularly the processes linking exposure to genotoxic pesticides with cancer risk, including dose-response relationships fundamental to establishing causality and identifying subpopulations at risk [167,309,310]. Hopefully, such a comprehensive approach will directly affect public health providing effective preventive measures aimed at significantly reducing human exposure to pesticides.

Acknowledgments

The authors thank Drs. Giorgio Reggiardo, Paola Roggieri, and Melanie Vitello for technical assistance.

REFERENCES

1. Forget, G. "Pesticides and the Third World," *J. Toxicol. Environm. Health(32)* 11 (1991).
2. *The Pesticide Manual.* A world compendium (C.R. Worthing, R.J. Hance, eds.), British Crop Protection Council, Farnham, Surrey, U.K., ninth ed., (1991).
3. Murphy, S.D. "Role of Metabolism in Pesticide Selectivity and Toxicity," *Toxicology of Pesticides: Experimental, Clinical, and Regulatory Perspectives* (L.G. Costa, C.L. Galli, S.D. Murphy, eds.), Springer-Verlag, Berlin, 19 (1987).
4. Shepard, H.H. *The Chemistry and Toxicology of Insecticides.* Burgess Publ. Co., Minneapolis, 383 (1939).
5. Hayes, W.J. "Introduction," *Handbook of Pesticide Toxicology,* Vol. 1, General Principles, chapt. 1 (W.J. Hayes, E.R. Laws, eds.), Academic Press Inc., San Diego, 1 (1991).

6. Ray, D.E. "Pesticides Derived from Plants and other Organisms," *Handbook of Pesticide Toxicology,* Vol. 2, Classes of Pesticides, chapt. 13 (W.J. Hayes, E.R. Laws, eds.), Academic Press Inc., San Diego, 585 (1991).
7. Smith, A.G. "Chlorinated Hydrocarbon Insecticides," *Handbook of Pesticide Toxicology,* Vol. 2, Classes of Pesticides, chapt. 15 (W.J. Hayes, E.R. Laws, eds.), Academic Press Inc., San Diego, 731 (1991).
8. Gallo, M.A., Lawryk, N.J. "Organic Phosphorus Pesticides," *Handbook of Pesticide Toxicology,* Vol. 2, Classes of Pesticides, chapt. 16 (W.J. Hayes, E.R. Laws, eds.), Academic Press Inc., San Diego, 917 (1991).
9. Eells, J.T., Watabe, S., Ogata, N., Narahashi, T. "The Effects of Pyrethroid Insecticides on Synaptic Transmission in Slices of Guinea Pig Olfactory Cortex," *Toxicology of Pesticides: Experimental, Clinical and Regulatory Perspectives,* (L.G. Costa, C.L. Galli and S.D. Murphy, eds.), Springer-Verlag, Berlin, 267 (1987).
10. Davidse, L.C. "Advances in Understanding Fungicidal Modes of Action and Resistance," *Pesticide Science and Biotechnology.* Proceedings of the Sixth International Congress of Pesticide Chemistry, Ottawa, Canada, 169 (1987).
11. Smith, E.A., Oehme, F.W. "A Review of Selected Herbicides and their Toxicities," *Vet. Hum. Toxicol. (33)* 596 (1991).
12. Bailey, G.W., White, J.L. "Herbicides: A Compilation of their Physical, Chemical, and Biological Properties," *Residue Reviews* (10) 97 (1965).
13. Duke, S.O. "Overview of Herbicide Mechanisms of Action," *Environ. Health Perspect. (87)* 263 (1990).
14. Gehring, P.J., Nolan, R.J., Watanabe, P.G., Schumann, A.M. "Solvents, Fumigants, and Related Compounds," *Handbook of Pesticide Toxicology,* Vol. 2, Classes of Pesticides, chapt. 14 (W.J. Hayes, E.R. Laws, eds.), Academic Press Inc., San Diego, 637 (1991).
15. Spear, R. "Recognized and Possible Exposure to Pesticides," *Handbook of Pesticide Toxicology,* Vol. 1, General Principles, chapt. 6 (W.J. Hayes, E.R. Laws, eds.), Academic Press Inc., San Diego, 245 (1991).
16. Coats, J.R. "Toxicology of Pesticide Residues in Foods", *Nutritional Toxicology,* Vol. II, chapt. 10, Academic Press Inc., 249 (1987).
17. Levine, R. "Recognized and Possible Effects of Pesticides in Humans," *Handbook of Pesticide Toxicology,* Vol. 1, General Principles, chapt. 7 (W.J. Hayes, E.R. Laws, eds.), Academic Press Inc., San Diego, 275 (1991).
18. Igbedioh, S.O. "Effects of Agricultural Pesticides on Humans, Animals, and Higher Plants in Developing Countries," *Arch. Environm. Health (46)* 218 (1991).
19. Wright, C.S., Doan, C. A., Haynie, H. C. "Agranulocytosis Occurring after Exposure to a DDT Pyrethrum Aerosol Bomb," *Am. J. Med (1)* 562 (1946).
20. Friberg, L., Masrtensson, J. "Case of Panmyelophthisis after Exposure to Chlorophenothane and Benzine Hexachloride," *Arch. Ind. Hyg. Occup. Med. (8)* 166 9 (1953).

21. Danopoulus, E. Melissinos, K. Katsas, G. "Serious Poisoning by Hexachlorocyclohexane," *AMA Arch. Ind. Hyg. (8)* 582 (1953).
22. Marchand, M. Dubiello, P. Goudemand, M. "Agranulocytosis in a Subject Exposed to Hexachlorocycloexane Vapors," *Arch. Malad. Profess. (17)* 256 (1955).
23. Moore, C. "Exposure to Insecticides, Bone Marrow Failure, Gastrointestinal Bleeding, and Uncontrollable Infections," *Am. J. Med. (10)* 274 (1955).
24. Conley, B.E. "The Present Status of Chlordane," Council on Pharmacy and Chemistry, Committee on Pesticide. *JAMA, J. Am. Med. Assoc. (158)* 1364 (1955).
25. Albahary, C. Dubrisay, J. Guerin, M. "Pancytopenic Reaction to Hexachlorocyclohexane," *Arch. Malad. Profess. (18)* 687 (1957).
26. Scott, J. Cartwright, J.E., Wintrobe, M.M. "Acquired Aplastic Anemia: An Analysis of 39 Cases and Review of the Pertinent Literature," *Medicine (38)* 119 (1959).
27. Gewin, M.H. "Benzene Hydrochloride and Cytoplastic Anemia," *J. Am. Med. Assoc. (171)* 1624 (1959).
28. Bar-Haim, I. Levy, Z. "Acute Pancytopenia, Probably due to Hexalon," *Harufah (59)* 72 (1960).
29. McFarland, W. Granville, N. Schwartz, R. "Therapy of Hypoplastic Anemia with Marrow Transplantation," *Arch. Intern. Med. (108)* 91 (1961).
30. Rankin, A.M. "A Review of 20 Cases of Aplastic Anemia," *Med. J. Aust. (2)* 95 (1961).
31. Hughes, D.W.D. "Acquired Aplastic Anemia in Childhood. A Review of 22 Cases," *Med. J Aust. (2)* 251 (1962).
32. Roberts, H.J. "Aplastic Anemia due to Pentachlorophenol and Tetrachlorophenol," *Souther Med. J. (56)* 632 (1963).
33. Sanchez-Mendal, L., Castenedo, J.P., Garcia-Rojas, F. "Insecticides and Aplastic Anemia," *N. Engl. J. Med. (269)* 1365 (1963).
34. Reeves, H.J. "Aplastic Anemia due to Pentachlorophenol and Tetrachlorophenol," *Souther Med. J. (56)* 632 (1963).
35. Loge, J.P. "Aplastic Anemia Following Exposure to Benzene Hexachloride (Unidane)," *JAMA, J. Am. Med. Assoc. (193)* 110 (1965).
36. West, I. "Lindane and Hematologic Reactions," *Arch. Environ. Health (15)* 97 (1967).
37. Stieglitz, R., Stobbe, H. Schuttmann, W. Hermann, H. Schmidt, V. "On the Pathomechanism of Panmyelopathies Induced by the Insecticide Lindane," *Folia Haematol. (91)* 293 (1969).
38. Muirhead, E.E., Groves, M., Guy, R., Halden, E.R., Bass, R.K. "Acquired Hemolytic Anemia, Exposure to Insecticide Preparations," *Vox. Sang. (4)* 277 (1973).
39. Murray, V., Saveeda, F. M., Navarez, G. M. "Environmental Contamination and the Health of Children," *Salud. Publica. Max. (15)* 91 (1973).

40. Vodopick, H. "Cherchez la Chienne: Erythropoietic Hypoplasia after Exposure to Gammabenzenehydrochloride," *J. Am. Med. Assoc. (234)* 850 (1975).
41. Furie, B., Trubavitz, S. "Insecticides and Blood Dyscrasias: Chlordane Exposure and Self-Limited Refractory Megaloblastic Anemia," *J. Am. Med. Assoc. (235)* 1720 (1976).
42. Jenkyn, I.R., Rudd, R.C., Fein, S.H. "Insecticide/Herbicide Exposure, Aplastic Anemia and Pseudotumor Cerebri," *Lancet (2)* 178 (1979).
43. Morgan, D.P., Stockdale, E.M., Roberts, R.J., Walter, A.W. "Anemia Associated with Exposure to Lindane," *Arch. Environ. Health (35)* 307 (1980).
44. Woodliff, H.F., Connor, P. M., Scopa, J. "Aplastic Anemia Associated with Pesticides," *Med. J. Aust. (1)* 628 (1986).
45. Wang, H.H., Grufferman, S. "Aplastic Anemia and Occupational Pesticide Exposure: A Case-Control Study," *J. Occup. Med. (23)* 364 (1981).
46. Princi, F., Spurbeck, G.H. "A Study of Workers Exposed to the Insecticides Chordane, Aldrin and Dieldrin," *Arch. Ind. Hyg. Occup. Med. (3)* 64 (1951).
47. Alvarez, W.C., Hyman, S. "Absence of Toxic Manifestations in Workers Exposed to Chlordane," *Arch. Ind. Hyg. Occup. Med. (8)* 480 (1953).
48. Fishbein, W.I., White, I.V., Isaacs, H.J. "Survey of Workers Exposed to Chlordane," *Ind. Med. Surg. (33)* 726 (1964).
49. Stein, W.J., Hayes, W.J. "Health Survey of Pest Control Operators," *Ind. Med. Surg. (33)* 549 (1964).
50. Samuels, A.J., Milby, T.H. "Human Exposure to Lindane: Clinical Hematological and Biochemical Effects," *J. Occup. Med. (13)* 147 (1971).
51. Morgan, D.P., Lin, L.I., Saikaly, H.H. "Morbidity and Mortality in Workers Occupationally Exposed to Pesticides," *Arch. Environ. Contam. Toxicol. (9)* 349 (1980).
52. Thomas, D., Stram, D., Dwyer, J. "Exposure Measurements Error: Influence on Exposure-Disease Relationships and Methods of Correction," *Annu. Rev. Publ. Health. (14)* 69 (1993).
53. Copeland, K.T., Checkoway, H., Holbrook, R.H., McMichael, A.J. "Bias due to Misclassification in the Estimate of Relative Risk," *Am. J. Epidemiol. (105)* 5 488 (1977).
54. Gullen, W.H., Bearman, J.E., Johnson, E.A. "Effects of Misclassification in Epidemiologic Studies," *Public Health Rep. (53)* 1956 (1968).
55. Checkoway, H., Pearce, N.E., Crawford-Brown, D.J. *"Research Methods in Occupational Epidemiology,"* Oxford University Press, New York, NY (1989).
56. Blair, A., Hoar Zahm, S., Pearce, N.E., Heineman, E.F., Fraumeny, J.F., Jr, "Clues to Cancer Etiology from Studies of Farmers," *Scand. J. Work Environ. Health (18)* 209– (1993).
57. Verluys, J.J. "Cancer and Occupation in the Netherlands," *Br. J. Cancer (3)* 162 (1949).

58. Guralnick, L. "Mortality by Occupation and Cause of Death," Washington, DC: Department of Health Education and Welfare. (DHEW [vital statistics special rep. no. 53/3]) (1963).
59. Williams, R.R., Stegens, N.L., Goldsmith, J.R. "Associations of Cancer Site and Type with Occupation and Industry from the Third National Cancer Survey Interview.," *J. Natl. Cancer Inst. (59)* 1,147 (1977).
60. Decoufle, P., Stanislawizyk, K., Houten, L., Bross, I.D.J., Viadana, E. "A Retrospective Survey of Cancer in Relation to Occupation," Cincinnati, OH: National Institute for Occupational Safety and Health. (DHEW (NIOSH) publ. no. 77–178) (1977).
61. Petersen, G.R., Milham, S. Jr. "Occupational Mortality in the State of California, 1959–1961," Rockville, MD: National Institute for Occupational Safety and Health (DHEW (NIOSH, NIH) publ. no. 80–104) (1980).
62. Burmeister, L.F. "Cancer Mortality in Iowa Farmers, 1971–1978," *J. Natl. Cancer Inst. (66)* 461 (1981).
63. Statistic Sweden, "Dodsfalls registret 1961–1970," Stockholm: Statistic Sweden (1981).
64. Milham, S. Jr. "Occupational Mortality in Washington State. 1950–1979," Cincinnati, OH. National Institute for Occupational Safety and Health (DHHS (NIOSH) publ.no. 83–116) (1983).
65. Howe, G.R., Lindsay, J.P. "A Followup Study of a 10% Sample of the Canadian Labor Force: I. Cancer Mortality in Males, 1965–1973," *J. Natl. Cancer Inst. (70)* 37 (1983).
66. Gallagher, R.P., Threlfall, W.J., Jeffries, E., Band, P.R., Spoinelli, J., Coldman, A.J. "Cancer and Aplastic Anemia in British Columbia Farmers," *J. Natl. Cancer Inst. (72)* 1311 (1984).
67. Walrath, J., Rogot, E., Murray, J., Blair, A. "Mortality Patterns among U.S. Veterans by Occupation and Smoking Status," Washington, D.C., U.S. Government Printing Office (NHI publ. no. 85-2756) (1985).
68. Delzell, E., Grufferman, S. "Mortality among White and Nonwhite Farmers in North Carolina, 1976–1978," *Am. J. Epidemiol. (121)* 391 (1985).
69. Schwartz, E., Grady, K. "Patterns of Occupational Mortality in New Hampshire, 1975–1985," Concord, N.H.: New Hampshire Division of Public Health Services (1986).
70. Olsen, J., Jensen, O.M. "Occupation and Risk of Cancer in Denmark: An Analysis of 93,810 Cancer Cases, 1970–1979," *Scand. J. Work Environ Health (13)* suppl. 11–91 (1987).
71. Notkola, V.J., Husman, K.R.H., Laukkanen, V.J. "Mortality among Male Farmers in Finland during 1979–1983," *Scand. J. Work. Environ. Health (13)* 124 (1987).
72. Saftlas, A.F., Blair, A., Cantorn, K.P., Haorahan, L., Anderson, H.A. "Cancer and Other Causes of Death among Wisconsin Farmers," *Am. J. Ind. Med. (11)* 119 (1987).

73. Starck, A.D., Chang, H., Fitzgerald, E.F., Riccardi, K., Stone, R.R. "A Retrospective Cohort Study of Mortality among New York State Farm Bureau Members," *Arch. Environ. Health (42)* 204 (1987).
74. Brownson, R.C., Reif, J.S., Chang, J.C. "Cancer Risk among Missouri Farmers," *Cancer (64)* 2,381 (1989).
75. Rafnsson, V., Gunnarsdottir, H. "Mortality among Farmers in Iceland," *Int. J. Epidemiol. (18)* 146 (1989).
76. Reif, J., Pearce, N., Fraser, J. "Cancer Risks in New Zealand Farmers," *Int. J. Epidemiol. (18)* 768 (1989).
77. Wigle, D.T., Semenciw, R.M., Wilkins, K., Riedel, D., Ritter, L., Morrison, H.I., et al., "Mortality Study of Canadian Male Farm Operators: Non-Hodgkin's Lymphoma Mortality and Agricultural Practices in Saskatchewan," *J. Natl. Cancer Inst. (82)* 575 (1990).
78. Louis, T.A., Finenberg, H. V., Mosteller, F. "Findings for Public Health from Meta-Analysis," *Ann. Rev. Public Health (6)* 1 (1985).
79. Greenland, S. "Quantitative Methods in the Review of Epidemiologic Literature," *Epidemiol. Rev. (9)* 1 (1987).
80. Hoar Zahm, S., Blair, A. "Pesticides and Non-Hodgkin's Lymphoma," *Cancer Research Suppl. (52)* 5484 (1992).
81. Davis, D.L., Hoel, D., Fox, J., Lopez, A. "International Trends in Cancer Mortality in France, West Germany, Italy, Japan, England and Wales, and the USA," *Lancet (336)* 474 (1990).
82. Devesa, S.S., Silverman, D.T., Joung, J.L., Pollack, E.S., Brown, C.C., Horn, J.W., et al. "Cancer Incidence and Mortality Trends among Whites in the United States, 1947–84," *J. Natl. Cancer Inst. (79)* 701 (1987).
83. Pickle, L.W., Mason, T.J., Howard, H., Hoover, R.N., Fraumeni, J.F., Jr., "Atlas of US Cancer Mortality among Whites: 1950–1980," US Department of Health and Human Services, DHHS Publication (NIH) 87-2900, Washington, DC, United States Government Printing Office (1987).
84. Wang, H.H., MacMahon, B. "Mortality of Pesticides Applicators," *J. Occup. Med. (21)* 741 (1979).
85. Axelson, O., Sundell, L., Andersson, K., Edling, C., Hogstedt, C., Kling, H. "Herbicide Exposure and Tumor Mortality: An Updated Epidemiologic Investigation on Swedish Railroad Workers," *Scand J. Work. Environ. Health (6)* 73 (1980).
86. Zack, J.A., Suskind, R.R. "The Mortality Experience of Workers Exposed to Tetrachlorodibenzodioxin in a Trichlorophenol Process Accident," *J. Occup. Med. (22)* 1 11 (1980).
87. Barthel, E. "Increased Risk of Lung Cancer in Pesticide-Exposed Male Agricultural Workers," *J. Toxicol. Environ. Health (8)* 1027 (1981).
88. Thiess, A.N., Frentzel-Beyme, R., Link, R. "Mortality Study of Persons Exposed to Dioxin in a Trichlorophenol-Process Accident that occurred in the BASF AG on November 17, 1953," *Amer. J. Ind. Med. (3)* 179 (1982).

89. Riihimaki, V., Asp, S., Hernberg, S. "Mortality of 2,4-Dichlorophenoxyacetic Acid and 2,4,5-Trichlorophenoxyactetic Acid Herbicide Applicators in Finland," *Scand. J. Work Environ. Health (8)* 37 (1982).
90. Blair, A., Grauman, D.J., Lubin, J.H., Fraumeni, J.F. Jr. "Lung Cancer and other Causes of Death among Licensed Pesticide Applicators," *J. Natl. Cancer Inst. (71)* 31 (1983).
91. Lynge, E. "A Followup Study of Cancer Incidence among Workers in Manufacture of Phenoxy Herbicides in Denmark" *Br. J. Cancer (52)* 259 (1985).
92. Merlo, F. "Adverse Health Effects in Human Population Exposed to TCDD in Seveso: An Update," In: *Dioxins in the Environment,* Kamrin, M.A., ed, Hemisphere Publishing Corporation, Washington, DC, 241 (1985).
93. Steineck, G., Wiklund, K., "Multiple Myeloma in Swedish Agricultural Workers," *Int. J. of Epidemiology (15)* 3 321 (1986).
94. Coggon, D., Pannett, B., Winter, P.D., Acheson, E.D., Bonsall, J. "Mortality of Workers Exposed to 2 Methyl-4 Chlorophenoxyacetic acid," *Scand. J. Work Environ. Health (12)* 448 (1986).
95. Kauppinen, T.P., Partanen, T.J., Nurminen, M.M., Nickels, J.I., Hernberg, S.G., Hakulinen, T.R., Pukkala, E.J., Savonen, E.T. "Respiratory Cancers and Chemical Exposures in the Wood Industry: A Nested Case-Control Study," *Br. J. Ind. Med. (43)* 84 (1986).
96. Puntoni, R., Merlo, F., Fini, A., Meazza, L., Santi, L., "Soft Tissue Sarcomas in Seveso," *Lancet (2)* 8505 525 (1986).
97. Wiklund, K., Dich, J., Holm, L.-E. "Risk of Malignant Lymphoma in Swedish Pesticide Appliers," *Br. J. Cancer (56)* 505 (1987).
98. Wiklund, K., Dick, J., Holm, L.-E. "Soft Tissue Sarcoma Risk in Swedish Licensed Pesticide Applicators," *J. of Occupational Medicine (30)* 10 (1987).
99. Bond, G.G., Wetterstroem, N.H., Roush, G.J., McLaren, E.A., Lipps, T.E., Cook, R.R. "Cause-Specific Mortality among Employees Engaged in the Manufacture, Formulation, or Packaging of 2,4-Dichlorophenoxyacetic Acid and Related Salts," *Br. J. Ind. Med. (45)* 98 (1988).
100. MacMahon, B., Monson, R.R., Wang, H.H., Zheng, T. "A Second Followup of Mortality in a Cohort of Pesticide Applicators," *J. Occup. Med. (30)* 429 (1988).
101. Wiklund, K., Dich, J., Holm, LE., Eklund, G. "Risk of Cancer in Pesticide Applicators in Swedish Agriculture," *Br. J. Ind. Med. (46)* 809 (1989).
102. Corrao, G., Calleri, M., Carle, F., Russo, R., Bosia, S., Piccioni, P. "Cancer Risk in a Cohort of Licensed Pesticide Users," *Scand. J. Work Environ. Health (15)* 203 (1989).
103. Bertazzi, P.A., Zocchetti, C., Pesatori, A.C., Guercilena, S., Sanarico, M., Radice, L. "Ten-Year Mortality Study of the Population Involved in the Seveso Incident in 1976," *Am. J. Epid. (129)* 6 1,187 (1989).

104. Bond, G.G., McLaren, E.A., Lipps, T.E., Cook, R.R. "Update of Mortality among Chemical Workers with Potential Exposure to the Higher Chlorinated Dioxins," *J. Occup. Med. (31)* 2 121 (1989).
105. Zober, A., Messerer, P., Huber, P. "Thirty-Four Year Mortality Followup of BASF Employees Exposed to 2,3,7,8-TCDD after the 1953 Accident," *Int. Arch. Occup. Environ. Health (62)* 139 (1990).
106. Alavanja, M.C.R., Blair, A., Masters, M.N. "Cancer Mortality in the U.S. Flour Industry," *J. Natl. Cancer Inst. (82)* 840 (1990).
107. Saracci, R., Kogevinas, M., Bertazzi, P.A., Bueno de Mesquita, B.H., Coggon, D., Green, L.M., Kauppinen, T., L'Abbé, K.A., Littorin, M., Lynge, E., Mathews, J.D., Neuberger, M., Osman, J., Pearce, N., Winkelmann, R. "Cancer Mortality in Workers Exposed to Chlorophenoxy Herbicides and Chlorophenols," *Lancet (338)* 8,774 1027 (1991).
108. Manz, A., Berger, J., Dwyer, J.H., Flesch-Janys, D., Nagel, S., Waltsgott, H. "Cancer Mortality among Workers in Chemical Plant Contaminated with Dioxin," *Lancet (338)* 8,773 959 (1991).
109. Fingerhut, M.A., Halperin, W.E., Marlow, D.A., Piacitelli, L.A., Honchar, P.A., Sweeny, M.H., Greife, A.L., Dill, P.A., Steenland, K., Suruda, A.J., "Cancer Mortality in Workers Exposed to 2,3,7,8-Tetrachlorodibenzo-p-dioxin," *N. Eng. J. Med. (324)* 212–218 (1991).
110. Coggon, D., Pannett, B., Winter, P. "Mortality and Incidence of Cancer at Four Factories making Phenoxy Herbicides," *Br. J. Ind. Med. (48)* 173 (1991).
111. Hansen, E.S., Hastle, H., Lander, F. "A Cohort Study on Cancer Incidence Among Danish Gardeners," *Am. J. Ind. Med.(21)* 651 (1992).
112. Abate, L., Basso, P., Belloni, A., Bisanti, L., Borgna, C., Bruzzi, P., Dorigotti, G., Falliva, L., Fanuzzi, A., Formigaro, M., Maggiore, G., Marni, E., Meazza, L., Merlo, F., Puntoni, R., Rosa, A., Stagnaro, E., Vercelli, M., Santi, L. "Mortality and Birth Defects from 1976 to 1979 in the Population Living in the TCDD Polluted Area of Seveso," *Chlorinated Dioxins and Related Compounds,* O. Hutzinger et al, eds. Pergamon Press Oxford and New York (1982).
113. Bertazzi, P.A., Zocchetti, C., Pesatori, A.C., Guercilena, S., Consonni, D., Tironi, A., Landi, M.T. "Mortality of a Young Population after Accidental Exposure to 2,3,7,8-Tetrachlorodibenzodioxin," *Int. J. Epidemiology (21)* 1 (1992).
114. Semenciw, R.M., Morrison, H.I., Riedel, D., Wilkins, K., Ritter, L., Mao, Y. "Multiple Myeloma Mortality and Agricultural Practices in the Prairie Provinces of Canada," *J.O.M. (35)* 6 557 (1993).
115. Sharp, D.S., Eskenazi, B. "Delayed Health Hazards of Pesticide Exposure," *Ann. Rev. Public Health (7)* 441 (1986).
116. Sterling, T.D., Arundel, A.V. "Health Effects of Phenoxy Herbicides: A Review," *Scand. J. Work Environ. Health (12)* 161 (1986).

117. Lilienfeld, D.E., Gallo, M.A. "2,4-D, 2,4,5-T, and 2,3,7,8-TCDD: An Overview," *Epidemiologic Reviews (11)* 28 (1989).
118. Johnson, E.S., "Association Between Soft Tissue Sarcomas, Malignant Lymphomas, and Phenoxy Herbicides/Chlorophenols: Evidence from Occupational Cohort Studies," *Fundamental Applied Toxicology (14)* 219 (1990).
119. Johnson, C.C., Feingold, M., Tilley, B. "A Meta Analysis of Exposure to Phenoxy Herbicides and Chlorophenols in Relation to Risk of Soft Tissue Sarcoma," *Int. Arch. Occup. Environ. Health (62)* 513 (1990).
120. Bailar, J.C. III, "How Dangerous is Dioxin?," *N. Engl. J. Med. (324)* 260 (1990).
121. Hardell, L., Eriksson, M., Axelson, O., Fredriksson, M. "Dioxin and Mortality from Cancer," *N. Engl. J. Med. (324)* 1810 (1991).
122. Johnson, E.S. "Important Aspects of the Evidence for TCDD Carcinogenicity in Man," *Environ. Health Perspective (99)* 383 (1993).
123. Lucier, G.W. "Humans are a Sensitive Species to Some of the Biochemical Effects of Structural Analogs of Dioxin," *Environ. Toxicol. Chem. (10)* 727 (1991).
124. Hardell, L., Sandstrom, A. "A Case-Control Study: Soft-Tissue Sarcomas and Exposure to Phenoxy-Acetic Acids or Chlorophenols," *Br. J. Cancer (39)* 711 (1979).
125. Hardell, L., Eriksson, M., Lenner, P., Lundgren E. "Malignant Lymphoma and Exposure to Chemical Substance, Especially Organic Solvents, Chlorophenols, and Phenoxy Acids: A Case Control Study," *Br. J. Cancer (43)* 169 (1981).
126. Eriksson, M., Hardell, L., Berg, N.O., Mollen, T., Axelson, O. "Soft-Tissue Sarcomas and Exposure to Chemical Substance: A Case Referent Study," *Br. J. In. Med. (38)* 27 (1981).
127. Hardell, L., Johansson, B., Axelson, O. "Epidemiological Study of Nasal and Nasopharyngeal Cancer and their Relation to Phenoxy Acid or Chlorophenol Exposure," *Am. J. Ind. Med. (3)* 247 (1982).
128. Cantor, K.P. "Farming and Mortality from Non-Hodgkin's Lymphoma: A Case-Control Study," *Int. J. Cancer (29)* 239 (1982).
129. Cantor, K.P., Blair, A. "Farming and Mortality from Multiple Myeloma: A Case-control Study with the Use of Death Certificated," *JNCI (72)* 251 (1984).
130. Smith, J.G., Pearce, N.E., Fisher, D.O., Giles, H.J., Teaghe, C.A., Howard, J.K. "Soft Tissue Sarcoma and Exposure to Phenoxyherbicides and Chlorophenols in New Zealand," *J. Natl. Cancer Inst. (73)* 1,111 (1984).
131. Pearce, N., Smith, A.H., Fisher, D.O. "Malignant Lymphoma and Multiple Myeloma Linked with Agricultural Occupations in New-Zealand Cancer Registry-Based Study," *Am. J. Epidemiol. (121)* 225 (1985).
132. Morris, P.D., Koepsell, T.D., Daling, J.R., Taylor, J.W., Lyon, J. L., Swanson, G.M., Child, M., Weiss, N.S. "Toxic Substance Exposure and Multiple Myeloma: A Case-Control Study," *J. Natl. Cancer Inst. (76)* 987 (1986).

133. Hoar, S.K., Blair, A., Holmes, F.F., Boysen, C.D., Robel, R.J., Hoover, R., Fraumeni, J.J., Jr. "Agricultural Herbicides Use and Risk of Lymphoma and Soft Tissue Sarcoma," *J. Am. Med. Ass. (256)* 1,141 (1986).
134. Woods, J.S., Polissar, L., Severson, R.K., Heuser, L.S., Kulander, B.G. "Soft Tissue Sarcoma and Non-Hodgkin's Lymphoma in Relation to Phenoxy Herbicide and Chlorinated Phenol Exposure in Western Washington," *J. Natl. Cancer Inst. (78)* 899 (1987).
135. Vineis, P., Terracini, B., Ciccone, G., Cignetti, A., Colombo, E., Donna, A., Pisa, R., Ricci, P., Zanini, E., Comba, P. "Phenoxy Herbicides and Soft-Tissue Sarcomas in Female Rice Weeders: A Population-Based Case-Referent Study," *Scand. J. Work Environ. Health (13)* 9 (1987).
136. Lowengart, R.A., Peters, J.M., Cicioni, C., Buckley, J., Bernstein, L., Presto-Martin, S., Rappaport, E. "Childhood Leukemia and Parents' Occupational and Home Exposures," *J. Natl. Cancer Inst. (79)* 39 (1987).
137. Musicco, M., Sant, M., Molinari, S., Filippini, G., Gatta, G., Berrino, F. "A Case-Control Study of Brain Gliomas and Occupational Exposure to Chemical Carcinogens: The Risk to Farmers," *Am. J. Epidemiol. (128)* 778 (1988).
138. Shu, X.O., Gao, Y.T., Brinton, L.A., Linet, M.S., Tu, J.T., Zheng, W., Fraumeni, J.F. Jr. "A Population-Based Case-Control Study of Childhood Leukemia in Shanghai," *Cancer (62)* 635 (1988).
139. Hardell, L., Eriksson, M. "The Association Between Soft Tissue Sarcomas and Exposure to Phenoxy Acids: A New Case-Referent Study," *Cancer (62)* 652 (1988).
140. Pearce, N.F. "Phenoxy Herbicides and Non-Hodgkin's Lymphoma in New Zealand: Frequency and Duration of Herbicide Use," *Br. J. Ind. Med. (46)* 143 (1989).
141. Boffetta, P., Stellman, S.D., Garfinkel, L. "A Case-Control Study of Multiple Myeloma Nested in the American Cancer Society Prospective Study," *Int. J. Cancer (43)* 554 (1989).
142. Reif, J., Pearce, N., Fraser, J. "Cancer Risks in New Zealand Farmers," *Int. J. Epidemiol. (18)* 768 (1989).
143. Donna, A., Betta, P.B., Robutti, F., Crosignani, P., Berrino, F., Bellingeri, D., "Ovarian Mesothelial Tumors and Herbicides: A Case-Control Study," *Carcinogenesis (5)* 7 941 (1984).
144. Reif, J., Pearce, N., Fraser, J. "Occupational Risks for Brain Cancer: A New Zealand Cancer Registry-Based Study," *J. Occup. Med. (31)* 863 (1989).
145. LaVecchia, C., Negri, E., D'Avanzo, B., Franceschi, S. "Occupation and Lymphoid Neoplasms," *Br. J. Cancer (60)* 385 (1989).
146. Donna, A., Crosignani, P., Robutti, F., Betta, P.G., Bocca, R., Mariani, N., Ferrario, F., Fissi, R., Berrino, F. "Triazine Herbicides and Ovarian Epithelian Neoplasms," *Scand. J. Work Health (15)* 47 (1989).

147. Hoar Zahm, S., Blair, A., Homes, F.F., Boysen, C.D., Robel, R.J., Fraumeni, J.F. Jr. "A Case-Control Study of Soft-Tissue Sarcoma," *Am. J. Epidemiology (130)* 4 665 (1989).
148. Brownson, R.C., Reif, J.S., Chang, J.C., Davis, J. R. "Cancer Risks Among Missouri Farmers," *Cancer (64)* 2381 (1989).
149. Buckley, J.D., Robinson, L.L., Swotinsky, R., Garabrant, D.H., LeBeau, M., Manchester, P., Nesbit, M.E., Odom, L., Peters, J.M., Woods, W.G., Denman Hammond, G. "Occupational Exposures of Parents of Children with Acute Nonlympocytic Leukemia: A Report from the Childrens Cancer Study Group," *Cancer Research (49)* 4030 (1989).
150. Wingren, G., Fredrikson, M., Brage, H.N., Nordenskjold, B., Axelson, O. "Soft Tissue Sarcoma and Occupational Exposures," *Cancer (66)* 806 (1990).
151. Eriksson, M., Hardell, L., Adami, H.O. "Exposure to Dioxins as a Risk Factor for Soft Tissue Sarcoma: A Population-Based Case-Control Study," *J. Natl. Cancer Inst. (82)* 6 486 (1990).
152. Pasqualetti, P., Casale, R., Collacciani, A., Colantonio, D. "Attività Lavorativa e Rischio de Mieloma Multiplo: Uno Studio Caso-Controllo," *La Medicina del Lavoro (81)* 4 308 (1990).
153. Morris Brown, L., Blair, A., Gibson, R., Everett, G.D., Cantor, K.P., Schuman, L.M., Burmeister, L.F., Van Lier, S.F., Dick, F. "Pesticide Exposures and Other Agricultural Risk Factors for Leukemia among Men in Iowa and Minnesota," *Cancer Research (50)* 6585 (1990).
154. Hoar Zahm, S., Weisenburger, D.D., Babbitt, P.A., Saal, R.C., Vaught, J.B., Cantor, K.P., Blair, A.A. "A Case-Control Study of Non-Hodgkin's Lymphoma and the Herbicide 2,4-Dichlorophenoxyacetic Acid (2,4-D) in Eastern Nebraska," *Epidemiology (1)* 349 (1990).
155. Cantor, K.P., Blair, A., Everett, G., Gibson, R., Burmeister, L.F., Brown, L.M., Shuman, L., Dick, F.R. "Pesticides and other Agricultural Risk Factors for Non-Hodgkin's Lymphoma among Men in Iowa and Minnesota," *Cancer Research (50)* 2447 (1992).
156. Smith, J.G., Christophers, A.J. "Phenoxy Herbicides and Chlorophenols: A Case Control Study on Soft Tissue Sarcomas and Malignant Lymphoma," *Br. J. Cancer (65)* 442 (1992).
157. Scherr, P.A., Hutchison, G.B., Neiman, R.S. "Non-Hodgkin's Lymphoma and Occupational Exposure," *Cancer Research* (suppl.) 52 5503 (1992).
158. Holly, E.A., Aston, D.A., Ahn, D.K., Kristiansen, J.J. "Ewing's Bone Sarcoma, Paternal Occupational Exposure, and other Factors," *Am. J. Epidemiol. (135)* 122 (1992).
159. Richardson, S., Zittoun, R., Garin, S.B., Lasserre, V., Guihenneuc, C., Cadiou, M., Viguie, F., Faust, I.L. "Occupational Risk Factors for Acute Leukemia: A Case-Control Study," *Int. J. Epidemiol. (21)* 6 1063 (1992).

160. Franceschi, S., Barone, F., Bidoli, E., Guarnieri, S., Serraino, D., Talamini, R., LaVecchia, C. "Cancer Risk in Farmers: Results From a Multi-Site Case-Control Study in North-Eastern Italy," *Int. J. Cancer (53)* 740 (1993).
161. Hoar Zahm, S., Weisenburger, D.D., Cantor, P.K., Holmes, F.F., Blair, A. "Role of the Herbicide Atrazine in the Development of Non-Hodgkin's Lymphoma," *Scand. J. Work Environ. Health (19)* 2 108 (1993).
162. Davis, J.R., Browson, R.C., Garcia, R., Bentz, B.J., Turner, A. "Family Pesticide Use and Childhood Brain Cancer," *Arch. Environ. Contam. Toxicol. (24)* 87 (1993).
163. Hardell, L. "Phenoxy Herbicides, Chlorophenols, Soft-Tissue Sarcomas (STS) and Malignant Lymphoma," *Br. J. Cancer (67)* 1,154 (1993).
164. Smith, J.G., Christophers, A.J. "Replay to the Letter from Professor Hardell," *Br. J. Cancer (67)* 1,156 (1993).
165. GU n. L 230, 19. 8. 1991 (dir. 91/414/CEE).
166. Dearfield, K.L., Quest, J.A., Whiting, R.J., Stack, H.F., Waters, M.D. "Characteristics of the U.S. EPA's Office of Pesticide Programs' Toxicity Information Databases," *Environm. Health Perspect. (96)* 53 (1991).
167. IARC, Monographs on the Evaluation of Carcinogenic Risk to Humans. "Occupational Exposures in Insecticide Application, and some Pesticides." Vol. 53 (1991).
168. U.S. Environmental Protection Agency (EPA), "Guidelines for carcinogenic risk assessment." Federal Register 51 33992 (1986).
169. U.S. Environmental Protection Agency (EPA), "Guidelines for mutagenicity risk assessment." Federal Register 51 34006 (1986).
170. IARC, Monographs on the Evaluation of Carcinogenic Risk to Humans. "Some Organochlorine Pesticides," Aramite. Vol. 5 39 (1974).
171. IARC, Monographs on the Evaluation of Carcinogenic Risk to Humans. "Occupational Exposures in Insecticide Application, and some Pesticides," Chlordane and Heptachlor. Vol. 53 115 (1991).
172. NCI, National Cancer Institute, "Bioassay of Chlordane for Possible Carcinogenicity," Carcinogenesis Tech. Rep. Ser. No 8, DHEW Publ. No. (NOH) 77-808. U.S. Govt. Printing Office, Washington, D.C. (1977).
173. IARC, Monographs on the Evaluation of Carcinogenic Risk to Humans. "Some Halogenated Hydrocarbons," Chlordecone. Vol. 20 67 (1979).
174. NCI, National Cancer Institute, "Report on Carcinogenesis Bioassay of Technical Grade Chlordecone (Kepone)." U.S. Govt. Printing Office, Washington, D.C. (1976).
175. IARC, Monographs on the Evaluation of Carcinogenic Risk to Humans. "Occupational Exposures in Insecticide Application, and Some Pesticides," DDT. Vol. 53 179 (1991).
176. IARC, Monographs on the Evaluation of Carcinogenic Risk to Humans. "Occupational Exposures in Insecticide Application, and Some Pesticides," Dichlorvos. Vol. 53 267 (1991).

177. IARC, Monographs on the Evaluation of Carcinogenic Risk to Humans. "Overall Evaluations of Carcinogenicity: An Updating of IARC Monographs." Vols. 1-42. Dimethylcarbamoyl chloride, Suppl.7 199 (1987).
178. IARC, Monographs on the Evaluation of Carcinogenic Risk to Humans. "Overall Evaluations of Carcinogenicity: An Updating of IARC Monographs." Vols. 1-42. Hexachlorocyclohexanes, Suppl.7 220 (1987).
179. NCI, National Cancer Institute, "Bioassay of Heptachlor for Possible Carcinogenicity," Carcinogenesis Tech. Rep. Ser. No. 9, DHEW Publ. No. (NIH) 77-809. U.S. Govt. Printing Office, Washington, D.C. (1977).
180. IARC, Monographs on the Evaluation of Carcinogenic Risk to Humans. "Some Halogenated Hydrocarbons," Mirex. Vol. 20 283 (1979).
181. IARC, Monographs on the Evaluation of Carcinogenic Risk to Humans. "Some Halogenated Hydrocarbons," Toxaphene. Vol. 20 327 (1979).
182. IARC, Monographs on the Evaluation of Carcinogenic Risk to Humans. "Occupational Exposures in Insecticide Application, and Some Pesticides," Captafol. Vol. 53 353 (1991).
183. IARC, Monographs on the Evaluation of Carcinogenic Risk to Humans. "Occupational Exposures in Insecticide Application, and Some Pesticides, Pentachlorophenol." Vol. 53 371. (1991).
184. IARC, Monographs on the Evaluation of Carcinogenic Risk to Humans. "Some Halogenated Hydrocarbons," 2,4,6-Trichlorophenols. Vol. 20 349. (1979).
185. NCI, National Cancer Institute, "Bioassay of 2,4,6- Trichlorophenol for Possible Carcinogenicity," Carcinogenesis Tech. Rep. No. 155, DHEW Publ. No. (NIH) 79-1711. U.S. Govt. Printing Office, Washington, D.C. (1979).
186. IARC, Monographs on the Evaluation of Carcinogenic Risk to Humans. "Overall Evaluations of Carcinogenicity: An Updating of IARC Monographs." Vols. 1-42. Ethylene Thiourea, Suppl.7 207 (1987).
187. IARC, Monographs on the Evaluation of Carcinogenic Risk to Humans. "Overall Evaluations of Carcinogenicity: An Updating of IARC Monographs." Vols. 1-42. Hexachlorobenzene, Suppl.7 219 (1987).
188. IARC, Monographs on the Evaluation of Carcinogenic Risk to Humans. "Overall Evaluations of Carcinogenicity: An Updating of IARC Monographs." Vols. 1-42. Sodium ortho- phenylphenate, Suppl.7 392 (1987).
189. IARC, Monographs on the Evaluation of Carcinogenic Risk to Humans. "Overall Evaluations of Carcinogenicity: An Updating of IARC Monographs." Vols. 1-42. Amitrole, Suppl.7 92 (1987).
190. IARC, Monographs on the Evaluation of Carcinogenic Risk to Humans. "Miscellaneous Pesticides," Nitrofen (technical-grade). Vol. 30 271 (1983).
191. IARC, Monographs on the Evaluation of Carcinogenic Risk to Humans. "Miscellaneous Pesticides," Sulfallate. Vol. 30 283 (1983).

192. NCI, National Cancer Institute, "Bioassay of Sulfallate for Possible Carcinogenicity", Carcinogenesis Tech. Rep. Ser. No. 115, DHEW Publ. No. (NIH)78-1370. U.S. Govt. Printing Office, Washington, D.C., (1978).
193. IARC, Monographs on the Evaluation of Carcinogenic Risk to Humans. "Overall Evaluations of Carcinogenicity: An Updating of IARC Monographs." Vols. 1-42. 1,2-Dibromo-3- Chloropropane, Suppl.7 191 (1987).
194. NTP, National Toxicology Program, "Carcinogenesis Bioassay of 1,2-Dibromo-3-chloropropane (CAS No. 96-12-8) in F344 Rats and B6C3F1 Mice (Inhalation Study," Tech. Rep. Ser. No. 206, DHHS (NIH) Publ. No. 82-1762. U.S. Govt. Printing Office, Washington, D.C., (1982).
195. IARC, Monographs on the Evaluation of Carcinogenic Risk to Humans. "Some Halogenated Hydrocarbons," 1,2-Dichloroethane. Vol. 20 429. (1979).
196. NCI, National Cancer Institute, "Bioassay of 1,2-Dichloroethane for Possible Carcinogenicity," Carcinogenesis Tech. Rep. Ser. No. 55, DHEW Publ. No. (NIH) 78-1361. U.S. Govt. Printing Office, Washington, D.C., (1978).
197. IARC, Monographs on the Evaluation of Carcinogenic Risk to Humans. "Some Halogenated Hydrocarbons and Pesticide Exposures," 1,3-Dichloropropene (technical grade), Vol. 41 113. (1986).
198. NTP, National Toxicology Program, "Toxicology and Carcinogenesis Studies of Telone II (Technical Grade 1,3- Dichloropropane) (CAS 542.75.6) Containing 1.0% Epichlorohydrin as a Stabilizer in F344/N Rats and B6C3F1 Mice (Gavage Studies)," Tech. Rep. Ser. No. 269, DHHS (NIH) Publ. No. 85-2525. U.S. Govt. Printing Office, Washington, D.C., (1985).
199. IARC, Monographs on the Evaluation of Carcinogenic Risk to Humans. "Overall Evaluations of Carcinogenicity: An Updating of IARC Monographs." Vols. 1-42. Ethylene Dibromide, Suppl.7 204 (1987).
200. IARC, Monographs on the Evaluation of Carcinogenic Risk to Humans. "Overall Evaluations of Carcinogenicity: An Updating of IARC Monographs." Vols. 1-42. Tetrachloroethylene, Suppl.7 355 (1987).
201. IARC, Monographs on the Evaluation of Carcinogenic Risk to Humans. "Overall Evaluation of Carcinogenicity: An Update of IARC Monographs." Vols. 1-42. Aldrin, Suppl.7 88 (1987).
202. IARC, Monographs on the Evaluation of Carcinogenic Risk to Humans. "Overall Evaluation of Carcinogenicity: An Update of IARC Monographs." Vols. 1-42. Arsenic and Arsenic compounds, Suppl.7 100 (1987).
203. IARC, Monographs on the Evaluation of Carcinogenic Risk of Chemicals to Humans. "Miscellaneous Pesticides": Chlorobenzilate, Vol. 30 73 (1983).
204. IARC, Monographs on the Evaluation of Carcinogenic Risk of Chemicals to Humans. "Miscellaneous Pesticides": Dicofol, Vol. 30 87 (1983).
205. IARC, Monographs on the Evaluation of Carcinogenic Risk to Humans. "Overall Evaluation of Carcinogenicity: An Update of IARC Monographs." Vols. 1-42. Dieldrin, Suppl.7 196 (1987).

206. IARC, Monographs on the Evaluation of Carcinogenic Risk to Humans. "Overall Evaluation of Carcinogenicity: An Update of IARC Monographs." Vols. 1-42. Hexachlorocyclohexanes, Suppl.7 220 (1987).
207. IARC, Monographs on the Evaluation of Carcinogenic Risk of Chemicals to Humans. "Some Organochlorine Pesticides": Terpene Polychlorinates, Vol. 5 219 (1974).
208. IARC, Monographs on the Evaluation of Carcinogenic Risk of Chemicals to Humans. "Miscellaneous Pesticides": Tetrachlorvinphos, Vol. 30 197 (1983).
209. IARC, Monographs on the Evaluation of Carcinogenic Risk of Chemicals to Humans. "Miscellaneous Pesticides": Captan, Vol. 30 295 (1983).
210. IARC, Monographs on the Evaluation of Carcinogenic Risk of Chemicals to Humans. "Miscellaneous Pesticides": Chlorothalonil, Vol. 30 319 (1983).
211. IARC, Monographs on the Evaluation of Carcinogenic Risk of Chemicals to Humans. "Some Organochlorine Pesticides": Quintozene, Vol. 5 211 (1974).
212. IARC, Monographs on the Evaluation of Carcinogenic Risk to Humans. "Occupational Exposures in Insecticide Application, and some pesticides," Ziram, Vol. 53 423 (1991).
213. IARC, Monographs on the Evaluation of Carcinogenic Risk to Humans. "Occupational Exposures in Insecticide Application, and some pesticides," Atrazine. Vol. 53 441 (1991).
214. IARC, Monographs on the Evaluation of Carcinogenic Risk of Chemicals to Humans. "Miscellaneous Pesticides": Diallate, Vol. 30 235 (1983).
215. IARC, Monographs on the Evaluation of Carcinogenic Risk to Humans. "Occupational Exposures in Insecticide Application, and some pesticides," Monuron, Vol. 53 467 (1991).
216. IARC, Monographs on the Evaluation of Carcinogenic Risk to Humans. "Occupational Exposures in Insecticide Application, and some pesticides," Picloram. Vol. 53 481 (1991).
217. IARC, Monographs on the Evaluation of Carcinogenic Risk to Humans. "Occupational Exposures in Insecticide Application, and some Pesticides," Trifluralin, Vol. 53 515 (1991).
218. NCI, National Cancer Institute, "Carcinogenesis Bioassay of Trichloroethylene," Carcinogenesis Tech. Rep. Ser. No. 2, DHEW Publ. No. (NIH) 76-802. U.S. Govt. Printing Office, Washington, D.C. (1976).
219. NTP, National Toxicology Program, "Toxicology and Carcinogenesis studies of Trichloroethylene (CAS No. 79.10.6) in four strains of rats (ACI, August, Marshall, Osborne-Mendel) (Gavage studies), NTP Tech. Rep. Ser. No. 273, DHHS Publ. No. (NIH) 88-2529, U.S. Govt. Printing Office, Washington, D.C. (1988).
220. Murphy, R.S., Kutz, F.W., Strassman, S.C. "Selected pesticides residues or metabolites in blood and urine specimens from a general population survey," *Environm. Health Perspect. (48)* 81 (1983).

221. Kutz, F.W., Cook, B.T., Carter-Pokras, O.D., Brody, D., Murphy, R.S., "Selected Pesticides Residues and Metabolites in Urine from a Survey of the U.S. General Population," *J. Toxicol. Environm. Health (37)* 277 (1992).
222. Tennant, R.W., Zeiger, E "Genetic Toxicology: Current Status of Methods of Carcinogen Identification," *Environm. Health Perspect. (100)* 307 (1993).
223. Garrett, N.E., Stack, H.F., Waters, M.D. "Evaluation of the Genetic Activity Profiles of 65 Pesticides," *Mutat. Res. (168)* 301 (1986).
224. Fowle, J.R., Sexton, K. "EPA Priorities for Biologic Markers Research in Environmental Health," *Environm. Health Perspect. (98)* 235 (1992).
225. EPA, Environmental Protection Agency, "Manual of Analytical Methods for the Analysis of Pesticides in Human and Environmental Samples," EPA-600/8-80-038, (1980).
226. Coye, M.J., Lowe, J.A., Maddy, K.J. "Biological Monitoring of Agricultural Workers Exposed to Pesticides: II. Monitoring of Intact Pesticides and their Metabolites," *J. Occup. Med. (28)* 628 (1986).
227. Vasilic, Z., Drevenkar, V., Frobe, Z., Stengl, B., Tkalcevic, B. "The Metabolites of Organophosphorus Pesticides in Urine as an Indicator of Occupational Exposure," *Toxicol. Environm. Chem. (14)* 111 (1987).
228. Tafuri, J., Roberts, J. "Organophosphate Poisoning," *Ann. Emerg. Med. (16)* 193 (1987).
229. Shafik, M.T., Sallivan, H.C., Enos, H.F. "A Method for the Determination of 1-Naphthol in Urine," *Bull. Environ. Contam. Toxicol. (6)* 34 (1971).
230. DeBernardinis, M., Wargin, W.A. "High-Performance Liquid Chromatographic Determination of Carbaryl and 1-Naphthol in Biological Fluids," *J. Chromatogr (246)* 89 (1982).
231. Edwards, I.R., Ferry, D.H., Temple, W.A. "Fungicides and Related Compounds," *Handbook of Pesticide Toxicology,* Vol. 3, Classes of Pesticides, chapt. 21 (W.J. Hayes, E.R. Laws, eds.), Academic Press Inc., San Diego, 1409 (1991).
232. Draper, W.M., "A Multiresidue Procedure for the Determination and Confirmation of Acidic Herbicide Residues in Human Urine," *J. Agric. Food Chem. (30)* 227 (1982).
233. Vanderlaan, M., Watkins, B.E., Stanker, L. "Environmental Monitoring by Immunoassay," *Environ. Sci. Technol. (22)* 247 (1988).
234. Mumma, R.O., Brady, J.F. "Immunological Assay for Agrochemicals," *Pesticides Science and Biotechnology,* London, 341 (1987).
235. Coye, M., Lowe, J., Maddy, K. "Biological Monitoring of Agricultural Workers Exposed to Pesticides: I. Cholinesterase Activity Determinations," *J. Occup. Med. (28)* 619 (1986).
236. Alexiou, N.G., Williams, J.F., Yeung, H.W., Husting, E.L. "Paradoxical Elevation of Plasma Cholinesterase," *Am. J. Prev. Med. (2)* 235 (1986)
237. Yeary, R.A., Eaton, J., Gilmor, E., North, B., Singell, J. "A Multiyear Study of Blood Cholinesterase Activity in Urban Pesticide Applicators," *J. Toxicol. Environm. Health (39)* 11 (1993).

238. Santella, R.M. "DNA Adducts in Humans as Biomarkers of Exposure to Environmental and Occupational Carcinogens," *Environ. Carcinog. & Ecotox. Revs. (9)* 57 (1991).
239. Wogan, G.N. "Markers of Exposure to Carcinogens: Methods for Human Biomonitoring," *J. Amer. College Toxicol.(8)* 871 (1989).
240. Roberts, D.W., Wayne Benson, R., Hinson, J.A., Kadlubar, F.F. "Critical Considerations in the Immunochemical Detection and Quantitation of Antigenic Biomarkers," *Biomed. Environm. Sci. (4)* 113 (1991).
241. Beach, A.C., Gupta, R.C. "Human Biomonitoring and the 32P-Postlabeling Assay," *Carcinogenesis, (13)* 1053 (1992).
242. Gorelick, N.J. "Application of HPLC in the 32P-Postlabeling Assay," *Mutat. Res. (288)* 5 (1993).
243. Weston, A. "Physical Methods for the Detection of Carcinogen-DNA Adducts in humans," *Mutat. Res. 19* (288) (1993).
244. Poirier, M.C. "Antisera Specific for Carcinogen DNA Adducts and Carcinogen-Modified DNA: Applications for Detection of Xenobiotics in Biological Samples," *Mutat. Res. (288)*31 (1993).
245. Pfeifer, G.P., Drouin, R., Homquist, G.P. "Detection of DNA Adducts at the DNA Sequences Level by Ligation-Mediated PCR," *Mutat. Res. (288)* 39 (1993).
246. Shuker, D.E.G., Bailey, E., Parry, A., Lamb, J., Farmer, P.B. "Determination of Urinary 3-Methyladenine in Humans as a Potential Monitor of Exposure to Methylating Agents," *Carcinogenesis (8)* 959 (1987).
247. Farmer, P.B., Bailey, E., Gorf, S.M., Tornqvist, M., Osterman-Golkar, S., Kautiainen, A., Lewis-Enright, D.P. "Monitoring Human Exposure to Ethylene Oxide by the Determination of Hemoglobin Adducts Using Gas Chromatography-Mass Spectrometry," *Carcinogenesis, (7)* 637 (1986).
248. Sabbioni, G., Newmann, H.-G. "Biomonitoring of Arylamines: Hemoglobin Adducts of Urea and Carbamate Pesticides," *Carcinogenesis (11)* 111 (1990).
249. Lohman, P.H.M., Morolli, B., Darroudi, F., Natarajan, A.T., Gossen, J.A., Venema, J., Mullenders, L.H.F., Vogel, E.W., Vrieling, H., van Zeeland, A.A. "Contribution from Molecular/Biochemical Approaches in Epidemiology to Cancer Risk Assessment and Prevention," *Environm. Health Perspect. (98)* 155 (1992).
250. Brandt-Rauf, P.W. "Advances in Cancer Biomarkers as Applied to Chemical Exposures: The Ras Oncogene and p21 Protein and Pulmonary Carcinogenesis," *J. Occup. Med. (33)* 951 (1991).
251. Wolff, S. "Biological Dosimetry with Cytogenetic Endpoints," *Prog. Clin. Biol. Res. (372)* 351 (1991).
252. Fenech, M., Morley, A.A. "Kinetochore Detection in Micronuclei: An Alternative Method for Measuring Chromosome Loss," *Mutagenesis (4)* 98 (1989).
253. Thierens, H., Vral, A., de Ridder, L. "Biological Dosimetry using the Micronucleus Assay for Lymphocytes: Interindividual Differences in Dose Response," *Health Physics (61)* 623 (1991).

254. Tucker, J.D., Auletta, A., Cimino, M.C., Dearfield, K.L., Jacobson-Kram, D., Tice, R.R., Carrano, A.V. "Sister-Chromatid Exchanges: Second Report of the GeneTox Program," *Mutat Res. (297)* 101 (1993).
255. Sorsa, M., Salomaa, S., Ojajarvi, A. "Human Biomonitoring for Somatic Chromosome Damage," Human Carcinogen Exposure. *Biomonitoring and Risk Assessment* (R. Colin Garner, P.B. Farmer, G.T. Steel, A.S. Wright, eds), IRL Press, Oxford University, 85 (1991).
256. Carrano, A.V., Natarajan, A.T. "Consideration for Population Monitoring Using Cytogenetic Techniques," *Mutat Res. (204)* 379 (1983).
257. Barrett, J.C. "Mechanism of Multistep Carcinogenesis and Carcinogen Risk Assessment," *Environm. Health Perspect. (100)* 9 (1993).
258. Yoder, J., Watson, M., Benson, W.W. "Lymphocyte Chromosome Analysis of Agricultural Workers during Extensive Occupational Exposure to Pesticides," *Mutat. Res. (21)* 335 (1973).
259. Van Bao, T., Szabo, I., Ruzicska, P., Czeizel, A. "Chromosome Aberrations in Patients Suffering Acute Organic Phosphate Insecticide Intoxication," *Humangenetik (24)* 33 (1974).
260. Rabello, M.N., Becak, W., De Almeida, W.F., Pigati, P., Ungaro, M.T., Murata, T., Pereira, C.A.B. "Cytogenetic Study on Individuals Occupationally Exposed to DDT," *Mutat. Res. (28)* 449 (1975).
261. Czeizel, A., Kiraly, J., Ruzicska, P. "Studies on Chromosomal Mutations in Workers Producing Organophosphate Insecticides," *Mutat. Res. (29)* 279 (1975).
262. Czeizel, E., Kiraly, J. "Chromosome Examinations in Workers Producing Klorinal and Buvinol," *The Development of a Pesticide as a Complex Scientific Task* (L. Banki, ed.), Medicina Press, Budapest, (1976).
263. Pilinskaya, M.A. "Results of Cytogenetic Examination of Persons Occupationally Contacting with the Fungicide Zineb," *Genetika (10)* 140 (1974).
264. Pilinskaya, M.A., et al., "The Significance of Cytogenetic Observations on Occupational Populations for the Genetic and Hygienic Evaluation of Pesticides," *Mutat. Res. (97)* 211 (1982).
265. Crossen, P.E., Morgan, W.F., Horan, J.J., Stewart, J. "Cytogenetic Studies of Pesticide and Herbicide Sprayers," *New Zealand Med. J.,* September 13, 192 (1978).
266. Kapp, R.W., Picciano, D.J., Jacobson, C.B. "Y-Chromosomal Nondisjunction in Dibromochloropropane-Exposed Workmen," *Mutat. Res. (64)* 47 (1979).
267. Kiraly, J., Szentesi, I., Ruzicska, M., Czeize, A. "Chromosome Studies in Workers Producing Organophosphate Insecticides," *Arch Environm. Contam. Toxicol. (8)* 309 (1979).
268. Nehez, M., Berencsi, G., Paldy, A., Selypes, A., Czeizel., A., Szentesi, I., Csanko, J., Levay, K., Maurer, J., Nagy, E. "Data on the Chromosome Examinations of Workers Exposed to Pesticides," *Regul. Toxicol. Pharmacol. (1)* 116 (1981).

269. Volnjanskaya, A.V., Vesileos, A.F. "Chromosome Aberration Level in Agricultural Workers," *Gig. Truda (12)* 47 (1981).
270. de Cassia Stocco, R., Becak, W., Gaeta, R., Rabello-Gay, M.N. "Cytogenetic Study of Workers Exposed to Methyl-Parathion," *Mutat. Res. (103)* 71 (1982).
271. Linnainmaa, K. "Sister Chromatid Exchanges among Workers Occupationally Exposed to Phenoxy Acid Herbicides 2,4-D and MCPA," *Teratog. Carcinog. Mutag. (3)* 269 (1983).
272. Larripa, I., Matos, E., de Vinuesa, M.L., de Salum, S.B. "Sister Chromatid Exchanges in a Human Population Accidentally Exposed to an Organophosphorus Pesticides," *Rev. Brasil. Genet. (6)* 719 (1983).
273. Dulout, F.N., Pastori, M.C., Olivero, O.A., Gonzalez Cid, M., Loria, D., Matos, E., Sobel, N., de Bujan, E.C., Albiano, N. "Sister-Chromatid Exchanges and Chromosomal Aberrations in a Population Exposed to Pesticides," *Mutat. Res. (143)* 237 (1985).
274. Paldy, A., Puskas, N., Vincze, K., Hadhazi, M. "Cytogenetic Studies on Rural Populations Exposed to Pesticides," *Mutat. Res. (187)* 127 (1987).
275. Dulout, F.N., Pastori, M.C., Gonzalez Cid, M., Matos, E., von Guradze, H.N., Maderna, C.R., Loria, D., Sainz, L., Albiano, N., Sobel, N. "Cytogenetic Analysis in Plant Breeders," *Mutat. Res. (189)* 381 (1987).
276. Rita, P., Reddy, P.P., Venkatram Reddy, S. "Monitoring of Workers Occupationally Exposed to Pesticides in Grape Gardens of Andhra Pradesh," *Environm. Res. (44)* 1 (1987).
277. Steenland, K., Carrano, A., Ratcliffe, J., Clapp, D., Ashworth, L., Meinhardt, T. "A Cytogenetic Study of Papaya Workers Exposed to Ethylene Dibromide," *Mutat. Res. (170)* 151 (1986).
278. Shane, B.S., Scarlett-Kranz, J.M., Shaw Reid, W., Lisk, D.J. "Mutagenicity of Urine from Greenhouse Workers," *J. Toxicol. Environm. Health (24)* 429 (1988).
279. Desi, I., Palotas, M., Vetro, G., Csolle, I., Nehez, M., Zimanyi, M., Ferke, A., Huszta, E., Nagymajtenyi, L. "Biological Monitoring and Health Surveillance of a Group of Greenhouse Pesticide Sprayers," *Toxicol. Lett. (33)* 91 (1986).
280. Nehez, M., Boros, P., Ferke, A., Mohos, J., Palotas, M., Vetro, G., Zimanyi, M., Desi, I. "Cytogenetic Examination of People Working with Agrochemicals in the Southern Region of Hungary," Regul. Toxicol. Pharmacol.,8 37 (1988).
281. Rupa, D.S., Reddy, P.P., Reddi, O.S. "Chromosomal Aberrations in Peripheral Lymphocytes of Cotton Field Workers Exposed to Pesticides," *Environm. Res. (49)* 1 (1989a).
282. Rupa, D.S., Reddy, P.P., Reddi, O.S. "Frequencies of Chromosomal Aberrations in Smokers Exposed to Pesticides in Cotton Fields," *Mutat. Res. (222)* 37 (1989b).

283. Rupa, D.S., Reddy, P.P., Reddi, O.S. "Analysis of Sister-Chromatid Exchanges, Cell Kinetics and Mitotic Index in Lymphocytes of Smoking Pesticide Sprayers," *Mutat. Res. (223)* 253 (1989c).
284. Jablonicka, A., Polakova, H., Karelova, J., Vargova, M. "Analysis of Chromosome Aberrations and Sister-Chromatid Exchanges in Peripheral Blood Lymphocytes of Workers with Occupational Exposure to the Mancozeb-Containing Fungicide Novozir Mn80," *Mutat. Res. (224)* 143 (1989).
285. See, R.H., Dunn, B.P., San, R.H.C. "Clastogenic Activity in Urine of Workers Occupationally Exposed to Pesticides," *Mutat. Res. (241)* 251 (1990).
286. Garry, V.F., Griffith, J., Danzl, T.J., Nelson, R.L., Whorton, E.B., Krueger, L.A., Cervenka, J. "Human Genotoxicity: Pesticides Applicators and Phosphine," *Science (246)* 251 (1989).
287. Carbonell, E., Puig, M., Xamena, N., Creus, A., Marcos, R "Sister Chromatid Exchange in Lymphocytes of Agricultural Workers Exposed to Pesticides," *Mutagenesis (5)* 403 (1990).
288. Rupa, D.S., Reddy, P.P., Reddi, O.S. "Clastogenic Effect of Pesticides in Peripheral Lymphocytes of Cotton Field Workers," *Mutat. Res. (261)* 177 (1991).
289. De Ferrari, M., Artuso, M., Bonassi, S., Bonatti, S., Cavalieri, Z., Pescatore, D., Marchini, E., Pisano, V., Abbondandolo, A. "Cytogenetic Biomonitoring of an Italian Population Exposed to Pesticides: Chromosome Aberration and Sister-Chromatid Exchange Analysis in Peripheral Blood Lymphocytes," *Mutat. Res. (260)* 105 (1991).
290. Gomez-Arroyo, S., Noriega-Aldana, N., Osorio, A., Galicia, F., Ling, S., Villalobos-Pietrini, R. "Sister-Chromatid Exchange Analysis in a Rural Population of Mexico Exposed to Pesticides," *Mutat. Res. (281)* 173 (1992).
291. Dulout, F.N., Lopez Camelo, J.S., von Guradze, H.N. "Analysis of Sister Chromatid Exchanges (SCE) in Human Population Studies," *Rev. Brasil. Genet. (15)* 169 (1992).
292. Kourakis, A., Mouratidou, M. Kokkinos, G., Barbouti, A., Kotsis, A., Mourelatos, D., Dozi-Vassiliades, J. "Frequencies of Chromosomal Aberrations in Pesticide Sprayers Working in Plastic Green Houses," *Mutat. Res. (279)* 145 (1992).
293. Bolognesi, C., Parrini, M., Bonassi, S., Ianello, G., Salanitto, A. "Cytogenetic Analysis of a Human Population Occupationally Exposed to Pesticides," *Mutat. Res. (285)* 239 (1993).
294. Pruett, S.B., Ensley, D.K., Crittenden, P.L "The Role of Chemical-Induced Stress Responses in Immunosuppression: A Review of Quantitative Associations and Cause-Effect Relationships between Chemical-Induced Stress Responses and Immunosuppression," *J. Toxicol. Environ. Health (39)* 163 (1993).
295. Luster, M.I., Rosenthal, G.J. "Chemical Against and the Immune Response," *Environm. Health Perspect. (100)* 219 (1993).

296. Thomas, P.T., Busse, W.W., Kerfvliet, N.I., Luster, M.I., Munson, A.E., Murray, M. et al. "Immunologic Effects of Pesticides," *The Effects of Pesticides on Human Health* (S.R. Baker, C.R. Wilkinson, eds.), Princeton Scient. Public., Princeton, NJ, (Advance Modern Environm. Toxicol., XVIII), 261 (1990).
297. Newcombe, D.S. "Immune Surveillance, Organophosphorus Exposure and Lymphoemogenesis," *Lancet (339)* 539 (1992).
298. Green, T. "Species Differences in Carcinogenicity: The Role of Metabolism in Human Risk Evaluation," *Carcinog. Mutagen. (10)* 103 (1990).
299. Calabrese, E.J. "Animal Extrapolation: A Look Inside the Toxicologists' Black Box," *Environ. Sci. Technol.(21)* 618 (1987).
300. Calabrese, E.J. "Comparative Biology of Test Species," *Environ. Health Perspect. (77)* 55 (1988).
301. Knudson, A.G. "Overview: Genes that Predispose to Cancer," *Mutat. Res. (247)* 185 (1991).
302. Schull, W.J. "The Segregation of Cancer-Causing Genes in Human Populations," *Mutat. Res. (247)* 191 (1991).
303. Nebert, D.W. "Role of Genetics and Drug Metabolism in Human Cancer Risk," *Mutat. Res. (247)* 267 (1991).
304. Kawajiri, K., Fujii-Kuriyama, Y. "P450 and Human Cancer," *Jpn. J. Cancer Res. (82)* 1325 (1991).
305. Guengerich, F.P. "Roles of Cytochrome P-450 Enzymes in Chemical Carcinogenesis and Cancer Chemotherapy," *Cancer Res. (48)* 2,946 (1988).
306. Li, W.F., Costa, L.G., Furlong, C.E. "Serum Paraoxonase Status: A Major Factor in Determining Resistance to Organophosphates," *J. Toxicol. Environ. Health (40)* 337 (1993).
307. Schroeder, K.R., Hallier, E., Peter, H., Bolt, H.M. "Dissociation of a New Glutathione S-Transferase Activity in Human Erythrocytes," *Biochem. Pharmacol. (43)* 1671 (1992).
308. Billings, P., Beckwith, J. "Genetic Testing in the Workplace: A View from the USA," *Trends in Genetics* (8) 198 (1992).
309. Hattis, D. "The Use of Biological Markers in Risk Assessment," *Stat. Sci. (3)* 358 (1988).
310. Perera, F., Brenner, D., Jeffrey, A., Mayer, J., Tang, D., Warburton, D., Young, T-I., Wazneh, L., Latriano, L., Motykiewicz, G., Grzybowska, E., Chorazy, M., Hemminki, K., Santella, R. "DNA Adducts and Related Biomarkers in Population Exposed to Environmental Carcinogens," *Environ. Health Perspectives (98)* 133 (1992).

CHAPTER 29

MEDICAL WASTE MANAGEMENT

Lella Kolachalam

SciTech Technical Services, Inc.
457 Highway 79
Morganville, NJ 07751

CONTENTS

INTRODUCTION

Medical waste can be defined as any solid waste generated in the diagnosis, treatment (e.g., provision of medical services), or immunization of human beings or animals, in research pertaining thereto, or in the production or testing of biologicals, such as the following:

- Cultures and Stocks:
  Cultures and stocks of infectious agents and associated biologicals, including: cultures from medical and pathological laboratories; cultures and stocks of infectious agents from research and industrial laboratories; waste from the production of biologicals; discarded live and attenuated vaccines; culture dishes and devices used to transfer, inoculate, and mix cultures.
- Pathological Wastes:
  Human pathological wastes, including tissues, organs, and body parts, and body fluids that are removed during surgery, autopsy, or other medical procedures, and specimens of body fluids and their containers.
- Human Blood and Blood Products:
  Liquid waste human blood; products of blood; items saturated and/or dripping with human blood; items that were saturated and/or dripping with human blood that are now caked with dried human blood including serum, plasma, and other blood components, and their containers, which were used or intended for use in either patient care, testing and laboratory analysis or the development of pharmaceuticals; and intravenous bags.
- Sharps:
  Sharps that have been used in animal or human patient care or treatment or in medical, research, or industrial laboratories, including hypodermic needles, syringes (with or without the attached needle), pasteur pipettes, scalpel blades, blood vials, needles with attached tubing, and culture dishes (regardless of presence of infectious agents). Also included are other types of broken or unbroken glassware that were in contact with infectious agents, such as used slides and cover slips.
- Animal Waste:
  Contaminated animal carcasses, body parts, and bedding of animals known to have been exposed to infectious agents during research (including research in veterinary hospitals), production of biologicals, or testing of pharmaceuticals.
- Isolation Waste:
  Biological waste and discarded materials contaminated with blood, excretion, exudates, or secretions from humans or animals isolated to protect others from certain highly communicable diseases.
- Unused, Discarded Sharps:
  Hypodermic needles, suture needles, syringes, and scalpel blades.

Concern with medical waste has increased following public exposure to discarded blood vials, needles, empty prescription bottles, and syringes, particularly along the nation's beaches and in streets and landfills. Hospitals and other health care facilities are caught in the middle of a major environmental as well as emotional issue.

OPERATING PROCEDURES

In the management of medical wastes, a plan should be established to ensure protection of public health and the environment. The plan should incorporate a cradle-to-grave approach to infectious medical wastes, like that which has been part of RCRA wastes since 1978. This includes the adoption of standard operating procedures (SOPs) which address:

- The generation of wastes,
- Segregation of wastes,
- Collection of wastes,
- Containerization and storage of wastes,
- Wastes treatment,
- Wastes handling and transportation,
- Waste disposal, and
- Contingency planning.

In addition to establishing SOPs, the management plan should outline the required training, refresher training, and medical monitoring requirements for staff dealing with medical wastes.

Procedures must be established by qualified personnel to ensure that operating personnel subject each waste load to the required treatment. Actions which, when followed, will ensure effective treatment of all proposed waste. The following issues should be addressed:

- **Type of waste:** Characteristics of each waste load must be determined to evaluate which treatment method is most appropriate. The potential hazard of each waste also must be examined so operating personnel take appropriate action in handling the waste.
- **Variations in waste stream:** Operating personnel should be aware of inconsistencies in the waste stream and should ensure that the residence time is adjusted appropriately to adequately treat the waste.
- **Limitation of treatment apparatus:** Operating personnel must be familiar with the limitations of each apparatus. Waste may not be effectively treated and may require secondary treatment or alternative treatment to remove biological hazards.

Generation of Medical Wastes

The primary purpose of the medical waste management plan is to address the generation of wastes within the respective facility. Medical wastes include pathological, hazardous, and infectious and noninfectious wastes. It is essential that individuals responsible for implementation of the management plan identify all sources from each department. An environmental audit by an independent party is probably the best means of isolating all sources within the facility.

The EPA estimates that the generation rate of hospital waste is at the rate of 13 lb/bed/day, but some independent estimates put the generation range between 16 and 23 lb/bed/day. The composition of hospital waste also may typically include radioactive waste and pharmaceuticals; hazardous waste such as cytotoxic agents used in chemotherapy; mercury and other heavy metals; waste chemicals; infectious wastes such as human blood and blood products; contaminated sharps; anatomical wastes; and pathological wastes. Sharps are objects with the potential of penetrating the skin. Typically included in this category are hypodermic needles, syringes, Pasteur pipettes, broken glass, and scalpel blades.

Segregation

Noninfectious medical wastes can constitute as much as 70% of wastes generated at medical facilities. This portion can be disposed of as sanitary waste, at lower cost and without treatment. It is essential in preparing a waste management plan that steps to segregate the infectious from the noninfectious waste be clearly outlined. This has been proven to be a cost-effective method of handling waste.

It is preferable that medical waste be segregated at the source: Staff handling the material can readily identify the hazards of each. It is important that the responsibility is never conferred on anyone who is untrained in the medical field, unless explicit instructions are given for specific wastes. The segregation of waste at the point of generation reduces the amount of subsequent exposure staff receive in handling the waste. It also serves to curtail the spread of pathogens by concentrating the infectious waste.

While wastes must be segregated into infectious and noninfectious as minimum, care also should be taken to subdivide the infectious waste according to the degree of hazard. Waste with multiple hazards, for example a contaminated sharp, require more than one form of treatment prior to disposal. Wastes with multiple hazards should be identified at the source and appropriately segregated, thus avoiding multiple treatments of large quantities of waste. Multiple hazard wastes should be placed in appropriate receptacles to await treatment. Reusable items, although not strictly disposable, must be identified as a waste and should be segregated at the source for appropriate treatments. Generally, different waste categories can be identified as follows:

- Sharps—needles, blades etc. (disposable),
- Surgical—pathological and animal (disposable),
- Soiled linens (reusable),
- Rubbish or mixed refuse (disposable),
- Patient care items (reusable),
- Noncombustible glass, metals and ashes (disposable),
- Garbage (nongrindable) (disposable), and
- Food service items (reusable).

Waste Collection System

Hospital waste collection systems fall into three general categories: Carts, gravity chutes, or pneumatic tubes. Wastes are typically bagged and may be segregated to separate microorganism-bearing items for special treatment. Collected wastes must be transported from the point of generation to collection points for processing and to appropriate disposal.

The hospital cart has become a fixture in the medical care delivery system. These carts are used to transport all types of medical devices, supplies, and eventually wastes. Wastes carried in carts represent a health hazard because of the physical contact involved. Bagging of wastes reduces exposure, but bags are easily broken or torn open, thus releasing the contents. Automated carts allow movement of materials with less personal contact, but exposures still can occur.

Gravity chutes offer a means of transporting wastes vertically through the hospital. This method avoids the use of elevators for vertical transportation, but research has shown that chutes may release pathogens from floor to floor. Gravity chute systems should have locking doors to prevent inadvertent opening, especially when the chute is in use from another floor. Whenever containers of wastes are falling through the chute, they generate a positive pressure than can exhaust contaminated air into hallways if the door arc left open during use. Some hospitals generate positive air pressure in hallways to prevent contaminated airflow. Chutes can produce severe problems if wastes become jammed or break while dropping or upon hitting the bottom.

Pneumatic tubes, while expensive, offer the best overall waste disposal performance. Some advantages include:

- High-speed movement,
- Movement in any direction,
- A completely closed system under negative pressure,
- Minimal intermediate storage of wastes awaiting disposal, and
- Minimal possible jamming of wastes.

Often, separate pneumatic tubes and gravity chute systems are used for both linen and waste disposal. Obviously, mixing of laundry and waste must be avoided. This problem can be typically solved through careful design and appropriate markings.

Because pneumatic tubes move waste at high speeds, some items may break and bags may burst. Glass waste is usually not permitted in pneumatic systems and must be disposed of separately. Advantages of pneumatic systems typically outweigh the disadvantages, and such systems are finding wider use in both new and old facilities. Such systems minimize human contact with wastes and allow transportation of waste directly into incinerators or other intermediate collection points.

Containerization and Storage

All waste is subject to containerization and storage between the time of generation and time of destruction. The medical waste management plan must address the containerization throughout this time. It also must outline requirements for storage areas.

The containerization of medical wastes provides for the environmentally safe handling, transportation, storage, and treatment of waste to the point of disposal. Therefore, it must be compatible with the type of waste, the handling and transportation, the storage, and the proposed treatment method.

In evaluating the type of waste, the waste management plan must determine hazardous and aesthetic considerations to be addressed during the "active life" of the waste. Contaminated sharps should be placed in sealed, puncture-proof containers at the point of generation. This reduces the potential for injury and spreading of infection during the handling and transportation of waste. Liquid waste should be contained in a leak-proof can or jar. It should have a removable bung or cap to allow the waste to be decanted prior to or after treatment. Unsightly pathological waste such as limbs and organs should be placed in opaque plastic bags and sealed prior to removal from the point of generation. All containers and bags preferably should be red and labeled "INFECTIOUS" or with the international biohazard symbol.

Wastes that must be transported to another facility for storage or treatment must be contained in packaging of sufficient integrity to withstand the journey. Wastes that have been single-bagged should be double-bagged and placed inside a semi-rigid box for shipment. At no time should waste be subject to any handling that would jeopardize the integrity of the packaging. Wastes should be loaded and unloaded manually and never compacted. This would defeat the purpose of containerization by assisting in the spread of pathogens.

The selection of containers and packaging materials should address the proposed treatment method. The waste management plan for treatment should outline compatible materials for each form of treatment. If the waste is to be incinerated, packaging does not present a problem. However, with steam sterilization it is advisable not to use high density polyethylene plastic bags. This impedes steam penetration and hinders treatment. For successful results, waste should be held in a single, light-duty plastic bag and placed in a dish. This will contain the waste if the bag should fail.

Whether treatment is performed on-site or not, a storage facility always should be available for the storage of waste should the existing system fail. It is normal, however, to use a storage facility to gather wastes over a short period of time. Storage of wastes should be outlined explicitly in the waste management plan and limits should be set on the type of waste, the duration of storage periods, temperature of facility, and operational criteria. State laws outline the maximum periods of time allowed between generation and destruction of wastes. It is expected that storage facilities will operate at cool if not cold temperatures to prevent the propagation of infectious agents and spreading of disease during storage. Wastes should be segregated to ensure and homogenous feed for treatment. They should be arranged chronologically to prevent violation of state storage periods. Finally, the storage facility should have limited access to personnel, and all entrances should be clearly marked by the international biohazard symbol and word "INFECTIOUS."

Handling and Transportation

While some waste may be treated at the source, for example, utensils may be autoclaved, the majority of waste generated at medical facilities must be transported to an on-site or off-site treatment facility prior to disposal. It is very important that the medical waste management plan address the handling and transportation of this waste.

Waste must be appropriately packaged for handling and transportation. Single-bagging is insufficient for this purpose. Any solid or semi-solid waste should be double-bagged and placed within a semi-rigid container. This is usually constructed of cardboard. Wastes must then be placed by hand onto a trolley container for transportation to the loading bay. This trolley should also be capable of containing any potential spills of waste, including liquids. It would be prudent to suggest that it contain either 10% of the volume being carried or the volume of the largest container, whichever is larger. Trolleys should be used solely for the purpose of transporting medical wastes and nothing else. They should be subject to frequent decontamination and be labeled with the word "INFECTIOUS" and the international biohazard symbol. All packages should be loaded and unloaded manually and without the aid of any mechanical devices that might compromise their integrity.

It should be strongly advised that the handling and transportation of medical waste occur during nighttime hours. At this time here is less likelihood of encountering patients, staff, and visitors. This limits the potential spread of disease within the facility. During off-site transportation, all waste should be containerized to prevent any leakage while in transit. Most trucks are not leak-proof and would require waste hoppers to contain the waste. Hoppers are easily loaded and unloaded. The truck should be labeled "INFECTIOUS" in three-inch high letters, accompanied by the international biohazard symbol. All infectious medical waste should treated prior to disposal to render it harmless to public health and the environment.

Disposal

Disposal of medical waste is a controversial aspect of waste management. This can be the result of illegal disposal and the health risks to the general public, but more often than not, it involves the aesthetic concerns. Proper disposal of treated medical wastes must be clearly outlined in the waste management plan.

In some states, medical waste is regulated irrespective of the quantity involved. This entails treatment and then disposal at a licensed facility. Some wastes still may be treated on site and then ground and discharged to the sanitary sewer. However, we cannot recommend the discharging of any untreated waste to sanitary sewers. Some states have initiated a chain-of-custody procedure for medical wastes, similar to those currently in effect for RCRA wastes. The treatment facility must acknowledge destruction of the waste within two weeks of generation. If the generator receives no satisfaction from the treatment facility within two weeks, the state will take responsibility for the case. The proper implementation of medical waste management plans should eradicate the illegal disposal of medical waste, which in recent years has become endemic of the profession.

Contingency Planning

In dealing with medical waste management we must realize that the individual responsible for implementation will not always be present to implement emergency responses or make crucial decisions. It is of the utmost importance that in emergency situations, standard operating procedures clearly outline the staff action to be undertaken. An emergency does not constitute a disaster, but it must be dealt with quickly. This usually refers to a waste spill, an accidental staff injury, or the failure of a waste container. The cleanup and disposal of waste and the precautions for affected staff shall be followed precisely. This will ensure that the individual responsible for implementation has control over an emergency situation whether he is present or not.

TREATMENT METHODS

In the majority of states, infectious waste is required to be treated prior to disposal. Exemptions do exist due to the quantities generated in some facilities. This section discusses various treatment methods presently available that eliminate the pathogenic hazard from hospital waste prior to discharging or disposal.

This treatment methods are:

- Incineration,
- Steam sterilization,
- Hydropulping,
- Thermal inactivation,
- Gas/Vapor sterilization,
- Radiation sterilization, and
- Chemical disinfection.

To ensure the effective treatment of each waste load, operating personnel must be familiar with both the waste and the treatment method. Monitoring of treatment is recommended to ensure that the apparatus is operating correctly and waste loads are receiving adequate treatment.

Incineration

Incineration refers to the combustion of wastes under excess air conditions to form ash, noncombustible residues, and off-gases with the elimination of pathogens. Historically, this has consisted mainly of pathological wastes, which for cosmetic as well as health-related reasons require a comprehensive means of pretreatment. An added benefit of incineration is a reduction in the amount of disposal volume involved. It is not uncommon to achieve as much as 95% reduction in disposal volume with incineration. The essential mechanism of a small-capacity incinerator consists of:

- A primary combustion chamber,
- A secondary combustion chamber, and
- An air pollution control system.

Primary Combustion Chamber

The primary combustion chamber is designed to heat the waste, dry it, and release any constituent volatiles and moisture. It is recommended that the primary combustion chamber operate at a temperature between 1,500°F and 1,800°F to achieve good burnout and maximize the kill of infectious agents present in the waste. Additional air also should be incorporated to minimize the formation of products of incomplete combustion (PICs), generated as a result of the high plastic content of medical waste (20% to 30%).

Unreacted or partially reacted combustion products, which may range from benzene to the chlorinated isomers of dibenzo-p-dioxin (CDD) and dibenofuran (CDF), can result from incomplete combustion. There are various proposed mechanisms for the formation of CDD and CDF compounds, but the one most plausible mechanism for hospital waste is that chlorinated plastics (PVC compounds) in the waste may produce precursor compounds in the oxygen-starved regions of the primary combustion chamber. These precursor compounds can subsequently combine to form CDD and CDF compounds. This formation of unreacted or partially reacted compounds also occurs in the combustion of other organics and may lead to the formation of lower-molecular weight compounds and CO.

Secondary Combustion Chamber

The secondary chamber is required to thermally destroy the airborne contaminants from the primary chamber. It is the secondary chamber that is largely responsible for meeting the air pollution-control requirement for the incinerator. The secondary chamber should be operated at temperatures between 1,800°F and 2,000°F. The retention time for gases in the chamber should be between 1 and 2 seconds. Oxygen availability in the secondary combustion chamber is also an important consideration. Finally, it is essential to maintain strong turbulence in conjunction with the aforementioned requirements to assure complete destruction of organic compounds and infectious agents. Turbulence helps to promote good heat transfer and oxygen availability, which is especially important because the combustor may become stratified and pockets of partially oxidized organics may pass through the secondary combustor. Good mixing minimizes the release of organic compounds from the main combustor.

Air Pollution Control

To ensure stack emissions are within state and federal standards, it is recommended that, in addition to proper operation of the incinerator, the stack should incorporate a baghouse or scrubber to lower the particulate content. Past air emission control requirements for infectious waste incinerators were directed toward the control of particulates, opacity, and odors. Regulations, as proposed and adopted by state agencies, have been directed toward:

- More stringent control of particulate,
- Control of acid gases, especially hydrochloric acid (HCl), and
- Control of carbon monoxide (CO).

The more stringent control of particulates mainly has been due to the availability of improved control technology. The control of acid gases has been due to the amount of chlorinated plastics contained in hospital waste. Hospital waste may contain 20% to 30% plastics. The plastics are made up of primarily polyethylene, polypropylene, and polyvinyl chloride, with the polyvinyl chloride being approximately 45% chlorine. This large quantity of chlorine may cause combustion products that include HCl and toxic organic compounds.

The control of CO has been directed toward the control of other organic compound emissions with CO serving as an indicator. The minimization of CO emissions indicates the maximization of combustion efficiency and also the minimization of other organic compound emissions.

Historically, hospital incinerators have been operated in a less-than-desirable fashion. It is essential that in the future, the operational constraints of the incinerator will be understood by operating personnel. This will ensure the proper treatment of all medical waste. To achieve high burnout and removal of infectious agents, the burner should be operated with a continuous waste stream and with a steady fuel and air feed rate. Overloading should be avoided at all times to prevent the formation of PICs. The content of the waste should be monitored continually to ensure that temperature surges are avoided. With the high proportion of plastic in medical wastes, this not uncommon. Incineration of infectious or pathological wastes at the start or end of a burn cycle also should be avoided to prevent the possible release of pathogens through the stack or to the ash residues.

As with all treatment methods, monitoring is essential to maintain desired results. It is advisable to perform test burns prior to startup operation. The EPA recommends monitoring test results by seeding the waste with a *Bacillus subtillus* variety *niger*. The stack emissions should be monitored to determine the number of viable spores surviving the burn. The operating parameters would then be adjusted to achieve required removal efficiencies.

Steam Sterilization

Steam sterilization is probably the most common from of sterilization in use in health care and related industries today. It involves the treatment of infectious waste with saturated steam at a sufficiently elevated temperature to kill infectious agents. This process is performed in an enclosed vessel more commonly known as autoclave. It is most commonly used to treat dressings, surgical instruments, and supplies. There are two types of autoclaves commercially available: the gravity displacement model and the pre-vacuum model. The gravity displacement autoclave expels the chamber air through an exhaust valve as it fills with pressurized steam. This is less effective than the pre-vacuum model, as some dilution is experienced in filling the chamber which leads to less penetration of waste. The pre-vacuum model, as the name implies, removes the air from the autoclave chamber prior to the addition of steam. The addition of pressurized steam elevates the temperature within the waste sufficiently to inactivate or kill the infectious agents present.

The choice of steam is pertinent to success of treatment. Steam may exist in any of three forms at a single temperature. With dry saturated steam, all water present is in the vapor phase. The heat content of dry steam would be in equilibrium with water at the same temperature. Steam is referred to as "wet" when the heat content is insufficient to maintain all water in the vapor phase. As a result, some aqueous liquid in present. Superheated steam is water, entirely in the vapor phase, but at a higher temperature than dry steam.

Dry steam is commonly used in steam sterilizers. Wet steam is unpopular for use in autoclaves because the total heat content is lower than dry steam. The introduction of aqueous liquid also requires subsequent frying. Superheated steam is avoided because the heat content is not readily available until the steam has chilled and subsequently reached saturation. Infectious agents such as spores are also more susceptible to moist heat than to dry heat. The decontamination of infectious waste occurs during autoclave primarily with the transmission of steam through the waste. Heat conduction through the media is of little additional benefit to the overall treatment. To operate the autoclave to maximum benefit, the following considerations should be taken into account:

- All excess air should be easily removed from the main chamber. Trapped air dilutes the steam and lowers penetration of waste. It is therefore recommended that all bottle caps be loosened prior to placement in an autoclave and all heat-resistant packages be opened.
- The type of wastes suitable for autoclaves are specific. Low density waste is easily penetrated by steam and requires a short residence time. Body parts however, are not easily penetrated and require a prolonged residence time. It is essential that waste be maintained at the required temperature and specific residence time for adequate treatment.

The operator should be familiar with the specific wastes and the operational limitations of the autoclave to ensure the effective treatment of all wastes.

Periodic monitoring of equipment and treated wastes is essential for effective quality control. Pressure gauges and thermometers should be monitored throughout each treatment cycle and any peculiarities noted. Treated waste should be periodically tested to ensure treatment is effective. The United States Pharmacopeia recommended *Bacillus stearothermophilus* for effective monitoring of treated wastes.

The major advantage of steam sterilization is that is has been used for many years in hospitals for small quantities of waste and sterilization of instruments and containers. Hospitals are familiar with the operation of the units. The second advantage is waste can be properly sterilized if it is processed correctly. A disadvantage is that the waste does not change in appearance or volume. Also, it can be difficult to ensure the time/temperature relationship have been met in the unit.

Hydropulping

Medical wastes can be ground and sewered, effectively removing it from the environment, as well as greatly reducing storage, transport, and handling costs. However, sewage is an inherently infectious material. If this treatment procedure is used, sewage facilities will have to disinfect the sewage before a safe effluent can be discharged back into the environment. Hydropulping is a system in which waste is ground in the presence of an oxidizing fluid, such as hypochlorite solution. The waste is fed into the top of the unit where it is pulped with a hammer mill while it is sprayed with the solution. An extractor then removes the liquid from the pulp. The pulp can be disposed of at a sanitary landfill, and the liquid can be directly discharged to a public sewage treatment facility. However, if the discharge requirement for solid, chlorine and pH are such, the liquid waste may have to be pretreated prior to discharge to the sewer.

An advantage of the hydropulping system is that waste can be reduced in size and disinfected within the same system for additional treatment or disposal. Another advantage is that the pulped waste is innocuous and may be more acceptable to the public. Public perception of waste may be such that it will not be a problem taking it to a sanitary landfill or a municipal incinerator for further processing.

Disadvantages associated with the hydropulping system include fugitive emissions. The water from the extraction system is high in chlorides, so discharge to the sewer may not be allowed. Many sewer authorities are not willing to accept this material without pretreatment. Another disadvantage of the system is difficulty in conducting microbiological tests on the pulp. This is necessary to determine if the waste is fully oxidized and all organic matter and infectious organisms from the waste have been destroyed.

Thermal Inactivation

Thermal inactivation or dry heat sterilization is a process whereby infectious or pathogenic wastes are rendered sterile by exposure to heat. It is similar in nature to steam sterilization with the notable absence of steam in the main chamber. This protects sharps and other steam-sensitive materials from corrosion during treatment. Materials normally sterilized by dry heat, but not steam, include sharp surgical instrument, oils, powders, and greases.

Thermal inactivation is currently performed on both solid and liquid medical wastes. Liquid wastes are normally treated by a coil or heat exchanger prior to distillation and subsequent discharge. Solid wastes, however, are treated in an oven chamber. Both wastes are subject to sufficient heat to inactivate any pathogenic organisms present.

The dry heat sterilizer for treatment of wastewaters may be a batch or continuous-feed process. The batch sterilizer heats the liquid waste to a predetermined temperature, at which it is maintained for a specified period, to produce the desired kill of pathogens. This is followed by distillation and cooling in a heat exchanger before the treated waste is discharged to an appropriate waste stream. The continuous-feed apparatus consists of a similar preheating tank from which the waste is fed through heat-retention tubing to maintain the desired treatment temperature. This discharge allows the continuous feed of waste to the preheating tank and does not disrupt the generation of liquid waste. Again, the waste is distilled and passed through a heat exchanger, which may be used to reheat incoming discharge while cooling discharge waste simultaneously. Thermal inactivation is less efficient than steam sterilization because steam does not penetrate the waste. It is essential to designate treatment temperatures and treatment cycles for various medical wastes.

Gas/Vapor Sterilization

Gas/vapor sterilization is a treatment process whereby medical waste is sterilized by a vaporized chemical in a sealed chamber. Ethylene oxide (ETO) and formaldehyde are most commonly used, both of which are probably carcinogens. While the most common use of this form of sterilization is in the decontamination of surgical instruments in both hospitals and industry, it can also be applied to the treatment of medical waste. Other forms of treatment are equally effective without involving the hazards of ETO or formaldehyde, and treatment may expose staff to a greater risk than the untreated waste itself. The gas/vapor sterilizer has various forms including the vacuum chamber, autospheric chamber, ampule/line bag, Steri-jet system, and tent fumigation. Units vary in size from small (less than 4 cu. ft) to large (100 to 3,000 cu. ft).

In using the vacuum chamber model, materials and equipment to be sterilized are placed in the airtight chamber and a vacuum pump removes the entrapped air. Unit dose cartridges or a gas mixture of ETO is then drawn by vacuum into the chamber. The load is exposed to the sterilizer for a predetermined length of time. On termination, the vacuum pump draws filtered air into the chamber. It is recommended that sterilized equipment and materials be stored in a vented area to prevent leaching of the potentially hazardous gases.

While it is recognized that the gas/vapor sterilizer has potential uses in the treatment of contaminated equipment, it must be stressed that alternative treatment methods be carefully examined before deciding to use this method.

Radiation Sterilization

Radiation of materials for sterilization purposes has various applications in the treatment of medical waste. Irradiation kills infectious agents and pathogenic bacteria and prevents replication. It has been the basis of presterilization of medical products such as surgical sutures and bandages, and it would seem to have potential for treating medical waste prior to discharge or disposal.

The principal forms of irradiation in use today include gamma irradiation, ultraviolet (UV) irradiation, electron beam (EB) irradiation and infrared irradiation. Electron beam is limited in its treatment of medical waste as it has poor penetration ability. Infrared sterilization has been used extensively to sterilize air ducts in central air circulating systems but is also limited in its application to medical waste treatment. However, gamma irradiation an UV irradiation have been used extensively in the treatment of waste and wastewater.

Gamma irradiation is the bombardment of materials by electron particles. The most common source material for gamma rays is ^{60}cobalt, a product of nuclear reactors. The rays have a penetration capacity of several meters, which is ideal for the treatment of waste sludges and packages. Gamma irradiation is advantageous for the treatment of medical waste, as it has very low electricity requirements and does not heat-treat material. Consequently, treated wastewater does not require cooling prior to discharging or disposal. In the treatment of wastes, the ^{60}cobalt source is passed through a maze-like structure to the exposure chamber for subsequent treatment. After the required cycle time, the waste may be safely discharged without cooling. When using radioactive source material, it is recommended that the source strength be monitored because it diminishes with time. The exposure time may be varied by controlling the flow of wastewater or the speed of the conveyor belt.

While UV radiation does not have the penetration power of gamma radiation, it has been considered in the treatment of wastewater as a means of tertiary treatment prior to discharge. During treatment, wastewater passes through a channel, across which are arranged a series of UV bulbs. The UV light, which is emitted at the 254-nanometer wavelength, is close to the peak germicidal wavelength. Exposure at this wavelength kills the pathogenic organisms present in the wastewater. Monitoring of irradiation may be performed by seeding waste spores of bacillus bubbles. The EPA recommends monitoring of treatment at least once every two weeks.

Chemical Disinfection

Chemical disinfection is the inactivation of waste by the addition of limited quantities of chemicals. Disinfection may be performed by adding hydrogen peroxide, acids, alcohols, quaternary ammonium compounds, or ketones. Disinfection is normally performed on surfaces, utensils, and medical supplies on a widespread basis. It is not recommended for treatment of waste or wastewater. The quantity of chemical additive varies with the variability of waste contaminants and constituents. The use of chemical additives would require the monitoring of individual treatment batches, which is neither recommended nor economical.

SEGREGATION AND TREATMENT

Isolation wastes are generated in health care facilities by patients whose illness is of such severity or contagiousness as to require isolation from the general public. In recent years, the AIDS crisis has given rise to increasing numbers of patients of this class. While sanitary wastes from isolation wards may be discharged directly to the sanitary sewer, all other wastes generated as a result of patients care must be treated prior to disposal. EPA recommends that these wastes be incinerated or subject to steam distillation. It also is recommended that isolation wastes be segregated from other waste to be treated. This will result in a more homogenous waste feed, which improves the efficiency of the treatment.

Cultures and Stocks of Etiologic Agents

Cultures and stocks of etiologic agents are generated principally in laboratories. They constitute a health risk to the general public due to the high concentrations of pathogens present. The recommended treatment for these wastes is steam sterilization. Because autoclaves are common in most laboratories, source treatment thereby removes any potential for the spreading of contamination during handling and transportation. If an autoclave is unavailable, then incineration and dry heat sterilization are acceptable methods of treatment.

Blood and Blood Products

In light of today's AIDS and hepatitis crisis, it is prudent to assume that all blood and blood products are sources of contamination. It is essential that all blood and related materials receive treatment prior to discharge or disposal. Both steam sterilization and incineration are appropriate methods of inactivating infectious agents present in blood and should be utilized for all related products.

While the disposal of blood into sanitary sewers has been practiced in the past, this method of disposal is not recommended. The EPA suggests limiting blood discharges into sanitary sewers where secondary treatment is assured. This practice is not recommended under any circumstance for the following reasons:

- Blood coagulation may cause blockage of plumbing and allow exposure of personnel during clearing.
- Personnel who dispose of blood supplies may be exposed to potential contamination during spillage by aerosols and splashes.
- In combined sewer systems, sanitary waste sometimes bypasses treatment facilities with stormwater flows. This creates an exposure potential for the general public.

While it is recognized that some blood will be discharged to the sanitary sewer during gross cleaning of surgical instruments and utensils, the treatment of such equipment by steam sterilization or dry heat sterilization prior to cleaning is strongly recommended. This will ensure that off-site discharges will not expose the general public to potential health risks.

Pathological Wastes

Pathological wastes consists of any body parts removed from a person or animal in a health care or veterinary facility. It is essential in the management of medical waste that personnel address both treatment and disposal of each waste. This is especially true in the case of pathological waste. All body parts should be treated as blood products and subject to steam sterilization or incineration to inactivate any potentially infectious agents present. Incinerator ash may be safely disposed of in a sanitary landfill. However, it is recommended that pathological waste such as limbs or organs that have been sterilized be disposed of in an unrecognizable form. This may consist of incineration following treatment or grinding with subsequent disposal to a sanitary sewer. Because incineration alone is a complete treatment, we suggest that sterilization and incineration not be performed, as dual treatment would be redundant. While present practice may allow the grinding of body parts to sanitary sewers where secondary treatment is available, we cannot recommend this practice.

Waste from Surgery and Autopsy

Waste generated from surgery and autopsy consist of body parts, body fluids, feces, blood and blood products, and contaminated utensils. All wastes should be treated as potentially infectious and subject to treatment prior to disposal or discharge with the use of steam sterilization or incineration for this purpose. For cosmetic reasons, all body parts should be incinerated or subject to grinding as a matter of policy. Waste residues from either treatment method may be safely disposed of at a solid waste disposal facility.

Contaminated Laboratory Wastes

Laboratories, both small-scale and commercial, generate wastes that require treatment prior to disposal. Under EPA guidance, these wastes should be segregated into groups based on the type, virulence, or quantity of infectious agent present. Wastes which are of lesser potency (biosafety levels 1,2 and 3) should be treated by steam sterilization or incineration. Their residues may be safely disposed of at a solid waste disposal facility. Wastes of greater potency (biosafety level 4) should be treated in a pass-through double-door steam sterilizer prior to their removal from the laboratory. Wastewaters from miscellaneous laboratory sources should be subject to dry heat sterilization, irradiation (if available), or possibly chemical disinfection. Chemical disinfection should not be adopted as laboratory practice because there is greater unreliability in killing pathogens, and it also requires testing of individual loads.

Contaminated Sharps

Needles, syringes, and blades, more commonly known collectively as "sharps," represent a dual hazard in medical waste management. Primarily, the sharp represents a source of potentially contaminated blood and requires treatment. Secondly, the sharp has a potential to cause injury and transmit infection to the bloodstream of personnel handling the equipment. Both aspects of disposal must be addressed. It is essential that sharps be collected for treatment in a puncture-proof container. Typically, these containers are made of thick polyethylene and come in a variety of sizes. The containers also can be subject to treatment that ensures a minimum of handling of exposed sharps. Many companies have successfully instituted segregation of waste at the source and greatly reduced the number of injuries (sticks) as a result.

Steam sterilization or incineration is the treatment method. If steam sterilization is utilized, the sharps must still be subject to treatment to remove their points or edges. Heat treatment can melt syringes, needles, and sharps into a homogenous block. Grinding will remove sharp points and edges. Needle clipping in an enclosed system is also approved. However, under no circumstances should staff practice clipping of needles outside a closed system. This exposes personnel to an increased risk of infection from clippings and aerosols.

Dialysis Unit Wastes

Dialysis unit wastes consist of body fluid, sharps, catheters, and disposable accessories. Just as in the treatment of blood and blood products, the waste must be considered to be contaminated with AIDS or hepatitis viruses. Testing of individual waste loads is not recommended. Steam sterilization or incineration of all dialysis unit wastes is the recommended method.

Disposable Biologicals

"Biological" is a term broadly applied to bottles of medication normally given intravenously to patients. While they are not sources of pathogens, they must be treated to destroy the biological activity remaining in the bottles. Steam sterilization, dry heat sterilization, or incineration are recommended for effective results. The purpose of treatment is to destroy the content of the vials, not the vials themselves. Molten glass would cover the waste and diminish its effective destruction. A lower temperature and longer residence time are recommended for the treatment of biologicals. Treated waste from any of the aforementioned treatment processes may safely be disposed of with other solid wastes in a sanitary landfill.

Miscellaneous Wastes

The principal forms of medical waste have been discussed. Other medical wastes requiring treatment include:

- **Animal carcasses and body parts.** Due to their size, incineration is strongly recommended. Autoclaving is possible but requires extensive duration times.
- **Animal wastes.** Due to its low heat conductivity and low permeability, steam sterilization is not recommended for these wastes. Incineration is the preferable treatment.
- **Contaminated equipment.** Contaminated equipment should be treated similarly to utensils. Incineration or steam sterilization is recommended. Larger machinery should be subject to *in situ* vapor sterilization.

Radioactive Medical Waste

Radioactive medical wastes are regulated by the Nuclear Regulatory Commission (NRC), the Department of Transportation (DOT), the Environmental Protection Agency, the state Department of Environmental Protection (DEP), and the state Department of Health (DOH). They are not regulated by the EPA, DEP, or DOH as RCRA wastes, as are all other medical wastes.

Radioactive supplies have been used for diagnostic purposes since the early 1900s in the form of X-rays. However, today their use includes therapy, diagnosis, and radio-pharmaceuticals, with the recent development of nuclear medicine. The form of source material varies from vials and ribbons to beads. The quantity of radiation to which each patient is exposed is referred to as "a dose." In radiation therapy, a patient is exposed to doses of radioactive rays in order to kill a cancerous growth or curtail its spreading about the body. Currently, ^{125}iodine is used for the treatment of prostate cancer. It is obtained by the hospital in a sealed container and is decayed in-house to the point at which it may be used. ^{137}Cesium and ^{192}iridium are also used in this manner to treat cancer. ^{32}Phosphorous is obtained in solution form and is injected into the patient for therapy.

In nuclear medicine, patients generally receive doses of radiation orally or by injection. Isotopes are used for diagnosis, therapy, and as radiopharmaceuticals. ^{99}Technetium is the most commonly used isotope and is administered to the patient by injection. With its half-life of just six hours, it can be decayed in-house. ^{123}Iodine is ingested as a capsule and is used in the treatment of thyroid cancer. ^{131}Iodine also is taken in capsular form, but at much higher doses, for thyroid cancer ablation. ^{67}Gallium is taken in liquid form to localize abscesses, cancers, or neoplasms. ^{201}Thallium is injected in liquid form to enhance cardiac imaging. Doses of radioactive isotopes usually are supplied to hospitals individually. Any remaining material is either returned to the supplier for subsequent disposal or is disposed of by hospital staff.

The packaging and transportation of radioactive wastes is regulated by the NRC and the DOT. Radioactive waste generally includes solidified liquids, liquid scintillation vials, absorbed liquids, biological waste, in vitro wastes, and animal carcasses.

BIBLIOGRAPHY

Carlile, J. "Finding Disposal Options For Medical Waste," *American City & County*, November 1989.

Cheremisinoff, P. N., and Shah, M. K. "Hospital Waste Management," *Pollution Engineering,* April 1990.

Cross, F. L. and Rykowski, P. K. "Infectious Waste Incinerator Retrofit," *Pollution Engineering,* June 1989.

Cross, F. L. "Siting a Medical Waste Treatment Facility," *Pollution Engineering,* September 1990.

Cross, F. L. *Incineration For Hospital And Medical Wastes,* SciTech Publishers, Inc., N.J. 1990.

Dine, D. D, "Strict Laws Aimed at 2 States' Disposal Woes," *Modern Healthcare,* September 1988.

Hall, S. K. "Infectious Waste Managements, A Multifaceted Problem," *Pollution Engineering,* August 1989.

Hershkowitz, A., "Without A Trace, Handling Medical Waste Safely," *Plastics World,* February 1990.

Holthous, D. "States Seek Tighter Rules On Infectious Waste," *Hospital,* September 1988.

Hudson, T. "Hospital Adjust To New State Pollution Regulations," *Hospital,* July 1990.

Marks, C. H., "Burn Or Not To Burn," *Pollution Engineering,* November 1988.

Meaney, J. G. and Cheremisinoff, P. N. "Medical Waste Strategy," *Pollution Engineering,* October 1989.

Nelson, S. "Infectious Hospital Waste Disposal A Troublesome, Costly Problem," *Modern Healthcare,* January 1987.

Tessitore, J. and Cross, F. L. "Incineration of Hospital Infectious Waste," *Pollution Engineering,* November 1988.

Tokarski, C. "Public Outcry Forces Hospitals To Confront Medical Waste Issue," *Modern Healthcare,* September 1988.

Tokarski, C. "Needle Scare Provokes Push For Federal Law Governing Disposal Of Medical Wastes," *Modern Healthcare,* July 1988.

Tokarski, C. "Hospital Brace For Waste-Tracking Costs," *Modern Healthcare,* April 1989.

Tokarski, C. "Debating The Benefits Of Regulation," *Modern Healthcare,* August 1990.

CHAPTER 30

INSTRUMENTATION FOR REMOTE WATER QUALITY MONITORING

Kenneth D. Smith

Naval Air Warfare Center
Trenton, NJ 08628, USA

CONTENTS

INTRODUCTION

A study was performed on remote monitoring sensor technology and instrumentation for "real-time" measurement of water quality. The information was compiled from a search of available literature and information on commercially available monitoring products and also those under development. This report attempts to accurately reflect the state of the art in remote water quality monitoring. The overall concept of remote water quality measurement is very extensive. This chapter focuses on remote sensing technologies that may be used for wastewater and storm water discharges as well as groundwater monitoring. Although an attempt was made to identify as many techniques and applications as possible, it is recognized that some devices may have been inadvertently overlooked. Opinions on the actual status of the development or application of these remote devices are suggested only with regard to the technology employed in general and are not intended to endorse or question the performance of any particular product.

The need for remote water quality monitoring in lakes, rivers, reservoirs, and aquifers is becoming increasingly important. The lack of efficient, cost-effective remote monitoring capability hampers the measurement of water quality parameters needed to access the condition of water resources in "real time." Conventional sampling and analysis techniques currently used for water quality determination typically rely on manpower-intensive methods of direct sampling and laboratory analysis. For some types of analysis, however, no other methods are available that provide accurate results. Direct sampling methods can fail to give adequate coverage of large areas in a timely manner to protect public health and the ecosystem because of inherent delays in analysis.

Regulations requiring water quality monitoring for pollutant releases and transport are expected to be continually expanded by the U.S. Environmental Protection Agency as well as by state environmental authorities. "Best available technology" standards are being required for effluent discharge permits. Current methods of collection and analysis of samples for determination of water quality requires a vast array of expensive equipment and highly trained personnel as well as extensive coordination with the lab performing the analysis. Automated remote monitoring instruments can greatly reduce or eliminate the time delays between sampling and analysis results. Environmental regulations are expanding the number of chemicals that will have to be determined in the future. As these monitoring programs are implemented, associated sampling and analytical costs will escalate considerably beyond existing high costs unless new technology is developed to remotely monitor water quality in real time. The development of *in situ* chemical sensors and the refinement of existing instrumentation will help reduce analytical costs and provide timely information on water quality and pollution migration in rivers, lakes, oceans, and groundwater.

REGULATORY OVERVIEW

The regulatory environment requiring the development and installation of remote water quality sensing instrumentation falls under four major environmental regulations:

- Safe Drinking Water Act (SDWA),
- Resource Conservation Recovery Act (RCRA),
- Water Quality Act (WQA), and
- Comprehensive Environmental Response, Compensation, and Liability Act (CERCLA).

Specific requirements are found in subsections of these major environmental programs.

Safe Drinking Water Act (SDWA)

The Safe Drinking Water Act of 1974 was promulgated to protect groundwater resources used for drinking water supplies. In 1986, the SDWA amendments established a list of priority contaminants, the drinking water priority list, that occur in drinking water and have adverse health effects. The first list, published in 1988, contained 53 contaminants. The list was to be updated every three years. A second list, published in 1991, is provided in Table 1 and contains 77 contaminants and contaminant groups. The SDWA of 1986 also established both enforceable maximum contaminant levels (MCL) and non-enforceable recommended maximum contaminant levels (RMCL). Inorganic chemicals regulated under the National Priority Drinking Water Regulation (NPDWR)(40CFR141,1986) are listed in Table 2, which lists both primary and secondary MCLs for the inorganic chemicals. The MCLs on the secondary list are unenforceable but provide water quality guidelines (40 CFR 143, 1986).

Table 3 lists the organic chemicals from NPDWR. These chemicals include unleaded gasoline additives, cleaning solutions, solvents, and pesticides, all of which commonly are detected in groundwater drinking supplies. The pesticides are inherently toxic and enter the groundwater either through direct application or from precipitation runoff.

The lists are subject to frequent changes. Chemicals may be added or deleted, but in any event, monitoring requirements and analytical costs both will significantly increase in years to come.

Resource Conservation and Recovery Act (RCRA)

The Resource Conservation and Recovery Act (RCRA) of 1976 (40 CFR 261) provided for the "cradle-to-grave" management of hazardous waste and changed

(text continued on page 765)

Table 1
1991 Drinking Water Priority List

Inorganics	
Aluminum	Cyanogen chloride
Boron	Hypochlorite ion
Chloramines	Manganese
Chlorate	Molybdenum
Chlorine	Strontium
Chlorine dioxide	Vanadium
Chlorite	Zinc
Pesticides	
Asulam	Metalaxyl
Bentazon	Methomyl
Bromacil	Metolachlor
Cyanazine	Metribuzin
Cyromazine	Parathion degradation product (4-nitrophenol)
DCPA (and its acid metabolites)	Prometon
Dicamba	2,4,5-T
Ethylenethiourea	Thiodicarb
Fomesafen	Trifluralin
Lactofen/Acifluorfen	
Synthetic Organic Chemicals	
Acrylonitrile	Dichloroacetonitrile
Bromobenzene	1,3-Dichlorobenzene
Bromochloroacetonitrile	Dichlorodifluoromethane
Bromodichloromethane	1,1-Dichloroethane
Bromoform	2,2-Dichloropropane
Chlorination/chloroamination by-products such as haloacetic acids, haloketones, chloral hydrate, MX-12 [3-chloro-4-(dichloromethyl)-5-hydroxy-2(5H)-furanone], and N-organochloramines	1,3-Dichloropropane
	1,1-Dichloropropene
	1,3-Dichloropropene
	2,4-Dinitrophenol
	2,4-Dinitrotoluene
	2,6-Dinitrotoluene
Chloroethane	1,2-Diphenylhydrazine
Chloroform	Fluorotrichloromethane
Chloromethane	Hexachlorobutadiene
Chloropicrin	Hexachloroethane
o-Chlorotoluene	Isophorone
Dibromoacetonitrile	Methyl ethylketone
Dibromochloromethane	Methyl isobutylketone
Dibromomethane	Methyl-t-butylether

Synthetic Organic Chemicals (cont.)	
Nitrobenzene	Naphthalene
Ozone by-products such as aldehydes, epoxides, peroxides, nitrosamines, bromate, iodate	1,1,2,2-Tetrachloroethane
	Tetrahydrofuran
	Trichloroacetonitrile
1,1,1,2-Tetrachloroethane	1,2,3-Trichloropropane
Microorganisms	
Crytosporidum	

Table 2
National Primary Drinking Water Standards for Inorganics

Contaminant	MCL, mg/L	MCLG, mg/L
Primary Regulations:		
Arsenic	0.05	N[a]
Asbestos	7 million fibers/L	7 million fibers/L
Barium	2	2
Cadmium	0.005	0.005
Chromium	0.1	0.1
Copper	Treat if ≥1.3 mg/L	1.3
Lead	Treat if ≥0.015 mg/L	Zero
Fluoride	4	4
Mercury	0.002	0.002
Nitrate[b]	10 (as N)	10 (as N)
Nitrite	1 (as N)	1 (as N)
Selenium	0.005	0.05
Secondary Regulations:		
Aluminum	0.05–0.2	N
Chloride	250	N
Fluoride	2.0	N
Iron	0.3	N
Manganese	0.05	N
Silver	0.1	N
Sulfate	250	N
Zinc	5.0	N

[a]*MCLG for total nitrate and nitrite is 10 mg/L (as N)*
[b]*MCLG not established*

Table 3
National Primary Drinking Water Standards for Organics

Contaminant	MCL, mg/L	MCLG, mg/L
Volatile Organic Chemicals		
Benzene	0.005	Zero
Carbon tetrachloride	0.005	Zero
o-Dichlorobenzene	0.6	0.6
p-Dichlorobenzene	0.075	0.075
1,2 Dichloroethane	0.005	Zero
1,1 Dichloroethylene	0.007	0.007
c-1,2 Dichloroethylene	0.007	0.007
t-1,2 Dichloroethylene	0.1	0.1
1,2 Dichloropropane	0.005	Zero
Ethylbenzene	0.7	0.7
Monochlorobenzene	0.1	0.1
Styrene	0.1	0.1
Tetrachloroethylene	0.005	Zero
Toluene	1	1
1,1,1 Trichloroethane	0.2	0.2
Trichloroethylene (TCE)	0.005	Zero
Trihalomethanes	0.1	N[a]
Vinyl chloride	0.002	Zero
Xylenes (total)	10	10
Synthetic Organic Chemicals		
Acrylamide	Treatment Technique[b]	Zero
Alachlor	0.002	Zero
Aldicarb	0.003	0.001
Aldicarb sulfone	0.003	0.002
Aldicarb sulfoxide	0.003	0.001
Atrazine	0.003	0.003
Carbofuran	0.04	0.04
Chlordane	0.002	Zero
1,2-Dibromo-3-chloropropane (DBCP)	0.0002	Zero
2,4-D	0.07	0.07
Endrin	0.0002	N
Epichlorohydrin	Treatment Technique	Zero
Ethylene dibromide (EDB)	0.00005	Zero
Heptachlor	0.0004	Zero
Heptachlor epoxide	0.0002	Zero
Lindane	0.0002	0.0002

Contaminant	MCL, mg/L	MCLG, mg/L
Methoxychlor	0.04	0.04
Polychlorinated biphenyls	0.005	Zero
Pentachlorophenol	0.001	Zero
Toxaphene	0.003	Zero
2,4,5-TP (Silvex)	0.05	0.05

[a]*MCLG not established*
[b]*No analytical methods available; treatment techniques used.*

(text continued from page 761)

the management criteria for other solid waste materials as well. The Hazardous and Solid Waste Amendments (HSWA) of 1984 also established Underground Storage Tank (UST) regulations as part of RCRA.

Under RCRA, wastes are considered hazardous that exhibit the characteristics of ignitability, corrosivity, reactivity, toxicity, or are designated from lists of 450 other chemicals and substances. Chemicals listed in Table 4 are substances that contain these materials that can not be landfilled if the substance fails the toxic leachate characteristic leaching procedure (TCLP) test.

In 1987 USEPA published a list of chemicals, Appendix IX (40CFR 264 & 270), for which groundwater monitoring will be required at hazardous waste treatment, storage, and disposal facilities (TSDF). As part of the RCRA permit requirements (Part B), owners and operators of TSDFs are required to conduct one of three types of groundwater monitoring programs:

- Detection Monitoring Programs,
- Compliance Monitoring Programs, or
- Corrective Action Programs.

The detection monitoring program is designed to determine if anything from Appendix IX has leaked from the TSDF. If the TSDF permit indicator parameters, such as total organic carbon (TOC), are exceeded, compliance monitoring must be initiated. If contamination from Appendix IX chemicals is shown to be increasing, USEPA may require corrective action to remediate the site.

Underground storage tanks (USTs) have a high potential for leaks. As part of UST regulations, all underground storage tanks require upgrades to protect against leaks and/or must be replaced. Even new tanks with double-wall construction require leak-detection sensing devices.

Water Quality Act

The Federal Water Pollution Control Act of 1972 was amended in 1977 as the Clean Water Act, which was subsequently amended in 1987 as the Water

Table 4
Regulatory Limits for TCLP Analytes

Metals			
Arsenic	5.0	Lead	5.0
Barium	100	Mercury	0.2
Cadmium	1.0	Selenium	1.0
Chromium	5.0	Silver	5.0
Volatile Organics			
Benzene	0.5	1,1-Dichloroethylene	0.7
Carbon tetrachloride	0.5	Methyl ethyl ketone	200
Chlorobenzene	100	Tetrachloroethylene	0.7
Chloroform	6.0	Trichloroethylene	0.5
1,2-Dichloroethane	0.5	Vinyl chloride	0.2
Semi-Volatile Organics			
o-Cresol	200	Hexachloroethane	3.0
m-Cresol	200	Nitrobenzene	2.0
p-Cresol	200	Pentachlorophenol	100
1,4-Dichlorobenzene	7.5	Pyridine	5.0
2,4-Dinitrotoluene	0.13	2,4,5-Trichlorophenol	400
Hexachlorobenzene	0.13	2,4,6-Trichlorophenol	2.0
Pesticides/Herbicides			
Chlordane	0.3	Lindane	0.4
2,4-D	10	Methoxychlor	10
Endrin	0.02	Toxaphene	0.5
Heptachlor (and its hydroxide)	0.008	2,4,5-TP (Silvex)	1.0

Quality Act of 1987. Under the Water Quality Act, the discharge of any pollutant into a navigable water is prohibited unless authorized by a permit. The National Pollution Discharge Elimination System (NPDES) permit specifies the limits for each discharge parameter. The NPDES permit is usually enforced through self-monitoring and reporting programs. NPDES parameters and typical limits are shown in Table 5. Additional monitoring in addition to those listed in Table 5 may be required depending on the discharger's industrial category.

Storm water discharge permits also are issued under the NPDES program. Storm water permits require monitoring for basically the same parameters as the NPDES discharge permit shown in Table 5.

Industrial facilities that do not discharge directly to navigable water often pretreat wastewater prior to discharge to publicly owned treatment works

Table 5
NPDES Permit Monitoring Parameters

Parameter	Discharge Limits	Type Sample
Biological Oxygen Demand (BOD)	30 ppm	composite
Chemical Oxygen Demand (COD)	50 ppm	composite
Total Suspended Solids (TSS)	50 ppm	composite
Total Organic Carbon (TOC)	10 ppm	composite
Ammonia (as N)	10 ppm	composite
Temperature	30 C	grab
pH	6 to 9	grab
Oil and Grease	10 ppm	grab

(POTW). Such indirect dischargers are required also to monitor their discharges under an industrial pretreatment program. Under the pretreatment program, a facility is regulated according to industry-specific categorical standards.

Comprehensive Environmental Response, Compensation, and Liability Act (CERCLA)

CERCLA, better known as superfund, was enacted in 1980 to expedite the cleanup of contaminated groundwater from abandoned dumps and hazardous wastes sites. Monitoring requirements for superfund sites primarily involve the EPA Contract Lab Program (CLP). The list of 125 analytes, which includes volatile organic compounds, semivolatile organics, pesticides and metals, is called the target compound list (TCL) or alternately the priority pollutant list (see Table 6). Remote sensing technology is desired to monitor contaminant migration in the groundwater from superfund sites and also to document progress in the cleanup programs.

APPLICATIONS FOR REMOTE MONITORING SENSORS

The most prominent force in the development and installation of remote monitoring sensors and remote data collection technology is the need to reduce the high costs incurred in conventional sampling and monitoring methods. Increased monitoring requirements from environmental regulatory agencies are expected to continue along with the increasing number of chemicals to be monitored with continually decreasing acceptable discharge limits for all parameters. Conventional sampling techniques and analytical methods require manual sampling in many cases followed by wet chemical analysis in laboratories, which are often located a considerable distance from the sampling site.

Table 6
Priority Pollutant List

Number	Chemical Name
1	Chloromethane
2	Bromomethane
3	Vinyl chloride
4	Chloroethane
5	Methylene chloride
6	Acetone
7	Carbon disulfide
8	1,1-Dichloroethene
9	1,1-Dichloroethane
10	1,2-Dichloroethene (total)
11	Chloroform
12	1,2-Dichloroethane
13	2-Butanone
14	1,1,1-Trichloroethane
15	Carbon tetrachloride
16	Bromodichloromethane
17	1,2-Dichloropropane
18	cis-1.3-Dichloropropene
19	Trichloroethene
20	Dibromochlormethane
21	1,1,2-Trichloroethane
22	Benzene
23	trans-1,3-Dichloropropene
24	Bromoform
25	4-Methyl-2-pentanone
26	2-Hexanone
27	Tetrachloroethene
28	Toluene
29	1,1,2,2-Tetrachloroethane
30	Chlorobenzene
31	Ethylbenzene
32	Styrene
33	Xylenes (total)
34	Phenol
35	bis (2-Chloroethyl) ether
36	2-Chlorophenol
37	1,3-Dichlorobenzene
38	1,4-Dichlorobenzene
39	1,2-Dichlorobenzene
40	2-Methylphenol
41	2,2'-oxybis (1-Chloropropane)
42	4-Methylphenol
43	N-Nitroso-di-n-propylamine
44	Hexachloroethane
45	Nitrobenzene
46	Isophorone
47	2-Nitrophenol
48	2,4-Dimethylphenol
49	bis (2-Chloroethoxy) methane
50	2,4-Dichlorophenol
51	1,2,4-Trichlorobenzene
52	Naphthalene
53	4-Chloroanaline
54	Hexachlorobutadiene
55	4-Chloro-3-methylphenol
56	2-Methylnapthalene
57	Hexachlorocyclopentadiene
58	2,4,6-Trichlorophenol
59	2,4,5-Trichlorophenol
60	2-Chloronaphthalene
61	2-Nitroaniline
62	Dimethylphthalate
63	Acenaphthylene
64	2,6-Dinitrotoluene
65	3-Nitroaniline
66	Acenaphthene
67	2,4-Dinitrophenol
68	4-Nitrophenol
69	Dibenzofuran
70	2,4-Dinitrotoluene
71	Diethylphthalate
72	4-Chlorophenyl-phenylether
73	Fluorene
74	4-Nitroaniline
75	4,6-Dinitro-2-methylphenol
76	N-Nitrosodiphenylamine
77	4-Bromophenyl-phenylether
78	Hexachlorobenzene
79	Pentachlorophenol
80	Phenanthrene
81	Anthracene
82	Carbazole

Number	Chemical Name	Number	Chemical Name
83	Di-n-Butylphthalate	104	Heptachlor epoxide
84	Fluoranthene	105	Endosulfan I
85	Pyrene	106	Dieldrin
86	Butylbenzylphthalate	107	4,4'-DDE
87	3,3'-Dichlorobenzidine	108	Endrin
88	Benzo [a] anthracene	109	Endosulfan II
89	Chrysene	110	4,4'-DDD
90	bis (2-Ethylhexyl) phthalate	111	Endosulfan sulfate
		112	4,4'-DDT
91	Di-n-Octylphthalate	113	Methoxychlor
92	Benzo [b] fluoranthene	114	Endrin ketone
93	Benzo [k] flouranthene	115	Endrin aldehyde
94	Benzo [a] pyrene	116	alpha-Chlordane
95	Indeno [1,2,3-c,d] pyrene	117	gamma-Chlordane
96	Dibenz [a,h] anthracene	118	Toxaphene
97	Benzo [g,h,i] perylene	119	Aroclor-1016
98	alpha-BHC	120	Aroclor-1221
99	beta-BHC	121	Aroclor-1232
100	delta-BHC	122	Aroclor-1242
101	gamma-BHC (Lindane)	123	Aroclor-1248
102	Heptachlor	124	Aroclor-1254
103	Aldrin	125	Aroclor-1260

Manual sampling requires assembling and transporting to the sample site numerous sample containers, chemicals to stabilize the sample, refrigeration (ice chests), and portable monitoring equipment such as a pH meter. Additional disadvantages of conventional sampling include inconsistent sampling techniques, time-consuming preparation of sampling equipment, delays in analysis, and exposure of personnel to potentially dangerous sampling locations and violent weather conditions.

Another problem worthy of mention with conventional sampling methods is the threat of liability and financial penalties that could be imposed if a leak or discharge into a navigable water from a facility were to continue undetected for an extended period of time. Adverse publicity and possible civil and criminal charges could result from undetected spills.

Areas of Application

Sensors are widely available for remote monitoring of physical measurements. Most physical sensors are well developed and acceptable for most remote monitoring requirements where only trends or pollution indicators are required. Some commonly used sensors for measuring the physical characteristics in water are:

- Temperature
- Conductivity
- Pressure
- pH
- Oxidation/Reduction (Eh)
- Depth

Inorganic and organic chemical sensors are developed to detect those contaminants listed in environmental regulations concerned with hazardous wastes, groundwater, surface water, and drinking water. There are several chemical sensors commercially available, but most are still in the research and development stage. Additional information concerning sensor capabilities is discussed later.

The potential applicability for remote sensing capability in the area of water quality measurement is almost limitless because many techniques still are being developed. Specific applications for remote sensors include:

- Storm water outfall monitoring,
- Wastewater treatment plant effluents,
- *In situ* groundwater monitoring,
- Agricultural runoff monitoring,
- Aboveground and underground tank and pipeline leak detection monitoring, and
- Monitoring of lakes, rivers, bays, and the oceans for oil spills, algae blooms, and sediment transport.

In general, remote monitoring instruments have the potential to provide real-time measurements in critical areas that have limited accessibility and where immediate knowledge of the water quality is required. Protection against accidental discharges of spills and leaks is a growing concern for industrial facilities. The ability of the sensors to communicate with centralized control and data acquisition systems is also a key element in the application of remote monitoring technology.

Additional applications for remote water quality sensors include:

- Ammonia detection in streams and rivers,
- Detection of microorganisms in drinking water,
- Detection of halogenated hydrocarbons,
- Oil slick and oil spill monitoring,
- On-line total organic carbon, and
- Heavy metals detection.

Criteria for Sensor Selection

When considering sensors for remote monitoring applications, several characteristics must be considered. These characteristics are:

- Detectability and sensitivity,
- Selectivity,
- Response time,
- Reversibility,
- Field calibration,
- Ruggedness and reliability,
- Data handling and transmitting capability,
- Power source availability, and
- Cost effectiveness.

The sensor must be capable of detecting the parameter at the sensitivity level required by the regulations. The sensor also must be selective for a specific contaminant and ignore interferences. A sensor must be reversible. In other words, a reversible sensor can be left in a monitoring location and respond to varying concentrations of the analyte over an extended period of time. An irreversible sensor is capable of only one measurement without being recalibrated. Sensors must have a fast response time so that immediate use can be made of the information collected. The sensing instrument must be rugged enough to withstand physical shock as well as extremes in weather conditions. Power sources are also an important consideration for remote locations. Hard wiring for power at remote locations and the transmission of data through land lines may be impractical or impossible. Batteries, solar, and wind energy sources must be relied on. Therefore, instruments must have low power requirements.

REMOTE MONITORING INSTRUMENTATION

Water quality is the condition of a water resource such as a lake, river, or groundwater used for recreation, drinking water, or navigation. The condition of the water can be described by a variety of physical and chemical parameters such as color, clarity, salinity, pH, temperature, dissolved oxygen, microorganism population, metals, and organic chemicals. Physical parameters are the most convenient measurements to make using remote sensing techniques. A significant amount of development work is taking place to improve the technology of fiber-optic based chemical sensors that use a variety of complex optical spectroscopy techniques. Spectroscopy, the study of spectra, has been used to identify atoms, molecules, and their associated structures and environments. Spectroscopy is performed by analyzing the wavelength and intensity of the radiation emitted, absorbed, or scatted by atoms and molecules under electromagnetic conditions. These techniques that are characterized by reflected energy in the visible, infrared, or microwave frequencies include:

- Absorbance,
- Reflectance,
- Refractive index, and
- Raman scattering

Older methods of remote sensing employed techniques that measured the backscatter of solar energy. These methods are photography, radiometry, and spectoradiometry. These techniques were able to measure simple water quality parameters such as turbidity plumes, temperature, and color. Typically, these measurements were taken from aircraft or satellites. Several remote sensing technologies and sensing devices will be explored.

Today the term "remote sensing" is used more to describe the art of aerial and satellite photography to visually assess worldwide environmental conditions. This type of remote sensing is valuable for tracking oil spills, algae blooms, forest fires, weather, and some water quality parameters such as turbidity and temperature. "Remote monitoring," on the other hand may be the term that more accurately describes methods for direct, real-time measurement of water quality. Remote monitoring sensors also can transmit their data to satellites or aircraft. This is accomplished by reflected light back to the satellite sensors that convert the energy to digital images.

Turbidity and Suspended Solids

Turbidity is an optical property of water used to describe its clarity. It is caused by light scattering from particles. Suspended solid concentrations are defined in terms of mass and volume. There is no direct correlation between turbidity and suspended solids. Transmissometers, nephelometers, or a combination of the two are used to measure turbidity.

Transmissometers measure the attenuation of light caused by the absorption, scattering, and blockage by suspended and dissolved particles in water when a sample is passed between a light source and a detector. Figure 1 shows the

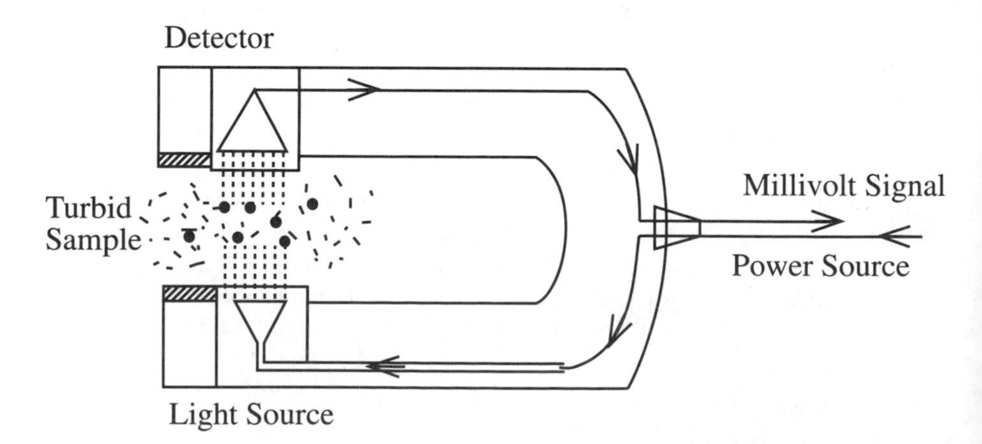

Figure 1. Transmissometer with photoelectric cell.

principle of operation of the transmissometer where a photo cell measures the amount of light that passes through the water sample from a light source.

Nephelometers or back-scattering meters measure the amount of light that is scattered from the particles in the water. Suspended particles scatter light rays in all directions. The amount of light that is scattered depends on the size, shape, refractive index, quantity, and composition of the particles. To measure the quantity of light lost by the scattering, the scattering must be measured at all angles within a certain window. The most common angle is 90 degrees, although some instruments can measure scattering up to 140 to 160 degrees.

The most common light source for the nephelometer is an infrared light-emitting diode (IRLED). Figure 2 demonstrates the turbidity measurement technique. The IRLED sends infrared light into the turbid water where it is scattered by the particles. The photodetector receives light scattered at 90 degrees from the light source. The amount of light detected by the photodetector is converted to an electrical signal (millivolts) that can be transmitted from remote locations.

Turbidometers can monitor drinking water and wastewater quality. Stormwater outfalls, non-point source run-off from agricultural lands, and sediment transport

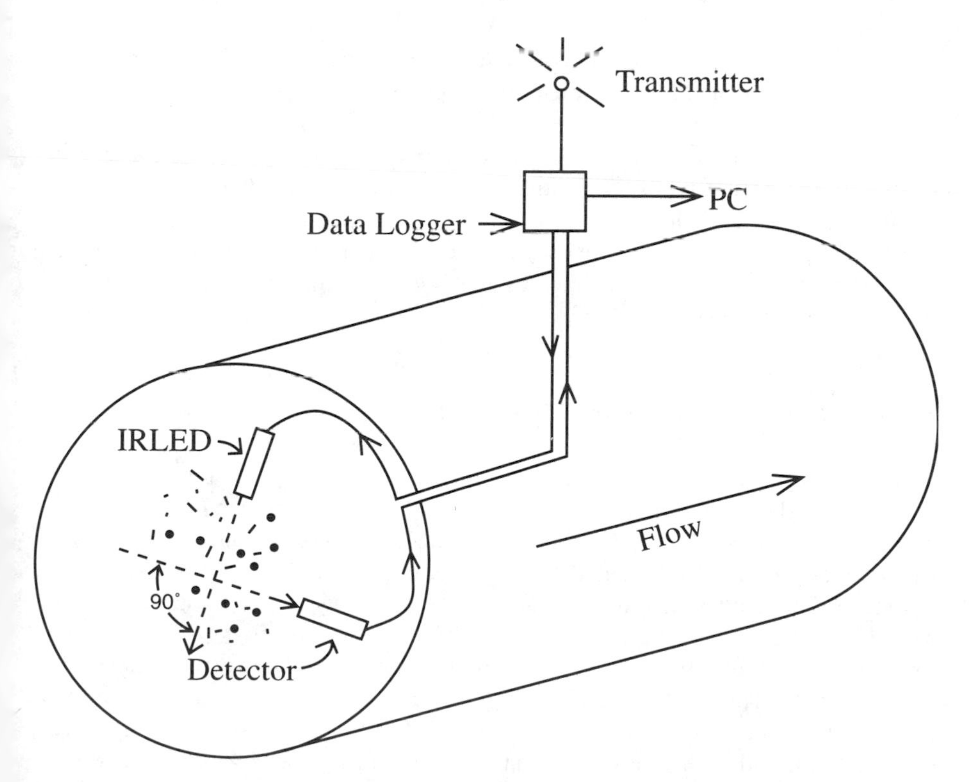

Figure 2. Turbidimeter with infrared light emitting diode (IRLED).

also can be monitored. Although turbidity and suspended solids are not the same thing, newly developed turbidity monitors can convert turbidity to suspended solid concentrations.

Measurement of turbidity and suspended solids are important in streams, rivers, and lakes because most solids eventually settle as sedimentation. Suspended solids and sediments have been shown to be an important means of pollutant transport. The transport mechanism of hydrocarbons, trace metals, and heavy metals, particularly mercury, has been determined to be directly associated with the concentration of suspended solids. Absorption of mercury on suspended solids is directly proportional to the surface area of the particles. Finer particles will increase surface area and therefore pollution absorption is greater. Oil is another substance that is transported largely by suspended solids. Because finer particles settle slowly and are disturbed by minor turbulence, their potential to transport pollutants is high. Therefore, turbidity as a water quality indicator is an important measurement.

Measurement of Conventional Physical Parameters

Basic information about groundwater in remote locations can be obtained with multisensor detectors. Measurement parameters such as temperature, specific conductance, pH, dissolved oxygen, oxidation-reduction potential, depth, and level can be incorporated into a single multiparameter probe. The probe can be placed in monitoring wells, at effluent outfalls or in the middle of a lake. The compact designs have low power requirements, miniaturized circuitry and data transmitting ability. Figure 3 shows how a multiparameter remote monitoring station might be placed in a lake. Figure 4 demonstrates an application at a groundwater monitoring well.

Storm Water and Wastewater Monitoring

Monitoring discharges at stormwater and wastewater outfalls traditionally has required cumbersome sampling techniques. Wet chemistry analysis for some parameters such as BOD and COD still is required, but continuing advances in the state of the art of automated sampling is easing the burden considerably.

One of the most challenging monitoring requirements is stormwater sampling. Monitoring regulations require two types of samples. The first is a "first-flush grab sample" to be obtained from the outfall during the first 30 minutes of a "storm event." The second type of sample is actually a series of flow-weighted composite samples taken over the duration of the storm event. A qualifying storm event is a rainfall of 0.1 inch and a storm duration of one half to one and one half times the length of the average storm in the area.

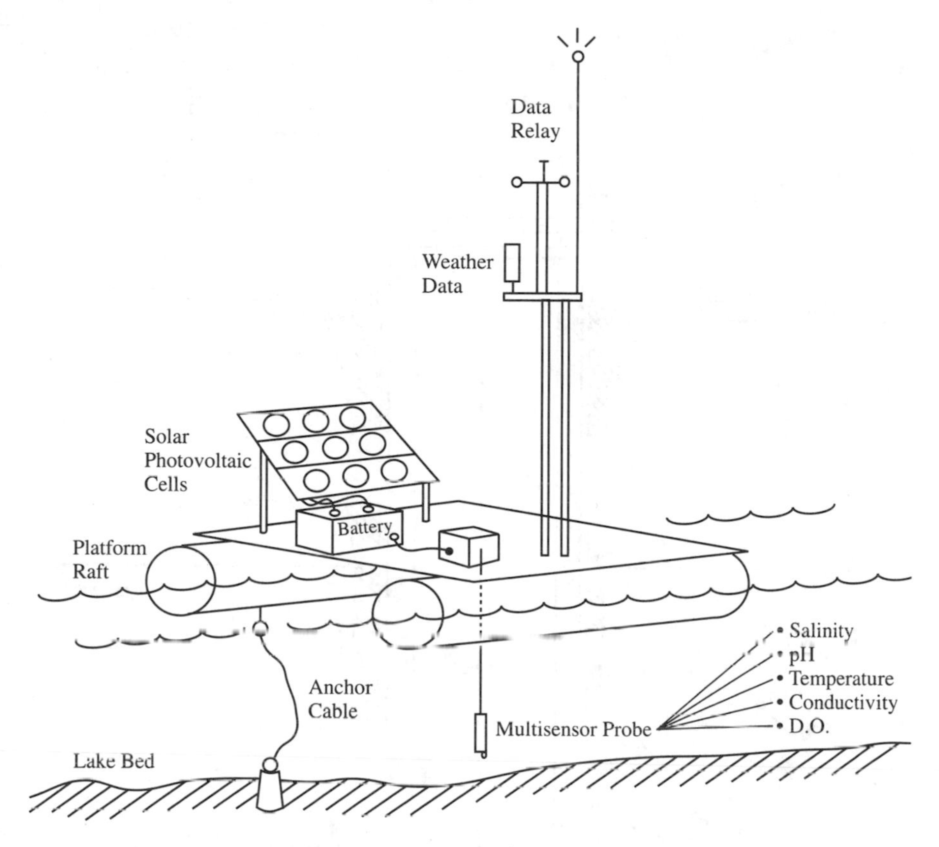

Figure 3. Multiparameter remote monitoring station.

Prior to automatic samplers, a facility had to scramble available personnel to manually take the required samples. A large facility with many outfalls also would need many people involved in the sampling program. Often brief showers did not qualify as storm events, leading to further frustration. The automatic sampler consists of a number of glass and plastic sample containers into which samples of the storm water are pumped. The sampler is initially activated by an automatic rain gauge. An open-channel flow meter is also necessary to control the length of time the sample pump operates to satisfy the flow-weighted composite sample requirement. Flow sensor techniques typically use either ultrasonic, submerged pressure transducer, or bubbler technologies. The type of flow sensor used depends on the specific application, but the ultrasonic design offers noncontact and nonfouling features. Figure 5 shows a typical equipment configuration for remote storm water sampling applications. Automatic samplers are routinely used for monitoring outfall discharges for NPDES permit compliance.

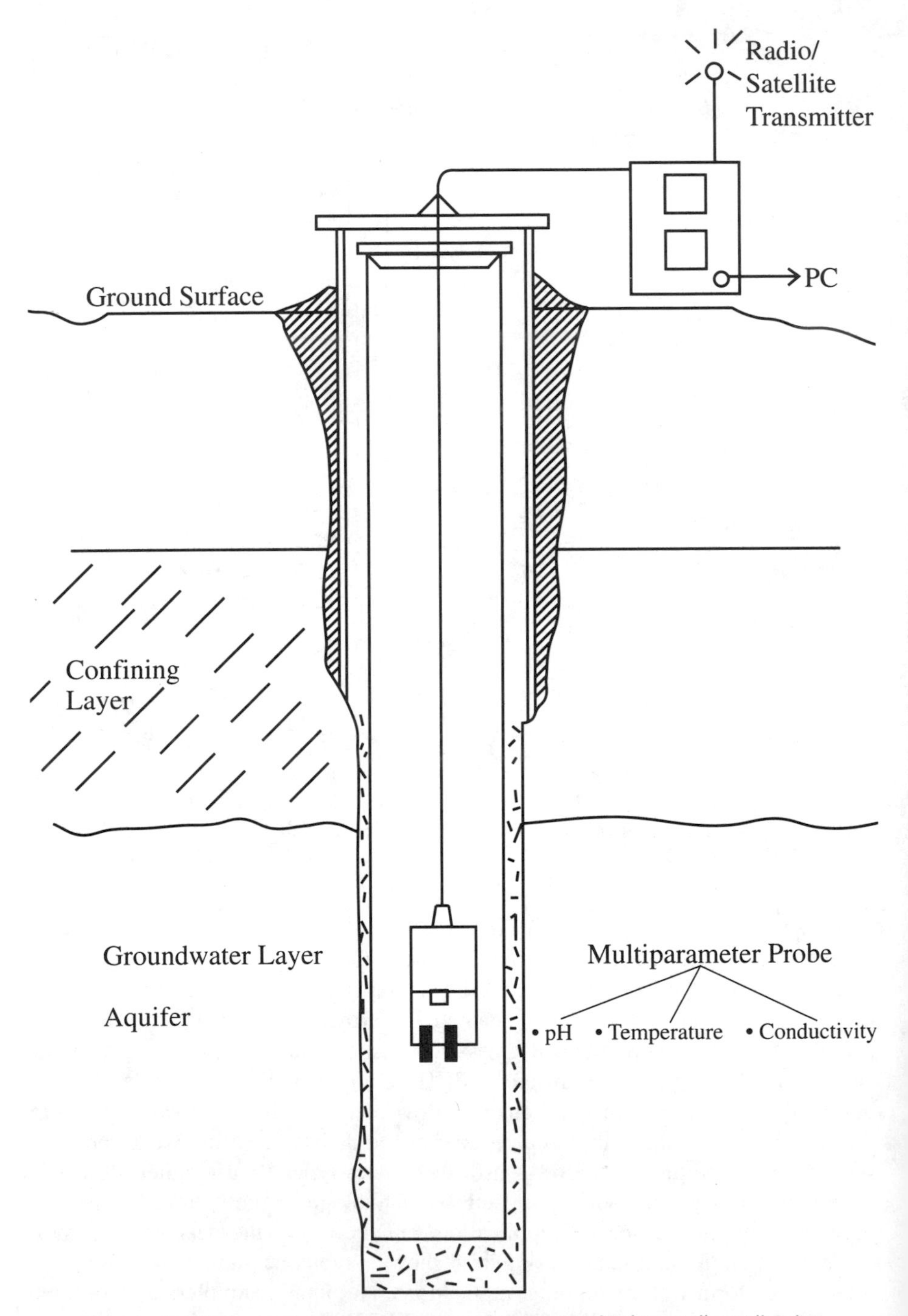

Figure 4. Multiparameter remote groundwater monitoring well application.

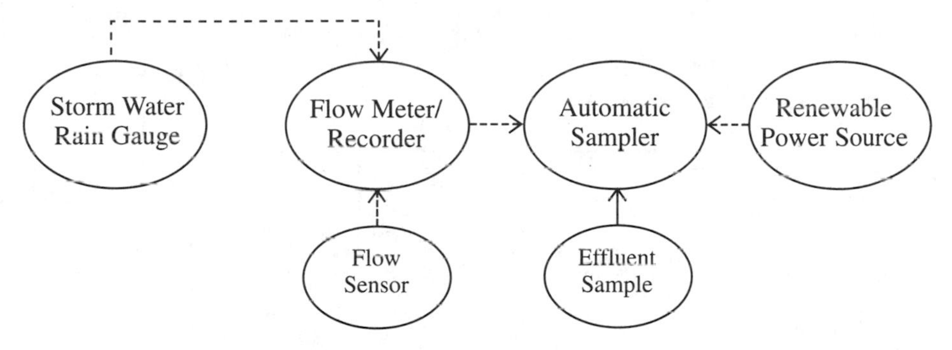

Figure 5. Equipment configuration for automatic sampling.

Fiber Optic Chemical Sensors

Significant development work is taking place in the field of fiber optic chemical sensors. To date, however, most fiber optic chemical sensor systems designed for environmental monitoring require substantial further development. Fiber optics technology originally was developed for the communications industry.

Optic spectroscopy techniques operate by modulation of a light beam. The light is transmitted in optical fibers 5 to 10 microns in diameter. The optical fibers typically are made from plastic or glass. The general configuration of a fiber optic sensing system involves three regions. The first region is a light source. The second is the sensing area where the measurement takes place. The third region is the detector that normally consists of either a photomultiplier, photodiode, or a diode array.

Two types of fiber optic chemical sensors for detecting chlorinated hydrocarbons have been developed. One type is irreversible and one is for continuous measurements. Both sensors use basic pyridine as a chemical reagent that changes to an intense red color when exposed to certain chlorinated compounds. Sensitivity has been demonstrated comparable to gas-chromatography for detecting trace levels of trichloroethylene and chloroform. The sensor is a modification of a previously developed fluorescence-based sensor. Figure 6 shows sketches of the two types of chemical sensor.

Groundwater contaminants can be detected *in situ* by making spectroscopic measurements through fiber optics. One technique of the fiber optic chemical sensor technology is being developed for use with the cone penetrometer. The cone penetrometer is a 1.25" diameter probe that provides a cost-effective alternative to conventional sampling methods. A cone penetrometer equipped with chemical sensors is driven into the ground to provide *in situ* real-time monitoring of contaminated groundwater plumes. The cone penetrometer is being developed by the U.S. Department of Energy and the U.S. Army Corps of Engineers.

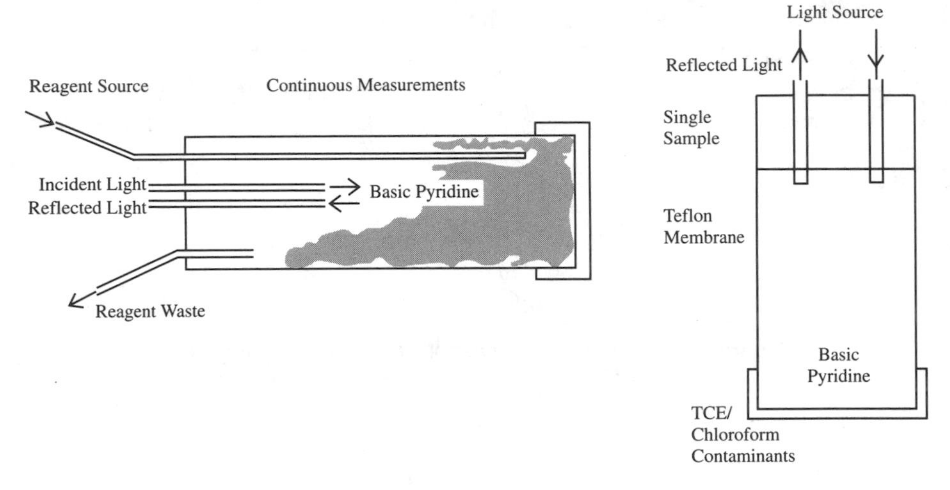

Figure 6. Fiber optic chemical sensors.

Current fiber optic chemical sensing techniques usually employ one of eight chemical sensing methods. These methods are:

- Optical fluorescence,
- Colorimetry,
- Optical scattering,
- Spetrochemical emission,
- Infrared absorption,
- Attenuated total internal reflection,
- Refractometry, and
- Chemical induced mechanical perturbations.

The main limitation for most fiber optic chemical sensors so far is that they are primarily qualitative, not quantitative. Fiber optic chemical sensors are being designed for detection of chlorinated organics and gasoline in groundwater.

Electrochemical Detectors

These detectors are used to monitor for specific chemicals, metals, or ions. Most require three electrodes—reference, working, and auxiliary—mounted in a single cell. A current flows between the working and auxiliary electrodes. The reference electrode and electrical interferences are disadvantages for electrochemical devices when compared to fiber optic sensors.

Oil Pollution Detection

Oil, in terms of quantities spilled or dumped into water resources, is the most prevalent water pollutant. Early detection is the most effective way to prevent damage to ecological systems, property, and treatment plants. The condition of oil in water may be either dissolved, undissolved, or emulsified. Most oil, however, tends to float on water. Several physical/chemical parameters may affect the proportion of each condition:

- Type of oil,
- Temperature,
- Contact time,
- Turbulence,
- pH, and
- Salinity.

A good oil detection sensor therefore should be unaffected by the oil/water conditions or the physical/chemical parameters.

Three techniques can be used to detect oil in water:

- Turbidimeters,
- Fluorescence meters, and
- Infrared reflectance.

Turbidimeters can measure both undissolved and emulsified oil and can detect oil down to 0.005 mg/l. Turbidimeters cannot measure dissolved oil in water, and the technique measures all particles in the water, not just the oil.

Fluorescence meters can measure dissolved, undissolved, and emulsified oil in water. Fluorescence meters measure the ability of the oil to absorb energy from an ultraviolet light source, then emit the energy as light. The emitted light is proportional to oil in the sample. Figure 7 shows the emitted light intensity versus concentration for three different hydrocarbons.

A newer method of oil detection is an electro-optical instrument designed for remote locations. Oil detection is based on changes in infrared reflectance. This method is qualitative only but detects an oil film on water down to a thickness of 0.1 microns. The advantage of this system is that there is no liquid contact and therefore no fouling of the optical surface due to oil in the water. Figure 8 demonstrates the infrared reflectance technology. This technology is new and relatively expensive.

All of the oil detection instruments can be remotely located and configured to interact with alarms. The instruments would typically be installed at facility storm water and wastewater outfalls. Early spill or leak warning devices are becoming increasingly important to eliminate cleanup liability, regulatory financial penalties, and bad public relations.

On-Line Continuous Monitors

NPDES permit monitoring at storm water and wastewater outfalls requires sample analysis for BOD and COD. Remote automatic samplers can ease the cost by taking the samples, but analysis still must be performed in the laboratory.

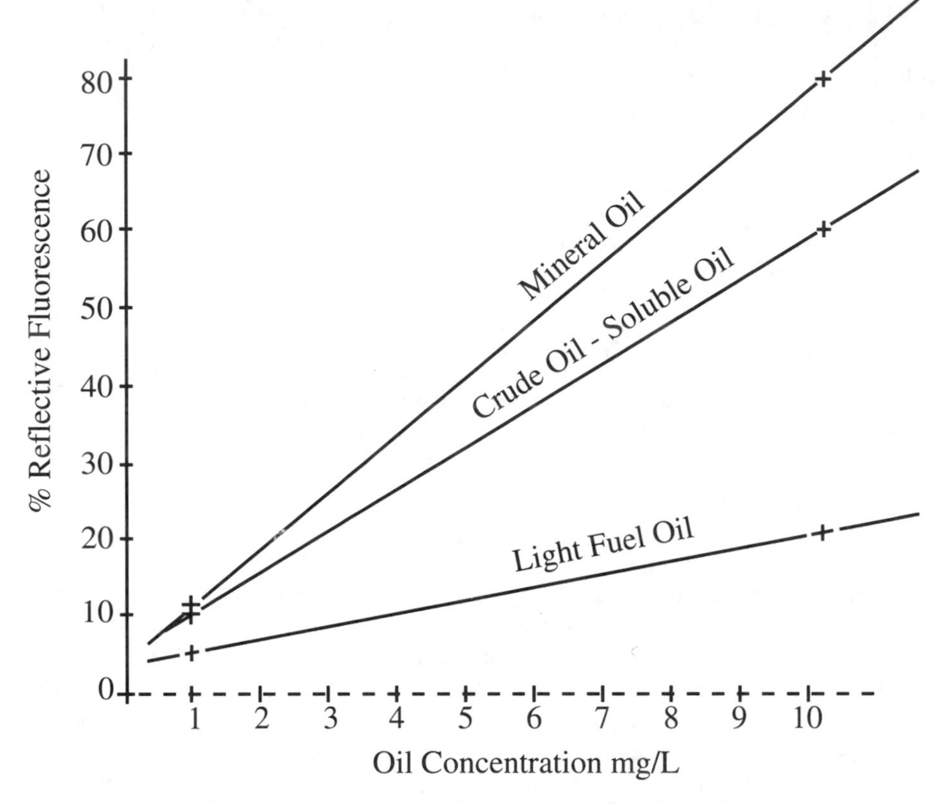

Figure 7. Approximate fluorescence of various oils.

On-line continuous monitors that perform total organic carbon (TOC) and gas chromatograph (GC) analysis are currently available.

The TOC analyzer can be used in applications where non-specific compound identification is required as with BOD/COD analysis. Samples are injected into a combustion tube where carbon is converted to carbon dioxide. A non-dispersive infrared (NDIR) detector measures the carbon dioxide, which is directly proportional to the organic carbon in the sample. The TOC analyzer also provides real-time analyses for total carbon (TC) and total inorganic carbon (TIC). The TOC system, which also includes a printer, sampler pump, and automatic carousel sampler, requires a 115-volt AC power supply, so remote is a relative term defined by the nearest available power source. Cost of the system could be about $20,000.

The gas chromatograph operates on the principle of purge and trap. The GC provides continuous automatic measurement of volatile organic compounds (VOCs). Unlike TOC, the GC can qualitatively and quantitatively measure for a variety of compounds in the low parts per billion range. The on-line GC is also expensive and has 115-volt AC power requirements.

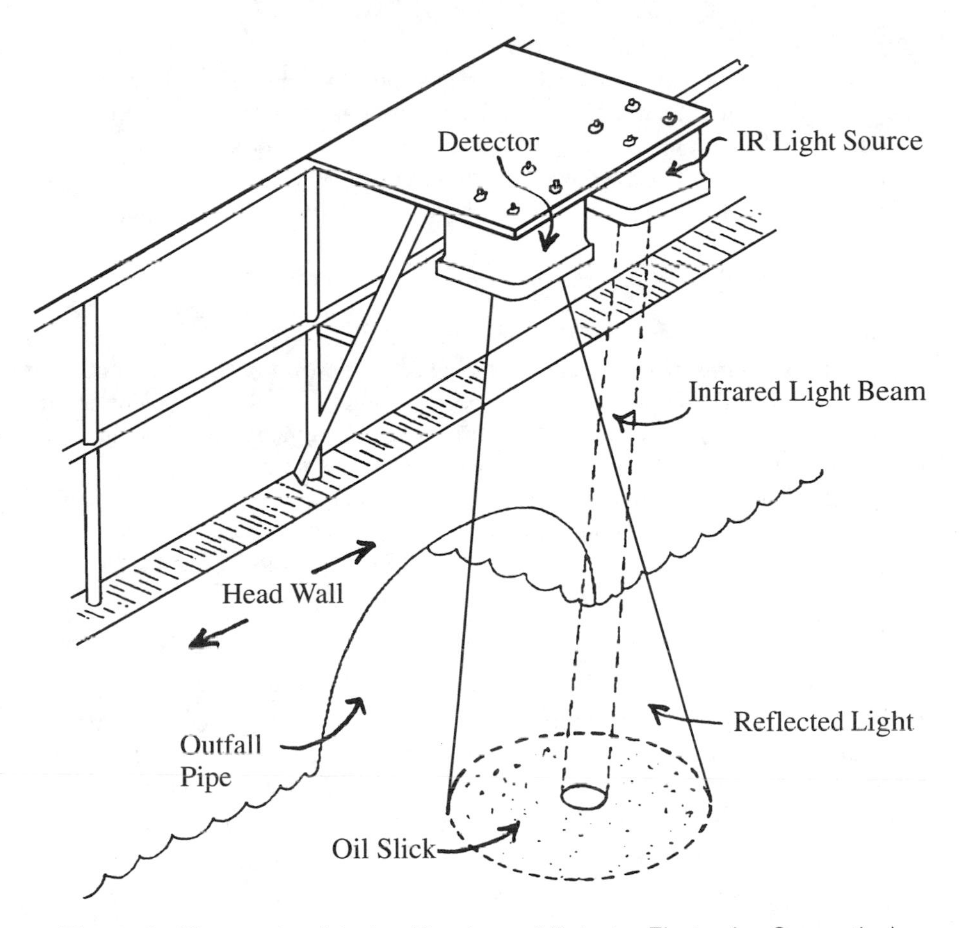

Figure 8. Oil on water detector (Courtesy of Detector Electronics Corporation).

More stringent, precise, and timely monitoring requirements eventually will make on-line TOC and GC monitors more cost effective. These instruments do not meet the ruggedness or low power requirement criteria for remote sampling beyond the practice range of a 115-volt AC power source. A special enclosure and security for this equipment is also necessary.

PRACTICAL APPLICATIONS FOR REMOTE MONITORING

Remote Sensing/GIS

Remote monitoring sensors have the capability of communicating with many information transmitting, receiving, and management systems. Among these systems are computers, modems, radios, and satellite telemetry transmitters.

"Remote sensing" via satellite is being integrated with remote monitoring systems on the ground and with a new system called the geographical information system (GIS). Satellite imagery, aerial photographs, and ground-based *in situ* remote monitors are being integrated with GIS to develop groundwater pollution plume maps, to map oil and chemical spills, and to track sedimentation transport in rivers. GIS is relatively new but is growing rapidly in popularity.

Automatic Sampling

Automatic samplers can be programmed to initiate and end storm water and wastewater sampling when levels of selected analytes are within limits defined by the facility with the aid of analyte-specific sampling technology. The ability to identify contaminants in intermittent discharges is a valuable wastewater discharge management tool. Microprocessor-controlled samplers and sensors can be programmed to operate under upset-only conditions. This would alert facility personnel to process malfunctions, spills, or leaks. Alarm signals from the sensors can be used to close sluice gates on outfalls to contain contaminants on site.

Leak Detection

Leak detection sensors are particularly useful for monitoring above-ground tanks and pipelines. Indirect methods of tank monitoring are level alarms and indicators. Continuous-level monitoring provides an instantaneous inventory record. Coupled with high- and low-level alarms, leaks and overfills can be detected rapidly. Figure 9 shows a typical tank level indicator/alarm configuration.

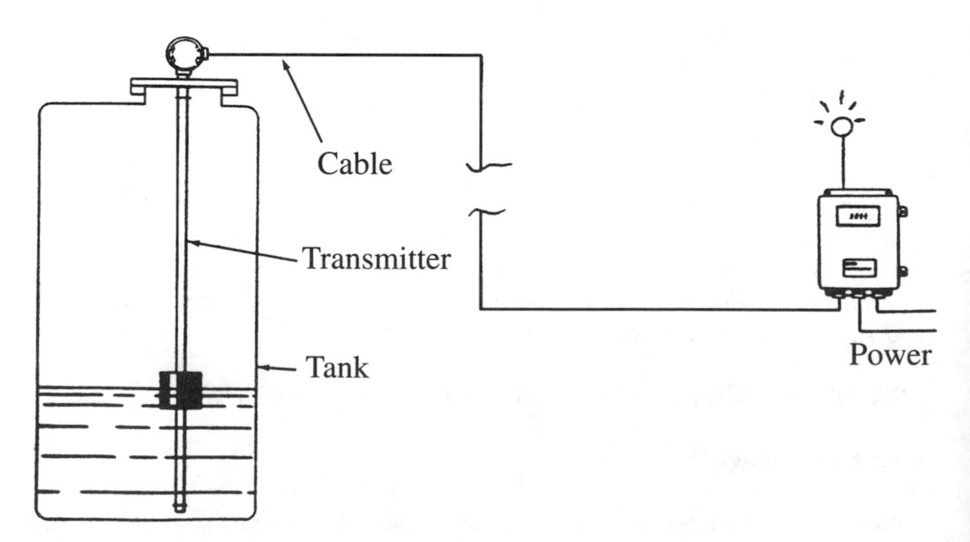

Figure 9. Tank level indicator/alarm (Courtesy of IMO Industries, Inc.).

New containment standards require double-walled piping for underground and preferably also above-ground storage tanks. Leak sensing cables that can detect fuels, solvents, acids, bases, and other conductive fluids can be installed in double-walled pipe between the inner fluid-carrying pipe and the surrounding containment pipe. The cables also can be used for continuous monitoring of tank containments and interstitial space in double-walled tanks. When a leaking liquid contacts the cable, an alarm is activated and the leak location is pinpointed by computing the length of the cable from the control panel to the leak. Figure 10 shows three examples of how the cables can be installed.

DATA MANAGEMENT

Remote Environmental Monitoring Systems

The heart of the remote monitoring system is the programmable remote terminal unit (RTU), which gathers and stores data from remote sensors and controls output devices. Data can be downloaded to a laptop personal computer from

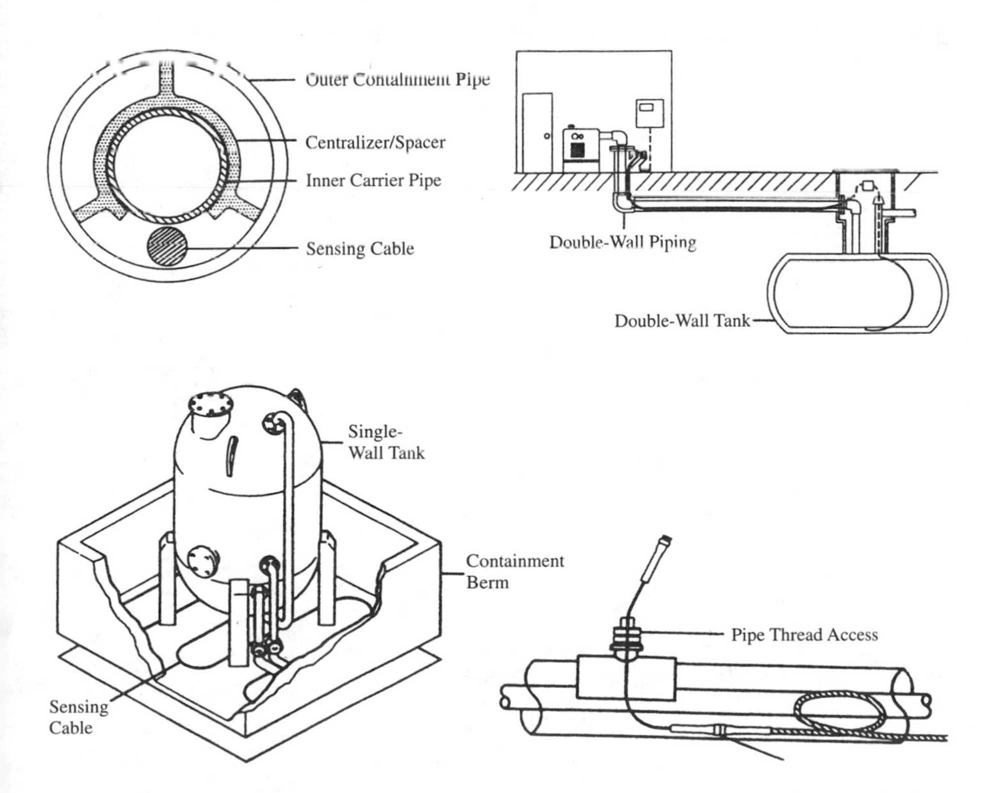

Figure 10. Leak sensing cable applications (Courtesy of Raychem Corporation).

the remote data logger. This data can then be integrated with modeling software and GIS. Data also can be accessed via satellite UHF/VHF radio, telephone modem, and cellular phones. Remote sensors input hydrology, meteorological, water quality, soil contamination, air quality, and radiation data to the remote terminal unit. Remote terminal units can be battery (12VDC) operated in remote areas with a solar panel. Figure 11 describes how the remote terminal unit interfaces with other peripherals.

Expert Systems

Environmental expert systems provide a new tool for professionals responsible for solving environmental problems. Environmental expert systems are only about eight years old, even though the technology for expert systems has existed for more than 20 years. Development of environmental expert systems has been rapid, and more than 68 systems exist today. Early systems were limited to sewage treatment plant operations. Today, systems are focusing on more complex environmental issues.

An expert system is defined as man and machine systems with specialized problem-solving expertise. The slow emergence of environmental expert systems is twofold. First, many environmental problems are not well understood, and few

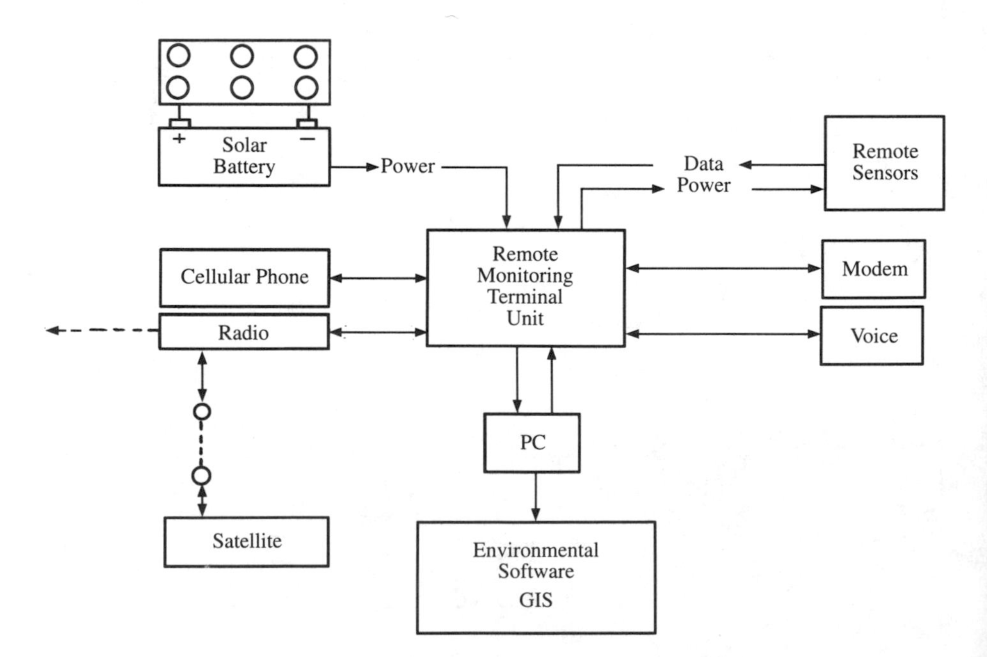

Figure 11. Remote monitoring terminal units.

environmental methods are absolutely agreed upon. Second, environmental problems are complex and cannot be solved by any single area of expertise. As remote sensor technology is perfected, the data generated by remote monitoring will be used in expert systems to understand complex environmental problems. Some of the areas where expert systems are currently being applied are:

- Hazardous waste-site cleanup technology selection,
- Characterization of waste sites using the MITRE hazard ranking system,
- Selection of treatment/recycle facilities,
- Selection of waste site remedial action technologies, and
- Geographic information systems.

CONCLUSIONS

This chapter has assessed the state of the art in remote environmental water quality monitoring. The escalating cost of sampling and analyzing and an ever-increasing list of analytes is the driving incentive to develop remote monitoring sensors and equipment that are reliable and capable of a high degree of both qualitative and quantitative measurement. Tables 7, 8, and 9 provide a partial list of commercially available water quality sensors and monitoring equipment.

Sensors are commercially available to monitor physical water quality parameters such as temperature, pH, dissolved oxygen, oxidation/reduction potential, and turbidity that are cost effective. The status of the sensors in the physical measurement category is sufficiently developed to provide accurate real-time detection of existing and developing water-quality deterioration.

Sensors to monitor chemical water quality parameters to determine concentrations of specific chemical pollutants and dissolved substances are, in general, not well developed. Fiber optic chemical sensors (FOCS) offer unique capabilities and versatility not available in other chemical sensors. FOCS technology is, however, still in the development stage and must demonstrate field reliability for remote monitoring applications. A major emphasis is in the direction of optical sensing methods. Technology also is being rapidly developed in the area of advanced aerial and satellite remote monitoring capabilities that can be integrated with expert systems such as the geographical information system (GIS).

Cost-effective and regulatory-acceptable technology needs to be developed to at least supplement the current manpower- and equipment-intensive methods used for direct sampling and laboratory analysis. The sheer number of environmental samples submitted for analysis has overwhelmed the analytical laboratory system. Delays in receiving results from laboratory analysis compromise efforts to recognize and correct potential environmental threats. Real-time remote water-quality measurement technology is the direction for future environmental information gathering. Existing

(text continued on page 791)

Table 7
Organic Chemical Sensors for Water Quality Monitoring

Class of Compounds Detected	Specific Analytes	Manufacturer	Detection Method	Support Equipment
Aliphatic and Aromatic hydrocarbons	Benzene Octane Hexadecane	In-Situ, Inc. Laramie, WY	Chemical Fuse Trips, Mechanical switch	Remote Alarm Unit
Aromatic and Chlorinated Hydrocarbons	Organics Gasoline Fuels	Raychem Corp. Menlo Park, CA	Liquid Sensor	Sensing Cables
Aromatic Hydrocarbons, Volatiles	Gasoline	Fiberchem, Inc. Las Vegas, NV	Chem. Reaction, Fluorescence, Refractive Index	Fiber Optic Detector, Argon Laser
Various Organics	Polycyclic Aromatics, Chlorophyll	Turner Designs Sunnyvale, CA	Fluorescence	UV lamp, Fluorometer
Various Hydrocarbons	Hydrocarbons	Warrick Controls Royal Oak, MI	Resistance Change due to Adsorption	Electronic

Various Hydrocarbons	Hydrocarbons	Emhart Elect. Group Indianapolis IN	Resistance Change due to Adsorption	Controller, Meter
Various Hydrocarbons	Hydrocarbons	Leak-X Corporation	Relay Closure when Cables Detect Hydrocarbons	Cables, Controller
Various Organics	Organic Carbon TOC	Astro International League City TX	NDIR Detection of CO from Combustion of Organics	115 VAC, Portable, Autosampler, Computer, Printer
Various Organics	Volatile Organics (VOCs)	Sentex Systems, Inc	Gas Chromatography	115 VAC, Portable, Computer, Carrier Gas

Table 8
Physical/Chemical Water Quality Monitoring Sensors and Equipment

Parameters Measured	Manufacturer	Detection Method	Support Equipment
pH, Temperature, Salinity, Dissolved Oxygen, Depth, Specific Conductance, Oxidation/Reduction	Hydrolab Corp. Austin, TX	Electrochemical, Multiparameter Probe	Submersible Battery Pack, Communication Software, Cables
Weather Station: Rainfall, Wind Speed, Barometer	Solus Systems, Inc. Beaverton, OR	Anemometer, Humidity Meter, Rain Gauge	Remote Terminal, Battery, Solar Panel, Cables
Weather Station: Rainfall, Wind Speed, Air/Soil Temp., Conductivity, Radiation	Handar Div. of TSI Incorporated Sunnyvale, CA	Ultrasonics, Electro-chemical, Electrical Resistance	Data Logger, Solar Panel, Telemetry Modem, Radio
Outfall Monitoring and Control	Control Microsystems Kanata, Ontario, Canada	Electrical Inputs from Remote Sensor	Remote Terminal Unit, 115 VAC, Back-up Battery, Alarms
Turbidity	Monitek Technologies Hayward, CA	Forward Light Scatter	In-line Fitting, Receiver/Transmitter
Suspended Solids	Monitek Technologies	Photoelectric Cell, Transmissometer	In-line Pipe Fitting, Light Source/Detector Probe, Receiver

Turbidity, Suspended Solids	D&A Instrument Co. Port Townsend, WA	Backscatter Light Radiation	Underwater Housing, Cables, Data Logger
Oil on Water, Oil Slick	Detector Electronics Corporation Minneapolis, MI	Infrared Reflectance Non-contact	Light Source, Detector Mounting Bracket
Dissolved Oil in Water	Great Lakes Instruments, Inc. Milwaukee, WI	Fluorescence, Scattered UV Light	Fluorescence Meter, Ultraviolet Light
Turbidity	BTG, Inc. Decatur, GA	Infrared Light Scatter	Data Logger/Transmitter, 115VAC, 24 VDC
Stormwater/Effluent Samples	American Sigma Medina, NY	Rain Gauge and Flow Meter Activate Sampler	Automatic Sampler, Battery, Sample Containers, Recorder
Stormwater/Effluent Samples	ISCO Environmental Division Lincoln, NE	Rain Gauge and Flow Meter Activate Sampler	Automatic Sampler, Battery, Sample Containers, Recorder
Open channel Flow	Monitek Technologies, Inc. Hayward, CA	Ultrasonic Non-contact	Parshall Flume or Weir

Table 9
Miscellaneous Water Quality Monitoring Sensors and Equipment

Class of Compounds Detected	Specific Analytes	Manufacturer	Detection Method	Support Equipment
Various Chemicals, *In Situ* Monitoring	Custom Applications	Guided Wave/ Perstorp El Dorado Hills, CA	Optical Spectroscopy	Spectrometer, Fiber Optics, Computer
Various Chemicals	Custom Applications	NIRS Systems, Inc. Silver Springs, MD	Near Infrared Spectroscopy	Spectrometer, Fiber Optics, IR Source
Various Chemicals	Custom Applications	Volpi Mfgr. USA Auburn, NY	Optical Spectroscopy	Light Source, Detector, Fiber Optics
Various Chemicals	Custom Applications	Guided Wave/ Perstorp El Dorado Hills, CA	UV-VIS-NIR Analysis Scanning Spectro-Photometer	Fiber Optics
Cooling Water Contaminants	Chlorine, Phenol, Oil	Teledyne Analytical Instruments San Gabriel, CA	UV/Visible Electro-Magnetic Photo-metric Analysis	115 VAC,
Oil	Oil in Water	Teledyne Analytical	Single Beam Dual-Wavelength UV Photometric Analyzer	115 VAC, Rack Mount

(text continued from page 785)

and forthcoming legislation and a growing public environmental awareness will drive the development of advanced water quality sensors and instrumentation.

BIBLIOGRAPHY

40 CFR 122,1990, National Pollutant Discharge Elimination System, Federal Register, November 16, 48062.

40 CFR 141, 1986, National Primary Drinking Water Regulations, Federal Register, July 3, 24328.

40 CFR 141,1991, Priority List of Substances Which May Require Regulation Under the Safe Drinking Water Act, Federal Register, January 14, 1470–1474.

40 CFR 141 and 142, 1987, National Primary Drinking Water Regulations—Synthetic Organic Chemicals; Monitoring for Unregulated Contaminants; Final Rule, Federal Register, July 8, 25690-25717.

40 CFR 141 and 142, 1991, Maximum Contaminant Level Goals and National Drinking Water Regulations for Lead and Copper; Final Rule, Federal Register, June 7, 26460–26564.

40 CFR 141, 142, and 143, 1991, National Primary Drinking Water Regulations, Final Rule, Federal Register, January 30, 35253597.

40 CFR 143, 1986, National Secondary Drinking Water Regulations, Federal Register, July 3, 24328.

40 CFR 261, 1990, Identification and Listing of Hazardous Waste, Federal Register, November 13, 47329–47334.

40 CFR 264 and 270, 1987 List (Phase 1) of Hazardous Constituents for Ground-Water Monitoring; Final Rule (Appendix IX), Federal Register, July 9, 25942-25953.

42 USC 300f, 1974, The Safe Drinking Water Act, Title 42, United States Code, Part 300f.

42 USC 6924, 1980, RCRA Subtitle C, Part 3004, Title 42, United States Code, Part 3004.

Baxter, D. and Goodwin, W., August 1992, "Remote Surface Water Monitoring," Environmental Technologies, Golden, CO.

Bentz, H., June 1993, "Turbidity Measuring Technology Tested," *Water Environment & Technology,* Washington, DC, p. 12–13.

Bogue, R.W., September 1991, "Environmental Monitoring: The Way Forward for Advanced Optical Sensors?" *Measurement Science and Technology,* p. 889–890.

Cheremisinoff, P. N., November 1988, "A Special Report: Measurement of Toxic Air and Water Pollutants," *Pollution Engineering,* p. 69–79.

Civil Engineering, August 1992, "Remote Sensing Zeros in on River Spill," p. 20.

Coursin, K., November 1988, "Effluent Monitoring for Oil in Water," *Pollution Engineering,* p. 100–102.

Detector Electronics Corporation, 1993, Oil Slick Detection, sales literature.

Executive Enterprises, Optrode Research, Summer 1991, *Federal Facilities Journal,* p. 225–226.

Friling, L., January 1, 1993, "Monitoring Storm Water Runoff," *Pollution Engineering,* p 36–39.

Griffin, J.W., and OLSEN, K.S., January 1991, "A Review of Fiber Optic and Related Technologies for Environmental Sensing Applications," Pacific Northwest Laboratory, Richland, Washington, DOE Report, DE-AC06-76RLo 1830.

Hushon, J. M., April 1990, "Overview of Environmental Expert Systems," American Chemical Society, Washington, DC.

IMO Industries Inc., GEMS Sensors Division, November 1991, GEMS Continuous Liquid Level Indicators, p. 17.

ISCO, 1992, "Storm Water Runoff Guide," ISCO Environmental Division, Lincoln, NE.

Milanovich, F. P., January 1991, "A Fiber Optic Sensor for the Continuous Monitoring of Chlorinated Hydrocarbons," Lawrence Livermore National Laboratory, DOE Report UCRL-LR-105199

Newton, J., July 1989, "Groundwater Investigation and Monitoring," *Pollution Engineering,* p. 66–73.

Nudo, L. and Simpson, C. A., November 15, 1992, "Analyzing the Environment," *Pollution Engineering,* p. 34–38.

Parker, H.D. and Pitt, G.D., 1987, *Pollution Control Instrumentation for Oil and Effluents,* Graham and Trotman, London.

PL 94-580, Resource Conservation and Recovery Act of 1976, Federal Public Law 94-580, October 31.

PL 98-616, 1984, The Hazardous and Solid Waste Amendments of 1984, Federal Public Law 98-616, November 8.

PL 99-339, 1988, The Safe Drinking Water Act Amendments of 1986, Federal Public Law 99-339, June 19.

PL 100-4, 1987, The Water Quality Act, Federal Public Law 100-4.

Pollution Engineering, July 1, 1992, "Remote Sensors Track Remediation Site Data," p. 88.

Power, September, 1992, "Water Pollution Monitors and Controls," p. 5,456.

Raychem Corporation, June 1989, "Trace Tek Leak Detection and Location Systems," Double-Containment Design Guide, p. 7,15,25,26.

Rossabi, J., April 1992, "Fiber Optic Sensors for Environmental Applications: A Brief Review," Westinghouse Savannah River Company, DOE Report, WSRC-RP-92-471.

Schabron, J.F., Niss, N.D., and Hart, B.K., September 1991, "Application and State of Development for Remote Chemical Sensors in Environmental Monitoring—

A Literature Review," Morgantown Energy Technology Center, Morgantown, West Virginia, DOE Report DOE/MC 11076-3063.

Seitz, W. R., August 1990, "*In-Situ* Detection of Contaminant Plumes in Ground Water," U.S. Army Corps of Engineers, Special Report 90–27.

Woltz, D., Tuttle, D., Pivinski, J., and Woodin, T., March 1, 1992, "Measuring, Sampling, and Analyzing Storm Water," *Pollution Engineering,* p. 50–55.

INDEX